BASIC HUMAN GENETICS

Elaine Johansen Mange

Arthur P. Mange

UNIVERSITY OF MASSACHUSETTS

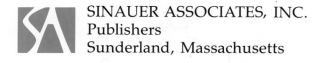
SINAUER ASSOCIATES, INC.
Publishers
Sunderland, Massachusetts

THE COVER

The Farmers, original batik 44″ × 54″
by Ugandan artist Paul Nzalamba.

BASIC HUMAN GENETICS
Copyright © 1994 by Sinauer Associates Inc.
All rights reserved. This book may not be reproduced
in whole or in part without permission
from the publisher. For permission address
Sinauer Associates, Inc.
Sunderland, Massachusetts 01375 U.S.A.

Library of Congress Cataloging-in-Publication Data

Mange, Elaine Johansen
    Basic human genetics / Elaine Johansen Mange, Arthur P. Mange.
        p.   cm.
    Includes bibliographical references (p.   ) and index.
    ISBN 0-87893-495-2
    1. Human genetics.   I. Mange, Arthur P.   II. Title.
QH431.M2785   1993
573.2′1—dc20   93-17808
                                                   CIP

Printed in U.S.A.

5   4   3   2   1

To our children
Jennifer Anna Mange
Steven Arthur Mange
Paul Franklin Mange

Contents

WORKING WITH DNA

GENETIC DISEASE

POPULATION PERSPECTIVES

GENETIC PRACTICES AND PROSPECTS

BOXES

Preface

This book is designed for a one-semester course in human genetics for students with diverse interests and no prior college biology course. Readers with high school biology and chemistry should have little difficulty with the material. We present the general principles of genetics, show how they operate in humans, and discuss their implications for individuals and for society. In writing this book we have tried to be as up-to-date, accurate, and clear as possible. We have introduced some "user-friendly" features: a number of actual case histories in our discussions of genetic disorders, boxes containing interesting sidelights, and the judicious use of a second color in our illustrations.

The text is divided into seven parts, totaling 20 chapters:

Part	Topic	Number of chapters
One	Essential Ideas	6
Two	Extensions of Essential Ideas	3
Three	Cytogenetics	2
Four	Working with DNA	2
Five	Genetic Disease	3
Six	Population Perspectives	2
Seven	Genetics and Society	2

Part One introduces some fundamental aspects of genetics; it touches on readers' prior experience and prepares them for more advanced material in later parts. Throughout the book we incorporate the ideas of molecular biology and their myriad applications in basic and clinical genetics. For readers who may need review, we present some basic chemical concepts in an appendix. Other learning aids include chapter summaries, lists of key terms, many questions, answers to virtually all of the questions, an extensive glossary, and a detailed index.

Much of the material in this book has been carefully distilled from our longer and more advanced text, *Genetics: Human Aspects*, 2nd Edition, published in 1990. But new findings accumulate so rapidly that it is hard to stay current from one month to the next, and a number of very recent developments appear here for the first time. These include new discoveries or significant new insights about the following topics: genetic anticipation; expanding genes (e.g., fragile X syndrome and Huntington disease); uniparental disomy; imprinting; genetics of personality traits; origin of nondisjunction causing Down syndrome; fluorescence in situ hybridization; DNA fingerprinting; tumor suppressor genes; development of colon cancer; monoclonal antibodies; ecogenetics; mitochondrial DNA and molecular evolution; human evolution; problems of screening for cystic fibrosis; and gene therapy.

But science is more than just technologies, facts, and concepts. Every science has an interesting history and progresses toward an unpredictable future through the collective efforts of many creative people. For these reasons, we provide historical material where appropriate and carefully selected references to past and current scientific literature. The citations in the text, giving the author's name and year of publication, refer the reader to a listing of bibliographic data at the back of the book. For the most part we have tried to choose references that are readable by non-science students.

We hope that our readers will gain not only factual knowledge but also the ability to better appreciate and critically evaluate reports in the various media about recent genetic discoveries. We also hope that readers will gain some understanding of the way this knowledge is acquired and of the limitations in its applications. At this point, however, we should make a confession related to the following comment:

> Some students reading biological science courses at university seem happy in the belief that the stories they read in books are true. (Luzio 1992)

The problem with virtually all science textbooks, including this one, is that they do not tell the whole truth. Instead, they give shortened, sanitized, and therefore unrealistic accounts of how science operates. But telling the whole truth would take vastly more space than we have been allotted, and reading the whole truth would take much more time than the average student can devote to a one-semester course.

Real science rarely follows a direct path. Instead, it usually meanders—sometimes without precise goals or maps or guides—and may wind up in swamps or on dead-end trails. Miscalculations, mirages, and misinterpretations can lead investigators far astray. But just as travellers occasionally stumble into fascinating out-of-the-way places that provide new experiences and insights, scientists may find some amazing things during their wanderings. Only rarely, however, does a particular scientific experiment provide a missing piece of the big picture or blaze a useful and important new trail. Scientists can never be sure, in advance, of where their journeys will lead or what they will find along the way. For many of them, this uncertainty is what makes the trip worthwhile. We hope that our short account of basic human genetics nevertheless has captured some of both the uncertainty and the wonder of science.

A number of our colleagues, teachers, and students contributed to the development of this book. We are grateful to Emanuel Hackel (Department of Medicine, Michigan State University) and to Joan K. Burns (Clinical Genetics Center, University of Wisconsin, Madison), who reviewed the entire manuscript. Their pithy comments and suggestions for improvement were extremely helpful. We also thank the following persons, who read and offered useful suggestions for certain parts of this manuscript: L. Herbert Bruneau (Oklahoma State University), Patricia A. DeLeon (University of Delaware), David Klingener (University of Massachusetts, Amherst), and Alan Swedlund (University of Massachusetts, Amherst). And we owe a continuing debt of gratitude to the 20 other experts who reviewed particular sections of our 1990 book. The responsibility for any errors that remain, however, is ours.

Finally, we are pleased to acknowledge the help and support of the personnel of Sinauer Associates, especially Andrew D. Sinauer, Carol J. Wigg, Joseph Vesely, Janet Greenblatt, Janice Holabird, and Dean Scudder. Their professionalism, patience, and attention to fine detail have contributed greatly to the finished product.

Elaine Johansen Mange
Arthur P. Mange

A Frame for Genetics

The year was 1934 and the place was Norway. Mrs. E., a sad but very persistent young mother, had consulted many doctors, hoping that somebody could explain why her two children were profoundly retarded and why they gave off such an odd, mousy odor. Both her seven-year-old daughter and four-year-old son had seemed completely normal at birth but within a few months had begun to show serious developmental problems. The girl did not start to walk until she was 22 months old. She could only speak in single words and was only partly toilet trained. She was hyperactive, always moving about aimlessly. She showed some muscle rigidity and had a skin rash, but otherwise her physical health was relatively good. The boy could not walk, sit without support, speak, feed himself, be toilet trained, or focus his eyes. All he could do was cry, smile, gurgle unintelligibly, and play. Aside from some urinary problems, irregular muscle spasms, and rough skin, his physical health was also fairly good.

Finally, the mother brought her children to Professor Asbjörn Fölling, a biochemist and physician. He noted the severe retardation and peculiar odor and also a somewhat stooped posture and quite fair skin in both children. While analyzing their urine, he discovered that it turned green after he added a chemical called ferric chloride—a very unusual reaction. When he asked the mother to bring in some more samples, she supplied him with about five gallons of the children's urine. In two months he isolated and purified the unexpected compound that was responsible for the green color.

Fölling then searched for more cases of what he suspected was a newly discovered disease associated with severe retardation. After testing the urine from several hundred patients in institutions for the retarded or mentally ill and in homes for the aged, he found eight more people (including another brother-sister pair) whose urine turned green with the addition of ferric chloride. When the survey was expanded to cover much of the country, the total came to 34 cases from 22 families, and Fölling, with the help of a Norwegian geneticist, was able to show that the condition was inherited in a simple way. Although the mother and father were themselves normal, they both carried the same hidden (recessive) disease-causing hereditary factor (gene) and passed it on to their children. The children would thus have that gene in double dose and be affected.

In 1938 Fölling and a colleague reported that all of the patients had too much of a substance called phenylalanine in their urine and blood, and hypothesized that these patients suffered from a defect in the way their bodies handled phenylalanine. But the exact cause was not known until 1947, when the American scientist George Jervis reported that the culprit was a specific enzyme that in normal people converts phenylalanine to a substance called tyrosine:

$$\text{phenylalanine} \xrightarrow{\text{normal enzyme}} \text{tyrosine}$$

An **enzyme** is a protein molecule that is specialized to greatly increase the rate of a certain chemical reaction without itself being changed by that reaction. Enzymes are made by cells under the control of genes. The reaction described here (the conversion of phenylalanine to tyrosine) did not occur in people whose enzyme was either missing or defective. So they had too little tyrosine in their bodies and far too much accumulated phenylalanine—some of which was converted to another substance and excreted in their urine. There it was detected by the ferric chloride reaction. The excess of phenylalanine (and perhaps some derivatives) or the lack of tyrosine (and its derivatives), or some combination of these, must cause the brain damage seen in all these patients.

Fölling and many others continued to study this disease, which later was named **phenylketonuria**, or **PKU**. (The name refers to *phenyl ketones* —conversion products of excess phenylalanine—found in the urine.) It turns out that most of the phenylalanine present in our bodies comes from the breakdown of the proteins we eat. Thus, the scientists reasoned, maybe the disease could be cured or at least made less serious if most of the protein and phenylalanine were removed from patients' diets. This strategy was tried in the 1950s—and, amazingly, it worked. If PKU was detected within a few weeks of birth, babies who were fed a specially concocted low-phenylalanine diet escaped brain damage. If kept on the restrictive diet until their teen years, they usually developed intelligence and behavior patterns within the range shown by normal people.

But the special diet was ineffective if started later than one or two months after birth, so early identification of affected babies was the key to preventing brain damage. In families known to harbor the PKU gene, it was possible to detect (by somewhat increased levels of phenylalanine in their blood) those people who, although normal themselves, ran a risk of producing affected offspring. By 1963 a simple test had been devised that would detect excess phenylalanine in a few drops of blood. Soon after, entire populations of babies were being tested for PKU within a few days of birth.

Clearly, there was much to rejoice about. For the first time, an inherited disease was being conquered: Its tragic effects could be prevented by a change in the environment, namely, dietary therapy. Those who were born with a double dose of the PKU gene variant could be detected. Indeed, thousands of young people who previously would have been doomed by PKU have been rescued and allowed to live normal lives.

Unfortunately, the story does not end here. When the young women who were spared the ravages of this condition reached child-bearing age, it was found that their babies were often irreversibly retarded. The reason is this: The mothers had been allowed to stop their restrictive diets some years before, and with a regular protein-rich diet, the levels of phenylalanine in their blood increased dramatically. This situation did not seem to seriously affect the young mothers, but it was devastating for the babies they bore. Unless these women return to the restrictive low-phenylalanine diet before they become pregnant, the medical triumph over PKU will be largely undone. So the struggle to conquer PKU continues.

GENETICS AND SOCIAL ISSUES

The case of phenylketonuria illustrates how far the science of human genetics has progressed in the past half-century. It also shows how the solution to one medical/scientific/social problem can sometimes quite unexpectedly give rise to other medical/scientific/social problems that are equally complex and difficult to solve.

Only during the past several decades has the genetics of humans become a thriving discipline with ever-improving techniques for the diagnosis and treatment of many inherited conditions. Along with the new possibilities for reducing human suffering, however, have come many social, political, and legal problems. Consider the following examples:

> What happens if a woman who was born with PKU (and rescued by dietary therapy) refuses to return to the low-phenylalanine diet before or just after she becomes pregnant? Should she be held legally responsible for "causing" mental retardation in her offspring? Does society have any right to interfere in this individual's decision-making process? Does society have any obligation to help pay the increased costs for the care of the retarded child?
>
> Should parents be allowed to choose the sex of their children? In the long run, couldn't this lead to a change in the near 1:1 sex ratio?
>
> Should certain populations be screened to detect medically normal individuals who carry certain detrimental genes, such as the sickle-cell gene in blacks or the Tay-Sachs gene in Jews?
>
> Should families be told when a child is born with a poorly understood genetic abnormality? Will the benefits of studying and perhaps later helping such children be offset by the possibility of stigmatizing them?
>
> Should society be asked to pay for expensive treatments for genetic disease? If so, should society have a say in whether such affected individuals ought to be born, perhaps preventing their birth through prenatal diagnosis and selective abortion?
>
> If it becomes possible to change our genetic destiny through techniques such as gene replacement, how should these methods be controlled?

Tidy answers to such questions do *not* appear at the back of this book, or anywhere else for that matter. The facts and their interpretations often turn out to be quite slippery, changing as scientists ask new questions and develop new techniques for answering them. However unsatisfying (and perhaps risky) it may be, we usually have to make do with incomplete information. And because public policy decisions on complex social issues must be based on much more than the facts alone, they should involve the thoughtful participation of *all* citizens, and not just the scientists.

SCIENCE AND SCIENTISTS

If the public is to have a say in developing guidelines for dealing with certain scientific problems, it should first have a clear sense of what science can and cannot be expected to do. This, in turn, depends on some understanding of how research is carried out. Unfortunately, great scientific achievements in many fields have led the public to expect more of science than it can possibly deliver: Too little food? Not enough fuel? Too many people? Not to worry—science will take care of everything! Then, when science fails to solve all our old problems and sometimes even creates some new ones, disillusionment and hostility set in (LaFollette 1990). But withdrawing support for science can be as harmful to public welfare as ignoring the ethical problems that arise when science begins to impinge on certain human rights and values.

3

JUDGING THE MEDIA

We know we can't believe everything we see or hear or read. But how do we figure out which medical stories to ignore and which ones to take seriously? There is no foolproof method for separating the reliable from the dubious, but it helps to ask the following questions.

Who did the study and where was it done? Was the work carried out at a university or a research institution? At a drug or biotechnology company?

Who paid for the research? A government agency? A not-for-profit foundation or organization? A drug company or other industry? Are desired results likely to be profitable to the sponsor or to the scientists? The quality of research is not necessarily dictated by the type of sponsor, but we should always consider the possibility of vested interests when we assess the latest scientific breakthroughs.

What and how large were the good effects of a drug or treatment? What and how large were the bad effects? Rather than accepting an unspecified "statistically significant" difference, which could be as little as 5% (and medically unimportant), look for a specific percentage difference in effects between the treated group and the untreated controls. Also look for large benefits and small risks.

What organism was studied? Humans? Rats? Fruit flies? Yeast? For practical and ethical reasons, many preliminary studies cannot be done on humans. And because biochemical processes in other organisms resemble those in humans—often to an astonishing degree—it makes sense to use other organisms first. But we should be skeptical about applying those results to ourselves until scientists have actually carried out the same studies in people.

Was the human research done on cells in test tubes or on whole people? Reports of test tube cures may raise our hopes, but we can't really be sure that a drug will work until it has actually been swallowed by or injected into or rubbed on intact bodies.

How large were the numbers? 20? 200? 2,000? *And how long did the tests last?* A month? A year? A decade? Sometimes the good effects are only temporary. Sometimes bad side effects occur in only a very small proportion of test subjects, or they develop only after some period of time has elapsed. Thus, the bigger the numbers and the longer the tests are, the more precise the results and the more reliable the benefit-to-risk estimate will be.

Was a simultaneous control group used for comparison? Control organisms must be very similar to the tested subjects in every way except that they did not undergo the experimental treatment in question. This is always important in biological research, but critical for testing new drugs or other medical treatments.

How were the test subjects and the controls chosen, and how carefully were they monitored? If humans were used, were they volunteers or nonvolunteers? Did they represent the general population or a special subgroup? Might they have been healthier or less healthy than average? More motivated or less motivated? Did supervised treatment take place in a hospital or clinic, or were subjects left alone and just assumed to have followed the prescribed regimen?

Were the studies double-blind? In a double-blind study, the subjects do not know whether they belong to the test group or the control group, and the researchers also do not know (until after the results are tallied) which group any subject belongs to. The *placebo effect*—an improvement reported by people who think they are taking medication but who are actually given fake pills or treatment—is well documented. Likewise, especially if differences are subtle, even the most honest investigators may unconsciously bias individual judgments if they know whether a subject is in the test group or the control group. (Even Mendel has been accused of generating data that are "too good"!)

Did the researchers look at the future or at the past? The better studies are *prospective* (looking forward and actually observing what happens to the subjects during an experimental period) as opposed to *retrospective* (looking backward and relying on possibly incomplete or faulty memories and past records of the subjects).

Has the research been published in a reliable scientific journal, or has it only been announced at a meeting or to the media? Be especially wary of "science by press conference," that is, announcing results to the media before they have been reviewed by peers and published in a scientific journal.

Have the experimental results been confirmed by other researchers? Have the same or similar studies been done by other groups, and if so, how do they compare? Although it does sometimes happen that true pioneers are not taken seriously, we should be cautious about accepting the news of a one-of-a-kind breakthrough that has not been verified by others. Look for statements by other researchers or by institutions (such as the National Academy of Sciences or the National Institutes of Health) that periodically review large amounts of research.

If there are testimonials by medical authorities or celebrities, were they paid for by the research sponsor? If so, such opinions may be biased or even unreliable.

How long will it take for the experimental advance to be parlayed into actual medical treatment? It usually takes years to repeat research results in test animals, develop a drug or treatment for humans, test the drug or treatment in humans, get government approval for the new drug or treatment, and produce it for public consumption. Thus, the premature announcement of new cures based on preliminary test results may raise false hopes for millions of people.

It would help if scientists tried harder to explain their everyday work to the public. On the other hand, premature publicity—for example, divulging research results to the press before publishing them in scientific journals—can create problems too, especially if the results do not hold up to further scientific scrutiny (Maran 1991). If gee-whiz type media announcements of new discoveries were judged against broad-ranging questions, a more balanced picture would emerge (Box A). But science is usually viewed with awe, not to mention a little fear or suspicion; it is perceived as a realm that most people cannot understand, let alone hope to enter. Such stereotypes, hopes, and fears can be dangerous if they affect public policies concerning the use of science. For these reasons, and because this topic is rarely treated in textbooks, we wish to elaborate a bit on the popular images of scientists.

The people who "do" science are, like any other group, quite varied. Yet they tend to be stereotyped and are often misunderstood. In cartoons and comic strips, "mad scientists" abound (Figure 1.1). In movies and books, scientists are often pictured as brilliant but out of touch with reality, or as totally objective and dedicated to their work, or as coldly sinister (Tudor 1989). Usually, they are male. Seldom are they treated as just plain folks capable of pettiness or insecurity. Moreover, the social aspects of science—the daily collaboration with co-workers in the laboratory, the informal exchange of ideas at meetings and at parties, the personal interactions that can speed up or slow down the solution to a problem—are rarely if ever mentioned.

Textbooks (including this one) usually describe only the successes; rarely do readers learn about the intervening false starts, dead ends, misinterpretations, or outright failures. As Harvard scientist Stephen Jay Gould (1986) points out, the same criticism can be made of scientific

"Crack out the liquid nitrogen, dumplings . . . we're on our way."

Figure 1.1
Misconceptions about genetics. This cartoon about the potential outcome of recombinant DNA research, although whimsical and amusing, has no basis in reality. Nevertheless, many people believe that such results are possible. For information on this type of research, see Chapters 12 and 13. (Reprinted courtesy of The Boston Globe.)

journals, whose standard format (introduction, materials and methods, results, discussion, conclusions) makes scientific research look more orderly than it really is. Gould notes, "The false starts are in the wastebasket, not [in] the *Science Citation Index.*" Francis H. C. Crick, codiscoverer of the DNA double helix, puts it this way (Horgan 1992): "Exploratory research is really like working in a fog. You're just groping. Then people learn about it afterwards and think how straightforward it was."

Truth be told, scientific research quite often yields confusing or negative or equivocal results, even at the hands of the world's greatest scientists. Furthermore, even successful research involves a lot of tedium, and were it not for the hope of finding something exciting now and then, some scientists might at times wish they had chosen another line of work. So why do people do science? Some are strongly motivated by practical or humanitarian goals—such as finding new and better ways of solving old problems, or developing new commercial products, or expanding the space program, or conquering diseases, or feeding the hungry, or solving environmental problems.

Others, however, are simply curious about the world around them and get tremendous pleasure and satisfaction from finding out how things work and what basic laws govern our universe. If (as sometimes happens) their research ends up having some practical or humanitarian value, well and good—but that is not the main reason they do it. In an essay on science and scientific attitudes, astronomer S. Chandrasekhar (1990) points out that curiosity was the motivating force behind the ancient Greeks' fascination with geometrical curves:

> They sought curves that would encompass the geometrical properties of straight lines and circles in a more beautiful and harmonious synthesis; and they discovered this harmony in the curves obtained as sections in a cone: the ellipse and the hyperbola. Their curiosity with respect to these curves was not motivated by any physical fact they discerned. . . . Apollonius of Perga wrote eight monumental volumes devoted to these curves. . . . But it did not occur to him, or to the other Greek mathematicians, that the curves, which they studied so earnestly for their intrinsic beauty, had any relevance to the real physical world.
>
> Yet some eighteen centuries later, when Kepler was analysing the orbits of the planets on the copernical system, he discovered that the very curves that the Greek mathematicians had studied for their intrinsic mathematical beauty were exactly those needed to represent the orbits of the planets.

A more recent example of a scientific project undertaken purely for the sake of curiosity was the 1990 space launch of the Hubble telescope. This instrument was designed to see ten times further into the past than any earthbound telescope—perhaps even as far back as the Big Bang that may have produced our universe. As astronomer Timothy Ferris (1990) comments:

> It has no practical purpose at all. . . . Hubble is an instrument of discovery; its paternity lies as close to the Niña, the Pinta and the Santa Maria as to Pasteur's test tubes or Einstein's notebooks; and it is intrinsic to discovery that one cannot accurately predict what will be discovered. . . . The Hubble Space Telescope is a machine for subjecting our conception of the wider universe to an ordeal by fire.*

---

* The failure of its giant lenses to focus properly owing to faulty production and inspection procedures—one of the most stupendous, costly, and highly publicized disasters of modern science—has also subjected its manufacturers to an ordeal by fire. But these errors do not really take away from the original intent of this project and have not prevented the Hubble from making a number of important discoveries. Furthermore, it may yet be repaired.

THE POSTMATURE DISCOVERY

Certain basic facts and technologies may exist for years before anybody uses them to produce important new findings that are quickly understood and accepted by colleagues. Two researchers who made such postmature discoveries have written about their experiences.

Joshua Lederberg shared a Nobel prize for his demonstration (as a 21-year-old medical student) that bacteria can reproduce by sexual union rather than only by solitary fission—a finding that opened up the vast field of bacterial genetics. He did this by crossing bacterial strains, each containing at least two different nutritional mutations that prevented growth in unsupplemented medium. Then he looked for normal progeny, which could grow in unsupplemented medium and must therefore have resulted from a sexual recombination of the parental traits. "It took about six weeks, from the first serious efforts at crossing in mid-April 1946, to establish well-controlled, positive results" (Lederberg 1986). True, he was extremely lucky in choosing his bacterial strains, since only about 5% of randomly chosen strains of *E. coli* would yield positive results in his experiments. Yet as social scientist Harriet Zuckerman and Lederberg (1986) point out: "In principle, the investigation was techni-

cally feasible by 1908 . . . [H]ad it been demonstrated experimentally, it would have been understood and appreciated by geneticists and possibly even by bacteriologists." Why several decades passed before scientists tried or succeeded in crossing bacteria remains a mystery.

A more recent postmature discovery is described by Cetus Corporation scientist Kary B. Mullis (1990), who tells how in 1983 he "stumbled across a process that could make unlimited numbers of copies of genes, a process now known as the polymerase chain reaction (PCR)." Details of this revolutionary but simple method, which allows molecular biologists to do many experiments much more quickly and easily than in the past, will be presented in Chapter 13. Here we will only mention that it makes use of a key DNA-copying enzyme, called *DNA polymerase*, to zero in on the specific piece of DNA to be repeatedly copied. With this procedure, over 100 billion replicas of the desired DNA segment can be made in just a few hours in a single test tube.

Mullis came up with the method late one night while thinking about another DNA problem:

> For the next few weeks I described the idea to anyone who would listen.

No one had heard of its ever being tried; no one saw any good reason why it would not work; and yet no one was particularly enthusiastic about it.

When Mullis tried the experiment, he found that it did indeed work. Shortly after, he presented his exciting results in the form of a poster at a scientific meeting:

> Yet nobody seemed to be interested in my poster, and I felt increasingly anxious. . . . Finally, I noticed Joshua Lederberg . . . and I snared him into looking at my results. Josh looked the poster over carefully . . . and we talked for a long time. . . . I think that Josh, after seeing the utter simplicity of the PCR, was perhaps the first person to feel what is now an almost universal first response to it among molecular biologists and other DNA workers: "Why didn't I think of that?" And nobody really knows why; surely I don't. I just ran into it one night.

Postmature discoveries may cause some anguish to the established scientists who didn't think of them. But for aspiring young researchers, it should be comforting to know that not everything that *could* be discovered in a given period *has* been discovered.

In the course of solving practical problems and satisfying human curiosity, the pursuit of science has usually involved several commonly recurring themes. One is the *codiscovery* of a principle by several researchers working independently. In some cases codiscoveries are not surprising, because the exchange of ideas in scientific journals and at meetings may draw many people to work on an interesting problem, especially in hot new areas of inquiry. In other cases, and particularly in recent times, the problems have become so complex that solutions require teamwork by several laboratories, each specialized in a different type of research.

A second theme in the history of science is the so-called *premature discovery*, which, although correct, is neglected for a long time by other scientists. There are several striking examples of this in genetics, starting with the work of Gregor Mendel. On the other hand, we also know of a few *postmature discoveries*, which are long past due and swiftly accepted by other scientists (Box B).

A third theme is the critical role that *new technology* plays in the development and testing of scientific ideas. The process is never-ending:

Better microscopes, new ways of staining cells and cell parts, improvements in growing and handling cells, new techniques for isolating and analyzing genetic material, increasingly powerful and accessible computers, and many other technological advancements were all necessary for knowledge to progress to its current level. These research tools do not by themselves generate, prove, or disprove hypotheses. People do that, often alone but rarely in complete isolation.

Of all the world's scientists who ever existed, most are alive today. A tour of their laboratories would reveal a wide range of intellects and personalities—from geniuses to plodders, from people with many talents to those with no outside interests, from the showy to the shy, from highly competitive workers to those with modest aspirations, from generous and caring individuals to selfish boors.

Very few of even the most intelligent and ambitious scientists will ever be mentioned in a textbook. Indeed, some of their published papers will never be cited anywhere, because scientific reports are now being turned out at such a rate that nobody can keep track of more than a tiny fraction of them. Browsing through scientific journals, we find a broad and amazing array of articles whose titles are so specific that they may sound silly to lay people—especially to those who feel that science should only solve practical problems and not bother with apparently trivial pursuits. But the gradual accumulation of unexciting bits of knowledge and experimental techniques by many workers often sets the stage for the great discoveries or insights of a few, usually in ways that could never have been predicted.

GENETICS IN AGRICULTURE, MEDICINE, AND SOCIETY

From the questions and comments in this chapter, from feature stories in the popular media, and perhaps from personal experience, most of you are probably aware that genetics now plays an important role in medicine and in certain social issues. Its key role in agriculture goes much further back.

For thousands of years, people have been puzzled and intrigued by

Figure 1.2
Dog breeds. Dogs (Canis familiaris), probably domesticated from the gray wolf (Lupus) over 10,000 years ago, now exist as more than 100 different breeds that range in size from Great Danes (30 inches shoulder height, 140 pounds) to tiny Chihuahuas (5 inches shoulder height, 4 pounds). All, however, still belong to one species and can interbreed. Perhaps the first wolf-to-dog changes that occurred were an upcurved tail and smaller teeth. Other traits that have been selected by humans include short legs, foreshortened face, drooping ears, and various coat colors and textures. Behavioral variations are seen, too, in dogs bred for different tasks, such as herding, hunting, guarding, pulling, guiding, and household companionship.

the similarities and differences seen within families of humans, other animals, or plants. Sometimes they were able to develop special breeds of crop plants or domesticated animals by selective mating. But their successes were hit-and-miss, depending partly on luck and partly on past observations, but with no real understanding of the biological principles behind them.

From anthropological findings, written records, and ancient art, we know that agriculture itself began at least 10,000 years ago. Although experts disagree on the details, it now appears that there were several centers of origin. These include the Near or Middle East, Asia, Africa, and Central America. The earliest herders and farmers selectively bred both animals and plants, using as parental stock those individuals or strains possessing the most desirable traits and culling out the offspring with the least desirable traits. In this way they gradually improved on or developed animal breeds (such as goats, sheep, oxen, camels, dogs, cattle, and horses) and crops (such as wheat, rice, and maize). Figure 1.2 shows how the process of selection, carried out for thousands of years, can take advantage of the tremendous genetic diversity within a given species.* Trade and migration patterns also played important roles in the evolution of domesticated animals and plants.

Only during the past century have clear explanations of such hereditary phenomena emerged. By experimenting with many plants, animals, and microorganisms, scientists have learned (1) how traits are passed on from one generation to another, (2) how genetic information is chemically "decoded" and expressed during each individual's development, and (3) how the genetic variability that arises among the members of groups can account for the gradual evolution of populations. These three major areas of inquiry are known, respectively, as *transmission genetics, molecular* (and *biochemical*) *genetics,* and *population genetics.*

Statistical techniques devised in the early 1900s also improved agricultural practices. By even more intensive selective breeding and interspecies crosses, agronomists have developed crops that are higher yielding, more uniform, more vigorous, and more resistant to diseases, insect pests, and weed killers (Figure 1.3). The results are seen around the world as the *Green Revolution.* Using statistical techniques coupled with technological advances such as *artificial insemination* and *embryo transfer,* breeders have also developed livestock with more uniform and commercially desirable traits, such as higher (or lower) yield and butterfat content of milk in cows and better quality and yield of wool in sheep.

The 1980s brought significant advances in tissue culture and in the new molecular techniques for cloning individual genes and implanting them into foreign species. These methods may revolutionize crop and livestock development, but close attention must be paid to the biological, environmental, and social effects of such programs.

Is it possible to significantly change or "improve" the human species through genetics? Not really, if human values and rights are also to be preserved. Sir Francis Galton coined the term *eugenics* to describe the hoped-for improvement of our own genetic endowment. It reflects a naive optimism about applying methods of artificial selection, so successful in agriculture, to humans: Just encourage "superior" individuals to be fruitful

---

* There are limits, however, to the selectability of some traits, including the speed of thoroughbred horses. Despite improved breeding, better training, and excellent health care, American and English racehorses today are running no faster in the 2.5-minute "classic" races than they did four or five decades ago. Yet the speeds of American quarter horses running 20-second sprinting races have shown small increases. The limiting physiological factor may be the accumulation of lactic acid in muscles, which occurs in the longer races but not in short sprints. Humans, on the other hand, are still improving all their track records —not by selective breeding, but by better training, health care, and equipment.

and discourage or prevent "inferior" people from reproducing at all. Unfortunately, the eugenics movement attracted demagogues along with idealists, and distortion of the meaning of Darwinian selection contributed to the genocidal Holocaust in Nazi Germany. With its emphasis on eliminating the "bad" genes of paupers, criminals, feebleminded individuals, and immigrants, this brand of eugenics fell into disrepute.

We now realize that even the most honorable attempts at improving the human gene pool create major problems. First of all, the simplest equations of population genetics show that preventing people with rare recessive disorders from reproducing will *not* significantly reduce the frequency of these detrimental genes. Moreover, the criteria used to determine whether reproduction should be encouraged or discouraged are fuzzy at best and highly susceptible to social and political mischief.

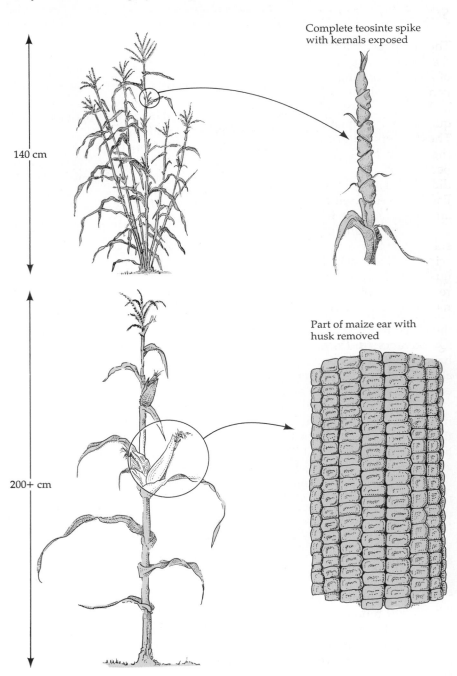

Complete teosinte spike with kernals exposed

140 cm

Part of maize ear with husk removed

200+ cm

Figure 1.3
Corn evolution. Modern maize (Zea mays) probably evolved from a wild grass called teosinte (Zea mexicana). The male flowers (tassels) in the two species are similar, as is the general form of the stem and leaves. Teosinte, however, has many stems and numerous small female flowers (spikes), whereas maize has one stem and one or two large female flowers (ears). The small spike of teosinte consists of a single row of six to ten kernels, each enclosed in a hard, triangular fruit case. Beadle (1980) and Galinat (1977) suggest that the fruit case of ancient teosinte was modified by successive mutations during domestication into the large cob structures of modern maize. Maize is probably the most highly variable cultivated crop. Several dozen distinct varieties are recognized in Mexico alone, and several dozen quite different varieties in Peru.

10

But human genetics can be employed humanely at the personal level. Increasingly sophisticated prenatal testing and genetic counseling are able to avert family tragedy or reassure those individuals who turn out not to be at risk for certain hereditary disorders. Great caution, care, and sensitivity are required, however, when considering or carrying out the large-scale testing of populations that are known to be at increased risk for certain conditions. Even for those who oppose testing, there is still the problem of just maintaining our current genetic endowment, as when, for example, medical advances preserve detrimental genes that would otherwise be eliminated by the death of these individuals. The possible long-term genetic and social consequences of medical practices will be debated for years to come.

SOME QUESTIONS ASKED BY GENETICISTS

The discoveries reported in this book took place between the 1860s and the 1990s. Our goal, however, is not to give a highly detailed description of all these findings. Rather, we hope to define the broad scope of human genetics while stressing the important experimental role of other species in learning about our own. We try to show how answers to scientific questions rest on layers of knowledge accumulated by many investigators. These layers do not form at a constant rate. Indeed, for certain periods, there may have been no progress at all in some disciplines, perhaps because scientists did not ask the right questions or perhaps because they did not have the right technologies available to answer their questions. Each age sees itself as the modern age; but in science, as elsewhere, final answers are rarely final.

Before plunging ahead with some of the answers, we will set the stage by listing a number of the major questions that genetically inclined scientists have asked at various times for over a century. Some of these questions have been answered, but others continue to challenge today's researchers. We hope that this short preview suggests a broad and fascinating field of inquiry.

Questions in Transmission Genetics

What are the rules for predicting how simple traits are passed on to succeeding generations?

What is the relation between invisible hereditary factors, the *genes*, and the microscopically visible cellular structures called *chromosomes*?

How are genes arranged on chromosomes?

How is sex determined?

What is the pattern of inheritance of genes present on the sex-determining chromosomes?

How many chromosomes do humans have?

How many genes do humans have?

What happens if chromosomes break or misbehave during cell division?

Can human genes be assigned to specific places on our chromosomes?

Are all genetic traits inherited in a simple way?

What human traits are inherited in complex ways?

To what extent do environmental differences play a key role in bringing about the variability we see among individuals of the same species?

For any complex trait determined in part by the joint action of many genes, can we measure the relative contributions of genetic factors (nature) and environmental factors (nurture)?

Questions in Molecular and Biochemical Genetics

What are chromosomes made of?

What are genes made of?

What is the physical structure of DNA (*deoxyribonucleic acid*), the genetic material?

What does a gene actually do?

How do genes make the gene products called *polypeptides*?

How does a change in a gene lead to a change in a polypeptide?

How does the genetic code work? That is, which groups of DNA subunits (called *bases*) specify which polypeptide subunits (called *amino acids*)?

What causes genes to operate only at certain times and in certain cells?

Are there special kinds of genes that control the early development of an organism?

Can genes from one species function in another species?

What kinds of genetic changes occur in cancer cells?

Is cancer inherited?

How does the immune system work?

How can a body defend itself against so many different foreign agents?

Why do organ transplants often get rejected?

Questions in Population Genetics and Evolution

Do species change over time—and, if so, how?

In genetic terms, how does Darwin's "natural selection" work?

How frequent are gene variations (*mutations*)?

Can mutations be produced artificially?

How do mutations spread through a population?

How do factors like migration patterns and population size affect evolutionary processes?

What are the genetic "rules" that help explain and quantify evolutionary processes?

Does inbreeding play any role in evolution?

Would it be possible to eliminate some extremely serious inherited disorders by discouraging affected individuals from reproducing? If not, why not?

Human Genetics: Some Special Problems

Human beings are not the best subjects for genetic study. Researchers are hampered by our long generation time, our small families, our highly varied genetic and environmental backgrounds, and the absence or incompleteness of our family records. Human geneticists also have to search the world's populations for those few families that provide some information on inheritance patterns rather than relying on the experimental matings that are the stock in trade of all other geneticists. Yet human biochemistry is better understood than that of any other complex species, and research on humans *has* led to new discoveries—such as the first insight into the relationship between human genes and enzymes and the precise effect of gene mutations on protein structure in humans. Obviously, the use of experimental organisms has been essential for discovering and testing fundamental genetic phenomena. But because we are so curious about ourselves, scientists have been willing to cope with the special problems of human genetics, or to find ways around them. What is difficult today may be less difficult tomorrow with new ideas and new technology.

FURTHER READING

We encourage you to delve further into the literature of genetics. Pertinent articles and books are mentioned throughout the text and at the end of each chapter. Most of these should be understandable to most readers. All citations are by author and date of publication. A single alphabetical listing of these works, with bibliographic information, appears at the end of the book.

A number of readings are appropriate to Chapter 1. Bauer (1992) very engagingly discusses the roles of (and misconceptions about) science and technology in modern life. Heussner and Salmon (1988) discuss how the media inform and misinform the public about scientific and medical news. Among the best histories of early genetics are those by Sturtevant (1965) and Judson (1979). Articles on the history and development of human genetics include Dunn (1962), McKusick (1975), Motulsky (1978), Jacobs (1982), Caskey (1986), and Bodmer (1986). Kevles (1984a–d), in a very readable series of magazine articles, describes the development and evolution of the eugenics movement.

2

Cells and Chromosomes

With the invention of the compound microscope around 1600, a whole new world of beauty and knowledge opened up. In 1655 Robert Hooke described the porelike *cells* seen in thin slices of cork. Between then and 1879, when Walther Flemming discovered the darkly staining threads (*chromosomes*) present in every cell and described their remarkable dance-like behavior during cell division, the basic outlines of cell theory were defined.

This period included many other major landmarks. In 1838–1839 the German researchers Matthias Schleiden and Theodor Schwann put forth their *cell theory*, stating that the cell is the underlying unit of structure in all living organisms. And in 1858, Rudolf Virchow proposed that all cells come from preexisting cells. From these ideas arose the conception of

> . . . an unbroken series of cell-divisions that extends backwards from our own day throughout the entire past history of life. . . . It is a *continuum*, a never-ending stream of protoplasm in the form of cells, maintained by as-similation, growth and division. The individual is but a passing eddy in the flow which vanishes and leaves no trace, while the general stream of life goes forwards. (Wilson 1925)

The study of cells is known as **cytology**. The kinds of cells that cytologists usually study—whether taken from multicellular organisms (plants, animals, and fungi) or from single-celled organisms (such as pro-tozoa, yeasts, and some algae)—have a very complex structure (Figure 2.1). Most of their genetic material resides in *chromosomes* located in a **nucleus**. The nucleus is separated by a double membrane from the rest of the cell's contents, which make up the **cytoplasm**. Organisms with this kind of cell structure are called **eukaryotes**. Their cellular complexity con-trasts with that of single-celled **prokaryotes** (bacteria and blue-green al-gae), which lack nuclei and have a primitive type of chromosome.

Multicellular eukaryotes contain two main kinds of cells. **Somatic cells** make up all the body parts: the skin, bones, muscles, brain, and internal organs of animals, as well as the leaves, stems, and roots of plants. **Germ cells** are reproductive cells—the eggs and sperm of animals, the egg cells and pollen grains of plants—that unite sexually to produce new individuals in the next generation. In this chapter we will focus on the structure of somatic cells only.

CELL STRUCTURE

The structures and substances present in a somatic cell do not just slosh around at random. They are distributed among several different compart-ments,* each separated from the others by at least one selectively perme-

* Compartments include the nucleus, cytosol, lysosomes, peroxisomes, mitochondria, en-doplasmic reticulum, and Golgi complex.

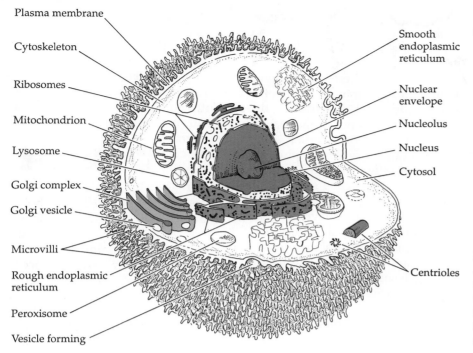

Plasma membrane

Cytoskeleton

Ribosomes

Mitochondrion

Lysosome

Golgi complex

Golgi vesicle

Microvilli

Rough endoplasmic
reticulum

Peroxisome

Vesicle forming

Smooth
endoplasmic
reticulum

Nuclear
envelope

Nucleolus

Nucleus

Cytosol

Centrioles

Figure 2.1
A three-dimensional representation of an idealized animal cell, sectioned to show internal structures as they appear in the electron microscope. The proportions of these structures vary from one type of cell to another.

able membrane. In every nondividing cell an intact nucleus is the central compartment. Its **nuclear envelope** consists of two membranes that contain many pores, through which materials are exchanged with the surrounding cytoplasm. Within the nucleus are at least one dark-staining **nucleolus** and many finely dispersed chromosomes. (The latter, not shown in Figure 2.1, are visible under a light microscope only during cell division, when the nucleus breaks down and the chromosomes condense and appear as darkly stained threads.)

Another compartment is the jellylike **cytosol**, the matrix in which organelles and intricate membrane systems are suspended. Each cell type has a somewhat different proportion and arrangement of these various components. The cytosol also contains thousands of *enzymes*, protein molecules that speed up chemical reactions. Proteins, the end products of gene action, are synthesized on tiny particles called **ribosomes**, which number over a million per cell. Some ribosomes lie free in the cytosol, while others are bound to membranes. Supporting and giving shape to the cytosol is the **cytoskeleton**, a network of filaments and tubules made up of specialized proteins. The tubules, called **microtubules**, also act as an internal transport system upon which cytoplasmic components can be shuttled to wherever they are needed in the cell. Lying just outside the nucleus in animals cells is a pair of tiny **centrioles**, which function as an organizing center for microtubules.

The remaining compartments, which form narrow channels in some areas and large cavities or small vesicles in others, provide the huge surface area needed for biochemical processes. They are also involved in the distribution of substances to different regions of the cell. Attached to these membranes or to the cell's outer membrane and acting as "door openers" are numerous **receptor proteins**, each specialized to recognize and admit (or bind to) only a certain type of molecule. On the cell surface, for example, there may be 500 to 100,000 receptors for a given substance. Each such complex triggers a specific reaction inside the cell. The mal-

15

functioning or absence of receptors can have very serious consequences. One example is the genetic disorder *familial hypercholesterolemia* (Chapter 5), which accounts for 5% of all heart attacks in patients under the age of 60.

Enzyme systems are everywhere. Some are embedded in or loosely bound to membranes, while others lie free in the cytosol. Processes of construction and destruction are constantly going on in cells—but in a highly ordered and integrated fashion, with each cellular structure specialized to carry on a certain activity. For example, all worn-out cell parts and invading organisms plus many imported materials end up in the **lysosomes**, the membrane-bounded vacuoles that act as the cell's garbage disposal and recycling systems. Inside the lysosomes, all large molecules are broken down into their smaller components by powerful enzymes. Fortunately, these enzymes function only in the innards of lysosomes; they would destroy the entire cell if they were released into the surrounding cytoplasm! Several dozen hereditary conditions called *lysosomal storage diseases* are known, each caused by the absence or malfunctioning of a specific lysosomal enzyme. Most of these disorders are very serious, involving skeletal and nervous system abnormalities, and some are lethal (Chapter 14).

Other tiny membrane-bounded structures called **peroxisomes** contain enzymes that use oxygen to detoxify certain substances, including ethanol. The hydrogen peroxide generated as one of the intermediates is then converted to water by another peroxisomal enzyme called *catalase*. Of those rare people who lack catalase owing to an autosomal recessive disorder called *acatalasia*, most show few if any ill effects. But the malfunctioning or absence of peroxisomes can cause very serious illnesses (Chapter 14).

The digestion products released from the lysosomes, as well as many other substances coming into the cell, are gradually degraded, releasing the energy needed for doing the work of the cell. An oxygen-requiring stage, called *cellular respiration*, takes place inside **mitochondria**. These miniature power plants—between 300 and 600 per cell—are bounded by two membranes and contain hundreds of mitochondrial enzymes.* The energy released is used to form new molecules, both large and small, and to drive many other metabolic reactions.

The **endoplasmic reticulum**, a compartment continuous with the outer membrane of the nuclear envelope, is a large and ever-changing network of tubules and cavities of various shapes and sizes. It is involved in the production of all membranes and secretory proteins, nearly all lipids (fatty substances), and certain complex carbohydrates. The *rough endoplasmic reticulum* is studded with ribosomes, whereas the *smooth endoplasmic reticulum* has none. Most proteins destined to be secreted by the cell have chains of sugars attached to them as they travel through the rough endoplasmic reticulum on their way to the Golgi complex.

The **Golgi complex** is the cell's "shipping agent," receiving, sorting, tagging, and directing many newly made proteins and membrane parts to their final destinations inside or outside the cell. It abuts the endoplasmic reticulum on one face (often near the nucleus) and the nuclear membrane on the other, and it is made up of saucerlike arrays of smooth membrane sacs. Enzymes in the Golgi complex modify or add sugar groups to the proteins and lipids passing through. The resulting molecules, surrounded by sections of membrane, form secretory or storage *vesicles* that later fuse with the plasma membrane and release their contents

---

* Mitochondria (and in plant cells, chloroplasts too) are partially self-replicating organelles. They contain some unique molecules that are much more like those found in prokaryotes than in eukaryotes. For these and other reasons, mitochondria are thought to have arisen from free-living, aerobic bacteria that invaded and were retained by primitive, anaerobic eukaryotic cells.

outside the cell. Alternatively, these molecules may become a structural part of the plasma membrane itself.

Enclosing the cytoplasm is the **plasma membrane**. In some kinds of cells its entire surface is fuzzy with slender cytoplasmic extensions (*microvilli*) that protrude into the surrounding space. The plasma membrane contains many specialized membrane proteins. It acts as the "gatekeeper" that controls which substances will pass in and out of the cell. This function is accomplished by means of specialized receptor proteins and by molecule-specific channels that pass specific ions and molecules through the membrane. *Cystic fibrosis*, an inherited disease, results from defects in a chloride channel protein.

The plasma membrane "swallows" certain outside materials (such as fluids, large molecules, bacteria, and parts of degenerating cells) by engulfing them and pinching off internally to form vesicles of various sizes. The new vesicles may fuse with other intracellular structures (e.g., lysosomes), to be degraded, or they may be stored, or they may return to the plasma membrane and release their contents elsewhere. The release of ingested materials and the secretion of cell products occur through a reverse process whereby certain vesicles bud off from the Golgi complex and fuse with the plasma membrane. They then open to expel their contents from the cell surface.

CHROMOSOMES

Walther Flemming, a German cytologist, was among the first to devise ways of staining cells so that their internal features could be seen clearly. What he then observed in stained nuclei was "a very delicately interconnected basket-work of winding threads of uniform thickness" (Flemming 1879). He described cyclical changes in these threads, which were later named **chromosomes** (Greek *chroma*, "color"; *soma*, "body"), and recognized their central role in cell division.

A few years earlier, Friedrich Miescher had developed ways of separating intact nuclei from the cytoplasm that surrounds them. From these isolated nuclei he extracted an acidic substance, rich in both phosphorus and nitrogen, that was "not comparable with any other group at present known" (Miescher 1871). He called it *nuclein* and speculated that some day the several types of nuclein would be considered as important as the proteins.

These two lines of research, one on the structure and behavior of chromosomes and the other on the chemistry of the hereditary material, did not actually converge until around 1950. Now they are very closely interwoven. In this section we present a brief overview of what is known about the physical and chemical nature of chromosomes a century after the inquiries began.

Only in dividing cells do chromosomes appear as compact bodies (Figure 2.2). At other times, when cells are going about their metabolic business, the chromosomes exist as ultrafine threads of **chromatin** dispersed throughout the nucleus. Despite these striking changes in form, which result from different degrees of compaction, each chromosome retains its structural continuity and individuality throughout successive cell cycles. With few exceptions, the number of chromosomes is the same for all somatic cells in an organism and indeed within a given species. Humans, for example, have 46 chromosomes; dogs, 78; carp, 104; red ants, 48; and fruit flies, 8. Among plants, potatoes have 48 chromosomes; broad beans, 12; and white oaks, 24. The relation, if any, between chromosome number and size or complexity of the organism is not clear.

(A) Chromosomes during
cell division (metaphase)

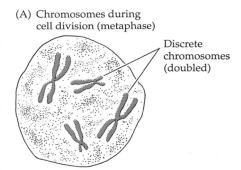

Discrete chromosomes (doubled)

(B) Nondividing cell

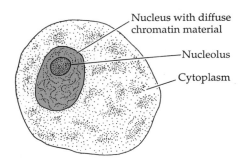

Nucleus with diffuse chromatin material

Nucleolus

Cytoplasm

Figure 2.2
(A) Dividing and (B) nondividing cells. These drawings show the chromosomal material as it would appear in a light microscope; the cells are not drawn to exactly the same scale. It is assumed that this species has two pairs of chromosomes.

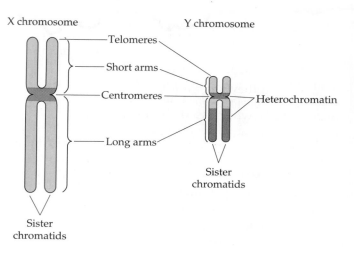

Figure 2.3
Drawing of human X and Y chromosomes during division to show the heterochromatin (color) and the euchromatin (gray). Most genes that control the production of proteins are located in euchromatic regions. The chromosome tips of both arms are called telomeres.

Gross Structure of Chromosomes

During cell division, individual chromosomes can usually be identified by size, shape, and banding patterns produced by various biological stains (Figure 2.3). These properties remain constant from cell generation to cell generation. Every chromosome has an indented region, the **centromere**, which is necessary for movement during cell division. The position of the centromere is constant for a given chromosome, dividing it into two **arms** of specific lengths. The chromosome tips are known as **telomeres**. Some staining procedures reveal unique **banding patterns** for each chromosome; others show that each centromere is flanked by darkly staining blocks of highly condensed material called **heterochromatin**. Smaller regions of less condensed heterochromatin are dispersed throughout all the chromosomes. Heterochromatin contains few known genes, and its function is not well understood. The more lightly staining material, called **euchromatin**, contains most of the genes. Surprisingly, however, these genetically active regions of DNA make up only about 10% of the total DNA in most cells.

Chromosomes duplicate themselves before the onset of cell division. After replication, each chromosome consists of two identical copies known as **sister chromatids**, lying side by side and still connected at the centromere region. After cell division begins, these tremendously long and threadlike chromosomes gradually shorten and thicken, becoming more rodlike (as shown in Figure 2.3). During their most condensed period (a brief stage called *metaphase*), the chromatids are partly separated; consequently, each chromosome—depending on the relative lengths of its arms—takes on a somewhat X-like or V-like shape (as seen also in Figure 2.8).

By counting and analyzing the shapes of these metaphase figures, cytologists can describe each species by its distinctive array of chromosomes—known as the **karyotype** (Greek *karyon*, "nut" or "nucleus"). Deviations from the normal karyotype are often associated with abnormalities or even death of the organism (Chapters 10 and 11). In most species the chromosomes of all cells (except eggs and sperm) occur in pairs, the two members of a pair looking alike and carrying the same genes. For example, human cells have 23 pairs; carp cells, 52 pairs; and white oak cells, 12 pairs. The normal number of chromosomes in a nucleus is called the **diploid** (or *2n*) number. During the production of gametes, this number is halved in such a way that an egg or sperm contains only one member of each chromosome pair; such cells are said to contain the **haploid** (or *n*) chromosome number. This special reduction process prevents the dou-

bling and redoubling of the chromosome number in successive generations. Instead, the union of two haploid (*n*) gametes restores the diploid
(2*n*) chromosome number in the zygotes of each generation.

Some early cytologists suspected that chromosomes played a role in
heredity, but they had no firm evidence for this until after Mendel's work
was uncovered. The 1903 **chromosome theory of heredity** described the
parallel behavior of genes and chromosomes and visualized the chromosome as a structure carrying many genes in linear order.

Sex Chromosomes

In many plants and animals the sexes are separate, and the mode of
inheritance of certain traits can differ according to which parent possesses
them. For such organisms, the biological basis of sex determination was a
mystery until research on insects and other animals showed that sex is,
with few exceptions, an inherited trait. Specifically involved is a pair of
identical-looking **sex chromosomes**, called the **X chromosomes**, or two
different sex chromosomes, one X and one **Y chromosome**. All the remaining chromosomes are called **autosomes** (Greek *autos*, "same"). Human males, for example, have an XY pair of sex chromosomes and 22
pairs of autosomes.

How sex is determined varies from one organism to another, but a
common pattern is the one found in humans. Individuals with two X
chromosomes are female, and those with one X chromosome and one Y
chromosome are male. As is the case with many organisms, human X and
Y chromosomes are structurally and functionally quite different. The medium-sized X is about 2.5 times as long as the tiny Y chromosome. Because
of differences in chromosomal organization, the two also look quite different when stained and viewed under a microscope. In later chapters we
discuss in some detail the sex chromosomes, sex determination, inheritance patterns of the genes on sex chromosomes, and the syndromes
associated with sex chromosome abnormalities.

The Components of Genes and Chromosomes

In 1953, half a century after the chromosome theory of heredity was
formulated, James Watson and Francis Crick, researchers at Cambridge
University, proposed a structural model for the stuff that genes are made
of—**deoxyribonucleic acid (DNA)**. Two other workers, Rosalind Franklin
and Maurice Wilkins, at the University of London, had been analyzing
the crystalline structures of DNA. Their X-ray photographs led Watson
and Crick to deduce that the molecule consists of two chains wound about
each other to form a **double helix**. This structure can be crudely pictured
as a twisted ladder (Figure 2.4), the vertical support or **sugar-phosphate
backbone** of each chain being a monotonous series of alternating sugar
and phosphate groups. Projecting inward from each sugar, to form the
horizontal rungs, are pairs of **bases**.

Four different bases occur in DNA: two larger ones, called **purines**,
and two smaller ones, called **pyrimidines**. The purines are *adenine (A)* and
guanine (G); the pyrimidines are *thymine (T)* and *cytosine (C)*. Each rung of
the DNA ladder consists of a purine connected to a pyrimidine: either A
with T (forming an AT or TA base pair) or G with C (forming a GC or CG
base pair). Any other base pair combination leads to instability in the
regular overall structure of the double helix. Thus, if you know the sequence of bases on one chain of the helix, you automatically know the
sequence of bases on the opposite chain. The two chains are said to be
complementary.

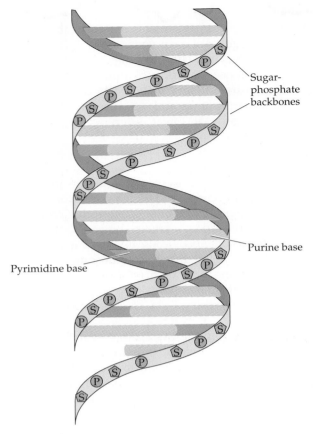

Sugar-
phosphate
backbones

Purine base

Pyrimidine base

Figure 2.4
A highly simplified view of the DNA double helix. Two twisted sugar-phosphate back-bones are linked together by purine-pyrimidine base pairs.

The particular sequence of bases along the length of the molecule can and does vary enormously. Indeed, it is the lengthwise base sequences that provide all the genetic information within DNA molecules. Like any written language, a sequence is a coded message. A *gene* is a segment of the DNA molecule with a unique sequence of thousands of bases that usually tells the cell how to make a particular protein. Later chapters, especially Chapters 12 and 13, provide details about the nature of genes and how they function.

After Watson and Crick's discovery, it became immediately obvious that the general properties of DNA could explain the behavior of genes. First of all, both DNA and genes can scrupulously *reproduce* themselves. Although genes and DNA usually copy themselves perfectly, both are known to exhibit rare mistakes. These **mutations**, if not repaired, are perpetuated when copies are made. Thus arises new material for evolution.

In addition to replicating and mutating, genes and DNA must *store the information* that is transmitted from mother to daughter cells or from parents to offspring. The stored information in DNA resides within the sequence of millions of bases along the DNA molecules—a developmental blueprint that is converted into each organism's unique set of characteristics through the process of protein synthesis. A gene is said to be *expressed* when its information is used by a cell to make a specific protein, this protein consisting of a special sequence of units called *amino acids*. How it all happens will be described in Chapter 12.

Fine Structure of Chromosomes

Eukaryotic chromosomes consist of more than DNA. Their other main ingredients are collections of proteins known as histones and nonhistones. The best understood of these are the histone proteins.

Histones are small molecules that regulate the degree of chromosomal compaction of the DNA double helix during cell division. They also play a role in regulating genes, that is, turning them on and off as gene products are needed in different cells at different times. Produced in the cytoplasm, histones immediately migrate into the nucleus and become tightly bound to newly replicated DNA. This DNA-histone complex forms the elementary structural unit of chromosomes. Histones have scarcely changed throughout evolution; indeed, those taken from organisms as different as calves and pea plants are virtually identical! Such extreme conservation of sequence probably means that (1) histones have the same function in all organisms and (2) almost every amino acid in the molecule is functionally important.

Nonhistone proteins (NHPs) are a mix of several hundred kinds of chromosomal proteins that are not histones. Unlike histones, NHPs seem to vary, in amount and composition, with the degree of metabolic activity in different cell types, or in the same cell at different times, or in differentially active regions of the same chromosome. Found in this mix are enzymes, gene regulators, and structural components of the chromosomes.

Determining how all these molecules are organized and packed into functional chromosomes has for decades been an area of very active research, speculation, and debate. We now know that each chromosome contains just one very long molecule of DNA. Stretched out, the total DNA in a single human egg or sperm nucleus would measure just under 1 meter (1 m = 3.28 feet).* During cell division, the packing process involves repeated coiling and folding of the basic structural unit of the chromosome, the DNA-histone fiber. This process is very hard to study because, as shown in Figure 2.5, whole chromosomes under the electron microscope look a lot like piles of spaghetti. Although it is impossible to follow the incredibly tortuous path of a single DNA-histone fiber through its entire length, much has been learned from electron microscopic studies combined with biochemical and physical analyses.

But whatever the approach, all researchers must struggle with the following problem: After chopping up and chemically treating chromosomes (or other cell parts), can they be sure that what they finally analyze in the test tube or under the microscope bears a reasonably close resemblance to what exists in functional cells? Might their treatments change or destroy some of the cell constituents they hope to study or, conversely, induce the formation of new constituents (artifacts) that simply do not exist in the native material?

The elementary chromatin fiber in all higher organisms is made up of repeating units called **nucleosomes** (Figure 2.6). Each nucleosome includes a total length of about 200 base pairs of DNA. When organized into nucleosomes, a DNA double helix undergoes a sevenfold shortening or compaction to form a fiber 11 nanometers (nm) in diameter. This fiber is further coiled and folded to form the 30-nm-wide chromatin fiber shown in Figure 2.5. Then additional levels of coiling and folding yield a human metaphase chromatid whose DNA double helix has been compacted 8,000- to 10,000-fold.

Certain nonhistone proteins also play an important structural role in chromosomal compaction. If all histones are removed from tightly con-

---

* See Appendix 1 for some common measurements.

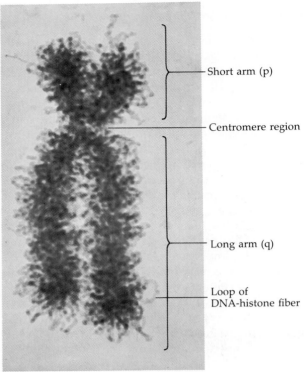

Figure 2.5
Electron micrograph of human chromosome 12 during division. Note the two chromatids joined at the centromere (constricted region). Each chromatid consists of a single DNA-histone fiber that is coiled and folded. The thinnest fiber seen in this photo is about 30 nm wide. Shown about 25,000 times actual size. (From DuPraw 1970.)

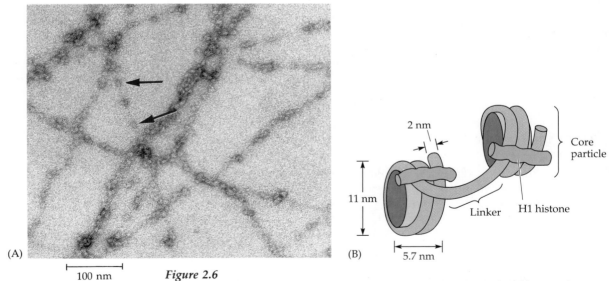

Figure 2.6
Nucleosome structure. (A) Chromatin from chicken blood cells. In this electron micrograph (×150,000), individual nucleosomes (arrows) look like tiny, clear beads on a string. (Courtesy of Christopher Woodcock.) (B) Two nucleosomes. A 146-base-pair length of DNA (color) is wrapped 1.8 times around a "spool" of eight histone molecules (two each of four kinds) to form each disk-shaped core particle. The remaining stretch of DNA, the linker, is associated with a fifth type of histone molecule called H1.

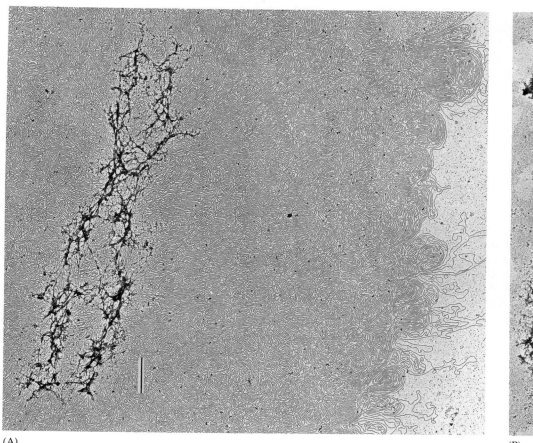

(A)

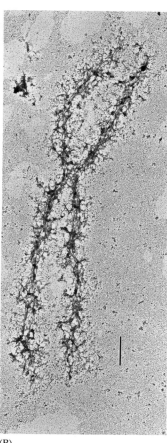

(B)

Figure 2.7
Human chromosomes, each with a dark-staining central scaffold of nonhistone protein outlining the shape of two chromatids joined at the centromere region. (A) After all histones are removed from this chromosome, an immense and tortuous maze of DNA loops out from the scaffold. (B) With both histones and DNA removed, all that remains is the central scaffold. The bar in each photo represents 1 μm. (Courtesy of Ulrich K. Laemmli.)

densed human chromosomes, the DNA becomes greatly unraveled but still keeps its identifiable chromosomal shapes. This is because each histone-free chromosome contains a central **scaffold** made of some nonhistone proteins. The DNA is attached to this scaffold in loops (Figure 2.7). When the DNA is also removed from these histone-depleted chromosomes, the scaffolds remain intact. How these structures arise is not known, but researchers think that they persist in some form throughout the entire cell cycle.

HUMAN CHROMOSOMES

It would be unfair to suggest that cytologists before the 1950s were careless in examining human chromosomes; rather, the techniques then available were simply not up to the task. In 1923 the American cytologist Theophilus Painter published unequivocal drawings of 48 human chromosomes. He wrote that he was able to do this because, unlike previous workers, he used fresh material and because, in his own "modern era of cytology," he utilized better techniques of fixation, sectioning, and staining. Although he mentioned that the counting error "might well amount to one chromosome" and although he had previously written that the best preparations showed 46 chromosomes, Painter nonetheless settled on 48. The question of the number of human chromosomes thus seemed to be settled, and for more than three decades cytologists and geneticists deferred to Painter's drawings or to those of others who also found 48. Actually, even the best examples show that accurate counting was hampered by vague outlines and touching or overlapping chromosomes.

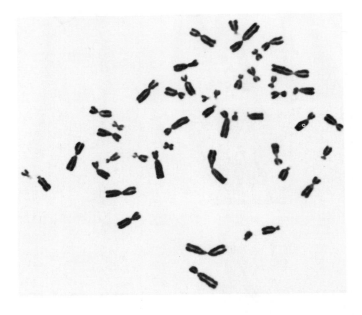

Figure 2.8
The chromosomes at metaphase (middle of cell division) from a white blood cell of a man (×2,000). The chromosomes are stained with Giemsa, which here colors all parts roughly the same. Each chromosome appears somewhat X-shaped because it has replicated, but the replicas remain attached at the centromere region. How many are there? (Courtesy of Irene Uchida.)

In 1956 Joe-Hin Tjio and Albert Levan, working with more advanced techniques, reported that they could find only 46 chromosomes in the lung tissue from four human embryos. Within months this finding was verified in other human cells. Tjio and Levan's experience teaches us that objective reexamination and reevaluation of "established facts" are very important for scientific progress. Figure 2.8, showing the chromosomes from a white blood cell, provides more evidence that the chromosome number in our species is 46.

Chromosome Banding

The original techniques for staining chromosomes used dyes that produced uniform coloration throughout each human chromosome. Thus, the homologues in a metaphase spread could be only crudely matched by size and shape. But starting in 1970, new staining procedures were devised in which the dyes produced **bands** of euchromatin interspersed with regions of partly condensed heterochromatin (Figure 2.9). Because each chromosome in a haploid set has a unique banding pattern, every pair of homologues can now be distinguished from other chromosomes of similar size and shape. Figure 2.10 shows chromosome 1 after preparation by several different banding methods.

The original banding methods could produce up to 320 metaphase bands in a haploid set of human autosomes. This technique allowed cytogeneticists to detect many new chromosomal anomalies in humans and to study the evolutionary relations among sets of normal chromosomes from various primate species. Later, some additional staining tricks expanded the power of such analyses many times, to up to 2,000 bands. The newer methods gave rise to many high-quality preparations—not only in the stage (called *metaphase*) during which chromosomes are most condensed, but in the preceding less-condensed stages (called *prophase* and *prometaphase*) as well (Figure 2.11). With this new **high-resolution banding** technique for examining prophase chromosomes, researchers were able to detect tiny chromosome defects that were previously unidentified in some patients with congenital disorders. Moreover, similar analyses have allowed investigators to identify chromosomal defects in the

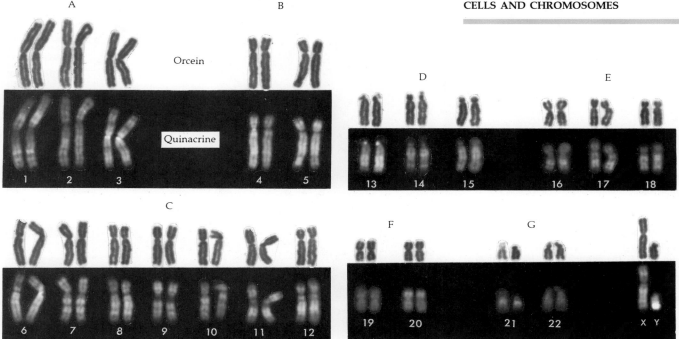

Figure 2.9
On the black background are quinacrine-stained human metaphase chromosomes arranged in homologous pairs according to size and similarities in their banding patterns. Above each pair, on the white background, are the same chromosomes stained with orcein after the quinacrine treatment. Although orcein is a good stain for giving defined chromosomal outlines, quinacrine or some other banding stain is necessary for matching up many homologous pairs. (Note: Human autosomes are arranged in groups A–G according to size and centromere position and are numbered 1–22 according to size. This pictorial representation of a person's chromosomes is called a karyotype.) (Courtesy of Irene Uchida.)

cancerous cells from many more patients than would have been possible with the older banding methods. Yet despite these improvements, scientists still do not know much about how banding patterns relate to chromosome structure:

> When it comes to mechanisms that would explain chromosome banding, there are many contradictory reports and very few clear answers. . . . It is clear that the answers to the questions of chromosome structure and condensation will not come from the folks who stain chromosomes and stare at them in the microscope, but they will come from the genome sequencing effort and studies in cell biology and histochemistry. (Ute Francke 1992)

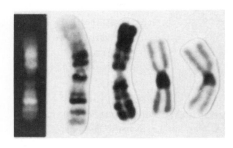

Figure 2.10
Human chromosome 1 prepared by four different methods. From left: Q-banding, G-banding, R-banding, and C-banding (two examples of C-banding, showing both homologues from a heterozygote). (From Therman and Susman 1992.)

25

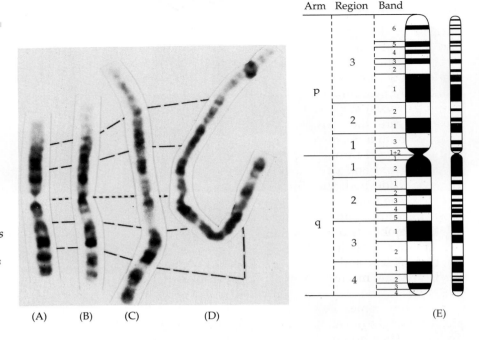

(A) (B) (C) (D) (E)

Figure 2.11

Human chromosome 1 as it looks at (A) mid-metaphase, (B) early metaphase, (C) just before metaphase, and (D) late prophase. In mid-metaphase the bands are fewer and thicker, while in progressively earlier stages the bands are more numerous and thinner. (E) In this diagram of chromosome 1, the chromatid on the left shows the mid-metaphase banding pattern, and the chromatid on the right shows the G-bands seen in late prophase. (From Yunis 1976.)

Fragile Sites

In 1977 Australian cytogeneticist Grant Sutherland described an interesting abnormality: a nonstaining constriction connected to a tiny knob at the tip of the long arm of the X chromosome (Figure 2.12). It was proposed that this **fragile site** is associated with heritable X-linked mental retardation in males. In the 1960s there had been a few reports of such X chromosome variants in families with retarded males, but they seemed to be rare and of limited significance. We now know, however, that the *fragile X syndrome* is, next to Down syndrome, the most common cause of mental retardation that can be specifically diagnosed (Chapter 7). About 20 heritable fragile sites have also been identified in the autosomes. These, however, seem to be correlated with certain kinds of cancer rather than with mental retardation.

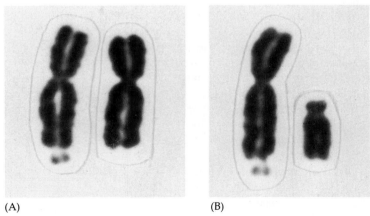

(A) (B)

Figure 2.12

(A) Sex chromosomes from a female with one normal X chromosome (on the right) and an X with a fragile site. (B) Sex chromosomes from a male with a normal Y chromosome (right) and an X with a fragile site. The fragile sites, located near the bottom of the X chromosomes on the left of (A) and (B), are the nonstaining regions that separate the tiny, stained tips from the rest of the chromosome. (From Richards and Sutherland 1992.)

Each method for preparing chromosome spreads has unique advantages and disadvantages. Since the techniques chosen depend on the type of analysis being undertaken, there is no single all-purpose protocol. Often, white blood cells are isolated from a blood sample and stimulated to divide in a culture medium. Treating the cells with *colchicine* (an extract from the autumn crocus) stops cell divisions at the metaphase stage, when chromosomes are the most highly condensed. The cells are then killed, dropped onto a microscope slide, and dried. The slides are then stained, and well-spread chromosomes are photographed. A karyotype is prepared by cutting out and pairing the chromosomes according to their banding patterns. If desired, the slide is destained and restained with a different dye, and the same nuclear spread is rekaryotyped, so the identical chromosomes can be compared by using different stains.

A major development is **automated analysis** of standard chromosome preparations. This involves methods for staining cells in culture (rather than on slides) and having the chromosomes automatically sorted and karyotyped by machines with special photometric and computer capabilities.

Chromosomes in the metaphase and immediately preceding (late *prophase*) stages of cell division have the unique property of appearing as doublets. They do not look this way during most of their cycle of metabolism and division. Before division they are virtually invisible, an indistinct tangle of vastly elongated threads; and in late division stages they appear as compact singletons. Only after their progressive shortening and thickening in the early part of cell division do the replicated sister chromatids of each chromosome become separately visible. Their continuing attachment at the centromere region provides an important means of identification, however. As shown in Figures 2.8 and 2.9, the resulting X and V shapes depend on the position of the centromere. The X-shaped chromosomes, with centromeres near the middle, are called **metacentrics**; the V-shaped chromosomes, with centromeres near one end, are called **acrocentrics** (Greek *akros*, "topmost" or "extreme"). Most chromosomal shapes fall somewhere in between and are called **submetacentric**.

Twenty-two of the 23 pairs of chromosomes occur in the cells of both sexes. These *autosomes* are numbered from 1 to 22 according to length (Table 2.1). The two remaining chromosomes, the sex chromosomes, are not numbered. The larger, metacentric one is the X chromosome, and the smaller, acrocentric one is the Y chromosome. Recall that male cells normally contain one X and one Y chromosome, whereas female cells contain two X chromosomes.

Although the absolute length of a given chromosome may differ somewhat from one metaphase spread to another, its **relative length**—that is, its length as a percentage of the total combined length of a haploid set of 22 autosomes—remains fairly constant from cell to cell. Thus, relative rather than absolute lengths are usually given. As shown in Table 2.1, the longest chromosome (chromosome 1) constitutes about 8.4% of the length of all the autosomes.

Note that the relative lengths of human chromosomes decrease with increasing chromosome number—except for the last two, 21 being shorter than 22. This misordering is due to a historical fluke involving the extra chromosome present in Down syndrome individuals, which was originally named 21 in the mistaken belief that it was the *next-to-shortest* autosome. Later, with improved techniques, cytogeneticists discovered that it was really the *smallest* autosome and should be called 22. But by this time the

Table 2.1
Characteristics of human somatic cell metaphase chromosomes.

Group	Number	Diagrammatic Representation	Relative Length[a]
LARGE CHROMOSOMES			
A	1		8.4
	2		8.0
	3		6.8
B	4		6.3
	5		6.1
MEDIUM CHROMOSOMES			
C	6		5.9
	7		5.4
	8		4.9
	9		4.8
	10		4.6
	11		4.6
	12		4.6
D	13		3.7
	14		3.6
	15		3.5
SMALL CHROMOSOMES			
E	16		3.4
	17		3.3
	18		2.9
F	19		2.7
	20		2.6
G	21		1.9
	22		2.0
SEX CHROMOSOMES			
	X		5.1
	Y		2.2

Source: Data from ISCN 1985.

[a] Percentage of the total combined length of a haploid set of 22 autosomes.

error was so entrenched in the scientific literature that it was easier to accept the inconsistency in relative length than to correct the number.

Before human autosomes were distinguishable by banding methods, they could only be arranged according to length and centromere position into seven groups (A–G). As shown in Table 2.1, within each of these groups the centromeric position is quite similar. For example, groups A and F contain the four most metacentric chromosomes; groups D and G consist of acrocentric chromosomes; and the chromosomes in groups B, C, and E are submetacentric.

In addition to the centromere, some chromosomes have other pinched-in sites called **secondary constrictions**. When appropriately treated and magnified, for example, the five pairs of chromosomes in groups D and G exhibit secondary constrictions near the tips of the short arms. These **nucleolar organizer regions (NORs)** give rise to the *nucleoli*, which contain the special substances destined to form ribosomes in the cytoplasm. The sites and number of NORs are constant for a given species; in human diploid cells there are ten. Nucleoli tend to fuse together, however, so that fewer than ten may be seen in an interphase nucleus. Group D and G chromosomes may also have tiny knobs of chromosomal material (called *satellites*) at the very tips of their short arms.

Within and among the chromosomes of normal individuals, the two members of a homologous pair of autosomes may often be distinguishable by slight differences in banding patterns. These observed variations were at first thought to be artifacts induced by the treatment procedures. But recent evidence indicates that some chromosome variations are inherited, unaccompanied by any known abnormalities. The best example of this is seen in the Y chromosome, whose long arm is unusually long in some males and unusually short in others.

It is often convenient to describe an individual's karyotype with cytogenetic shorthand rather than by a picture. At the simplest level this description is merely the total number of chromosomes and the sex chromosomal complement; thus, a normal female is designated 46,XX and a normal male 46,XY. To describe a site on a particular chromosome (Figure 2.13), we first list the chromosome number, then the arm (**p** designating the short arm and **q** the long arm), then the region number within an arm, and finally the specific band within that region. For example, 7p13 refers to chromosome 7, short arm, region 1, band 3—a narrow, light band. The fact that the bands appear much fuzzier in a photograph (Figure 2.9) than in a diagram (Figure 2.13) points up the need for great technical expertise in analyzing karyotypes.

To describe abnormalities in chromosome structure and number, cytogeneticists use standardized symbols. For example, a boy with typical Down syndrome is designated 47,XY,+21. He has 47 chromosomes, including normal X and Y sex chromosomes, with the forty-seventh chromosome being an extra chromosome 21—that is, three 21s altogether. The karyotype of a female with Turner syndrome is 45,X—which means that she lacks one sex chromosome. Other symbols and some interspecies comparisons will be presented in later chapters.

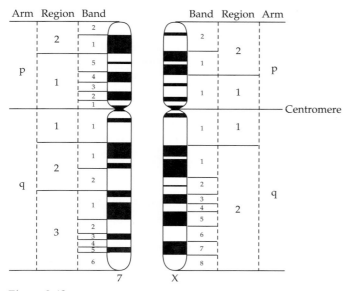

Figure 2.13
The quinacrine and Giemsa banding patterns (Q- and G-bands) of chromosomes 7 and X at metaphase. The dark bands are those that fluoresce brightly with quinacrine or stain darkly with Giemsa. These two chromosomes have almost the same length (p = short arm, q = long arm) and centromere position, but they can be distinguished by their different banding patterns. (From ISCN 1985.)

1. The cell is the underlying unit of structure in all living organisms. Eukaryotic cells have a nucleus, which contains the chromosomes, plus the surrounding cytoplasm, which contains many structural components specialized to carry out the varied activities of the cell.

2. Chromosomes, which carry hereditary determiners called genes, are the genetic link between generations. They occur in homologous pairs in body cells, one member of each pair derived from the female parent and the other from the male parent.

3. Chromosomes can best be seen during cell division, when they are tightly coiled. Each replicated chromosome at first consists of two identical chromatids joined by an undivided centromere.

4. DNA is a long molecule composed of two complementary chains wound about each other to form a double helix. Genetic information is encoded in the sequence of bases in DNA. This information is decoded in the cytoplasm to form specific proteins. When rare mutations occur, they are copied during DNA replication.

5. Chromosomes of higher organisms contain DNA, histone proteins, and nonhistone proteins. The basic chromosomal fiber is a DNA-histone complex, and chromosomal structure is provided by some nonhistone proteins.

6. Each chromosome can be unequivocally identified in metaphase or late prophase by its size, shape, and banding pattern. The human karyotype consists of 44 autosomes (22 pairs) plus 2 sex chromosomes (1 pair). Males are designated 46,XY and females 46,XX. In addition, the karyotypes of normal individuals sometimes have special features, such as variations in Y chromosome length.

KEY TERMS

acrocentric	double helix	metacentric	plasma membrane
autosome	endoplasmic reticulum	mitochondrion	receptor protein
banding pattern	euchromatin	nonhistone proteins (NHPs)	ribosome
base	fragile site	nuclear envelope	sex chromosomes
centromere	Golgi complex	nucleolar organizer region	sister chromatid
chromosome	haploid (*n*)	(NOR)	somatic cell
chromosome arm	heterochromatin	nucleolus	submetacentric
cytoplasm	high-resolution banding	nucleosome	telomere
cytoskeleton	histone	nucleus	X chromosome
deoxyribonucleic acid (DNA)	karyotype	peroxisome	Y chromosome
diploid (2*n*)	lysosome		

QUESTIONS

1. Distinguish between the terms in each of the following sets:

somatic cell—germ cell
cytoplasm—cytoskeleton
nucleus—nucleolus—nucleosome
plasma membrane—nuclear membrane
mitochondria—endoplasmic reticulum—Golgi complex
lysosome—ribosome—peroxisome
autosome—sex chromosome
haploid—diploid
chromosome—chromatid

centromere—telomere
euchromatin—heterochromatin
acrocentric—metacentric—submetacentric
X chromosome—Y chromosome
p—q
unbanded chromosome—banded chromosome
receptor protein—enzyme
DNA—histone protein—nonhistone protein

2. Suppose that along one-half of a double helix, a tiny segment of DNA includes the following sequence of bases (where A = adenine, C = cytosine, G = guanine, and T = thymine): -C-G-G-A-T-G-T-A-A-C-C-C-T-. What is the order of bases at the same nucleotide positions on the complementary half of this double helix?

3. Using "shorthand" symbols, give karyotypes for the following syndromes, which will be described in Chapter 10. (a) Males with Klinefelter syndrome have an extra X chromosome. (b) Females with Turner syndrome are missing one X chromosome.

4. Although the presence of an extra autosome is usually lethal before birth, some cases survive to birth. (a) In Down syndrome, there is an extra chromosome 21. (b) Trisomy 13, with an extra chromosome 13, is rare; it leads to gross malformations and early death. (c) Also rare and characterized by serious deformities and a very short life span is trisomy 18, caused by an extra chromosome 18. Describe these karyotypes using the standard shorthand symbols and assuming that (a) is female, (b) is male, and (c) is female.

5. If two breaks occur in a chromosome, the intervening section may be turned around 180 degrees before the broken ends heal, a process producing an *inversion*. Within the inversion, gene order is reversed and the location of the centromere may be altered (e.g., *a b c* ♦ *d* can change to *a* ♦ *c b d*). For each of the following hypothetical inversions, indicate (1) the new gene order, centromere (♦) position, and banding pattern; (2) whether the inversion would be detectable in unbanded spreads; and (3) whether it would be detectable in banded spreads.

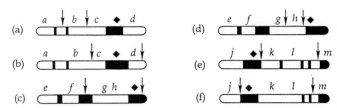

FURTHER READING

Much of the information on cell and chromosome structure was gleaned from three superbly written and illustrated books: Alberts et al. (1989), Darnell, Lodish, and Baltimore (1987), and Watson et al. (1987). De Duve (1984) gives an interesting, scaled-up cell tour. For very readable accounts

of the history of human cytogenetics, see Jacobs (1982) and Hsu (1979). Human chromosomes and cytogenetic techniques are described by Therman and Susman (1992), Wagner et al. (1992), and Macgregor and Varley (1988).

3 Gametes and Cell Division

Now that you are familiar with the general structure of eukaryotic cells and their chromosomes, we can describe how they reproduce themselves.

> In each of our bodies there are molecular choreographers programming a minuet in which chromosomes appear from obscurity, line up with their partners, separate, rejoin, and then disperse. That minuet is called the cell cycle, and it must proceed according to certain rules and cadences if we are to lead normal lives. In embryonic cells, the cycle must frequently proceed very rapidly, in some adult cells more slowly, and in some neural tissue not at all. If the cycle fails in growing cells, death results. If it goes incorrectly in mature cells, cancer is caused. (Koshland 1989)

GAMETES AND FERTILIZATION

Humans practice **sexual reproduction**. This scheme involves the alternation of a *diploid (2n)* generation of cells, which characterizes virtually all of the familiar plant and animal forms, with a *haploid (n)* generation of cells, the often microscopic eggs and sperm.

With the union of one haploid egg and one haploid sperm, another diploid life begins. This new cell, the **zygote**, immediately divides and redivides by a process called *mitosis*. In humans the zygote replicates about 50 times in about 38 weeks, forming a marvelously complex newborn of more than a trillion cells. Continued divisions occur throughout life, replacing the millions of body cells that are always dying for one reason or another. Within the sex organs, certain cells also undergo the special process of *meiosis* to produce eggs or sperm. Thus the human sexual cycle is complete.

The Mature Egg

Roughly five times the diameter of an average human somatic cell, the mature **egg** measures about 0.1 millimeter (mm), or 100 micrometers (μm), across (Figure 3.1). When ready for fertilization, the mammalian egg is actually suspended partway through cell division; thus it lacks a well-defined nucleus. At the stage shown in Figure 3.1, the egg's nuclear envelope has disintegrated into bits and pieces that will later be reused. The chromosomes, having already participated in one cell division and part of the next, are lined up on a **spindle**, awaiting the signal to complete a second cell division.

In the space just outside the egg membrane lies the tiny *first polar body*, a product of the first meiotic division (see page 45). It carries the same number of chromosomes found on the egg spindle, but very little cytoplasm. Enveloping both the first polar body and the egg is the **zona pellucida**, a dense, jellylike layer secreted by the egg and whose surface

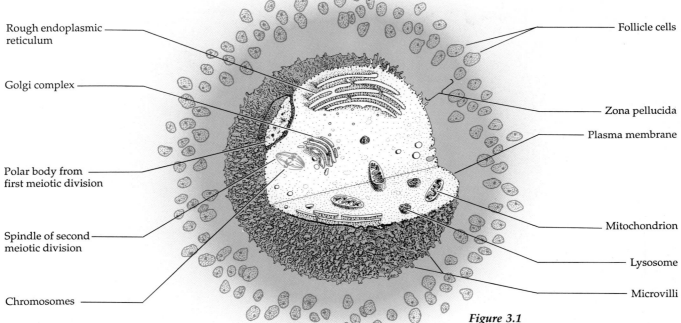

Rough endoplasmic reticulum

Golgi complex

Polar body from first meiotic division

Spindle of second meiotic division

Chromosomes

Follicle cells

Zona pellucida

Plasma membrane

Mitochondrion

Lysosome

Microvilli

Figure 3.1
Egg cell (idealized and sectioned) ready for fertilization. The various inclusions shown here are suspended in the cytosol. At this stage of development, the nuclear membrane has broken down. Drawn about 500 times actual size.

is thickly peppered with species-specific sperm receptors. This layer protects the egg and early embryo and prevents the entry of foreign sperm. Surrounding the zona are follicle cells, which by the time of fertilization have begun to decompose and separate from each other.

The Mature Sperm and Fertilization

The **sperm**, extremely tiny compared with the egg, is a cell stripped down to bare essentials—basically a *nucleus* enclosing the chromosomes, plus *mitochondria* for energy and a *tail* for locomotion (Figure 3.2). The total length of a human sperm cell is about 0.06 mm (60 μm), 93% of this being the tail. The head contains the nucleus with very densely packed DNA. Hugging the upper part of the nucleus is a thin **acrosomal cap** filled with several kinds of enzymes. A short neck, composed of a **centriole** surrounded by segmented columns, attaches to the long tail. The core of the tail (see the cross section in Figure 3.2) arose from a second centriole. Mitochondria are spirally wrapped around the core of the upper part of the tail, providing the energy needed for movement; and a tough, fibrous sheath covers most of the remaining length. A plasma membrane encloses the entire sperm.

One ejaculate from a fertile male contains at least 250 million sperm, of which up to 100 may find and attach to the material surrounding the egg (Box A). Sperm are swept to the fertilization site mainly by rhythmic contractions of the female reproductive tract, and they need to swim only during their final approach. Whether they are attracted to the egg or simply collide with it is not clear. In any event, some will attach to and penetrate the egg's coatings (Figure 3.3), thereby reaching the plasma membrane, but normally only one sperm enters the egg and forms a zygote.

The binding of the first sperm to the egg's plasma membrane triggers two events. First, electrical changes in the egg membrane and cytosol render it relatively impervious to additional sperm for a short while. As a second line of defense, the underlying *cortical granules* expel their contents into the surrounding space. This reaction alters the receptors in the zona pellucida and permanently blocks any more sperm from adhering to the egg.

33

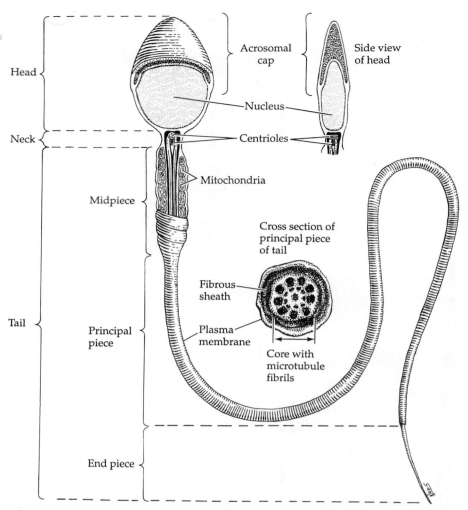

Figure 3.2
Human sperm cell (idealized and sectioned). After fertilization, all but the centrioles and the chromatin in the nucleus will disintegrate inside the egg cytoplasm.

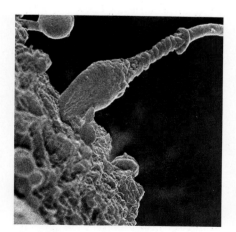

Figure 3.3
Human sperm touching the surface of an oocyte. (From Nilsson 1977. Copyright Lennart Nilsson.)

Following fusion of its plasma membrane with the egg's plasma membrane, the entire sperm is engulfed by microvilli and pulled into the egg's cytoplasm. The mitotic apparatus of the egg is suddenly activated and completes the second meiotic division. The egg chromosomes separate into two groups, one of which is pinched off to form the *second polar body*. The remaining egg chromosomes decondense, and a newly constituted nuclear envelope encloses them to form the **female pronucleus**.

Meanwhile, the sperm's tail, midpiece, and nuclear envelope disintegrate. Its chromatin material becomes more swollen and dispersed, and the special proteins that were present in the sperm nucleus are replaced by histones produced by the oocyte. Formation of a nuclear envelope around this chromatin completes the **male pronucleus**. The two pronuclei approach each other but do not fuse (Figures 3.4 and 3.11). Within each pronucleus, the chromosomes duplicate, condense, and (following breakdown of the two nuclear envelopes) arrange themselves on the first cleavage spindle. Maternal and paternal chromosomes are not enclosed within the same nuclear envelope until the two-cell state. In humans, the first cleavage division occurs 24 to 36 hours after fertilization.

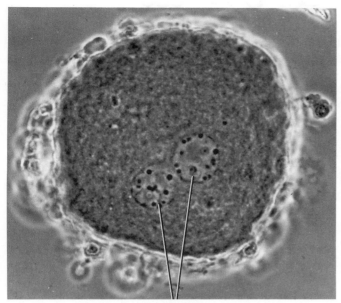

(A)

Male and
female pronuclei

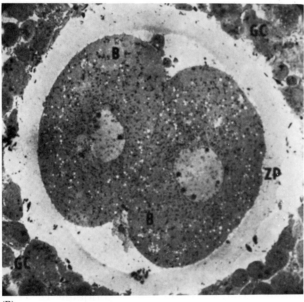

(B)

Figure 3.4
(A) The male and female pronuclei lying side by side, shortly after penetration of the sperm has triggered completion of the second meiotic division in the egg. The first mitotic division of this human zygote will follow. Shown about 300 times actual size. (From Dickmann et al. 1965.) (B) Human two-cell embryo. n, nucleus; zp, zona pellucida; c, cumulus cells outside zona. (From Dvořák et al. 1982.)

THE CELL CYCLE AND MITOSIS

Adult humans, like all multicellular organisms, are formed by the proliferation of a single cell first into 2 cells, then into 4, 8, 16, and so on, followed by gradual differentiation of the resultant cellular mass into tissues, organs, and organ systems. Whatever their developmental fate, all of these cells exhibit the same basic behavioral pattern: a period of

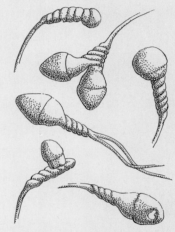
35

growth and metabolism, known as **interphase**, alternating with a period of cell division called **mitosis**. These two delicately integrated functions constitute the **cell cycle**, whose orderly course includes four distinct phases. The first three, G_1, S, and G_2, make up interphase; the final phase, M, is the mitotic division. Most cultured cells, including those of mammals, complete the entire cycle in 16 to 24 hours, of which only 1 to 2 hours are spent in division. This is summarized in the following diagram:

G_1	S	G_2	M
5–9 hours	6–9 hours	2–5 hours	1–2 hours
—I–N–T–E–R–P–H–A–S–E→			DIVISION

Each cell type seems to have its own precise protocol, which is not easily altered by experimental means.

Interphase

The first phase, **G_1**, is the initial time gap between cell "birth" at division and the beginning of DNA synthesis. All the cytoplasmic activities described in Chapter 2 are carried on during this period. By far the most variable in duration among different cell types, this phase may be too brief to measure in rapidly dividing cells, or it may last for weeks, months, or years in cells that seldom or never divide. G_1 in the "average" dividing cell takes 5 to 9 hours. Late in this period is the **R** (resting or restriction) **point**, at which time a cell may slow down or stop dividing if conditions are unsuitable. R is a "point of no return," too, because cells that pass it will almost always proceed through S, G_2, and M—regardless of local conditions. Also at the end of G_1, the cell's centriole pair (see page 38) begins to replicate. (Exactly how centrioles duplicate, and whether they contain their own nucleic acid, is still a mystery.)

Then a molecular signal at the end of G_1 triggers the onset of the **S** (synthesis) period, during which nuclear DNA replicates. Meanwhile, histones are being synthesized in the cytoplasm and transported through pores of the nuclear envelope into the nucleus. There they combine with replicating DNA to form new nucleosomes and chromatin fibers (Figure 3.5). After the DNA wraps around histone proteins to form nucleosomes, the nucleosome string is compacted into a 30-nm wide chromatin fiber. This fiber gets folded into **looped domains** that are attached to the chromosomal scaffold and stabilized by a network of nonhistone proteins. Besides being important in chromatin compaction, looped domains are thought to have an important genetic function. Each one contains a specific group of genes regulated as an independent unit and protected by the domain boundaries from regulation by surrounding domains.

How does DNA in a chromatin fiber manage to replicate? Rather than beginning at just one point on each chromosome, the unwinding of nucleosomes and replication of DNA starts at one specific spot (the **origin of replication**) in each looped domain and continues in both directions at a rate of about 50 base pairs per second. It takes 15 minutes to replicate the average domain. The mammalian diploid nucleus contains about 30,000 domains, all of which replicate once (and only once) within an S period. The initiations of replication are highly ordered; in general, euchromatin replicates early and heterochromatin replicates late in the S period. But certain regions may replicate early or late, depending on whether their tissue-specific genes are active or inactive. It is not well

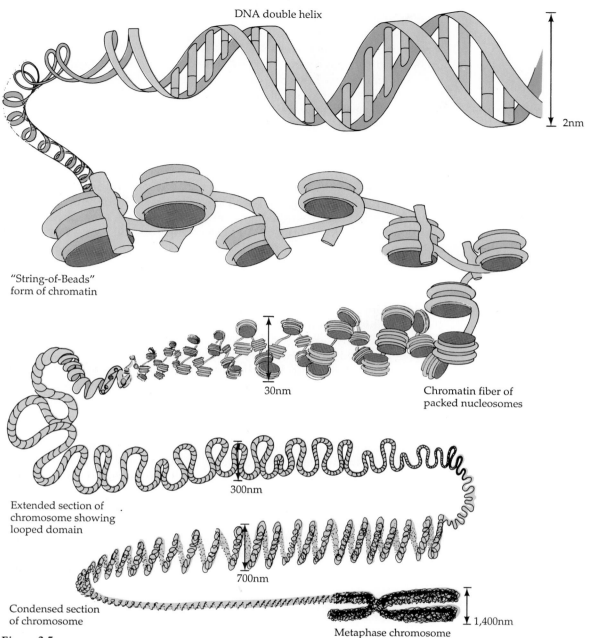

DNA double helix

2nm

"String-of-Beads" form of chromatin

30nm

Chromatin fiber of packed nucleosomes

Extended section of chromosome showing looped domain

300nm

700nm

Condensed section of chromosome

1,400nm

Metaphase chromosome

Figure 3.5
The different orders of chromatin packing assumed to give rise to a metaphase chromosome.

understood how the timing of replication is controlled and how DNA associated with nucleosomes replicates.

The two copies of each chromosome, when they eventually become visible during mitosis, are called sister chromatids. The centrioles also complete their replication during the S period, after which a second gap of 2 to 5 hours, G_2, occurs before the nucleus is somehow stimulated to undergo mitosis—during **M**.

Some cells, especially those of blood-forming and epithelial tissues, continue to divide regularly throughout the lifetime of the individual. Other cells, including those of nerve and muscle tissues, do not normally divide at all after birth. Such *noncycling* cells, when removed from the body, can often be induced to divide in tissue culture; thus, the capacity for cycling has not been lost. Conversely, cancerous cells are those whose cycling controls have run amok; as a result, they divide uncontrollably. If scientists can learn how the division cycle is regulated in normal cells, they might be able to use this knowledge to prevent the growth of cancer cells and to promote wound and tissue healing. Thus, in addition to its scientific value, the study of the cell cycle has immense practical value.

Mitosis

In the 1870s, with the development of special dyes and improved microscopes, European cytologists first discovered how cells divide. Since then, the process has been studied in ever more exquisite detail by focusing with more and more sophisticated microscopes on all kinds of cells and cell parts that have been treated with increasingly complicated physical and biochemical techniques. Yet, strange as it may seem to beginning students, we still do not fully understand the most basic events of cell division: exactly how chromosomes replicate and condense; how spindles form and operate; how chromosomes attach to the spindle; how daughter chromosomes separate and move to opposite poles of the spindle; how the cytoplasm divides; what limits the number of cell divisions in normal cells (as opposed to transformed or cancerous cells); and finally, how all these processes are controlled and integrated.

Along with these more familiar problems, we know very little about a less obvious but equally important aspect of normal differentiation called **programmed cell death**, whereby certain types of cells are eliminated at certain stages of development. This orderly interplay between cell proliferation and cell death controls the shape and functioning of organs in both the developing embryo and the adult organism.

All higher plants and animals follow, with some variations, the same general routine during nuclear replication and cell division. Although it is a continuous process, cytologists have for the sake of convenience separated it into several stages: prophase, prometaphase, metaphase, anaphase, telophase, and cytokinesis. These stages are shown in Figure 3.6.

The longest stage, **prophase**, begins as the nucleus swells and its chromatin fibers, previously stretched out to their ultimate interphase length, start to condense (Figure 3.6A). Heavy phosphorylation of H1 histones is associated with this process. The looped domains that are attached to nuclear scaffold proteins also gradually shorten or coalesce. Finally, each gossamer thread is densely packed into a short, thick, coiled chromosome whose looped domains radiate out in all directions from a central scaffold. Its two identical sister chromatids, replicated in the preceding S phase, become visible by late prophase. The dispersal of the nucleoli (containing ribosomal RNA and proteins) is completed by late prophase.

In the cytoplasm the two pairs of **centrioles**—replicated during the previous interphase and lying just outside the nuclear envelope—separate and start to migrate in opposite directions around the nucleus. As they do this, spindle microtubules arise from the diffuse material surrounding the centrioles. (A centriole pair plus this surrounding material is called a **centrosome**.) Most of these fibers will form the *spindle*. The remaining microtubules make up the two *asters*, whose role in cell division is not clear.

The next division stage, **prometaphase**, begins when the nuclear envelope starts to undulate in the polar regions (Figure 3.6B). Ruptures

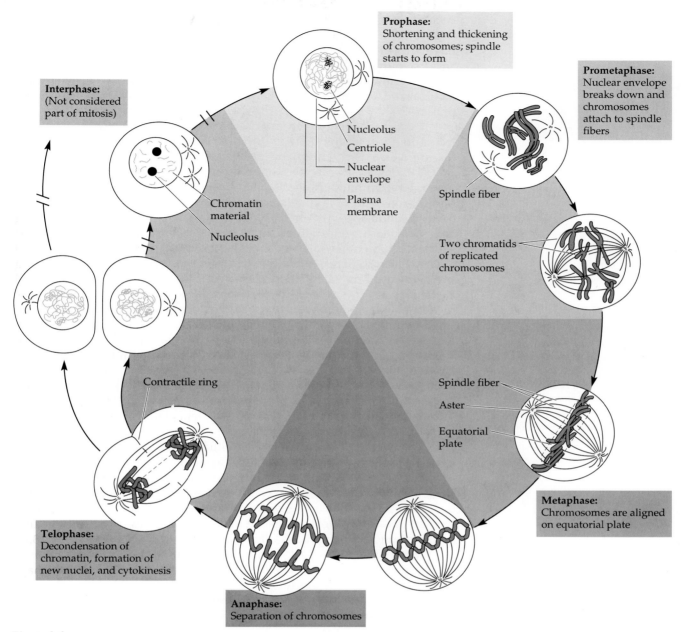

Interphase:
(Not considered part of mitosis)

Prophase:
Shortening and thickening of chromosomes; spindle starts to form

Nucleolus

Centriole

Nuclear envelope

Plasma membrane

Prometaphase:
Nuclear envelope breaks down and chromosomes attach to spindle fibers

Spindle fiber

Chromatin material

Nucleolus

Two chromatids of replicated chromosomes

Spindle fiber

Aster

Equatorial plate

Contractile ring

Metaphase:
Chromosomes are aligned on equatorial plate

Telophase:
Decondensation of chromatin, formation of new nuclei, and cytokinesis

Anaphase:
Separation of chromosomes

Figure 3.6
Chromosome behavior during mitosis. Only six chromosomes (three homologous pairs) are shown, with centromeres drawn as small circles rather than as constrictions. One member of each pair came from the female parent and one came from the male parent. Two nucleoli are shown; they and the nuclear envelope disappear in prophase and reappear in telophase. Note that the partitioning of this diagram does not reflect the relative amount of time needed for each mitotic stage, or for the much longer interphase period.

then appear, and the whole envelope rapidly breaks down into small vesicles that remain scattered just outside the developing spindle. Meanwhile, the centriole pairs have become stationed at opposite poles of the spindle, which has taken its final shape. The centromeres of replicated chromosomes, which have started to converge toward the midregion of the spindle, develop platelike **kinetochores**. These proteinaceous structures, which bind on one side to special sequences of centromeric DNA,

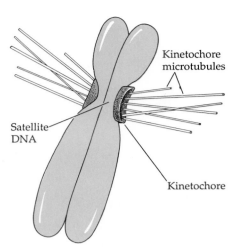

Kinetochore
microtubules

Satellite
DNA

Kinetochore

Figure 3.7
Diagram of the mammalian centromere, showing the primary constriction, the platelike kinetochore associated with the outer edges of the centromere, and kineto-chore microtubules attached to the kineto-chore. A special type of repeated DNA (called satellite DNA) and several special-ized types of centromere proteins have been detected in the primary constriction region. (From Willard 1990.)

become attached on their free surface to bundles of *kinetochore microtubules* (Figure 3.7). Each chromatid pair gets violently jerked back and forth as one chromatid's kinetochore attaches to a bundle that goes to one pole and its counterpart finally gets connected to the opposite pole. The key rule governing proper chromatid segregation is that the two members of a chromatid pair must never be connected to the same pole at prometa-phase.

By **metaphase** the chromosomes, now short and thick, have all col-lected midway across the spindle in a region called the **equatorial plate** (Figure 3.6D). Each chromatid pair is suspended between the two poles by the equal tension exerted on each centromere through the kinetochore fibers. Keep in mind, however, that during mitosis the two members of a homologous pair of chromosomes may end up close together or far apart on the spindle.

Anaphase begins when the centromeres of each sister chromatid pair suddenly jump apart. Then the kinetochore fibers appear to tow the attached centromeres, with the two chromosome arms trailing behind them, toward opposite poles (Figure 3.6E and F). At this point, each sister chromatid is now called a chromosome. Exactly how chromosomes get moved on the spindle is not well understood, but cytologists have iden-tified two clearly separable components. First, the spindle elongates, ap-parently as the polar microtubules of the two half-spindles lengthen and push apart in the equatorial region. Second, the kinetochore microtubules shorten, thereby moving the chromosomes to the separating poles. It has even been suggested that in certain organisms, kinetochores contain mo-lecular "motors" that help propel chromosomes along the microtubules. The net result is that each side of the elongated dividing cell gets one copy of each chromosome.

The events of **telophase** are essentially the reverse of what happens during prophase and prometaphase. Chromosomes begin to decondense, and nuclear envelopes reaggregate around the two identical groups of chromosomes. Chromosomes continue to decondense, becoming individ-ually indistinguishable. Nucleoli, produced by special ribosomal DNA (rDNA) sites on nucleolar organizer chromosomes, appear inside the two daughter nuclei (Figure 3.6G and H).

Meanwhile (actually starting in late anaphase), a **contractile ring** of actin and myosin forms just under the plasma membrane, in the equatorial plane midway across the long axis of the cell. Like a tightening belt, it gradually pinches in until the cytoplasm and spindle are split into two. This process of bisection, called **cytokinesis**, is independent of nuclear division. The cytoplasm in cells whose nuclei have been experimentally removed can still divide; conversely, some nuclei can undergo mitosis without any accompanying cytoplasmic division, thereby yielding multi-nucleate cells.

The end result of mitosis is the production, from one original cell, of two daughter cells with identical nuclei. *One interphase chromosome doubling (S period) has been followed by one division (M); so chromosome number and composition remain constant from one cell generation to the next.* The cytoplasms of daughter cells are usually about the same, too; but because there is no mechanism for the precise distribution of cytoplasmic components, differ-ences may arise between the two cells from a mitotic division. This vari-ation is particularly true of the early divisions of the fertilized egg (zygote) whose organelles and inclusions, such as yolk, may be present at different concentrations in various regions. In some organisms (such as frogs), such gradients result in the formation of daughter cells with greatly disparate cytoplasms.

All the cells descended from one zygote generally have the same nuclear genotype, yet they somehow manage to differentiate into many different types of cells, each with a specialized function. Among these specialized cells are the **germ cells**, which will divide by a special process called meiosis to give rise to eggs and sperm.

Cell Cycle Mutants

As far as we know, the cell cycle progresses by a series of ordered steps, each of which is usually required before the next step can proceed and is presumably catalyzed by one or more specialized proteins. Thus, the cell cycle must itself be under genetic control, and the *cell cycle genes* must be subject to occasional mutation. Although work on cell cycle mutants is difficult and slow, researchers have managed to isolate and study several hundred cell cycle genes in fruit flies, in yeast cells, and in mammalian cells in tissue culture.

Most cell cycle mutants isolated so far have been recessive and include suppressors as well as activators. Cytological studies have revealed abnormalities in chromosome organization and stability, chromosome condensation, spindle formation, DNA replication, chromosome segregation, nuclear division, or splitting of the cytoplasm. Other cell cycle mutants involve genes for proteins called growth factors and growth factor receptors.

Virtually all mitotic mutants exhibit similar defects in the comparable stages of meiosis (the cell divisions leading to the formation of eggs and sperm). In addition, geneticists have isolated many meiotic mutants that affect specialized functions, such as premeiotic DNA synthesis and various aspects of meiotic chromosome behavior.

What can we say about cell cycle mutants in humans? Not a lot yet. But some recent high-tech cloning experiments have shown that cell cycle mutants isolated from yeast cells are comparable to genes found in human cells! Indeed, the normal counterpart of a cell cycle mutant from a yeast species could function in human cells, and vice versa. Thus, it appears that all cells have very similar cycling controls. This very exciting prospect has stimulated renewed research efforts and new approaches to the genetic analyses of cell division.

MEIOSIS AND GAMETOGENESIS

Every egg that a human female will ever produce is already present, although immature and in a state of "suspended animation," at birth; thus any egg is as old as its host is at the time of the egg's release from the ovary. Although ovulated at the rate of one (or perhaps a few) per month for the 35 to 40 years between puberty and menopause, these eggs will not fully complete their meiotic divisions unless fertilized. It has been suggested that the increased risk of offspring with birth defects among older mothers may be somehow related to the longer storage time of their eggs.

Although estimates vary widely, a female produces perhaps several million immature eggs, of which only a few hundred are matured and ovulated during her lifetime. A male, on the other hand, probably makes and releases several trillion sperm in his lifetime. Sperm production does not begin until puberty, however, and usually continues for at least 50 years. From start to finish, a human sperm takes only about 3 months to mature, and it is released immediately.

Egg formation (**oogenesis**) and sperm formation (**spermatogenesis**) occur by a special pair of cell divisions known as **meiosis**, during which **diploid (2n)** cells give rise to **haploid (n)** cells. In humans this means that cells containing 46 chromosomes (23 pairs) produce gametes with just n = 23 chromosomes. Unlike the mitotic cycle, in which one chromosome replication is followed by one division, meiosis involves a single replication (**premeiotic S period**) followed by two divisions, called **meiosis I** and **meiosis II**. Along the way, the maternal and paternal members of each pair of chromosomes become separated from each other. This event results from a unique feature of meiosis, the pairing of homologous chromosomes, and from the exchange of parts during prophase of the first meiotic division. These processes, called *synapsis* and *crossing over*, usually ensure that the resultant daughter cells contain one member of each chromosome pair. Both meiotic divisions include the stages observed in mitosis, but by far the longest and most complex of these—occupying up to 90% of the total time—is prophase of the first division, or **prophase I**.

First let us follow the developmental fate of some primordial germ cells in a female. These cells appear in the undifferentiated gonadal tissue of the fetus during the fifth week of development. By the twelfth week, they have undergone repeated mitotic divisions to form daughter cells called **oogonia** in the developing ovaries of the females. Some of these oogonia continue dividing mitotically to produce more oogonia, but others begin to divide by meiosis. The behavior of the latter cells is illustrated in Figure 3.8. The first sign of change is the enlargement of an oogonium, which is then called a **primary oocyte**. Like the chromosomes of any cell in interphase, the 46 chromosomes appear diffuse and nearly invisible in stained cells.

Meiosis I

Because prophase I is such a long and important stage of meiosis, cytologists have divided it into five substages: leptotene, zygotene, pachytene, diplotene, and diakinesis. To understand meiosis, we must keep in mind that *maternal and paternal chromosomes are associated as homologous pairs during the first division.*

As a primary oocyte enters **leptotene** of prophase I, its chromosomes become visible as delicate threads that are indistinguishable from one another (Figure 3.8A). Although already doubled during the previous S period of interphase, they are not visibly double when viewed through a light microscope. In some species their *telomeres* (chromosome tips) are attached to the inside of the nuclear envelope. As in mitosis, the progressive shortening and thickening of chromosomes during prophase is due to folding and coiling of the fine DNA-histone fibers and to coalescence of the scaffold proteins.

Figure 3.8 ►
Chromosome behavior during the meiotic divisions of gametogenesis. In females, the smaller cells represent polar bodies; in males (nomenclature in parentheses), the four products of meiosis are the same in size and function. Only two sets of homologous chromosomes are shown, one member of each being shown in color and the other in black. It is convenient to assume that the color chromosomes were received from one parent and their black homologues from the other. Only one crossover is indicated, and the very brief prophase II is omitted. The nuclear envelope and nucleoli are also omitted. Note that (1) the segregation of maternal and paternal chromosomes at anaphase I could equally well have been both black chromosomes to one pole and both color ones to the other, and (2) each of the four end products contains one of the four chromatids that make up each tetrad.

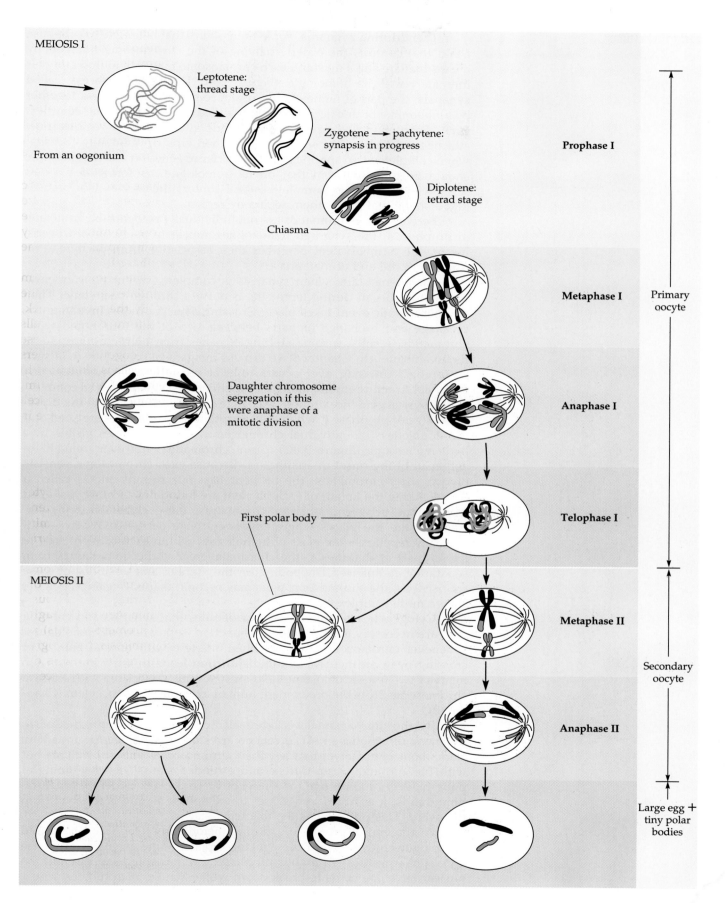

MEIOSIS I

From an oogonium

Leptotene: thread stage

Zygotene → pachytene: synapsis in progress

Diplotene: tetrad stage

Chiasma

Daughter chromosome segregation if this were anaphase of a mitotic division

MEIOSIS II

First polar body

Prophase I

Metaphase I

Anaphase I

Telophase I

Metaphase II

Anaphase II

Primary oocyte

Secondary oocyte

Large egg + tiny polar bodies

Then, during **zygotene**, something occurs that is unique to prophase I and that controls the orderly halving of the chromosome number. By means that are still a mystery, each chromosome manages to find and pair intimately with its homologue (Figure 3.8B). During this process, called **synapsis**, the pair of homologous chromosomes becomes bound together by a ribbonlike proteinaceous structure called the **synaptonemal complex**. Each synaptonemal complex begins to form at the telomere regions (now attached to the nuclear envelope) and then zips up gradually. Pairing, when complete, is so specific that if the order of genes on one chromosome should be different from that on its homologue, as sometimes occurs (Chapter 11), the two homologues will contort themselves like pretzels into a shape that aligns them region by region.

Because sister chromatids are not individually recognizable until later in prophase I, the synapsed homologues appear to be a pair of threads; hence, the term **bivalent** describes their association from pachytene on through to the end of metaphase I.

Pachytene starts when synapsis is complete; chromatin condensation, which began during leptotene, continues through pachytene. Then another meiotic event takes place. **Crossing over**, a process involving the breakage and exchange of parts between two of the four homologous chromatids, results in some shifting of genes from each chromosome to its homologue. (In Chapter 9 we discuss the genetic consequence of such swapping.) Crossing over occurs only at **recombination nodules**, tiny spherical structures found along the length of each synaptonemal complex.

Physical evidence of these exchanges is not observed until the **diplotene** stage of prophase I, when the synaptonemal complexes are shed and the doubleness of individual chromosomes finally becomes visible in a light microscope (Figure 3.8C). Sister chromatids, however, remain together until the onset of anaphase I. Because each chromosome consists of two sister chromatids, the bivalent has four strands and is called a **tetrad**. Along the length of each bivalent are found one or more X-shaped connections between nonsister chromatids; these **chiasmata** (singular, chiasma) are the places where crossovers have occurred. When—as if mutually repelled—the pairs of homologues within each bivalent start to move apart at **diakinesis**, the chiasmata prevent the homologues from separating completely. Depending on the number and positions of these connections, the chromosome pairs may become shaped like an X, a figure 8, or a more complex form.

Cytogeneticists have known for decades that synapsis and crossing over are necessary for the proper disjunction (separation) of homologues. Whenever chromosomal abnormalities interfere with normal pairing or crossing over, an increase in **nondisjunction** leads to offspring with too many or two few chromosomes. Down syndrome, associated with an extra chromosome 21, is the preeminent human example of this problem (Chapter 10).

By the time a female is born, all her oocytes have progressed to diplotene of prophase I. They remain arrested at this stage for anywhere from about one to five decades, their chromosomes still in bivalents but with the chromatin in a diffuse and extended state. Then beginning at puberty, every month one (or a few) of these oocytes undergoes **maturation**; that is, it enlarges, progresses to metaphase II, and is released from the ovary. This process has the following stages.

At the end of prophase I, during meiotic prometaphase, the paired homologues move toward the equatorial plate. Some events in prophase I are similar to those in mitotic prophase: The nucleoli disappear and the nuclear envelope starts to break down while spindle microtubules arise.

During **metaphase I** (Figure 3.8D), the spindle is completed. Lined up on it are 23 bivalents, their homologous chromosomes now held together by chiasmata, their centromeres ready for departure in opposite directions. During **anaphase I** (Figure 3.8E), the two members of each chromosome pair finally separate (disjoin), moving to opposite poles. Unlike mitotic anaphase—during which the sister centromeres separate, thus bringing about the separation of their two chromatids—we see here the *disjunction of whole chromosomes*, each still composed of two chromatids that are held together at their centromere regions. At **telophase I** (Figure 3.8F), the chromosomes at one pole, together with a small amount of cytoplasm, are extruded and pinched off to form the tiny **first polar body**. The remaining chromosomes plus the bulk of the cytoplasm constitute a **secondary oocyte**, which begins to divide again immediately.

Meiosis II

Meiosis II involves the separation of sister centromeres and thus the separation of the two chromatids of each chromosome. It is like a mitotic division, but with 23 rather than 46 chromosomes. During the very brief **prophase II**, a new spindle arises just under the surface of the egg. In **metaphase II**, the stage at which mature human eggs are expelled from the ovary, chromosomes are lined up in the equatorial region (Figures 3.8G and 3.9A) and remain so until the penetration of a sperm stimulates them to continue to **anaphase II** (Figure 3.8H). At the onset of anaphase

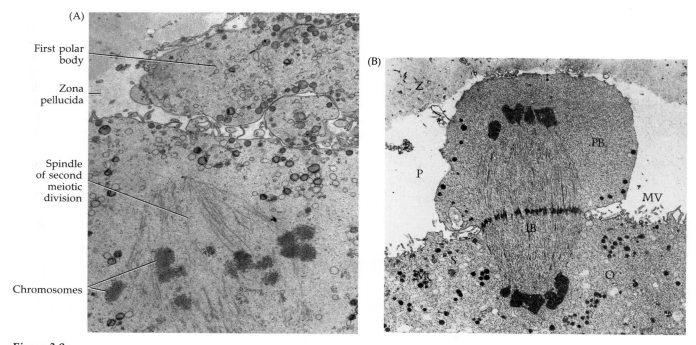

Figure 3.9
(A) An unfertilized human egg at metaphase II. The egg has just been released from the ovary. Its chromosomes are lined up on the equatorial plate of the meiotic spindle and will remain at this stage until the egg is fertilized. The first polar body (just above and to the right of the spindle) is a product of the first meiotic division. (From Baca and Zamboni 1967.) (B) Telophase II in a just-fertilized egg, with the second polar body (PB$_2$, top) being formed. The two groups of separated chromosomes lie at opposite poles of this second meiotic spindle. Cytokinesis has begun, and the dense material (interbody, IB) in the equatorial region of the spindle determines where it will be pinched in two. Microvilli, MV; ovum or egg, O; perivitelline space, P; smooth endoplasmic reticulum, S; zona pellucida, Z. (From Sathananthan et al. 1986.)

45

II, the cohesion of sister chromatids finally lapses, and they move to opposite poles. At **telophase II** (not shown in Figure 3.8, but seen in Figure 3.9B), one of the haploid chromosome groups is expelled, again with only a small glob of cytoplasm, to form the **second polar body**. By this time the first polar body might also have divided into two tiny cells. Thus, meiosis in the female leads to the formation of four haploid cells: one **egg** and three very small, nonfunctional polar bodies (Figures 3.8I and 3.10).

Overview

The whole process from oogonium to egg takes from 12 to 50 years (Figure 3.11). Its most important feature is the accurate reduction of chromosome number from diploid to haploid, but the genetic shake-up that occurs may also be important. Because the orientation of chromosome pairs on the metaphase I spindle is the result of a random process (i.e., it is equally likely that a maternal or a paternal centromere points toward a given pole), the chromosomes found at telophase I poles are mixtures of those originally derived from both the mother and the father. In each oocyte under-

Figure 3.10

The overall scheme of mitotic and meiotic divisions in the two sexes, from zygotes to gametes and back to zygotes. The haploid number of chromosomes found in the gametes is designated n. In males, the spermatogonia present at puberty divide about 20 times per year to provide a continuing supply of primary spermatocytes. In females, the oogonia present at birth differentiate without division into the primary oocytes.

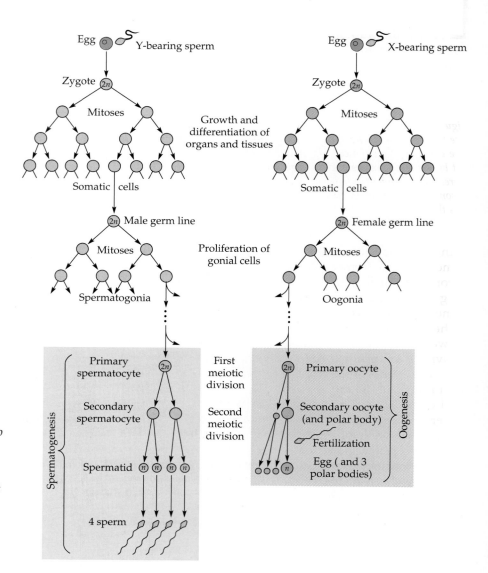

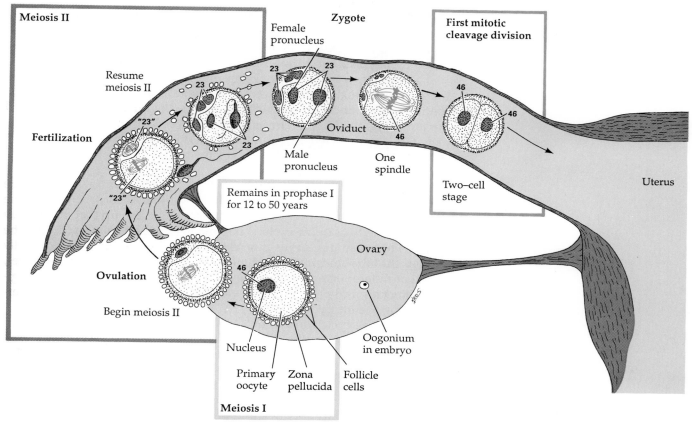

Figure 3.11

The location and timing of oogenesis, fertilization, and cleavage. The larger arrows indicate the completion of a cell division. The number of chromosomes per nucleus is shown, but between the two meiotic divisions the "23" refers strictly to the number of centromeres seen through the light microscope. Counting chromosomes in this "in-between" cell is somewhat ambiguous, because after meiosis I each chromosome consists of two chromatids that become chromosomes after meiosis II.

going meiosis, different combinations of maternal and paternal chromosomes will be formed. In addition to this **independent assortment** of chromosome pairs, the process of crossing over causes a further reshuffling of genes. Superimposed on the number of different whole chromosome combinations possible from a single individual (2^{23}, or over 8 million) is the additional variability produced by exchanges of chromosome parts between homologues. Hence, no two eggs (or sperm) produced by an individual are ever expected to be identical.

The events of meiosis in the male are the same as those in the female with regard to the nucleus, but they begin at puberty (rather than during fetal development) and proceed without interruption once started. After puberty, some diploid **spermatogonia** (descendants of primordial germ cells) in the walls of seminiferous tubules in the testes continually enter meiosis as **primary spermatocytes** (Figures 3.8 and 3.10). From one primary spermatocyte, meiosis I produces two **secondary spermatocytes** of equal size. Each of these in turn forms two **spermatids** after meiosis II. Division of the cytoplasm at telophase I and telophase II is equal, however, so that four haploid cells of equivalent size and functional ability are formed (Figure 3.10). There follows a complex developmental process, **spermiogenesis**, whereby each spermatid becomes an extremely specialized **spermatozoon** (sperm). Much cytoplasm is lost. Of the remainder, the Golgi complex produces the acrosome, mitochondria are incorporated into the midpiece of the tail, and centrioles form the neck and fibrils of the tail (Figure 3.2). Chromosomal proteins are replaced by protamines, which tightly pack the chromosomes into the sperm head.

Human spermatogonia develop into mature sperm in roughly nine weeks, of which about three weeks are spent in meiosis. Some men may continue to produce sperm throughout their lifetime, but in most males the fertilizing ability of sperm is greatly reduced by the time they are in their 60s. As depicted in Figure 3.11, fertilization restores the diploid chromosome number to the newly created zygote.

SUMMARY

1. In the human life cycle, a haploid egg and a haploid sperm, each with 23 chromosomes, unite at fertilization to make a diploid zygote with 46 chromosomes. Repeated mitotic divisions form diploid body cells that differentiate into all tissues and organs. Diploid oocytes and spermatocytes undergo meiosis to yield haploid eggs and sperm, thus completing the life cycle.

2. The cell cycle consists of a long interphase, during which protein synthesis and DNA replication occur, alternating with a short mitotic division, during which the replicas are distributed to daughter cells. Hormones and growth factors are important cell cycle regulators.

3. During the cell cycle, the dimensions of a chromosome vary, but its basic structure remains constant. One DNA double helix combines with histones to make a long string of nucleosomes, which then coils into a chromatin fiber. This fiber folds into a series of looped domains, which attach to scaffold proteins. During cell division the fiber continues to condense; by metaphase the length of the DNA helix has undergone about a 10,000-fold compaction.

4. Mitosis has four main stages. Chromosomes shorten and thicken in prophase and line up on the spindle in metaphase; at anaphase the centromeres separate and pull identical sister chromatids (now called chromosomes) to opposite poles; in telophase, chromosomes uncoil, nuclear envelopes re-form, and the cytoplasm splits. Thus, two genetically identical cells are formed from one. No pairing of homologous chromosomes occurs during mitosis.

5. The cell cycle consists of an ordered series of genetically controlled steps that can be studied by isolating cell cycle mutants. Many of the genes that control the cell cycle are probably the same or similar in all species.

6. The formation of haploid eggs and sperm by meiosis involves one replication and two successive cell divisions. Meiosis I results in the separation of pairs of homologous chromosomes, each individual chromosome consisting of sister chromatids held together in their centromere regions. Meiosis II results in the separation of sister chromatids.

7. A primary oocyte forms one egg and up to three nonfunctional polar bodies; a primary spermatocyte yields four functional sperm. Oocytes are in prophase I by the time of birth and in metaphase II at ovulation and will complete the second meiotic division only if fertilized. Spermatocytes do not begin to develop until after puberty.

8. Two key meiotic processes—the independent assortment of nonhomologous chromosomes and crossing over between homologous chromosomes—result in a great reshuffling of gene combinations.

KEY TERMS

anaphase	germ cell	pachytene	sexual reproduction
bivalent	independent assortment	primary oocyte	spermatid
cell cycle	interphase	primary spermatocyte	spermatogonium
centriole	kinetochore	programmed cell death	spindle
centrosome	leptotene	prometaphase	synapsis
chiasma	meiosis	pronucleus	synaptonemal complex
crossing over	meiosis I and II	prophase	telophase
cytokinesis	metaphase	R point	tetrad
diplotene	mitosis	S	zygote
equatorial plate	nondisjunction	secondary oocyte	zygotene
first polar body	oogenesis	secondary spermatocyte	
G_1 and G_2	oogonium	second polar body	

QUESTIONS

1. Review oogenesis in human females by listing the meiotic stages that occur during each of the following periods: embryo/fetus until birth; birth until start of oocyte maturation; oocyte maturation until ovulation; fertilization until first cleavage division. Give the amount of time that elapses for each period.

2. Explain why a metacentric chromosome has an X shape at mitotic metaphase and a V shape at mitotic anaphase. Explain why an acrocentric chromosome has a nearly V shape at mitotic metaphase and a sort of J shape (or nearly rod shape) at mitotic anaphase.

3. Consider these metaphase homologues:

(a) After one division, the two daughter cells had the following chromosomes. Was the division mitosis or meiosis?

(b) After one division, the two daughter cells had the following chromosomes. Was the division mitosis or meiosis?

(c) What would the homologous set look like (diagrammatically) before and after the second meiotic division?

4. Consider a spermatogonium with three sets of homologues: 1, 1′; 2, 2′; X, Y. List, in a systematic fashion, the eight different kinds of sperm that can be formed, considering the three pairs together and assuming no crossing over within the intervals we are following. Choose one of these possibilities and reconstruct the meiotic divisions that produced it.

5. Consider two sets of homologues: 1, 1′; 2, 2′. (a) If a secondary oocyte contains the centromeres of 1 and 2, what does the first polar body contain? (b) If an egg cell contains the centromeres of 1 and 2, what does the second polar body contain?

6. During interphase and before DNA replication, a diploid human cell contains 46 chromosomes, and the term *chromatid* is inapplicable. How many chromosomes and chromatids should be visible in this cell at metaphase? (Recall that at anaphase, chromatids become the chromosomes of incipient daughter cells.)

7. How many human chromosomes and chromatids (if applicable) are present at these stages of meiosis: (a) a spermatogonium before the S period? (b) a primary spermatocyte at metaphase I? (c) a secondary spermatocyte at metaphase? (d) a spermatid? (e) a sperm?

8. Among spontaneously aborted human fetuses, many are triploid ($3n$), and some are tetraploid ($4n$). How many chromosomes do these fetuses have in their somatic cells?

9. How many chromosomes has a mule whose mother (a horse) had 64 and whose father (a donkey) had 62?

FURTHER READING

The literature on chromosomes, cells, and cell division is vast. Much of the information in this chapter was gleaned from three superbly written and illustrated books: Darnell, Lodish, and Baltimore (1990), Alberts et al. (1989), and Watson et al. (1987). On gametes and fertilization, see articles by Wassarman (1987, 1988), Grobstein (1979), and Epel (1977, 1980) and a book by Longo (1987).

Various aspects of the cell cycle and cell division are described in Murray and Kirschner (1991), Chandley (1988), Stewart (1990), Manuelidis (1990), Strange (1992), Prescott (1987), Baserga (1981), and Mazia (1974, 1987).

4

The Rules of Inheritance

The Austrian monk Gregor Mendel experimented for nine years to confirm theories of inheritance that—by some remarkable insight—he might already have formulated. Far from being a shy recluse who labored in secrecy and obscurity, Mendel was a broadly trained, energetic, and highly respected scientist and teacher. He was also a chairman of the Moravian Mortgage Bank, a founder of the Austrian Meteorological Society, and very active in the agricultural societies organized by local sheep, apple, and grape growers to further their interests in animal and plant breeding and to keep the city of Brünn a thriving commercial center. Mendel tackled the problem of heredity at the suggestion of his superior, Abbot Napp, an enlightened administrator who fostered scientific research at the monastery in Brünn and who even had a greenhouse built for some of Mendel's peas (Orel 1984). Despite their confidence in Mendel and their interest in the hybridization problem, however, his colleagues failed to grasp the significance and generality of his findings (Box A).

MENDEL'S FIRST LAW: SEGREGATION OF A PAIR OF ALLELES

In this chapter and the next one, we answer the first and most basic question in transmission genetics: What are the rules for predicting how simple traits are passed on to succeeding generations? Gregor Mendel solved the basics of this problem of heredity in 1865, but nobody realized it until 1900. His work, which marked the beginning of quantitative biology, still stands as a model of careful experimental design. Drawing on his training in mathematics, physics, and chemistry, Mendel then applied mathematical principles to the analysis of his biological studies. He noted the different classes of offspring that resulted from specific matings, as well as how many individuals occurred in each class. He calculated the ratios among these classes and showed that they fit the terms of simple algebraic expansions. He recognized that "the greater the numbers, the more are merely chance effects eliminated" (Mendel 1865).

Mendel's experiments were designed to test a specific hypothesis about inheritance. Rather than worrying about how *all* traits are inherited in *all* organisms, he focused on the inheritance of easily distinguishable forms of a *few* traits in a *single* organism—the garden pea. This plant was chosen for three reasons: (1) It had a number of clearly defined variations, such as tall versus short plants, round versus wrinkled seeds, inflated versus constricted pods, and white versus gray seed coats. (2) Its flower structure, tightly surrounding both male and female parts, normally guarantees *self-fertilization*. But *cross-fertilization* can be arranged by removing the male parts from each flower before the pollen matures and later dusting the female parts with pollen from a different line. (3) The hybrids between two lines are as fertile as the parental lines, so large numbers of progeny can be studied in every generation.

Mendel started with **pure-breeding** lines, in which all the progeny from self-fertilization resemble the parents, generation after generation. A pure-breeding line has little or no genetic variation within it. When making crosses between two different lines, he always set up **reciprocal matings**: females of line A mated with males of line B, as well as males of line A mated with females of line B. In this way he could see whether the inheritance of a trait was influenced by the sex of the parents. Mendel was careful to keep the plant generations separated from each other and to study the descendants of each hybrid line through at least four generations of self-fertilization. He seemed to recognize the importance of keeping environmental conditions as constant as possible. The need for every one of these precautions is obvious today, but apparently was not understood in Mendel's time.

Mendel's Experiments

With each of seven pairs of traits, Mendel did the same series of experiments and got the same kinds of results. Here we present data for only one pair: *tall* plants (about 200 centimeters, or 6.5 feet) versus *short* plants (about 30 cm, or 1 foot). (We now know that this difference in height is due to the presence or absence of a growth hormone.) All the crosses are diagrammed in Figure 4.1. Each generation represents one year's work.

First, Mendel made many reciprocal matings between a pure-breeding tall line and a pure-breeding short line. These make up the **parental generation (P)**. Because the parents differed from each other in just one characteristic, the mating is called a **monohybrid cross**. The progeny of these matings, called F_1 **hybrids**, were all tall. Thus, Mendel reasoned, the tall plant character must be **dominant** in expression to the short plant character, which was designated **recessive**. For the other six pairs of traits, likewise, only one of the two alternative forms was expressed in the F_1 hybrid.

Next, Mendel allowed the F_1 hybrid plants to self-fertilize, collected the resulting seeds, and sowed them the following year. Among this F_2 **generation**, 787 plants were tall and 277 were short, a ratio of about 3 to 1 (3:1). Comparable F_2 ratios were obtained for the other six pairs of traits. No plants of medium height, or with any characteristic intermediate between the two parental lines, ever appeared.

The third step was to again allow self-fertilization, sow the seeds from individual F_2 plants, and look at the F_3 generation. The plants with the recessive trait were always found to be pure-breeding: short plants yielded only short progeny. But the progeny of the F_2 plants with the dominant trait were of two types. About 1/3 of the F_2 tall plants were pure-breeding, giving rise to only tall plants after self-fertilization. But the remaining 2/3 of the tall F_2 plants, after self-fertilization, yielded both tall and short progeny in a ratio of 3:1. Thus, Mendel concluded, they must be hybrids like the F_1. As shown in Figure 4.1, he continued this procedure for several generations and found that this pattern repeated in all cases.

These experimental results did not fit prevailing ideas about heredity, which held that the progeny of two different lines would all be intermediate in appearance as a result of a permanent blending of their characteristics. According to this hypothesis, the F_1 plants should have been all of medium height, rather than all tall. From the blending hypothesis it also followed that the original parental traits (in this case, tall and short) should never be recovered in future generations, having contaminated

BOX A

GREGOR MENDEL, 1822–1884

"A huge concourse of mourners attended the funeral train from the monastery church to the Brünn central cemetery . . . . [M]any notables were present: professors, teachers, members of the Catholic clergy, but also the Protestant pastor and the Jewish rabbi, representatives of the numerous societies in which Mendel had worked and to which he had subscribed, deputies from the Heinzendorf fire brigade, and so on—but above all, many of the poor, to whom he had always been so kindly and so helpful, often enough their only refuge in time of trouble. But though hundreds realised that they had lost a good friend, and hundreds more came as in duty bound or from curiosity to witness the interment of a dignitary, not a single one of those present on the occasion knew that a great scientific investigator, a man of imperishable reputation had passed away." (Description of Mendel's funeral from Iltis 1924.)

Generation	Matings	Type of fertilization

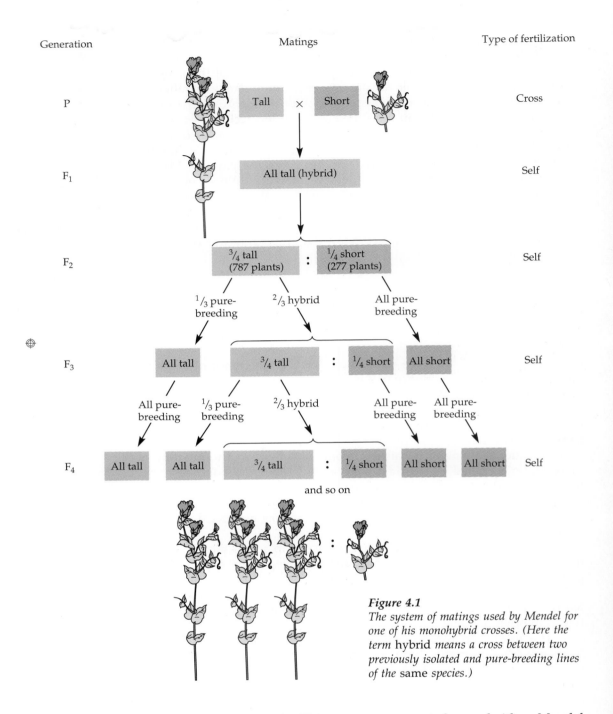

Figure 4.1
The system of matings used by Mendel for one of his monohybrid crosses. (Here the term hybrid *means a cross between two previously isolated and pure-breeding lines of the same species.)*

each other in the hybrids. This outcome was not observed either. Mendel found that each F_1 hybrid line gave rise to F_2 progeny resembling one or the other of the two original parents (i.e., tall or short) but never intermediate in appearance. Thus, *no blending or contamination of the two alternative traits occurred while they were combined in the hybrids.*

Mendel's First Law

Mendel's interpretation of these data is called the **law of segregation**. From the ratios described, he reasoned that each observable trait (e.g., plant height) is determined by a pair of cellular factors. During the formation of **gametes** (reproductive cells), the *two members of each pair of factors* *are separated (segregated)*, so an egg cell or a pollen grain contains one or the other, but not both, of the original factors.*

Because reciprocal crosses gave the same results, Mendel assumed that the male and female parents contribute equally to the formation of offspring. Thus, as shown in Figure 4.2, each F_1 hybrid inherits one "tall" factor (*A*) from its tall parent plus one "short" factor (*a*) from its short parent, and its factor composition can be designated as *A/a*. Although the recessive *a* factor is not expressed in the *A/a* hybrid, it remains unchanged and may be expressed in future generations.

When the hybrid *A/a* plants form gametes, half of them will contain the *A* factor and the other half will carry the *a* factor.

---

* Although chromosomes had not yet been discovered, the cell theory propounded by Schleiden and Schwann in 1838–1839 was fully accepted in Mendel's time, as were reports that eggs and sperm were single cells. Cytological proof that an egg is fertilized by just one sperm did not come until 1879, however.

Figure 4.2
Genetic interpretation of Mendel's monohybrid crosses; compare with Figure 4.1. The purebreeding tall plants (A/A) are represented by boxes of solid gray, the pure-breeding short plants (a/a) by boxes of solid color, and the hybrid tall plants (A/a) by boxes with stripes. In the F_2, F_3, and F_4 generations the 1/4 A/A and 2/4 A/a are summed to give 3/4 A/−. The latter notation indicates that at least one dominant factor (in this case A, representing tall) is present.

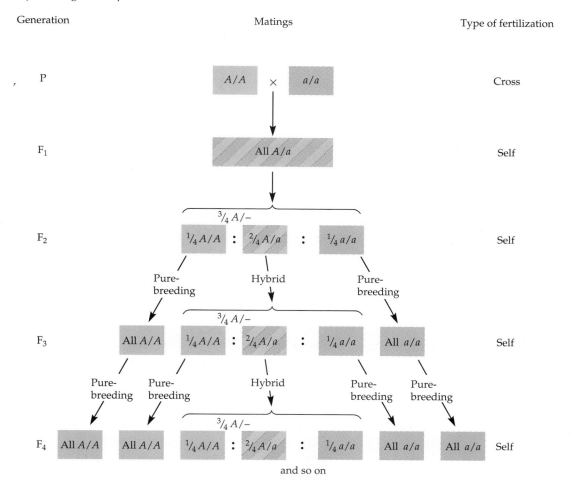

It remains, therefore, purely a matter of chance which of the two sorts of pollen will become united with each separate egg cell. However, according to the law of probability, it will always happen, on the average of many cases, that each pollen form, *A* and *a*, will unite equally often with each egg cell form, *A* and *a*. (Mendel 1865)

The four equally likely fertilization combinations are as follows:

Type of egg cell		Type of pollen grain		Type of seed and plant		
1/4	*A*	×	*A*	→	*A/A*	
1/4	*A*	×	*a*	→	*A/a* } 1/2 *A/a* } 3/4 tall
1/4	*a*	×	*A*	→	*a/A*	
1/4	*a*	×	*a*	→	*a/a*	1/4 short

Hence, the recovery, in the F_2 generation, of 1/4 *A/A* : 1/2 *A/a* : 1/4 *a/a* plants. Since the tall trait is dominant, however, it is impossible to distinguish the *A/A* plants from the *A/a* plants just by looking at them. What one sees is 3/4 tall plants and 1/4 short plants.

Modern Nomenclature

Several terms were coined after Mendel's time to describe these situations. The factors *A* and *a* are **alleles** of each other; they are alternative forms of a **gene** that influences plant height. We can thus refer to *A* as the dominant allele of the gene determining plant height and to *a* as its recessive allele. The terms *gene* and *allele* are sometimes used interchangeably, but in most contexts only one term is correct. It is proper to say that we are discussing the gene (not the allele) for plant height and that this gene exists chemically in different allelic forms, one form determining taller growth than the other. (Most genes, by the way, exist in more than two allelic forms, but a given individual carries only two alleles of a particular gene.)

Genotype refers to the precise allelic composition of a cell, such as *A/A* or *A/a* or *a/a*. **Phenotype** refers to what is actually seen in an organism, that is, tall plants or short plants. The phenotype *short* is due to the genotype *a/a*. The phenotype *tall* is due to either the genotype *A/A* or the genotype *A/a*. A **homozygous** genotype is one in which the two alleles are the same (*A/A* or *a/a*); in a **heterozygous** genotype, the two alleles are different (*A/a*). When we are not sure whether a phenotypically tall plant is genotypically homozygous (*A/A*) or heterozygous (*A/a*), we write the genotype *A/−*, the dash representing an allele that could be either *A* or *a*. The designation *A/−* can also be used when we are not interested in distinguishing between the dominant homozygote and the heterozygote.

Genes are present on chromosomes inside the nucleus of a cell in higher organisms (Chapters 2 and 3). Mendel was unaware of the existence of chromosomes. But just as Mendelian factors (alleles) are paired, so, too, are chromosomes. In fact, the two alleles of the same gene occupy the same position on **homologous chromosomes**. The particular site or location where a gene is found along the length of a chromosome is the **locus** of that gene. For example, in a heterozygous tall plant, *A/a*, the allele *A* occupies a particular locus on a particular chromosome. At a corresponding position on the homologous chromosome is found its allele, *a*:

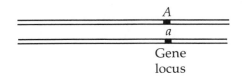

Gene
locus

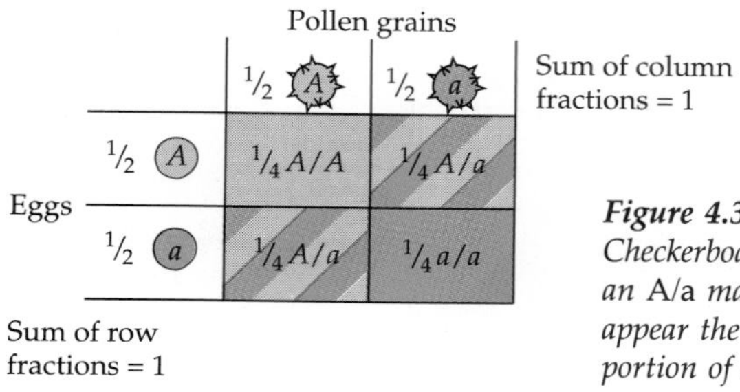

Pollen grains

Sum of column
fractions = 1

Eggs

Sum of row
fractions = 1

Figure 4.3
Checkerboard depicting Mendel's F_1 cross of an A/a female with an A/a male. In the boxes within the body of the checkerboard appear the genotypes of the F_2 progeny generation, and the proportion of each genotype (the gamete fraction).

A Checkerboard Method for Predicting Offspring

The time-honored *Punnett square* helps to visualize simple crosses involving just one or two gene loci. A variant, which we call the *checkerboard method*, includes fractions as well as the genotypes of gametes and offspring. The steps, pictured in Figure 4.3, are as follows:

1. Label the rows of a checkerboard with the various kinds of eggs that the female parent can make. There should be one row for each *different* kind of egg. The labeling of rows should include, in addition, the fraction represented by each egg type—what we call the *gamete fraction*—and these egg gamete fractions should together add up to exactly 1.
2. Label the columns of the checkerboard in a similar fashion with the different kinds of male gametes. Include the gamete fractions of each different kind. If the number of kinds of pollen or sperm is different from the number of kinds of eggs, you will end up with a rectangle rather than a square.
3. Fill in the body of the checkerboard with the *offspring genotypes* by combining the headings of the corresponding row and column. The fraction of each offspring genotype is the *product* of the fractions heading the corresponding row and column (see also Chapter 9). In Figure 4.3, for example, 1/2 A × 1/2 A gives 1/4 A/A for the upper left-hand box.

Summing the two identical genotypes on the ascending diagonal boxes (lower left + upper right) gives a total of 2/4 A/a. Phenotypically, A/A and A/a are tall (3/4), and a/a is short (1/4). In this checkerboard the egg and pollen types are the same because the cross was achieved by self-fertilization, but with other types of crosses this will probably not be true. Also, in this case the fractions heading the rows (or columns) are both 1/2, reflecting Mendel's law of segregation; the alleles separate randomly into the gametes. We will now be coming to situations in which the different kinds of gametes are not formed in equal proportions (see also Chapter 9).

Progeny Testing and Matings

Mendel figured out the genotypes of the tall F_2 plants by **progeny testing**. He took ten seeds from each self-fertilized F_2 plant, grew them, and observed the F_3 phenotypes (Figure 4.2). Homozygous tall F_2 plants (A/A) can produce only A gametes; thus, when self-fertilized, they yield only tall plants. But heterozygous tall F_2 plants (A/a) produce equal numbers of both A and a gametes (1/2 A and 1/2 a). Like their hybrid F_1 parents, they yield both tall and short progeny after self-fertilization. What Mendel

(A) *Heterozygous* tall × short: *A/a* × *a/a*

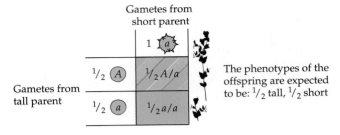

The phenotypes of the offspring are expected to be: ¹/₂ tall, ¹/₂ short

(B) *Homozygous* tall × short: *A/A* × *a/a*

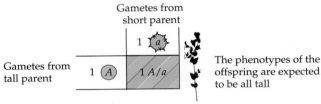

The phenotypes of the offspring are expected to be all tall

(C) *Homozygous* tall self-fertilized: *A/A* × *A/A*

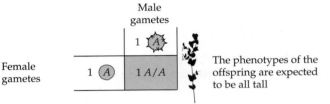

The phenotypes of the offspring are expected to be all tall

Figure 4.4
Checkerboards for one gene, two alleles, with A dominant to a. The phenotype of A/− is tall; the phenotype of a/a is short. (The phenotypes of progeny arising from the self-fertilization of a heterozygous tall plant are shown in Figure 4.3.)

found among the tall F₂ plants, by these breeding tests, was a 1:2 ratio of homozygotes to heterozygotes. This he expressed as 1/3 pure-breeding to 2/3 hybrid. (Recall that a ratio is converted into fractions by dividing each member of the ratio by the sum of the members.)

Actually, allowing a tall plant to self-fertilize is not the best way of determining whether it is homozygous or heterozygous. To maximize the probability of detecting the recessive allele in a heterozygote, we should cross each tall plant with a short plant (*A/−* × *a/a*). As shown in Figure 4.4A, heterozygous tall plants, when crossed with short plants, will yield a ratio of 1 tall to 1 short progeny, rather than the ratio of 3 tall to 1 short plant recovered after self-fertilization (Figure 4.3). Note in Figure 4.4B and C that homozygous tall plants produce all tall progeny, whether mated with short plants or self-fertilized. Note also in Figure 4.4 that when all the gametes from a parent are the same, they are represented by the fraction 1 and by just one row or one column in the checkerboard.

Consider that Mendel planted only ten seeds from each self-fertilized F₂ plant. From heterozygous tall plants (*A/a* × *A/a*), he could expect to recover two or three short plants among the progeny ($1/4 \times 10 = 2.5$). Of course, as a result of random variation, he observed sometimes more and sometimes fewer than this number of short plants. The occurrence of even one short plant among the ten progeny would be enough to identify its parent as a heterozygote. But what if, by chance, *A/a* × *A/a* yielded ten tall and no short progeny, instead of the expected seven or eight tall and two or three short plants? In such a case the parent plant would *mistakenly* be called a homozygote rather than a heterozygote. We can even calculate how often this might happen, and it is indeed likely that Mendel missed a few heterozygotes this way.

Now, if ten seeds from the cross $A/a \times a/a$ are planted, we would expect to find five short plants among the progeny ($10 \times 1/2 = 5$). Here, too, we would observe sometimes more and sometimes fewer than five short plants because of random fluctuation. But the occurrence of ten tall and no short plants from this cross is much less likely than in the case of self-fertilization.

For this reason, geneticists will, whenever possible, cross a dominant phenotype to a recessive phenotype to determine whether the former is homozygous or heterozygous. Such a mating is called a **testcross**. A testcross is one of the six matings possible for a single pair of alleles, A and a. That is, there are six different ways that the three genotypes (A/A, A/a, and a/a) can be combined in pairs, as shown in Table 4.1. In the course of describing Mendel's experiments, all but one of these has been considered. We suggest that you verify for yourself the outcome of each of these matings, using checkerboards if necessary, and then simply commit them to memory.

Mendel did many additional experiments to see what happens when more than one pair of characters are considered simultaneously, for example, the cross $A/a\ B/b \times A/a\ B/b$. Here, however, we abandon Mendel's pea patch and turn our attention to some human traits.

An Example from Humans

Mendel's law of segregation is fully applicable to genes present on human autosomes. **Alkaptonuria**, for example, is one of the "charter group" of heritable metabolic diseases first proposed as Mendelian recessives in the early 1900s. An English physician and biochemist, Archibald Garrod, correctly suggested that the basic defect in this rare, relatively mild condition is the absence of a specific enzyme that normally converts a substance called *homogentisic acid* to something else (Chapter 14). Homogentisic acid, having nowhere to go biochemically, then accumulates and is excreted in the urine of affected individuals. In some (but not all) cases the urine turns black after long exposure to air. A more reliable symptom is dark pigmentation of certain connective tissues (the outer layer of the eye, the cartilage of the external ear) and even of perspiration in young adults. Older patients also suffer from arthritis and may exhibit a higher than usual incidence of heart disease. The earliest known case was detected in an Egyptian mummy, dated about 1500 B.C., whose X-rayed bones showed the typical signs of alkaptonuria.

Table 4.1
Matings for one pair of alleles, A and a.

Parental Genotypes[a]	Genotypic Ratios among Progeny	Phenotypic Ratios[b]
1. $A/A \times A/A$	all A/A	
2. $A/a \times A/a$	1/4 A/A:1/2 A/a:1/4 a/a	3/4 $A/-$:1/4 a/a
3. $a/a \times a/a$	all a/a	
4. $A/A \times A/a$	1/2 A/A:1/2 A/a	all $A/-$
5. $a/a \times A/a$	1/2 a/a:1/2 A/a	
6. $A/A \times a/a$	all A/a	

[a] In the first three, the genotypes of the two parents are the same. In the last three, the genotypes of the two parents are different, but here we do not care to specify which sex has which genotype. If reciprocal matings were considered separately, then there would be a total of nine different matings, rather than the six listed here.
[b] Where different from genotypic ratios and assuming A is dominant to a.

The two alleles of the gene can be symbolized K and k, and the genotypes and phenotypes are as follows:

Genotype	Phenotype
K/K	Unaffected
K/k	Unaffected
k/k	Affected with alkaptonuria

Affected children are almost always born to unaffected parents who are heterozygous ($K/k \times K/k$) and who do not know that they carry the unusual allele until they have produced a child with alkaptonuria. By Mendelian rules, we know that 1/4 of any future children are expected to be affected.

MULTIPLE ALLELES

So far, we have only considered genes with two alleles. But there is nothing magical about this number. The various forms of a gene—the alleles—are generated by the process of **mutation**. This is the chemical change of one allele—say, the "standard," or "wildtype," one—into another—the "variant," or "mutant," one—or vice versa. Mutations could theoretically generate as many different alleles as there are chemical changes possible in a gene. Indeed, some genes have hundreds of known alleles.

To describe the various allelic forms of a gene, called **multiple alleles**, geneticists use sequential superscripts (or subscripts): A^1, A^2, A^3, . . . , A^N, where N is the highest number in the allelic series. The important thing to remember about any series of multiple alleles is that any one individual still has only two alleles (or *one*, for X-linked genes in a human male). These can be selected from a larger number, N, existing in the population as a whole.

A good example of a multiple allelic series is the gene that codes for the beta (β) polypeptide chain of the hemoglobin molecule (Chapter 12). Over 100 alleles of this gene are known; among these, the most common are Hb^A, Hb^S, Hb^E, and Hb^C. The superscript A refers to the normal allele, and S, E, and C refer to allelic forms of the gene that lead to various abnormal β polypeptide chains. Phenotypically, homozygotes for Hb^S have the disease sickle-cell anemia and are often quite ill. Homozygotes for Hb^E and homozygotes for Hb^C have only a mild or moderate anemia. Individuals who are heterozygous for the normal allele combined with either Hb^S, Hb^E, or Hb^C are unaffected, or nearly so.

Consider a gene with three alleles: B^1, B^2, and B^3. For each allele there is a homozygous genotype: B^1/B^1, B^2/B^2, and B^3/B^3. For each pairing of different alleles selected from the three available, there is a heterozygous genotype: B^1/B^2; B^1/B^3, and B^2/B^3.

With four alleles of a gene—C^1, C^2, C^3, and C^4—there are ten genotypes: four homozygotes and six heterozygotes. One methodical way for writing down all the genotypes is to first list all the alleles, as we have done for these four. Then combine the first allele (C^1) with itself and sequentially with each allele to the right of it in the list, here yielding a subtotal of four genotypes (C^1/C^1, C^1/C^2, C^1/C^3, C^1/C^4). Then combine the second allele (C^2) with itself and sequentially with each one to the right of it, here yielding a subtotal of three genotypes (C^2/C^2, C^2/C^3, C^2/C^4). Continue this process until you have run out of combinations, in this case as far as C^4/C^4, a subtotal of one genotype.

In general, for a gene with N alleles—A^1, A^2, A^3, . . . , A^N—this listing and combining will yield all possible genotypes, albeit in a tedious fashion. Readers proficient in algebra can derive a formula for the number of possible different genotypes for a gene with N alleles:

$$\frac{N(N + 1)}{2}$$

Of this number, N are homozygous and the rest are heterozygous. Using this formula, confirm the number of possible different genotypes given earlier for $N = 2$, 3, and 4 alleles. As the number of alleles of a gene increases, the number of possible different genotypes increases dramatically: With 10 alleles there are 55 different genotypes; with 20 alleles, 210 different genotypes. Although all the possible genotypes may occur in some populations, some genotypes may be missing—that is, not present in any individual now living, or even once living. In any event, we emphasize again that any individual genotype is made up of only two alleles for each gene.

SOME PROBABILITY

Up to this point, we have used checkerboards to predict the offspring of crosses involving the alleles of one gene. This method could be expanded to deal with several genes, but there are quicker, less tedious ways of arriving at the same predictions: using explicit rules of probability. Although some aspects of this kind of thinking might be new to you, much of the following material you have probably used intuitively.

When we flip a coin, we can ask, "What is the *chance* or *likelihood* of a head?" When we do several repetitions of flipping the coin 100 times, we might ask how often the *proportion* or *frequency* of heads will be greater than 80 (or some other number). These terms are roughly synonymous with **probability**. To be specific, however, probability is always written as a number between zero and one, expressed as a simple fraction (like 1/3) or a decimal fraction (like 0.333).

Definition

More rigorously, we can think of doing an experiment like picking a card, rolling a die (plural, dice), or selecting a gamete. In each case, there is a certain number of equally likely outcomes: 52 for a deck of cards, 6 for a die, and 2 for the case of a person heterozygous at a given locus. Before doing the experiment, we ask about some event A occurring; its probability can be written

$$P(A) = \frac{\text{number of outcomes "favorable" to A}}{\text{total number of equally likely outcomes}}$$

The notation $P(A)$ should be read "the probability of event A occurring," and its definition is a commonsense one. In the case of drawing a card randomly from a deck, there are 52 equally likely outcomes (the denominator). Of these, 13 are in a suit "favorable," say, to the event "drawing a heart" (the numerator). Therefore, the probability of drawing a heart is

$$P(\text{heart}) = 13/52 = 1/4 = 0.250$$

Using the definition again, note that for a die, fairly balanced and thrown,

$$P(\text{even number}) = 3/6 = 1/2 = 0.500$$

$$P(\text{at least a 3}) = 4/6 = 2/3 = 0.667$$

The phrase "at least a 3" means throwing a 3 or a higher number: a 3 or 4 or 5 or 6, a total of four favorable outcomes. For a B/b heterozygote,

$$P(B \text{ gamete}) = 1/2 = 0.500$$

$$P(b \text{ gamete}) = 1/2 = 0.500$$

If an event is *impossible*, there will be *no* outcomes favorable to it, so the numerator is 0. On the other hand, if an event is *inevitable*, then *every* outcome is favorable to it, and the numerator is the same as the denominator. Thus,

$$P(\text{impossible event}) = 0$$

$$P(\text{inevitable event}) = 1$$

For example, the probability of rolling a 7 with one die is 0, and the probability of rolling at least a 1 is 1.

If the probability of event A occurring is some number x, what is the probability of A *not* occurring? Something must happen; it is inevitable that either A occurs or it does not occur, and so the probabilities of events A and not-A must add to 1. Therefore, if

$$P(\text{A occurring}) = x$$

then

$$P(\text{A not occurring}) = 1 - x$$

For example, if the probability of snow tomorrow is 1/10, then the probability of no snow is 9/10.

We can generalize the foregoing a bit by saying that the probabilities of a set of events must add to 1 if every possible outcome is included. Thus, the sum of probabilities of a set of exhaustive events equals 1. This statement provides a useful check on arithmetic. For example, from the mating $B/b \times B/b$, the probability of heterozygous offspring plus the probability of homozygous offspring must equal 1.

Two Rules for Combining Probabilities

Sometimes the probability of interest is not a simple one that can be gotten directly from the definition. For example, the probability of a straight in poker (five cards in sequence regardless of suit) is far from obvious.* What we need are some rules for combining several simple probabilities to give us the probability of a more complex event. Here we present, in a fairly intuitive way, two such rules.

The Addition Rule ("Or" Rule). Consider an experiment that can result in the events A, B, C, D, ... Our interest is in not just one of these events, but in, say, A *or* B occurring. Event A could be "picking a heart" and B, "picking a spade." In shorthand notation, we are interested in $P(\text{A or B})$. Intuitively,

* The probability is 768/216,580, or about 0.35%.

$$P(A \ or \ B) = P(A) + P(B)$$

or, for more than two events,

$$P(\textbf{A} \ or \ \textbf{B} \ or \ \textbf{C} \ or \dots) = P(\textbf{A}) + P(\textbf{B}) + P(\textbf{C}) + \dots$$

In words, if we want to know the probability of one event *or* another *or* yet another, and so on, we *add* their individual probabilities together. This is the **addition rule**.

Whenever the conjunction *or* is used in everyday language, the addition rule is usually appropriate. For example, for a gene with two alleles, *B* dominant to *b*, what is the probability of a person having the dominant phenotype? Note that the dominant phenotype arises if a person is either *B/B* or *B/b*. Therefore,

$$P(\text{dominant phenotype}) = P(B/-) = P(B/B) + P(B/b)$$

There is a restriction in the use of the addition rule: *The events in question must be mutually exclusive.* This means that the occurrence of one event excludes the other(s) from happening. You will discover that in genetics, this is usually the case. If an organism has the genotype *B/B*, it does not have the genotype *B/b* or any other genotype. If a card is a heart, it is not a spade or any other suit; thus,

$$P(\text{heart } or \text{ spade}) = 1/4 + 1/4 = 1/2$$

$$P(\text{heart } or \text{ spade } or \text{ diamond } or \text{ club}) = 4(1/4) = 1 \text{ (inevitably)}$$

To illustrate events that are not mutually exclusive, let's say that A is "picking a heart" and B is "picking a queen." Because it is entirely possible for a card to be both a heart and a queen at the same time,

$$P(\text{heart } or \text{ queen}) \neq 13/52 + 4/52$$

The sign $\neq$ means "does not equal." Since we cannot use the addition rule here, the probability P(heart or queen) must be obtained some other way. It can be calculated directly from the definition of probability as 16/52 (i.e., 13 hearts plus 3 other queens out of the 52 cards in the deck).

The Multiplication Rule ("And" Rule). Consider an experiment that has as possible results the events A, B, C, D, ... If the experiment is repeated several times, we might be interested in the joint outcome of successive trials. For example, the joint probability of A on the first trial and B on the second is

$$P(A \ and \ B) = P(A) \times P(B)$$

or, for more repetitions,

$$P(\textbf{A} \ and \ \textbf{B} \ and \ \textbf{C} \ and \dots) = P(\textbf{A}) \times P(\textbf{B}) \times P(\textbf{C}) \times \dots$$

In words, if we want the joint probability of one event *and* another *and* another, and so on, we just multiply their individual probabilities together. This is the **multiplication rule**.

With two decks of cards before you, what is the probability of drawing an ace from the first *and* another ace from the second? By the multiplication rule, we obtain

P(ace from first *and* ace from second decks) $= (1/13)(1/13) = 1/169$

Note that each factor of 1/13 is itself obtained either from the addition rule $(1/52 + 1/52 + 1/52 + 1/52)$ or directly from the definition of probability: 4 favorable outcomes (aces) out of 52 (cards).

The restriction in the use of the multiplication rule is that *the events must be independent*. This means that the occurrence of one event does not affect in any way the probability of the other events occurring. Selections from separate decks are independent events because handling one deck of cards cannot affect the other.

If we draw two cards from the same deck, however, the outcomes are *not* independent. Having drawn one card, ace or not, the probabilities applying to the second draw are changed. To compute the probability of picking two aces from the same deck *without replacing the first card*, we need to take note of the altered probability:

4/52 (ace on first draw) $\times$ 3/51 (ace on second draw) $= 1/221$

Questions of independent events arise in very different situations from questions of mutually exclusive events, a condition of the addition rule. Mutually exclusive events can never be independent because excluding one event affects drastically the probability of another event's occurrence.

Whenever the conjunction *and* is used in everyday language, the multiplication rule is usually appropriate. For example, what is the probability that four siblings will all be girls? This can be interpreted as

$$P(4 \text{ girls}) = P(\text{1st born a girl } and \text{ 2nd a girl } and \ . \ . \ .)$$

$$= P(\text{girl}) \times P(\text{girl}) \times P(\text{girl}) \times P(\text{girl})$$

$$= \text{approximately } (1/2)^4$$

$$= 1/16 = 0.0625$$

Based on actual birth records, the likelihood of a liveborn girl is slightly, but significantly, less than 1/2—about 0.48 or 0.49, depending on which real population is sampled. So, a slightly more accurate answer might be

$$P(4 \text{ girls}) = (0.49)^4 = 0.0576$$

As an example of using both the addition and multiplication rules, consider the offspring from $A/a \times A/a$. From a checkerboard, the genotypic results are

$$P(A/A) = 1/4 \qquad P(A/a) = 1/2 \qquad P(a/a) = 1/4$$

Here are the rules of probability that lead to these same results:

$$P(A/A) = P(A \text{ egg } and \ A \text{ sperm}) = (1/2 \times 1/2) = 1/4$$

$$P(A/a) = P(A \text{ egg } and \ a \text{ sperm, or } a \text{ egg } and \ A \text{ sperm})$$

$$= (1/2 \times 1/2) + (1/2 \times 1/2) = 1/4 + 1/4 = 1/2$$

$$P(a/a) = P(a \text{ egg } and \ a \text{ sperm}) = (1/2 \times 1/2) = 1/4$$

$$P(\text{dominant phenotype}) = P(A/A) + P(A/a) = 1/4 + 1/2 = 3/4$$

There are some other basic rules of probability that answer questions such as, "What is the probability of getting two *or more* girls in a family of five?" These rules are based on expansion of the binomial equation, however, and will not be presented here. Interested readers can find them in Mange and Mange (1990) or in any textbook on general genetics.

MENDEL'S SECOND LAW: INDEPENDENT ASSORTMENT OF TWO PAIRS

We now illustrate Mendel's second law, not with traits in garden peas, but by following simultaneously the inheritance of two traits in humans: a type of dwarfism and the ABO blood type. Short stature has many causes. One kind stems from a lack of growth hormone, which is normally produced by the pituitary gland (a tiny but very important structure surrounded by the brain). Affected individuals exhibit *pituitary dwarfism*. Those with type I, the most common form (Figure 4.5), have normal body proportions, pinched facial features with high foreheads, and high-pitched voices. Letting *d* represent the recessive allele for pituitary dwarfism, affected persons are *d/d* and unaffected persons are *D/D* or *D/d*.

The ABO blood type, expressed as antigens on the surface of red blood cells, depends on multiple alleles, I^A, I^B, and I^O, of a gene. The alleles I^A and I^B are codominant to each other and cause antigens A and B, respectively, to be present. These two alleles are both dominant to I^O. (A discussion of the gene and antigens is presented in Chapter 16.) We can summarize the genotype/phenotype correspondences as follows:

Full genotype	Shorthand genotype	Phenotype (blood type)
I^A/I^A or I^A/I^O	*A/A* or *A/O*	A
I^B/I^B or I^B/I^O	*B/B* or *B/O*	B
I^A/I^B	*A/B*	AB
I^O/I^O	*O/O*	O

Figure 4.5

Charles S. Stratton (3 feet, 2 inches) and Lavinia Bump (2 feet, 8 inches), both healthy and intelligent, at their marriage in 1863. Their short stature apparently resulted from homozygosity for recessive genes that interfered with the production of pituitary growth hormone. Stratton's parents were first cousins, and Bump's were third cousins. Stratton (better known as General Tom Thumb), Bump, her sister, and her sister's husband (both also of short stature) were affiliated with the circus showman P. T. Barnum. For more information, see McKusick and Rimoin (1967). (Photograph courtesy of Circus World Museum, Baraboo, WI.)

The shorthand genotypes, which we use throughout this chapter, omit the basic locus designation, I, leaving just the superscripts.

The Proportions of the Gametes

Consider a doubly heterozygous male, $D/d\ A/B$, who then produces four kinds of gametes:

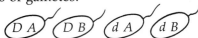

In what proportions are these gametes formed? Mendel's law of segregation states that 1/2 the gametes are expected to carry D and 1/2 to carry d. The same law states that 1/2 the gametes are expected to carry A and 1/2 to carry B. The question is this: How are these two pairs of alleles distributed into gametes with respect to each other? Mendel was able to show that alleles of different genes are assorted into gametes independently: *The segregation of one pair of alleles in no way alters the segregation of the other pair of alleles.* Each gene pair is acting alone, not interacting with the other, so the various kinds of gametes have random frequencies. Thus, we calculate the probability of any kind of gamete by multiplying together the probabilities for each allele it contains.

A *tree diagram* is useful to determine the different kinds of gametes from a double heterozygote. Recalling the "and" rule for probabilities, here is the diagram for our example:

Dwarfism alleles	ABO alleles	Gamete constitution
1/2 D	1/2 A →	1/4 (D A)
	1/2 B →	1/4 (D B)
1/2 d	1/2 A →	1/4 (d A)
	1/2 B →	1/4 (d B)

This diagram is the essence of Mendel's second law, the **law of independent assortment**.

Offspring Types for Two Genes

Let us now determine the offspring phenotypes from the mating

$D/d\ A/B$ male $\times$ $D/d\ O/O$ female

In a checkerboard, rows and columns are labeled with the various kinds of gametes and their probabilities. The four different kinds of sperm cells have already been given in the tree diagram. For the eggs, the two types are derived from a similar diagram, except that it does not branch because the female is homozygous for the ABO gene:

Dwarfism alleles	ABO alleles	Gamete constitution
1/2 D	all O	1/2 (D O)
1/2 d	all O	1/2 (d O)

Thus, the offspring checkerboard becomes:

		Egg types	
		1/2 (D O)	1/2 (d O)
Sperm types	1/4 (D A)	D/D A/O 1/8 unaffected type A	D/d A/O 1/8 unaffected type A
	1/4 (D B)	1/8 unaffected type B	1/8 unaffected type B
	1/4 (d A)	1/8 unaffected type A	1/8 affected type A
	1/4 (d B)	1/8 unaffected type B	1/8 affected type B

Each box in the checkerboard corresponds to a probability of 1/8, the product of 1/2 and 1/4, the gamete fractions heading the rows and columns. Using the genotypes in the first row as examples, fill in the remaining genotypes. Since pituitary dwarfism is recessive, each offspring type with a capital *D* is unaffected; this includes the top two rows and the first column. Summing similar boxes, we get the following offspring phenotypic expectations:

$$\left.\begin{array}{l} 3/8 \text{ unaffected, type A} \\ 3/8 \text{ unaffected, type B} \\ 1/8 \text{ affected, type A} \\ 1/8 \text{ affected, type B} \end{array}\right\} \text{ sum} = 1$$

The first result, 3/8 unaffected, type A offspring, comes from combining probabilities according to the "or" rule:

$$\left.\begin{array}{l} \text{top row, first column: } 1/8 \; D/D \; A/O \\ \text{third row, first column: } 1/8 \; D/d \; A/O \\ \text{top row, second column: } 1/8 \; D/d \; A/O \end{array}\right\} \text{ 3/8 unaffected, type A}$$

Notice that among unaffected, type A offspring there is a 1:2 ratio of *D/D* to *D/d*. Thus, 1/3 of unaffected, type A are homozygous unaffected and 2/3 are heterozygous unaffected for growth hormone. The unaffected, type B offspring are constituted in the same way.

Gene × Gene Method

The checkerboard could be called the **gamete × gamete method** for solving genetics problems, because first the various types of gametes are determined and then their proportions are multiplied together. Another procedure, called the **gene × gene method**, is especially useful if many genes are considered simultaneously and just one or a few of the possible combinations are of interest. We illustrate the gene × gene method using the same mating: *D/d A/B* male × *D/d O/O* female. Considering one pair of alleles at a time, we know that Mendel's first law predicts the following:

D/d × *D/d* genotypes: 1/4 *D/D*, 1/2 *D/d*, 1/4 *d/d*
             phenotypes: 3/4 unaffected, 1/4 affected
A/B × *O/O* genotypes: 1/2 *A/O*, 1/2 *B/O*
             phenotypes: 1/2 type A, 1/2 type B

Now we make use of the fact that the two genes are inherited independently and so multiply the individual expectations. From the phenotypic results, we obtain

P(unaffected, type A) = P(unaffected) × P(type A) = (3/4)(1/2) = 3/8
P(unaffected, type B) = P(unaffected) × P(type B) = (3/4)(1/2) = 3/8
P(affected, type A)   = P(affected) × P(type A)   = (1/4)(1/2) = 1/8
P(affected, type B)   = P(affected) × P(type B)   = (1/4)(1/2) = 1/8

These gene × gene results are, of course, the same as the gamete × gamete results (and in either case we might have considered genotypes rather than phenotypes).

We now apply the gene × gene method to the offspring of parents who are *both* doubly heterozygous, for example, *D/d A/O* × *D/d A/O*. The growth hormone alleles, considered alone, yield 3/4 unaffected and 1/4 affected with pituitary dwarfism. The *ABO* alleles, considered alone, yield 3/4 type A and 1/4 type O. Multiplying, gene × gene, we get the joint phenotypic expectations

P(unaffected, type A) = (3/4)(3/4) = 9/16
P(unaffected, type O) = (3/4)(1/4) = 3/16
P(affected, type A)   = (1/4)(3/4) = 3/16
P(affected, type O)   = (1/4)(1/4) = 1/16

This is the 9:3:3:1 ratio that Mendel observed in pea plants when he studied the joint inheritance of two independent traits.

The two methods of analysis, gamete × gamete and gene × gene, involve about the same amount of work when we are considering just two genes. But consider a situation involving more than two genes: What proportion of the offspring from the following cross is expected to show the dominant phenotype for all genes? (Assume that each uppercase allele is dominant to the corresponding lowercase allele.)

C/c D/d E/e F/F × *C/c D/d e/e F/f*

For this, the checkerboard is quite tedious, but the gene × gene analysis is easy:

Mating by gene	*P*(dominant offspring)
C/c × *C/c*	3/4
D/d × *D/d*	3/4
E/e × *e/e*	1/2
F/F × *F/f*	1

Multiplying the fractions together, we get 9/32, the proportion of offspring who have the dominant phenotype for all genes. Solving this problem by means of a checkerboard would require an 8 × 8 diagram. Should you care to try it, the different kinds of gametes from each parent can be obtained by a tree diagram that branches for each heterozygous gene.

In summary, we can always set up a checkerboard or a tree diagram, but the gene × gene method may be much less work.

MENDEL'S LAWS AND MEIOSIS

By the time Mendel's work was uncovered, the basic facts of cell reproduction were already known. Thus, the striking parallels between the behavior of his "factors" and the behavior of chromosomes in cell division were almost immediately recognized. The parallels between Mendel's "factors" and nuclear chromosomes were synthesized as the **chromosome theory of inheritance**.

Cytological proof for the chromosome theory of inheritance came a bit later. Of course, it cannot be shown that one pair of chromosomes segregates independently of another if the two members of a homologous pair look exactly alike—as they usually do. A direct microscopic proof required *two pairs* of visibly different homologues. In 1913 E. Eleanor Carothers found in the grasshopper not only a sex chromosome difference but also an additional set of homologous chromosomes in which one member of the pair had an extra piece of chromosomal material stuck to it. Thus, she was able to see differences between the homologues of two sets. In microscopic examinations of testes, Carothers found the four types of secondary spermatocytes in roughly equal numbers. This showed conclusively that chromosome behavior underlies both Mendelian rules of transmission.

Examples in Humans

Occasionally, a human chromosome carries a distinctive cytological feature that is found in several generations of a family. Such cases are known as **marker chromosomes**. If this family is also segregating for a trait whose genetic locus is present on the unusual chromosome, it should be possible to observe their joint inheritance. This is in fact how the first assignment of an autosomal gene to a specific human chromosome came about. In 1968, while karyotyping members of his own family, a graduate student at Johns Hopkins University discovered a peculiar "uncoiled" region in the centromeric heterochromatin of the long arm of chromosome 1 (Figure 4.6) Chromosomes with this property are considerably longer than their normal homologues, but carriers of this variant chromosome are phenotypically normal. Roger Donahue noted that his family also showed segregation patterns for the Duffy blood group (*Fy*) that coincided with the inheritance of the extra-long chromosome. Specifically, family members who carried the marker chromosome also had the Fy^a allele for the Duffy

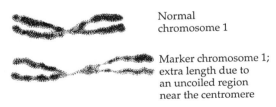

Normal chromosome 1

Marker chromosome 1; extra length due to an uncoiled region near the centromere

Figure 4.6
An "uncoiled" region in the centromeric heterochromatin of human chromosome 1. Joint inheritance of this chromosome 1 marker and Duffy allele a in one family suggested that they were physically linked together. This 1968 finding led to the first assignment of a human autosomal gene to a specific human autosome.

67

blood group, while those without the marker chromosome had the Fy^b allele. Statistical analyses of several such pedigrees in which *uncoiler* and *Fy* were segregating, as well as other studies, placed the Duffy locus very close to the heterochromatin on the long arm of chromosome 1. Other kinds of cytologically detectable chromosomal variants and their use in genetic studies will be discussed in later chapters.

Now let us reconsider the joint inheritance of the two human traits discussed earlier, pituitary dwarfism and ABO blood group, whose genes are known to be borne on different pairs of homologous chromosomes. The gene causing pituitary dwarfism is near the middle of the long arm of chromosome 17, and the ABO gene is located near the tip of the long arm of chromosome 9. The genotype and karyotype of the doubly heterozygous father, *D/d A/B*, are roughly diagrammed as follows:

Since *D* and *d* are alleles of each other, they occupy exactly corresponding places (loci) on homologous chromosomes. The same is true for alleles *A* and *B*. This male inherited one chromosome 9 and one chromosome 17 from his mother, and one each from his father. When he makes sperm, the maternal and paternal chromosomes 9 segregate at the first meiotic division independently of the segregation of maternal and paternal chromosomes 17. As shown in Figure 4.7, the genetic result is equivalent to the diagram shown in an earlier section.

SEX DETERMINATION AND SEX LINKAGE

Mendel found no differences among the progeny from reciprocal matings. The same phenotypic classes occurred, in the same proportions, whether a given trait was inherited through the male or through the female parent. Pea plants are not distinguishable on the basis of sex; both male and female structures appear in the same individual, a situation that characterizes some animals as well. In other plant and animal species, however, the sexes are separate, and the mode of inheritance of some traits does differ with reciprocal matings. For such organisms, sex is (with few exceptions) an inherited trait.

Sex Determination

Recall that among the sperm produced by a male, half carry an X chromosome and the other half carry a Y chromosome. All the eggs produced by a female, on the other hand, contain an X chromosome. So it is the father, rather than the mother, whose gamete determines the sex of their offspring. If an X-bearing sperm fertilizes the egg, a female is produced; the union of a Y-bearing sperm with an egg gives rise to a male. This system, diagrammed in Figure 4.8, perpetuates a presumed 1:1 sex ratio at the time of fertilization, generation after generation. Actually, the sex ratio at birth differs slightly from 1:1, perhaps as a result of a differential mortality of the sexes during intrauterine life.

We should emphasize that sexual development is an immensely complex process controlled by many genes and susceptible to alteration by environmental conditions at any stage from before birth through adulthood. In many species (including humans), some sexual characteristics can be modified, or perhaps even reversed, by hormonal treatments,

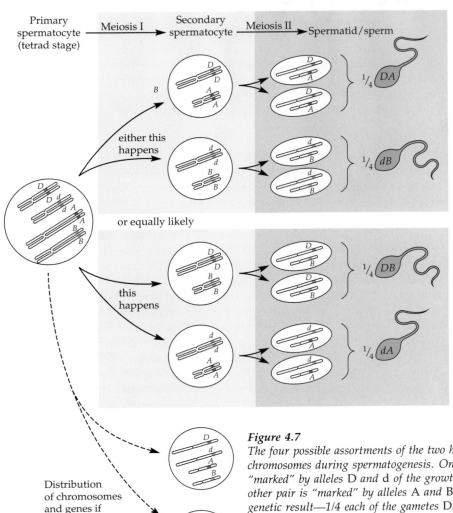

Figure 4.7
The four possible assortments of the two homologous pairs of chromosomes during spermatogenesis. One pair of homologues is "marked" by alleles D and d of the growth hormone gene. The other pair is "marked" by alleles A and B of the ABO gene. The genetic result—1/4 each of the gametes DA, DB, dA, and dB—arises because of independent assortment of different homologous sets at the first meiotic division. The results of oogenesis would be the same, except that only one of the four products of any one meiosis is a functional gamete.

injury, surgery, and many other factors. So chromosomal sex and phenotypic sex are usually, but not always, the same (Chapter 6). Anyway, what concerns us here is the behavior of the sex chromosomes and the genes they carry rather than the development of sex per se.

Sex Linkage

The X chromosomes of most species bear roughly as many genes as do autosomes of similar size. The human X chromosome is medium-sized; it carries over 160 definitely identified loci, with another 175 loci possible but not yet fully validated. These genes are said to be **X-linked** or *sex-linked*, because their inheritance is coupled with that of the X chromosome. Keep in mind, however, that X-linked genes are no different in kind from those found on autosomes. They affect characteristics and systems as diverse as blood groups, vision, hearing, the nervous system, the muscular system, teeth, skin, and glucose metabolism. Indeed, most X-linked genes have nothing to do with sexual differentiation. Conversely, there are a

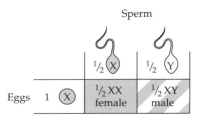

Figure 4.8
Checkerboard for the chromosomal sex-determining system that operates in humans and many other organisms. All the eggs from an XX female (fraction 1) contain an X chromosome; the sperm from an XY male are 1/2 X and 1/2 Y. Thus, a 1:1 sex ratio at fertilization is maintained generation after generation.

69

number of autosomal genes known to play a role in sexual development. Thus, *X-linkage is a matter of gene geography, not gene function.*

The human Y, on the other hand, is one of the smallest chromosomes and carries very few genes. Nevertheless, these two chromosomes do possess a tiny homologous region at the tips of their short arms, and during the cell division that leads to the production of gametes, they form a homologous pair that obeys Mendel's first law.

X Linkage in Humans

A familiar example of X-linked inheritance in humans is **color blindness.** The term is a misnomer, however, because most "color-blind" people see a range of colors—especially among the blues and yellows. What we really mean is *partial* color blindness or *color defectiveness*, which is mainly a problem of discriminating between reds and greens. Although pedigree studies as early as 1911 indicated that the gene resides on the X chromosome, complete understanding of its inheritance had to await molecular studies carried out in the mid-1980s. Research by Jeremy Nathans and colleagues at Stanford University showed that there are actually *two* color vision loci on the X chromosome; they lie very close to each other and differ by only 2% of their DNA base sequences. Indeed, one locus probably arose as a duplication of and subsequent divergence from the other locus.

Each of the two loci controls the production of a light-absorbing protein, one making up part of the *red-sensitive* color vision pigment and the other making up part of the *green-sensitive* pigment. These two pigment types occur separately in color-sensitive cells (cones) within the retina of the eye, as does a third, blue-sensitive pigment (autosomally inherited and not under consideration here). Four types of X-linked color vision defects are known: At each of the two loci, one mutant allele is associated with an anomalous pigment and the other mutant allele with a missing pigment. At each locus the normal (third) allele is dominant to either mutant, and the mutant with defective pigment is dominant to that with missing pigment.

Altogether, about 8% of males of western European extraction are color-blind. Here we will only consider the most common defect, *deuteranomaly* (which we call *green-shift*), occurring in about 5% of white males. (For reasons that will become clear later, all these conditions are rare among females.) Affected individuals have all three pigment types, but their green-sensitive pigment is anomalous. In certain ranges they see fewer hues than normal: Reds are perceived as reddish browns, brighter greens as tans, and olive greens as indistinguishable from browns. (Individuals with the more serious condition of *deuteranopia*—which occurs with a frequency of 1% in white males—cannot tell reds from greens at all, nor can they distinguish either of these from yellow. Basically, what they see are yellows, blues, blacks, whites, and grays.)

Recall that a female inherits two sets of X-linked genes, one from her mother and one from her father. Since the green-shifted allele (call it *g*) is recessive to the normal *G* allele, a female must be homozygous *g/g* to express the trait. Heterozygous *G/g* females, with one normal allele and one green-shifted allele, are called **carriers,** and their own color vision is usually normal or near normal. Homozygous *G/G* females have normal color vision.

A male's single X chromosome can only be inherited from his mother. Whatever genes are carried on his X chromosome will be expressed, whether dominant or recessive, because his other sex chromosome is a gene-poor Y received from the father. Thus, we cannot use the terms *homozygous* or *heterozygous* with respect to X-linked traits in a male. Rather, he is said to be **hemizygous** (Greek *hemi*, "half") for *G* (i.e., phenotypically normal) or hemizygous for *g* (i.e., phenotypically green-shifted). The corresponding genotypic description is *G*/(Y) or *g*/(Y).

Compare the results of reciprocal crosses between normal and green-shifted individuals. First consider the most common case, a homozygous normal mother and a green-shifted father (Figure 4.9A). All the mother's eggs (fraction = 1) have a normal *G* gene, so all her sons are normal. The father produces two kinds of sperm; one kind carries the Y chromosome, which is passed on to sons, and the other kind carries the X chromosome with the *g* gene. Thus, all daughters will be carriers.

What can be expected from matings between heterozygous normal mothers and green-shifted fathers? As shown in Figure 4.9B, the eggs are of two types, *G* and *g*. The sperm are of two types, *g* and Y. The progeny will include four possible genotypic combinations, which are also phenotypically distinct: 1/4 carrier females, 1/4 green-shifted females, 1/4 normal males, and 1/4 green-shifted males. Among the offspring of either sex, half will be normal and half will be green-shifted.

Finally, consider the reverse of the preceding, in which the father is normal but the mother is green-shifted (Figure 4.9C). The father must be *G*/(Y) and the mother must be *g*/*g*. The mother's eggs all carry *g*; of the father's sperm, half carry *G* and half carry the Y chromosome. Thus, all sons will be green-shifted and all daughters will be carriers. Note that in this case, the pattern of color vision in the male and female parents is "crisscrossed" in their offspring.

Now you can see why deuteranomaly occurs much more frequently

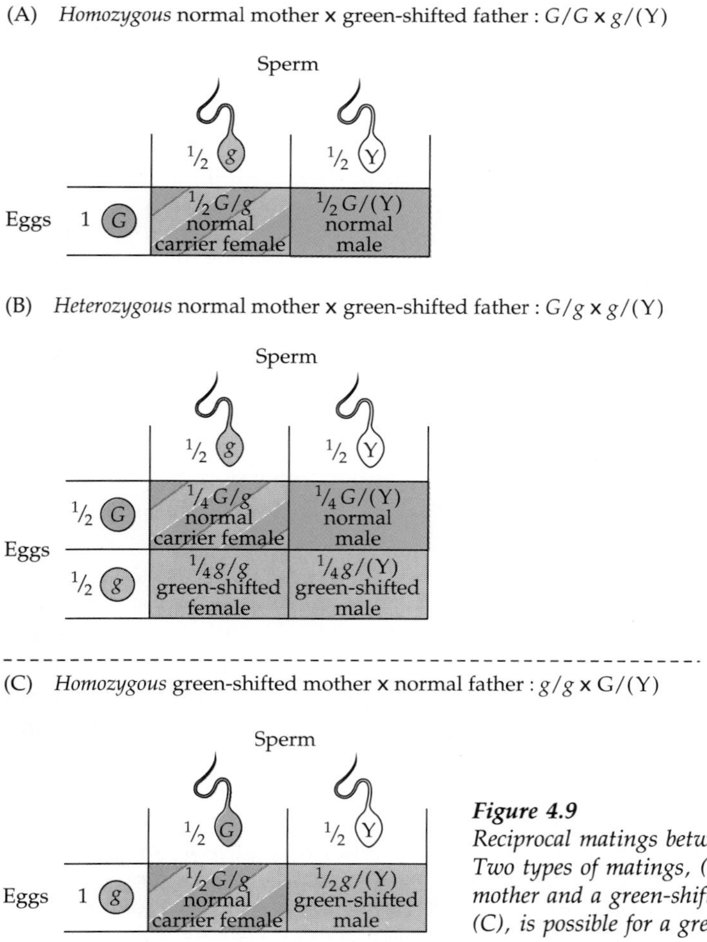
(A) Homozygous normal mother x green-shifted father : G/G x g/(Y)
Sperm
1/2 g
1/2 Y
Eggs
1 G
1/2 G/g normal carrier female
1/2 G/(Y) normal male
(B) Heterozygous normal mother x green-shifted father : G/g x g/(Y)
Sperm
1/2 g
1/2 Y
Eggs
1/2 G
1/2 g
1/4 G/g normal carrier female
1/4 G/(Y) normal male
1/4 g/g green-shifted female
1/4 g/(Y) green-shifted male
(C) Homozygous green-shifted mother x normal father : g/g x G/(Y)
Sperm
1/2 G
1/2 Y
Eggs
1 g
1/2 G/g normal carrier female
1/2 g/(Y) green-shifted male

Figure 4.9
Reciprocal matings between normal and green-shifted individuals. Two types of matings, (A) and (B), are possible for a normal mother and a green-shifted father. Only one type of mating, (C), is possible for a green-shifted mother and a normal father.

among males than among females. A male needs only one mutant gene, inherited from either a carrier or a green-shifted mother, to express the trait; the genotype of the father, who contributes only a Y chromosome, is irrelevant. A green-shifted female, however, must have not only a carrier or a green-shifted mother, but a green-shifted father as well. The low probability of such parentage leads to about a 20:1 ratio of green-shifted males to affected females in the general population.

Because one mutant gene guarantees the expression of *X-linked recessive* traits in males, whereas two are needed for the expression of *autosomal recessive* traits in either sex, it is much easier to detect the former than the latter in pedigree studies. This is not true for *dominant* mutations, which are expressed in single dose whether X-linked or autosomal. Since most mutations are recessive, however, there is a striking difference between the number of genes assigned to the X chromosome and the number known for any one autosome. It is unlikely that more genes reside on the X chromosome than on autosomes of comparable size. Rather, X-linked mutants are just easier to detect and verify. But with the advent of molecular genetic techniques, which do not favor X-linked over autosomal DNA, this distinction will disappear.

X-Linked Lethal Genes and Shifts in Sex Ratio

So far we have discussed genes that produce normal phenotypes and genes that may cause conditions detrimental to the organism. There is a third class of genes, called **lethal genes**, that disrupt processes absolutely essential to life. Like other genes, lethals can be dominant or recessive and can exist on any chromosome. They can also take effect at various stages of the life cycle, from the earliest embryonic stage through adulthood. Some spontaneous abortions, for example, are due to lethal genes acting during fetal development.

Suppose that a recessive lethal mutation, causing fetal death, occurs on the X chromosome. Any males that inherit it would die before birth. Females who inherit one lethal X chromosome are usually unaffected, but as shown in Figure 4.10, half of their sons are not expected to survive. Thus, the sex ratio among their live-born offspring will be *2 females to 1 male*, rather than the usual 1 female to 1 male. Of course, to distinguish an abnormal sex ratio from random fluctuations about a normal sex ratio, large numbers of progeny from extended human families must be observed. X-linked lethals are rare in human populations, however, and so do not appreciably affect the overall sex ratio.

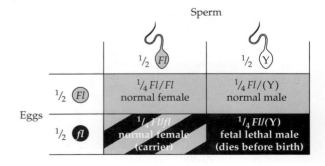

Figure 4.10
Progeny expected from mating between a female heterozygous for an X-linked recessive lethal and a normal male. (We let fl *represent the X-linked recessive fetal lethal allele and* Fl *represent its dominant allele for normal fetal development.) When the lethal gene is expressed during fetal development, only half the sons conceived will survive to birth. Thus, among the live-born progeny there will be twice as many females as males: 2/3 females and 1/3 males.*

X-linked recessive lethals have the same effect in fruit flies, where females are also XX and males XY. In the 1920s H. J. Muller used this shift in sex ratio to estimate the frequency of X-linked recessive lethal mutations in natural populations of fruit flies. By comparing this "spontaneous" mutation rate with what happened after X-ray treatments, he was the first to show that X-radiation increases the number of lethal mutations and other types of mutation as well. Evaluating the effect of irradiation in humans by screening large populations for slight changes in sex ratio is extremely difficult and expensive. One such study attempted to determine whether or not the atomic bombs dropped on Japan in 1945 led to an increase in the mutation rate there (Chapter 12).

SUMMARY

1. In modern terminology, Mendel's law of segregation states that the two alleles of a gene separate during the formation of gametes, so each gamete receives one of the two alleles at random.

2. The law of segregation and the randomness of fertilization allow predictions of the genotypic and phenotypic ratios among the offspring of any mating. A convenient way to set up these predictions is to construct a checkerboard whose rows and columns are labeled with the various kinds of eggs and sperm and with gamete fractions representing their relative abundance.

3. The testcross is the most informative mating for determining whether an organism with the dominant phenotype is homozygous or heterozygous.

4. More than two (i.e., multiple) alleles may exist for one gene. The number is determined by the chemical changes (mutations) that lead to recognizable phenotypes. Although one individual has only two alleles for each gene, multiple alleles are the basis of a large number of possible genotypes in the general population.

5. The notion of probability is similar to intuitive ideas of chance or frequency. Probabilities are expressed as numbers between 0 and 1 inclusive.

6. The way probabilities are combined depends on the kind of question and the nature of the events. When *or* is used in describing mutually exclusive events, the combined probability is obtained by addition. When *and* is used for independent events, the joint probability is obtained by multiplication.

7. Mendel's second law, called the law of independent assortment, deals with the inheritance of two or more pairs of traits. It states that the segregation of one pair of alleles in no way alters the segregation of the other pair of alleles.

8. The X chromosome carries many genes, most of which have nothing to do with sexual development. Very few genes are present on the tiny Y chromosome.

9. A father transmits his X chromosome to daughters and his Y chromosome to sons; a mother transmits an X chromosome to both sexes. A son receives a Y chromosome from his father and his single X chromosome from his mother; being hemizygous, he expresses all his X-linked genes. A daughter receives an X chromosome from both parents.

10. X-linked recessive lethal genes that act during fetal development can alter the sex ratio to favor females.

QUESTIONS

Classical albinism is caused by the lack of an enzyme necessary for the synthesis of melanin pigments. Enzyme production requires the dominant allele, C, so C/C or C/c (which can be abbreviated $C/-$) represents normal pigmentation, and c/c represents albino.

1. What progeny are expected, and in what proportions, from a normally pigmented woman who has an albino husband and an albino father?

2. A normally pigmented woman and an albino man have nine normally pigmented children and one albino child. What is the best guess for the woman's genotype? Is any other genotype possible?

3. A normally pigmented woman and an albino man have ten normally pigmented children and no albinos. What is the best guess for the woman's genotype? Is any other genotype possible?

4. Two parents are heterozygous for albinism. Assume that they will have five children.

    (a) What is the probability that the children will have the following phenotypes in the order stated: first born = unaffected, second = albino, third = unaffected, fourth = albino, fifth = unaffected?

    (b) What is the probability that all five will be unaffected?

    (c) What is the probability that *at least* one child will be albino?

    (d) Suppose that the first three children are albino. What is the probability that the fourth child will be albino?

5. Consider a gene with two alleles, B and b. (a) List all the matings (the parental genotypes) that could produce a heterozygous child. (b) Which mating in your list gives the greatest proportion of heterozygous offspring?

6. (a) Again considering alleles B and b, list all the matings whose offspring can be of only one genotype. (b) List all the matings whose offspring can be of two, and only two, genotypes. (c) What one mating is not on either of these lists?

7. For some rare autosomal dominant diseases, it is possible that the dominant homozygote has never been seen. One reason is that the fetus with this genotype may die early in development and be spontaneously aborted. Consider a situation in which

    Q/Q dies as an early fetus
    Q/q lives to reproduce but is affected
    q/q lives and is unaffected

What genotypes are possible among the live-born children of the mating $Q/q \times Q/q$? Among these live-born children, what is the proportion of each genotype?

8. What kinds of gametes, and in what proportions, can be made by a person who develops from an egg carrying genes A and B and a sperm carrying a and b, if the two loci are on different (nonhomologous) chromosomes?

9. List all the kinds of eggs that can be formed by a woman who is heterozygous for the three independently inherited genes for brachydactyly, pituitary dwarfism, and sickle-cell anemia: $B/b \; D/d \; Hb^A/Hb^S$. What is the proportion of each kind of gamete?

10. If the woman in question 10 were heterozygous for a fourth independent gene, Xg^a/Xg, how many different kinds of gametes could she make?

11. What phenotypes, and in what proportions, are possible from the following mating involving the growth hormone and ABO genes:

    $D/d \; B/O \times D/d \; B/O$

12. Consider these three independent genes and their alleles:

    B = brachydactyly (short fingers) and
    b = its normal recessive allele
    P = polydactyly (extra fingers) and
    p = its normal recessive allele
    S = syndactyly (joined fingers) and
    s = its normal recessive allele

Consider the mating $B/b \; P/p \; S/s \times B/b \; P/p \; S/s$. (a) What is the one most likely *genotype* among the offspring? What is its probability? (b) What is the one most likely *phenotype* among the offspring and its probability? (c) What is the probability of a child with completely normal fingers?

13. A good example of a *multiple allelic series* is the gene that codes for the β polypeptide chain of the hemoglobin molecule. Over 100 alleles of this gene are known; among these, the most common are Hb^A, Hb^S, Hb^E, and Hb^C. List methodically all the genotypes possible when considering these four alleles of the hemoglobin β chain gene. (Your work will be easier if you omit the basic gene designation, Hb, and just use the superscripts.) How many genotypes are homozygous? How many are heterozygous?

14. Give the genotypes and probabilities for the offspring of the following matings. The basic gene designation, Hb, has been omitted.

    (a) $A/A \times A/S$ (b) $A/C \times A/S$ (c) $A/E \times C/S$

15. Consider the three-allele, X-linked locus that controls the green-sensitive pigment for color vision. Recall that the normal allele G is dominant to both mutant alleles; the deuteranomaly (green-shifted) allele, which we will call g^1, is dominant to the deuteranopia (missing pigment) allele, which we will call g^2. List the offspring genotypes, phenotypes (including sex and carrier status), and proportions of each, expected from the following matings:

    (a) g^1/g^1 female $\times$ $G/(Y)$ male
    (b) G/g^2 female $\times$ $G/(Y)$ male
    (c) G/g^1 female $\times$ $g^2/(Y)$ male
    (d) female heterozygous for deuteranomaly $\times$ unaffected male

FURTHER READING

For biographies of Mendel, see Orel (1984) and Iltis (1924). One translation of Mendel's paper is in Stern and Sherwood (1966), which also reprints an interesting statistical analysis of his experimental results. Corwin and Jenkins (1976), Srb, Owen, and Edgar (1970), Peters (1959), and L. Levine (1971) reprint research papers important to the early development of general genetics. For comments on Mendel's contribution to human genetics, see Stern (1965). Textbooks by Crow (1983), Strickberger (1985), and Levitan (1988) contain good, short accounts of probability.

Human Pedigrees

Stormie Jones was born in May 1977, in Cumbry, Texas. The blond, brown-eyed, freckle-faced little girl seemed normal except for one thing. Beginning at age three months, yellow bumps appeared on her buttocks, knees, elbows, knuckles, and toes. Not until she was six years old, however, did any doctor recognize these fatty deposits as a sign of life-threatening illness.

Stormie had **familial hypercholesterolemia (FH)**, a defect of fat metabolism characterized by excess cholesterol in the blood, yellow fatty deposits in the skin and tendons, and coronary heart disease. This condition, inherited as an autosomal dominant, is now known to be one of the most common Mendelian disorders in humans. Most affected individuals are heterozygotes. But Stormie, having inherited one defective allele from each of her heterozygous parents, was a rare homozygote for FH. When finally diagnosed, she had a blood cholesterol level about nine times that expected for a normal child, and her arteries were already severely blocked. By the age of 6½ years, she had already suffered two heart attacks and undergone two coronary artery bypass operations.

Soon it became clear that only a liver transplant could save Stormie's life,* but her own heart was too weak to withstand such a trauma. So at age six years and nine months, Stormie made medical history as the first person to get a transplanted heart and liver in one operation (Figure 5.1). The procedure was successful: Within a few weeks, her cholesterol level had dropped almost fivefold, and she was able to return to a fairly normal home and school life. Six years later, however, she needed a second liver transplant. Then in November 1990, at age 13, she suddenly took ill and died from a totally unexpected event: rejection of the heart she had received almost seven years before.

During Stormie's short lifetime, researchers learned a tremendous amount about the physiology and genetics of FH. A few details will be presented later in this chapter, when we focus on the inheritance pattern of FH. But first we need to explain how geneticists collect and organize pedigree data.

PEDIGREE CONSTRUCTION

One might think that the recent explosion of molecular techniques would make old-fashioned pedigree analyses unnecessary. Not so. Even "high-tech" geneticists often need to know the exact interrelationships among the sources of DNA molecules they analyze, while clinicians and genetic

---

* Normally, liver cells have the major responsibility for clearing cholesterol from the bloodstream; but being homozygous for FH, Stormie's liver cells were incapable of performing this function.

counselors need to use all possible genetic tools, both new and old. Thus, they still summarize, in diagrammatic form, information from family studies. These **pedigrees** are drawn up in a fairly standardized way. Figure 5.2, for example, represents a hypothetical pedigree in which affected persons possess a dominant autosomal gene.

Separate generations occupy separate horizontal lines, with the most ancestral at the top, and are numbered from top to bottom by Roman numerals. The individuals within a given generation are numbered from left to right by Arabic numerals. Note that consecutive individuals (such as II-2 and II-3 in Figure 5.2) may be genetically unrelated. Within each sibship (group of brothers and sisters), the birth order, from oldest to youngest, is arranged from left to right.

Males are designated by squares, females by circles. If sex is unknown, a diamond is used; this symbol, with a number enclosed, may also designate unaffected individuals whose sex is irrelevant. Stillbirths, miscarriages, and abortions are indicated by tiny symbols, like that for II-7 and III-6; deceased individuals are indicated by a symbol with a diagonal slash. An arrow points to the **propositus** (or **proband**), the individual through whom the pedigree was discovered (IV-15) and who must be accounted for in statistical analyses.

Parents who are unaffected and unrelated, such as the mate of III-5, may be omitted from the diagram. Likewise, unaffected sibs may be lumped into a single diamond (IV-11 through IV-13), circle, or square,

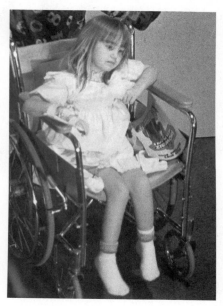

Figure 5.1
Stormie Jones at age six. (U.P.I./The Bettmann Archive.)

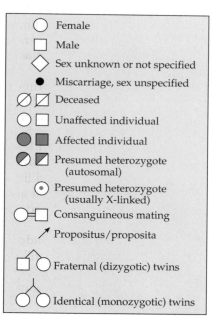

◯	Female
☐	Male
◇	Sex unknown or not specified
●	Miscarriage, sex unspecified
⊘ ⊠	Deceased
◯ ☐	Unaffected individual
● ■	Affected individual
⊘ ⊿	Presumed heterozygote (autosomal)
⊙	Presumed heterozygote (usually X-linked)
◯—☐	Consanguineous mating
↗	Propositus/proposita
☐◯	Fraternal (dizygotic) twins
◯◯	Identical (monozygotic) twins

Figure 5.2
Hypothetical pedigree of a rare autosomal dominant trait, illustrating the symbols most commonly used by American geneticists. The gene is not fully penetrant in individual III-13, who therefore does not express the trait.

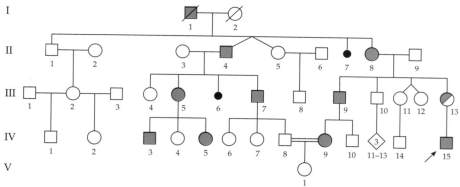

with a number inside to indicate how many individuals are included. To save space, huge pedigrees are sometimes drawn in circular or spiral form rather than in rectangular form.

Two parents are joined by a horizontal line, from which drops an inverted T, to which their offspring are attached by short vertical lines. Individuals III-4 through III-7, for example, are sibs, the children of II-3 and II-4. Note that III-2 had two mates, III-1 and III-3; thus, IV-1 and IV-2 are half sibs. A single child (such as III-8) is attached by a long vertical line directly to its parents' horizontal "mating line." Twins, such as II-4 and II-5, attach at the same spot along the inverted T if nonidentical. If identical, like III-11 and III-12, they may branch from a short vertical line or be connected by a line. (If the type of twinning is unknown, a question mark appears between them.)

Individuals affected with the trait under consideration (e.g., I-1, II-4, II-8, and so on) are represented by solid symbols; unaffected individuals are represented by open symbols. For dominant disorders, affected individuals may be shown as "half solid," because they are usually heterozygotes. Alternatively, carriers of autosomal genes, including those showing nonpenetrance, may be represented by half-solid symbols. (Individual III-13, for example, did not express her dominant allele.) Partial expression of a trait may be indicated by lighter shading. A dot inside the symbol is usually used to identify a heterozygous carrier of an X-linked recessive gene. (It may sometimes be used to identify any individual who has a gene but does not express it, such as the carrier of an autosomal recessive allele or even of an unexpressed dominant allele.) When two or more traits are being studied simultaneously, more complex symbolism must be used.

Consanguineous matings (matings between related individuals) are important to note. Individuals IV-8 and IV-9 are related as second cousins, because they have one set of great-grandparents (I-1 and I-2) in common. The mating line between IV-8 and IV-9 (doubled to indicate consanguinity) closes a genetic loop that also passes through ancestral individuals III-7, II-4, II-8, and III-9. The daughter of IV-8 and IV-9, although inbred, is normal with respect to the trait considered in the figure.

Data Collection

If you think that the "low-tech" collection of pedigree information is a fairly simple task, you are mistaken. Living family members may be widely scattered, and because few people keep careful written records, it is often difficult to get reliable information about deceased relatives. Recollections may be hazy or exaggerated, especially about medical histories; also, adoptions, abortions, artificial inseminations, and illegitimate births may go unmentioned.* Another problem is the reluctance to discuss family members who are in some way abnormal. Indeed, close relatives may be unaware of such individuals if they died young or were institutionalized. Stillbirths and miscarriages are even less likely to be remembered. Given all these difficulties, most geneticists would say that looking carefully at all the members of just a few generations is usually better than relying on incomplete or faulty information about generations further back.

Also complicating most pedigree analyses is the fact that the phe-

---

* Indeed, the results of one large study indicated that 2.3% of tested children were apparently the result of infidelity, concealed adoption, or some other such event. Estimates from some other studies have ranged much higher. Wettke-Schäfer and Kantner (1983) comment: "Reviewing the literature, we found a remarkable lack of information regarding [spontaneous] abortions and stillbirths. . . . Often the occurrence of abortion is unknown, or there is not enough pedigree information. When abortions are mentioned, . . . often there is no information regarding the time of abortion or regarding the sex of the aborted child."

notypic expression of a trait often varies among individuals (Chapter 7). This variation can be due to differences in mutant alleles at a given locus, differences in background genotypes (including modifying genes), and differences in environmental conditions. Although they rarely encounter the ideal research situation, human geneticists have learned to make the most of what they find. The few populations (such as the Amish, the Hutterites, and the Mormons) that produce large families, live in well-defined areas, marry almost exclusively among themselves, maintain a uniform life-style for generations, and also keep good records have provided unique opportunities for genetic study (Chapter 17).

By analyzing all kinds of pedigrees and medical data, geneticists have, over many decades, determined the mode of inheritance for many phenotypic traits (Box A). At first they could only separate these phenotypes into two groups, X-linked and autosomal, without knowing which of the 23 autosomes harbored any given trait. But starting in 1968 with the assignment of the Duffy blood group trait to chromosome 1, new cytogenetic and molecular techniques have allowed geneticists to map many genes to their exact locations. McKusick's 1990 catalog of simply inherited human phenotypes includes 2,656 "proven" traits: 70% autosomal dominants, 24% autosomal recessives, and 6% X-linked traits.* Of these, over 60% of the autosomal loci have been assigned to specific chromosomes, and most of the X-linked loci have been mapped.

Gene Function

For most of the disorders listed in McKusick's catalog, the exact map locations and gene products are unknown. Yet from what is known, an interesting rule of thumb has emerged: *Almost all disorders caused by enzyme deficiencies are recessive*, whereas those *disorders caused by defects in nonenzymatic (i.e., structural) proteins are often inherited as dominants*. The reason given for this distinction is as follows: Although heterozygotes make only half as much of the normal gene product as homozygous unaffected individuals do, this is usually still enough to get the job done in the case of enzyme reactions, where the catalyst is needed in only small amounts. But in the case of structural proteins, such as collagen, the abnormal gene product may actively interfere with normal cell functions.

To a surprising extent, gene structure and function are *conserved* in closely related and even between distantly related species.† Thus, researchers often make interspecies and interclass comparisons of phenotypes, genotypes, and genes. What they find are astonishing similarities between humans, mice, flies, worms, yeast, and bacteria. For inherited human disorders in particular, animal models are both instructive and practical, allowing many kinds of crosses and experiments that could never be undertaken on people.

And now that traditional pedigree studies can be combined with powerful new techniques for analyzing differences in the structure of DNA itself, a new brand of genetics is emerging. From Mendel's time until the mid-1970s, allelic differences could be detected and studied only if they gave rise to phenotypic variation. But with the ability to isolate, sequence, and even insert specific mutations into small segments of DNA (Chapter 13) comes the possibility of studying and using genes whose allelic variants

---

* In addition, it lists 2,281 "provisional loci," which have not yet been fully validated or identified.

† One of the most unusual examples is the gene for *hemoglobin*, the oxygen-carrying molecule of animal blood cells. A very similar gene is found in some plants, prompting the suggestion that one might be able to get blood from a turnip after all.

INHERITANCE OF SOME "REGULAR" HUMAN TRAITS

"What are the chances that our child will have brown eyes? Red hair? Do you think it might be left-handed like me? Tall like my husband? Will its nose be large or small? Will its skin color be light or dark? Is it likely to inherit my father's baldness or his bushy eyebrows? My mother's big ears?" Sometimes the questions come *after* the fact: "My son and his wife both have blue eyes, but their child has brown eyes. How could this happen?"

In studying the inheritance of human traits, geneticists have quite understandably concentrated on diseases and physical abnormalities rather than on the ordinary phenotypic variants that may interest us. Still, it often comes as a surprise that we know so little about the inheritance of hair color, height, and other run-of-the-mill physical characteristics. Furthermore, much of the data that have been reported on such traits are not as extensive or reliable as we would like, so they need to be taken with a grain of salt.

Consider, for example, one genetic "fact" that everybody seems to remember from their biology course: Brown eyes are always dominant to blue eyes. Right? Wrong! It turns out that eye color is a complex rather than a simply inherited trait, and—although alleles for darker colors tend to be dominant over alleles for lighter colors—several loci are probably involved, and it is possible for two blue-eyed parents to produce a brown-eyed child. Indeed, the author of the "bible" of human genetics (McKusick 1990) writes, "My monozygotic twin brother and I, brown-eyed, had blue-eyed parents and blue-eyed sibs."

Also probably *not* inherited as single-gene traits are hair color, skin color, height, nose shape, earlobe type, tongue curling, handedness, hand clasping pattern, and production of excess intestinal gas. Another common misconception about simply inherited traits is that the dominance or recessiveness of a trait determines its frequency in a population. This is not true. As will be explained in Chapter 17, dominant traits can be rare and recessive traits can be common.

Confirmed Traits

Name of trait	Description
Achoo syndrome	Sneezing in response to bright light
Camptodactyly	Palmward bending of a finger, usually fifth
Cleft chin	Chin dimple
Clinodactyly	Shortened and radially curved fifth finger
Distichiasis	Two rows of eyelashes
Ear malformation	Tightly curled, cup-shaped external ear
Ear pits	Pit in front of ear opening
Earwax, wet type	Sticky (as opposed to dry) earwax
Epicanthus	Vertical fold of upper eyelid skin covering inner corner of eye, as present in Asians
Extra nipples	Extra nipples/breasts in males or females
Hairy palms/soles	Hairy areas near wrist and on arch of foot
Male pattern baldness	Gradual loss of hair from top of scalp
"Michelin tire baby" syndrome	Multiple deep skin creases on limbs
Midphalangeal hair	Hair on top of middle segment of fingers
Nevus flammeus, nape of neck	Port wine stains on nape of neck
Odd-shaped teeth	Peg-shaped teeth
Piebald trait	White forelock, and lack of pigment in midline extending from face to abdomen
Prognathism	Hapsburg jaw; prominent lower jaw, often with thickened lower lip and flat cheekbones
PTC tasting	Ability to taste phenylthiocarbamide
Amylase enzyme variants	Variations in the breakdown of starch and glycogen
Trembling chin	Trembling triggered by anxiety or upset
Tune deafness	Inability to distinguish different notes
Uncombable hair	Longitudinal groove present in each hair

We have selected some normal or benign morphological and behavioral traits from McKusick's 1990 listing of autosomal dominant genes. For these *confirmed* traits, the evidence for autosomal dominant inheritance is strong and the phenotype is known to be determined by a distinct locus.

may show no obvious phenotypic differences whatsoever, or even genes whose phenotypes and gene products are unknown.

We begin, however, with the standard approach. Summarized in this chapter are the three types of single-gene inheritance most often found by traditional pedigree studies: autosomal dominant, autosomal recessive, and X-linked recessive inheritance. We will also look briefly at X-linked dominant, Y-linked, and mitochondrial inheritance. (The kinds of traits determined by the *interaction* of numerous genes will be discussed in later chapters.)

Some Warnings

In trying to determine how a given disorder is inherited, genetic sleuths must keep several things in mind. Although a single pedigree may be used to rule out certain possibilities, it may not be enough to prove a specific genetic hypothesis. For example, some pedigrees for X-linked recessive inheritance would also fit an autosomal gene acting as a dominant in males and a recessive in females. (The gene for male pattern baldness, discussed in Chapter 7, may behave this way.) Thus, it may be necessary to pool together many studies and analyze the data by more sophisticated mathematical techniques.

On the other hand, because mutations at different loci may occasionally give rise to very similar phenotypes, some inherited disorders do have multiple modes of transmission. McKusick's 1990 catalog includes several dozen phenotypes with more than one pattern of inheritance (Table 5.1). Some of these phenotypes can be inherited in three different ways: as an autosomal dominant, an autosomal recessive, or an X-linked trait! In such cases the initial diagnosis of a given disease does not always predict its gene location or its mode of inheritance, and a very careful clinical and genealogical follow-up is needed to determine which form it really is and to provide accurate genetic information to the family.

For these reasons, geneticists who are armchair physicians or physicians who are armchair geneticists should be very cautious about giving advice to families with suspected genetic conditions. It is better to refer such patients to specialists who are trained in both medicine and genetics. These genetic counselors are more likely to be familiar with new developments in diagnosis and treatment and with the various problems—psychosocial as well as medical and genetic—encountered by such families. Some of these problems will be touched on in this chapter as we discuss several simply inherited human disorders.

AUTOSOMAL DOMINANT INHERITANCE

Stormie Jones died because her body's cells lacked the ability to remove cholesterol from her bloodstream. Instead, much of this fatty substance remained trapped there, while some got deposited as lumps in her tendons and under her skin (Figure 5.3) and as plaques that clogged her arteries and caused her heart attacks. Before discussing the exact cellular defect involved, however, we should describe the phenotypes and present a pedigree of familial hypercholesterolemia.

Unaffected newborn babies have serum cholesterol levels of about 30 milligrams per deciliter (mg/dl). In unaffected adults raised on a low-fat diet, serum cholesterol levels are 50–80 mg/dl. But in adults raised on high-fat diets in Western societies, "ideal" serum cholesterol levels are 130–190 mg/dl.

FH heterozygosity is quite common, with a worldwide frequency of 1 in 500, accounting for 5% of all heart attacks in patients under the age of 60. Heterozygotes are born with cholesterol levels at least twice the normal adult levels. Heart attacks are 25 times more frequent than in unaffected

Table 5.1.
Some human disorders with known multiple modes of inheritance and gene loci or with one mode of inheritance and two or more gene loci.

Name and Inheritance Mode(s)[a]	Brief Description	Chromosome Loci[b]
ALL THREE MODES: AD, AR, AND X		
Cataracts	Clouding of the eye lens or of its capsule	1, 2, 16, X
Charcot-Marie-Tooth disease	Progressive weakness and atrophy of leg and arm muscles	1, 17, X
Cleft lip/palate	Opening in midline of upper lip and/or palate[c]	1, 6, X
Craniosynostoses	Premature fusion of skull bones	7, X
Ehlers-Danlos syndrome	Loose joints, stretchy and fragile skin, skin scarring	2, 7, 17, X
Hypoparathyroidism	Low or no parathyroid hormone, leading to low calcium and to muscle spasms	11, X
TWO MODES		
Diabetes insipidus (AD, X)	Excess thirst and urination due to disorder of a pituitary hormone	20, X
Diabetes mellitus (AD, AR)	Faulty sugar metabolism due to lack of pancreatic insulin	6, 11, 19
Epidermolysis bullosa (AD, AR)	Occurrence of large blisters, spontaneously or with mild trauma	8, 11
Glycogen storage disease (AR, X)	Accumulation of glycogen in cells, due to errors in metabolism	1, X
Hypothyroidism (AD, AR)	Deficient thyroid hormone, causing cretinism or mental and physical slowing	1, 8, 22
Osteogenesis imperfecta (AD, AR)	Brittle, easily broken bones	7, 17
Severe combined immunodeficiency disease (SCID) (AR, X)	Impaired or absent immune response	20, X
Spherocytosis (AD, AR)	Abnormal spheroid red blood cells, causing chronic anemia	1, 8, 14
Von Willebrand disease (AD, X)	Deficiency of clotting factor VIII, causing gastrointestinal, urinary, and uterine bleeding	12, X
ONE MODE, TWO OR MORE LOCI		
Adrenal hyperplasia (AR)	Ambiguous sex organs due to steroid hormone imbalances	1, 6, 8, 10
Elliptocytosis (AD)	Elliptical red blood cells, mild anemia due to red blood cell destruction	1, 14
Hyperphenylalaninemia (AR)	Excess phenylalanine in urine, other PKU and PKU-like symptoms, some mild	4, 12
Porphyria (AD)	Abnormal porphyrin metabolism, affecting skin, liver, and brain	1, 9, 11, 14
Tuberous sclerosis (AD)	Epilepsy, mental retardation, skin tumors and depigmented spots, bone cysts	9, 11
X-linked mental retardation (X)	Several phenotypically distinct types (all with X-linked loci)	X

[a] AD = autosomal dominant, AR = autosomal recessive, and X = X-linked.
[b] Relatively few human gene loci have been mapped to their exact locations. Thus, for any given disorder, the chromosomal sites listed here are not exhaustive.
[c] Cleft lip and/or palate is a complex trait, often not inherited in a simple, predictable fashion; also, over 200 syndromes include cleft lip and/or palate as a feature. Thus, extreme care must be taken in diagnosis and in counseling families about the chances of recurrence.

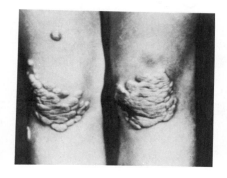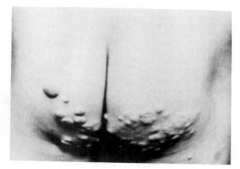

Figure 5.3
Lumpy deposits of cholesterol found in people (especially homozygotes) with hypercholesterolemia. (From Goldstein and Brown 1989.)

relatives and begin to occur at about age 35; 50% of males and 15% of females die by age 60. Surprisingly, FH patients usually have a slim body build and normal blood pressure. Rare *FH homozygotes* such as Stormie occur only about once in a million births. They are born with serum cholesterol values about four to six times the normal levels; this excess is limited to the **low-density lipoprotein (LDL)** fraction of cholesterol (discussed shortly). Heart attacks are common from age 5 on and often cause death before age 20. Thus, homozygotes rarely reproduce.

Pedigree Analysis: Familial Hypercholesterolemia

The pedigree in Figure 5.4 shows all the characteristics of **autosomal dominant inheritance** when a gene is fully, or almost fully, penetrant—that is, expressed phenotypically wherever it occurs (Chapter 7). Note the following characteristics:

1. *All affected individuals have at least one affected parent.* Also, affected individuals are usually found in every generation of a sizable pedigree.
2. *Males and females are affected in about equal numbers* (here 19 females, 20 males), which makes X linkage less likely and rules out Y linkage.
3. *Both males and females transmit the trait.* Note that females I-1, II-1, II-14, and II-18 and males II-12, II-16, and II-20 each produced affected progeny of both sexes. Male-to-male transmission rules out X linkage, and transmission by a female rules out Y linkage.
4. *Matings of heterozygous affected with unaffected individuals produce about 1/2 affected and 1/2 unaffected offspring.* Here, when all the progeny from such matings are added together, the totals are 33 known affected, 23 known unaffected. The phenotypes of 3 additional offspring were unknown.
5. *The trait does not appear in the descendants of two unaffected parents* unless the dominant allele is reintroduced by matings with unrelated affected individuals.
6. *The trait may be more severe or extreme in homozygotes than in heterozygotes, or it may even be lethal.* Here, the only mating that might possibly have produced homozygotes is II-18 × II-19.

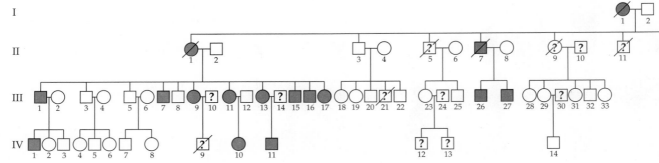

Figure 5.4
Pedigree of familial hypercholesterolemia, an autosomal dominant trait. All affected persons are heterozygotes. (Redrawn from Schrott et al. 1972.)

Metabolic and Genetic Control of Cholesterol

Despite its bad press, cholesterol is needed in our bodies. It is a major component of cell membranes and also the raw material for bile acids and steroid hormones. But here's the rub:

> The very property that makes it useful in cell membranes, namely its absolute insolubility in water, also makes it lethal. For when cholesterol accumulates in the wrong place, for example within the wall of an artery, . . . its presence eventually leads to the development of an atherosclerotic plaque . . . If cholesterol is to be transported safely in blood, its concentration must be kept low. (Brown and Goldstein 1986)

Our bodies get their cholesterol in two ways: from what we eat and from what our cells (mostly liver cells) make from scratch. These two pathways are interconnected by means that were discovered by Michael S. Brown and Joseph L. Goldstein of the University of Texas Health Science Center at Dallas—research that won them the Nobel prize in 1985.

Within our intestines, fats are broken down into cholesterol and fatty substances called triglycerides, which then move into the bloodstream and travel throughout the body. But being insoluble in water, they cannot be transported "as is"; instead, they are packaged into *lipoprotein particles*. The major carrier of cholesterol is the low-density lipoprotein (LDL) fraction. Each LDL particle has a cholesterol core protected by an outer coat and topped by a special protein molecule.

How is LDL-packaged cholesterol removed from the bloodstream? Brown and Goldstein (1984) found that cell surfaces are studded with binding sites for the "topping" protein. These **LDL receptors** are made of protein with some sugars chains attached, and their numbers increase or decrease according to a cell's need for cholesterol. After latching onto the LDL receptors, LDL particles get pulled into the cytoplasm and processed in various ways. But these regulatory mechanisms cannot control cholesterol levels *outside* cells when large amounts of fats and cholesterol are present in the bloodstream. Blocked from entering the liver, the excess cholesterol remains trapped in the bloodstream and ultimately gets deposited inside artery walls, causing atherosclerotic plaque formation and heart disease.

In genetically normal individuals with low-fat diets, the LDL pathway works very effectively. But the high-fat diets common in Western societies can swamp the LDL transport and metabolism pathways, leaving dangerous levels of cholesterol circulating in the bloodstream.

In the 1980s, researchers purified the LDL receptor molecule and determined the sequence of its 839 amino acids. They also isolated the

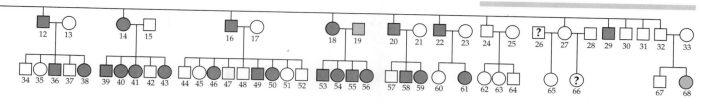

○ □ Normal
● ■ Possibly affected
● ■ Affected
? ? Not sampled
⊘ ⊠ Deceased

LDL receptor gene, a medium-sized locus residing on the short arm of chromosome 19. About 20 mutations of this gene have been found. Some produce no LDL receptors at all, while others produce LDL receptors that are nonfunctional for one reason or another. These different mutations are not randomly scattered throughout the peoples of the world; instead, particular defects tend to be found in high frequency among particular populations.

In certain self-contained populations—especially Lebanese, South African Afrikaners, and French Canadians—FH is extremely frequent. Among South African Afrikaners, for example, FH heterozygotes are 1 in 100 and homozygotes are 1 in 30,000, and two particular mutants account for nearly all affected individuals. This situation is most likely due to the **founder effect**—the rapid multiplication in a small population of just one or two mutant *FH* alleles (Chapters 17 and 18). Similarly, there is good evidence for a founder effect among the ancestors of present-day French Canadians and Lebanese. In such cases the original mutation(s) can sometimes be traced back to one or two persons in a group of immigrants to a new territory.

From detailed molecular studies it has also become clear that *many affected individuals, previously thought to be homozygotes, are actually heterozygous for two different LDL receptor mutant alleles.* This finding is true not only of FH, but of virtually every important genetic disorder that has been carefully analyzed in recent years. It helps to explain why the phenotypes of these disorders can be quite variable.

Low-fat diets are the first approach to reducing cholesterol levels in both FH heterozygotes and genetically unaffected individuals. With proper treatment, FH heterozygotes should be able to lead reasonably normal lives, even though their one normal allele of the LDL receptor gene can make only about half as many LDL receptors as do unaffected individuals. But FH homozygotes, having no normal alleles, cannot make receptors under *any* conditions. Thus, their only treatment possibilities are repeated plasma exchanges and/or surgery, including liver-heart transplants from organ donors with normally functioning LDL receptor genes. In families or populations where FH is especially common, the detection of FH homozygotes by prenatal diagnosis may be desirable.

Another autosomal dominant trait with tragic consequences is described in Box B. For Huntington disease, however, the lethal effects occur late rather than early in life. Some additional examples of human disorders inherited as autosomal dominants are listed in Table 5.2.

AUTOSOMAL RECESSIVE INHERITANCE

The time was October 1971, and the place was Redding, Connecticut. The arrival of baby Alexandra—the first girl born into the Deford family since 1904—was greeted with great joy. Although underweight, she was de-

HUNTINGTON DISEASE

Huntington disease (HD) is an autosomal dominant neurological disorder that killed folksinger Woody (Woodrow Wilson) Guthrie (1912–1967). His mother Nora, an intense and very musical person, was affected, and two of his eight children have also died from the disease.

Among the first to clearly describe this tragic and incurable condition was Johan Christian Lund, a physician in rural Norway. His 1860 report (translated by Ørbeck 1959) includes four-generation pedigrees of two families (both males and females affected) in which "chorea St. Vitus," his name for the condition, recurred as a hereditary disease in Saetersdal:

> It is commonly known as the 'twitches'. . . . It usually occurs between the ages of 50 and 60, generally starting with less obvious symptoms, which at times only progress slowly . . . but more often after a few years they increase to a considerable degree, so that any form of work becomes impossible and even eating becomes difficult and circuitous. The entire body, though chiefly the head, arms, and trunk, is in constant jerking and flinging motion. . . . A couple of the severely affected patients have become demented.

Lund's obscure report was rediscovered in 1914, but attracted very little attention. Indeed, a few American physicians had written even earlier descriptions that were likewise ignored.

What did attract attention was a report (still considered a classic) by George Huntington, an American physician whose father and grandfather were also family doctors on Long Island, New York. His first encounter with this disorder—and the beginning of his medical education—took place when he was a young boy accompanying his father on professional rounds. In 1872, drawing on the 78 combined

This 1943 photo shows Woody Guthrie at age 31, about a decade before he began to show the symptoms of Huntington disease. (Courtesy of Woody Guthrie Publications, Inc.) For more information about Woody Guthrie and his family, see Klein 1980.

years of the three physicians' experiences in that area, the 22-year-old Dr. Huntington described a

> . . . hereditary chorea . . . confined to certain and fortunately a few families transmitted to them, an heirloom from generations away back in the dim past. . . . It is attended generally by all the symptoms of common chorea, only in an aggravated degree, hardly ever manifesting itself until adult or middle life, and then coming on gradually but surely, increasing by degrees, and often occupying years in its development, until the hapless sufferer is but a quivering wreck of his former self. . . .
>
> Of its hereditary nature. When either or both the parents have shown manifestations of the disease, . . . one or more of the offspring almost invariably suffer from the disease, if they live to adult age. But if by any chance these children

go through life without it, the thread is broken and the grandchildren and great-grandchildren of the original shakers may rest assured that they are free from the disease. . . . Unstable and whimsical as the disease may be in other respects, in this it is firm, it never skips a generation to again manifest itself in another; once having yielded its claims, it never regains them. (Huntington 1872)

Note that this description of autosomal dominant inheritance in humans followed the publication of Mendel's laws by just seven years! Today, the protein product and the primary action of this mutant gene are still unknown. But their result, the progressive degeneration of certain groups of brain cells, is well documented. Age of onset is highly variable, the average being about 38 years. There is no effective treatment, and death usually comes 10 to 25 years after the onset of symptoms. The frequency of HD is roughly 1 in 20,000 in most Western countries, but much lower among Asians and blacks.

In 1993, after an intensive and frustrating ten-year search, researchers finally found and cloned the HD gene, which lies near the tip of 4p (Chapter 13). This discovery should make it possible to determine with virtual certainty whether at-risk relatives possess the HD allele (Morrell 1993). The decision about whether to take this test is an agonizing one, however. The possible relief of knowing that one is most likely free of the disease must be balanced against the grief and loss of hope that might overwhelm those who learn that they most likely have the mutant allele. Indeed, genetic counselors worry about the potential for suicides in the latter group. And for a variety of reasons, most members of HD families have thus far chosen not to take the test.

clared healthy at birth. Yet from the very beginning, she was "sick all the time":

> She ate voraciously, but food went right through her and she couldn't put on any weight. Besides, she always seemed to have colds and ear infections. The pediatrician finally said it was "failure to thrive." (Deford 1986)

Table 5.2.
Some human disorders inherited as autosomal dominants, and their chromosomal locations.

Name	Brief Description	Site
Angioneurotic edema, hereditary	Allergy-caused itching and swelling, often involving the airways	11p
Dentinogenesis imperfecta	Teeth abnormally shaped, blue-gray or brown, and opalescent	4q
Hyperthermia of anesthesia	Abnormal/lethal response to anesthesia; high fever, muscular rigidity	19q
Mandibulofacial dysostosis (Treacher Collins syndrome)	Abnormal jaw/facial bones, ears; down-slanting eyes, other defects	5q?
Marfan syndrome	Tall, loose-jointed; long, thin limbs, hands, feet; eye and heart/artery defects	15q
Myotonic dystrophy	Difficulty in relaxing contracted muscles; muscle wasting, cataracts, defects in gonads, heart, mental retardation	19q
Nail-patella syndrome	Poorly developed or absent kneecaps and nails; defects of joints, kidneys, eyes	9q
Neurofibromatosis	Defects in nerves, muscles, bones, skin, and eyes; pigmented spots and soft tumors all over body	17q
Polyposis, adenomatous intestinal	Precancerous growths in lining of large intestine (usually colon)	5q
Retinoblastoma	Tumor(s) of retina of the eye(s), with predisposition to bone cancer	13q
Ventricular hypertrophy, hereditary	Overgrowth of tissue between heart ventricles, obstructing blood flow	14?
von Hippel-Lindau syndrome	Blood-vessel-forming tumors in the retina and cerebellum	3p
Waardenburg syndrome	Wide-set inner corners of eyes, wide nose, white forelock, hearing loss	2q

Every day Alex got weaker. At age four months, scrawny and extremely ill with pneumonia, she was given a "sweat test" to measure the amount of salt in her sweat. The results showed that she had **cystic fibrosis (CF)**—such a serious case, in fact, that she could die at any time. Her parents, themselves healthy and having no known relatives with CF, were shocked and devastated by the news.

But Alex survived, becoming a bright, lively, and extraordinarily plucky child who charmed and amazed all those who knew her (Figure 5.5). She attended nursery school and then elementary school, for a few years managing to live a fairly normal life except for occasional hospital stays. But she also had to endure a daily routine of long and painful treatments: inhaling decongestants to loosen the thick mucus that clogged her lungs; having her chest pounded and pinched to force the mucus out; taking antibiotics to prevent infections; and taking enzyme preparations to help with digestion.

Cystic fibrosis, her parents learned, is the most common lethal autosomal recessive disorder among Caucasians and a leading cause of childhood death. Frequency estimates vary from about 1 in 500 to 1 in 3,800 among different white populations; but overall, about 1 in 2,000 newborns are affected in the United States. CF is rare among American blacks (1 in 17,000 births), however, and rarer among Hawaiian Asians (1 in 90,000). About 1 in 20 to 25 people are heterozygous for this gene, but most carriers never know they harbor a lethal allele until they produce an affected child.

The disease is variable, but its most common and serious form has three major manifestations: extremely salty sweat* and serious digestive and respiratory problems. These and other effects are caused by decreased chloride and sodium reabsorption in the ducts of glands that release their

Figure 5.5
Photo of Alex Deford. (Courtesy of Frank Deford.)

---

* An old German folk saying states that the baby who is kissed at birth and tastes salty will have a short life.

products into tubules (rather than into blood or lymph vessels). As a result, thick, sticky secretions clog up all these tubules, causing irreversible damage and malfunctioning of the lungs, pancreas, liver, sweat glands, and reproductive organs. Clinically, there are two main subgroups among CF patients: 15% show *pancreatic sufficiency* and 85% show *pancreatic insufficiency*. In the latter group, the pancreatic enzymes fail to reach the intestines; thus, digestion and absorption are incomplete and intestinal complications are common. Growth may be retarded in children with CF, and they all have a greatly increased (and often fatal) susceptibility to bacterial infections, especially pneumonia.

In 1950 few patients survived past infancy; but now the median age of survival is about 24 years, with over 25% of patients living into their 30s. Most adult CF males are sterile because of damaged tubules in the testes, and fertility is greatly reduced in females. But many CF adults marry, and most lead active, productive lives. Despite vast improvements in diagnosis and treatment over the last few decades, however, many CF patients still die within a year or two of birth. There is still no cure for this chronic, relentless disease, and the suffering it causes is monumental:

> For CF families the divorce rate is several times the national average, and the incidences of separation, alcoholism, desertion, wife beating, and so on are all comparably high. And the physically healthy siblings, no less than the parents, are often crippled in many emotional ways. (Deford 1986)

Alex was lucky to have a strong, loving family. But despite their best efforts, at age five her health began to deteriorate. She coughed and gagged, struggling to breathe—a result of her severe mucus blockage. Her appetite waned and she lost weight. She tired easily, growing too weak to play at recess time. Her fingertips became pale and clubbed. She spit up blood. Her heart weakened. She had liver problems, a regular high fever, and even developed arthritis. Bouts with pneumonia grew more serious, and her hospital stays increased in length and frequency.

Alex attended the first day of third grade, but then had to return to the hospital for several months. She was in constant pain; her breathing became labored, and she often needed oxygen. A lung collapsed three times, requiring the insertion of tubes through her chest wall. Alex knew she was dying and talked freely about it. Because nothing more could be done for her, she returned home just before Christmas. A few weeks later, eight-year-old Alex died in her parents' arms.

Since Alex's death, scientists have made phenomenal progress in unraveling the mysteries of cystic fibrosis: determining the location of the CF gene on chromosome 7; isolating it and identifying many of its mutations; analyzing the phenotypic variants; isolating the CF protein; and beginning to study its functions in the cell membrane. These and other findings have greatly improved the accuracy of prenatal diagnosis and made possible, for the first time, carrier detection in families with cystic fibrosis—and, to a certain extent, in the general population. Some details will be presented in later chapters; here we will focus on the inheritance pattern of CF.

One major puzzle is how such a lethal gene can be so common, especially when so few affected individuals reproduce. Perhaps the mutation rate of this gene is unusually high. Perhaps the CF mutations persist because of some small beneficial effect—such as an increase in fertility or a resistance to diarrhea caused by bacterial toxins—among unaffected carriers. Another possibility is a founder effect.

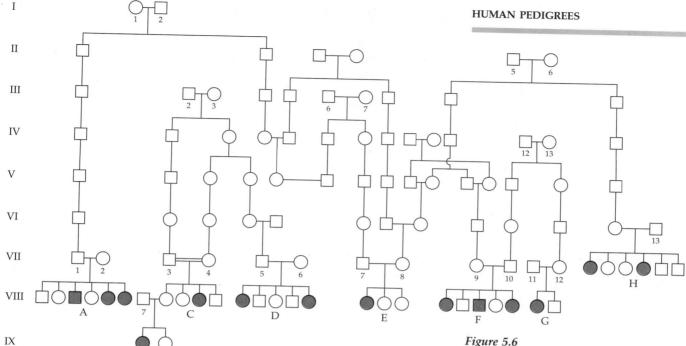

Figure 5.6
Autosomal recessive inheritance of cystic fibrosis in eight related families (A–H) from the department of Finistere in northwestern Brittany. By interviewing family members and examining birth, death, and marriage registers, geneticists were able to trace these families back six to eight generations, within which there are at least five ancestral couples: I-1 and I-2, II-5 and II-6, III-2 and III-3, III-6 and III-7, and IV-12 and IV-13. Note that there is only one known consanguineous marriage, in generation VII. (Redrawn from Bois et al. 1978.)

Pedigree Analysis: Cystic Fibrosis

The pedigree shown in Figure 5.6 is taken from a genealogical survey of a small area in northwestern Brittany where cystic fibrosis is common (1 in 377 live births) and the population is quite isolated and stable. It exhibits some, but not all, of the features of **autosomal recessive inheritance**:

1. *Most affected individuals have two unaffected parents*, so that the trait tends to show up erratically and unpredictably. In fact, such a gene is often passed on invisibly for many generations, through unaffected heterozygotes. Only when two carriers produce an affected child is the condition manifested. In this pedigree, for example, all 14 affected individuals have unaffected parents, and there are no recorded cases of CF for five to six consecutive generations.

2. *The trait is expressed in both sexes and transmitted by either sex to both male and female offspring in roughly equal numbers.* Here we see 12 affected females and only 2 affected males—clearly an extreme sex ratio, but perhaps due to chance variation, because another study of a larger area in northern Brittany turned up 16 boys and 18 girls. The trait is obviously carried by heterozygotes of both sexes.

3. *Matings between unaffected heterozygotes yield about 3/4 unaffected and 1/4 affected offspring.* Here we see eight sibships (A–H) produced by matings between known heterozygotes; they contain a total of 33 offspring, of which 14 are affected. At first glance there appear to be far too many CF offspring, that is, 14 rather than about 8. But in pedigrees like this, the *numbers are always biased in favor of affected offspring, because we only identify those heterozygous parents who have at least one affected child!* Moreover, fam-

ilies having a larger number of affected members are more likely to come to medical attention. *Heterozygous parents who by chance produce no affected children will escape detection*; such sibships with heterozygous parents and all unaffected children are wrongfully excluded from our CF totals. On the other hand, all CF homozygotes are quite readily detectable.

4. *With rare traits, the unaffected but heterozygous parents of an affected individual may be related to each other.* When this occurs, the recessive allele that each parent carries has usually originated in an ancestor that they have in common (Chapter 17). In this pedigree, sibship C resulted from a marriage between two distant cousins (VII-3 and VII-4), each of whom probably inherited a CF gene from the same common ancestor (III-2 or III-3).

5. *Matings between an affected and an unaffected parent will usually yield all unaffected offspring*; only if the unaffected parent is heterozygous can affected children be produced. CF pedigrees usually lack such matings because of the sterility of males and the shortened life span of affected individuals, but a few pregnancies of CF females have been reported.

6. *When both parents are affected, all their children are affected.* (This outcome contrasts with that for a dominant autosomal trait, where two affected parents can produce unaffected children.) For reasons already mentioned, no such matings occur in CF pedigrees.

The Cystic Fibrosis Gene and Protein

Attempts to map the CF gene involved years of effort and numerous detours down blind alleys. But in 1989 a research team led by Lap-Chee Tsui at the Hospital for Sick Children in Toronto finally found it near the middle of the long arm of chromosome 7. Although a giant gene, its known mutations are restricted to a relatively small region.

In 1990 researchers isolated the CF protein, **cystic fibrosis transmembrane regulator (CFTR)**, which contains 1,480 amino acids. Its structure is similar to that of several proteins known to transport small molecules across cell membranes. How it regulates the passage of chloride through epithelial cells is not well understood. Recent studies suggest that the normal CFTR molecule may be unusual in having two separate functions: to provide a chloride channel (a pore in the plasma membrane through which chloride ions pass) and to act as a regulator, opening and closing that pore—that is, to be both the gate and the gatekeeper.

The most common mutation, found on 70–75% of all CF chromosomes worldwide, is a deletion of three bases that results in the loss of the amino acid phenylalanine (abbreviated F) at position 508 in the CF protein. Thus this mutation is called *ΔF508*. The ΔF508 protein is improperly processed and gets hung up in the Golgi complex rather than migrating to the cell surface. Over 200 other CF mutations have been reported, but each of them is rare and together they still do not account for all of the remaining 25–30% of CF chromosomes.

Some studies suggest a correlation between CF mutational genotypes and clinical phenotypes. Patients homozygous for the *ΔF508* mutation are often diagnosed at an earlier age and have a much greater frequency of pancreatic insufficiency (99%) than do patients with any other CF genotype. The pancreatic-sufficient patients who are not homozygous for *ΔF508* may show milder and more heterogeneous symptoms. Oddly enough, a few patients with a greatly shortened CFTR protein (owing to mutations that produce a lopped-off gene product) exhibit milder symptoms than those with the most common mutant protein, which lacks just one out of 1,480 amino acids! The reasons for this unexpected finding are not understood.

Identification of the CF gene and its CFTR protein have greatly improved prospects for the development of drugs or antibodies to treat the disease. In addition, some researchers are now considering the possibility of *gene therapy* for CF—that is, getting normal copies of the gene into those CF epithelial cells that line the lungs and airways. Indeed, a few experimental steps in this direction have already been taken. But it is a long way to go from curing CF cells in a laboratory dish, or inserting human genes into rat lung cells, to curing CF in people. In Chapter 20 we will discuss gene therapy in greater detail.

Detection of CF Homozygotes and Carriers

For any disease, geneticists must consider several levels of urgency when devising ways of detecting homozygotes (before or after birth) and heterozygotes (before or after producing an affected child):

1. Top priority is given to *high-risk families*, that is, those in which an affected child has already been born. Here, depending on the mode of inheritance, the carrier status of one or both parents has thus been revealed, and the major task is to detect any additional affected offspring before or after they are born.
2. Next priority is given to the *close relatives* (i.e., sibs, aunts, uncles) of high-risk families, whose carrier status is unknown and who have not yet produced affected offspring. Here the double challenge is to identify carrier parents before they produce affected children and to detect the affected children before birth, if possible.
3. For inherited conditions that are lethal (or very serious) and also relatively common—and especially if an effective treatment is available—the screening of entire populations might be undertaken to detect both carriers and future affected individuals.

Thus far, cystic fibrosis testing has progressed only to the second level. Although the third level is to some extent possible, it is also very controversial, and only a few trial population screenings have been undertaken.

In families known to be at risk, there are several ways to detect CF prenatally. One procedure measures the levels of certain intestinal enzymes in amniotic fluid samples from pregnant women known to have a 1-in-4 chance of producing a baby with CF. But the most reliable indicators involve DNA analysis together with enzyme tests. In families where DNA samples can be obtained from a CF child, molecular techniques allow the direct prenatal detection of future affected offspring. Carriers can also be directly detected in families where DNA is available from a CF patient. But when a CF child has died and a DNA sample is *not* available, the probability estimates for prenatal diagnosis and carrier status are more complicated and depend partly on whether DNA is available from an unaffected sib.

One promising approach to mass screening of newborns is a test for blood levels of the pancreatic enzyme *trypsin*, which tends to be higher in babies with CF, but even this test misses a small percentage of the cases. Another goal is to find a simple and reliable test for detecting heterozy-

BOX C

SICKLE-CELL DISEASE

Another example of autosomal recessive inheritance is sickle-cell (or hemoglobin S) disease, which is most common (up to 1 in 50) among West African blacks. Several centuries ago, tribes there gave it specific names—*ahotutuo* (Twi tribe), *chwechweechwe* (Ga tribe), *nwiiwii* (Fante tribe), and *nuidudui* (Ewe tribe)—and recognized that it ran in families. They described two clinical types, "severe" and "not-so-severe," the latter type being essentially normal and vastly more frequent than the severe type. They even realized that matings between "severe" and "not-so-severe" types (which we would call homozygote × heterozygote) gave rise to a much greater proportion of severely affected children than did matings between two "not-so-severe" parents (Konotey-Ahulu 1991).

In 1910, Chicago physician J. B. Herrick published a report (the first in Western medical literature) about "peculiar elongated and sickle-shaped red blood corpuscles in a case of severe anemia" in a black dental student from Grenada, West Indies. Twelve years later, another case report and a review of previously published cases finally aroused widespread interest in this condition. Mason (1922) comments:

> The patients all complained of weakness and poor health since early childhood. All had swelling of the ankles associated with leg ulcers. . . . Two patients had had repeated attacks of abdominal pain associated with fever and jaundice. . . . The blood picture does not resemble that seen in any of the more common anemias, and it is possible that the disease represents a clinical entity. If that is true, it is of particular interest that up to the present the malady has been seen only in the negro.

Later studies established that sickle-cell anemia also occurs in India, the Middle East, the Mediterranean area, and the Caribbean. Individuals affected with *sickle-cell disease* are homozygous for a defective hemoglobin S allele. Heterozygotes are unaffected under ordinary circumstances, but in unusual situations of oxygen deprivation they may show some clinical symptoms. Thus, heterozygotes are said to have the *sickle-cell trait*. Further discussions of sickle-cell disease appear in Chapters 7, 12, 18, and 20.

gotes in the general population before they produce affected offspring. Identification of the CF locus has raised the possibility of doing more widespread screening for both homozygotes and heterozygotes. About 85% of CF carriers possess the most common mutant allele, *ΔF508*, but the remaining 15% of carriers have various rare mutant alleles for which good tests are not yet available. For this and other reasons (Chapter 19), there are currently no large-scale plans for mass screening to detect the 7 to 10 million unsuspecting CF carriers in the United States.

Box C discusses another serious but fairly common autosomal recessive disorder: sickle-cell anemia. For examples of additional human disorders inherited as autosomal recessives, see Table 5.3.

X-LINKED RECESSIVE INHERITANCE

All muscular dystrophies are characterized by gradual weakening and wasting of muscle tissue (*dys*, "abnormal"; *trophy*, "growth"), and all are inherited; but the symptoms, probable metabolic defects, and modes of inheritance vary considerably.

The most common and deleterious of these syndromes is **Duchenne**

Table 5.3.
Some human disorders inherited as autosomal recessives, and their chromosomal sites.

Name	Brief Description	Site
Adenosine deaminase (ADA) deficiency	Severely impaired immunity, skeletal and neurological abnormalities	20q
α-1 antitrypsin deficiency	Lack of protease inhibitor (Pi), causing severe lung and liver disease	14q
Ataxia telangiectasia	Poor muscle coordination, dilation of capillaries in eyes and skin, immune defects	11q
Albinism, "classic"	Absence of melanin, affecting eyes, hair, skin, and hearing	11q
Disaccharide intolerance I	Diarrhea due to inability to absorb sucrose that has not been split into glucose and fructose	3q
Friedreich ataxia	Degeneration of cerebellum and spinal cord; defects in limb coordination, tendon reflexes, speech, cardiac function	9q
Galactosemia	Inability to digest milk sugar; liver enlargement and damage, cataracts, mental retardation	9p
Gaucher disease	Blood abnormalities, enlarged spleen, bone and neurological defects, skin pigmentation	1q
Hemochromatosis	Excess absorption/storage of iron; liver damage, diabetes, darkened skin, heart failure	6p
Homocystinuria	Displaced eye lens, defects of skeleton and blood vessels, mental retardation	21q
Hurler syndrome	Mental retardation, enlarged liver and spleen, bone defects, coarse features	22q
Pyruvate kinase-1 deficiency	Hemolytic anemia (loss of red blood cells), liver damage	1q
Tay-Sachs disease	Developmental retardation, paralysis, dementia, blindness, red spot on retina	15q
Thalassemias, α- (inactivation of 1-4 α-globin alleles at the two loci)	Varies from no symptoms to mild or moderate anemia to stillbirth (from lethal anemia, heart failure, fluid accumulation)	16p
Thalassemias, β- (defects or inactivation of one or two β-globin alleles)	More than 70 different mutations known; varies from no symptoms to thalassemia major (severe anemia, defects of bone, liver, spleen; ultimately fatal)	11p
Wilson disease	Defects in copper metabolism; damage to liver and central nervous system; corneal changes, tremors, emotional changes	13q
Zellweger syndrome	Absence or defects of peroxisomes, leading to defects of skull, face, ears, eyes, hands, feet, liver, and kidneys	7q

muscular dystrophy (DMD), which was first described in 1868 by French neurologist Guillaume Benjamin Amand Duchenne. In 1879, 21 years before the rediscovery of Mendel's Laws, it was reported that affected males inherited this condition through unaffected mothers. DMD is now known to be a fully penetrant, X-linked recessive condition occurring in about 1 in 3,500 to 4,000 newborn males. Like cystic fibrosis, DMD remains at a high frequency in the population. About one-third of all cases appear to be new mutations; that is, the mothers are not carriers.

"Usually the onset . . . occurs before age 6 years, and the victim is chairridden by age 12 and dead by age 20" (McKusick 1990). Most patients show normal intelligence, although some are mildly retarded and about 20% are seriously retarded. The earliest signs are developmental delays in walking and talking. Then come weakness and gradual wasting of the thigh and pelvic muscles, leading to unsteadiness, difficulty in walking stairs or rising from chairs, and a cautious, waddling gait. Although weak, the calf muscles are noticeably thickened. As the shoulder, trunk, and back muscles gradually weaken and degenerate, the child develops a "swayback" posture (Figure 5.7), has trouble maintaining balance, and

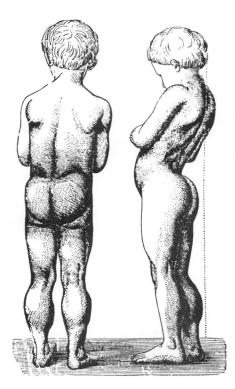

Figure 5.7
Duchenne muscular dystrophy.
Thickened calves and swayback posture
seen in young patients. (From Duchenne
1868.)

falls a lot. With gradual contraction of the Achilles tendons (heel cords), the child also begins to walk on his toes. Later, tendons connected to the hips, knees, and elbows will also shorten, and curvature of the spine (scoliosis) will squeeze and interfere with the functioning of internal organs (notably the lungs). Heart problems are the rule, sometimes leading to sudden death. Usually, however, patients die of respiratory infections or of respiratory failure when diaphragm muscles become affected.

Under the microscope, Duchenne dystrophic muscles show a significant loss of muscle fibers and variable sizes of those that remain, along with considerable infiltration of fat and connective tissue. Muscle physiology goes awry, and a muscle enzyme called *creatine phosphokinase (CPK)* is found in the serum at levels 50 to 100 times higher than normal. Because CPK levels may be normal before birth, however, this diagnostic test is not reliable enough for prenatal use.

A similar but rarer X-linked condition is **Becker muscular dystrophy (BMD)**. Although Becker muscular dystrophy seems identical to Duchenne muscular dystrophy with regard to muscle tissue abnormalities, its overall phenotype is more benign and variable, with a tendency to later onset, slower progression, little or no intellectual impairment, and a much longer life span. The genetic relation between the DMD and BMD mutants—whether they are alleles of the same gene or found at independent loci on the X chromosome—was not determined until 1986 (discussed shortly).

There is no cure for either DMD or BMD, and no highly effective drugs are available. All that can be done is to try to slow the progressive deformity of joints and the loss of muscle strength, thereby prolonging the patient's ability to sit and walk. Treatment includes the use of lightweight splints and braces, physical therapy, pulmonary therapy, and perhaps orthopedic surgery.

Pedigree Analysis: Duchenne Muscular Dystrophy

Here are the characteristics of **X-linked recessive inheritance**, many of which are illustrated in Figure 5.8:

1. *If the gene is not common, many more males than females are affected.* Here, for example, we find five affected males and no affected females. These males are *hemizygous* for the abnormal allele.
2. *Father-to-son inheritance is never seen. Instead, affected males always get their abnormal allele from their mothers,* who are usually unaffected but may have affected fathers, brothers, or uncles. Here all five mothers of affected males are unaffected, and four of them (II-4, II-8, III-7, and IV-19) have known affected brothers and/or uncles. Because males with DMD do not usually reproduce, we cannot say anything about their progeny; but with other X-linked recessive conditions (e.g., red-green color blindness), affected males produce only unaffected sons and carrier daughters.
3. *Among the sons of carrier mothers, about half are affected and half are unaffected.* Among live births in this pedigree, known carrier mothers (I-1, II-4, II-8, III-7, and IV-19) produced a total of six unaffected and five affected sons. These numbers are too small to provide strong evidence, but they do fit the expected pattern. (When geneticists pool data for analysis, they must take into account the bias introduced by the failure to detect those carrier mothers who produce no affected sons.)

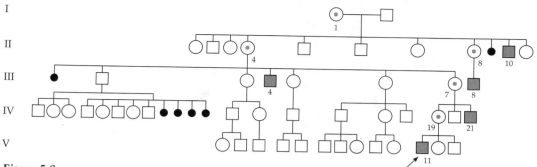

Figure 5.8

One of 33 Duchenne muscular dystrophy kindreds discovered in Utah. It illustrates X-linked recessive inheritance, with sterility of affected males. Note, however, that unaffected males do reproduce but fail to transmit the disorder to any of their offspring. Instead, carrier females (circles with dots inside) are the sole transmitters. (Redrawn from Stephens and Tyler 1951.)

4. *All the daughters from the mating of a carrier female with an unaffected male are unaffected, but half will be carriers.* Because carriers may be identified only *after* they have produced an affected son or a carrier daughter, classification of female offspring is iffy, and we cannot even guess about childless females. (This is not true, however, when heterozygotes are identifiable by laboratory tests.) Counting only those females who had at least one son, we see that among the daughters of known carriers, four (II-4, II-8, III-7, and IV-19) were carriers and three (III-3, III-5, and III-6) were possible noncarriers. Adding the figures from six additional pedigrees collected by Stephens and Tyler (1951), the grand totals are 16 carrier daughters and 19 possible noncarriers.

5. *The mating of an affected male with a homozygous unaffected female produces no affected children, but all their daughters will be carriers.* Because males with DMD rarely reproduce, we cannot illustrate this principle here. But with other X-linked conditions, such as hemophilia or red-green color blindness, it has been amply demonstrated.

6. *Occasionally, a carrier female may manifest some symptoms of a recessive X-linked condition.* Indeed, the enzyme CPK is elevated in about 70% of known female carriers, and about 20% of carriers show some muscle weakness. These effects are due to a phenomenon called **lyonization** (Chapter 7), a normal process whereby one or the other X chromosome in every female cell is rendered inactive. If by chance a female ends up with many cells whose active X chromosome carries the mutant allele (in tissues affected by this disorder), she may be partly or (in very rare cases) fully affected by a recessive condition that usually strikes only males. In fact, all females heterozygous for X-linked traits are **mosaics**; their tissues contain various proportions of cells in which one or the other of the two alleles is expressed.

Identification of the DMD Gene

Detection of the DMD gene itself became a possibility in the early 1980s. Investigators took several different approaches. In analyzing a few rare cases of females with full-blown Duchenne muscular dystrophy, geneticists found that all of them carried a chromosomal abnormality involving the breakage and fusion of parts of two nonhomologous chromosomes

(Chapter 11). Each female had a different autosome joined to an X chromosome, but in all cases the X chromosome had broken at the same spot, the so-called Xp21 band of the short arm. Along the way, an amazing case turned up: a male patient who had Duchenne muscular dystrophy plus four other X-linked conditions and an X chromosome deletion at Xp21 (Chapter 11). These discoveries pinpointed the DMD gene to the Xp21 region, which makes up about 20% of the short arm.

Zeroing in on the new reference point and using many clever tricks of molecular biology, scientists set out to identify the genes in that region. By sharing their DNA samples and information about their patients, an international group of 25 research teams amassed and analyzed data on 1,346 males with Duchenne or Becker muscular dystrophy.

This study, published in 1986 by Louis Kunkel, of Harvard University, and 76 coauthors, showed for the first time that the DMD locus is gigantic—over ten times larger than any other known human gene! Perhaps because of its huge size, mutations of the DMD gene vary from patient to patient. Rather than being caused by a *base substitution* (mutation of a single base) within the DNA sequence that constitutes the gene, in at least 60% of cases a DMD mutation results from the *deletion* of a long stretch of bases within that region of the DNA. Becker muscular dystrophy patients, too, often show deletions in their DNA.

Quite surprisingly, it also turns out that the size of a DMD deletion is not correlated in any obvious way with the severity of the disease. DMD patients with large deletions often have phenotypes similar to those of DMD boys with no detectable deletions, and the same is true for BMD patients.

Dystrophin

By 1988 Kunkel's team had found the product of the normal allele of the *DMD/BMD* gene. This previously unknown protein, dubbed **dystrophin**, exists at very low levels in the muscle cell membranes of unaffected individuals. It is thought to act as a strengthener, helping to protect the fibers (especially of skeletal muscle cells) from damage due to repeated contraction. It is present at even lower levels in a dozen other normal tissues (including brain, lung, and kidney tissues), where its function is unknown. Dystrophin is missing from the muscles of nearly all DMD patients, but present in altered form or amounts in BMD patients.

A mutant strain of mice carries a mutation called *mdx*. These mice exhibit some traits comparable to those of DMD: They have deletions in the same X-linked gene; they lack dystrophin in their muscle cells; and they show the same kind of tissue damage seen in human dystrophic muscle cells. Furthermore, there is great similarity between the DNA sequence of the normal alleles of the *DMD* and *mdx* genes, confirming this mutation's status as a genetic counterpart to DMD. Yet, for unknown reasons, these mice do not express a severe DMD-like phenotype; indeed, they are fairly normal and seem to manage quite well without dystrophin. Obviously it would be desirable to find some animals that do develop full-blown DMD, and a few rare cases have been found in dogs.

In recent years, scientists have made great progress in their attempts to understand this devastating illness. However,

> . . . the discovery of the gene and gene product for Duchenne's dystrophy is not the end of the story, because we now face a new set of problems that cannot be solved only by seemingly magical tricks with DNA. Other methods in cell biology, biochemistry, physiology, and developmental biology will be

needed to ascertain the function of dystrophin, how it is synthesized and placed in the membrane, how lack of dystrophin causes Becker's or Duchenne's muscular dystrophy, what role dystrophin may have in other diseases, how Becker's and Duchenne's dystrophies differ, why the mdx mouse escapes disability, and whether this information can be used to define effective therapy. (Rowland 1988a)

Detection of Carriers and Hemizygotes

Until 1982, the only way of identifying DMD carriers was by measuring the levels of the enzyme creatine phosphokinase (CPK) in their blood: levels exceeding the normal 95th percentile appear in about two-thirds of definite carriers. In addition, up to 80% of definite carriers show thickening of the calf muscles, about 8% exhibit some muscle weakness, and some show other serum and muscle abnormalities.

Unfortunately, elevated CPK levels do not show up reliably in affected males before birth. Thus, in families known to be at risk for DMD, about the only option before 1982 was to test for sex and then abort all male fetuses—even though 50% would be unaffected. Now it is possible, by using molecular techniques described in Chapter 13, to detect carrier mothers and affected offspring with 95–100% accuracy in most known DMD families.

But it will never be possible to detect all cases of DMD in advance through prenatal testing of carriers' offspring because about one-third of all cases represent new mutations. That is, the mothers of these individuals do not carry the DMD gene. Why such a high mutation rate? Nobody knows for sure, but one factor might be the colossal size of the DMD gene, which presents a big target for mutation and in addition allows for a great deal of intragene recombination (exchange of homologous parts) during the formation of gametes in carrier mothers. There is also some evidence that at least one small area within the DMD locus is a "hot spot," especially prone to breakage and deletions.

Because of the high mutation rate, the testing for DMD in all newborn infants (rather than just those in known high-risk families) might be worthwhile if reliable and cost-effective tests were available. It is possible, for example, to detect CPK in the tiny amounts of dried blood (on filter paper) that are routinely obtained from all newborns to test for certain other inherited conditions. Although no treatment yet exists for the affected babies that might be found through such screening programs, their families could be offered genetic counseling, giving them the information and the choice of what to do about future pregnancies.

The pedigree for yet another X-linked recessive disease is presented in Box D. For examples of additional human disorders inherited as X-linked recessives, see Table 5.4.

OTHER MODES OF INHERITANCE

Here we describe three much less common ways by which genes can be transmitted: X-linked dominant inheritance, Y-linked inheritance, and mitochondrial inheritance.

X-Linked Dominant Inheritance: Hypophosphatemia

X-linked dominant inheritance is relatively rare—perhaps in part because when families are small, it may be impossible to distinguish X-linked

HEMOPHILIA

Another example of X-linked recessive inheritance is hemophilia, a failure of blood clotting. Although hemophilia is rare, its mother-to-son inheritance was recognized over 2,500 years ago. The Talmud states that if a woman produces two "bleeders," any additional sons shall be excused from circumcision. If as many as three sisters bear sons who are bleeders, the sons of any other sisters are not to be circumcised; but no such prohibition applies to the sons of their brothers.

There are two non-allelic X-linked recessive forms of this disorder: hemophilia A (the "classical" type, or factor VIII deficiency) and hemophilia B (Christmas disease, or factor IX deficiency). Hemophilia A appeared in several interrelated royal families of Europe, apparently arising from a mutation in one parent of Queen Victoria of England (1819–1901). The pedigree shown here presents the ten known males with hemophilia (genotype *h*/(Y) who are descended from Queen Victoria. McKusick (1965) gives brief sketches of their lives; most died from uncontrolled bleeding at the ages indicated in parentheses. The pedigree is incomplete in many respects, and children indicated by the multiple-person symbols are not necessarily in the birth orders indicated. But the pedigree does include all the sibs of the affected males.

Probably the most famous victim of hemophilia was Victoria's great-grandson Tsarevitch Alexis, heir to the throne of Russia (Hartl 1985; Massie 1985). His illness caused his parents to put their faith, and considerable political power, into the hands of the "mystic" and charlatan Rasputin. Rasputin's evil doings may have hastened the onset of the Russian Revolution, to be followed by the murder of 14-year-old Alexis and his family by Bolshevik revolutionaries.

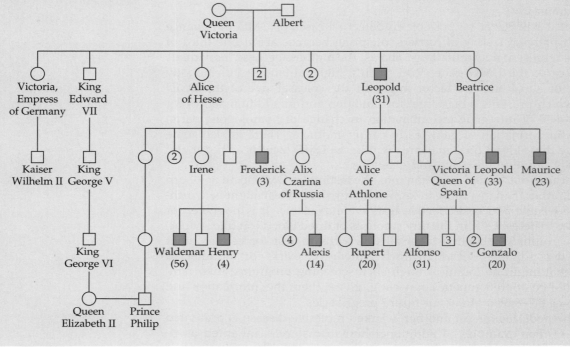

dominant from autosomal dominant inheritance. Only a few disorders are known to be inherited as X-linked dominants. In at least three of these (incontinentia pigmenti, oral-facial-digital syndrome I, and focal dermal hypoplasia), the trait is lethal before birth in hemizygous males.

Hypophosphatemia (Greek *hypo*, "less than"; Greek *emia*, "in the blood") is characterized by abnormally low levels of phosphorus in the blood, probably due to defective reabsorption of phosphate in the kidneys. Deficient reabsorption of calcium in the intestines also occurs, leading to softening of the bones and rickets (bowlegs) in virtually all affected males

Table 5.4.
Some human disorders inherited as X-linked recessives, and their chromosomal locations.

Name	Brief Description	Site
Fabry disease	α-galactosidase A deficiency; kidney damage, corneal changes, vascular skin growths	Xq
G6PD deficiency (favism, primaquine sensitivity)	Acute hemolytic anemia brought on by certain substances, such as fava beans and more than 340 drugs (primaquine, aspirin, sulfas, etc.)	Xq
Hunter syndrome	Heparitin sulfate in urine; milder skeletal and mental symptoms, but more deafness, than seen with Hurler syndrome	Xq
Lesch-Nyhan syndrome	Defect in purine metabolism; mental retardation, spastic cerebral palsy, involuntary movements, uric acid urinary stones, self-mutilation (biting fingers and lips)	Xq
Ornithine transcarbamylase (OTC) deficiency	Excess ammonia in blood, mental deterioration, extreme irritability, vomiting	Xp
Testicular feminization (androgen insensitivity)	XY individuals with testes but cells unresponsive to androgen, so develop female external genitals, breasts, blind vagina	Xq
X-linked mental retardation with fragile site	Mental retardation; large testes, ears, and jaw; high-pitched voice, jocular speech	Xq

and about 50% of affected females. Because affected individuals may be bowlegged despite the intake of normal-to-high doses of vitamin D, this condition is also known as **vitamin D–resistant rickets**. Its frequency is roughly 1 in 60,000.

Those individuals with bone disease have abnormally short legs and short stature, along with certain other skeletal abnormalities that can vary considerably within a family. Pain due to softening or overgrowth of the bones is common, as are tooth abnormalities. Onset is early, usually between 6 and 12 months of age. There is no cure, but treatment with four or five daily doses of oral phosphate plus vitamin D, from early childhood through early adulthood, can lead to marked improvement.

Much has been learned about the hypophosphatemic phenotype from a comparable mutation, also inherited as an X-linked dominant, in the laboratory mouse. In humans, the gene has been mapped to the distal part of the X chromosome short arm. Thus far, it is not possible to diagnose hypophosphatemia prenatally.

Figure 5.9 is a pedigree for hypophosphatemia. Among the following features of this inheritance pattern, note that affected males and their offspring and parents provide the best clues:

1. There is no male-to-male transmission of the trait.
2. Affected males (when mated to unaffected females) produce all unaffected sons and all affected daughters.

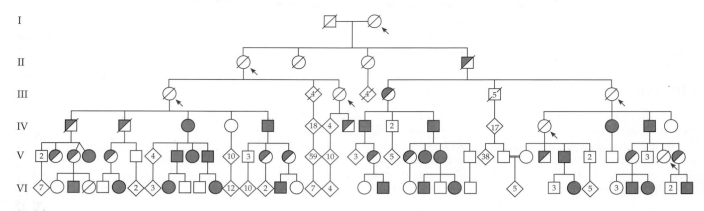

● ■ Hypophosphatemia and rickets

◐ ◪ Hypophosphatemia without rickets

⊘ ⬓ Deceased

Figure 5.9

A pedigree for X-linked dominant hypo-phosphatemia, also known as vitamin D–resistant rickets. Note that (1) the trait was observed in every generation except the first one; (2) there is no father-to-son transmission; and (3) affected males produce daughters who are all affected. There are ambiguities among some deceased females, however. If the trait is indeed an X-linked dominant, then—because they have affected sons or are in the line of descent of the mutant gene—females I-2, II-1, III-1, III-6, III-17, IV-50, and V-161 (indicated by arrows) should have been affected. Perhaps they were only mildly affected and thus eluded detection. These kinds of problems are not uncommon in analyses of human pedigrees. (Redrawn from Winters et al. 1957.)

3. Affected females can produce affected sons and daughters as well as unaffected sons and daughters. Among these sons and daughters, the ratio of affected to unaffected is about 1:1.

4. Every affected individual has one affected parent; but for affected males, the affected parent is always the mother.

5. Males, being hemizygous, are usually more severely affected than females, who are heterozygous.

6. Owing to the phenomenon of X chromosome inactivation (Chapter 7), expression of the trait is also more variable among females.

7. If affected males and affected females reproduce at the same rate, in the population as a whole there will be twice as many affected females as males. This is because females have two X chromosomes, whereas males have only one.

Y-Linked Genes

Very few genes have been detected on the Y chromosome of most species. In humans, the Y is less than half the length of the X chromosome (which as of 1990 contained 160 confirmed and about the same number of provisional loci) and slightly longer than the smallest autosome (which contains 16 confirmed gene loci and 3 provisional loci).

Only three or four loci have been definitively assigned to the Y chromosome. Besides these, some factors for sperm production, skeletal height, and tooth size have also been provisionally assigned to the Y chromosome. We know that the Y is critical to the development of maleness and the prevention of femaleness because certain rare individuals possessing only one X chromosome but no Y are females (Chapter 10). More precisely, a particular region of the Y is needed for male development, and femaleness is what develops in its absence.

The tip of the Y chromosome short arm is actually homologous to a tiny region at the tip of the X chromosome short arm, and this region contains at least two simply inherited traits. But what concerns us here is the rest of the Y chromosome, the *nonhomologous* part. Any gene located on this latter part of the Y chromosome has a very striking mode of inheritance. Such a **Y-linked gene** is received, and usually expressed, by all the male descendants of an affected male, generation after generation. (And in the other direction, it should extend from any male through his paternal grandfather, his paternal great-grandfather, and so on, all the way back to the first human male.) Female descendants, lacking the nonhomologous Y region, could neither receive, express, nor transmit such a trait. In the past half century, claims of Y linkage have been advanced for over a dozen traits. On closer examination of the family histories, however, exceptions to the given requirements were nearly always discovered.

Mitochondrial Genes

In the 1960s scientists discovered that aerobic (oxygen-using) cells of animals contain two genetic systems: the main one (50,000 to 100,000 genes) in the nucleus and a very tiny one (about 35 genes) in the mitochondria, the "powerhouses" of the cell (Chapter 2). All the mitochondria present in our cells are derived exclusively from our mothers, and none from our fathers. This is because egg cells contain many mitochondria, but sperm cells contain very few mitochondria—and none of the sperm mitochondria survive inside the fertilized egg.* Thus, any trait associated with a **mitochondrial gene** must be transmitted by the mother to all of her children, both male and female. In the other direction, this lineage should extend from any individual through the maternal grandmother, the maternal great-grandmother, and so on, all the way back to the first human female! Clearly, this situation is an exception to Mendel's rules of inheritance.

Any mitochondrial mutations that cause diseases might be detected by their maternal pattern of inheritance. They must also be associated with defects in mitochondrial metabolism and affect those tissues—brain and spinal cord, skeletal muscle, heart, kidney, and liver—that have high energy requirements. In humans, there is evidence of at least two such types of conditions, both quite variable. One group of disorders, called *mitochondrial myopathy* (Greek *myo*, "muscle"; English *pathy*, "disease"), involves muscular weakness or wasting or both. Another, affecting vision, is *Leber's hereditary optic neuropathy*. This defect has been traced to a specific enzyme of the mitochondrial membrane involved in the production of the high-energy compound ATP; it is caused by the mutation of a single nucleotide base.

SUMMARY

1. Gathering reliable family data is not easy. Interpreting genetic data may be difficult, too, especially when dealing with rare traits.

2. Some inherited conditions show unique patterns of transmission that are recognizable by means of standard pedigrees. In humans, the most common types of simple inheritance are autosomal dominant, autosomal recessive, and X-linked recessive.

3. Autosomal dominant traits that are fully penetrant do not skip generations. They are expressed equally in both sexes and are transmitted by both sexes (including father-to-son transmission).

4. Autosomal recessive traits often do not appear in every generation, and affected individuals usually have unaffected parents. In cases of very rare traits, the unaffected parents of an affected child are usually related. The trait is expressed equally in both sexes and transmitted by both sexes (including father-to-son inheritance).

5. X-linked recessive traits skip generations, are more frequent in males than females, and are transmitted differently by the two sexes. Father-to-son inheritance is never observed; instead, males inherit the trait from carrier mothers, who may have affected male relatives and

* But in 1991 some scientists were astonished to find that the sperm of mussels *do* transmit mitochondria. Also, in the progeny of crosses between two different species of mice, very rarely (1 in 10,000) a mitochondrial DNA molecule from the sperm survives. (This situation was *not* found for crosses within the same species, however.) Whether the phenomenon ever occurs in other organisms is not yet known.

carrier female relatives. Because of the lyonization effect, cultured cells taken from unaffected carriers may consist of two phenotypic populations, and an occasional carrier female may be partly or fully affected.

6. Few X-linked dominant traits are known. Like autosomal dominants, they are expressed in every generation and are transmitted by both sexes. Unlike autosomal dominants, however, they show no male-to-male transmission, and the offspring of affected males consist of all affected daughters and all unaffected sons.

7. Some genetic disorders have multiple modes of inheritance or exhibit differences in the molecular basis of their mutations. For this and other reasons, genetic counseling should be done only by those trained in both clinical medicine and genetics.

8. Familial hypercholesterolemia is a common disorder, leading to increased risk of heart attacks. It is caused by autosomal dominant alleles on the short arm of chromosome 19. The normal gene product, the LDL receptor, removes excess cholesterol from the bloodstream.

9. Cystic fibrosis is a relatively common lethal condition caused by autosomal recessive alleles on the long arm of chromosome 7. The gene and its normal product, cystic fibrosis transmembrane regulator, have been isolated. The defect involves the regulation of chloride ion transport through the cells of affected organs.

10. Duchenne muscular dystrophy is caused by a recessive lethal gene located on the short arm of the X chromosome. About 60% of patients with Duchenne muscular dystrophy have deletions in this huge gene. The normal gene product, called dystrophin, is absent from the membranes of dystrophic muscle cells; its function is not well understood.

KEY TERMS

autosomal dominant
    inheritance
autosomal recessive
    inheritance
cystic fibrosis
    transmembrane
    regulator (CFTR)

consanguineous mating
dystrophin
founder effect
LDL receptor
low-density lipoprotein
    (LDL)

mitochondrial gene
mosaic
pedigree
proband
propositus

X-linked dominant
    inheritance
X-linked recessive
    inheritance
Y-linked gene

QUESTIONS

1. Why are the terms *homozygous* and *heterozygous* not applicable to X-linked genes in males?

2. (a) Construct, as completely as possible, a pedigree of your own family. (b) Ask as many relatives as possible to eat some asparagus and determine whether or not they smell a distinctive odor in their urine afterward. If feasible, suggest that they also try to detect this odor in the urine of other family members (without discussing their observations among themselves) and then report all the results back to you. Then see if you can figure out how this simple and harmless trait may be inherited.

3. What mode(s) of inheritance could explain the pedigree shown here?

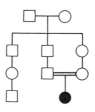

4. Galactosemia is an autosomal recessive disorder (genotype *gl/gl*) caused by an inability to metabolize the milk sugar galactose. The symptoms seen in untreated infants are lethargy, failure to thrive, severe vomiting, liver disease, cataracts, and mental retardation; early death may occur. But if newborns with galactosemia are placed on a diet free of lactose and galactose, symptoms regress, and development may even fall within the normal range. About 1 person in 150 is heterozygous (*Gl/gl*) and phenotypically unaffected.

(a) Two unaffected parents have a baby with galactosemia. What are the genotypes of the parents and baby?

(b) Two unaffected children are also born to this couple. What is the probability that their fourth child will have galactosemia? Would the probability be different if their first three children had been affected?

(c) If a treated individual reaches adulthood and is fertile, what are his or her chances of marrying a heterozygote and having children with galactosemia?

5. Refer to the pedigree of hemophilia in Box D. (a) Identify all females who must have been heterozygous, *H/h*. (b) What is the ratio of affected sons to unaffected sons among the children of known heterozygotes? (c) Is it possible that Queen Elizabeth II is a carrier?

6. What is the most likely mode of inheritance for the single condition illustrated in the following pedigrees? (From Mabuchi et al. 1978.)

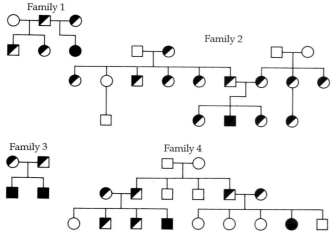

Family 1

Family 2

Family 3   Family 4

7. With regard to the X-linked gene for green-shift color blindness, females can have three different genotypes, *G/G*, *G/g*, or *g/g*; but males can have only two, *G/(Y)* and *g/(Y)*. The normal allele *G* is dominant to its green-shifted allele *g*.

(a) Systematically list the six possible matings for this X-linked gene, identifying the sex of each parent. For each mating, give the expected offspring. List daughters and sons separately.

(b) Which of these matings can produce a carrier daughter? In which of these are all daughters carriers?

(c) Which of these matings is expected to produce some sons with green-shift and some sons with normal color vision? What is the genotype of the mothers in these matings?

(d) What offspring phenotypes are possible from a green-shifted mother and a normal father? Identify offspring by sex. (Note that in this particular family there is a crisscross type of inheritance.)

(e) A couple has a green-shifted daughter and a son with normal color vision. What are the genotypes of the parents?

8. Around 1930, several published reports described males who lacked teeth and sweat glands and were partly bald. This condition had been noted much earlier by Charles Darwin (1875):

I may give an analogous case . . . of a Hindoo family in Scinde, in which ten men, in the course of four generations, were furnished . . . with only four small and weak incisor teeth and with eight posterior molars. The men thus affected have very little hair on the body, and become bald early in life. They also suffer much during hot weather from excessive dryness of skin. It is remarkable that no instance has occurred of a daughter being affected. . . . [T]hough the daughters in the above family are never affected, they transmit the tendency to their sons; and no case has occurred of a son transmitting it to his sons. The affectation thus appears only in alternate generations, or after long intervals.

How is this trait (now called ectodermal dysplasia) inherited?

9. If a male desires to maximize his contribution of genes to future generations, is he better off having daughters or sons? Why?

10. You could not have gotten a sex chromosome from one of your four grandparents. Which grandparent could not have transmitted, via your parents, a sex chromosome to you? (*Note:* The answer depends on your sex.)

11. Describe the expected pattern of inheritance for a rare, fully penetrant X-linked dominant trait that is lethal (before birth) in hemizygous males. How can it be distinguished from an autosomal dominant trait?

12. (a) What mode(s) of inheritance could explain the pedigree shown here, assuming that the trait is *rare*? (b) What mode(s) of inheritance could explain the same pedigree if the trait is *common*?

13. In the pedigree shown here, assume that the trait is fully penetrant and that none of the matings are consanguineous. Also consider whether the causative allele is rare or common in the general population. Why could the trait represented by solid symbols *not* be due to (a) a rare autosomal recessive, (b) a rare X-linked recessive, (c) a rare autosomal dominant, (d) an X-linked dominant, or (e) a common Y-linked gene? What type(s) of inheritance could explain this pedigree?

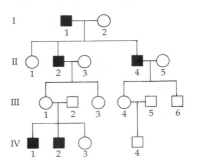

FURTHER READING

For simply inherited human traits, the "bible" is an encyclopedic catalog compiled and regularly updated by Mc-Kusick (10th ed., 1992); in addition to its descriptions of almost 5,000 phenotypes, the foreword and appendices are packed with information on human genetics. Milunsky (1989), written for the lay public, includes many interesting case descriptions and historical sidelights. Emery and Rimoin (1990) and Scriver et al. (1989) are two-volume reference works that present detailed clinical and genetic information on a vast number of human disorders. Krush and Evans (1984), an interesting nuts-and-bolts guide on how to design and conduct family studies, also discusses personal reactions and interactions between researchers and families.

Reproduction and Development

6

In 1953 John M. Morris, a professor at the Yale University School of Medicine, summarized 72 reports (published between 1817 and 1950) of a condition found in women who appear normal but who have testes in place of ovaries. He called the clinical syndrome **testicular feminization** and added his own case reports. One described a 44-year-old housewife, typically female-looking:

> The patient's history revealed that she had developed normally but had never menstruated. . . . She had been married for 20 years with normal coitus, libido, and orgasm, but had had no pregnancies. The patient was an only child. Her mother had three sisters who had never menstruated, two of them married, . . . and her maternal grandmother had one sister who had never menstruated. . . .
>
> The hands were rather large; breasts were well developed. There was no vestige of hair on the face, and no axillary [underarm] or pubic hair other than a trace of down on the vulva. . . . The vagina ended in a blind pouch 8 to 20 cm in depth.

A pelvic tumor contained "tubules of a testicular type." Morris suggested that this patient probably never had ovaries or any other female internal organs. Nevertheless, he pointed out, these patients are definitely female from the psychosexual aspect—with sex urges like those of other women and strong desires for childbearing.

These and many other studies have shown that testicular feminization is an X-linked recessive trait occurring in individuals who are chromosomally male (46,XY). The gene is located on the long arm of the X chromosome, near the centromere. Affected individuals produce plenty of male hormone (**androgen**), which circulates in their bloodstream. But the **androgen receptors** that normally allow the hormone to enter the appropriate cells—such as those that give rise to the external male genitals or produce beards and pubic hair—are absent or defective. Thus, unable to respond to the hormone's masculinizing effects, affected individuals develop as phenotypic females (Figure 6.1) who are sterile owing to the absence of internal female organs. The disorder, which occurs with a frequency of about 1 in 65,000 46,XY individuals, has also been called the *androgen insensitivity syndrome*. It is only one of many types of development errors that raise the question of what we mean by *male* and *female*.

USUAL SEXUAL DEVELOPMENT

Most of us have strong opinions about what constitutes maleness and femaleness, and our daily lives are permeated by personal and societal expectations for sexual development and behavior. In both plants and

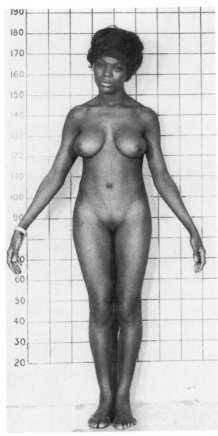

Figure 6.1
The testicular feminization syndrome in a 17-year-old 46,XY individual. Note the well-developed breasts, female-type body build, and absence of pubic hair. This woman is about 5 feet, 11 inches tall. (From Zourlas and Jones 1965.)

animals, there is great variability in modes of sex determination, systems of mating, and mating behavior. In most cases sex is determined genetically, but there are some fascinating examples—among fish, lizards, turtles, crocodiles, and birds—in which environmental factors (such as temperature or local sex ratio) play a major role, even causing sex reversal under some circumstances.

Some sexually reproducing plants and animals maintain both sexes in the same body but practice either self-fertilization or cross-fertilization. But for those plants and animals (including humans) in which the two sexes are normally maintained separately, individuals whose body structure or behavior cannot be classified as completely male or completely female have always been regarded with special interest. Indeed, the concept of male and female joined harmoniously in one body has been idealized in mythology and art (Box A). But in real life, the occurrence of both male and female physical traits in one body is not so ideal. We now know that some of these human sexual abnormalities result from variations in prenatal hormone levels or in sex chromosome constitution (Chapter 10). In the 1950s the first clinics dealing specifically with birth defects of the sex organs were established, and the first careful studies of human sexual behavior were undertaken. These sources of information, augmented by anthropologists' studies on widely divergent cultures throughout the world, have led to a deeper understanding of how maleness and femaleness normally develop in humans. The picture that emerges involves a series of steps, some occurring before birth and some after.

Prenatal Development: Genes and Hormones

The first step in sex determination occurs at the time of fertilization, when either an X-bearing or a Y-bearing sperm enters the egg. This establishment of **chromosomal sex** sets the stage for future changes, although the two sexes are physically indistinguishable at first. By the sixth week of development in humans, both XX and XY embryos have the same sex organs. These include a pair of "unisex" gonads called **gonadal ridges**, plus two sets of primitive duct systems: the male-type **Wolffian ducts** and the female-type **Müllerian ducts** (Figure 6.2A). The gonadal ridges will form the testes or ovaries, while the duct systems will eventually form the rest of the internal sex organs. Externally, as seen in the figure, both XX and XY embryos develop a knoblike **genital tubercle** and a **urogenital groove** surrounded by specialized folds and swellings of tissue.

Thus, in the beginning, the internal and external sex organs of any embryo are equipped to form either a male or a female. But unless male hormones intervene at later stages of fetal life, it will develop along female lines. Extremists on both sides of the battle of the sexes have seized on this developmental fact to "prove" either that females are more primitive than males or, conversely, that females represent the basic sex and males are an anomalous offshoot. We think that such leaps of fancy from biology to ideology are unwarranted and foolish.

In presumptive males, the second step occurs late in the sixth week of development, when a gene on the short arm of the Y chromosome produces a protein called **testis-determining factor (TDF)**. Apparently acting as a gene regulator, TDF triggers the inner parts of the gonadal ridges to begin developing into male gonads, the **testes**. Although necessary for differentiation of the mammalian testis, this gene by itself is not sufficient; autosomal and X-linked genes are also known to be involved.

The third step in male development occurs when the fetal testes begin to produce two different hormones. The **anti-Müllerian hormone**

BOX A

HERMAPHRODITES IN MYTH AND ART

The term *hermaphrodite* comes from Hermaphroditus, the son of the Greek god Hermes and the Greek goddess Aphrodite. Myth has it that while bathing, Hermaphroditus grew together with the nymph Salamacis, thus combining in one body both male and female characters. Hermaphroditic gods abound in the myths and art of India and Persia too; Ardhanarisvara, for example, represents the joining of the male god Siva with the female Sakti. And the creation stories of many other cultures (including one Hebrew version of Adam's creation) feature a hermaphrodite who becomes bisected into male and female sexes (Mittwoch 1986).

From the third century B.C., *a Greek statuette of a hermaphrodite, with breasts and male genitals. (Courtesy of the Museum of Fine Arts, Boston.)*

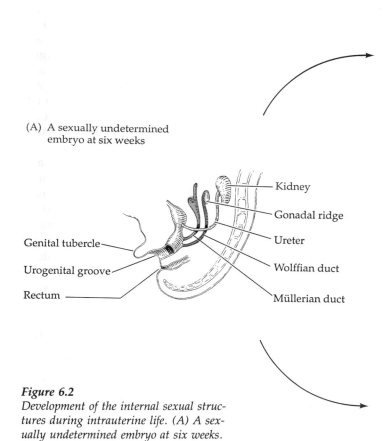

(A) A sexually undetermined embryo at six weeks

- Kidney
- Gonadal ridge
- Ureter
- Wolffian duct
- Müllerian duct
- Genital tubercle
- Urogenital groove
- Rectum

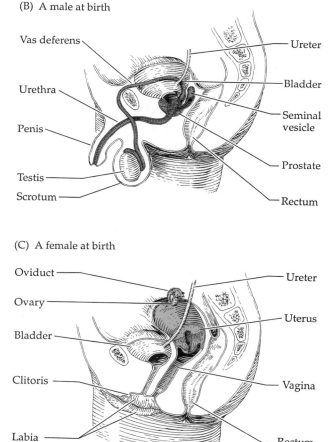

(B) A male at birth

- Vas deferens
- Urethra
- Penis
- Testis
- Scrotum
- Ureter
- Bladder
- Seminal vesicle
- Prostate
- Rectum

(C) A female at birth

- Oviduct
- Ovary
- Bladder
- Clitoris
- Labia
- Ureter
- Uterus
- Vagina
- Rectum

Figure 6.2
Development of the internal sexual structures during intrauterine life. (A) A sexually undetermined embryo at six weeks. (B) A male at birth. (C) A female at birth.

(AMH) causes the Müllerian ducts to regress. (The human *AMH* gene resides on chromosome 19q and is active only in the presence of a Y chromosome.) Meanwhile, the male sex hormone **testosterone** develops and maintains the Wolffian ducts. If the appropriate tissues are responsive to these hormones, the following changes occur by the seventh week. Internally, the Müllerian ducts begin to degenerate. Under the local influence of testosterone, the Wolffian ducts begin to form the *prostate gland*, *seminal vesicles*, and *vasa deferentia* (singular, vas deferens), tubes connecting the whole system (Figure 6.2B). In addition, neural pathways in the *hypothalamus* (part of the brain) that control menstrual cycling in females are altered in male embryos. Finally, a very potent derivative of testosterone (called *dihydrotestosterone*) organizes the shaping of external genitals. The genital tubercle and surrounding folds lengthen and fuse to form a **penis**, while nearby swellings develop and fuse to form the **scrotum**.

In presumptive females, the products of various genes (including female hormones) cause the outer parts of the genital ridges to begin to develop as **ovaries** around the twelfth week. Two X chromosomes, with genetic determinants on both arms, are required for ovarian maintenance. In addition, several autosomal loci are known to affect normal ovarian development. Quite independently, the Wolffian tubes (in the absence of testosterone secretion) regress, and the Müllerian tubes (in the absence of anti-Müllerian hormone secretion at an appropriate time) form the female ducts. The forward parts become the **oviducts** (Fallopian tubes), and the end parts fuse to form the **uterus** and most of the **vagina** (Figure 6.2C). The external genitals also develop: a **clitoris** from the genital tubercle, small and large **labia** (lips or folds) from the surrounding folds and swellings, and the lower part of the vagina from the urogenital groove. Thus, the sex of a fetus is clearly evident by the end of the fourth month.

Postnatal Development: Environmental Factors

On the basis of the appearance of its external genitals, a newborn's sex is assigned and recorded on the birth certificate. Sex-specific names and clothing usually follow, perhaps along with sex-specific treatment. Some (but not all) studies find that baby boys are cuddled and spoken to less frequently, but encouraged to be energetic and independent, whereas little girls are rewarded for cleanliness, passivity, and dependence. Responding to such cues, most young children begin to develop a sense of sexual identity by about 18 months of age.

Sex-specific toys and games are common, too, and by the age of five or six most children prefer companions of their own sex. Despite some recent changes, different roles and aspirations may be projected for the two sexes by certain political and religious philosophies, by the various media, and especially by advertisers. Males are usually portrayed as doers and rescuers—dependable, rational, and stoic; females are often portrayed as helpers, observers, or victims—unreliable, intuitive, and emotional.

How well do such assumed differences in behavior hold up under close scrutiny? A number of psychologists, after surveying and analyzing research on sex differences, have concluded that girls are not more social or suggestible, nor are they better at rote learning or worse at analytical thinking. Whether girls are more compliant and nurturing or less active and competitive is not clear. But there is considerable evidence that by the age of 12 or 13, girls excel in verbal ability, while boys excel in visual-spatial ability and mathematics and are more aggressive. Are the observed differences in these four traits related to differences in brain structure and/or function? This idea is still a matter of hot debate. But even if it were

GENDER IDENTITY AND SEX ROLES

The process by which a person comes to identify himself or herself as a male or a female is immensely complex. This private **gender identity**, the sum of one's feelings, is expressed publicly as a **sex role**, the sum of those actions that indicate a degree of maleness or femaleness to others.

Do boys and girls develop different patterns of behavior because they are treated differently, or are they treated differently because they show different behavioral patterns from the time of birth? Both physiological and environmental factors definitely play a role, but are behavioral differences due more to genetic differences between the sexes or more to differences in upbringing?

In birds and many mammals, including primates, the sexually stereotyped behaviors expressed in courting, mating, and caring for young are definitely controlled by hormones and neural pathways. Even so, normal females may occasionally exhibit mounting and thrusting behavior, and nor-

mal males may sometimes crouch to be mounted or display other aspects of female mating behavior. In humans, hormones also control the most basic differences between the sexes: gamete production, pregnancy, and lactation.

But much less is known about the relationship of mating behavior to hormones or brain circuitry in humans. How can these processes be studied? One approach is to look at the behavior patterns of people born with anatomical and hormonal abnormalities. These include 46,XX females who were masculinized during fetal life by hormone imbalance or by hormone treatment, but then surgically corrected, provided with appropriate hormone treatment, and raised as normal girls. Compared with control groups of females, the fetally masculinized group contained more "tomboys" and more late-marrying individuals. Observed heterosexual, homosexual, or bisexual tendencies vary from study to study, some showing no differences from controls and

others showing some difference.

Studies of hermaphrodites and other individuals with ambiguous genitals are also inconclusive. In many cases the chosen sex of rearing predominated well beyond puberty, regardless of the chromosomal sex or the development of inappropriate secondary sexual characteristics during puberty (e.g., breasts in "males" or virilization in "females"). These studies suggest that environmental factors are more important than genetic or hormonal factors. But other cases are known (mainly with 5-α-reductase deficiency) in which similar types of individuals who were raised as females abruptly changed their sexual identity and sex roles when virilization occurred at puberty. This change argues strongly in favor of nature over nurture. But the wide variety of outcomes from various kinds of studies suggests that gender identity does not develop automatically and unswervingly on the basis of chromosomal sex or even genital sex.

---

true, the *expression of these traits varies so widely within both sexes and shows so much overlap between the sexes that one could never predict, on the basis of sex alone, how any given male or female will behave.*

Anthropologists also find that sexual norms of behavior vary tremendously from one culture to another. Margaret Mead (1949) found a "great variety of ways, often flatly contradictory, in which the roles of the two sexes have been patterned." For example, traits such as emotional dependency, or the ability to do hard physical labor, or the tendency to gossip have each been assigned to different sexes in different societies. Such cultural attributes are superimposed, layer by layer, on a basic physiological distinction between the sexes: males inseminate, whereas females menstruate, gestate, and lactate. The developmental events of puberty that underlie these irreducible sexual functions are described in the following section. Some psychological factors are considered in Box B.

Postnatal Development: Physiological Factors

Between the ages of 10 and 15, the production of certain hormones sets off a chain of events that culminates in sexual maturity and the ability to reproduce. Triggered by genetic and environmental factors that are poorly understood, puberty starts when substances produced by the hypothalamus cause the tiny *pituitary gland* just beneath it to begin secreting **gonadotropic hormones**. These, in turn, stimulate certain cells in the gonads to make sex hormones, which bring about striking changes (secondary sex characteristics) in many sensitive tissues throughout the body. Both sexes produce the same hormones, but in much different proportions. (Actually, the different sex hormones are very similar to each other. They belong to a group of chemicals called *steroids*, which are derived from cholesterol.) Under the influence of nerve pathways set up in the hypothalamus during embryonic development, female hormone levels vary widely during the menstrual cycle. In adult males the cycling of gonadotropic hormone production is greatly reduced but not eliminated.

The two major gonadotropic hormones are named for their function in females. **Follicle-stimulating hormone (FSH)** activates a few egg follicles in the ovary to undergo maturation each month. With the increased production of **luteinizing hormone (LH)** in midcycle, one follicle is stimulated to release its egg and then to become a *corpus luteum* (yellow body). Before ovulation, the maturing follicle secretes the female sex hormone **estrogen**, which leads to development of the uterine lining.* Other effects of estrogen at puberty include increased growth of the sex organs and breasts, widening of the pelvis, and deposition of subcutaneous fat. Following ovulation, the corpus luteum secretes the hormones estrogen and *progesterone*. The latter stimulates further thickening of and increased blood supply to the uterine lining in preparation for implantation of a fertilized egg. Both sex hormones also act on the hypothalamus and pituitary gland to decrease production of FSH and LH as part of the menstrual cycle. A third gonadotropic hormone, *prolactin*, is secreted mainly after pregnancy and stimulates milk production; actually, small amounts of prolactin are secreted at all times by both sexes.

In males and females at puberty, the **adrenal glands** atop the kidneys make an androgen that is partly transformed into testosterone. In females, this hormone is responsible for the growth of pubic and underarm hair and the development of oil and sweat glands in the skin. An excess of adrenal androgens in females (say, due to adrenal tumors) can cause the abnormal development of male secondary sex characteristics, such as facial hair, large muscles, and deepening of the voice. It can also enlarge the clitoris.

In pubertal males, FSH and LH stimulate the maturation of germ cells into sperm in the *seminiferous tubules* of the testes. A little bit of estrogen is produced by the testes and the adrenal glands, but in the presence of normal amounts of androgens, it has no obvious effects. However, normal teenage males often show slight breast development at puberty, which usually regresses. When such tissues persist, excess estrogen may be the cause. Very high concentrations of testosterone are necessary within the testes for complete sperm maturation. At puberty, testosterone also brings about most of the masculine changes and acts on the pituitary and hypothalamus to inhibit the production of LH and FSH.

The steps leading to the development of normal males are summarized in Figure 6.3.

SOME ERRORS IN SEXUAL DEVELOPMENT

McKusick's catalog of inherited phenotypes (1992) lists several dozen aberrations in sexual development. For some of these, the modes of inheritance and the basic biochemical defects are well established, but for others the evidence remains scanty or ambiguous. It has been suggested that of about 50 known "sex genes," about 20 are X-linked, 2 are Y-linked, and the remainder are autosomal.

---

* At least six different estrogens are known, but we will not distinguish among them. Likewise, several different androgens (male hormones) exist, among which testosterone is the most important.

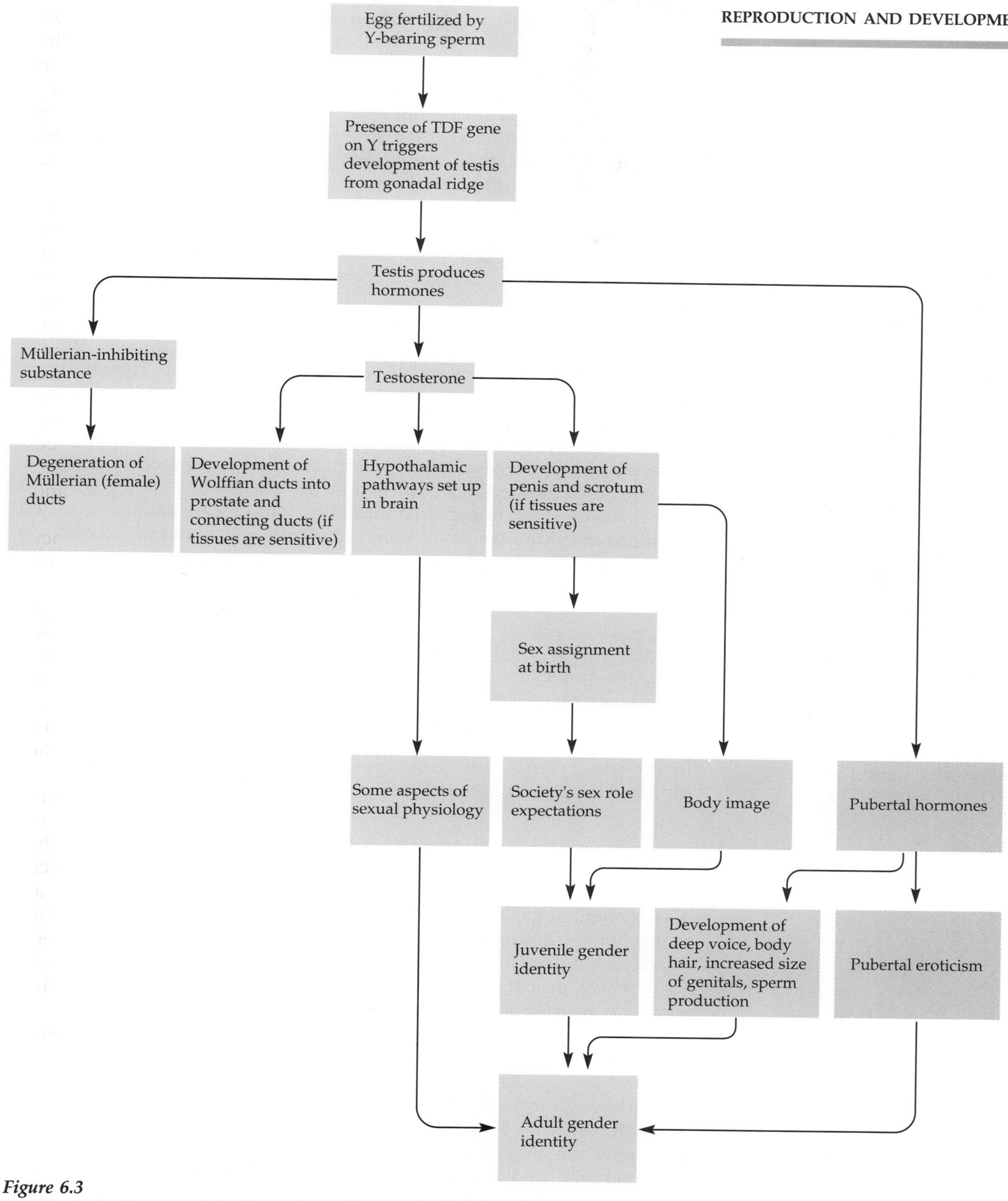

Figure 6.3
Steps necessary for normal development of maleness. An alteration at any step leads to weakening or failure of male development. (Adapted from Money and Ehrhardt 1972.)

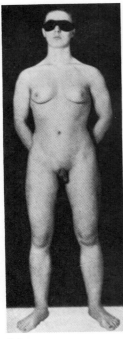

Figure 6.4
A person with true hermaphroditism, having both breasts and a penis. Many true hermaphrodites, however, have ambiguous or predominantly female-type external genitals. (From Overzier 1963.)

Usually, ovaries develop only in XX embryos, and testes develop only in XY embryos. But when neither or both occur, or when gonadal sex does not correspond to chromosomal sex, the result is a **hermaphrodite**, of which there are numerous types.

True hermaphrodites, with both ovarian and testicular tissue, are rare (Figure 6.4). Although their external genitals are often ambiguous, roughly two-thirds of them are raised as males, irrespective of their chromosomal sex. They often produce eggs, but rarely sperm, and they are almost always sterile. Most true hermaphrodites have a uterus, and some menstruate. Simpson (1990) reports that five 46,XX true hermaphrodites have become pregnant, "usually but not always after removal of testicular tissue." There are a few reports of apparent autosomal recessive mutations for true hermaphroditism, which indicates that genes on the sex chromosomes are not in full control of sexual development and differentiation.

Most true hermaphrodites are 46,XX, but 46,XY karyotypes and *mosaics* containing both 46,XX and 46,XY tissue are known too. How testicular tissue manages to develop in an XX individual is not clear, but one possibility is that during embryonic development, some XY cells were present but later disappeared or became very rare. In other cases, part of the father's Y chromosome might have been attached to another chromosome. In some cases, however, the cause remains unknown.

In contrast to true hermaphrodites, whose gonadal sex is mixed, **pseudohermaphrodites** have either testes or ovaries, but the external genitals are the opposite of the gonadal sex, or ambiguous, or abnormal in some other way. There are two main pseudohermaphroditic classes:

Type	Gonads	External genitals
Male pseudohermaphrodites	Testes	Femalelike or abnormal
Female pseudohermaphrodites	Ovaries	Malelike or abnormal

In the remainder of this section, we will describe several disorders that can produce pseudohermaphroditism.

Defects of Androgen Target Cells

Androgens produced by fetal testes normally permit and enhance the development of the Wolffian ducts into male-type structures, but problems arise when the tissues on which they act fail to respond because of receptor or enzyme defects. One of the most astonishing errors in sexual development was described at the beginning of this chapter. *Testicular feminization* (or *complete androgen insensitivity syndrome*) causes 46,XY persons with the usual levels of androgens to nevertheless develop into phenotypic females with breasts. Similar conditions have also been described in mice, rats, and cows; indeed, the mouse and human loci appear to be the same.

Although the testes are intact and functional inside the abdomen, no Wolffian structures are formed in the fetus. Nor are there oviducts, uterus, and upper vagina—which means that the response to Müllerian-inhibiting hormone produced by the fetal testes is appropriate. Because the external genitals are typically female and the short, blind lower vagina looks normal from the outside, the newborn is declared a girl and raised as such.

At the time of puberty, estrogens (converted from testosterone) stimulate breast growth and other female secondary sexual characteristics. Normal male levels of testosterone also circulate in the bloodstream of these women, but to no avail: The tissues that would usually be stimulated to maleness do not respond. Insensitivity to androgens also prevents the growth of sexual hair in the pubic and underarm areas, although hair on the head is often luxuriant and shows no male patterns of receding hairline. If affected individuals have not developed hernias during childhood (from the attempted descent of the testes into nonexistent scrotal sacs), what finally brings them to the doctor's office at the time of puberty is the failure to menstruate. In all other ways, including sexual relations, these *male pseudohermaphrodites* feel and act like women.

A similar but independent condition, called **incomplete androgen insensitivity**, is usually inherited as an X-linked recessive. The individual's tissues show a partial responsiveness to androgen and produce a more variable phenotype. These 46,XY individuals have undescended testes and normal male levels of testosterone; they also lack both Wolffian- and Müllerian-derived sex organs, are infertile, and at puberty develop breasts. But their external genitals, instead of being entirely female, are partly male, characterized by an enlarged clitoris and partially fused labioscrotal folds. At puberty these individuals become more hairy, but their phallus remains small. Thus, the preferred sex of rearing is female. Whether the assignment is female or male, the appropriate surgery should be done as early as possible, and hormone treatment should be provided at the time of puberty to maximize the affected person's adjustment to that sexual identity.

Deficiency of 5-α-Reductase

In normal male fetuses an enzyme called **5-α-reductase** catalyzes the conversion of testosterone to **dihydrotestosterone** in tissues destined to form the male external genitals. An autosomal recessive mutation has been described in a large, inbred kindred from the Dominican Republic; some members of the kindred exhibit *male pseudohermaphroditism* due to deficiency of 5-α-reductase and thus of dihydrotestosterone. It has no obvious effect in 46,XX females, who show normal sexual phenotype and fertility. Homozygous affected 46,XY males have normal internal male organs, but their external genitals are ambiguous at birth, and many are raised as females. At puberty, male patterns of facial and body hair growth do not occur, and the prostate gland fails to enlarge. Yet the scrotum becomes pigmented, the penis grows, the voice deepens, and height increases to normal male levels.

At puberty, 17 of the 18 Dominican subjects who were raised as girls adopted a male gender identity (i.e., they began to feel that they were male). This sudden change suggests to some researchers that gender identity may not be fixed in early childhood, as has been assumed. Instead, under the influence of sex hormones, it may remain flexible until the time of puberty. Other researchers, however, argue that—given the low status of females in some Latin societies—these individuals may prefer maleness for practical rather than biological reasons.

Congenital Adrenal Hyperplasia

Some cases of male and female pseudohermaphroditism have been traced to defects in the synthesis of testosterone in the testes or of androgens in the adrenal glands, or both. In these patients the adrenal glands grow excessively large in an attempt to compensate for the deficiency; hence the name **congenital adrenal hyperplasia (CAH)**. (*Hyperplasia* refers to excess production and growth of normal cells in a tissue or organ.) Recessive autosomal mutations affecting about half a dozen different enzymes are known, each blocking a specific step in the multibranched biochemical network. This metabolic pathway is summarized very simply as follows:

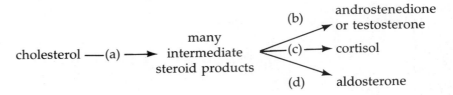

The hormones **cortisol** and **aldosterone** control sugar and salt balance in the body. Cortisol also acts as an inhibitor, regulating the amount of a substance that stimulates the adrenal cortex to produce precursor substances. Thus a deficiency of cortisol leads to an overproduction of hormone precursors and an accumulation of certain hormone products.

If step (a) is blocked in a 46,XY fetus, no testosterone is produced and a gonadal male with female genitals—a *male pseudohermaphrodite*—will result. These infants also suffer severe loss of salt and water because of the absence of cortisol and aldosterone, and they may die within a few weeks.

If the mutant block occurs somewhere in the intermediate steps of the pathway, there may be a biochemical detour around it that results in the production of some testosterone. Hence, the genitals of these male pseudohermaphrodites may be *partly* masculinized. In such cases, high blood pressure rather than extreme salt and water loss is usually observed. Early diagnosis is important so that the deficient hormones can be prescribed; following a decision on whether to raise the infant as male or female, the genitals can be surgically corrected.

When the enzyme block occurs at (c) or (d) in a 46,XY fetus, some of the backed-up intermediates are shunted into the testosterone branch, and an *excess* of male hormone is produced. The result is excess virilization in affected males and females. Over 90% of all cases of CAH are caused by a deficiency of the **21-hydroxylase** enzyme operating in one or both of these pathways. Four known forms of this disorder, all inherited as autosomal recessives, result from various combinations of several alleles at a duplicated locus on the short arm of chromosome 6.

Taken altogether, the worldwide frequency of **21-hydroxylase deficiency** is about 1 in 5,000 births. In the two rare *classic* forms (salt-wasting and simple virilizing) there is an early onset of virilization (Figure 6.5). With the other two (*nonclassic*) forms, individuals are born without symptoms. These cases are highly variable in phenotype and age of onset. Young children, for example, may show premature development of pubic hair. Later-onset cases may show no deficiency until late childhood or puberty. Symptoms may include advanced bone age (leading to short stature), severe acne, excess hairiness, and (in young women) infrequent menstruation, male pattern baldness, and ovarian cysts. The nonclassic forms are also much more frequent—found in up to 1% of all Caucasians and 3% of Jews of European origin. Indeed, nonclassic 21-hydroxylase deficiency may be the most common autosomal recessive disorder known in humans.

For families known to be at risk, it is now possible to detect heterozygote carriers as well as affected fetuses. Population screening of newborn babies has also been done, using the same drops of dried blood that were collected to test for PKU. With the more severe forms of this enzyme deficiency, treatment with the appropriate hormones may begin even before birth. Administration of the adrenal hormone *cortisone*, which in-

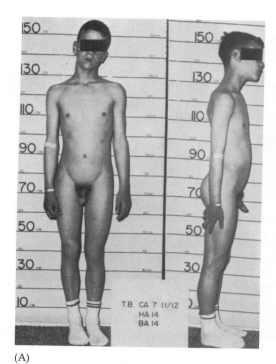

(A)

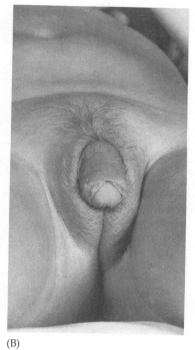

(B)

Figure 6.5
Virilized male and female, both with classic 21-hydroxylase deficiency. (A) This seven-year old 46,XY male has the type of congenital adrenal hyperplasia that leads to an excess of testosterone and precocious puberty (including pubic hair and enlarged penis), but no enlargement of the testes. His height is already about 155 cm (5 feet, 1 inch). (From Bongiovanni et al. 1967.) (B) This 26-month-old 46,XX female pseudohermaphrodite has ambiguous genitals (enlarged clitoris/phallus), pubic hair, and pigmented labio-scrotal folds. (From New and Levine 1973.)

hibits the action of testosterone, is effective in both female and male subjects. If detected and treated early enough with hormones and surgery, affected 46,XX females may be able to lead normal lives that include childbearing. Regarding the treatment of pseudohermaphrodites of any type, New et al. (1989) comment:

> Society sees phenotype, not genotype. In assigning a sex of rearing to a male or female pseudohermaphrodite, the genetic sex is of less consideration than the physiologic and anatomic character of the genitalia and their potential for development and function. Because of the wide individual variability in the presentation of ambiguous genitalia in these patients, there can be no all-inclusive rules for sex assignment based solely on genetic sex or type of enzyme deficiency.

Sex Reversal: XX Males and XY Females

About one in 20,000 male births is a phenotypic male with a 46,XX karyotype. Such *sex-reversed* males look entirely normal as boys, and often as men too. But all are sterile, and some adults show some of the clinical features associated with 47,XXY *Klinefelter males* (Chapter 10). These include small testes with abnormal tissue structure, decreased male hormones, and reduced male secondary sexual characteristics. Intelligence is normal, however, and mean height is shorter than 46,XY males but taller than 46,XX females.

Phenotypic females with a 46,XY karyotype are born with roughly the same frequency. These *sex-reversed females* also look normal as children, show normal intelligence, and are often taller than 46,XX females. They have Fallopian tubes, an underdeveloped uterus, and poorly differentiated (usually streak) gonads that render them sterile. Tumors may develop in the gonads. The absence of menstruation and secondary sexual characteristics can be treated fairly effectively with hormones. Both 46,XX males and 46,XY females may lead normal lives, including marriage.

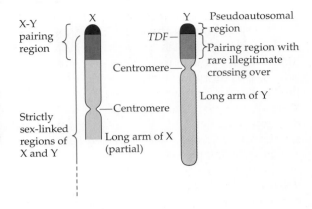

Figure 6.6
Fine structure of the human X chromosome (short arm) and Y chromosome. The tiny **pseudoautosomal region** *(black) carries a few homologous genes. During male meiosis, one obligatory crossover occurs here. The X-Y pairing region (black plus dark gray and dark red) includes a quarter of Xp and most of Yp. The dark gray and dark red regions carry different (nonhomologous) genes, but rare "illegitimate" crossovers can occur here too. The strictly sex-linked region (gray and red areas) makes up most of the X and Y.*

How do sex-reversed males and females come about? Many years ago it was proposed that some of them arise when a tiny but critical bit of Y chromosome is lost (from 46,XY females) or gained (by 46,XX males). The missing or extra piece was presumed to carry the gene for testis-determining factor (TDF). Indeed, 46,XX males and 46,XY females provided the key to geneticists' 32-year search for this elusive gene.

Recall that the human Y chromosome is about one-third the size of the X chromosome and carries fewer than 10 known genes, compared to over 160 known (plus 175 provisional) genes on the X. Yet despite their great dissimilarity, the X and Y chromosomes pair during meiosis. This is because the tips of their short arms share a tiny region of homology (Figure 6.6). During male meiosis, a synaptonemal complex and one chiasma always form in this region, resulting in a regular exchange of material between the tips of Xp and Yp. In 46,XY males the homologous region is present in *duplicate* (rather than singly, as with regular X-linked loci), and any genes within it are inherited as though they are autosomal—hence the term **pseudoautosomal** region. Two or three genes have been identified in the pseudoautosomal region.

Normal pairing between the X and Y, but *without* regular crossing over, also extends into the neighboring nonpseudoautosomal (and genetically nonhomologous) region, often just beyond the centromere of the Y chromosome. Very rarely an "illegitimate" crossover occurs in this latter region, resulting in the transfer of the *TDF* locus to the X chromosome and loss of that locus from the Y chromosome in that particular spermatocyte. If the resulting sperm fertilizes a normal egg, a 46,XX male or a 46,XY female will result (Figure 6.7).

Researchers focused their search for the *TDF* gene by analyzing and comparing the DNAs from several dozen 46,XX males (with only a tiny piece of a Y chromosome attached to one of their two Xs) and a few 46,XY females (with a tiny deletion of their Y chromosome). Both types of abnormalities were located in the same minute region of the Y chromosome where the gene for the testis-determining factor must reside.

In 1987 an international team of scientists reported that they had finally found the gene for testis-determining factor, but the excitement was short-lived. Further research indicated that this new gene, called *ZFY* (for Zinc Finger on the *Y*), did not meet all the necessary criteria. In 1990 a group of British investigators identified a nearby gene, located very close to the pseudoautosomal boundary, which they called **SRY** (Sex-determining Region of the *Y* chromosome). *SRY* codes for a small protein that closely resembles some DNA-binding proteins known to turn other genes on or off.

Additional evidence also points to *SRY* as the testis-determining factor: (1) Essentially the same Y chromosome DNA sequence is detected in all males, but not in the females, of a wide variety of mammals. (2) In XY embryos of mice, the *Sry* gene* is expressed only in cells of the gonadal ridges, and at a time just before testes should begin to form. (3) In mice, the *Sry* DNA sequence is present in the male-determining region of normal Y chromosomes, but missing in a mutant Y chromosome (with a tiny deletion) that has lost its male-determining ability. (4) If mouse *Sry* gene sequences are transplanted into XX zygotes, some (but not all) of these prospective females develop into males. The sex-reversed mice possess male rather than female sex organs and exhibit normal male copulatory behavior. (5) In humans, two sex-reversed 46,XY women were found to have mutations (rather than cytologically observable deletions) within a key part of their *SRY* genes.

Although they seem to have found the long-sought "master switch" that activates a cascade of steps leading to maleness, scientists sound a strong note of caution: Not all sex-reversed cases can be explained by these findings. Indeed, most 46,XY females possess the *SRY* gene, and the causes of their sex reversal are unknown. Likewise, some 46,XX males lack any detectable Y chromosome DNA. Certainly, non-Y-linked mutations must affect later steps in the cascade of events leading to the development or nondevelopment of gonads—but they have not yet been clearly identified. Thus, to fully understand the complex process of sex determination in both males and females, researchers will have to find and analyze the other genes—known to exist on both sex chromosomes and autosomes—that function in sexual development.

INFERTILITY AND THE NEW WAYS TO GET PREGNANT

About 8.5% of married couples who want children, or who want more children, are unsuccessful in producing them because of a variety of physical or emotional problems. For about 50% of infertile couples, the problem lies with the female; in about 30% of all cases, it lies with the male; and 20% of all cases involve problems with both partners. Besides surgery, other corrective treatments, and adoption, several techniques of "assisted reproduction" are now available. Although these reproductive techniques fall more within the realm of embryology than genetics, their use has generated tremendous interest, concern, and controversy. Thus, we present a brief overview here.

Artificial Insemination

Although not universally accepted on religious or moral grounds, **artificial insemination donor (AID)**† is a simple, inexpensive medical procedure dating back to the nineteenth century. But it has been plagued by secrecy, poor record keeping, and varying degrees of care in execution. Perhaps 6,000 to 10,000 children per year in the United States are now conceived through AID, using fresh or frozen donor semen. Employed primarily when a husband is infertile because of a low sperm count or poor sperm motility, it has also been used by couples when a fertile husband could transmit a deleterious allele. Unmarried women, too, have used AID.

Usually the doctor selects the donor, who may be a medical student matched as closely as possible to the husband for obvious physical char-

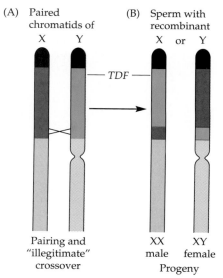

Figure 6.7
Results of "illegitimate" crossing over between the X and the Y chromosome short arms during spermatogenesis. Only one chromatid for each chromosome is shown. (A) An illegitimate crossover (indicated by chiasma) occurs in the non-pseudoautosomal X–Y pairing region. (B) The recombinant chromosome gives rise, after fertilization, to an XX male (with TDF) or an XY female (lacking TDF). Note: A recombinant chromosome is named according to the centromere it carries. Thus, in (B), the recombinant chromosome on the left has the X centromere (not shown) and is called an X. The one on the right has the Y centromere (shown) and is called a Y.

---

* The names of mouse genes that are equivalent to human genes often are written without full capitalization.

† Not to be confused with acquired immune deficiency syndrome (AIDS).

acteristics. Near the estimated day of ovulation, the doctor uses a syringe to deposit the donor's semen close to the woman's cervix. This procedure is repeated over as many months as may be necessary to attain fertilization. The donor and the "preadopting" couple remain unknown to each other, and a few doctors deliberately muddle the matter by mixing semen from several donors. The couple, the child, the doctor, and the donor bear legal relationships that are to some extent unresolved. Thirty states have laws stating that the AID child is legitimate and entitled to the same rights as a child conceived the old-fashioned way.

AID has been made safer and easier by means of frozen sperm, which can be stored for long periods of time and thawed just before use. The increased safety results from the greater time available for testing the donor and his semen for the bacteria or viruses that cause sexually transmitted diseases. Most sperm banks quarantine sperm while conducting follow-up tests, especially the test for HIV antibodies, which may show up months after the donor is capable of transmitting the AIDS virus through his semen. For this reason, current guidelines by several fertility associations recommend only frozen semen for AID, although pregnancy rates are somewhat lower.

In Vitro Fertilization and Embryo Transfer

The number of babies worldwide conceived by **in vitro fertilization and embryo transfer (IVF-ET)** is now in the thousands, but the pool of patients who might benefit from the procedure is thought to be in the millions. Over 100 centers in the United States now perform IVF-ET—but not all successfully, and none with federal funding.

The general procedure is to obtain an egg directly from the ovary, fertilize it with the mate's sperm in a culture dish, allow the zygote to undergo a few cleavage divisions, and transfer the developing embryo to the woman's uterus. There it might implant in the uterine wall and develop into a normal fetus. The various steps require several outpatient hospital visits for the prospective mother, numerous other examinations and tests, skilled medical and technical personnel, and a specialized laboratory.

The prospective mother is treated with gonadotropic hormones (or other fertility drugs) to induce her oocytes to begin to mature. At the appropriate time, the ovaries are viewed with a laparoscope (a lighted instrument with a magnifying lens and a collecting tube) inserted into the abdomen, and the contents of the mature follicles on their surface are vacuumed into a collecting chamber. Three or more eggs are put into a special culture medium and bathed with fresh sperm from the husband. If fertilization occurs, the resultant zygote cleaves to form two, four, and eight cells over several days (Figure 6.8). The developing embryo is transferred back to the female's uterus at the four- or eight-cell stage by means of a fine tube inserted through the cervical canal. Actually, to improve the chances of success, several fertilized eggs are usually transferred to the recipient's uterus. This practice has resulted in many multiple births— about 20–25%, compared to just 1% in the general population. The extremely high frequency of multiple births with IVF-ET causes an excess of premature deliveries and low birth weight babies, as well as stillbirth and infant mortality rates that are about double those in the general population.

Although there is much variation among clinics and among different subpopulations of infertile women, the overall success rate for IVF-ET remains disappointingly low. For each month that an attempt is made,

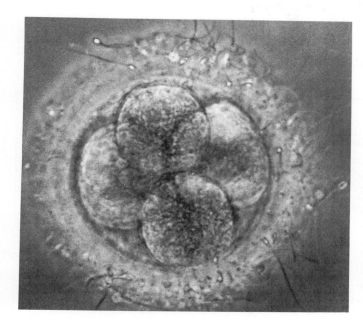

Figure 6.8
A four-cell human embryo derived from an egg recovered directly from a woman's ovary and fertilized in vitro. A few sperm can be seen in the zona pellucida surrounding the embryo. (Courtesy of R. G. Edwards, Cambridge University.)

only about 10% of starts eventually end in a live birth. The major stumbling block seems to be the failure of the embryo to implant after transfer to the recipient's uterus. Because the cost of this multistepped technological assist is about $6,000 per cycle (for as many monthly cycles as the process is attempted), the financial as well as emotional costs are extremely high. But the overall low rate of success should be compared with the natural situation. For normally fertile couples, it is estimated that only 30% of eggs exposed to sperm through sexual intercourse produce a viable offspring.

Researchers have now developed techniques for freezing and storing four- to eight-cell embryos. This practice reduces the need to repeat ovulation stimulation and laparoscopy to retrieve additional eggs. Subsequent menstrual cycles occur without drug treatment (because no more eggs need be collected), and this may improve the chances of achieving pregnancy. Freezing also reduces the risk of multiple births, because the physician can transfer just one or two fresh embryos, saving the remainder for successive attempts. This technique therefore sidesteps the ethical problem of discarding four- and eight-celled human embryos, which some in vitro centers do to avoid multiple pregnancies. Unfortunately, it also creates new problems.

Variations on a Theme

Other technological methods for dealing with the various causes of infertility are summarized in Figure 6.9. One spin-off of IVF-ET is a technique called **gamete intrafallopian transfer (GIFT)**. This technological assist might be helpful when the husband has defects of sperm concentration, motility, or morphology or when there are various other causes of infertility. However, a woman using GIFT must have at least one functional oviduct. The induction of ovulation and egg recovery with a laparoscope are similar to the operations used for IVF-ET. But instead of allowing fertilization to occur externally, eggs and sperm are introduced via a thin tube into the upper end of the oviduct. There fertilization occurs as it would normally. The rate of success seems to be better than for IVF-ET.

119

Name of technology	Sperm source	Egg source	Place of fertilization	Place of pregnancy
Traditional situation			Fallopian tube	Uterus
Artificial insemination donation (AID)			Fallopian tube	Uterus
			Fallopian tube	Uterus
Gamete intrafallopian transfer (GIFT)			Fallopian tube	Uterus
In vitro fertilization and embryo transfer (IVF-ET)			In vitro	Uterus
			In vitro	Uterus
			In vitro	Uterus
Egg transfer			Fallopian tube	Uterus
			In vitro	Uterus
Embryo transfer			Fallopian tube	Uterus
			Fallopian tube	Uterus
Traditional surrogacy			Fallopian tube	Uterus
Gestational surrogacy			In vitro	Uterus

Figure 6.9
Reproductive technologies that are currently available. Spouses' contributions are shown in gray; color represents a third-party contribution. For some techniques, several variants are possible.

When a woman does not produce viable eggs or when she chooses not to risk transmitting a dominant or X-linked recessive disorder to a child, two other techniques are sometimes used. In **egg transfer**, a donor is hormonally stimulated to produce one or more eggs, which are then obtained by laparoscopy. The eggs may then be fertilized within the wife, with the husband's sperm, using GIFT procedures. **Embryo transfer** may be used when both husband and wife are infertile. In one scenario, an egg donor is artificially inseminated by a sperm donor. The resulting embryo is flushed out of the donor's uterus for transfer to the wife, whose menstrual cycle must be synchronous. An obvious complication of this process is that failure to find and aspirate the developing embryo could result in unwanted pregnancy for the egg donor. The success rate of embryo transfer is usually low.

There is also **surrogate motherhood**, by which a husband and wife contract for the services of another woman, the surrogate. Surrogacy differs from the other technologies in that the fetus does not develop in the woman who intends to be the mother of rearing. Typically, the surrogate is artificially inseminated with sperm from the husband and is legally obligated to turn over the resulting newborn to the contracting couple. From 1984 to 1988, over 600 births occurred by this method in the United States alone. This unusual social situation has provoked strong public reactions and at least one sensational court case, when the surrogate (biological) mother changed her mind and decided to keep the baby rather than accept the fee and relinquish the child to the contracting couple (the biological father and his wife). Despite its legal and emotional pitfalls, as well as its high cost, this practice seems likely to continue.

In a new twist, called **gestational surrogacy**, the surrogate mother provides the uterus but not the egg and is usually completely unrelated to the child she bears. Instead, the child is conceived by IVF-ET, using eggs from the wife and sperm from the husband. Such a procedure allows a woman with a missing or nonfunctional uterus to nonetheless have a child who is genetically her own. The opportunities for misuse in nonmedical situations are obvious, however. In particular, many people worry about the "rent-a-womb" possibilities for women who are capable of bearing children but simply find it inconvenient for one reason or another.

HUMAN EMBRYONIC DEVELOPMENT

Whether conceived naturally or in a glass dish, normal human embryos go through the same developmental process inside the uterus.

Shortly after "unassisted" fertilization in the upper part of an oviduct, the zygote begins to divide as it slowly migrates down the oviduct to the uterus—yielding two cells, then four, eight, and so on, without any intervening periods of cell growth. During these **cleavage** divisions, the total amount of cytoplasm remains constant as it is partitioned among smaller and smaller cells. Metabolic needs are met by yolk and other materials originally laid down in the egg.

After about three days, a solid ball of approximately 16 cells enters the uterus. Further division and the absorption of fluid convert it by the fourth day into a hollow **blastocyst** of 100 to 200 cells. Among these, a tiny clump (the **inner cell mass**) destined to form the embryo proper can already be distinguished from the remaining cells, which will contribute to its surrounding membranes (Figure 6.10). After floating free for about two days, the blastocyst attaches to and begins to burrow into the spongy lining of the uterus, whose rich blood supply will nourish the developing embryo. By the end of the second week, **implantation** is complete. Inside the rapidly growing blastocyst, the embryo begins to absorb nutrients diffusing from maternal tissue.

By the third week, the inner cell mass has become a flat, pear-shaped embryo (Figure 6.11). Above a central rod called the **notochord**, primitive neural tissue arises, becomes grooved, and then folds together to form a **neural tube**, forerunner of the brain and spinal cord. Small cellular blocks called **somites** develop on both sides and ultimately differentiate into muscle, skeletal, and skin tissues. Blood vessels and primitive blood cells soon make up a simple **cardiovascular system** that distributes oxygen and nutrients throughout the embryo while removing waste materials. The cardiovascular system is the first organ system to become functional. It links the embryo with specialized maternal-fetal tissue called the **placenta**, which is embedded in the uterine wall. Here the embryonic and maternal bloods flow very close to each other but remain separated by cell membranes. The placenta also produces hormones that maintain pregnancy and prevent menstruation.

No signs of sexual development appear until the fourth week, when swellings that will ultimately form the external sex organs arise and **primordial germ cells** develop in the wall of a primitive cavity called the yolk sac (Figure 6.11B). These cells, the descendants of which are destined to become eggs or sperm, migrate during the fifth week into the newly forming internal gonadal ridges.

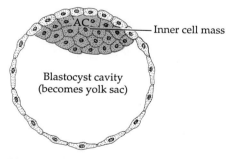

Figure 6.10

A human blastocyst (diagrammatic cross section) at about 4½ days. The inner cell mass (color) will form the embryo proper, and the outer circle of cells the chorion. The amniotic cavity will begin forming in the region marked AC at about eight days. The blastocyst consists of about 200 cells and is drawn here about 400 times actual size.

Figure 6.11
A human three-week embryo. (A) Top view, looking down from the amniotic cavity on what had been the inner cell mass. (B) Cross section of an embryo at the level indicated by the plane in (A). The yolk sac will get much smaller as the embryo grows and expands into the amniotic cavity. (Adapted from Moore 1988.)

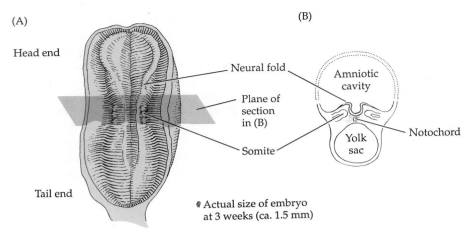

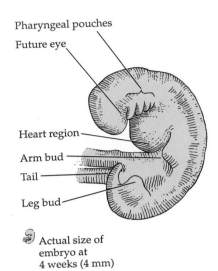

Actual size of embryo at 4 weeks (4 mm)

Figure 6.12
A human four-week embryo, without its surrounding membranes. (Adapted from Moore 1988.)

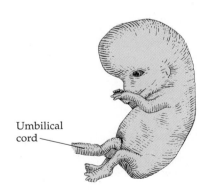

Figure 6.13
A human eight-week embryo without its surrounding membranes. All major organ systems are developing. The toes and fingers have separated, the webbing that once was between them having now disappeared. The embryo is shown actual size (30 mm). (Adapted from Moore 1988.)

The most crucial and sensitive developmental period spans weeks 4 through 7, a time when pregnancy may still go undetected. By the end of this stage, every major organ system has begun to take form; thus, the presence of drugs, viruses, or radiation may lead to malformations. Early in this period, the flat, disk-shaped embryo bends to form a C-shaped cylinder with a relatively large head (Figure 6.12). By the seventh week, rudimentary eyes, ears, nose, jaws, heart, liver, and intestines are distinctly visible, and the development of other structures (lungs, kidneys, sex organs, and skeleton) is also well under way.

By the end of the eighth week, the embryo is about 30 mm (1.2 inches) long and weighs roughly 5 grams (⅙ ounce). Its tail has disappeared. Limbs with fingers and toes are clearly formed, and the head constitutes almost half of its total length (Figure 6.13). The embryo is fully encased in two transparent membranes: an inner **amnion** pressed close to an outer **chorion**. Anchored to the placenta by an umbilical cord, the embryo moves quite freely in the fluid-filled amniotic cavity. The amniotic cavity acts as a protective shock absorber; it also prevents the embryo from forming adhesions with the surrounding membranes and promotes normal symmetry in development.

From the eighth week of development to birth at 38 weeks, the fetus will increase about 650-fold in weight (to 3,180 grams or 7 pounds) and about 17-fold in length. Meanwhile, development of all the organ systems continues (Figure 6.14). Because the respiratory system is the last to develop, lung function sets the limits of viability; brain development is crucial too. Actually, a 26-week-old fetus may survive if born prematurely, even though it has attained only 30% of the normal birth weight. With increased knowledge, improved technology, and extraordinary effort and expense, it is becoming possible to keep alive some babies born even earlier than this. Such heroic measures present an unresolved ethical dilemma, however, because most of the very premature survivors will suffer from serious problems for years to come.

GENETICS OF EMBRYONIC DEVELOPMENT

How does a single cell, the zygote, become transformed into an exquisitely complex organism made up of many distinctive cells, tissues, and organs? And how does each species develop unique patterns and shapes? These questions have fascinated and baffled scientists for more than a century.

Over many decades, embryologists have described in immense detail the changes and interactions of specific embryonic cells during their orderly course of differentiation. But what they found was not one grand scheme to fit all embryos; instead, they observed a bewildering array of developmental plans among the various groups of organisms studied. And although important discoveries and useful concepts helped to focus the search, clear-cut and universally applicable answers to these major biological problems continued to elude them.

In particular, neither embryologists nor geneticists could explain how such striking differences can arise among cells whose nuclei remain genetically identical. In the 1970s and 1980s the availability of some organisms in which many mutants of early development had been isolated and genetically analyzed, together with the use of new high-powered molec-

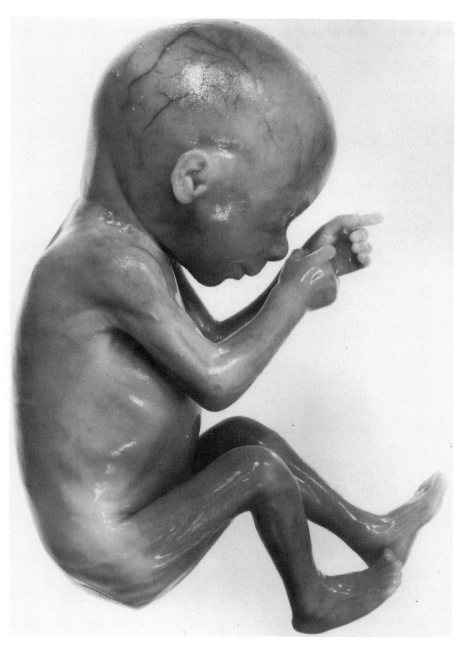

Figure 6.14
A human 17-week fetus, actual size. (From Moore 1988.)

ular techniques, brought new insights. Much of this recent knowledge in developmental genetics comes from intensive studies of a few representative animals, including free-living roundworms, fruit flies, and mice. Studies in fruit fly embryos indicate that three kinds of gene families control development: *maternal effect* genes, *segmentation* genes, and *pattern formation* genes. To a surprising extent, many similar genes operate in a wide variety of organisms.

Maternal Effect Genes

Long before the advent of molecular genetics, it was known that the egg contains stores of nutritive material (e.g., yolk) great enough to support the early embryo through at least its first few cleavage divisions. In addition, the egg was known to exert some genetic influence over the very early embryo. In snails, for example, the direction of shell coiling (left versus right) is controlled by the maternal genotype—which determines the spindle position and thus the planes of cleavage in the zygote and embryo—rather than by the snail's own genotype.

Later studies in a wide variety of organisms showed that maternal effects involve the stockpiling in the egg cell of various materials in addition to yolk. These include everything needed for cleavage divisions: nucleotides and enzymes for DNA replication, as well as tubulin for spindle formation. Another category of reserves includes constituents needed for protein synthesis, including ribosomes and messenger RNA molecules that encode genetic information transcribed from maternal DNA. No wonder the egg is so much larger than other animal cells!

The distribution of these various developmentally significant inclusions within an egg cell is not uniform, however. Some types of eggs show definite *gradients* of concentration, whereas others exhibit a *mosaic* partitioning of important substances. Either pattern eventually causes an unequal apportionment of certain substances among the various cells in a developing embryo, but neither pattern can by itself account for all the observed complexities of development; interactions between cells are critical too.

The first developmental step occurs when **maternal effect genes** in the mother set up important gradients (or other partitions) of substances in the egg cytoplasm. These gene products will give rise to the head-to-tail and (in some species) front-to-back axes in the developing embryo. How is this known? Scientists have analyzed many mutants of maternal effect genes and seen how such mutants can alter the fates of future embryonic cells. These mutants give rise to embryos with extra or missing heads, tails, dorsal structures, or ventral structures. In every case, the effects are best explained by the existence of one or more gradient systems.

Segmentation and Pattern Formation Genes

At some point the embryo's own genes become activated and take over from the maternal genes. (In humans this first occurs between the four- and eight-cell stages.) In fruit flies the establishment of the basic body plan is accomplished by two kinds of events. First the **segmentation genes** interact with one another to subdivide the embryo into regions—a linear series of progressively smaller but roughly similar segments. Then the **pattern-forming genes** assign a regional identity to each resulting domain (Figure 6.15). What finally results is a series of repeated patterns or units along the length of the embryo.

Such pattern formation occurs through the action of so-called **hom-**

eotic genes, which act by "naming," or specifying, the structures unique to each subunit. They were originally identified when flies carrying mutant alleles displayed bizarre effects, such as legs sprouting from antennal sockets or from the mouth region, wings emerging from the region where eyes belong, and so on (Figure 6.16). Thus, when the normal function of these genes is disrupted by mutation, the fate assigned to various regions is inappropriate to their location.

Using molecular techniques, geneticists isolated and sequenced the genes and gene products of some segmentation and homeotic loci. All were found to be complex and interrelated. In particular, they shared a highly similar sequence of about 180 DNA base pairs, which was more specifically dubbed the **homeobox**. Further analyses of numerous other maternal, segmental, and homeotic genes revealed that the homeobox was present in many of them.

Even more exciting was the finding that these master sequences also exist in a wide range of organisms: some mollusks, worms, amphibians, birds, and mammals (including mice and humans). Although their segmentation patterns are not so obvious as those of insects, these organisms are segmented in one way or another—including the structure of their central nervous systems, muscles, and (where they exist) backbones. The homeobox sequences have remained highly conserved throughout evolution. The similarity among such disparate organisms demonstrates that at least some developmental genetic mechanisms are widespread in the animal kingdom. Fruit flies have about 10 homeobox (*Hox*) genes that make up a single complex on chromosome 3. In both mice and humans, about 40 *Hox* genes have been identified, distributed among four different gene complexes on separate chromosomes.

In vertebrates, embryonic and regenerating limbs are good subjects for analyzing the homeobox proteins. Over 50 mutations in mice and at least 8 mutations in chickens are known to affect limb development. And in the embryonic forelimbs of *Xenopus* (a frog), there is an early molecular

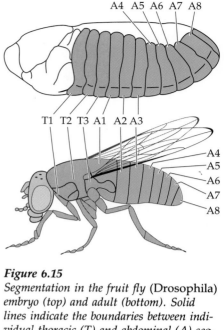

Figure 6.15
Segmentation in the fruit fly (Drosophila) *embryo (top) and adult (bottom). Solid lines indicate the boundaries between individual thoracic (T) and abdominal (A) segments. In the adult, each thoracic segment has a unique set of appendages: T1 has legs only; T2 has legs and wings; T3 has legs and halteres, which are small, stalklike balancing organs.*

Figure 6.16
Homeotic mutants of Drosophila. *(A) This four-winged fruit fly has several mutations that collectively change the third thoracic segment (T3) into another second thoracic segment (T2). As a result, it has wings where there should be halteres. (Photo courtesy of E. B. Lewis.) (B) This fruit fly has a mutation that causes normal legs to sprout from the area of the head where small, bristlelike antennae should be. (Photograph courtesy of J. Haynie.)*

(A)

(B)

difference—based on gradients of a particular homeobox protein—between presumptive arms and legs. Experiments with developing mouse limbs give similar results, suggesting that this gene (called *X1Hox 1*) is involved in forelimb development in many animals. Homeotic genes also affect the development of internal structures, such as the central nervous system (especially the hindbrain and spinal cord) as well as muscles and testes.

The effects of one *hox** mutation on very early mouse embryos seem to mimic a known syndrome in humans. Using some tricks of molecular biology, researchers produced some homozygous *hox-1.5⁻/hox-1.5⁻* offspring, all of which died at birth or shortly thereafter. These homozygotes have a fairly well defined set of developmental abnormalities. Their body shape is more rounded, with short necks and with heads tucked toward their chests. Thymus and other neck glands (including the thyroid) are missing or reduced in size. Other neck, throat, and tongue abnormalities occur, and the jawbones are shortened and misshapen. Serious heart and blood vessel defects are probably the cause of death. The airway tubes are smaller than normal, the esophagus lacks a ring of muscle, and the stomach and intestines are full of air.

Many of the defective tissues and organs seen in *hox-1.5⁻/hox-1.5⁻* mice develop from segmented embryonic structures called *pharyngeal arches and pouches* (shown in a human embryo in Figure 6.12), which give rise to structures in the lower face and neck. Particularly intriguing to researchers is the fact that the abnormalities appearing in *hox-1.5⁻/hox-1.5⁻* mice are strikingly similar to those seen in a human birth defect called *DiGeorge syndrome*. Babies with this disorder exhibit similar head and facial abnormalities, lack thymus and parathyroid glands, have too little thyroid tissue, and suffer from cardiovascular and immune system defects. Most die in infancy. So far, there is no clear-cut genetic correspondence between the two phenotypes, however. DiGeorge syndrome is inherited as an autosomal dominant, sometimes linked to breaks in human chromosome 22, whereas the counterpart of the mouse *hox-1.5* locus is known to be on human chromosome 7. Perhaps the DiGeorge phenotype results from mutations in different homeotic genes. Yet the remarkable similarities between these two conditions suggests a common developmental pathway.

Adhesion Molecules and Genes

The processes by which embryonic organs take shape and begin to produce tissue-specific proteins involve many complex interactions. Cells bind to each other, move, divide, differentiate, and die in precisely patterned ways. How are such interactions and movements coordinated and regulated? Research begun in the 1970s has uncovered key proteins called **adhesion molecules**. Those occurring on the cell surface are called **cell adhesion molecules (CAMs)**. They exhibit different binding specificities, but each will bind most tightly to similar CAMs on neighboring cells. Investigators showed that CAM-mediated cell-to-cell adhesions set up intricate patterns of chemical signals that cause cells to behave in specific ways. They have also identified the genes giving rise to several different CAMs and determined their DNA sequences. CAMs are quite widespread among animals, including fruit flies.

Cell adhesion molecules have also been implicated in *Kallmann syndrome*, an X-linked disorder. Affected males show an odd combination of two symptoms: small testes and no sense of smell. The defects result from a deficiency of an LH-releasing hormone in the hypothalamus and from

---

* Here, lack of capitalization distinguishes a mouse *hox* gene from its human *Hox* counterpart.

missing olfactory bulbs and tracts in the brain. It turns out, however, that there is an embryological connection between these traits. The cells that normally produce the LH-releasing hormone actually arise in the developing nose of the fetus and then normally migrate to the hypothalamus. The formation of olfactory bulbs and tracts is normally induced by the olfactory nerve, whose cells also arise in the developing nose and migrate to the brain.

In people with Kallmann syndrome, these two types of embryonic cells (secretory and nerve) never make it to their usual destinations in the brain and thus never function properly. Scientists have analyzed the DNA of several patients, and from the nucleotide sequence of the gene, they have deduced the structure of the protein product. Certain parts of its 680-amino-acid sequence are very similar to known cell adhesion molecules. Thus, it appears that the defects of embryonic cell migration that characterize Kallmann syndrome result from defects in CAM-like proteins and genes.

Programmed Cell Death and Suicide Genes

Odd as it may seem, embryos regularly produce many more cells than exist in the newborn organism. In fact, cell death is an important feature of embryonic development. We have already noted the normal loss of millions of primary oocytes in female fetuses, as well as the breakdown of Wolffian ducts in female fetuses and Müllerian ducts in male fetuses. Many other structures and systems also are affected by **programmed cell death**. For example, early in development our hands are paddle-shaped; but after the loss of parallel rows of cells, the remaining structures form our fingers. (Incomplete loss of such cells may leave webbing between the digits.) And in certain parts of the vertebrate brain and spinal cord, 40–85% of the nerve cells normally die before birth of the organism! More familiar examples in other creatures include the loss of a tadpole's tail as it becomes a frog and the loss of most of a caterpillar's tissues during its metamorphosis into a butterfly or moth.

Exactly how does this happen? Recent studies in several different organisms indicate that the "condemned" embryonic cells do not just die at random. Instead, it appears that by expressing specific **cell death genes**, they actively commit suicide. Mutations of these genes may block cell death, in which case the affected cells survive and function normally. In a well-studied roundworm, 17% of the embryonic somatic cells are normally destroyed before hatching. Over a dozen different *ced* (*cell death*) genes are known to direct several stages of this process, from actual death to engulfment by neighboring cells. Similar findings come from studies of moths—specifically, the quick death of certain abdominal muscles after moth pupae leave their cocoons and become adults. Here, too, several cell death genes have been identified. Researchers hope that the study of such genes will help us understand developmental processes that not only span the course from early development to aging, but include pathological conditions such as cancer and degenerative diseases as well.

Although scientists are still a long way from answering all the questions about how genes control development—including the question of what regulates the regulators—they have found some important clues and approaches to this major unsolved biological problem. It appears that at the molecular level, development in widely differing groups of animals operates by mixing and matching the same few kinds of genes and proteins. The challenge is to explain how a few types of genes can give rise to such different organisms.

GENOMIC IMPRINTING

Gregor Mendel reported that the outcomes of *reciprocal crosses* were identical. That is, the phenotypes and phenotypic ratios seen among the progeny were the same whether a particular trait was inherited through the male or the female parent. For almost a century, experiments with a wide variety of plants and animals have verified this concept.* For example, some invertebrates (such as male bees and wasps) regularly develop by **parthenogenesis**—that is, from an unfertilized egg—and thus have maternal chromosomes only. And although it is not their usual form of reproduction, parthenogenesis has been reported to occur occasionally in nonmammalian vertebrates as well. (Some birds, especially turkeys, can develop this way.) But there have never been any scientifically authenticated cases of such "virgin birth" occurring naturally among mammals. Biologists have always assumed, however, that—as with garden peas and parthenogenetically developed animals—the mammalian genome normally inherited from a female parent was functionally equivalent to that inherited from a male parent.

Why Mammals Need Both a Mom and a Dad

In the 1980s, researchers were amazed to discover that the two parental genomes are *not* always functionally equivalent in mammals. In mouse zygotes the two pronuclei (one male, one female) look somewhat different from each other and can be individually sucked out of the cell. By removing a pronucleus from one mouse zygote and putting it into another one, researchers can produce a zygote containing two female pronuclei or two male pronuclei. They can also fuse a second polar body nucleus with the egg pronucleus. But early embryonic development from such zygotes is abnormal. The two pronuclei come together and form the usual diploid number of chromosomes, just as normal fertilization would. Yet viable offspring are never produced. Purely female-derived embryos have poorly developed extraembryonic membranes, and purely male-derived embryos show highly retarded development of the embryo proper. So mammalian offspring need both a mom and a dad for normal development to occur.

In human zygotes the same situation apparently prevails. An occasional pregnancy produces, instead of a normally developing fetus, a uterine growth called a *hydatiform mole*. Such placental masses contain extraembryonic membranes but no fetal tissue, and their chromosomal composition is entirely paternal—presumably the result of the fertilization of an egg without a pronucleus and the subsequent doubling of the sperm chromosomes. Conversely, *ovarian teratomas*, which are embryonic tumors that lack placental tissue, are derived from non-ovulated oocytes and have two maternal sets of chromosomes. Human *triploids* possess three complete chromosome sets (Chapter 10) and rarely survive past early pregnancy. Triploid early abortuses with two sets of paternal chromosomes and one set of maternal chromosomes usually have a large placenta and some molelike features, whereas triploid abortuses with two sets of maternal chromosomes and one set of paternal chromosomes have an underdeveloped placenta.

Do mammalian zygotes need to have an *entire* genome from each parent, or will certain critical chromosomes, or chromosome parts, or groups of genes, or even single genes do the trick? Experiments with some mouse embryos that possess rearranged chromosome parts (Chapter 11) show that certain genes or groups of genes are active when contributed by one parent but not active when contributed by the other parent.

* With the exception of reciprocal crosses involving sex-linked genes.

This phenomenon, whereby progeny phenotypes differ according to whether a particular gene (or group of genes) came from the mother or the father, is called **genomic imprinting**. It can be considered a form of temporary gene inactivation. Researchers estimate that perhaps 10–20% of mouse genes are subject to imprinting, but only a few of these naturally imprinted genes have been identified so far. One encodes *insulin-like growth factor II (IGF-II)*, transmitted as an autosomal recessive. It is active when inherited from a male but "silent" when inherited from a female. IGF-II is necessary for normal embryonic growth. When males with a mutated form of this gene are mated to normal females, the heterozygous offspring are viable, fertile, and normally proportioned—but only 60% of normal size. The reciprocal cross, however, does not give rise to any "proportionate dwarfs"; instead, all progeny are phenotypically normal. In humans the *IGF-II* gene is inherited on chromosome 11; its behavior with regard to imprinting is unknown. (Some traits in fruit flies, maize, and yeast also seem to show imprinting.)

How does the imprinting process occur? Nobody knows for sure. But one possibility is that during gamete formation, certain genes are differentially altered, or "tagged," by the attachment of certain small molecular groups (methyl groups) to DNA. This kind of process is also thought to be involved with the *X chromosome inactivation* that occurs in cells of all female mammals and in a common type of mental retardation called *fragile X syndrome* (Chapter 7). But it may not be the only answer, or even the primary answer, to the riddle of imprinting. Some changes in the degree of chromosome coiling, for example, have been noted in association with gene activity and imprinting.

Whatever its nature, the imprinting procedure would have to be *erased* some time during the embryo's development, so that each individual could *re-imprint* its genes according to its own sex during gametogenesis (Figure 6.17).

Imprinting and Human Disorders

There is evidence of differential imprinting in a few dozen human disorders. The age of onset or severity or segregation ratios of these conditions seem to differ according to which parent contributed the gene. *Huntington disease (HD)*, an autosomal dominant neurological condition that is lethal in adulthood, provides the best example. Its age of onset is highly variable, the average being 30 to 40 years. But significantly more persons with juvenile- or adolescent-onset HD inherit the gene from an affected father than from an affected mother. And significantly more cases of late-onset HD (after age 50) can be traced to an affected mother than to an affected father.

The most astonishing example of imprinting in humans involves two rare disorders whose physical and genetic features in some ways present a mirror image. Both involve growth and behavioral defects. The *Angelman syndrome* (Figure 6.18A) was once called the "happy puppet" or (in French) "marionette joyeuse" syndrome because of the behavior and appearance of affected children. Despite severe mental and motor retardation and absence of speech, they have a happy disposition and laugh excessively. They tend to be hyperactive, with movements that are jerky, repetitive, and puppetlike. They have red cheeks, a large jaw, and a large mouth with a prominent tongue. They also show abnormal brain waves and suffer from seizures. Children with *Prader-Willi syndrome* (Figure 6.18B) have small hands and feet, underactive gonads and tiny external genitals, short stature, and some characteristic facial features. They are mentally retarded, slow-moving, obese, and compulsive overeaters.

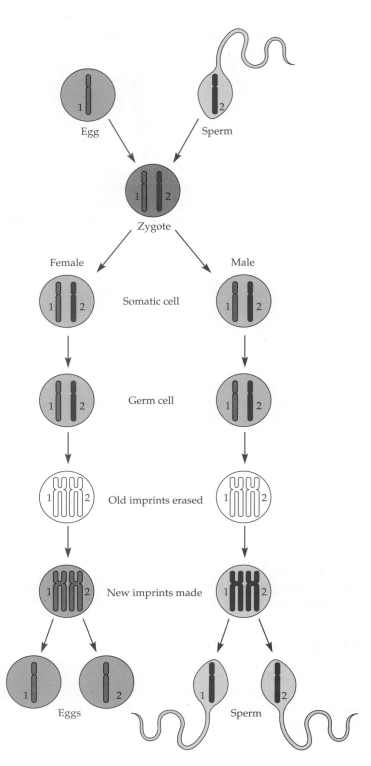

Figure 6.17
The processes of imprinting, erasure, and re-imprinting, presumed to occur between two generations of gametes. As genes pass from one sex to another (i.e., mother to son or father to daughter), the imprints (shown as colored in the egg, black in the sperm) will change even though the alleles remain the same. Consider what happens to two homologous autosomes if a particular zygote is female (left column); now compare that to what happens if the zygote is male (right column). (We have labeled the homologues 1 and 2.) In both sexes the imprinting is erased following chromosome duplication inside the gamete-producing cells. By the end of meiosis, however, the newly produced gametes are re-imprinted. The gametes formed by either sex contain the same chromosomes. But now one of the chromosomes — that labeled 2 in the egg and 1 in the sperm — is differently imprinted from when it entered the zygote. (Adapted from Hoffman 1991.)

The unusual situation is that both syndromes, in about 50% of cases, show tiny deletions in the q11–13 region of chromosome 15. Whether these deletions are identical is not clear, but they do overlap to a considerable extent. What makes all the difference, however, is the pattern of transmission of these deletions to heterozygous offspring (i.e., those with one normal and one deleted chromosome). When the deletion is inherited from the father, it results in the Prader-Willi syndrome; but when inherited from the mother, it results in the Angelman syndrome!

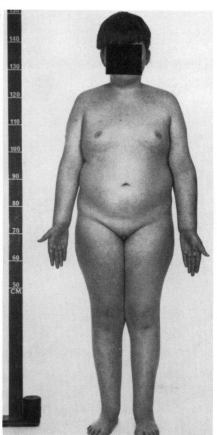

Figure 6.18
Two syndromes that result from differential imprinting of the same region on chromosome 15q. (A) Two brothers with Angelman syndrome, which results from a deletion of the region in the maternal chromosome 15. Facial features include a wide mouth, thin upper lip, pointed chin, and prominent jaw. (Photo courtesy of Devlin and Robb et al.) (B) A male with Prader-Willi syndrome has a deletion of the same region in the paternal chromosome 15. Note the obesity, tiny penis, and small hands and feet. (Photo from Connor and Ferguson-Smith 1987; courtesy of Blackwell Scientific Publications.)

Certain types of childhood cancers, during the course of their malignant growth, also show intriguing patterns of selective loss of either maternal or paternal chromosome segments. Some of these tumors have also exhibited patterns of inheritance or selective chromosome loss that suggest imprinting. Such studies of the relation of the parental origin of chromosomes to cancer development and growth provide a new twist to an extremely complex area of research (Chapter 15).

The various phenomena described in this chapter indicate that scientists have made some exciting beginnings in describing the genetics of normal development as well as developmental disorders. It is now clear that development is regulated by a changing pattern of gene expression. Although the techniques of molecular biology may never reveal *all* the answers to *all* questions of how differentiation is regulated, with new insights and ever-improving experimental techniques, the future should continue to bring new understanding.

SUMMARY

1. In very early development, the sex organs have the potential to form either a male or a female. Presence of a Y chromosome triggers the development of testes. The hormone testosterone, produced by the embryonic testis, stimulates the development of male sex organs.

2. The presence of two X chromosomes, along with the absence of a Y chromosome and male hormones, leads to the development of female sex organs.

3. Physical changes at puberty are regulated by two main types of hormones. Gonadotropins are secreted by the pituitary and act on the gonads. Sex hormones are secreted by the gonads and to some extent by the adrenal glands and act on many parts of the body. Both sexes secrete all these hormones, but in different proportions.

4. True hermaphrodites, which are usually 46,XX, have both ovarian and testicular tissue. Pseudohermaphrodites, either 46,XX with ovaries (female) or 46,XY with testes (male), have external genitals that are

ambiguous or characteristic of the opposite sex. Either type of individual may be raised as a male or female.

5. There are numerous types of male and female pseudohermaphroditism. The most common inherited forms are collectively known as congenital adrenal hyperplasia. Most cases of CAH are caused by a deficiency of the 21-hydroxylase enzyme. Different mutant alleles, in various combinations, lead to different forms of CAH.

6. Testicular feminization occurs when 46,XY individuals, with testes that produce normal amounts of testosterone, develop into apparently normal (but sterile) females. The absence of androgen receptors renders their tissues unresponsive to male hormone.

7. The tips of Xp and Yp are homologous and make up a tiny, pseudoautosomal region that (during male meiosis) regularly pairs and includes a crossover. Next to the pseudoautosomal region of Yp is a segment that harbors the gene for the testis-determining factor. Illegitimate crossing over in this segment may give rise to sex-reversed 46,XY females or 46,XX males.

8. New reproductive technologies are designed to help infertile couples have children, but they also raise ethical, legal, social, and financial questions. These methods include the use of artificial insemination donor (AID), in vitro fertilization, embryo transfer, gamete intrafallopian transfer of a woman's egg and a man's sperm (GIFT), similar manipulations with donated eggs or donated embryos, and surrogate motherhood with or without a donated egg.

9. The human zygote undergoes repeated divisions to form an embryo and surrounding membranes. The most critical and sensitive developmental period spans weeks 4 to 7, during which all the major organ systems are forming.

10. Several hierarchies of gene action control early development. These include the maternal effect genes of the mother, as well as the segmentation and pattern formation genes present in the zygote. Pattern formation (homeotic) genes contain a highly conserved region called the homeobox. Its protein product is thought to regulate groups of genes by binding to their DNA.

11. Cell movements and tissue differentiation that give rise to animal body form are regulated by various types of adhesion molecules. Another normal feature of development, programmed cell death, is orchestrated by a series of cell death genes.

12. Mammalian zygotes do not develop properly unless they contain both maternal and paternal genomes. In both sexes, some genes become differentially imprinted (silenced) during gamete formation and are expressed differently during development.

KEY TERMS

adrenal gland
5-α-reductase
amnion
androgen
androgen receptor
anti-Müllerian hormone (AMH)
blastocyst
cell adhesion molecule (CAM)
cell death gene

cleavage
egg transfer
embryo transfer
estrogen
genomic imprinting
gestational surrogacy
hermaphrodite
homeobox
homeotic gene
21-hydroxylase deficiency
maternal effect gene

Müllerian duct
ovary
oviduct
pattern formation gene
penis
placenta
pseudoautosomal region
pseudohermaphrodite
scrotum
segmentation gene
sex reversal

SRY gene
testis
testicular feminization
testis-determining factor (TDF)
testosterone
true hermaphrodite
uterus
vagina
Wolffian duct

QUESTIONS

1. Between 1968 and 1992, female athletes at the Olympic Games were required to take simple tests that indicate the number of X chromosomes and (later) the presence of Y chromosome DNA. This policy was intended to detect any male members of female teams. Could these test unequivocally accomplish their purpose? Explain.

2. A few cases have been described of males who possess oviducts and a uterus as well as the usual male internal organs. The condition is probably inherited as an autosomal recessive. What type of error in fetal development could account for this phenotype?

3. Why are males generally more vulnerable than females to errors in sexual development?

4. List all the exceptions to Mendel's laws—especially regarding reciprocal matings—that you can think of. (To do this, take into account what you have learned in previous chapters as well.)

5. What mode(s) of inheritance could explain the following pedigree (Bowen et al. 1965) for partial androgen insensitivity?

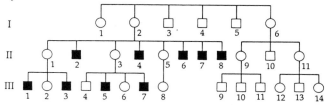

6. Voss et al. (1989) describe a cystic fibrosis patient who was diploid but who had two identical chromosomes 7 from his mother and none from his father. This phenomenon resulted in homozygosity for the maternal CF allele. In addition to having cystic fibrosis, however, he and one other such CF patient (with two identical maternal chromosomes 7 and no paternal chromosome 7) were both *very small* for their age. Can you suggest any reason(s) for the small stature?

Questions 7 to 9 are meant primarily to stimulate discussion, and we do not provide answers.

7. Do you approve or disapprove of the following practices that alter human reproduction? For yourself? For others? What conditions would you impose on their use? Which practices are the most feasible? The least feasible?

> Artificial insemination using a husband's sperm
> IVF-ET using a wife's egg and a husband's sperm
> Banking of eggs from distinguished and identified women
> Reproductive technologies to aid homosexual couples or single women
> Surrogate motherhood, in which a *nonrelated* surrogate provides both the egg and the uterus
> Surrogate motherhood, in which a *related* surrogate provides the both the egg and the uterus
> Surrogate motherhood, in which a nonrelated surrogate does not provide the egg
> Surrogate motherhood, in which a related surrogate does not provide the egg

8. It has been reported that a certain doctor, using AID, arranged 50 pregnancies from one sperm donor. Is there anything wrong with this?

9. When they were killed in a plane crash, a wealthy couple from Los Angeles left behind two frozen embryos conceived by IVF using the wife's egg and donor sperm. The couple had provided no instructions for the disposition of the embryos. Common law generally holds that those who have been conceived have the right to inherit if born alive. What would you do with the embryos?

FURTHER READING

Sexual differentiation in animals, including humans, is summarized by McLaren (1988), Mittwoch (1986), and Austin and Edwards (1981). Various disorders of sexual differentiation are described by Simpson (1990), New et al. (1990), and Griffin and Wilson (1989).

The new reproductive technologies are presented by Austin (1989), Seibel (1988), and Office of Technology Assessment (1988). Andrews (1991) and Edwards (1989) discuss the ethical and legal ramifications of in vitro fertilization, while Lasker and Borg (1987) focus on personal experiences with high-tech conception.

Prenatal human development is clearly presented by S. G. Gilbert (1989), Moore (1988), and Nilsson (1989). For information on the genetics of embryonic development, see S. F. Gilbert (1991). Genomic imprinting is discussed by Sapienza (1990), Hall (1990), and Reik (1989).

7

Complicating Factors

In the remote rain forests of New Guinea live a group of Melanesians who speak the Fore language. Until recently, these Fore people and some of their nearby neighbors practiced ritual cannibalism of dead relatives as a way of honoring them. But what attracted the attention of geneticists in the 1950s was not these cannibalistic practices. Rather, it was the high frequency and unusual transmission pattern of a strange and terrible disease that plagued the Fore people. They called it **kuru**, or laughing death.

Signs of brain damage—uncoordinated movements, trembling, loss of balance, inability to talk or swallow, emotional instability, and uncontrollable laughter—would suddenly appear. No treatment could ease or cure this quick-acting neurodegenerative disorder, and death came within a year after the onset of symptoms. About 75% of kuru victims were adult women, and the rest were mostly children of both sexes above the age of four (Figure 7.1). Kuru was rare among adult men and very young children and totally absent in outsiders who lived among the Fore. On the other hand, a few Fore people developed kuru while living (for up to six years) far from their villages.

Because it did not behave like a standard infectious disease, and because it seemed to run in families, some scientists suggested that kuru might be inherited. To account for its unusual age and sex distribution, they proposed an autosomal disease-causing allele that was dominant in females and recessive in males, with the onset of symptoms occurring early in homozygotes and late in female heterozygotes. The main problem with this scheme lay in explaining how a highly lethal allele could ever come to have high frequency. One view was that the kuru allele had become lethal only recently through some environmental change and that prior to this time it had been neutral or perhaps even beneficial.

IS THE TRAIT REALLY GENETIC?

Common sense tells us that a trait or disorder that runs in families could be attributed to genetic factors (simple or complex), or to shared environmental factors (diet, living habits, chemical pollutants, etc.), or to some combination of genetic and environmental factors. Gene interactions can be complex, and the methods for investigating genetic traits are not always straightforward. Thus, the first question we need to ask about any phenotype—even if it obviously runs in families—is whether it is genetically determined at all. At least two important criteria, taken together, are needed to establish a genetic causation:

1. *The trait must occur more often among genetic relatives of probands than among the general population.* If the trait has a one-gene mode of inheritance without complicating factors, the occurrence of clear-cut phenotypic ratios among the offspring of particular matings may clinch the genetic argument. But if two or more genes are involved or if complications occur, then it becomes increasingly difficult to compute expected ratios, and the genetic causation becomes less and less clear.

2. *The trait must not spread to unrelated persons exposed to similar environmental situations.* This criterion helps exclude infections, environmental poisons, nutritional deficiencies, and other conditions that may run in families or be prevalent in local populations. But some infectious agents may mimic biological inheritance. For example, syphilis and some viral diseases can be transmitted from mother to fetus across the placental barrier.

These two criteria are necessary but not always sufficient. Additional, and usually less forceful, observations supporting a genetic basis may apply in some situations:

3. *The trait appears at a characteristic age without any obvious precipitating event.* Many genetic disorders are present at birth, but so are developmental abnormalities not related to the genotype of the newborn. Other genetic disorders, such as Huntington disease, appear at a later age but with some variation in the age of onset. However, the absence of an obvious trigger may reflect ignorance of environmental agents rather than the operation of genes.

4. *The trait varies in frequency among separate populations.* Human populations differ in the frequencies of many alleles. Racial or ethnic characteristics may indicate genetic differences, but it is often difficult to disentangle genetic from environmental causations.

5. *Identical twins (who have identical genotypes) share the trait more often than do nonidentical twins (who, like ordinary siblings, have about 50% of their genes in common).* Also, if the trait is genetic, adopted twins will resemble their biological parents more than their adopting parents. Substantial difficulties arise in the interpretation of twin studies, however, as presented in Chapter 8.

6. *A human trait resembles one known to be inherited in experimental or farm animals.* Scientists can gain knowledge of a human genetic disease by studying animals that appear to have the same inherited defect.

Kuru

For about eight years after kuru was discovered by scientists, its pattern of transmission remained ambiguous, and in the end, the genetic hypothesis turned out *not* to stand the test of time. The first clue pointing to an environmental causation came when scientists noted the similarity of kuru symptoms to those of *scrapie*, a fatal infectious disease of sheep and goats. (Scrapie can also be experimentally transmitted to other mammals.) The progress of the infection in scrapie, kuru, and another human illness called *Creutzfeldt-Jakob disease* is similar.* An extremely long period separates the time of exposure and the onset of symptoms. (This latent period of up to several years first led scientists to call the disease agent a "slow virus.") Brain damage leads to loss of coordination and mental deterioration and usually ends in death. The immune system of infected animals does not respond effectively, and unusual fibrils can be found in infected brain tissue. The infectious agent of scrapie—called a **prion**—is very strange, consisting largely or entirely of protein with little or no nucleic acid. The

Figure 7.1
A Fore child incapacitated by kuru. Although still alert, the child could not stand, sit, or talk, and died a few months after this photograph was taken. (From Gajducek 1977.)

* Similarities to some characteristics of *Alzheimer disease* (a type of mental deterioration seen primarily in the elderly) are also intriguing.

infectious nature of kuru was finally confirmed by inoculating chimpanzees with the brain tissue of kuru victims. After year-long incubation periods, the animals developed a fatal disease closely resembling kuru.

An explanation for the unique age and sex distribution of kuru among the Fore tribes lay in the ceremonial cannibalism of close kin. Preparing for the rites, women and young children of both sexes handled the dead body, including the brain, which was packed into bamboo cylinders and steamed. The infectious agents probably entered the women and young children through breaks in their skin or through the mucous membranes of the eyes or nose. Men and older boys seldom took part in the ritual. The practice of cannibalism has since been abandoned by the Fore people, and the incidence of kuru has been much reduced.

In the remainder of this chapter, we will describe many of the difficulties encountered in trying to establish a genetic basis for various traits, as well as the problems of interpreting human genetic data. We will also present some recent molecular findings that shed new light on a few of these old problems.

VARIATIONS IN PHENOTYPIC EXPRESSION

From observations of gene transmission in plants and animals, it is known that possession of a gene does not guarantee that it will be expressed the same way in every individual. When a gene is "turned on," its RNA and polypeptide products interact with elements of the internal environment (including other gene products and metabolites) and external environment (including nutrition, medicines, climate, pollutants, exercise, and mental stress). Some genes, such as those determining the blood groups, are always expressed in essentially the same way, regardless of the "background" genotypes or outside conditions. Other genes, such as that for brachydactyly, may be expressed in various ways, or not at all, depending on the presence or absence of additional factors that are often not specifically identifiable.

Sex Differences

The sex hormones can exert a powerful influence on the internal environment. The product of any gene exists in one or the other of the two different chemical milieus that characterize the two sexes. Although this does not seem to affect how most genes are expressed, some inherited traits do differ between males and females. In general, we are considering here the effects of *autosomal* loci, whose pattern of inheritance would be distinct from sex-linked genes.

An example of such **sex-influenced inheritance** is *pattern baldness*, in which premature hair loss occurs on the front and top of the head but not on the sides (Figure 7.2). Its inheritance is difficult to study because hair loss may be caused by many factors, both genetic and environmental. In addition, hair loss begins at different ages and varies in extent. Pattern baldness affects mainly males; females who carry the same alleles may show only thinning of hair rather than complete loss. Male hormones, the androgens, are implicated in the expression of the trait. For example, a full head of hair, sometimes luxuriant, is seen in men who do not mature sexually because of testicular injury or castration prior to puberty. But after androgen treatment, these men sometimes lose their head hair. Likewise, females with tumors of the adrenal gland may secrete large amounts of androgens and become bald. Removal of the tumor in these cases again

(A)

(B)

(C)

(D)

Figure 7.2
Pattern baldness in the Adams family. Each man fathered the next one in this list. (A) John Adams (1735–1826) at about age 65, second U.S. president. (B) John Quincy Adams (1767–1848) at age 52, sixth U.S. president. (C) Charles Francis Adams (1807–1886) at age 41, diplomat. (D) Henry Adams (1838–1918) at about age 45, historian. (A, lithograph by Nathaniel Currier, courtesy of the New York Historical Society. B, engraving by Francis Kearny, courtesy of the New York Historical Society. C, lithograph by Nathaniel Currier, courtesy of the Library of Congress. D, photograph by his wife, Marian Hooper Adams, courtesy of the Massachusetts Historical Society.)

promotes growth of head hair. Thus, expression of this type of baldness requires *both* the possession of a specific allele (which appears to be common) and the presence of male hormones.

Father-to-son transmission (as in the Adams family) and other data tend to rule out X-linked inheritance in favor of autosomal inheritance, but the exact mode or modes of transmission are still not clear. It has been suggested that the autosomal gene responsible for pattern baldness, B_1, acts dominantly in males and recessively in females. If so, the genotype/phenotype correspondences are as follows:

Genotype	Female phenotype	Male phenotype
B_1/B_1	Pattern bald	Pattern bald
B_1/B_2	Nonbald	Pattern bald
B_2/B_2	Nonbald	Nonbald

The particular pedigree in Figure 7.3 is consistent with this view, but it is also consistent with other genetic hypotheses (see question 3).

Traits affecting structures or processes that exist in only one sex are called **sex-limited traits**. The manner of gene transmission is irrelevant here; it is the interaction of gene products with the internal environment of the body that results in the sex difference. For example, *hydrometrocolpos*, a condition characterized by the accumulation of fluid in the uterus and vagina, occurs in women homozygous for a certain autosomal recessive

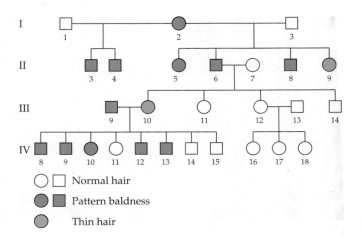

Figure 7.3
Part of a pedigree of pattern baldness, presumed to be dominant in males (B_1/—) but recessive in females (B_1/B_1). Female I-2 was reported to be "very bald on the top and front of the head" from an early age. Her sons, II-6 and II-8, and her great-grandsons, IV-8, IV-9, IV-12, and IV-13, became bald by age 25 in the same pattern. (The age of onset was not known for II-3 and II-4, her sons by a previous marriage.) If pattern baldness is recessive in females, then II-5 should not have been affected. Since she was reported to be chronically ill and did not become bald until rather late in life, other factors may have been responsible. (From Osborn 1916, with original numbering of the pedigree.)

gene. Obviously, males can never exhibit this symptom, even if homozygous. Still another example can be seen in dairy cattle, where both the quantity and quality of milk are known to be influenced by autosomal genes inherited equally through both parents. Because bulls cannot express these economically important genes, breeders must evaluate their genetic worth by assessing the milk production of their female relatives. (A prize bull can father thousands of offspring after his death through frozen sperm and artificial insemination.) Genes that influence human milk production undoubtedly exist, although little is known of their inheritance or specific effects.

Variable Expressivity

The expression of sexual traits is just one factor that interacts with gene products. More generally, the totality of both genotype and environment can modify the way a particular gene product is utilized. **Variable expressivity** refers to the differences in the observed effects of a given allele in different individuals—for example, the degree of severity of a certain inherited syndrome. Thus, expression may be represented as a *continuum* rather than as an all-or-none phenomenon. A sex-influenced trait can be considered a specialized case of variable expressivity.

In experimental organisms, a number of genes that have been studied affect the expression of other genes in clearly defined ways. **Suppressor mutations** reduce the effect of another mutant gene to a normal or near-normal phenotype. **Modifiers** are genes that change the degree of expression of other genes. In mice, mink, and other mammals, for example, many genes are responsible for coloring the fur various shades of black, brown, or yellow, but a modifier gene called *dilute* reduces their intensities, leading to faded grays, tans, or buffs.

Siamese cats and Himalayan rabbits present a familiar example of *environmentally caused variation* in a gene-controlled trait (Figure 7.4). These animals are homozygous for a mutant allele of a coat color gene that produces light or white fur all over the body except on the extremities. The paws, ears, nose area, and tail have dark fur, the particular color variation being determined by other coat color genes. The development of the dark fur in these animals is dependent on temperature; the lower the temperature, the darker the fur and the more extensive the dark areas. Apparently, the pigment-producing enzyme made by this mutant allele works best at the somewhat lower temperatures that occur farthest from the core of the body. If a patch of white fur is shaved from the side of a

(A)

(B)

Figure 7.4
Some animals possess a pigment-producing enzyme that works best at the somewhat lower temperatures of their extremities. (A) Siamese cats. (B) Californian rabbits, a commercial breed derived in part from the Himalayan breed. (A, copyright J. F. Glisson; courtesy of Hatcher R. Granville, Rich-Hat Cattery, Atlanta, and the Siamese Cat Society of America. B, Barry L. Runk, Grant Heilman Photography.)

Himalayan rabbit and an ice pack is applied (or if the rabbit is simply put in a cold environment), the new hair that grows back will be dark.

Variable expressivity is the rule in humans, but it is not usually possible to pinpoint specific modifying genes or specific environmental agents. Nearly all mutant phenotypes, especially those due to dominant genes, can be shown to vary in some respect from one affected person to another. A good example is the autosomal dominant disease *neurofibromatosis*, more accurately called *neurofibromatosis type 1 (NF1)* or *von Recklinghausen disease* to distinguish it from at least half a dozen related disorders. Although expressed to some degree in anyone who possesses the gene, the phenotype is extremely variable—between families, within families, and even in the progression of the disease within a single individual. Mild cases may involve only a few lightly pigmented areas called *café-au-lait spots* (which are the color of coffee with milk) and a few *skin neurofibromas* (benign but disfiguring tumors that arise from the fibrous coverings of nerves). These features are shown in Figure 7.5. Also usually present are so-called *Lisch nodules* (small lumps) in the iris of the eye. In more serious cases, however, patients may develop thousands of neurofibromas of various sizes and a wide range of additional symptoms. Many of the affected tissues are derived from the *neural crest*, a transient embryonic structure that lies atop the newly formed neural tube. Its cells migrate throughout the embryo and give rise to many different cell types, including certain pigment cells, nerve cells, hormone-producing cells, skeletal cells, and connective tissue cells.

Despite the relatively high frequency of NF1—about 1 in 3,000 to 4,000 births among all ethnic groups—and despite much research, physicians still cannot predict how it will be manifested in a given individual, nor can they treat the major symptoms effectively. In some cases, cosmetic surgery may lessen severe disfigurement and improve social interaction. In 1987 the gene was mapped to chromosome 17q. This was the first step toward analysis of the gene itself, identification of the gene product, prenatal diagnosis, and possible therapy.

Variable Age of Onset. Another aspect of gene expression is the age at which the phenotypic consequences first appear. Although many genetic conditions, such as albinism or polydactyly, are present at birth, others only appear at a later age. For example, even though the metabolic error in phenylketonuria can be diagnosed at birth by a simple urine test, there are no overt clinical abnormalities for a few months. The change of an

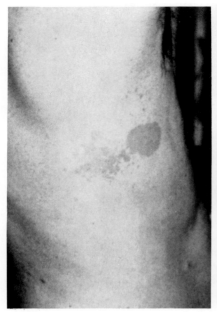

(A)

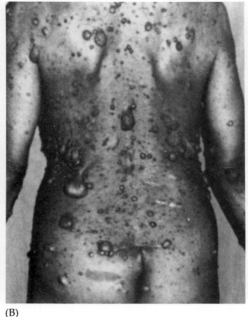

(B)

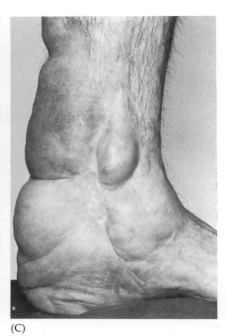

(C)

Figure 7.5
Some features of neurofibromatosis. (A) Café-au-lait spots. (B) Skin neurofibromas of various sizes on a mature woman. (C) Neurofibroma with tissue overgrowth on the lower left leg. (A and B, from Riccardi 1990. C, courtesy of Susan M. Huson.)

apparently healthy baby into a severely affected one (if untreated) occurs gradually over the first year of life, with variation from patient to patient in the **age of onset** of various symptoms.

Exceedingly variable in this regard is *diabetes*. The basic defect involves the hormone *insulin*, which is needed for proper use of the sugar glucose. In diabetic patients, glucose builds up in the blood and is excreted in the urine. (In fact, the disorder's full name is *diabetes mellitus*, from the Latin meaning "to pass sweetness.") There are two major forms of the disease: a juvenile, more severe form in which insulin is absent, and an adult, less severe form in which insulin is present but is not working correctly.* Each form is due in part to a different genetic system involving one or more genes.

The expression of symptoms seems to require the interaction of certain mutant alleles and particular dietary and health-related factors. These include obesity and viral or autoimmune diseases that damage the cells of the pancreas that manufacture insulin. Variation in the age of onset exists within each form, and various clinical features begin anytime from childhood to old age. Without treatment, diabetes leads to progressive debilities: blindness, kidney failure, nerve damage, gangrene in the feet, heart attacks, and strokes. Treatment can lessen the severity of the symptoms and allow a near-normal life-style in some cases. In people under 25 years of age, the frequency of diabetes in many populations is about 1 in 1,000, rising to about 1 in 25 in persons over the age of 60.

Huntington disease is transmitted as an autosomal dominant trait (see Chapter 5, Box B). The primary action of this mutant gene is unknown, but it leads to progressive degeneration of brain cells, which in turn causes severe muscle spasms and personality disorders. There is no effective treatment, and death comes 10 to 15 years after the onset of symptoms. Only occasionally do the first signs of disease appear in children; about 60% of cases are diagnosed between the ages of 35 and 50, with the remainder occurring equally before and after this age range. Because per-

---

* The juvenile-onset form is also known as type 1 or insulin-dependent diabetes mellitus (IDDM). The adult- or maturity-onset form is type 2 or non-insulin-dependent (NIDDM). About 20% of cases are type 1, with the remaining 80% type 2.

sons with the allele may reproduce before symptoms occur, their children live with the knowledge that they, too, may someday be stricken. DNA analysis can now reveal the presence or absence of the causative gene before symptoms arise. For persons at risk, however, deciding whether or not to have such a test can be an agonizing problem.

Incomplete Penetrance

An extreme form of variable expressivity is the total absence of expression in persons known to carry a particular allele. Such nonexpression of a gene that usually has a detectable phenotype is called **incomplete penetrance**. It may occur for many reasons, both genetic and environmental. Complete versus incomplete penetrance is an *all-or-none* phenomenon: Either a particular genotype is expressed in some form or another or it is not expressed. As diagnostic methods improve and the threshold for detecting abnormal phenotypes is lowered, however, the criteria by which we label a particular genotype as expressed or not expressed may also change.

In the case of a dominantly inherited trait, incomplete penetrance is manifested as "skipped generations." This situation in a pedigree of *polydactyly* (extra digits) is seen in Figure 7.6. Since the condition is fairly rare, it is highly likely that all six affected individuals possessed copies of precisely the same dominant allele. Individual II-6 must have received the allele from her father and transmitted it to her daughter, although she herself was not affected. A two-generation "skip" is seen on the right side of the pedigree. (Although dominant inheritance is fairly clear in this pedigree, a lower degree of penetrance might obscure the mode of inheritance.) Polydactyly also illustrates another aspect of variable expressivity. Figure 7.7 illustrates a case in which six well-formed digits appear on both hands and both feet, although each appendage reveals slightly different bone development.*

The precise cause of incomplete penetrance is usually unknown. But in some special cases, it is found that a specific allele at one locus masks the characteristic effects of an allele at another locus. This phenomenon is called **epistasis**. In the mouse, for example, the genotype *b/b* is phenotypically brown and *B/−* is phenotypically black, but neither brown nor black pigment appears in the fur if the mouse is also homozygous *c/c* at another

* References to a polydactylous giant are found in II Samuel 21:20 and I Chronicles 20:6: "And yet again there was war at Gath, where there was a man of great stature, whose fingers and toes were four and twenty, six on each hand, and six on each foot: and he also was the son of a giant."

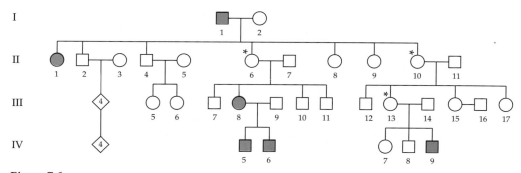

Figure 7.6
A pedigree of polydactyly. The three persons indicated by asterisks undoubtedly possessed the dominant gene for polydactyly but did not express it—examples of incomplete penetrance. (From Neel and Schull 1954.)

141

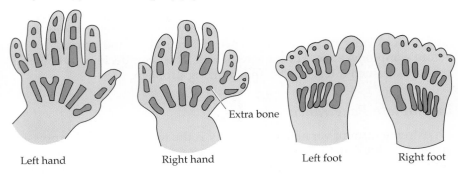

Left hand   Right hand   Left foot   Right foot

Figure 7.7
X-rays of a seven-day-old Norwegian boy (not from the pedigree in Figure 7.6) with polydactyly of both hands and both feet. The sixth digit is fairly well formed on each appendage, but in different ways. A fork in one of the bones of the palm of the left hand and in the left foot is clearly seen. Curiously, the extra small bone in the palm of the right hand also occurred in the right hand of the polydactylous mother. (From Sverdrup 1922.)

locus. The genotype *c/c* leads to an albino mouse, which develops *no* fur pigment, so color alleles at other loci are irrelevant. We say that the genotype *c/c* is epistatic to alleles at the *B* locus and other coat color loci. Epistasis should not be confused with dominance, which is the masking of one allele by another allele at the same locus.

One Gene with Several Effects

Given the complexity of embryonic development, it is not surprising that a single abnormal gene can have multiple, sometimes seemingly unrelated consequences. One gene product can be involved in many biochemical processes affecting different pathways of growth and maturation. Recall, for example, that males with *Kallmann syndrome* have an odd combination of symptoms: small testes and no sense of smell. These traits turned out to be embryologically related and probably due to a defect in a cell adhesion molecule. The resulting cascade of effects is called **pleiotropy**—literally, "many turnings." As with variable expressivity, pleiotropy is the rule rather than the exception, especially when a given phenotype is carefully examined at different levels.

Consider the multiple effects of insulin deficiency in *juvenile-onset diabetes*, as shown in Figure 7.8. Note that the excess utilization of fats for energy results in the production of acid substances (ketone bodies) in such high concentrations that the body is unable to neutralize them completely, a situation that can even cause coma or death.

One Phenotype from Different Causes

We have just described how one gene can have multiple effects. Conversely, any one of several different genes can result in similar effects. *Polydactyly* (Figures 7.6 and 7.7) is a case in point. Extra digits can be due to mutant alleles of any of about a half dozen different genes, usually showing dominant inheritance with incomplete penetrance. (We consider here only the phenotypes in which polydactyly is the sole observable malformation; other cases are known in which it is associated with more serious syndromes.) In one form, a reasonably functional digit appears on the little-finger side of the hand (or little-toe side of the foot). In a second form, an extra digit in the same position is much reduced in size. Other types of polydactyly involve an extra thumb, or a thumb with an extra

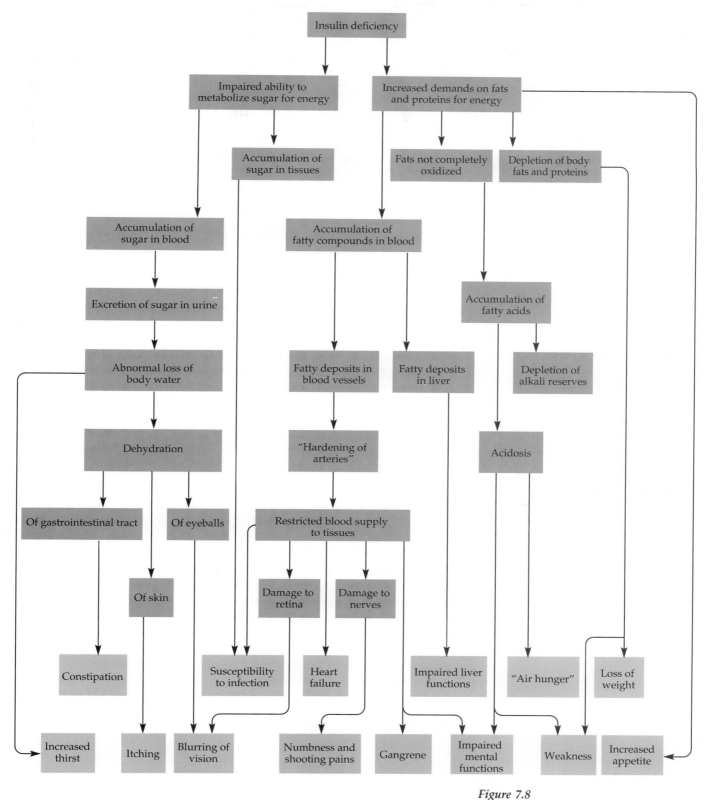

Figure 7.8
Pleiotropic consequences of insulin deficiency occurring in juvenile-onset diabetes mellitus. (Redrawn from Neel and Schull 1954.)

joint, or a duplicated index finger. The frequencies of the various types of polydactyly among newborns differ considerably, the extra little finger being the most common. Polydactyly is much more common in blacks than in whites.

Another feature of polydactyly is that not all cases are genetic. Errors of development brought about by slight changes in the prenatal environment can also produce bone malformations. Such environmental mimics of a characteristically genetic condition are called **phenocopies**. A person with polydactyly who has no affected relatives—a so-called sporadic case—*may* be a phenocopy, but other causes of sporadic cases are possible (see question 8).

The term used to described multiple genetic causes of the same, or nearly the same, phenotype is **genetic heterogeneity**. From the point of view of an affected person, determining a precise causation can be important. For example, *hemophilia A* results from a defect in so-called clotting factor VIII, whereas *hemophilia B,* with a similar phenotype, results from a defect in clotting factor IX. The symptoms of hemophilia A can be alleviated by using a concentrate of factor VIII; but this treatment would be of no use to a person suffering from hemophilia B.

Genetic heterogeneity is always expected for complex anatomical, physiological, or behavioral traits. The ability to hear, for example, depends on the proper functioning of perhaps hundreds or thousands of genes. *Deafness,* of a greater or lesser degree, can result when one of these genes mutates to an allele that impairs the development of the intricate structures of the middle or inner ear or the nerve connections to the brain. It has been estimated that 30–50% of all childhood deafness stems from such genetic factors, with the remainder—phenocopies—resulting from certain viral infections of the mother during pregnancy (measles and rubella especially), prematurity, or other types of environmentally imposed harm to the fetus or newborn. Based on audiological tests and associated pleiotropic effects, Nance and McConnell (1973) describe about 50 different genetic hearing loss syndromes. Roughly two-thirds of these genetic cases appear to be autosomal recessives, about one-third are autosomal dominants, and a few are X-linked. With refinements of testing procedures, these syndromes may be found to have multiple causations.

The most common example of inherited deafness, accounting for about 3% of all deaf children, is *Waardenburg syndrome.* This autosomal dominant disorder is seen most frequently in Kenya and South Australia. Its phenotype, which also involves skin abnormalities, is highly variable. Indeed, only 20–25% of affected individuals show a hearing loss. The main features are wide-set eyes, a wide bridge of the nose, eyebrows that meet in the midline, and a white forelock (Figure 7.9). Other defects may include patches of unpigmented skin, eyes that differ in color, prematurely gray hair, and white eyelashes. The normal allele, which is located on chromosome 2q, may somehow be involved with the migration of neural crest cells that become nerve and pigment cells. Waardenburg syndrome (WS) is homologous to the *splotch phenotype* in mice. Recent studies reveal that in many cases, both phenotypes are caused by the same homeobox-type gene—called *PAX-3* in humans, *Pax-3* in mice, and *paired box* in fruit flies. The gene product is a DNA-binding protein that is thought to regulate other genes during early development. Some cases of WS appear to involve other genes, however.

Homozygotes That Are Really Heterozygotes

So far, in discussing the genotypes and phenotypes associated with simply inherited human traits, we have usually considered one locus, two alleles (one normal and one mutant), three genotypes (homozygous normal, heterozygous, and homozygous mutant), and either two phenotypes (if one allele is completely dominant) or three phenotypes (if the two alleles are codominant). It is true we discussed *multiple alleles* in Chapter 4 and

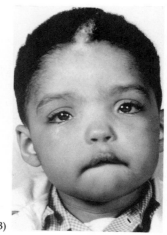

Figure 7.9
Three people (including a mother and daughter) with Waardenburg syndrome. Note the wide-set eyes, the wide bridge of the nose, the white lock at the hairline in the middle of the forehead, and (in B) growth of the eyebrows toward the midline. (A, from Partington 1959. B, reproduced by kind permission of Professor A. E. H. Emery.)

have occasionally mentioned that multiple alleles exist for certain disease-causing loci. But the implication has generally been that the different alleles are found in different populations and that a given mutant phenotype results from homozygosity for the *same* mutant allele.

But this scheme oversimplifies the true picture. In recent years, with rapid advances in molecular techniques, geneticists have been able to isolate and chemically dissect the mutant alleles associated with a number of human disorders. What they discovered is that some mutant phenotypes previously thought to result from homozygosity of *one* mutant allele at a given locus actually arise from the combination of *two* different mutant alleles at that locus. Thus some mutant phenotypes are genotypically heterozygous rather than homozygous. Such an individual is called a **compound heterozygote**.

This situation has been found, for example, with phenylketonuria, familial hypercholesterolemia, cystic fibrosis, and sickle-cell anemia—indeed, with most of the disorders we have discussed so far. Rather than trying to understand just a few genotypes and many phenotypes at a given locus, we are now presented with many genotypes, each giving rise to one or a few variant phenotypes. Certainly, it greatly complicates the study of genetic disorders. But knowledge of these multiple alleles and the resulting increase in genotypic combinations helps to explain much of the variable expressivity and incomplete penetrance noted years ago with many disease-causing genes.

Anticipation

In one of the earliest textbooks of human genetics, Curt Stern (1949) wrote:

> We have seen that the age of onset of Huntington's chorea is variable from one affected individual to another. Such variability is typical for many inherited diseases whose symptoms appear late in life. It is a widely held opinion among medical men, and some statistics seem to support it, that the age of onset of these diseases becomes earlier and earlier in successive generations. . . . The phenomenon is called "*anticipation*."... Furthermore, it is held that diseases whose severity varies increase in severity from one generation to the next. . . .

145

The concept of anticipation does not readily fit in with the system of genetic facts and interpretations which have proven fruitful in so many ways in the study of man and a great variety of experimental organisms.

Other examples of **anticipation** had also been noted. Starting in 1911, a number of reports were published on an inherited muscle-wasting disease called *myotonic dystrophy* (described later in this chapter). They suggested that the associated eye cataracts often appeared alone one or two generations before the muscle-wasting phenotype showed up in these pedigrees. Many geneticists, however, dismissed such findings, attributing them to biases in observation and data collection rather than to some novel kind of biological process. Nevertheless, many clinicians remained convinced that the data were reliable and that anticipation was a real phenomenon.

So the matter stood for about four decades. Then in the late 1980s, some remarkable discoveries about a few genes shed new light on the old observations. As we will see, the concept of anticipation is now alive and well. This is not to say, however, that biases in data collection and interpretation do not occur. In the next section, we discuss some of these pitfalls.

COLLECTING AND INTERPRETING GENETIC DATA

Even when the alleles of a single, fully penetrant gene give rise to uniform, easily recognized phenotypes (and no phenocopies exist), human geneticists still may not observe a clear-cut Mendelian ratio that would confirm a specific genetic hypothesis.

Biases in Collecting Data

Departures from standard ratios of a simple trait may occur even in large samples, because raw data can be biased by the methods of collection. For example, suppose that 100 engineers are asked to tabulate the sex of persons in their own sibships, including themselves. Assuming an average sibship size of three, typical data might be as follows:

Males	Females	Total	% males
189	111	300	63.0%

The percentage of males seems unusually high. But the puzzle is solved by realizing that the engineers themselves—the *probands*—are included in the ratio, and they are predominantly male. Assume, for example, that 90 of the engineers are men and 10 are women. Subtracting these probands from the raw data, leaving only their brothers and sisters, provides an unbiased sample for calculating the sex ratio:

	Males	Females	Total	% males
Raw data	189	111	300	63.0%
Probands	−90	−10	−100	—
Unbiased sample	99	101	200	49.5%

Looking at the raw data, we might propose some scheme to account for the seemingly unusual sex distribution. On more careful analysis, however, we see that the raw data are simply skewed by the method of collection: An excess of males was "forced" into the observations. This example illustrates **ascertainment bias**, the term *ascertainment* referring to the method used to discover and record data. The bias does not mean that the collected information is useless or wrong, only that it must be handled with care to avoid drawing unwarranted conclusions.

Ascertainment bias can be important in the case of autosomal recessive inheritance, for which we might reasonably expect to find 25% affected children from heterozygous parents. In fact, the reasonable expectation may *not* be 25% affected or even close to 25%. In trying to establish ratios that prove autosomal recessive inheritance, there are two main problems: dominance and small families. Incomplete data can also cause difficulties.

Dominance. If the normal allele is truly dominant, then a heterozygote cannot be distinguished from a normal homozygote by his or her phenotype alone. Without further analyses, a heterozygote can only be identified as the child or parent of an affected person (Figure 7.10). If the recessive phenotype is very obvious, then the cases that come to medical attention are for the most part affected children from unaffected heterozygous parents (Figure 7.10B). *But those heterozygous parents who happened not to produce any affected children are never found.* Yet the expected 25% affected is valid only if *all* children from *all* heterozygous parents are accounted for. In other words, geneticists often locate heterozygote × heterozygote parents indirectly (through their affected child) rather than directly (by looking at the parents themselves). This method leads to a bias, because at least one child must be affected before the A/a × A/a matings can be ascertained. Thus, some families—that is, the A/a × A/a matings with no affected offspring—are cut off from observations.

Small Families. The problem of overlooking A/a × A/a families who happen not to produce any a/a offspring is minimal in large sibships. For example, what is the probability that no a/a child will show up in a sibship of size 15? Simple calculations show that in the case of recessive inheritance among families with 15 offspring each, only about 1% of parents [i.e., $(\frac{3}{4})^{15}$]* would be missing by virtue of having all A/− children. But if A/a × A/a parents produce only *one* child, then 75% (rather than merely 1%) would be undetectable when the one and only child is unaffected. Geneticists use such probabilities in making statistical corrections for ascertainment bias.

Incomplete Data. In summary, for recessive inheritance in humans, where experimental matings are not possible and where the geneticist must make use of whatever data are available, the simple Mendelian ratios may not be the expected values. Furthermore, genetic surveys may be incomplete. Geneticists may use medical or dental records, death certificates, or other data that were collected for other purposes and with various degrees of thoroughness and care over an extended period. Questionnaires directed to practicing physicians may yield different degrees of completeness, depending on the phenotype in question: A serious condition that is easy to treat will bring affected persons to doctors more often than a mild condition that is difficult to treat. Reliance on cases reported in the scientific literature may also be biased: Families with one or a few affected persons may be missing because they are less interesting than families with many affected.

Thus, when we analyze patterns of inheritance, some fairly sophis-

* Refer to the discussion of probability in Chapter 4.

(A) Persons with a normal phenotype have an affected parent

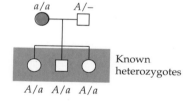

Known heterozygotes

(B) Persons with a normal phenotype have an affected child

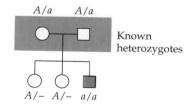

Known heterozygotes

○ □ Normal: A/A or A/a

● ■ Affected: a/a

Figure 7.10
The two ways of establishing heterozygosity for a strictly recessive trait. We assume that no aspect of phenotype allows the normal heterozygote, A/a, to be distinguished from the normal homozygote, A/A.

ticated statistics are often needed. Even the most refined mathematics, however, cannot correct for not knowing how affected persons first came to the attention of the data collector. Poor data yield unreliable conclusions regardless of how they are manipulated.

Interpreting Data and Testing Hypotheses

We have noted that observed data are usually not *exactly* the same as ideal expectations based on some hypothesis; there are **deviations** between the two of a greater or lesser amount. For example, Mendel counted 787 tall and 277 short plants out of 1,064 offspring from heterozygous parents (Chapter 4). But these numbers are not exactly 798 (¾ of 1,064) and 266 (¼ of 1,064) as predicted by the law of segregation:

Class	Observed number	Expected number	Deviation
Tall	787	798	−11
Short	277	266	+11
Total	1,064	1,064	0

The deviations of −11 in one phenotypic class and +11 in the other could be due to chance fluctuations, or they could be due to a wrong hypothesis. Because the deviations are "small," they are likely to be due to chance and therefore inconsequential. So we ignore them and state with some assurance that the observations are consistent with the Mendelian hypothesis. But had the deviations been "large," we would propose that the *expected numbers* are not appropriate and seek some other hypothesis to give a better fit.

Coin tossing provides another example. Suppose that you flip a penny 20 times, getting 15 heads and 5 tails. Here is the corresponding table:

Class	Observed number	Expected number	Deviation
Heads	15	10	−5
Tails	5	10	+5
Total	20	20	0

Here, the expectation of 10 heads:10 tails is calculated from the hypothesis of a true coin, fairly flipped. Are the deviations (−5 and +5) sufficiently small that we consider them mere chance? Or are the deviations sufficiently large that we are suspicious of the coin or of the way it was tossed?

Some rules regarding probabilities allow us to decide between these possibilities. Essentially, we want to use the size of the deviations to evaluate whatever hypothesis is set forth to explain the observations. As the deviations become larger, at what point do we begin to doubt the hypothesis? An answer of sorts is provided by a statistical tool (not described here) whose end result is a **probability P.*** The value of P decreases as the deviations increase. Conceptually, P reflects the credibility of the hypothesis: The smaller the value of P, the less plausible the hypothesis. We then make a decision based on the value of P.

If P is low, then we *reject* the hypothesis. We say that the observed results are not consistent with the hypothesis, that the hypothesis is not believable, or that the large deviations between observed and expected

---

* Details about the *chi-square method* can be found in most general genetics textbooks and upper-level human genetics texts.

are significant—that they are not likely to be due to mere chance. In many biological applications, "low" is defined as 0.05 or smaller. This is sometimes spoken of as the 5% level of significance.

If P is high, then we *accept* the hypothesis. We say that the observed results are consistent with the hypothesis, that the hypothesis seems like a good one, or that the deviations are not significant. "High" is defined as more than 0.05.

In summary, we want to know whether some idea we have is worth pursuing, and we gather some evidence to help us decide. The reasoning is similar to that in a court trial: The hypothesis is that the accused is innocent. If the evidence supporting the hypothesis is incredible—has low *P*—then the hypothesis of innocence is *rejected.* The jurors, of course, have no mathematical means for reaching a decision but must grapple with concepts like "reasonable doubt."

In the coin-tossing experiment, the *P* value turns out to be low. But this statistical procedure does not tell us *what* is wrong with the coin experiment or even that there *is* something wrong, only that we should be suspicious. A low probability value does not mean that the hypothesis is false, only that it is *likely* to be false. Mistakes can be made. We might reject a hypothesis that is really true, or accept a hypothesis that is really false. Good statistical procedures are designed to minimize these sorts of errors. They help us make judgments about evidence, just as the adversary system does for jurors in a court of law. And just as a "hung jury" may cause a new trial, so a borderline decision based on our statistical method may cause an investigator to gather new evidence.

Statistical analyses provide practical, standardized tests applicable to any discipline. For better or worse, statistics now pervades our world, from medical prognoses to weather forecasts to election-night predictions. In a sense, statistical thinking flowers when knowledge is scant and wilts as knowledge increases. But there is little doubt that probabilities help guide the direction of many human activities.

For about 60 years following the rediscovery of Mendel's laws, human geneticists struggled with the problems of collecting and analyzing data on simply inherited traits and on traits determined by the interaction of multiple loci. Just dealing with such "standard" genes in humans kept them fully occupied.

Meanwhile, other geneticists developed increasingly sophisticated statistical, chemical, and physical means of analyzing the behavior of genes and chromosomes in a wide variety of organisms. The more closely they looked, the more they began to find intriguing exceptions to standard genetic behavior. Some of these unexpected behaviors of genes and chromosomes were quickly recognized in a wide variety of organisms and provided explanations for unsolved genetic mysteries. But others seemed so bizarre, and were often so difficult to analyze, that they were thought to be oddities confined to one or a few experimental organisms. More and more often, however, some of these quirky types of genes and chromosomes are turning up in a broad array of organisms, including humans. In the following pages we describe several exceptional phenomena and suggest how they might explain some of the variations in gene expression discussed earlier in this chapter.

SEX CHROMATIN AND THE LYON EFFECT

Biologists have puzzled over the behavior, structure, and evolution of sex chromosomes ever since their discovery a century ago. Sex chromosomes

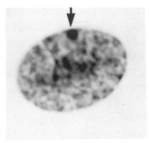 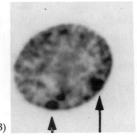

Figure 7.11
Sex chromatin (Barr) bodies, indicated by arrows, in nuclei of female cells. (A) Nucleus in a buccal smear from inside the cheek of a 46,XX female. One Barr body lies against the nuclear envelope. (B) Nucleus in a buccal smear from a 47,XXX female. Two sex chromatin bodies are visible. (Courtesy of Murray L. Barr, University of Western Ontario.)

are unique in several ways. For one thing, although small imbalances among *autosomal* genes and chromosomes are very poorly tolerated, causing severe defects or death, the sex chromosomes are much more flexible in this regard. Even in species where the X is very large, many organisms (including humans) can survive with too many or too few X chromosomes. Equally puzzling is the *normal* chromosomal makeup of females (who have two X chromosomes), compared with that of males (who have one X and a small, nearly geneless Y chromosome). Because females have twice as many X-linked alleles as males do, should not one or the other sex suffer from a relative excess or deficiency of X-linked gene products? By what mechanism do the two sexes compensate for this twofold difference in dosage of X-linked genes?

In mammalian somatic cells, it turns out that only one X chromosome is normally functional. Thus, there must be a way of silencing one X chromosome in females. The discovery of the silencing mechanism is a good example of serendipity in science—accidentally finding something new and different while looking for something else. During the 1940s two Canadian scientists, Murray Barr and Ewart Bertram, noted a darkly staining mass in the nuclei of some interphase cells in cat brains (Figure 7.11A). Investigating further, they discovered that interphase cells from females exhibited dark-staining spots in their nuclei, but cells from males did not. Examination of human nerve cells showed the same sex difference. They correctly speculated that each spot consists of one tightly condensed (heterochromatic) X chromosome.

Because of its correlation with sex, the spot was named the **sex chromatin**, although the term **Barr body** is also used. It can be seen in many types of somatic cells, usually lying against the inner surface of the nuclear membrane. Cells with one or more Barr bodies are said to be *chromatin-positive*, while those lacking any are *chromatin-negative*.

Detection of Sex Chromatin in Humans

The simplest and most commonly used method of checking for sex chromatin is the *buccal smear*: Cells are gently scraped from inside the cheek, spread on a glass slide, stained, and examined with a light microscope. Usually only 20–70% of the buccal cells from normal females contain a Barr body, although all cells are presumed to carry it. In female embryos it is first observed around the sixteenth day of development.

In the late 1950s a new twist to the sex chromatin story developed when researchers discovered that some humans have too many or too few sex chromosomes. People with *Klinefelter syndrome* have the karyotype 47,XXY and are male; conversely, people with *Turner syndrome* have the karyotype 45,X and are female. (Their phenotypes will be described in Chapter 10.) Quite surprisingly, these syndromes showed for the first time that the human Y chromosome plays a major role in sex determination. Until then it had been thought that the critical factor was (as in the fruit fly) the number of X chromosomes, with two X chromosomes leading to femaleness and one to maleness, and with the Y chromosome needed only for fertility in males.

What about the sex chromatin in people with these sex chromosome abnormalities? Cells from males with one extra X chromosome (47,XXY) are chromatin-positive, whereas cells from Turner (45,X) females are chromatin-negative. They are exceptions to the original "rule" that the sex chromatin is found only in females. Further investigations showed that more than one sex chromatin occurs in persons with additional X chromosomes, but not in persons with additional Y chromosomes. The more precise rule states that *the number of Barr bodies is one less than the number of X chromosomes:*

Number of Barr bodies	KARYOTYPE(S)	
	Female	Male
0	45,X	46,XY; 47,XYY
1	46,XX	47,XXY; 48,XXYY
2	47,XXX	48,XXXY; 49,XXXYY
3	48,XXXX	49,XXXXY
4	49,XXXXX	

The Inactive X Hypothesis

What happens to the chromosome that forms the Barr body? Are its genes still functional? Is the same X chromosome (maternal versus paternal) condensed in every cell? What determines whether or not a given chromosome will be condensed? Based primarily on studies with mice, the following **inactive X hypothesis** was proposed in 1961 by English geneticist Mary Lyon:

1. The condensed X chromosome found in cells of normal female mammals is genetically inactive.
2. Inactivation first occurs very early in embryonic development, during the blastocyst stage, at which time either the maternal or the paternal X chromosome is equally likely to be inactivated.
3. This inactivation event occurs independently and randomly in each blastocyst cell. But once the decision is made for a given cell, the same X chromosome will be inactivated in all its descendant cells, thereby producing a clone from each of the original cells.

The net effect of this whole process—called **lyonization**—is to equalize the phenotypes in males and females. This phenomenon is known as **dosage compensation**. Although normal females possess twice as many X chromosomes as males, they generally have the same number of active X-linked genes and the same amount of X-linked gene products.

Evidence for Lyonization

Evidence for lyonization was sought in females *heterozygous* for X-linked genes; these females should have clones of cells in which one allele or the other is active, but never cells with two active alleles. If entire clones

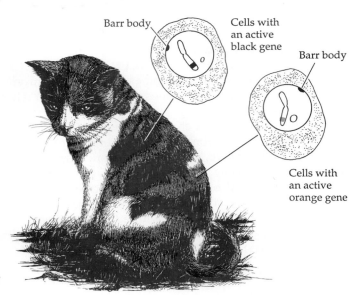

Figure 7.12
Calico cat. Many autosomal genes collectively affect the color and pattern (banded versus solid) of the individual hairs on a cat. Here we consider only one X-linked locus and its role in forming the light and dark patches of fur on a calico cat. This female is heterozygous for a dominant allele (O) that leads to orange (or yellow) hairs and a wildtype recessive allele (o) that produces black (or dark-speckled) hairs. In any one cell, either one allele or the other is active; the inactive allele is on the X chromosome forming the Barr body. The size of each patch of fur depends on how many cells descended from each cell present at the time of the random inactivation decision.

remain in place during development, good-sized phenotypic "patches" will result, each patch expressing only one X-linked allele of a gene. If cells from different clones intermingle during development, however, then an intermediate or finely variegated "salt-and-pepper" phenotype will be observed. One line of evidence given by Lyon involved the mottling or dappling on the fur of female mice heterozygous for a number of X-linked coat color genes. This effect is also seen in the familiar calico or tortoiseshell cats (Figure 7.12), whose patches of black or orange hair indicate which X-linked gene is active in each heterozygous cell. With rare exceptions, calico cats are female.

Because humans lack fur, we cannot observe calico people. Perhaps the next best trait, however, is phenotypic expression of the recessive X-linked trait *anhidrotic ectodermal dysplasia*. Some affected males were described by Charles Darwin as the "toothless men of Scinde"; they lack sweat glands, teeth, and most of their body hair. Studies of females heterozygous for this gene reveal considerable variability in phenotypic expression, taking the form of irregular skin patches with few or no sweat glands (Figure 7.13) and some regions of the jaw with missing or defective teeth. Note also in Figure 7.13 that the two pairs of identical twins in generations I and III show differing patterns of skin patches, a phenomenon called mosaicism. Many other human X-linked genes are also known to show mosaic expression in heterozygous females.

One of the best systems for studying lyonization makes use of an X-linked gene controlling production of the enzyme *glucose-6-phosphate dehydrogenase (G6PD)*. This enzyme, which occurs in nearly all plants and animals, acts in glucose metabolism. It turns out that human 46,XX females with *two* doses of the normal X-linked *G6PD* allele produce no more of the enzyme in their cells than do 46,XY males with *one* dose of the normal allele, although one might expect the former to produce twice as much. Furthermore, 47,XXY Klinefelter males and 47,XXX females produce no more G6PD than do normal males and females, suggesting that regardless of how many X chromosomes are present in a given cell, only one X is active.

Even more to the point were these observations: (1) A recessive allele *(gd)* of this gene produces a defective enzyme with a much lower level of activity than that of the normal allele *(Gd)*. When skin cells from females

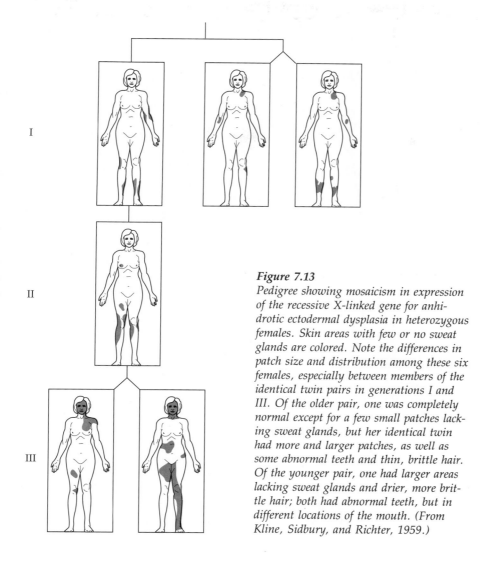

Figure 7.13
Pedigree showing mosaicism in expression of the recessive X-linked gene for anhidrotic ectodermal dysplasia in heterozygous females. Skin areas with few or no sweat glands are colored. Note the differences in patch size and distribution among these six females, especially between members of the identical twin pairs in generations I and III. Of the older pair, one was completely normal except for a few small patches lacking sweat glands, but her identical twin had more and larger patches, as well as some abnormal teeth and thin, brittle hair. Of the younger pair, one had larger areas lacking sweat glands and drier, more brittle hair; both had abnormal teeth, but in different locations of the mouth. (From Kline, Sidbury, and Richter, 1959.)

heterozygous for the normal and mutant G6PD alleles (i.e., *Gd/gd*) were individually cloned and then tested for enzyme activity, some clones expressed the normal allele (*Gd*) and others expressed the mutant (*gd*), even though all clones were genotypically identical (*Gd/gd*). (2) Individual red blood cells from heterozygous females also show high or low activity but never intermediate activity of G6PD, as is found in heterozygotes for enzymes produced by autosomal genes. (3) Similar types of studies with other X-linked enzyme-producing genes yielded similar results. The inescapable conclusion is that *for any X-linked gene in an XX mammalian cell, only one of the two alleles remains active.*

Interest in X chromosome inactivation is keen because this process might be the same or similar to the one(s) that control the selective turning on and off of autosomal genes and also parental imprinting. Scientists have found that the inactive X replicates later in the cell cycle than do the other chromosomes. Also, X inactivation starts and spreads out from an *X inactivation center* in the long arm near the centromere. (A gene called *XIST* has recently been found in this region.) Yet nobody knows exactly what happens there and what prevents *all* the X chromosomes in a cell from being inactivated. *Methylation* of cytosine bases in the DNA may be involved in maintaining X inactivation, but there are still many more questions than answers about this phenomenon.

Other Considerations

Studies of X chromosome inactivation have revealed some additional twists to the story. Here we present three of them.

1. *Part of the X chromosome is not inactivated.* If all but one X chromosome is completely inactivated in every individual, then why do Turner females differ from regular females, and why do Klinefelter males differ from regular males? To explain these abnormal phenotypes, it was suggested that some X-linked genes must always escape inactivation. These genes would be expressed doubly in regular females and Klinefelter males (both with two X chromosomes) and singly in regular males and Turner females (both with one X chromosome). We now know that this is indeed the case: A few X-linked genes near the tip of the short arm of the human X chromosome do *not* exhibit mosaicism in heterozygous females. These loci include the blood group gene *Xg*, the steroid sulfatase gene (*STS*) (whose mutant forms cause scaling of the skin), and gene *MIC2* (which codes for a cell surface protein). Unlike the late-replicating inactive genes, they reproduce early in the cell cycle. Their products, however, show no obvious connection to the phenotypic features of Turner or Klinefelter syndrome. So maybe there are additional X-linked genes (influencing stature and fertility, for example) that escape inactivation.

2. *Germ cells behave differently.* In female fetuses, future germ cells undergo lyonization along with somatic cells. But as the germ cells develop into oocytes and enter meiosis, their inactivated X chromosomes become *reactivated*. Thus, each egg cell ends up with an active X. In males, on the other hand, the X chromosome undergoes precocious condensation in all primary spermatocytes just as they enter meiosis. But this process is poorly understood and may not be true lyonization.

3. *Some genes get reactivated as mammals age.* Studies of female mice carrying a certain chromosomal abnormality have shown that some inactivated genes may not stay that way. This work involved a piece of autosome that was stuck into the middle of an X chromosome. Normally, autosomal genes do not lyonize. But when moved to the X chromosome, they get inactivated right along with the flanking X-linked genes. In this experiment the displaced piece of autosome carried a wildtype allele for the enzyme tyrosinase, which leads to normal pigmentation. Its recessive allele (which, when homozygous, causes albinism) was present on a normal autosome. Thus, the fur was mottled with colored and albino patches, similar to that of calico cats. But as these animals aged, they became more darkly pigmented—apparently because the displaced tyrosinase gene was reactivated in some cells. Other experiments with mice have given similar results.

X chromosome inactivation is perhaps the first and best-known example of unexpected phenomena that explain ambiguous patterns of gene expression. But it is definitely not the only one. In recent years, geneticists have been astonished to learn that some genes behave in a most unusual fashion.

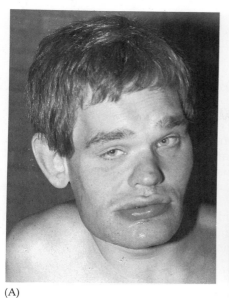
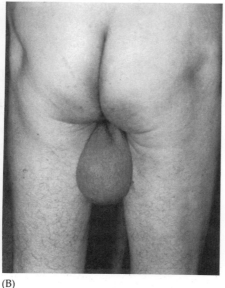

(A)   (B)

Figure 7.14
Fragile X syndrome in a 26-year-old male.
(A) Note the coarse features and the big
head with relatively large forehead, jaw,
and ears. (B) Extremely large testes. (From
De Boulle et al. 1993.)

THE FRAGILE X SYNDROME

Decades ago it was noted that males outnumber females by about 25% among the mentally retarded at institutions. In the 1940s several researchers suggested that the excess could be due to X-linked recessive inheritance. In 1969 a secondary constriction was found on the long arm of the X chromosome of four retarded males in one family. Since this cytological defect was associated with chromosome breakage, it was called a **fragile site**, and the related X-linked mental retardation was dubbed **fragile X syndrome**. Not until the late 1970s, however, was this syndrome recognized as the most common form of inherited mental retardation.* Exactly how the fragile site brings about retardation remains unknown. About 20 different heritable fragile sites exist on other chromosomes, but none of these seem to be associated with a particular phenotype.

Roughly 1 in 1,250 males exhibits a fragile site in band q27.3 of the X chromosome (Figure 2.18), in up to 50% of his tested cells.† Most of these males are moderately to severely retarded. But some are only mildly retarded, and a few males with this fragile site may show normal intelligence. Many affected males have huge testes and large, protruding ears (Figure 7.14). A long jaw and big head are common; stubby hands, lax joints, and a heart valve defect may occur too. Affected males tend to be tall as children but short as adults. Their speech is usually delayed in development and often high-pitched and repetitive in nature. Although some may be hyperactive or autistic as children, their behavior as teenagers and adults tends to be shy and nonassertive, but friendly. Few if any affected males are known to have produced offspring.

The frequency of heterozygous mentally affected females is roughly 1 in 2,250, and the percentage of their cells in which the fragile site can be seen (about 30%) is lower than that in affected males. Affected females, who are thought to represent about one-third of all carrier females, show mild retardation or some kind of learning disability. But beyond this they have no consistent physical abnormalities. Severely affected males and females rarely mate. But fragile X carriers are highly fertile, and for un-

* Down syndrome, the most frequent form of mental retardation in general, occurs randomly and does not usually run in families (Chapter 10).

† It was noted in Chapter 2 that the detection of fragile sites is rather tricky.

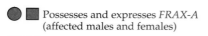

● ■ Possesses and expresses *FRAX-A*
(affected males and females)

⊙ ⊡ Possesses but fails to express *FRAX-A*
(transmitter males and carrier females)

○ □ Does not possess (or express) *FRAX-A*

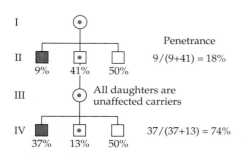

Figure 7.15
Inheritance and penetrance of fragile X syndrome, with emphasis on male members of a pedigree. Compare the frequencies of retarded males and transmitter males in generation II with those in generation IV. Note that the generation I and III carrier females should have the same fragile X allele, yet they give rise to greatly differing percentages of affected male progeny.

known reasons the frequency of twins among their children is two to four times the usual rate. Homozygotes for the fragile X defect are rarely if ever found.

Genetics of the Fragile X Syndrome

Something weird happens to the expression of fragile X syndrome across generations. In trying to make sense of it, geneticists assumed the following: There is a *familial mental retardation locus*—called **FRAX** for *fragile X gene*—located in or very close to band q27.3. Its abnormal allele (called *FRAX-A*) is expressed cytologically as the fragile site and phenotypically as the fragile X syndrome. Because carrier females produce 50% offspring who possess this allele and 50% who do not, the segregation of *FRAX-A* is normal. But unlike all other X-linked genes, *FRAX-A* shows a high degree of nonexpression (nonpenetrance) in males, who should express all their hemizygous genes. Even more surprising, affected offspring tend to cluster in certain generations.

As shown in Figure 7.15, the inheritance of fragile X syndrome mainly involves normal carrier females and **transmitter males**. The latter have the *FRAX-A* allele but do not express it: They are phenotypically unaffected and do not show the fragile X site in any of their cells. Likewise, their *carrier daughters* show no retardation and few or no fragile sites. The key fact of this pedigree is that in generation II, 41% of the males are transmitters and only 9% are mentally retarded, whereas in generation IV, 13% of the males are transmitters and 37% are mentally retarded. So among males who supposedly inherit the same allele, four times as many are affected in generation IV as in generation II. (Although not shown here, a similar difference is seen among heterozygous females: About 3.5 times as many are mentally subnormal in generation IV as in generation II.) Indeed, depending on the generational position of *FRAX-A* in a pedigree, its *penetrance* ranges from 18% to 74% in sons of carrier females and from 10% to 34% in daughters of carrier females.

Molecular Biology of Fragile X

A partial explanation of the unusual inheritance pattern of fragile X came in 1991, when the fragile X site was isolated and mapped by several groups of scientists.[*] They were amazed to find that the locus varied in length, not only among fragile X males, but even among other members of an affected family. This instability was traced to a *specific sequence of three nucleotide bases, CGG (cytosine, guanine, guanine), that are repeated a variable number of times.* People without the *FRAX-A* allele have 6 to 58 copies of the triplet; unaffected carriers usually have about 60 to 230 copies; and individuals with the fragile X syndrome usually have over 230 copies.

Researchers call an intermediate-sized allele a *premutation*; its carriers are unaffected, but their children and grandchildren are at high risk of being affected. The largest-size allele is called a *full mutation*. When transmitted through males, the sequence's copy number stays constant or decreases. Only with *transmission through females* do the two-step increases in triplet copy number (first from normal to premutation, then to full mutation) occur. The exact mechanism for these changes is not known; nor has any gene product been identified. But some early ideas about anticipation now have a molecular basis: In succeeding generations, this gene somehow grows longer and its abnormal phenotype becomes more severe.

* Team leaders included Grant Sutherland in Australia, Jean-Louis Mandel in France, and Stephen T. Warren and C. Thomas Caskey in the United States.

Before the discovery of the CGG amplification, counseling of fragile X families based on cytological findings was extremely difficult. Because only about 50% of all known carriers show the fragile site, it was hard to say which sisters of an affected male were at risk for producing affected children. And among females with the fragile site, there was no way of telling whether they would be retarded or unaffected. Male fetuses with the fragile site would usually be retarded, however.

Even now, with direct diagnosis by DNA analysis (i.e., determining the exact number of CGG repeats), the forecasts are not always clear-cut. Male fetuses with over 233 repeats will be retarded, and males and carrier females with fewer than 167 repeats will probably be mentally normal. But males with 167 to 233 repeats are borderline cases, and their phenotypes cannot be predicted with absolute certainty. Similarly, carrier females with over 167 repeats have about a 50% risk of being mentally retarded (Richards and Sutherland 1992).

OTHER UNEXPECTED VARIATIONS IN GENE STRUCTURE AND FUNCTION

As the preceding discussions indicate, scientists are beginning to correlate some variations in phenotypic expression of specific alleles with irregularities in DNA structure. Following are a few additional examples.

More on Expanding Genes

Three other disorders, two autosomal and one X-linked, have been traced to the expansion of certain triplet base sequences within their genes. These **expanding genes** show great variability in phenotype and age of onset and, like the fragile X syndrome, provide examples of *anticipation*.

The most common form of muscular dystrophy, called *myotonic muscular dystrophy*, occurs with a frequency of about 1 in 7,500. *Myotonia* is a defect of muscle fibers that results in abnormally prolonged contractions. Patients have trouble relaxing a movement after any vigorous effort, such as letting go of their grip after a handshake. *Dystrophy* refers to inadequate growth of some tissue, usually muscle. Here it involves progressive muscle wasting and weakness, especially in the face, jaw, neck, and limb extremities (Figure 7.16).

In affected individuals, the severity, age of onset, and range of affected tissues vary tremendously, from being almost undetectable, even in old age, to being lethal in infancy. In addition to the muscles, many other organs and tissues are affected. Central nervous system defects—mental retardation in affected children and unusual apathy and drowsiness in adults—are also common. Clinicians recognize three main phenotypic groups: a mild form with late onset, a severe (classical) form with adult onset, and a severe (congenital) childhood form inherited from a mildly affected mother.

The underlying cause of this disease—the defective gene product—remains a mystery, although a cell membrane abnormality of some sort is suspected. No new mutations are known; every case, when carefully studied, can be traced back through the family. Indeed, it may be that most known cases are descended from one original mutation. The gene, which is inherited as an autosomal dominant, was mapped to chromosome 19q in 1992. But in the course of zeroing in on its exact location, the three main groups of researchers independently made the same astonishing discovery: The myotonic dystrophy gene gets bigger, and gives rise to a more severe phenotype, as it passes from one generation to the next.

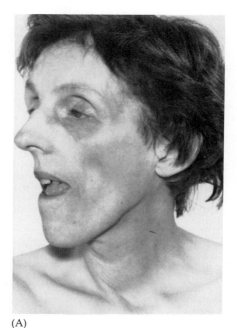

(A)

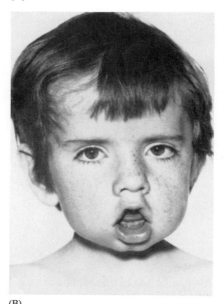

(B)

Figure 7.16
Facial features of myotonic dystrophy in two severely affected patients. (A) Note the drooping eyelids, facial weakness, and wasting of muscles of the jaw and neck. (B) Note the "tent mouth" and jaw weakness. (From Harper 1990.)

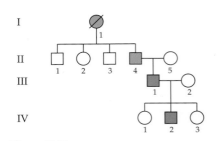

Figure 7.17
Pedigree of myotonic dystrophy. Individuals I-1 and II-4 each possess about 60 extra copies of the CTG triplet. They are mildly affected, showing few symptoms other than cataracts. Individual III-1, with at least 700 extra copies of the CTG triplet, had adult onset of the severe form of myotonic dystrophy. Individual IV-2, with 700 or more extra copies, had teenage onset of the severe disease. (Redrawn from Brook et al. 1992.)

Affected individuals have a normal allele and an abnormal allele. The normal allele contains a string of three nucleotide bases—CTG (cytosine, thymine, and guanine)—that are repeated, one right after the other, 5 to 27 times. The size of the abnormal allele varies between affected siblings. But people with the mild form of myotonic dystrophy have a string of at least 50 CTG triplets, and those with the severe forms have hundreds (perhaps up to 2,000) of copies of this triplet! The pedigree of one such family is shown in Figure 7.17.

As with the fragile X locus, there are many more questions than answers about this bizarre gene. What triggers its expansion? When and how does it occur? How does expansion of the CTG region affect the gene's function? How do the mutant gene products differ from normal? Scientists have determined that the normal gene encodes a type of *protein kinase*, an enzyme that adds phosphate groups to many other proteins and thereby modifies their behavior in the cell. Could there be a large number of undetected carriers who show few or no phenotypic effects? Perhaps the most difficult question of all is an ethical one. What should genetic counselors tell individuals who are minimally affected with this disorder but who run the considerable risk of producing grandchildren or great-grandchildren who are severely affected with myotonic dystrophy?

Many of the same kinds of questions are being asked about *X-linked spinal and bulbar muscular atrophy* (also called Kennedy disease). It is a rare, nonlethal, muscle-wasting disorder that is often associated with breast development and impotence in affected males. These individuals have a defect in their hypothalamus that leads to estrogen excess and androgen deficiency. Researchers have pinpointed the molecular basis of the disorder: an excessively repeated triplet of the bases CAG (cytosine, adenine, guanine) that leads to a defect in the cell membrane receptor for androgens. The normal allele contains about 21 CAG repeats, but that number is doubled in the mutant allele.

In 1993 an international research collaborative finally tracked down the gene for *Huntington disease*, whose exact location near the tip of chromosome 4p had eluded them for a decade. Upon cloning and sequencing the mutant gene, they were astonished to find that it also contained many repeats of the base triplet CAG (cytosine, adenine, guanine). Studies of this large gene in 75 HD families revealed that unaffected individuals have 11 to 34 copies of the triplet, whereas affected members have 42 to more than 66 copies. The more copies present, the more severe the symptoms and the earlier the onset, with the longest repeats being found in juveniles with HD. (The most extreme case was an *HD* allele with about 100 copies, from a child showing onset at age 2.) The number of triplet copies usually increases from one generation to the next, but occasional decreases have been noted. Even within sibships there is considerable size variation of the *HD* allele.

The predicted protein product, whose sequence of 3,144 amino acids is unlike that of any known protein, has not been isolated. But it should soon be possible to directly test the DNA of individuals at risk for HD to determine (1) whether they possess a mutant allele, and (2) when it is likely to take effect.

More on Genomic Imprinting

In Chapter 6 we introduced the phenomenon of *genomic imprinting*, whereby progeny phenotypes differ according to whether a particular gene (or group of genes) came from the mother or the father. We also described some developmental defects known to be associated with imprinting in the mouse and the human. Recall, for example, that the *Prader-Willi and Angelman syndromes* are associated with tiny deletions of the same small chromosomal region; but different phenotypes result, depending on whether the deletion is inherited from the father or from the mother.

We also pointed out that differential imprinting might explain the varying age of onset or severity of some human disorders, such as Huntington disease and juvenile diabetes. In a review of this topic, Hall (1990) lists about a dozen human diseases suspected on a clinical basis to be imprinted and whose genes reside in chromosomal segments homologous to known imprinted regions of mouse chromosomes. She also lists about 100 additional diseases that should be considered candidates for imprinting because their genes likewise reside in these chromosomal regions homologous to imprinted segments in mice.

First, some definitions. An **imprinted allele** is one whose expression is changed or silenced when it passes through a particular sex. Somehow, the DNA in this allele gets temporarily marked or tagged, altering its genetic information. An allele is said to be **paternally imprinted** if it is not expressed when inherited from the father; it is **maternally imprinted** if it is not expressed when inherited from the mother. Individuals who carry an allele in its imprinted or "silenced" form, and thus do not express it, are **nonmanifesting carriers**.

What do pedigrees of imprinted genes look like? They resemble standard inheritance patterns, but with phenotypic expression among children varying according to the sex of the transmitting parent. Consider the hypothetical case of a rare dominant allele, called *D*, that is imprintable. Figure 7.18 shows how the expression of this allele would vary from generation to generation, depending on whether it is (A) paternally imprinted or (B) maternally imprinted.

In the two pedigrees, the structures of generations II to IV are identical with regard to the number, sex, and birth order of offspring. They are also arranged so that the sister and brother II-1 and II-2 are both

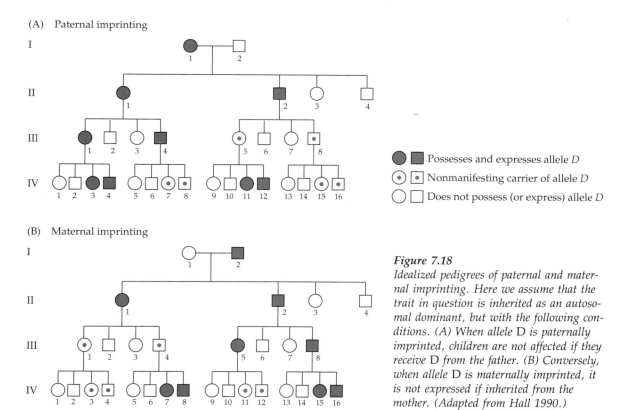

(A) Paternal imprinting

Possesses and expresses allele *D*

Nonmanifesting carrier of allele *D*

Does not possess (or express) allele *D*

(B) Maternal imprinting

Figure 7.18
Idealized pedigrees of paternal and maternal imprinting. Here we assume that the trait in question is inherited as an autosomal dominant, but with the following conditions. (A) When allele D *is paternally imprinted, children are not affected if they receive* D *from the father. (B) Conversely, when allele* D *is maternally imprinted, it is not expressed if inherited from the mother. (Adapted from Hall 1990.)*

159

affected in both pedigrees. But in generation I, the affected, transmitting parent is female in (A) and male in (B). And in generations III and IV, there is an interchange of affected versus nonmanifesting individuals between (A) and (B). For example, III-1 and III-4 are affected in (A) but nonmanifesting carriers in (B). Likewise, III-5 and III-8 are nonmanifesting carriers in (A) but affected in (B). We analyze this pedigree further by making the following observations:

1. Expression of the trait skips generations in a specific way, shown in generations II to IV and summarized as follows:

	(A) Paternal imprinting, affected male II-2	(B) Maternal imprinting, affected female II-1
If allele is inherited		
Children will be:	Unaffected	Unaffected
Grandchildren through sons will be:	Unaffected	Affected
Grandchildren through daughters will be:	Affected	Unaffected

2. In each generation we expect roughly equal numbers of affected males and females and roughly equal numbers of nonmanifesting males and females, regardless of which parental sex transmits the imprinted gene.
3. This type of pedigree is easily distinguishable from pedigrees of mitochondrial inheritance. In the latter, only the progeny of females—and never any descendants of males—express the particular trait.
4. Either pedigree (A or B) could be explained by a simple autosomal gene with incomplete penetrance. Thus, geneticists need more extended pedigrees to identify with certainty the phenomenon of imprinting.

Genes Within a Gene

Earlier in this chapter we said that type 1 neurofibromatosis (NF1), one of the most common autosomal dominant diseases, shows a high mutation rate and extremely variable expressivity. Doctors are unable to predict how severe and disfiguring it will be in any particular individual, nor can they treat it very effectively.

In 1990 the *NF1* gene, which resides on chromosome 17, was isolated by two research groups, one led by Francis Collins at the University of Michigan, the other by Raymond White of the University of Utah. The key molecular analyses needed to pinpoint its exact location took three years. As is often the case, chromosomal abnormalities helped to clinch the case. Researchers found two NF1 patients, each with a break on chromosome 17q that had reattached to a different chromosome. Since it was likely that the break had occurred within or near the *NF1* gene, the two groups focused their efforts on the stretch of DNA around and between the two breakpoints.

By analyzing the DNA from other NF1 patients—some with point mutations, some with deletions, and one with an extra piece of DNA in that region of chromosome 17—they determined that the *NF1* gene was huge (about 300,000 base pairs) and actually included three smaller genes. Although such **nested genes** were known to occur in lower organisms, this was only the second such example found in humans.* Whether the

---

* The other human example is the X-linked *factor VIII gene,* which causes hemophilia A and which contains a single embedded gene of unknown function.

nested genes play any role in the development of neurofibromatosis is not known.

What does the *NF1* gene do? Recent evidence suggests that it plays an important role in controlling cell division by acting as a *tumor suppressor* (Chapter 15). Interestingly, it also appears that most new *NF1* mutations are of paternal origin. Whether this effect is due to an increased mutation rate in older men, or to imprinting, or to some other phenomenon is not known.

Jumping Genes

Genes that don't stay put were first discovered and analyzed by corn geneticists in the 1940s. They found that certain genes had a tendency to move around within the genome. Wherever these so-called *controlling elements* landed, they caused instability in the form of new mutations, which were detected as colored spots on corn kernels or some other form of phenotypic variability. When the controlling elements took off again, however, the mutations in that chromosomal region—along with their associated unstable phenotypes—reverted back to the normal state.*

Although most mainstream geneticists accepted the presence of jumping genes in corn, they did not think that such exotic gene behavior occurred in other organisms. That viewpoint changed in the 1960s and 1970s with the discovery of similar genes—called **transposable elements** —in bacteria and fruit flies. We now know that jumping genes are quite widespread in plants and animals (Chapter 12).

The first known example of jumping DNA in humans that gives rise to a distinct phenotype involved hemophilia A. In 1988 Hugh Kazazian and colleagues at Johns Hopkins University analyzed the DNA sequences of two different mutations of the X-linked *clotting factor VIII gene*. They found that in both cases a piece of a special kind of repeated DNA— called *LINE-1* for *long interspersed elements*—was inserted from a different chromosome into the mutant gene. Both of the new insertions produced an enzyme called *reverse transcriptase*. This enzyme, quite surprisingly, allows RNA to be copied into DNA—the reverse of the usual situation (Chapter 12).

The second known example of a jumping gene in humans, reported in 1991, involved the *neurofibromatosis* gene. As if the presence of nested genes within this locus were not amazing enough, Francis Collins's research group also discovered that one *NF1* mutation occurred when a small piece of DNA jumped from its usual location into the middle of the *NF1* gene. This DNA insert was a type of repetitive DNA called *Alu* (Chapter 12) and does not itself produce a protein product. But its insertion into the *NF1* gene led to an abnormal NF1 protein product.

Nobody knows how common jumping genes might be in humans or if they always cause abnormal mutant phenotypes when they move around. Nor is it known whether the persistence of jumping genes in many organisms means that they play some important role in evolution.

In the beginning of this chapter, we discussed some problems in interpreting observations about traits that run in families but do not exhibit clear-cut phenotypes or patterns of inheritance. We have now described some unusual and newly discovered types of gene structure or behavior that might explain some of these old problems. Whether future molecular studies will unravel all such mysteries we cannot say. But it is clear that such exceptions to Mendel's standard rules, especially as they apply to human illnesses, will continue to provide a challenge to geneticists and physicians.

* For her pioneering research on jumping genes, Barbara McClintock of Cold Spring Harbor Laboratory in New York received the Nobel prize in 1983.

SUMMARY

1. A trait that is genetic will run in families and will not spread to unrelated persons in the same circumstances, but these conditions alone may not be enough to establish a genetic hypothesis.

2. Sex-influenced genes are expressed differentially in the two sexes.

3. Variable expressivity of a gene, including variable age of onset, results from variation in environments and in background genotypes.

4. Incomplete penetrance of a gene refers to the nonexpression of a characteristic phenotype. This outcome may be caused by epistasis, the masking of a phenotype by the action of a nonallelic gene.

5. A pleiotropic gene affects various aspects of an individual's phenotype.

6. Genetic heterogeneity refers to multiple genetic causes of nearly the same phenotype. A phenocopy is a nongenetic condition that mimics a genetic condition.

7. In traits involving multiple alleles, some phenotypes previously thought to be homozygous are found to be heterozygous for two different mutant alleles.

8. Indirect methods of data collection produce biases that require special methods of statistical analysis. The deviations of observed data from those expected on the basis of some hypothesis can be used to test the credibility of that hypothesis.

9. X-linked traits are usually expressed equally in both sexes, even though females have twice as many X-linked alleles as do males. In mammals the dosage of X-linked genes is equalized by the inactivation of one X chromosome in every female body cell. Thus, females heterozygous for X-linked genes are phenotypic mosaics.

10. A fragile site on the X chromosome is associated with familial mental retardation (fragile X syndrome). The disorder often increases in severity and degree of penetrance in successive generations of a family owing to an unstable mutant allele that grows in size.

11. The myotonic muscular dystrophy gene also tends to expand and get more dangerous as it passes from one generation to the next. Here, too, the culprit is a nucleotide triplet that is repeated in excess in mildly affected people and in great excess in severely affected people. The genes for Huntington disease and X-linked spinal and bulbar muscular atrophy also contain extra copies of a nucleotide triplet.

12. The expression of an imprinted allele is silenced or changed when it passes through one sex. Thus, in pedigrees of imprinted genes, the phenotype varies according to the sex of the parent.

13. Neurofibromatosis type 1 varies greatly in severity and age of onset. The *NF1* gene is huge and acts as a tumor suppressor. It also contains at least three small genes whose functions are unknown.

14. Jumping genes (transposable elements) are found in many plants and animals. At least two examples—both involving insertions of repeated DNA into the mutated allele—are known in humans.

KEY TERMS

anticipation	dosage compensation	genetic heterogeneity	modifier
ascertainment bias	epistasis	inactive X hypothesis	nested gene
Barr body	expanding gene	incomplete penetrance	phenocopy
compound heterozygote	fragile X syndrome	lyonization	pleiotropy

prion
sex chromatin
sex-influenced inheritance

sex-limited trait
suppressor mutation
transmitter male

transposable element
variable age of onset
variable expressivity

QUESTIONS

1. Assume that either the recessive genotype d/d (at the D locus) or the recessive genotype e/e (at the E locus) can produce deafness. If these loci are on different chromosomes, what is the probability of a deaf child from the mating of two unaffected individuals who are double heterozygotes, that is, $D/d\ E/e \times D/e\ E/e$?

2. Assume that pattern baldness is an autosomal trait that is dominant in males (bald = B_1/B_1 or B_1/B_2; nonbald = B_2/B_2) but recessive in females (bald = B_1/B_1; nonbald = B_1/B_2 or B_2/B_2). What are the expected genotypes and phenotypes of the children of a bald woman and a nonbald man?

3. Is the pattern of transmission of baldness in Figure 7.3 consistent with the hypothesis of an X-linked recessive? Here, assume that $b/(Y)$ = bald man; $B/(Y)$ = nonbald man; b/b = bald woman; $B/-$ = nonbald woman.

4. In mice, $B/-$ is black and b/b is brown. At another locus, $C/-$ is colored and c/c is colorless (albino). The albino genotype is epistatic to all genotypes at the B locus. What offspring phenotypes, and in what proportions, are expected from two doubly heterozygous black mice—that is, $B/b\ C/c \times B/b\ C/c$?

5. In mice, the recessive allele d (dilute) interacts with B genotypes in this way:

$B/-\ D/-$ = black $b/b\ D/-$ = brown
$B/-\ d/d$ = gray $b/b\ d/d$ = light chocolate

What offspring, in what proportions, are expected if a triple heterozygote is crossed with a completely recessive homozygote—that is, $B/b\ C/c\ D/d \times b/b\ c/c\ d/d$?

6. In rabbits, the three alleles C, c^b, and c give these phenotypes:

C/C or C/c^b or C/c = full color
c^b/c^b or c^b/c = Himalayan
c/c = albino

What are the genotypes of the parents if a litter of rabbits contains all three phenotypes?

7. In humans, a few families have been reported in which both parents are albino but all their children have normal pigmentation. Some families are also known in which both parents are deaf but all their children have normal hearing. What explanation is reasonable for these cases? Suggest genotypes for the albino parents and their offspring.

8. How many Barr bodies would you expect to find in cells from the following types of individuals: (a) karyotypically normal female; (b) karyotypically normal male; (c) female with Turner syndrome; (d) 46,XY female with pure gonadal dysgenesis; (e) 46,XY female with androgen insensitivity syndrome; (f) 46,XX male pseudohermaphrodite; (g) 46,XY female pseudohermaphrodite; (h) 46,XX sex-reversed male?

9. Why are phenotypic differences between members of identical female twin pairs likely to be greater than those between identical male twin pairs? (*Note:* Identical twins, being derived from a single zygote, have identical genotypes.)

10. Linder and Gartler (1965) studied certain benign uterine tumors up to several centimeters in diameter that were removed from women who were heterozygous A/B for these two alleles at the $G6PD$ locus. In a given individual, some tumors consisted of all A cells and others of all B cells, but none were mixtures of A and B cells. On the other hand, pieces of adjacent uterine tissue as small as 1 mm in diameter contained both A cells and B cells. What can be concluded about the origin of the uterine tumors?

11. In Figure 7.15, what offspring phenotypes would be expected if a transmitter male in generation IV married a normal, noncarrier woman?

12. In Figure 7.18, what offspring phenotypes would be expected if IV-4 married IV-9 (his second cousin)? Consider separately the situations of (a) paternal imprinting and (b) maternal imprinting.

FURTHER READING

Good summaries of genetic heterogeneity in humans appear in Skinner (1990) and in Beaudet et al. (1989). For more details on data collection and analysis, see Mange and Mange (1990) or any upper-level human genetics text. The story of kuru is told by Zigas (1990). Lyon (1988, 1992), Riggs and Pfeifer (1992), and Graves (1987) discuss X chromosome inactivation and dosage compensation. The fragile X syndrome is reviewed by Davies (1989), Hagerman and Silverman (1991), and Richards and Sutherland (1992a). Harper et al. (1992) and Sutherland and Richards (1992) discuss anticipation and myotonic muscular dystrophy. Hall (1990, 1992) and Solter (1988) review genomic imprinting, and Wallace and Collins (1991) describe neurofibromatosis.

8 Quantitative and Behavioral Traits

Soon after their births as identical twins, Jim Lewis and Jim Springer were adopted into different Ohio families. Thirty-nine years later they were reunited as part of a large, carefully designed twin study conducted by Thomas Bouchard and colleagues at the University of Minnesota (Franklin 1989). The twins discovered they had led astonishingly similar lives (Figure 8.1). Like their given names, some of the coincidences were probably the result of blind chance: Each was brought up with an adopted brother named Larry and a family dog named Toy; each married and divorced a woman named Linda and remarried one named Betty; their first sons were named James Allan and James Alan.

Other coincidences could have been influenced by inborn tendencies: In school, both liked math and disliked spelling; both enjoyed carpentry and mechanical drawing; both were trained in law enforcement and were deputy sheriffs in different Ohio towns; both drove similar blue Chevrolets and vacationed at the same stretch of Florida beach; they shared nearly identical patterns of drinking and chain-smoking. The twins thought that these events were weird, even downright spooky (but see Question 1). Bouchard noted: "If someone else brought this material to me and said, 'This is what I've got,' I'd say I didn't believe it" (Chen 1979).

The study of similarities and differences among identical twins raised apart brings into sharp focus questions related to heredity and environment. Our major concerns here are physical and behavioral characteristics that may reflect the functioning of the whole person. Such composite traits may have a partial genetic basis, perhaps a subtle basis not explained by the Mendelian rules of inheritance that apply to single genes. Rather than finding one gene with a major effect, researchers are finding that many genes with individually small effects appear to influence a particular behavior.

Some students may find it odd to bring together the concepts of *genetics* and *behavior* at all. Certainly, behavior genetics is a relatively new field of study. This is because early behavioral scientists were uninterested in genetics, and early geneticists could not easily study ill-defined behaviors. But in the 1930s, Robert Tryon, at the University of California, Berkeley, began systematic genetic studies of the behavior of rats. Starting from a *single* group of animals, Tryon was able to breed one line that made few errors in running a maze and another line that performed poorly (Figure 8.2). The ability to separate out maze-bright or maze-dull rats by selective mating over several generations demonstrated genetic variation for a behavioral trait.

In ensuing decades, behavioral geneticists experimented with many other animals, including fruit flies, bees, birds, mice, cats, dogs, and monkeys. Investigations of human conditions as diverse as normal personality traits and devastating disorders such as Alzheimer or Huntington

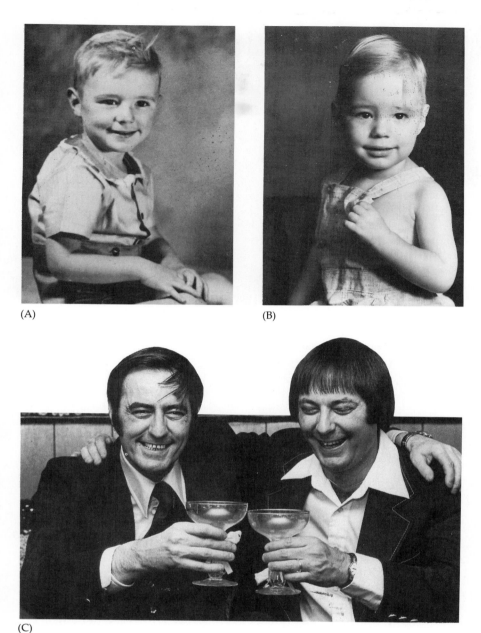

(A)

(B)

(C)

Figure 8.1
Identical twins who were separated when they were about one month old and brought up in different families. (A) Jim Lewis at age 3. (B) Jim Springer at age 2. (C) The 39-year-old twins at their reunion in 1979. (A and B, courtesy of Thomas J. Bouchard, Jr. C, Ira Berger/New York Times Pictures.)

disease leave no doubt that some of the behavioral differences among us depend on genetic differences. The relevant genes affect hormones, neurotransmitters, and cell membranes, as well as the form and function of sense organs, nerves, and muscles—systems that are also important in many physical characteristics. Of course, environmental factors also influence the same behavioral traits, and their modifying effects may be greater than the effects of genetic factors. We especially want to emphasize that evidence for genetic influence, even strong genetic influence, never rules

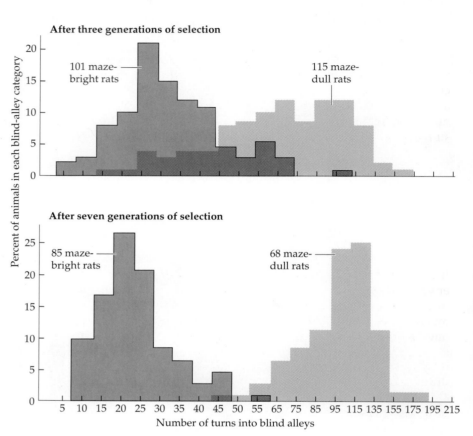

Figure 8.2

The effects of selective breeding on maze learning, represented as histograms (adapted from Tryon 1940). The horizontal scale along the bottom is total number of errors—turns into blind passageways— made by a rat in running a maze. For example, an animal that made 8 errors would be included in the first step of the histogram. The vertical scale represents the percentage of animals in each step. Three generations are represented. In the parental generation (top), error making by the rats spanned a wide range. From this group, the rats making the fewest mistakes were bred together, and among their offspring, the best were bred again to produce successive generations of maze-bright rats. Similarly, the worst performers were selected as parents for successive generations of maze-dull rats (colored histograms). After just three generations of selection (middle), but especially after seven generations (bottom), the averages for the two selected groups were markedly different.

out the possibility of environmental influence. A disease (such as phenyl-ketonuria) that is clearly genetic in origin but nevertheless treatable to some degree shows the relevance of both genes and environment.

GENETIC AND ENVIRONMENTAL VARIATION

Complex physical and behavioral traits are rarely "either/or" situations like many of the Mendelian characteristics that we have previously considered. Rather, they vary *continuously* and are measured by length, weight, time, patterns, color gradations, activity levels, test scores, or some other suitable scale. And when it comes to intelligence or personality, the definitions of the traits themselves may be imprecise and the resulting measurements correspondingly uncertain. In particular, studies of human intelligence have sparked considerable academic, social, and political controversy. Yet these difficulties have not stopped geneticists, anthropologists, psychologists, educators, and others from studying such **quantitative variation**. They have tried to measure how much of the variation depends on people being conceived with different genotypes and how much depends on people living in different environments.

This is the old nature/nurture question, which seeks to partition the underlying causes of observed differences. The question is asked because it deals with interesting or important human traits. Robert Plomin (1990), of Pennsylvania State University, writes:

> Some of society's most pressing problems, such as drug abuse, mental illness, and mental retardation, are behavioral problems. Behavior is also key in health as well as illness, in abilities as well as disabilities, and in the personal pluses of life, such as sense of well-being and the ability to love and work.

A person's phenotype unfolds during development and maturation, when genes and gene products interact with each other and with life's circumstances, which consist of all environmental factors both before and after birth and every influence, whether obvious or subtle, that may help make one person different from another. These happenings are sometimes summarized in a cause-and-effect diagram:

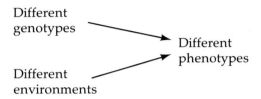

It is not proper to ask whether a trait is "due to genes" or "due to the environment," since a person cannot exist without both. Rather, the nature/nurture question is one of degree. We can ask, for example, how much of the observable phenotypic variation in hair color is due to people having different hair color alleles and how much is due to hair being exposed to different environments. In a sense, we are asking how boldly, relative to each other, the two arrows in the diagram above should be drawn.

Because the genes provide the initial guidelines for the development of a new person, this diagram is sometimes described in the following way: The genotype of an individual determines a *range* of possible phenotypes, and within that predetermined range, a specific phenotype is molded by environmental influences. The range determined by genes may be narrow or broad, depending on the trait under consideration. For example, one's ABO blood type is gene-determined within a very narrow range. In fact, we usually do not consider environmental influence at all. But researchers have noted that the strength of the blood-typing reactions may change with age, and very rarely, blood type B may be acquired later in life.

At the other extreme, infectious diseases are primarily caused by contact with external agents and are seemingly independent of a person's genotype. Still, we all know people who, through all adversity, never get the colds we do. Studies of twins and adoptees reinforce the suggestion that disease resistance has a genetic component, expressed through the manifold cells and reactions of the immune system. In trying to discover one characteristic that is largely, or only, affected by environmental variables, Plomin has suggested "niceness," the degree to which a person is sympathetic, trusting, and cooperative. Still, many personality traits are increasingly found to be influenced to lesser or greater degree by genetics. Thus, for no characteristic of interest are differences between people entirely of genetic or entirely of environmental origin.

The analysis of quantitative traits makes use of **biometry**, which is

167

SIR FRANCIS GALTON

Sir Francis Galton (1822–1911) was the originator of many statistical concepts and the founder of *eugenics*, the study of ways to improve the genetic endowment of humankind (Chapter 20). He read at an early age, studied medicine and mathematics, explored southwest Africa for the Royal Geographical Society, and established a widespread network of self-recording weather stations. The first to utilize twins for studies of stature and intelligence, he also began the classification of fingerprints. Measuring was his passion.

In his autobiography (1909), Galton wrote:

> Many mental processes admit of being roughly measured. For instance, the degree to which people are bored, by counting the number of their fidgets. I not infrequently

tried this method at the meetings of the Royal Geographical Society, for even there dull memoirs are occasionally read. . . . I have often amused myself with noticing the increase in that number as the audience becomes tired. The use of a watch attracts attention, so I reckon time by the number of my breathings, of which there are fifteen in a minute. They are not counted mentally, but are punctuated by pressing with fifteen fingers successively. The counting is reserved for the fidgets.

Galton shared and appreciated Mendel's zeal for counting things, and he felt a sentimental bond because of their common birth year (1822). Galton, however, never recognized the fundamental importance of Mendel's work.

statistics applied to biological variation. Although the computations are beyond the scope of this book, we will describe why the nature/nurture problem is a statistical concept. The question was first seriously addressed by the inventive and many-sided Sir Francis Galton (Box A), a contemporary of Gregor Mendel and a half first cousin of Charles Darwin. Galton measured many physical and behavioral characteristics in humans and inquired into their inheritance.

Heritability

A useful way to describe the origins of variability is by the algebraic equation

$$V_P = V_G + V_E$$

This equation can be understood on two levels. Most simply, it expresses that the total phenotypic variation (V_P) that is observed for a given trait in a given population has two components: variation in the genotypes (V_G) and variation in the environments (V_E). The V's can also be taken to represent a more precise measure called *variance*. This estimate of variability is the one that statisticians usually compute to express how diverse a group is. The more widespread, or extreme, the values, the greater the variance. This is illustrated by the histograms in Figure 8.2. The variance that is represented (but not calculated) in these histograms relates to the number of mistakes made by rats in running a maze. The parental generation has the greatest variance, since the rats in this group span the entire range of maze-running ability. Within the selected group of maze-bright rats, the variation in error rate becomes less and less in successive generations. Similarly, the maze-dull rats show progressively less variation. Histograms like these can be a starting point for estimating the separate genetic and environmental components of variation.

The formula $V_P = V_G + V_E$ assumes that genetic and environmental sources of variation are independent of each other. This assumption is not always true, however, because *interactions* may occur. For example, environmental factors reinforce genetic factors whenever artistic or musical youngsters are sent to special schools for art or for music. A second example of genetic-environmental interaction involves the effect of cigarettes on death from diseases of the heart and blood vessels. The cardiovascular death rate in smokers that also have a family history of heart attack is much greater than the summed risks of smoking, on the one hand, and family history, on the other. In cases such as these, corrective terms can be included in the variability equation.

A quantity called **heritability in the broad sense (H_b)** expresses the proportion of phenotypic variation that is due to the genotypic differences among members of a population. Mathematically, it is the genetic variance divided by the total phenotypic variance:

$$H_b = V_G/V_P$$

Suppose that all members of a population have the same genotype, as is approximately true in an inbred line of mice. Then $V_G = 0$, and any observed variability in a trait must be attributed to environmental variation; thus, the heritability in this restricted case would be zero. Conversely, suppose that all members of a population grow up under precisely controlled environmental conditions (like plants in a greenhouse), in which case $V_E = 0$. Here, any variability is attributed to genotypic differences, and the heritability would be 100%.

In human populations, it is not usually possible to control either the genotypes or the environments, and values for V_G and V_E must be estimated indirectly. This calculation can be done very roughly by measuring the *correlations* between the phenotypes of relatives. The correlations measure the similarities between, say, parents and offspring, or between pairs of siblings. Twins, either identical or fraternal, can also be used. Within given populations, heritability values for, say, height and fingerprint patterns are in the range of 80–90%, while typical heritabilities for weight, blood pressure, and IQ scores are in the range of 40–60%. Although the heritability of a particular trait is a rough measure of the importance of genetic differences between people in comparison with modifiability by environmental factors, this interpretation must be substantially qualified. We will return to this topic later in the chapter.

Another quantity, **heritability in the narrow sense (H_n)**, is the proportion of variation in a population that is due to just a *part* of the genetic variation—the so-called additive part, symbolized V_A. Because V_A is less than (or possibly equal to) V_G, H_n is less than (or equal to) H_b. The narrow-sense heritability provides animal breeders with the best prediction of offspring phenotypes from knowledge of parental phenotypes. For example, breeders can more easily alter through selective mating the butterfat percentage of cow's milk (H_n for this trait is 60%) than they can increase the total yield of milk ($H_n = 30\%$). To the extent that a desired phenotype is due to a favorable environment, or to the extent that the effect of a desirable or undesirable allele is masked by genetic dominance or epistasis, artificial selection can make no predictable headway.

THE GENETIC COMPONENT OF VARIATION

During the first decade of this century, the Danish botanist Wilhelm Johannsen calculated the separate contributions of genetic and environmental variables to the weight of seeds of the garden bean. (In doing so, he also coined the commonly used terms *gene, genotype,* and *phenotype*.) Later, investigators in Sweden and in the United States showed that Mendelian rules were adequate to explain continuous variation in the seed color of wheat and in the dimensions of tobacco plant flowers. The explanations supposed that several genes affected the trait in question, each by a small amount. This type of heredity was called **multiple gene inheritance** or **polygenic inheritance**. Note that the idea of multiple *genes* is quite different from that of multiple *alleles*. The former refers to many genes affecting one phenotypic characteristic, the latter to many alleles of one gene. We will now present a polygenic model for the inheritance of any quantitative trait, ignoring, for the moment, environmental influence.

An Additive Model

As far as variability in a quantitative trait can be traced to genetic differences between individuals, let us assume the following:

1. The trait is affected by, say, three genes (*G, H, I*), each with two alleles: *G,g; H,h; I,i*.
2. Each allele indicated by an uppercase letter acts similarly, the combined effects simply adding together.

Suppose, for example, that the genotype *g/g h/h i/i* has the phenotype 40 units (in combination with a standard environment and a constant genetic background at other loci). Let each uppercase allele add 5 units above this. Thus, each of the seven phenotypes is specified by noting the number of uppercase alleles (Table 8.1). For example, a genotype with two uppercase alleles has the phenotype 50 units, regardless of whether the uppercase alleles are both at the same locus (e.g., *G/G h/h i/i*) or at different loci (e.g., *G/g H/h i/i*). Altogether there are 27 possible genotypes generated by combining any genotype at *G* (3) with any at *H* (3) with any at *I* (3).

Using either the gamete-by-gamete (checkerboard) method or the gene-by-gene method for solving genetic problems, we can use the information in Table 8.1 to predict offspring phenotypes from any set of parental genotypes. For example, we can ask what phenotypes, and in what proportions, are expected among the progeny of two triple heterozygotes, *G/g H/h I/i*. Since the same offspring phenotype may result from several different genotypes, some simplifications are possible. For example, a triple heterozygote can make eight types of gametes, but we need not

Table 8.1
An additive model of polygenic inheritance involving three genes.

Number of Uppercase Alleles	Phenotype (Arbitrary Units)	Representative Genotype	Number of Different Genotypes
0	40	*g/g h/h i/i*	1
1	45	*G/g h/h i/i*	3
2	50	*G/G h/h i/i*	6
3	55	*G/g H/h I/i*	7
4	60	*G/G H/H i/i*	6
5	65	*G/G H/H I/i*	3
6	70	*G/G H/H I/I*	1
Total			27

(A)

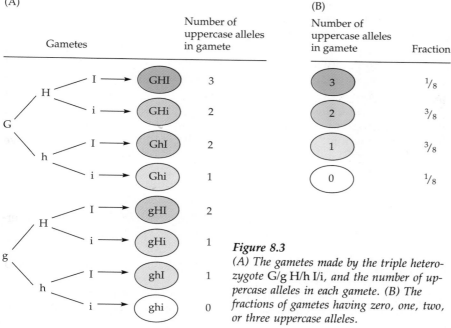

Figure 8.3
(A) The gametes made by the triple hetero-
zygote G/g H/h I/i, and the number of up-
percase alleles in each gamete. (B) The
fractions of gametes having zero, one, two,
or three uppercase alleles.

resort to an 8 × 8 checkerboard. The gametes have either zero, one, two, or three uppercase alleles, so a 4 × 4 checkerboard will do. The fractions of the gametes are worked out in Figure 8.3, and the relevant checkerboard is set up in Figure 8.4.

You may also wish to work out offspring distributions for some of the other matings involving three genes with additive alleles. You will discover that the average of the offspring phenotypes is always the same as the average of the two parents. This is characteristic of genetic situations based on additive alleles and provides important information to breeders. To the extent that V_P is due to V_A, they can be assured of producing, on the average, offspring as superior as their hand-picked parents.

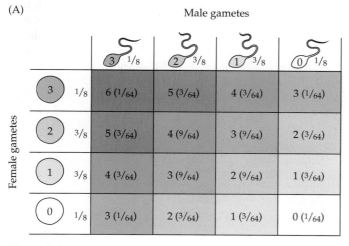

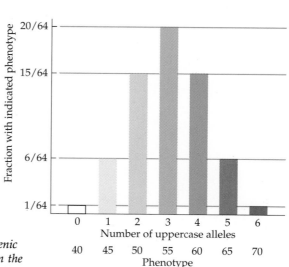

Figure 8.4
The offspring from G/g H/h I/i × G/g H/h I/i under the additive model of polygenic inheritance. (A) Checkerboard. The numbers of uppercase alleles are given within the gamete symbols. The total number of uppercase alleles in the zygotes and the correspond-ing fractions (product of row and column fractions) are given within the body of the checkerboard. (B) Summary of expected phenotypes. Zygotes with the same number of uppercase alleles are combined; the corresponding phenotypic value is obtained from Table 8.1. Note that the average offspring phenotype is 55, the same as that of the parents.

The additive model of gene action underlying quantitative traits may involve any number of genes. A more general formula would involve more than two alleles per locus and perhaps genes with one allele dominant to the others at that locus. There are further refinements that could be added to the model: gene interactions, such as epistasis, or gene linkage (next chapter), or major and minor genes making different contributions to the phenotype.

If we assume that a trait is also influenced by environmental factors, then the histogram in Figure 8.4B would become smoothed out, and intervening phenotypic values (such as 46 to 49 units) would be filled up. The distribution would be broadened at the left and right by individuals with all lowercase or all uppercase alleles living in correspondingly extreme environments. Thus, as few as three additive genes, accompanied by environmental variations, can account for quite a range of phenotypes among the offspring from a particular mating. Even more variation would be expected among the members of a large population.

Skin Color

The degree of skin darkness in people seems to have evolved in response to varying environmental conditions, thereby producing populations that are fairly homozygous for different genes controlling the rate of production of skin pigments called melanins (Chapter 14). The mechanism of action of the control genes is not well understood, nor is the nature of the environmental agents responsible for selection. One suggestion, however, involves the ability of ultraviolet light to convert precursor substances into vitamin D under the surface layers of the skin. Lighter surface layers are conducive to greater vitamin D synthesis, compensating peoples living in far northern or far southern latitudes for weaker sunlight. Since vitamin D is needed for bone formation and its deficiency leads to rickets, strong selective pressures could exist if dietary vitamin D were restricted.

The American geneticist Charles B. Davenport studied skin color inheritance in the families of black-white matings in Louisiana, Bermuda, and Jamaica. He suggested that *two* genes with additive alleles were involved. (Table 8.2). His measurements of skin darkness were obtained by matching a skin patch to a continuous color scale, but his cut-off points between phenotypic classes were necessarily arbitrary. (You will also recognize that "white" skin is not the white of this page, nor is "black" the color of this ink.) Nevertheless, Davenport was able to show generally good agreement between his observations and the predictions of a simple two-gene model.

More recent studies have used a spectrophotometer to measure the

Table 8.2
Davenport's hypothesis for skin color variation in black-white crosses.

Number of Uppercase Alleles	PHENOTYPE		
	Percentage Black	Shorthand Description	Possible Genotypes
0	0–11	White	*aa bb*
1	12–25	Light	*Aa bb, aa Bb*
2	26–40	Medium	*Aa Bb, AA bb, aa BB*
3	41–55	Dark	*AA Bb, Aa BB*
4	56–78	Black	*AA BB*

percentage of light reflected from skin: The darker the skin, the less light reflected. In one study, the analyses of averages and variances in a British population suggested that more than two genes were involved, probably three or four. The investigators were able to roughly calculate the components of variance: V_E was about 35% of the total variance in skin color; V_G (also V_A in this case) was about 65%. Thus, in this population, in either the broad or narrow sense, the heritability of skin color was about 65%.

In summary, in the simplest cases the genetic contribution to variation in quantitative traits may be explained on the basis of several genes inherited according to Mendelian rules, with each gene having a small, additive, phenotypic effect.

TWINS IN GENETIC RESEARCH

Occurring in somewhat more than 1% of pregnancies, twins are always a source of family interest and general curiosity. More often than with single births, they also cause pregnancy complications for their mothers and are subject to health-related problems as newborns (often premature) and as infants. Because identical twins share the same genotype, they provide human geneticists just a little of what a highly inbred line gives to animal or plant geneticists. Twin data provide the starting point for a more detailed analysis of the nature/nurture question.

The Biology of Twinning

There are two types of twins: identical, or **monozygotic (MZ)**, and fraternal, or **dizygotic (DZ)**.

Monozygotic twins derive from a single zygote (one egg fertilized by one sperm) that divides into two separate cell masses within the first two weeks of development. MZ twins are the same sex, their genotypes being identical except for possible somatic mutation (as is thought to happen with some regularity in the development of the cells of the immune system; Chapter 16). *Conjoined* twins may arise in those very rare instances when the cell masses remain partially joined.*

Dizygotic twins result from two zygotes (two eggs separately fertilized). DZ twins are like-sexed about half the time, and they have only half their alleles identical, on the average—the same as for sibs born at different times. Although DZ twins or sib pairs could share *all* their genes or *none* of their genes, depending on the vagaries of chromosomal segregation and crossing over, these extreme events are virtually impossible. We all know sib pairs, nonetheless, who are very similar or very different in some aspects of their phenotype.

The diagnosis of MZ versus like-sexed DZ twins on the basis of physical appearance and mannerisms is usually, but not always, reliable. Visual appearance can be supplemented with information on blood groups and other traits for which one or both parents are heterozygous. Just *one* genetic difference between twins is enough to establish that they are DZ. When no genetic difference is found, the twins may be either MZ or DZ. The more traits examined without finding a difference, however, the more

* Conjoined twins are often called *Siamese twins* after a famous exhibition pair, Chang and Eng, who were joined in the lower chest region by a tough, flexible ligament 3–4 inches thick and extending 5–6 inches between them. Born in Siam (Thailand) in 1811, the twins were bright, resourceful, and wry. At one New York performance, they refunded half the admission fee to a one-eyed man because they said he could see only half of what other viewers could. They eventually settled in North Carolina as farmers, married sisters, fathered a total of 21 children, and died within hours of each other at age 62 (Wallace and Wallace 1978).

Table 8.3

Approximate twinning rates in different populations, ranked by DZ rate.

Population	TWIN-PAIRS PER 1,000 PREGNANCIES[a]	
	MZ	DZ
AFRICAN		
Nigerians	5	40
South African blacks	5	22
U.S. blacks	4	12
CAUCASIAN		
Italians	4	9
Swedes	3	9
U.S. whites	4	7
ASIANS		
Koreans	5	6
Chinese	5	3
Japanese	5	3

[a] Data are average values primarily from tables in Bulmer (1970).

likely it is that the twins are MZ. Since there are so many polymorphisms that can be examined, including DNA markers (Chapter 13), monozygosity can be established with little doubt, if deemed necessary.

The pattern of occurrence of MZ versus DZ twinning can be contrasted in four ways (Bulmer 1970): by race, by maternal age, by family, and by time.

Racial Variation. The frequency of MZ twinning is remarkably constant throughout the world; with about 4 MZ twin pairs per 1,000 pregnancies. On the other hand, the frequency of DZ twinning varies widely (Table 8.3), being greatest among African groups and least among Asians. This suggests that MZ twinning results from a random accident of early development equally likely to occur in any embryo. DZ twinning, however, is influenced by whatever environmental and genetic differences exist between different racial groupings.

Maternal Age Variation. Nearly constant MZ rates but quite variable DZ rates are also observed as a function of maternal age. In the Italian population depicted in Figure 8.5, mothers between 30 and 40 years have a higher rate of DZ twinning than those who are older or younger. It is evident that to produce DZ twins, mothers must ovulate two eggs in the same menstrual cycle. Ovulation is under the control of gonadotropic hormones secreted by the pituitary gland. Gonadotropin levels increase from adolescence through the entire reproductive period, paralleling the increase in DZ twin rates up to about age 37. The sharp drop after this age seems to be due to a general failing of ovarian function as menopause approaches.

A further indication of the pivotal role of gonadotropins on reproduction comes from women with certain infertility problems. Treatment with gonadotropins (or some other drugs) increases the number of eggs that are ovulated and that could be fertilized during any one menstrual cycle. Especially striking is the high frequency (10–40% in different studies) of twins, triplets, quadruplets, quintuplets, and even higher multiplicities that result from the use of various ovulation stimulants.

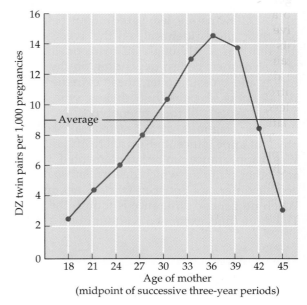

Figure 8.5

The dizygotic twinning rate by age of mother for Italian births, 1949 to 1965. The monozygotic twinning rate was almost constant, at about 4 per 1,000 for all maternal age-groups. (Adapted from Bulmer 1970.)

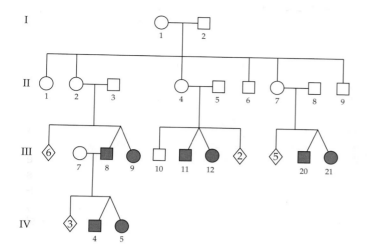

Figure 8.6

An unusual pedigree of unlike-sexed DZ twins (shown as solid symbols). Each of three sisters in generation II (individuals 2, 4, and 7) had a set of DZ twins. Hard to understand is the set of DZ twins fathered by III-8. He may have received heritable factors that foster DZ twinning from his mother and passed them on to his children. But he could not have influenced the apparent multiple ovulation by his wife that led to his DZ twin children. Thus, the twin set in generation IV is probably coincidental. The diamond symbol with number inside indicates children of either sex. (Redrawn from Gedda 1961.)

Family Patterns. MZ twinning shows no tendency to recur in the same mother or to run in families. A woman with one set of DZ twins, however, is somewhat more likely to have another set of DZ twins than a "non-twinning" mother. Furthermore, the twinning rate among the close female relatives of a DZ-producing mother is slightly increased, suggesting a small genetic component of DZ twinning, but not of MZ twinning (Figure 8.6). The nature of the genes involved in DZ twinning is unknown, but they may act by increasing the level of gonadotropins. The genes are expressed only in females but can be transmitted by either sex.

Temporal Patterns. Worldwide, dizygotic twinning rates decreased significantly from the 1950s through the 1970s—as much as 40% in England and Wales. The reasons are unknown.

The Analysis of Twin Data

If geneticists assembled inbred lines of mice, each line consisting of just two animals, they would not have particularly good material for quantitative genetic analysis. Yet this is the type of situation that human geneticists must deal with in twin studies. Some information of interest can be obtained, but the relatively small numbers of twins limit the reliability of the data. Further difficulties arise because human geneticists have little control over the environmental variables they might wish to evaluate. Indeed, it is not always clear what environmental factors are relevant to variation in complex traits such as intelligence.

The analysis of twin data for a quantitative trait, such as adult height, can proceed in the following way: For each twin pair, the difference in height is obtained. Then the data from all MZ twins (or from all DZ twins, like-sexed for a proper comparison) can be combined into a single figure that is the average of these differences (Table 8.4). Alternatively, we can compute the variance statistic: V_{MZ} (or V_{DZ}). Both types of calculation express how much two MZ (or DZ) twins differ from each other, on the average. The application of twin data to the nature/nurture question involves these quantities and how they are related to the components of variability V_E and V_G in a typical human population.

Because identical twins have the same genotype, whatever variability exists between them must be due to environmental differences. Therefore,

175

Table 8.4
Average differences between twin pairs, or sib pairs, for three quantitative traits.

	MZ TWINS		DZ Twins (50 pairs)	Sibs (50 pairs)
	Raised Together (50 pairs)	Raised Apart (9 pairs)		
Height	1.7 cm	1.8 cm	4.4 cm	4.5 cm
Weight	4.1 lb	9.9 lb	10.0 lb	10.4 lb
Stanford-Binet IQ	5.9	8.2	9.9	9.8

Source: Newman et al. 1937.

the quantity V_{MZ} is related to V_E. These variances are not equal, however, since environmental influences on identical twins are quite different from those affecting random, unrelated individuals, the basis of the V_E statistic. The same family environment and similar experiences based on being the same age suggest that V_{MZ} is less, to an unknowable degree, than V_E.

These difficulties are minimized but not eliminated when we examine identical twins who have been raised apart. Four studies have focused on such twin pairs: an ongoing study of personality by Thomas J. Bouchard, Jr., and colleagues at the University of Minnesota, with about 50 pairs; an older study by Horatio Newman and colleagues at the University of Chicago, with 19 pairs; a British study by James Shields, with 44 pairs; and a Danish study by Niels Juel-Nielsen, with 12 pairs (Rose 1982). (We ignore the data in a fifth study by Sir Cyril Burt; see Box B.) But even for identical twins raised apart, the differences between them would not be expected to be a very good estimate of V_E. For example, over half of the separated MZ twins studied by Shields were raised in families related to each other. In addition, adopting families generally tend to be at least middle class in socioeconomic level. Thus, although the individual case histories of twins raised apart make fascinating reading (Rosen 1987), the collected data are not extensive and consequently do not provide very precise conclusions.

The difference in variances, $V_{DZ} - V_{MZ}$, that one can calculate using twins is somewhat related to V_G. The relationship is based on the fact that DZ twins differ in both genotype and environment, whereas MZ twins differ in environment only. You might think that the subtraction of one from the other should cancel out the environmental variation, leaving just the genetic component. But there are good reasons why this difference is not equal to V_G, the genetic variance between *random* persons. One is that the genetic difference between fraternal twins or sibs is clearly not as great as that between unrelated persons. In addition, the environmental influences on the two types of twins may not be equal, so the subtraction would not cancel that source of variability. For example, parents and friends sometimes treat MZ twins more uniformly than they treat like-sexed DZ twins, exposing MZ twins to more similar experiences.

Recognizing the difficulties just outlined, we can obtain from twin data a quantity similar to heritability in the broad sense. For example, if we use Newman's data, the broad-sense heritability for height is about 84%. Twin data applied to many quantitative traits confirm that most of them are influenced by variation in both genotype and environment. For the traits in Table 8.4, it appears that the genetic component of height is greater than that of weight, which is greater than that of IQ; and this conclusion is borne out by many other sets of data as well. In summary, we can, at the least, state the following: To whatever extent MZ twins

DID BURT CHEAT?

The reputation of English psychologist Sir Cyril Burt (1883–1971) has had its ups and downs. Burt had a lifelong interest in childhood education, in theories of intelligence, and in the role of genes in determining mental processes. His studies supported the idea that intelligence is a highly heritable trait. During his lifetime he was much admired for his keen intellect, extraordinary knowledge, mathematical brilliance, and clinical skill, especially with children.

That bright view of Burt was darkened soon after his death. First, Princeton psychologist Leon Kamin discovered in 1974 that some figures in a series of Burt's papers were very suspicious. For example, correlation values for twins showing about 80% heritability of IQ remained the same despite increasing sample sizes. Then, a 1976 front-page story in the *London Sunday Times* reported that two of Burt's associates may not have existed. Finally, there came the biography of Burt by English psychologist L. S. Hearnshaw (1979). The book recounted how Burt could be not only charming and generous, but also cantankerous and jealous. He was fond of creating pseudonyms to write articles supporting his own work or attacking that of others, which he then published in his own journals. Hearnshaw documented how Burt had spoiled his own distinguished career—perhaps as the result of mental instability and personal setbacks later in life—by perpetrating several scientific frauds, including fabricated research on MZ twins raised apart.

More recent books (Fletcher 1991; Joynson 1989) have turned the matter upside-down again. They present critiques of Hearnshaw's biography, providing alternative explanations that do not involve Burt in fraudulent activity. For example, at least some of the constant correlations appear to be related to the partial reanalysis of old data that had been misplaced during the bombing of England in the early 1940s. Some twin data were only gradually rediscovered and reported in the 1950s and 1960s. As for the missing associates, they may have been prewar, unpaid social workers; Burt credited them with the work in postwar years, even though they were no longer around.

The arguments by Joynson and Fletcher may or may not be true, but their defense of Burt seems to be as plausible as the idea that a distinguished researcher would crudely make up data to support a firmly held belief. If he did invent some twins, would he not also change correlation values to render them more believable? Chances are the truth of the matter will never be known.

In any event, it is certainly true that the methods that Burt used were poor by today's standards for evidence or even by the standards of his own time. Details of his intelligence-testing methods never appear in any of his dozens of papers on the subject. Clearly, his results on IQ heritability are inadequately supported; but vagueness is not the same as fraud. Furthermore, although it is important to clarify the actions of a scientist who may have been unjustly maligned, the matter little affects the science of behavioral genetics. Removing Burt's data from the literature on IQ does not alter the conclusion of numerous other studies showing that genetic factors play a significant role in the development of mental functions.

---

differ, environmental variation is a factor; to whatever extent DZ twins differ more than MZ twins, genetic variation is a factor.

Threshold Traits. Some phenotypes have the following attributes:

1. They are discontinuous, being either present or absent.
2. They appear to have a genetic component.
3. They do not show simple Mendelian inheritance.

Examples of such traits include a number of birth defects, such as *cleft lip/cleft palate* (separately or together); behavioral abnormalities, such as *schizophrenia*; and metabolic disorders, such as *diabetes mellitus* (Table 8.5). As in the case of quantitative traits, it is likely that multiple genes as well as many environmental influences are involved. The discontinuity in expression—the trait being either present or absent rather than continuously variable—is explained by the concept of a **threshold**. This explanation implies that an essentially normal phenotype results from many different gene-environment combinations; beyond a critical accumulation of (usually unknown) genetic and environmental stresses, however, the phenotype is abnormal.

Table 8.5
Three threshold traits.

Trait	CONCORDANCE (%)		Phenotype
	DZ	MZ	
Cleft lip/cleft palate	5	35	A variable abnormality affecting the formation of the upper lip and the roof of the mouth during gestation. As a consequence, there is a vertical fissure in the midline of the lip (25% of cases), palate (25%), or both (50%). Modern surgery can usually correct the defect functionally and aesthetically. Although this condition often accompanies other syndromes, its frequency by itself is about 1 per 1,000 births.
Schizophrenia	15	45	A variable psychiatric disorder sometimes accompanied by delusions and hallucinations. Patients may lose all interest in their surroundings or may make illogical and inappropriate responses to persons or happenings. The diagnostic criteria vary, and frequency estimates range from about 1/80 to 1/200 among Americans. Biochemical and structural differences between the brains of schizophrenics and normal persons are being actively investigated.
Diabetes mellitus			Both types result in high levels of glucose in the blood. In type 1, affecting young people, an insulin deficiency prevents glucose from entering cells. The deficiency stems from destruction of the insulin source, the beta cells of the pancreas. The frequency is about 1/400. In type 2, affecting older people, insulin is present, but other proteins needed to transport glucose into cells do not seem to function properly. Obesity is an important predisposing factor in type 2, and the frequency is about 1/150. Both types may be characterized by constant thirst, frequent urination, vision difficulties, slowly healing sores, and coma.
Type 1	5	50	
Type 2	25	95	

The genetic component in threshold traits is indicated, in part, by twin studies. Twins are said to be **concordant** if they both have the trait in question and **discordant** if one has the trait and the other does not. For the traits in Table 8.5, the concordance among MZ twins is 35–95%, whereas the concordance among DZ twins is 5–25%. Although some investigators have disputed the role of genes in the occurrence of schizophrenia, data on adopted children support a genetic component in this disease also. For example, adopted children are more likely to develop mental disorders if their biological mother was schizophrenic than if their biological mother was not. For the two types of diabetes noted in Table 8.5, different MZ twin concordances have confirmed that the diseases' underlying genetic systems are quite different (Hrubec and Robinette 1984).

BEHAVIORAL TRAITS

Trying to sort out genetic and environmental influences on human behavioral traits, such as schizophrenia or intelligence, is especially difficult. One problem, noted previously, is inadequate knowledge of, and control over, important sources of environmental variation. In addition, some behavioral traits lack objective standards of measurement that all investigators can agree on.

A further difficulty is the ideological frame in which the nature/ nurture question is often put. Studies of IQ, for example, are construed to support either a *hereditarian* or an *environmentalist* outlook on society and then perhaps used to influence political or social policy. Hereditarians emphasize the role of genes in fixing the limits of behavioral characteristics, whereas environmentalists see the newborn as much more pliable, to be shaped in large degree by societal forces. Because the debate is sometimes bitter, both inside and outside the academic community, it is important to understand that hereditarians are not necessarily, or even often, racists, nor are environmentalists necessarily, or even often, utopian dreamers. Common sense suggests that neither dogma can be 100% true. Both sides recognize that human achievement can be enhanced by good parenting and solid education, regardless of any genetic predispositions.

A Genetic Component to Human Behavior

Behavior is what people do and how they do it. The mental or muscular responses by which such actions are expressed depend on environmental inputs to our sense organs. The various stimuli are coordinated and controlled by immensely complicated networks of nerve impulses and by hormonal messengers. Our activities are therefore based on aspects of anatomy and physiology that we know can be affected by gene-encoded proteins. It should therefore be expected that genetic variation can lead to behavioral variation.

We have already noted several simply inherited traits that have a behavioral component as part of the phenotype. One of the most distressing is the *Lesch-Nyhan syndrome*, caused by an X-linked recessive gene (Chapter 14). The normal allele controls the production of an enzyme needed in the metabolism of DNA components. Boys hemizygous for a mutant allele of this gene compulsively bite and mutilate their lips and fingers. How the enzymatic abnormality brings about the bizarre behavioral component is still a mystery.

Another example of a single-gene behavioral phenotype involves *porphyria*, which may have affected King George III of England (1738–1820). The unusual abnormalities of this rare autosomal disorder stem from a defect in the synthesis of heme, the iron-containing part of the hemoglobin molecule. The primary physical symptoms are abdominal pain, constipation, vomiting, and wine-red urine. (The word *porphyria* derives from the Greek, meaning "purple.") In the case of George III, neurological symptoms—visual disturbances and restlessness—were first noted when he was age 50. Then after three weeks he became delirious, with convulsions and a prolonged stupor. The king recovered from this bout but suffered from several subsequent attacks of what was then diagnosed as insanity (Macalpine and Hunter 1969). The illness of George III provided the early impetus for research in psychiatry.

Recall also from the last chapter that the mental retardation associated with *fragile X syndrome* stems from a peculiar mutated region near the tip of the long arm of the X chromosome. At this site, the more copies of a trinucleotide (CGG), the more severe the retardation. It is not known, however, how the repetitions of CGG bring about the phenotype. In general, retardation is not an uncommon characteristic of serious single-gene diseases—for example, *phenylketonuria* (Chapter 1). The mental deterioration (progressive loss of memory, confusion, anxiety) of *Alzheimer disease* also has clear genetic components. As noted in Chapter 10, a diagnostic feature is deposits in the brain of a short protein called β-amyloid, which is derived from a longer protein coded by a gene on chromosome 21. Other genes (and environmental factors) are also implicated in Alzheimer disease.

The reading problems known collectively as *dyslexia* may also have genetic components (Vellutino 1987). Although the criteria of definition vary, dyslexic children generally have unexpected difficulties in learning to read despite normal intelligence, customary instruction, and typical patterns of social and cultural support. One form of dyslexia that clearly runs in families may be influenced by a dominant gene near the centromere of chromosome 15. Most forms of dyslexia, however, probably emerge from a combination of biological and environmental factors that are not individually identifiable.

If single-gene mutations can affect behavior in obvious ways, there is no reason to doubt that multiple genes with subtle effects can affect various mental processes, even if specific loci and their products cannot be identified. In addition to gene mutations, gene imbalance (as occurs in chromosomal aberrations) can also have important effects on behavior. For example, Down syndrome patients have distinguishing personality features, such as friendliness, joviality, love of music, and ability as mimics. Patients with Turner syndrome also exhibit an interesting behavioral trait: Although the distribution of their IQ scores is not unusual, they tend to do very poorly on tests that require them to visualize forms in space. In school they often do poorly in arithmetic and mathematics despite otherwise normal or superior performance.

In summary, there are real genetic contributions to variations in behavioral traits, expressed on a continuing basis throughout life. The specific relationships between genotypes, environments, and the resulting phenotypes are often obscure, but possible neurological pathways for gene action are increasingly active areas of investigation.

Intelligence

All of us have a general idea of the meaning of intelligence (a dictionary synonym is "mental acuteness") by which we sometimes judge ourselves and others. Many psychologists emphasize abstract reasoning ability, which includes thinking rationally, solving problems, understanding the basics of a complex situation, and responding effectively to new environments. These meanings are sufficiently broad and applicable to such a wide variety of human endeavors that no obvious yardstick for measuring intelligence immediately presents itself. Some psychologists would even broaden the definition further to include such attributes as musical and artistic talent, physical ability, common sense, insight into oneself, and skill in interpersonal relationships (McKean 1985; Winn 1990). Apart from trying to identify and measure the various aspects of intelligence and determining the relative importance of nature and nurture, there is much interest in the relevant neural and biochemical networks, as well as in the evolution of intelligence in human and other primate species.

Galton, with his passion for measuring, appears to have made the first attempts to test mental abilities. At the 1884 International Exposition in London, he set up a booth where visitors paid threepence to be scored in several ways. Some of Galton's tests dealt with sensory perception (e.g., accuracy in discriminating different weights) or with quickness of reaction, which he felt correlated with a person's intellect.

In Paris, beginning in 1904, Alfred Binet and his colleagues at the Sorbonne were given the task of identifying subnormal schoolchildren who might need special education to improve their position. Binet's easily administered tests proved fairly successful in predicting how well a child performed in subsequent schooling. The test items were verbal, numerical, or pictorial problems of increasing difficulty for older children. Although Binet hoped to measure "general intelligence" rather than knowledge accumulated through schooling and other experiences, he recognized that test construction depended on trial-and-error standardization rather than on a theory of intelligence. Psychologists at Stanford University modified the Binet procedure to evaluate white, middle-class American children. Test questions were selected so that, overall, boys and girls performed equally well and scaled to average 100 for each age group. Gould (1981) gives a detailed history of IQ testing, including accounts of its misuse by the American eugenics movement.

Do IQ tests measure intelligence in any of its various meanings? No one doubts that the scores do predict reasonably well the future scholastic success of the children against whom the tests are standardized. That, of course, is what they were originally designed to do. In addition to the generalized Stanford-Binet, Wechsler, and Scholastic Aptitude Tests (SATs),* other quite different types of tests purport to measure one or another aspect of mental ability. Educational psychologists point to the considerable degree of correlation between many of these tests and suggest that some common capability is being examined. The factor common to these various forms of measurements is what some psychologists call intelligence. This conception of intelligence is derived from the statistical analysis of tests. It might be more satisfying if the flow of ideas were reversed so that test construction could be derived from an analysis of intelligence. Intelligence may be too complex a trait, however, to be defined by one or a few test scores.

We emphasize that intelligence is not a genotype; it is a *complex phenotype* that develops under the influence of genes and the experiences of a lifetime. An IQ test provides a highly simplified assessment of an aspect of that phenotype. Although IQ measurements are related to what it means to be intelligent, the results are not comprehensive measures.

Heritability of IQ. In this section we ask: To what degree do IQ variations *within* middle-class whites (the group on whom the tests were standardized) arise from genetic differences? It is, of course, the measured IQ scores themselves that are at issue, not intelligence in its broader context. A quite different question is raised in the next section: To what degree does the difference in average IQ *between* black and white Americans arise from genetic differences? The answers to the "within" and "between" questions need not be the same. In any event, numerical results such as heritability values are only starting points in assessing how social and educational changes can help improve intellectual development.

Many populations have yielded values for the heritability of IQ, although some individual studies are based on small numbers. A 1981 summary of 111 studies that met certain criteria is given in Figure 8.7. The data are presented as the correspondences (correlation coefficients) in scores between pairs of individuals on a variety of tests that purport to measure an aspect of intelligence. The pairs of individuals were raised either together or separately; they shared none of their genes (top line), or an increasing proportion of their genes (middle lines), or all of their genes (bottom two lines).

The statistical significance of these data is hard to assess, partly because the results in any one line are so heterogeneous. Overall, it appears that neither genetic nor environmental influences can be neglected. For example, if environmental influences were absent, then zero

---

* The various Wechsler scales, in addition to verbal tasks, include performance components (such as assembling an object) to judge aspects of mechanical ability. The Wechsler Adult Intelligence Scale (WAIS) is the most widely used IQ test for adults in the United States, having been standardized against age-groups from 16 to 75. Many of you will be familiar with the SATs, which are taken by more than a million college-bound students every year (Jensen 1981).

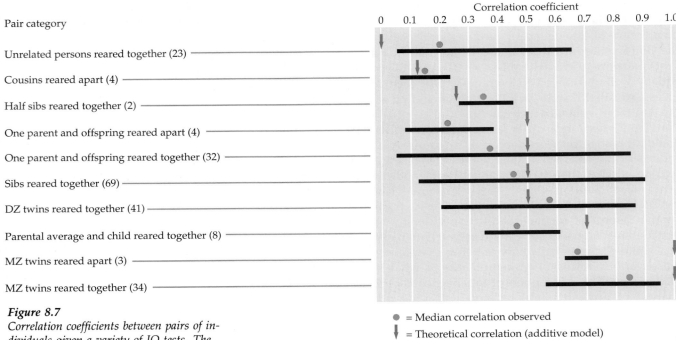

Pair category	Correlation coefficient

Pair category

Unrelated persons reared together (23)

Cousins reared apart (4)

Half sibs reared together (2)

One parent and offspring reared apart (4)

One parent and offspring reared together (32)

Sibs reared together (69)

DZ twins reared together (41)

Parental average and child reared together (8)

MZ twins reared apart (3)

MZ twins reared together (34)

● = Median correlation observed

↓ = Theoretical correlation (additive model)

Figure 8.7
Correlation coefficients between pairs of individuals given a variety of IQ tests. The number of studies of each type of pairing is given in parentheses, and the range of correlation values is given by the extent of the bar. The colored dot above a bar denotes the median correlation observed, and the colored arrow denotes the correlation that would be expected on the basis of a simple polygenic model with additive alleles. In rough terms, the similar trend of dots and arrows suggest the existence of genetic factors. That the dots and arrows are not superimposed suggests the existence of environmental factors. (Redrawn and abridged from Bouchard and McGue 1981.)

correlation should be found between unrelated persons reared together (first line); but the 23 studies summarized show that there *is* such an association. The last two lines in Figure 8.7 lead to a similar conclusion: The MZ twins raised together or apart do differ, and that difference must necessarily arise from *environmental* factors. On the other hand, the influence of *genotypic* factors on IQ is strongly suggested by the higher correlation between MZ twins reared together (about 0.84) than between either DZ twins reared together (0.57) or sibs reared together (0.45).

Estimating heritability values from correlations is complicated and involves assumptions about the study population that may be only approximately true. Typical calculations for children and adolescents show that heritability in the broad sense is about 50% (e.g., see Scarr and Carter-Saltzman 1983). For adults, the heritability values are higher (Bouchard et al. 1990). Researchers ascribe some of the remaining IQ variability to purely environmental variations, and some to associations between genotypes and environments. This latter source of variation arises, for example, when parents with "favorable" genotypes not only transmit favorable alleles to their children, but also (because of the favorable genotypes) provide their children with an enriched environment conducive to achieving high IQ scores.

We wish to reiterate the limits in the meaning of heritability. These cautions have been set forth by many persons, including University of California psychologist Arthur Jensen in the 1969 paper that sparked heated arguments over the race-IQ issue (discussed shortly).

1. *Population versus individual.* Heritability is not defined for an individual. It would make no sense to say, for example, that 60 points of a person's IQ score are due to his genes and 40 points to his environment. Rather, heritability is a measure of the genetic variability of persons within a population. In groups with relatively homogeneous environments, the heritability of a trait will tend to be larger than for groups in heterogeneous environ-

ments. Perhaps this tendency is one reason why heritability estimates among British populations are often higher than those estimated for the more diverse cultural and educational settings of the United States.

2. *Known versus unknown genes.* Heritability values do not depend on knowing the metabolic actions of whatever genes and alleles are involved. Although it would be nice to understand the relevant biochemistry, the heritability values do not depend on this knowledge. Undoubtedly, many genes—perhaps hundreds—with individually small effects influence the type of mental processes involved in IQ measurements.

3. *Constant versus changeable IQ.* Many times we have noted that an environmental change can modify the expression of a gene-influenced phenotype. The significant heritability of IQ does not rule out environmental modifications of IQ. Although there is a high correlation between IQ scores for the same person taken at different ages, the correlation is not perfect. Furthermore, the correlation applies to existing circumstances of schooling and middle-class values, and novel circumstances might affect the IQ phenotype more. No one is predestined to have and maintain a particular IQ value.

Racial Differences in IQ. In this area of investigation, the facts do not always speak for themselves. The same set of data may be interpreted by different scientists to mean different things, and these viewpoints may be oversimplified or exaggerated by journalists seeking eye-catching copy on scientific matters that impinge on public policy. The major undisputed fact about the IQ scores among white and black children is that the average racial difference is about 15 points. Nevertheless, there is considerable overlap, so the IQ score of a particular individual is no guide to his or her race, and vice versa. The reasons for the average IQ difference between blacks and whites are difficult to assess, and most investigators think that the underlying causes are unknowable on the basis of currently available data. The many interrelated environmental variables that could affect test scores include the following:

1. *Standardization.* Because blacks were not included in the development or standardization of the Stanford-Binet tests, cultural differences could lead to lower scores for blacks.

2. *Discrimination.* Social and educational opportunities that are available to whites of any socioeconomic level have often been denied to blacks. Predominantly black schools, for example, tend to be inferior to predominantly white schools. Also difficult to measure are the effects of hundreds of years of racial prejudice, both blatant and subtle, that are rooted in slavery but continue to exist.

3. *Socioeconomic status.* One measure of socioeconomic status (SES) combines, into a single index, data on father's and mother's education, father's occupation, number of siblings, and reading materials and appliances in the home. Studies show that blacks in general have lower SES ratings than whites, and people with low SES ratings typically get lower IQ scores.

4. *Language.* This particular cultural difference between blacks and whites is especially important, because IQ tests are predominantly verbal. Designing tests to eliminate language and other cultural differences is difficult, and most investigators believe that it has not yet been accomplished.

5. *Motivation.* What factors allow a test taker to do as well as possible? How is performance affected by a subject's self-esteem and by the individual's perception of how the results will be used? How important are the attitudes and expectations of teachers, friends, and parents? These elements have been shown to play significant roles (Goleman 1988).

These variables suggest that some or all of the average IQ difference between blacks and whites could be due to *environmental* differences between the two groups.

The argument that at least some of the average IQ difference could be due to *genetic* differences flows from the concept of races: subdivisions of humankind that have come to differ, to a greater or lesser degree, in the frequencies of some of the alleles they possess. Blacks and whites do differ in traits controlled by single genes—for example, hemoglobin variants and blood types. These differences have occurred through the processes of evolution, such as mutation and selection. There is no reason to suspect that alleles that affect behavioral processes could not also come to have different frequencies. But the idea that genetic factors for IQ performance *could* differ between races certainly does not mean that they *do* differ.

Jensen suggests that most of the black-white difference in IQ is due to genetic factors. His argument is that even when black and white children are matched for environmental factors, most of the IQ difference remains. Others believe that the matching can never be fairly done (Bodmer and Cavalli-Sforza 1970; Gould 1981). In short, the arguments for a genetic contribution to the average black-white difference in IQ are weak.

In any event, the possibility or reality of an average genetic difference—or any other kind of difference—between groups should be inconsequential in a democratic society. Although many of the conclusions of Jensen (1969) have been divisive, he does state that the full range of human talents is represented in all races and that it is

> unjust to allow the mere fact of an individual's racial or social background to affect the treatment accorded to him. All persons rightfully must be regarded on the basis of their individual qualities and merits.

Personality

It has been noted with amusement that parents of single children tend to be environmentalists, while parents of more than one—seeing the striking differences that may develop even in a steady environment—tend to be hereditarians. For example, many parents take the credit—perhaps rightly so—for providing the proper environment that produces a delightful first child: a warm home life, enriching toys, good schools, and so on. Yet a second or third child that they bring up in substantially the same way often turns out to have a completely different personality. One child may be outgoing and talkative, another shy and quiet, a third uniquely a worrywart or a big risk-taker or a slob.

The systematic study of genetic influences on personality traits has expanded over the last decade or two. A large, ongoing investigation is being done by Thomas Bouchard and colleagues at the University of Minnesota (Holden 1987). They have assembled and compared four groups of twins: MZ twins raised together, MZ twins raised apart, DZ twins raised together, and DZ twins raised apart. This design allows heritability values to be calculated in different ways. (One remarkable set of MZ twins raised apart, among the group of about 50 such sets, was described at the beginning of this chapter.) All twins were given batteries of tests having thousands of items, including one self-appraisal called the Multidimensional Personality Questionnaire. This questionnaire allowed investigators to evaluate such personality traits as social potency (a tendency toward leadership or charisma), social closeness (the need for intimacy, aid, and comfort), and traditionalism (respect for authority and rules).

From this study and from several others with similar intent, researchers have concluded that many different personality traits have sizable genetic components, with heritabilities in the range of about 40–60% (Table 8.6). Thus, on the average, about 50% of observed personality diversity can be ascribed to genetic diversity. Furthermore, and surprisingly, very little of the remaining 50%—the environmental influence—seems to be due to children sharing a common family environment. Rather, many of the environmental influences that affect the development of personality traits tend to come randomly from outside the home, that is, the non-shared environment. Even allegedly shared family experiences may be perceived differently by each child and produce nonshared responses. For example, strict discipline by parents might affect quite differently a naturally shy child with many personal interests and an outgoing sibling with many social interests (Bower 1991). Behavioral geneticists have indeed noted that MZ twins brought up in the same family are not much more alike than MZ twins brought up in different families.

In short, parents can take some credit or blame, but not a lot either way, for providing suitable environments that influence the development of personality traits in their children.

Alcoholism

We conclude this chapter with a brief look at alcoholism, which people view variously as a biological problem, a psychosocial abnormality, or a moral weakness. Occurring in about 10% of American men and 4% of American women at some time during their lives, alcohol addiction can have disastrous consequences. The annual toll in the United States includes more than 50,000 deaths and $100 billion in costs (Desmond 1987).

Alcohol intoxication causes depression of the central nervous system, producing uninhibited behavior, poor judgment, and mood changes, as well as incoordination, slurred speech, blackouts, and coma. In advanced stages of addiction, alcoholics are obsessed with alcohol to the exclusion of everything else—family, friends, work, food—all the while denying they have a problem. Long-term alcohol abuse is associated with many degenerative diseases, especially of the liver, but also of the stomach, esophagus, pancreas, heart, and brain. In affected women who drink heavily during pregnancy, alcohol diffusing across the placenta can affect the fetus, producing *fetal alcohol syndrome*, with symptoms ranging from mild to profound physical and mental abnormalities.

The development of an alcoholic is influenced by numerous personal, social, and cultural factors that interact with each other. In addition, the existence of some genetic variation for alcoholism is supported by two lines of evidence: (1) higher concordance in MZ twins than in DZ twins in some (but not all) studies and (2) increased risk for children of alcoholics even when adopted into nonalcoholic families (Goodwin 1991).

Three separate adoption studies were undertaken in the late 1970s in Denmark, Sweden, and the United States. The results were generally consistent with each other. The Swedish study (Table 8.7), for example,

Table 8.6
Heritabilities of some personality traits.

Personality Trait	Heritability
Well-being	0.48
Social potency	0.54
Achievement	0.39
Social closeness	0.40
Stress reaction	0.53
Alienation	0.45
Aggression	0.44
Control	0.44
Harm avoidance	0.55
Traditionalism	0.45
Absorption	0.50
Positive emotionality	0.40
Negative emotionality	0.55
Constraint	0.58

Source: Tellegen et al. 1988.

Table 8.7
Alcoholism among male adoptees

Biological Parent	Number in Sample	Percentage of Adopted Sons[a] Who Were Alcoholic	
Alcoholic father	89	39.4	avg = 34.0%
Alcoholic mother	42	28.6	
Nonalcoholic father	723	13.6	avg = 14.6%
Nonalcoholic mother	1,029	15.5	

Source: Bohman 1978.
[a] Swedish males born to unwed mothers between 1930 and 1949, adopted as infants by nonrelatives, and followed through 1972. Alcoholism was determined by official records of alcohol-related fines, arrests, clinic visits, hospitalizations, and so on.

investigated persons who were born to single mothers and who were adopted at a very early age by nonrelatives. The table shows that when adopted males had an alcoholic biological parent, they were 2.3 times more likely to develop alcoholism than when they had nonalcoholic biological parents. A lower rate of alcoholism among females (included in other investigations) is not well understood but may be due in part to societal roles, to a lower rate of identification, or to biological factors related to the metabolism of alcohol.

Researchers have tried to find physiological or neurological characteristics that would identify those at increased risk for alcohol abuse. Enzymes that have been examined include two in the liver, alcohol dehydrogenase and aldehyde dehydrogenase, which break down alcohol via the following pathway:

$$\text{alcohol} \xrightarrow{\text{alcohol dehydrogenase}} \text{acetaldehyde} \xrightarrow{\text{aldehyde dehydrogenase}} \text{acetate}$$

No consistent differences have been demonstrated between alcoholics and nonalcoholic controls with regard to these enzymes, but an interesting racial trait has come to light. More than 80% of Asians possess a variant, slow-acting form of the second enzyme, aldehyde dehydrogenase. It is thought that this difference (leading to a buildup of acetaldehyde) accounts for the marked facial flushing, accelerated heart rate, and other symptoms of distress that many Asians experience when drinking even small amounts of alcoholic beverages.

More recently, specific genes that influence alcoholism have been sought. In a 1990 study, researchers discovered a significant genetic difference between a group of 35 alcoholics and a control group of 35 nonalcoholics. The gene in question, on the long arm of chromosome 11, encodes a protein that acts as a receptor site for a *neurotransmitter*. A neurotransmitter is a small chemical that is released from the end of one nerve cell and then binds to receptor molecules at the beginning of the next nerve cell in a neural network (Figure 8.8). In this way, the neurotransmitter sets in motion chemical changes that advance (or sometimes retard) electrical impulses along the nerve path. Many drugs that alter mood or behavior act in the brain by enhancing or inhibiting neurotransmitters or their receptors.

The receptor protein investigated in this study specifically binds the neurotransmitter *dopamine*. This substance is known to be present in regions of the brain that are involved in pleasure seeking, including regions influenced by cocaine dependence. Using molecular probe techniques dis-

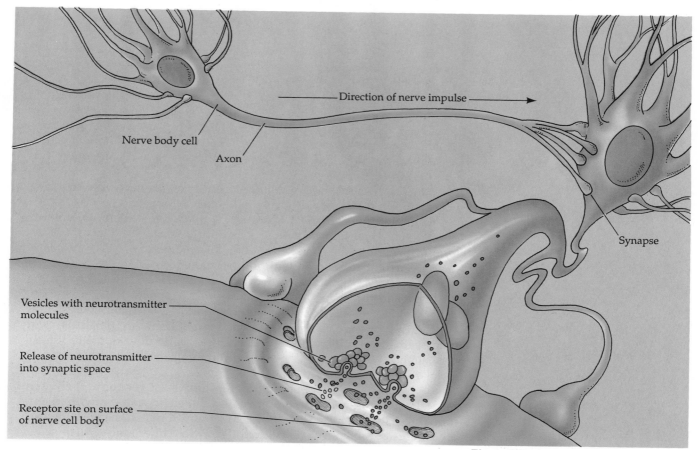

Nerve body cell

Axon

Direction of nerve impulse

Synapse

Vesicles with neurotransmitter
molecules

Release of neurotransmitter
into synaptic space

Receptor site on surface
of nerve cell body

Figure 8.8
Nerve cells and the connection—the synapse—between them. When the electrical impulse gets to the synapse at the end of the axon, neurotransmitter molecules from vesicles in the knob are released into the space separating the two cells. The binding of the transmitter molecules to the receptor molecules causes additional chemical reactions that may initiate further electrical conduction through the cell body and axon of the next cell.

cussed in Chapter 13, the researchers identified two alleles, *A1* and *A2*, of this **dopamine receptor gene**. The *A1* allele was present in 69% of the alcoholics, but in only 20% of the nonalcoholics. These data show that *A1* is neither necessary nor sufficient to cause alcoholism, and what the *A1* allele might do differently than *A2* is not known. Because susceptibility to alcoholism is affected by many biological, psychological, and social factors, it was, in fact, surprising that alleles of this one gene could discriminate so large a proportion of affected versus nonaffected persons.

Indeed, this striking difference has not been uniformly supported by additional studies, which are about equally divided—positive or negative—in associating alcoholism with the *A1* allele of the dopamine receptor gene (Horgan 1992). One problem in comparing different studies is the different criteria used to define alcoholism. In the initial 1990 study, DNA for analysis was taken from the brains of deceased persons. Thus, the diagnosis of alcoholism was made only after death, on the basis of medical and other treatment records. In other investigations the diagnosis of alcoholism is made on the basis of direct, standardized interviews of subjects. The age of onset and the severity of illness needed to define alcoholism may also vary from study to study. In addition, some researchers may exclude certain types of alcoholic individuals (opium users, those with a history of psychoses or certain organic diseases, etc.) that are included by other researchers. Also, differences exist in the way control groups are defined and collected.

In spite of these difficulties, genetic investigations of alcoholism are continuing and might cast light, not just on this serious disease, but also on broader aspects of dependencies, neurological disorders, and brain physiology.

1. The physical and behavioral differences between people are influenced by both genetic and environmental factors. The traits may be quantitative in nature, varying continuously over a broad range.

2. The heritability of a trait, in the broad sense, is the proportion of phenotypic variation that is due to all genetic differences. Narrow-sense heritability measures just the so-called additive portion of the genetic variability. Narrow-sense heritability is the best gauge of how effective selection can be in altering the average phenotype of a population over the course of generations.

3. A value for the heritability of a trait applies to the particular population in which it was measured and may not hold for the same population at other times or for other populations. A high heritability estimate in a specific population does not rule out the possibility of altering phenotypes by novel environmental changes.

4. The genetic component of quantitative variation involves the cumulative effects of several (or many) genes that are inherited according to Mendelian rules.

5. Monozygotic (MZ) twins are genetically identical, whereas dizygotic (DZ) twins are equivalent to sibs born at the same time. MZ twinning seems to result from random events of early development. On the other hand, DZ twinning tends to run in families and vary between racial groups.

6. Twin data provide only rough estimates of genetic and environmental contributions to quantitative and threshold traits. Environmental influences are suggested whenever MZ twins differ. Genetic influences are suggested whenever (like-sexed) DZ twins differ among themselves more than do MZ twins.

7. Variations in many human behavioral traits are influenced by genetic differences, although specific genes cannot always be identified and the pathways between the gene products and the behavioral phenotypes are usually obscure.

8. IQ tests were originally developed in France to identify schoolchildren who needed help, and IQ testing continues to be useful for that purpose. IQ tests are standardized by trial-and-error methods rather than by theories of intelligence. Mental abilities are probably too varied and complex to be adequately represented by one or a few scores.

9. Within white, middle-class populations (on whom most tests have been standardized), heritability of IQ, in the broad sense, is about 50%. These calculations are based on correlations in IQ between pairs of individuals of various degrees of relatedness and involve assumptions that may be only partly true.

10. Some or all of the difference in standardized IQ scores between blacks and whites could be due to environmental and cultural differences between races. Some could reflect genetic differences, but the evidence for this is weak.

11. Personality differences have substantial genetic and environmental components, the latter due primarily to influences outside the family environment.

12. Alcoholism also develops under the varied effects of both environmental and genetic factors. One genetic factor, the gene for dopamine receptors, may be important in predisposition to alcoholism.

biometry
concordant twins
discordant twins
dizygotic (DZ) twins
dopamine

dopamine receptor
dopamine receptor gene
heritability in the broad sense
heritability in the narrow sense

monozygotic (MZ) twins
multiple gene inheritance
neurotransmitter
polygenic inheritance
quantitative variation

threshold trait
variance

QUESTIONS

1. The list of coincidences that bonded the Jim twins (see beginning of chapter) becomes less weird when we recognize that the list is *highly selective*. What does this mean?

2. Try to think of a human trait that is *gene-independent*; that is, it varies among people solely as a result of different environmental influences.

3. The two selected groups of rats, maze-bright and maze-dull, depicted in Figure 8.2, did not diverge much more from each other after seven generations of selection. Why do you suppose that this was true?

4. Which population would tend to show the higher heritability for a particular trait: (a) One that is relatively homozygous for genes affecting the trait or one that is more heterozygous? (b) One that is in a relatively uniform environment for factors affecting the trait or one in a more heterogeneous environment?

5. Many of Galton's views on inheritance were vague or incorrect, but his law of filial regression is often observed for quantitative traits. Using weight as an example, this law states that offspring of very heavy parents are, on the average, not so heavy, and the offspring of very light persons are not so light; that is, a portion of the filial generation that might be expected to be extreme is closer to the population average. Assuming that weight is affected by both genetic and environmental factors, how would you explain filial regression?

6. In Table 8.1, what six genotypes have phenotype 60 (four uppercase alleles)?

7. Although the inheritance of eye color has not been worked out, it certainly involves more than one gene. Galton distinguished eight eye colors; by combining his two darkest categories, we can fit his scheme to the model of polygenic inheritance in Table 8.1:

Galton's classification	Assumed number of uppercase alleles
Light blue	0
Blue	1
Blue-green	2
Hazel	3
Light brown	4
Brown	5
Dark brown and black	6

(a) What are the genotypes of the lightest-eyed parents who could produce a child with brown eyes (five uppercase alleles)?
(b) Assume that anyone with zero, one, or two uppercase alleles is said to be blue-eyed and anyone with three or more uppercase alleles is said to be dark-eyed. (Eye color thus becomes a threshold trait.) What is the darkest-eyed child possible from two "blue-eyed" parents?

8. Under Davenport's model of skin color inheritance (Table 8.2), are these statements true or false?
(a) When one parent is white (zero uppercase alleles), the progeny can be no darker than the other parent.
(b) When one parent is light (one uppercase allele), the progeny can be no darker than the other parent.

9. It has been postulated that an egg and a polar body may very rarely be separately fertilized by two sperm to yield an unusual type of twin pair. Comment on the degree of similarity of these twins compared with regular DZ or MZ pairs. Assume that one sperm fertilized the mature egg and the second sperm fertilized the second polar body (sister to the egg). Also assume, for simplicity, no crossing over during meiosis.

10. Below are the Stanford-Binet IQ scores of the 19 pairs of MZ twins reared apart (Newman et al. 1937). The average of all the IQ scores is 95.7, and the average difference within pairs is 8.2 points. What conclusions are warranted?

106	95	91	85	101	90	93	127	102	116
105	94	90	84	99	88	89	122	96	109

102	88	115	97	78	92	106	96	116
94	79	105	85	66	77	89	77	92

11. *Very* low IQ can be a consequence of homozygosity for a single recessive allele (as in untreated PKU) or the result of a developmental birth defect. On the other hand, *moderately* low IQ reflects, to some extent, polygenic inheritance. Generally, the sibs of the severely retarded have higher IQ scores than the sibs of the moderately retarded. Explain.

12. Assume that in country A, all schools are *equally* bad, learning is *never* encouraged, and all other environmental factors relevant to taking IQ tests are *constantly* unfavorable. Thus, $V_E = 0$ with regard to IQ.

Assume that in country B, all schools are *equally* good, children are *impartially* encouraged in mental tasks, and all other environmental factors relevant to IQ are *constantly* good. Here, too, $V_E = 0$.

What is the heritability of IQ within A and B? Comment on the cause of the likely difference in IQ scores between the two countries.

13. Koshland (1987) writes that "the debate on nature and nurture in regard to behavior is basically over. Both are involved, and we are going to have to live with that complexity to make our society more humane for the individual and more civilized for the body politic." Do you agree or disagree?

FURTHER READING

A more detailed account of quantitative inheritance is included in Bodmer and Cavalli-Sforza (1976). A nontechnical account of behavioral genetics is provided by the Research and Education Association (1982), and a short summary is provided by Plomin (1990). The text by Anastasi (1988) provides background information on psychological testing, and Lord (1989) is an interesting summary on the twinning phenomenon.

Material on intelligence testing and its interpretation is voluminous, varied, and sometimes acrimonious. Sample both Jensen (1980, 1981) and his detractors (Gould 1981; Lewontin, Rose, and Kamin 1984). We also recommend the technical but clearly written article by Bouchard et al. (1990). Winn (1990) has written a popular article on some recent theories of intelligence, and Scarr (1987) includes a personal account of her involvement in a controversial field. Cloninger (1991) reviews recent genetic studies of alcoholism, and McHugh and McKusick (1991) edit a broad collection of articles on genetic studies of the brain.

Gene Mapping

Dr. Julia Bell, Fellow of the Royal College of Physicians and a pioneer in human genetics, retired from her research position at age 86 and died in 1979 at age 100. She had been a 21-year-old graduate student in mathematics when Mendel's work was rediscovered in 1900, and at age 41 she also earned a medical degree—all this at a time when few women entertained scientific careers. A perfectionist in gathering family data and a skilled statistician in analyzing the assembled pedigrees, she was well known for work on muscular dystrophy and hereditary diseases of the eye. Moreover, she and J. B. S. Haldane were the first to establish the chromosomal positions of two human genes with respect to each other, determining that the X-linked gene for color blindness was several "map units" from the gene for hemophilia. They said that it was important "to demonstrate that the principles of linkage which have been worked out for other animals also hold good for man" (Bell and Haldane 1937). Here we set out the basic ideas behind linkage analysis—also called **gene mapping**—and note why this sort of puzzle-solving activity is fundamental to other kinds of genetic research.

THE CONCEPT OF RECOMBINATION

In Chapter 4 we considered the joint inheritance of two independent genes. For example, consider the mating, $T/T\ U/U \times t/t\ u/u$, where the T locus (with allele T or t) and the U locus (with allele U or u) are on nonhomologous chromosomes. The gametes made by the first parent are all $\boxed{TU}$ and by the second parent all $\boxed{tu}$. The fertilization of the one gamete by the other leads to the doubly heterozygous offspring $T/t\ U/u$. Recall that Mendel's law of independent assortment predicts the following gametes from this **double heterozygote**:

  1/4 $\boxed{TU}$ 1/4 $\boxed{Tu}$ 1/4 $\boxed{tU}$ 1/4 $\boxed{tu}$

The first and last gametes from the double heterozygote, $\boxed{TU}$ and $\boxed{tu}$, are the same as the two gametes that formed the double heterozygote in the first place. They are called **parental gametes**. The two middle gametes $\boxed{Tu}$ and $\boxed{tU}$, however, contain new assortments of genes, different from those in the immediately preceding gamete generation; they are called **recombinant gametes**. This process is outlined in Figure 9.1, which shows two ways in which a double heterozygote can come about. In case 1, already presented here, the fertilizing gametes have either two uppercase

Figure 9.1

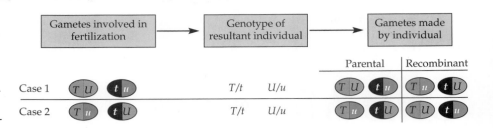

Figure 9.1
*The meaning of parental and recombinant gametes. Notice that the same doubly heterozygous genotype can be formed in two different ways. In either case 1 or case 2, the recombinant gametes made by the doubly heterozygous individual are **different** from the gametes at the preceding fertilization. As a consequence, the parental gametes from one case are the same as the recombinant gametes from the other. (In this figure, we have not distiguished between an egg and a sperm.)*

alleles or two lowercase alleles. In case 2, by contrast, each of the gametes giving rise to the double heterozygote carries one uppercase allele and one lowercase allele. The resulting double heterozygotes and the gametes they make are identical in the two cases, but the parental and recombinant types are reversed. The designation of a gamete as parental or recombinant depends only on how the nonallelic genes were combined in the first place.

The rest of this chapter deals with this question: What are the proportions of the four kinds of gametes made by a double heterozygote? As you already know, these proportions are 1/4 each in the case of independent genes—a total of 50% parental and 50% recombinant. But this result is not observed if the two genes happen to be relatively close to each other on the *same* chromosome.

CROSSING OVER

The example we have chosen to use here involves two loci relatively close to each other on the X chromosome, because X-linked genes are easier to deal with than are autosomal genes. Because homologous loci are absent from the Y, all the genes on a male's single X chromosome, including recessives, are directly expressed in his phenotype. Furthermore, the precise genetic makeup of a female's two X chromosomes can usually be inferred from the phenotypes of her father and sons. For its historical significance, we consider the two X-linked genes studied by Bell and Haldane:

G = dominant allele for normal color vision
g = recessive allele for green-shift (deuteranomaly)

H = dominant allele for normal blood clotting
h = recessive allele for hemophilia A (classical type)

First, consider the genotypes of two females, Jan and Fay, in Figure 9.2. Jan's father, who was both green-shifted and hemophilic, transmitted his X chromosome with *g* and *h* to his daughter. Because the pedigree shows that Jan has a normal phenotype, she must carry both a dominant *G* gene and a dominant *H* gene on her other X chromosome, the one received from her mother. Her two X chromosomes can be represented as follows:

$$\text{genotype of Jan} = \begin{array}{cc} g & h \\ \rule{3cm}{0.4pt} \\ \rule{3cm}{0.4pt} \\ G & H \end{array} \begin{array}{l} \text{(paternal X)} \\ \text{(maternal X)} \end{array}$$

She is doubly heterozygous, with the two nonallelic recessive genes, *g* and *h*, on the same chromosome; this is often written *g h/G H*.

On the other hand, Fay received a paternal X chromosome carrying g and H. Being phenotypically normal, Fay must have G on her other, maternal X chromosome. This chromosome must also carry the hemophilia gene, h, since Fay transmitted h to her son. Therefore,

$$\text{genotype of Fay} = \frac{\underline{g \qquad H}}{\underline{G \qquad h}} \quad \begin{array}{l}\text{(paternal X)}\\\text{(maternal X)}\end{array}$$

She, too, is doubly heterozygous, but the two nonallelic recessive genes, g and h, are on "opposite" homologous chromosomes ($g\,H/G\,h$).

These two females differ in what is called the **phase** of the double heterozygote. The term **coupling** (or *cis*) is sometimes used to refer to the two recessive nonallelic genes when they are on the same member of a homologous pair, as seen in Jan's genotype. The term **repulsion** (or *trans*) may be used when they are on opposite members, as seen in Fay's genotype. The terms derive from the work of Punnett (of checkerboard fame) and others who tried unsuccessfully to explain departures from Mendel's independent assortment by elaborate systems of attractions and repulsions. (*Note*: Codominant alleles do not fit this system of nomenclature.)

The Proportions of the Gametes

As a result of the work of Bell, Haldane, and others, a number of females like those in Figure 9.2 have been identified, and the phenotypes of their sons have been tallied. A completely normal son from a woman like Jan must have inherited his mother's X chromosome carrying the genes G and H; a doubly affected son must have inherited her other X chromosome, the one carrying the genes g and h. Both her $\underline{G\,H}$ and $\underline{g\,h}$ sons are derived from parental gametes. These are not the only types of sons possible from the double heterozygote in coupling phase, however. Two other types, involving combinations of the mother's X chromosomes, are shown in the left-hand column of Figure 9.3. The two additional types of sons are (1) those with green-shift and normal blood clotting and (2) those with normal color vision and hemophilia. They arise from gametes carrying X chromosomes represented by $\underline{g\,H}$ and $\underline{G\,h}$, as shown at the bottom left of the figure. These are *recombinant* gametes, since the alleles of the two genes have recombined to form a "new" assortment. The two recombinant gametes in this case are much less frequent (1.5% each, 3% total) than the two parental gametes (48.5% each, 97% total).

Figure 9.3 also shows the gametes produced by the female whose recessive genes are in repulsion. The percentage of recombination is the same (3% total), but the assortment of genes in a recombinant gamete is different. A recombinant gamete from a repulsion phase is a parental gamete from a coupling phase, and vice versa. The percentage of recombination is a function of the loci involved, not of the particular alleles that happen to be at those loci.

In summary, the genes for green-shift and for hemophilia A are seen to exhibit 3% recombination, a figure derived from data on the phenotypes of sons from doubly heterozygous females. Had we chosen other genes to investigate, the percentage recombination would probably be some other value. It will be demonstrated in the following sections that the percentage recombination depends on the physical distance separating the two genes along the length of the chromosome. Genes on the same chromosome that are fairly close to each other are said to be **linked**. Although

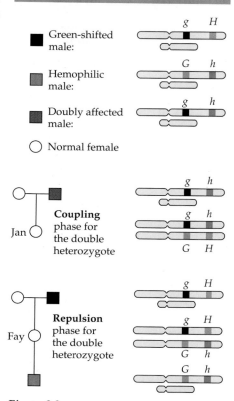

Figure 9.2

Pedigrees showing the two ways a person can be a double heterozygote for linked genes. Jan has the two recessive (nonallelic) genes, g and h, on one of her X chromosomes. Fay has the two recessive genes on "opposite" X chromosomes. The Y chromosome does not carry the genes in question.

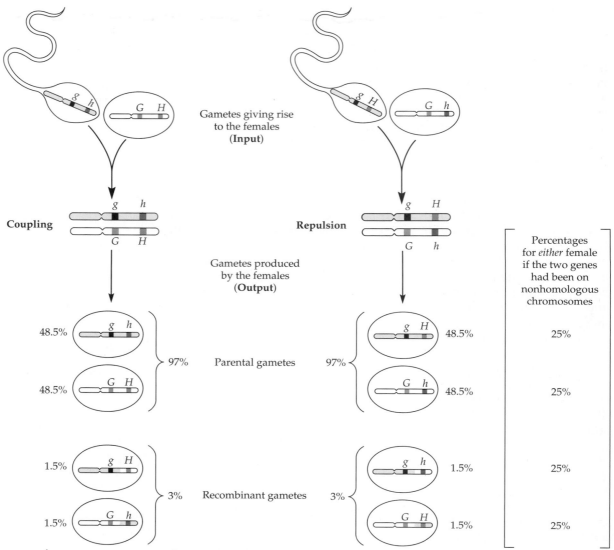

Figure 9.3
The approximate percentages of different gamete types made by females who are doubly heterozygous for green-shift and hemophilia A. The genetic constitutions of the gametes produced by the females were inferred from the phenotypes of their sons, who are hemizygous for X-linked genes. The two types of females correspond to Jan and Fay in Figure 9.2.

genes physically located on the same chromosome tend to be inherited together, they can recombine with their alleles on the homologous chromosome by the remarkable process described next.

The Physical Basis of Recombination

Genes nearby on the same chromosome are connected to each other like two knots spaced along a rope, or two links in a chain, or two rings on a raccoon's tail. The genes (or knots, or links, or rings) need not be next to each other; they are still physically bound together through the intervening sections. The problem then becomes: Why are linked genes not *always* inherited together? How do genes on one chromosome ever recombine with the genes on the homologous chromosome? How can the percentage of recombination be a number other than 0% (linked genes) or 50% (independent genes)?

The solution to this problem came with microscopic observations, made early this century, of interconnections between synapsed chromosomes during meiosis. As we noted in Chapter 3, these **chiasmata** (singular, chiasma) form "swaps" between homologous chromosomes, resulting in cross-shaped configurations at points along the chromosomes at the tetrad stage of meiosis I. As shown in Figure 9.4, cytologists can see chiasmata in stained meiotic cells from different organisms. Note that there may be several chiasmata per tetrad. Unfortunately, they are difficult to make out clearly in the meiotic cells of most mammals, including humans, partly because these chromosomes tend to be small. Very little information is available on human females, but in human males the tetrads of auto-

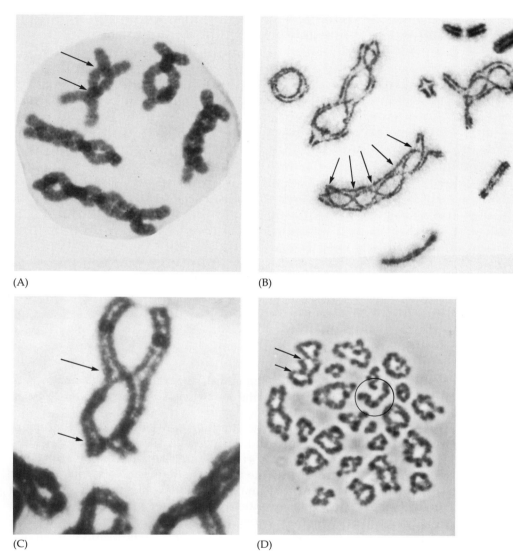

(A)

(B)

(C)

(D)

Figure 9.4

Photographs of chiasmata. These cross-shaped configurations (some indicated by arrows) are seen in late prophase of the first meiotic division, when homologous chromosomes start to move apart from each other. (A) During pollen formation in the trillium plant. (B) During sperm formation in a grasshopper; five chiasmata can be seen in one tetrad. (C) During sperm formation in a salamander; the centromeres and interchange points show particularly clearly. (D) During sperm formation in a man; the circled X and Y chromosomes appear to be synapsed end to end. (A, photograph by R. F. Smith; courtesy of A. H. Sparrow, Brookhaven National Laboratory. B, courtesy of Bernard John, The Australian National University. C, courtesy of James Kezer, University of Oregon. D, courtesy of Paul Polani, Guy's Hospital, London.)

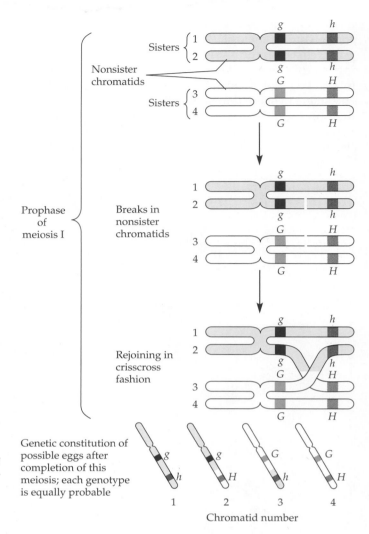

Figure 9.5

Diagram of the occurrence and consequence of a single crossover along the length of the X chromosome that joins the loci for green-shift and hemophilia. We assume that the woman is doubly heterozygous in coupling. Although the crossover is shown as occurring between chromatids 2 and 3, the same results come from a crossover between strands 1 and 3, or between 1 and 4, or between 2 and 4.

somes show from one to four (sometimes five, rarely six) chiasmata, with an average of about two chiasmata per chromosome pair. The XY pair in human males seems to have none.

Whatever the organism, each chiasma represents a place where **nonsister chromatids** have broken and rejoined in crisscross fashion, as shown in Figure 9.5. This exchange of corresponding segments of nonsister chromatids is called a *crossover*, and the resulting strands are known as *crossover chromatids* or *crossover strands*. The process itself, called **crossing over**, is a normal, relatively frequent event. It occurs early in meiosis, before the chromosomes are condensed enough for microscopists to make out individual chiasmata. The idea that alleles of linked genes can be separated from each other in this fashion was due to Thomas Hunt Morgan, one of the first geneticists to win a Nobel prize (Box A). More recent research has begun to unravel the mechanism of the amazingly exact exchange of chromatid parts at the level of DNA molecules.

Figure 9.5 depicts crossing over and its genetic consequences in a woman who is doubly heterozygous for green-shift and hemophilia in the coupling phase. Strands labeled 1 and 2 are sister chromatids and exact replicas of each other; similarly, strands 3 and 4 are sister chromatids and exact replicas of each other. For the sake of clarity, the crossover is shown as occurring between nonsisters 2 and 3, but a 1-3 crossover or a 1-4 or 2-

MORGAN'S FLY ROOM

Thomas Hunt Morgan (1866–1945) was born in Kentucky of an aristocratic family that included Confederate General John Hunt Morgan (uncle), composer Francis Scott Key (great-grandfather), and financier J. P. Morgan (remote cousin). From 1904 to 1928, as Professor of Experimental Zoology at Columbia University, he assembled a remarkable group of scientists who crowded together in a small space that was affectionately dubbed the Fly Room.

Early on, Morgan invited two undergraduates, Calvin Bridges and Alfred Sturtevant, to join the Fly Room; later, a graduate student named H. J. Muller joined the group. The lab atmosphere was remarkably relaxed, with students treated as equals. The four of them (and visiting colleagues) discovered many of the details of basic genetics: multiple alleles, sex-linked genes, crossing over and mapping, recessive lethal genes, sex determination, chromosomal aberrations, and nondisjunction. They established the fruit fly, *Drosophila*, as the premier organism for genetic research and wrote the first authoritative textbook of genetics, *The Mechanism of Mendelian Heredity* (Morgan et al. 1915).

The physical setting of the

laboratory—with its haphazard and make-do simplicity—was unconventional. The culture medium, which was poured into a miscellaneous assortment of glass milk bottles, contained mashed bananas.

The resulting smell was fierce and drew constant complaints from the rest of the biology department. . . . Near the entrance of the room a stalk of bananas hung conspicuously,

serving as a center of attraction for the numerous fruit flies that had escaped from their milk bottles. . . . Morgan was by nature a careless and sloppy man, but he also delighted in shocking others, in playing the imp. . . . More than once Morgan was mistaken for the janitor. (Shine and Wrobel 1976)

Describing the work in the Fly Room, Sturtevant (1959) wrote:

Each carried on his own experiments, but each knew exactly what the others were doing, and each new result was freely discussed. There was little attention paid to priority or to the source of new ideas or new interpretations. What mattered was to get ahead with the work. . . . There can have been few times and places in scientific laboratories with such an atmosphere of excitement. . . . This was due in large part to Morgan's own attitude, compounded of enthusiasm combined with a strong critical sense, generosity, open-mindedness, and a remarkable sense of humor.

Morgan was awarded the 1933 Nobel prize in physiology or medicine, but he did not attend the regular award ceremony in Stockholm. Perhaps he was too busy, or perhaps he just disliked fancy-dress occasions.

---

4 crossover (each a nonsister pairing) is equally likely and gives the same results. (In real life all four chromatids are touching one another in three dimensions rather than lying flat on a book page.) As a consequence of the crossover, the future egg may contain, on its X chromosome, the *parental* associations of genes, $g\,h$ or $\underline{G\,H}$, or the *recombinant* associations, $g\,\underline{H}$ or $\underline{G}\,h$. The choice of which of these four chromosomes is included in the functional egg is random; the other three chromosomes end up in polar bodies.

Three aspects of crossing over need to be mentioned. The first points up the difference between the terms *crossing over* and *recombination*. If crossing over occurs as depicted in Figure 9.5 but in a person who is *not* a double heterozygote, then crossing over cannot lead to recombination. Crossing over is independent of what particular alleles occupy the genetic loci, but the *detection* of a particular crossover requires that the exchange occur between loci that are both heterozygous. You can verify that nothing genetically different from parental associations can result when a person is homozygous (*g/g* or *G/G* or *h/h* or *H/H*); this nonrecombination is true whether a crossover occurs between the *G* locus and the *H* locus or anyplace else.

A second consideration is that chiasmata are seen to occur fairly randomly along the length of chromosomes. In a given meiosis, a crossover and the resultant chiasma may or may not occur in the length of chromosome between the locus for green-shift and that for hemophilia A. If a crossover does *not* occur in the region, then all products of this meiosis must necessarily be parental with respect to the *G* and *H* loci. A crossover to the left of the *G* locus or to the right of the *H* locus cannot recombine alleles of *G* with those of *H*.

The third point follows from the second. If crossing over is random along a chromosome, then the likelihood of recombination between two spots depends on the length of chromosome between them. That is, the closer together the two genes are, the less likely it is that a random crossover will occur somewhere within the *specific segment* of chromosome separating them. At the limit of closeness, the genes are adjacent on the chromosome, there is virtually no opportunity for an intervening crossover, and recombination is zero.

For two genes that are neither very close together nor very far apart on the same chromosome, the percentage recombination depends on their separation and falls between 0% and 50%. The genius of Morgan and his students, especially Alfred H. Sturtevant, was to equate percentage recombination with physical distance along the chromosome. A 3% recombination, as in the particular case of green-shift and hemophilia A, is taken to mean that the loci are 3 map units apart. More generally, we simply define:

x % recombination = x map units

That this statement is a logical interpretation of the genetic system has been amply verified by extensive mapping of genes in many species.*

When genes are very far apart, two or more crossovers may occur along the length of the chromosome between them. Remarkably enough, the effect of several crossovers is not different from the effect of one crossover, because the two strands involved in a second crossover are *random* with respect to a first crossover. A second (or third) crossover is as likely to "undo" the recombination resulting from an initial crossover as it is to make more recombination. Because more than one crossover has no additional effect on the percentage recombination, we can accurately measure only those distances within which multiple crossing over is negligible. In practice, this means restricting our attention to genes that are less than about 20 or 25 map units apart.

MAPPING MANY GENES ON THE X CHROMOSOME

All the genes on a chromosome are attached in series, like successive kilometer sections on a highway. It is a very long and narrow chromosomal highway: If the DNA double helix of a human X chromosome were magnified to stretch the 5,000 km from New York to Los Angeles, it would be about 10 cm (4 inches) wide. In theory, each gene can be mapped with respect to its near neighbors, and those neighbors with respect to additional neighbors, to yield the relative positions of the perhaps 2,000 X-linked genes. Like distances along a highway, the distances along the chromosome are also additive. Consider a chromosome with genes *a*, *b*, and *c*, *in that order*. If genes *a* and *b* are found to be 4 map units apart, and *b* and *c* are 8 map units apart, then *a* and *c* will be 12 map units apart.

---

* Many researchers call the unit of distance a *centimorgan* in honor of T. H. Morgan. We will, however, use the original straightforward term *map unit*; 1 map unit = 1 centimorgan.

Adding Red-Shift to the Map

We have seen that the distance between *g* (green-shift) and *h* (hemophilia A) is 3 map units. By finding mothers who are heterozygous for one of these two genes and also heterozygous for a third nearby gene, we can extend the map left or right. The pedigree in Figure 9.6 is a clear-cut example involving the joint inheritance of both green-shift (*g*) and red-shift (*r*) in a family from Sardinia. Recall that these two traits are separable entities leading to somewhat different defects in color vision (Chapter 5).

You can write out for yourselves the X-linked genotypes for all the males. For example, I-1 carries the X chromosome *G r*, and he transmitted this to his daughter, II-2. She received from her mother the X chromosome *g R*, a conclusion we can draw by noting that two of her brothers (II-3 and II-4) also carry *g R*. Thus, female II-2 is doubly heterozygous in repulsion.

Each of the three sons of II-2 has a different phenotype. Son III-2, the oldest, is both green-shifted and red-shifted; III-4, the middle son, is green-shifted only; III-5, the youngest, is red-shifted only. The oldest son's phenotype reflects recombination between the two color vision loci, whereas his younger brothers' phenotypes represent the two parental (noncrossover) chromosomes. Thus, in this particular sibship, we observe 33% recombination. (The phenotypes of the two daughters are uninformative.) Combined with all other pedigrees that have provided information, however, the best estimate for the distance between the two color vision genes turns out to be about 7 map units.

Mapping studies that combine the information from many families are complicated, and we omit the details. The usual type of analysis, called the *lod score method*, was invented by the mathematical geneticist Newton Morton. The term *lod* stands for "*lo*garithm of the o*d*ds favoring linkage." Researchers make a ratio of the logarithms of two probabilities for the inheritance patterns. The numerator assumes linkage between the genes; the denominator assumes the genes are independent. This likelihood ratio must be greater than 3 to provide some evidence for linkage. Fortunately, the calculations are now handled largely by computer programs.

The next step in building up our map of the X chromosome is to decide the relative positions of the three genes *g*, *h*, and *r*. Knowing just the *g–h* distance (3 map units) and the *g–r* distance (7 map units) does not provide a unique map; two alternatives are possible:

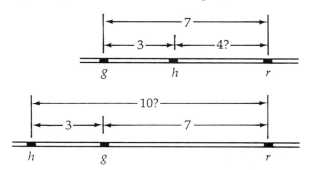

To choose between the two possibilities, we need information about the *h–r* distance directly. Is it 4 map units (the difference between 7 and 3) or 10 map units (the sum of 7 and 3)? Not enough families have been found to provide reliable information directly; such pedigrees would have to include sons from women doubly heterozygous for hemophilia A and red-shift. Nevertheless, each of these genes has been mapped with respect to additional loci in the immediate neighborhood, and it is likely that the upper alternative is the correct one.

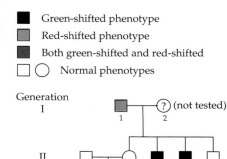

■ Green-shifted phenotype
▨ Red-shifted phenotype
■ Both green-shifted and red-shifted
□ ○ Normal phenotypes

Figure 9.6
A pedigree from Sardinia showing the joint inheritance of green-shift and red-shift controlled by linked genes on the X chromosome. (Data from Siniscalco, Filippi, and Latte 1964.)

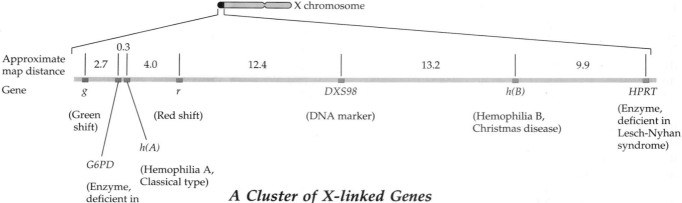

Figure 9.7
The approximate map of some gene loci near the tip of the long arm of the X chromosome. In addition to two loci that affect color vision (g and r) and two loci that affect blood clotting h(A) and h(B), the map includes three other loci. G6PD codes for an enzyme (glucose-6-phosphate dehydrogenase) that is present in most cells, but especially in red blood cells. A deficiency of this enzyme occurs in relatively high frequency in some Mediterranean, Asian, and African people. Although the deficiency produces a severe anemia whenever these individuals eat fava beans (favism) or are treated with a malarial drug (primaquine sensitivity), they are not usually adversely affected otherwise. DXS98 is one of about 1,000 DNA markers on the X chromosome, which are easily mapped by laboratory tests and which are not usually associated with any phenotype (Chapter 13). HPRT codes for an enzyme (hypoxanthine-guanine phosphoribosyltransferase). This enzyme is deficient in boys with the Lesch-Nyhan syndrome (Chapter 14). (Data largely from Keats, Ott, and Conneally 1989.)

A Cluster of X-linked Genes

Some additional genes in the g–h–r region that have been described and mapped near the tip of the long arm of the X chromosome are depicted in Figure 9.7. Note that two separate genes can lead to hemophilia. Hemophilia A or classical hemophilia due to h(A) (previously designated h) and hemophilia B or Christmas disease due to h(B). These map distances are not known with great accuracy, however, because they are based on limited family studies. Although derived by sophisticated statistical methods, the results can be no better than the raw data on which they are based. Another gene of interest in this region is the so-called *fragile X* site that predisposes to mental retardation (Chapter 11). Other clusters of X-linked genes and autosomal genes are also known.

Further Considerations

We noted that long distances along a chromosome cannot be measured directly because of inaccuracies introduced by multiple crossing over. But long distances can be calculated by adding together the shorter distances between intervening genes. In the map in Figure 9.7, for example, gene loci r and h(B) are 25.6 map units apart. If we wanted to predict the types of sons from women who are, say, of the genotype r h(B)/R H(B), we could make use of the knowledge that these loci recombine 25.6% of the time. Thus, the eggs made by these women would have the following approximate genetic compositions:

25.6% recombinant eggs { 12.8% (r H(B))
{ 12.8% (R h(B))

74.4% parental eggs { 37.2% (r h(B))
{ 37.2% (R H(B))

Note that the two recombinant eggs add up to 25.6% and are expected to be equally frequent (12.8% each), and the two parental eggs add up to the remainder, 74.4%, and are expected to be equally frequent (37.2% each). The phenotypes of their sons with regard to the red-shift phenotype and the hemophilia B phenotype together are predictable from the frequencies of these egg types. (Can you give their phenotypes?) These percentages are, of course, very different from the expectations we would have if the red-shift gene and hemophilia B gene were on nonhomologous chromosomes (25% each).

The DNA double helix within the X chromosome, if stretched out, is estimated to be about 10 cm long. To measure the length—not in physical units but in recombinational map units—would require that we know genes at opposite ends of the X chromosome and other genes at short intervals along the entire length. This array is not currently available, but a guess for the genetic length of the X chromosome can be based on the counts of chiasmata in meiotic cells, each chiasma representing 50 map units. Unfortunately, the meiotic cells of females are difficult to observe, and males do not have two X chromosomes between which chiasmata may form. Investigators can guess that the chiasmata frequency in the female X chromosome is similar to that in male autosomes. This procedure is prone to error, however, because the frequency of crossing over in females is usually greater than that in males—by about 50% in some regions of chromosome 1. (Such sex-related differences in recombination rates are common, but poorly understood. In male fruit flies, for example, there is no crossing over at all!) For these reasons, a guess of about 250 map units for the recombinational length of the X chromosome is tentative.

SIGNIFICANCE OF GENE MAPPING

> There is a certain satisfaction in knowing where a piece of a puzzle fits and what its relationship is to neighboring pieces. So it is with the "puzzle" of how the vast amount of genetic information is packaged and distributed throughout the genome. (Conneally and Rivas 1980).

The construction of the human gene map can, indeed, be likened to the assembly of a jigsaw puzzle. Initially there is disorder. Then, just as border pieces are separated from inside pieces, the X-linked genes are distinguished from autosomal genes and arranged in order wherever possible. Just as small groups of connected puzzle pieces are centers for adding others, previously mapped clusters of genes acquire new members, or two previously separated clusters on the same chromosome are joined by the discovery of intervening genes. Puzzle solvers may also find that a previously positioned piece—or gene—has to be moved to another site because it did not fit exactly.

Beyond the delight that some take in mapping for its own sake, however, there are practical applications in genetic counseling and medicine. Locating a gene is often a prerequisite to isolating and cloning it, which in turn can lead to information about the encoded protein. Knowledge of a human disease gene and its protein product brings us that much closer to early diagnosis and treatment. Similar kinds of research on crops and farm animals can also lead to greater agricultural productivity. In addition, the evolution of the genetic architecture in more or less closely related organisms can only be understood by dissecting and analyzing its component parts.

Building and Using Gene Maps

It is tedious to study human linkage by searching for recombination in pedigrees with some doubly heterozygous individuals. Some of the tedium is now being alleviated, however. In the 1970s, for example, entirely new methods of locating genes based on *somatic cell genetics* eliminated the need for any kind of family data. Instead, cells from various sources could be grown and manipulated in laboratory cultures. Such studies yield a different type of information: They assign a gene to a particular chromosome or, often, to a specific segment of a chromosome. If two genes are assigned to the same small region, linkage is established or at least suggested.

The 1980s contributed several additional mapping aids that involve

Table 9.1
Some genes mapped to chromosome 1 (with positions shown in Figure 9.8).

Standardized Gene Symbol	Name of Gene
APOA2	Apolipoprotein A-II
AT3	Antithrombin III
D1Sxx	DNA markers on chromosome 1 (*xx* is a number)
F13B	Clotting factor XIII (B polypeptide)
FUCA1	Fucosidase (α-L-1)
FY	Duffy blood group
GDH	Glucose dehydrogenase
HF	Complement component H
NGFB	Nerve growth factor (β polypeptide)
PGM1	Phosphoglucomutase-1
REN	Renin
RH	Rhesus blood group
UMPK	Uridine monophosphate kinase

manipulations of nucleic acids. It is now possible, by chemically analyzing a person's DNA, to determine heterozygosity for nucleotide sequences called **DNA marker sites.** Although no change in one's phenotype is necessarily associated with a change in a DNA marker, the heterozygous locus can nonetheless be used as a Mendelian site in family linkage studies. Thus, the newer somatic cell and molecular technologies, which we discuss in detail in Chapter 13, have come to augment and complement the older mapping methods. About 2,500 "standard" genes (and over 10,000 DNA marker sites) have been mapped by one means or another. Thus, if humans possess, say, 50,000 standard genes, researchers have mapped just 5% of them.

Exceedingly detailed descriptions of human gene maps can be found in scientific journals that publish proceedings of international human gene mapping workshops held annually at various places around the world. Recent information about chromosome 1, for example, includes about 250 standard genes and about 450 DNA markers. A small sampling of these genes and marker sites is presented in Table 9.1. By convention, each gene symbol consists of one to five characters—capital letters or Arabic numbers. Although a few of the genes in Table 9.1 may be familiar to you (e.g., the rhesus blood group gene), most of them will be new to you. Some of the genes code for cellular enzymes (often identified by a name ending in -ase). Others are blood proteins involved in clotting (antithrombin III and clotting factor XIII), transport of cholesterol (apolipoprotein A-II), or immune reactions (complement component H).

A map of recombinational distances among these chromosome 1 loci is presented in Figure 9.8, which includes the numbered dark and light chromosome bands and corresponding regional assignments. We also include several of the DNA markers on chromosome 1 (labeled D1S followed by a number). There are no inconsistencies between the order of genes determined by recombinational mapping (left column) and the regional assignments determined by somatic cell methods. Thus, the genetic map is said to be *colinear* with the cytological map.

Here are three ways in which this type of linkage information can be useful in genetic counseling:

1. To test fetuses in early pregnancy to determine whether they are affected with a genetic disorder. This can be done by analyzing amniotic fluid cells for the products of the alleles of an innocu-

ous gene linked to a disease gene (see also Chapter 20). For example, the dominant gene for *myotonic dystrophy*, a severe muscle-wasting disease, is closely linked on chromosome 19 to the so-called *secretor* locus, which has no medical significance. Until recently, myotonic dystrophy could not be detected prenatally, but the phenotypes associated with the secretor locus can. In some pedigrees the secretor status of the fetus can reveal with high probability (but not certainty) the presence of the gene for myotonic dystrophy.

2. To test persons at risk for an autosomal dominant disorder before the time when symptoms of the disease usually appear, to see whether the person has the gene. This information—good or bad—might affect personal decisions regarding treatment and childbearing. *Huntington disease* is a severely debilitating condition that afflicted folksinger Woody Guthrie. Combining extensive pedigree studies with DNA analyses, investigators showed in 1983 that the Huntington gene, although itself not (yet) detectable, is linked to a particular restriction enzyme site that can be identified by molecular techniques. (In 1993, researchers identified the Huntington gene itself.)

3. To test the normal sibs of persons affected with an autosomal recessive disorder to determine whether they are carriers. Here, too, the test results might affect reproductive decisions.

Gene mapping can also help detect or explain multiple causes of the same or similar genetic diseases. For example, abnormal blood clotting occurs for many different reasons related to mutant genes at different X-linked and autosomal loci. Proper therapy requires knowledge of the exact cause of the defect. Although both the hemophilia A (classic) and hemophilia B (Christmas disease) genes are X-linked, they map to distinctly different positions. (Note that Bell and Haldane did not distinguish between hemophilia A and B, nor did they distinguish between green-shift and red-shift. Their value of about 5 map units is thus sort of an average between a hemophilia locus and a color vision locus.)

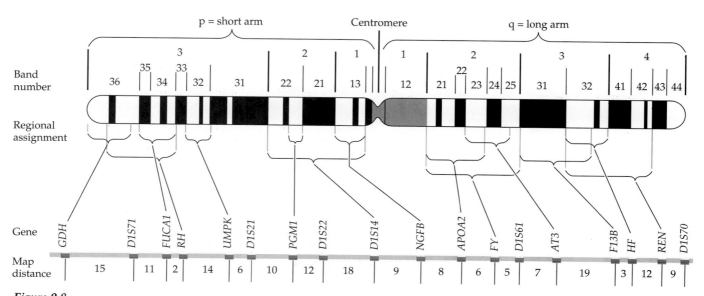

Figure 9.8
Some genes and DNA markers (designated D1S) on chromosome 1, showing both approximate recombinational map distances for males and regional assignments to the cytological map. The genes on this map are described in Table 9.1. (Data mostly from Keats, Ott, and Conneally 1989; Keats, Sherman and Ott 1990.)

203

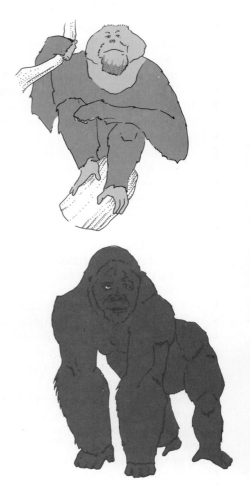

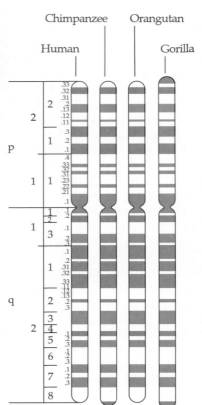

Figure 9.9
Comparison of X chromosomes from four primates: the human, chimpanzee, gorilla, and orangutan. The diagrams of the four X's represent Giemsa-stained chromosomes, not from metaphase, but from late prophase, when about 1,000 bands per haploid set can be resolved. The similarities in banding for the X chromosome are striking, and the similarities for autosomes (not shown here) are also remarkable. (Chromosomes from Yunis and Prakash 1982.)

Gene Mapping and Evolution

Comparative genome mapping is an active and exciting area of research. A great deal has been learned about the evolutionary histories of related species by looking at their chromosomes and gene maps. For example, for humans and mice, researchers have discovered about 50 similar chromosomal segments that contain about 390 homologous genes—350 autosomal and 40 X-linked. The more closely related two species are, the more similar their chromosomes and gene maps are expected to be. On these principles, a variety of evolutionary trees have been constructed.

The study of primate chromosomes is a case in point. Most similar to the human karyotype are those of the chimpanzee, gorilla, and orangutan, all having 48 chromosomes. Comparisons of banding patterns among these four species reveal virtually identical X chromosomes and remarkable similarity of the autosomes as well (Figure 9.9). Only the Y chromosomes vary considerably from species to species.

Chromosome 2 of humans seems to represent a fusion of two ancestral acrocentric autosomes, a conjecture that explains our having 22 rather than 23 autosomes per haploid set. Other differences among the four species can be accounted for by inverted regions within chromosomes and changes in the regions near centromeres. This approach suggests that humans are related most closely to the chimpanzee (13 seemingly identical chromosomes), and then to the gorilla (9 close matches) and orangutan (8 close matches). The gibbon is a much more distant relative; except for the X chromosome, there is very little similarity between its chromosomes and those of the other four primates (Figure 9.10). The same evolutionary relationships are obtained when we compare DNA sequences by molecular techniques (Lewin 1984). The precise branching of the evolutionary tree of the great apes and humans is, however, still in dispute, especially the question of whether humans are more closely allied to chimpanzees or to gorillas.

Considerable progress has been made in mapping primate genes, especially those of the chimpanzee. The production by humans and chimpanzees of hemoglobin chains and some other polypeptides that are identical means that the structural genes for these proteins must also be virtually identical. So similar are the karyotypes, gene maps, and proteins of humans and chimpanzees that it has been proposed that the obvious phenotypic differences between the two species must reside in differences in their genetic regulatory mechanisms rather than in the structural genes themselves (King and Wilson 1975). Even Down syndrome traits have been described in chimpanzees, associated with trisomy of a small acrocentric chromosome that is the counterpart of human chromosome 21.

The X chromosome, in particular, is remarkably constant in size, banding patterns, and gene content, not only among primates, but also among all mammals that have been studied. For example, the *G6PD* gene is X-linked in most mammals, including (but not limited to) the rat, mouse, cat, dog, cow, sheep, pig, mink, rabbit, kangaroo, donkey, and horse, as well as in seven primates that have been surveyed. Why the X chromosome has been preserved intact throughout mammalian evolution is an intriguing question. In Chapter 7, we saw that inactivation of one of the X chromosomes in the cells of females occurs in *all* mammalian species. This mechanism compensates for the double dose of X-linked genes in the female compared to the male. Susumu Ohno (1969), of the City of Hope Medical Center in California, proposes that once X inactivation evolved early in the mammalian line, genes on the mammalian X were stuck there. Transfer of a part of the X to an autosome would remove the genes on the transferred piece from inactivation, and the consequent dosage change would result in a serious disadvantage to the organism.

We will return to the subject of mapping in Chapter 13, after introducing the methods for detecting and analyzing DNA markers and for assigning a gene to a specific segment of a chromosome. There we discuss what is known as the Human Genome Project, a long-term, "big-ticket" scheme to locate sites (either standard genes or DNA markers) every 1 map unit along all human chromosomes and to determine the sequence of every nucleotide between these sites. Since the total length of the 23 pairs of human chromosomes is about 3,300 map units, and each map unit is about 1 million nucleotides long, the project will seek to determine the order of 3 billion nucleotide units.

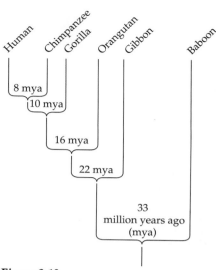

Figure 9.10
An evolutionary tree of humans and closely related primates based on an analysis of DNA. The numbers at branch points are estimates of the time of divergence in millions of years ago. (Based on data of Sibley and Ahlquist 1984.)

SUMMARY

1. For two genes on *different* chromosomes, Mendel's law of independent assortment applies (Chapter 4). For an individual heterozygous for two such genes, four sorts of gametes are expected in equal proportions: 50% parental and 50% recombinant. A parental gamete has the same genetic constitution as one of the gametes that formed the double heterozygote. A recombinant gamete has a new assortment of the two genes.

2. For two genes nearby on the *same* chromosome, a double heterozygote can still make four types of gametes, but their proportions will not be 1/4 each. The proportions of parental and recombinant gametes from such linked genes will depend on (a) the physical distance separating the two genes along the chromosome and (b) whether the double heterozygote is in the coupling or the repulsion phase.

3. The further apart two linked genes are, the greater is the opportunity for recombination due to crossing-over. Microscopically visible chiasmata represent the positions of the crossing-over events, exact exchanges of pieces of nonsister chromatids. Crossing over provides a measure of physical distance along a chromosome. This equivalence is reasonably accurate as long as recombination is under about 25%, in which case x% recombination is defined as x map units.

4. We can map genes on the human X chromosome by looking at the phenotypes of the sons of doubly heterozygous mothers. The phenotype of each son reflects the genetic constitution of the egg from which he arose. Investigators have identified several clusters of X-linked genes, one including two color vision and two hemophilia loci.

5. Chromosomes can be mapped in recombinational units or located to specific stained bands by family linkage studies or by somatic cell and molecular techniques (Chapter 13). Mapping information provides fundamental knowledge of genome organization and has practical applications in genetic counseling, medicine, and agriculture.

6. Comparisons of mammalian gene locations and chromosome banding patterns provide information on evolutionary sequences.

KEY TERMS

chiasma(ta)	double heterozygote	map unit	recombinant gamete
coupling	gene mapping	nonsister chromatids	repulsion
crossing over	linked genes	parental gamete	sister chromatids
DNA marker site	lod score method	percent recombination	

QUESTIONS

These questions are arranged roughly according to difficulty. Using the right terminology for linked genes where appropriate will take you a long way toward the solutions.

1. What kinds of gametes, and in what proportions, can be made by a person who develops from an egg carrying genes A and B and a sperm carrying a and b, if the two loci are (a) on different (nonhomologous) chromosomes, (b) on the same chromosome adjacent to each other? What is the percentage recombination in each case?

2. What kinds of gametes, and in what proportion, can be made by a double heterozygote in coupling ($A B/a b$) if the A and B loci are (a) 2 map units apart, (b) 8 map units apart, (c) 22 map units apart?

3. Why is a double heterozygote needed to map two genes on the same chromosome by observations on recombination?

4. Assume that the dominant genes E and H are 18 map units apart on chromosome 2. What progeny phenotypes

(dominant or recessive for E and for H) are expected from the following matings?

(a) $\underline{E\ H}/e\ h \times \underline{e\ h}/e\ h$
(b) $\underline{E\ h}/e\ H \times \underline{e\ h}/e\ h$

5. Assume two X-linked genes, with A dominant to a and B dominant to b. Max had both recessive traits and fathered many daughters, all with both dominant traits. These daughters, in turn, had a total of 40 sons (Max's grandsons): 15 had both recessive traits, 17 had both dominant traits, and 8 had the A dominant trait but not B, or vice versa. What is the genotype of Max's daughters? How far apart are the two genes?

6. A female with normal color vision is heterozygous in coupling for the X-linked genes for green-shift (G/g) and also for red-shift (R/r). The loci are about 7 map units apart. What percentage of her sons are expected to have normal color vision?

7. A husband and wife have the following genotypes with respect to green-shift and red-shift: $g\,r/G\,R \times g\,r/(Y)$. What percentage of their daughter's sons are expected to have normal color vision? (For this to happen, the daughters must have both G and R.)

8. Repeat the previous question for the case of a wife who is a double heterozygote in repulsion: $g\,R/G\,r \times g\,r/(Y)$.

9. Assume that the dominant genes D and R are 20 map units apart on an autosome. What progeny phenotypes (dominant or recessive for D and for R) are expected from the mating $D\,R/d\,r \times D\,R/d\,r$? Note that you will have to consider 20% recombination in both parents.

10. The nail-patella syndrome is an autosomal dominant (Np) characterized by absence or malformation of the fingernails, toenails, elbows, and kneecaps. Kidney disease may also be present. The gene locus is about 10 map units from the ABO locus on chromosome 9. A man with the nail-patella syndrome and blood type AB married a normal woman (np/np) of type O. Three children were type A and normal; one was type B and affected.

(a) What is the most likely phase of the doubly heterozygous parent?
(b) The woman is pregnant with her fifth child. A few red blood cells from the fetus reveal that it is type A. What is the probability that it will develop normally (without the nail-patella syndrome)? Assume the phase specified in part (a).
(c) What would be the probability of normality had the fetus been type B?

11. (a) What is the genotype of each person in the following pedigree? Be as specific as possible, giving alternative genotypes if necessary.
(b) Which person is definitely a recombinant?
(c) For individual III-3, what are the probabilities of being H/H? H/h? h/h? (These numbers must add to 1.) Assume that green-shift and hemophilia A are 3 map units apart. Note also that III-3 is known to be homozygous g/g.

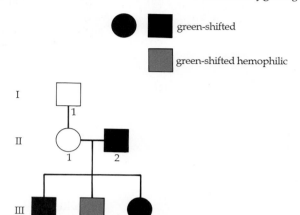

12. Crossing over seems to be less frequent in males than in females. To establish the fact that two autosomal genes are linked, would you rather find double heterozygotes who are male or female? Think about this a bit before answering. (Assume that all other pedigree parameters are the same.)

FURTHER READING

We have been able to find only a few nontechnical readings on gene mapping. An interesting historical tidbit on the first chromosome map is by Crow (1988). This paper contrasts with two articles on the present status of the human gene map: a feature story by Prakash (1988) and introductory comments to the Tenth International Workshop on Human Gene Mapping by Ruddle and Kidd (1989). Two books by committees of the National Research Council (1988) and the Office of Technology Assessment (1988) include easily understood explanations on the methods of mapping. Older articles that are not overly technical are by Francke (1983) and McKusick and Ruddle (1977). Articles by Lewin (1984) and Diamond (1984) are interesting comments on the human-chimpanzee-gorilla connections. Textbooks on evolution—Avers (1989) or Futuyma (1986), for example—provide good background information on making evolutionary trees.

10 Chromosome Nondisjunction

In 1866, just one year after Mendel published his rules of inheritance for garden peas, the English physician John Langdon Down reported some interesting observations. As medical superintendent of an asylum for the severely retarded, he had noticed that about 10% of the residents resembled each other and could be easily distinguished from the rest of the patients: "So marked is this, that when placed side to side, it is difficult to believe that the specimens compared are not children of the same parents." This condition, originally called *mongolism*, is now known as **Down syndrome**.

In addition to being mentally retarded, people with Down syndrome show some combination of the following traits (Figure 10.1): All parts of the body are shortened owing to poor skeletal development. The face is broad and flat with a small nose, irregular teeth, and abnormally shaped ears. The eyes may be close-set with narrow, slanting eyelids; a large, furrowed tongue may protrude from a mouth framed by rather thick lips. Hipbones are abnormally shaped and aligned, and the feet often display a sizable gap between the first and second toes; the little finger is often short and curved inward. Some highly unusual *dermatoglyphic* (Greek *derma*, "skin"; *glyph*, "carving") patterns of hand creases, fingerprints and footprints are also associated with Down syndrome.

Weak reflexes, loose joints, and poor muscle tone render these individuals rather limp and floppy. Defects of the heart, digestive tract, kidneys, thyroid gland, and adrenal glands are also common. Males have poorly developed genitals and are invariably sterile. In females, ovarian defects and irregular menstruation are the rule. But fertility is possible, and over two dozen live births have been recorded. (About 40% of these babies had Down syndrome.)

Many babies with Down syndrome die within a year, often from heart abnormalities. Susceptibility to infection (especially pneumonia) is also common, probably owing to defects in the immune system. Leukemia is 15 to 20 times more frequent in people with Down syndrome than in the general population, and other types of cancer are more common too. But antibiotics and other medical improvements have extended the mean life expectancy to over 30 years, with over 25% of affected individuals surviving into their 50s.

Motor development is slow, and bladder and bowel control may take years to develop. Many children learn to talk, but speech is usually thick and harsh sounding, perhaps in part because of hearing defects. Lively, cheerful, and very affectionate, people with Down syndrome are noted for their impishness, flair for mimicry, and enjoyment of music and dance. Although clumsy with their hands, they respond well to early intensive stimulation and training for simple tasks. Although IQs seldom exceed 60, some people with Down syndrome do learn to read and write, attend regular schools, and find employment.

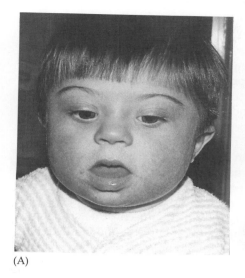
(A)

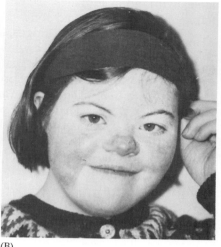
(B)

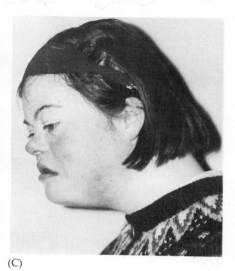
(C)

Figure 10.1
Down syndrome children. (A) One-year-old boy. (B, C) Sixteen-year-old-girl. (Photographs courtesy of Paul Polani, University of London.)

What causes the basic defects? Dr. Down could not possibly have guessed that they resulted from the presence of an extra chromosome, because his paper preceded the discovery of chromosomes by about 15 years. Indeed, 90 years passed before geneticists determined the correct human chromosome number, and not until 1959 did they realize that people with Down syndrome have a triple dose of the smallest chromosome. By then, decades of work—mainly with fruit flies—had shown how departures from the normal diploid number might come about.

This section of the text deals with such *changes in quantity* of otherwise normal genetic material, changes involving hundreds or even thousands of genes. This type of variation contrasts with most of the genetic variants we have discussed up to now, which arose from *changes in quality* (mutations) of individual genes in organisms that had a normal chromosome number. Cells with normal chromosome sets have **euploid** karyotypes (Greek *eu*, "good"; *ploid*, "set"). For humans, haploid gametes and diploid zygotes are the euploid conditions. In this chapter we discuss **aneuploid** organisms, those with unbalanced sets of chromosomes due to an excess or deficiency of individual chromosomes. It is the *imbalance* among genes, rather than the nature of individual genes, that causes the observed phenotypic consequences or even death of affected organisms.

Aneuploidy can arise by a variety of mechanisms. The first such phenomenon discovered was **nondisjunction**, the failure of two homologous chromosomes or sister chromatids to separate (disjoin) during cell division. If it happens during *meiosis*, a gamete may end up with one too many or one too few chromosomes. Union of this aneuploid gamete with a euploid one leads to aneuploidy in the zygote and its descendant cells. During a *mitotic* division, nondisjunction occurs when the two sister chromatids of a chromosome fail to separate, and this can lead to a mixture of euploid and aneuploid cells within one individual. Experiments with a vast range of species have shown that nondisjunction is a major cause of aneuploidy. It is also known to be the source of much grief to humans.

MISTAKES IN CELL DIVISION

Nondisjunction was first discovered in crosses between red-eyed and white-eyed fruit flies. This particular white-eye mutant is X-linked and recessive to its wildtype red-eye allele. Thus, matings of white-eyed females with red-eyed males, w/w x $+/(Y)$, should yield all red-eyed daughters, $+/w$, and white-eyed sons, $w/(Y)$.

Around 1913, Calvin Bridges, an undergraduate researcher at Columbia University, found a few white-eyed daughters and red-eyed sons among the F_1 generation. Bridges proposed that the unexpected white-eyed females developed when an egg with two X chromosomes was fertilized by a Y-bearing sperm, and microscopic analyses verified that these females were indeed XXY. Exceptional red-eyed males, on the other hand, had one X chromosome and no Y, so they must have come from the fertilization of a no-X egg by an X-bearing sperm. (Unlike humans, fruit flies with one X and no Y are male.)

How were the XX and no-X eggs formed? Bridges suggested that they could have arisen during the first meiotic division when the two X chromosomes failed to disjoin (i.e., separate) from each other. He coined the term *nondisjunction* to describe this phenomenon. Nondisjunction apparently occurs when the centromeres of homologous X chromosomes move to the same pole at anaphase. The discovery of nondisjunction was a great genetic breakthrough, an exception that proved the rule of Mendelian inheritance and provided direct physical evidence that genes exist on chromosomes. Later research showed that nondisjunction can occur in any organism, in any dividing cell, and can involve any pair of chromosomes. Consequences vary widely, depending on where and when the misdivision takes place.

Meiotic Nondisjunction

Consider the most common situation, in which only one pair of chromosomes misbehaves. For simplicity's sake, we will focus on spermatogenesis, in which all four products of meiosis are functional. Standard terminology is as follows: For a gamete, **nullisomic** ($n - 1$) and **disomic** ($n + 1$) mean, respectively, the presence of *no* homologous chromosomes and of *two* homologous chromosomes rather than one.* For a zygote, **monosomic** ($2n - 1$) and **trisomic** ($2n + 1$) refer, respectively, to the presence of *one* homologous chromosome and of *three* homologous chromosomes, rather than the usual pair.

Nondisjunction may happen during either the first or second meiotic division. As Figure 10.2 shows, the timing makes a difference. If nondisjunction occurs during *meiosis I*, *all* the sperm derived from that primary spermatocyte will be abnormal. Specifically, half of them will contain *neither* member of the given chromosome pair, and half will carry *both*. In humans, for example, these sperm will end up with a total of 22 or 24 chromosomes rather than the usual 23. It follows that after the fertilization of normal eggs, the resultant zygotes will have either 45 or 47, rather than 46, chromosomes.

If nondisjunction occurs in a secondary spermatocyte undergoing *meiosis II*, then only *two* of the four sperm will be abnormal. As before, these two will bear either one too few or one too many chromosomes. Note also that the disomic sperm from the two types of nondisjunctions are different. If a male is heterozygous A/a (as indicated), then the abnormal disomic sperm from *first*-division nondisjunction will have one chromosome with A and another with a. On the other hand, the disomic sperm from *second*-division nondisjunction will have both chromosomes with A or both chromosomes with a.

Normal development of the zygote in all diploid species depends on the presence of exactly two of each chromosome type, any deviation from this pattern causing abnormalities or death. Nondisjunction of sex chro-

* Recall that n = the haploid number of chromosomes, 23 in humans.

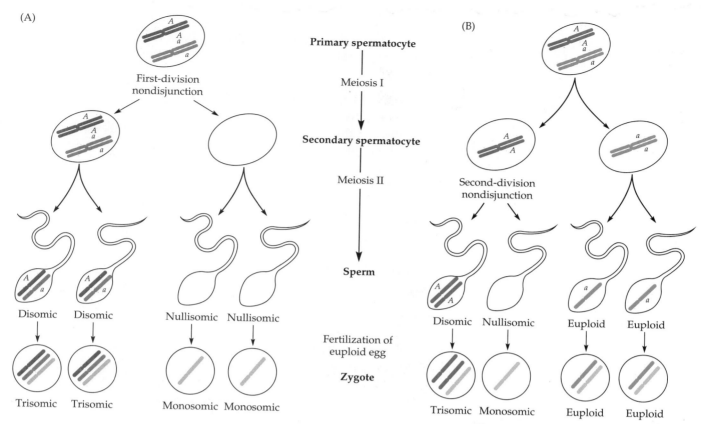

Figure 10.2
Results of (male) meiotic nondisjunction involving a single pair of chromosomes with a heterozygous gene. Assume that all other chromosome pairs segregate normally. (A) At the first meiotic division, nondisjunction involves homologous chromosomes moving to the same pole at anaphase. The eventual gametes are two disomic, two nullisomic, and no euploid. (B) At the second meiotic division (in just one of the two secondary spermatocytes), nondisjunction involves sister chromatids moving to the same pole at anaphase. The resulting gametes are one disomic, one nullisomic, and two euploid. Male chromosome A is shown in dark color and male chromosome a in medium color. In zygotes, the egg chromosome is shown in light color.

mosomes is better tolerated than that of autosomes, however, partly because of the Lyon effect. Recall from Chapter 7 that only one X chromosome is fully active in any cell. Because all additional X chromosomes are inactivated, they have less effect on phenotype than does the presence of too many (or too few) autosomes.

In addition, an *excess* of chromosomes is tolerated better than a *deficiency*. Thus, among autosomal aneuploidies, trisomies are more viable than monosomies. But only those trisomies involving the smallest or most heterochromatic (i.e., least gene-rich) chromosomes are able to survive at all. Among these, the greater the imbalance among autosomal genes, the more abnormal the phenotype. In humans, the only truly viable autosomal trisomy is Down syndrome, which involves the smallest chromosome.

What causes the phenotypic abnormalities seen in aneuploid organisms? Is it the small effects of many genes added together, or is it just a few scattered genes that are highly deleterious or lethal when present in anything but two doses? In fruit flies (and probably other organisms too) it is the former situation. Also, the effects of having a gene (or group of genes) in triplicate are less severe than having just one copy.

Mitotic Nondisjunction

In the cases discussed so far, an abnormal chromosome number is established in the zygote stage, the result of fertilization involving an aneuploid gamete. Consequently, *all* the cells of the body are expected to be aneuploid. In mitotic nondisjunction, however, the zygote may be normal, with aneuploidy occurring sometime during its later development. If the resultant cells survive and continue to divide, they give rise to descendant line(s) of aneuploid cells. The individual then has a mixture of cells with different chromosome numbers and is called a **mosaic**. This anomaly can occur in somatic cells or in cells destined to give rise to gametes. The latter case, called **germinal mosaicism**, can lead to aneuploidy among offspring as well.

The earlier that nondisjunction takes place, the larger the proportion of aneuploid cells that might be found in the mosaic. If nondisjunction occurs during the *first* cleavage division of the zygote, as shown in Figure in 10.3A, then *all* the daughter cells will be aneuploid; half will be monosomic and half trisomic. If it occurs during the *second* cleavage division (Figure 10.3B), only *half* of the resultant cells will be aneuploid. Nondisjunction in the *third* cleavage division will initially make one-fourth of the

(A) Mitotic nondisjunction during the first cleavage division of a zygote. All daughter cells are unbalanced.

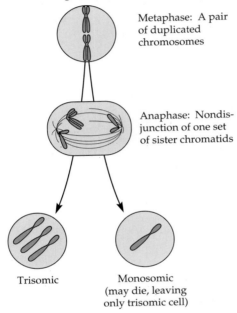

Metaphase: A pair of duplicated chromosomes

Anaphase: Nondisjunction of one set of sister chromatids

Trisomic

Monosomic (may die, leaving only trisomic cell)

(B) Mitotic nondisjunction during the second cleavage division. If the monosomic cell dies, the normal and trisomic lines remain.

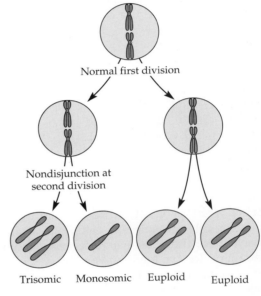

Normal first division

Nondisjunction at second division

Trisomic   Monosomic   Euploid   Euploid

(C) Chromosome loss as euploid cell divides

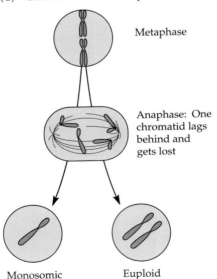

Metaphase

Anaphase: One chromatid lags behind and gets lost

Monosomic   Euploid

(D) Chromosome loss as trisomic cell divides

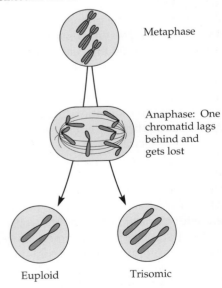

Metaphase

Anaphase: One chromatid lags behind and gets lost

Euploid   Trisomic

Figure 10.3
How mosaicism arises from mitotic nondisjunction (A and B) or from chromosome loss (C and D).

cells aneuploid. And so on. If nondisjunction occurs late in development, only a tiny group of aneuploid cells will be formed, and they may escape detection.

Actually, the proportion of aneuploid cells in an individual after birth does not necessarily reflect what happened in the embryo before birth, because aneuploid cells have a lower probability of survival than do normal cells. Indeed, embryos with high proportions of abnormal cells may not survive at all, as shown by the high frequency of aneuploidy in spontaneously aborted fetuses (Chapter 11).

Mosaics can also arise through other processes. The most frequent of these is **chromosome loss**, whereby a chromosome lags so far behind during anaphase that it is not included in either daughter nucleus. Consequently, one of the daughter cells will contain one less chromosome than the other. If the original cell was euploid, one of the daughter cells will be monosomic (Figure 10.3C). If the original cell was *trisomic*, then following the loss of the extra chromosome, one daughter cell will be normal rather than trisomic (Figure 10.3D). Clearly, this event could lead to mosaicism in an individual who started out as a trisomic; indeed, it has been documented in several Down syndrome mosaics. Like nondisjunction, chromosome loss may occur during any cell division, mitotic or meiotic, and can involve autosomes or sex chromosomes.

In any given cell division, more than one chromosome may undergo nondisjunction or loss, but the result is usually lethal. Those few cases that do survive beyond birth generally involve either mosaicism or autosomal aneuploidy in conjunction with a sex chromosome aneuploidy. We will see in Chapter 15, however, that specific single or multiple aneuploidies characterize certain types of cancer.

NONDISJUNCTION OF AUTOSOMES

Long before the cause of Down syndrome was known, studies of this condition in twins suggested a genetic role. (*Identical* twins arise from a single zygote and are expected to have identical genotypes; *fraternal* twins develop from two fertilized eggs and so are genetically equivalent to sibs.) Such studies are usually done by finding one affected member of a twin pair, then determining (1) the type of twinning and (2) whether or not the other member is also affected. Among identical twins it was found that if one member had Down syndrome, then almost always the other one did too.* In contrast, if one member of a pair of fraternal twins had Down syndrome, the other was in almost all cases unaffected.

Because the pattern of transmission did not fit the usual models for single-gene inheritance, however, it was suggested in the 1930s that Down syndrome might result from a chromosomal abnormality. Cytological methods then available were too primitive to test this hypothesis, but just three years after the normal human chromosome number was established came the exciting discovery that somatic cells of people with sporadic Down syndrome contain 47 chromosomes. The extra chromosome was one of the G group (Figure 10.4); at first researchers thought it was the next-to-smallest—that is, chromosome 21—so **trisomy 21** became another name for Down syndrome. Improved techniques later showed that the extra chromosome was actually the *smallest* member of the G group, but rather than trying to change the name, geneticists live with the inconsistency.

---

* It can happen that, after the splitting of an embryo to form identical twins with trisomy 21, one or the other twin loses a chromosome 21. Alternatively, nondisjunction occurring in the first mitotic division after separation of karyotypically normal twins could give rise to Down syndrome in one twin.

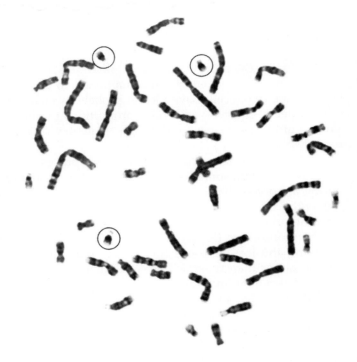

Figure 10.4
G-banded metaphase spread from an individual with Down syndrome, showing three chromosomes 21 (circled). (From Epstein 1989.)

Saying that people with Down syndrome have an extra chromosome does not really explain how their abnormalities come about. Does chromosome 21—which contains only about 1.7% of the total DNA in humans—carry genes involved with the development of the brain, skeleton, and other organ systems that are affected? More specifically, does chromosome 21 carry genes for those enzymes, hormones, blood groups, and so on, that might be expressed in excess in people with trisomy 21? Unfortunately, not enough is known about networks or cascades of biochemical events in normal human development to say how an extra chromosome interferes with these processes. Certainly, there must be complex interactions among the gene products of many chromosomes.

Not all of chromosome 21 has to be present in triplicate to produce Down syndrome. In a few rare individuals, the only extra chromosomal material is the distal half of the long arm. As geneticists use cytogenetic and molecular techniques to dissect and characterize chromosomes, some additional clues are beginning to emerge (Box A). Of the 1,000 or so genes thought to be on chromosome 21, only about 20 have been mapped. Among these, several lie near the tip of the long arm in bands q22.1 through q22.3—the region that, when triplicated, is most often associated with Down syndrome.*

Molecular studies reveal that the long arm of chromosome 21 has about two dozen special DNA marker sequences called RFLP sites (Chapter 13), as well as a few cytological markers. These studies allow researchers to start answering some other key questions about Down syndrome: In which parent did the nondisjunction occur? Does the phenotype of the affected child vary with parental origin of the extra chromosome? Are

---

* A 1993 report links one of these genes with *familial amyotrophic lateral sclerosis (FALS)*, or *Lou Gehrig's disease* (which also afflicts physicist Stephen Hawking), in about a dozen families. Its normal product, the enzyme *copper/zinc superoxide dismutase (SOD1)*, converts toxic superoxide ions to oxygen and hydrogen peroxide. Dominant mutations of the *SOD1* gene somehow damage the nerves carrying impulses from the brain to skeletal muscles, thus leading to paralysis and death. This finding accounts for only a fraction of FALS cases, however, and the familial cases make up only 10% of all cases of amyotrophic lateral sclerosis (ALS). The other 90% are sporadic, and may or may not have similar causes.

A PRELIMINARY PHENOTYPIC MAP OF CHROMOSOME 21

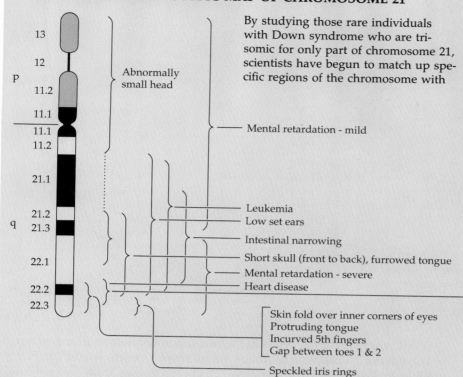

By studying those rare individuals with Down syndrome who are trisomic for only part of chromosome 21, scientists have begun to match up specific regions of the chromosome with specific features of Down syndrome. After analyzing the detailed cytogenetic and clinical descriptions for each patient with partially trisomy, they combine the data. The diagram shown here, which summarizes 17 case histories, indicates some possible associations within bracketed subregions. The number, identity, and exact location of the critical genes within each subregion are not yet known, but it is hoped that molecular studies will greatly refine the process of phenotypic mapping.

some parents at high risk for nondisjunction, and if so, can they be identified in advance? What is the correlation, if any, between crossing over and nondisjunction on chromosome 21? Is the maternal age effect (discussed shortly) due to increased nondisjunction in older women, or is it due to decreased destruction of their aneuploid embryos? Is the Down syndrome phenotype associated with just a few key genes on chromosome 21, or (as with fruit flies) is it mostly due to a generalized imbalance of genes?

Frequency and Origin of Nondisjunction Events

Down syndrome occurs in about 1 out of every 800 to 1,000 live births in all ethnic groups.* It afflicts families of wealth or intellectual achievement as well as families of the poor or uneducated. Nearly all cases occur just once within a given family and thus are called **sporadic Down syndrome**. The 1–2% of cases that run in families are discussed in Chapter 11.

Over a century ago it was noted that babies with Down syndrome are often the last-born members of a sibship. This observation suggested that birth order, maternal age, paternal age, or a combination of these might be implicated. In the 1930s, a **maternal age effect** was shown to be the critical factor in most cases: Women over 35 years of age produced over half of all babies with Down syndrome, although they accounted for only about 15% of all births. Of course, older women have older husbands and more children, but statistical analyses can disentangle such closely correlated factors and measure their individual effects.

* This is a decrease from the 1 in 700 frequency that occurred before prenatal testing was possible.

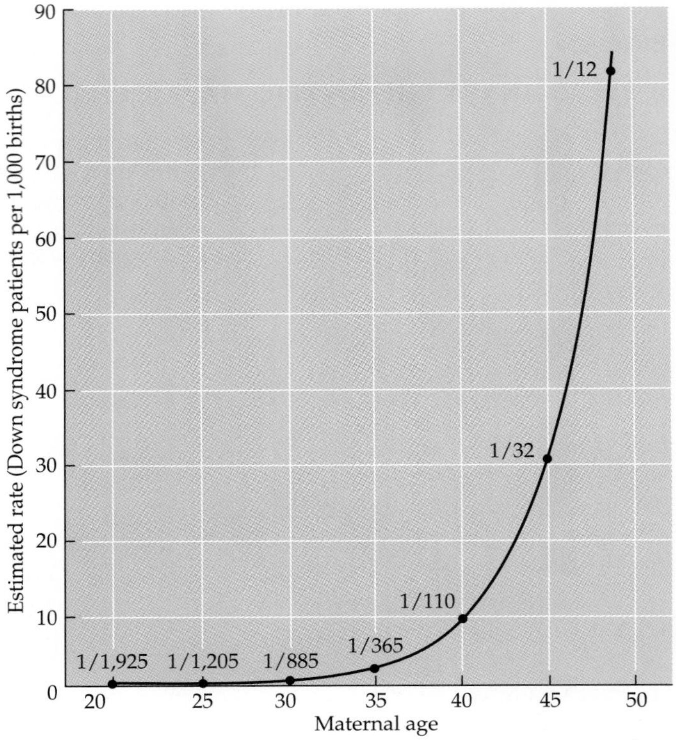

Figure 10.5

Estimated rate of occurrence of Down syndrome for different maternal ages. The number 1/1,925 for 20-year-old mothers, for example, means one Down syndrome offspring per 1,925 births, and so on. (From Hook and Chambers 1977.)

But where maternal age is not a factor, nondisjunction in *fathers* must be considered. Molecular methods often reveal variations that make it possible to distinguish the maternal 21 chromosomes from the paternal 21s. Such studies have recently shown that only about 4–5% of sporadic cases of Down syndrome are paternal in origin. Among these, most of the nondisjunction events occur during meiosis II. Among the 91–92% of cases caused by maternal nondisjunction, however, the situation is reversed: Most are meiosis I errors. The remaining 4–5% of cases can be traced to mitotic nondisjunction in somatic cells of the early embryo.

Figure 10.5 shows how the frequency of Down syndrome births varies among different maternal age-groups. Note that beyond age 30, the frequency rises dramatically, so that a 45-year-old woman is about 45 times more likely to produce a child with Down syndrome than is a 20-year-old woman. Specifically, the risks are about 1 in 2,000 for maternal ages 20–34, about 1 in 300 at age 35, and about 1 in 45 by the age of 45. (Some researchers report an increased incidence of Down syndrome among very young mothers, too, but others suggest that the difference may not be statistically significant.) Studies of spontaneous abortions have shown that the incidence of this disorder is at least four times higher at conception than among live births.

With the advent of prenatal testing in the late 1960s, the routine screening of pregnant women over the age of 35 became possible. As a result, fewer babies with Down syndrome have been born to older women, and the frequency of Down syndrome among live births has dropped somewhat. Whereas "over-35" mothers used to produce a large proportion of babies with Down syndrome, they now account for only about 20% of them. The remaining 80% of babies with Down syndrome are now being born to women under 35 years of age, who, although their risk factor is very much lower, vastly outnumber older mothers.

Although it has not yet been feasible to do prenatal tests on *all* women under the age of 35, researchers have been looking for some additional clues to help them identify those younger mothers who are at increased risk of producing babies with Down syndrome. In 1984 it was noted that the concentration of a substance called **α-fetoprotein (AFP)** is often lower than average in the blood serum of mothers who give birth to babies with Down syndrome.* However, this screening procedure missed about two-thirds of the Down syndrome pregnancies in one large study, which has led some people to question whether AFP alone is a useful screening test for Down syndrome. Still, when considered along with the levels of two particular hormones in maternal serum, it can help lead to the detection of about 60% of Down syndrome fetuses in younger mothers (Chapter 19).

Down Syndrome and Alzheimer Disease

Ironically, delayed development of people with Down syndrome is usually followed by premature aging—often including, among those who survive beyond 30 years, dementia of the type suffered by people with **Alzheimer disease (AD)**. Dementia involves memory loss, confusion, anxiety, loss of ability to perform simple tasks like dressing or feeding, and ultimately the loss of all functional capabilities. Alzheimer disease is a very serious personal and public health problem, afflicting about 10% of people who are in their 70s and about 30% of people in their 80s.

Recent studies have uncovered some striking similarities in the brain abnormalities associated with AD and Down syndrome (Figure 10.6): **senile plaques** (extracellular deposits of protein from the degeneration of nerve cell endings) and **neurofibrillary tangles** (clusters of protein filaments in the nerve cell bodies). The relations among these defects are not clear, however. Although it appears that all people with Alzheimer disease and all adults with Down syndrome over age 30 have plaques and tangles in their brains, only 25–40% of the latter group seem to have dementia. Perhaps, rather than a simple cause-and-effect connection, there is a certain *threshold* number of plaques and tangles must accumulate before dementia sets in.

Other poorly understood similarities between the neurological defects in Down syndrome and Alzheimer disease include the following: severe loss (20–50%) of nerve cells in the *hippocampus* (the center for receiving, storing, and sending messages to and from various brain areas) and in certain parts of the cerebral cortex; fewer and smaller contacts (synapses) between nerve cells; large numbers of vacuoles in the nerve cells; deficiencies in certain enzymes and neurotransmitters; and possible excess of nerve growth factor.

The protein present in the plaques and tangles of both AD and older people with Down syndrome is called **β-amyloid**. It is the short (43-amino-acid) degradation product of a larger (about 740-amino-acid) protein known as **amyloid precursor protein (APP)**. APP is found in the cell membranes of most tissues, but its normal function is unknown. The APP gene is located on chromosome 21q. So, too, is a gene for one dominantly inherited, early-onset form of AD called **familial Alzheimer Disease (FAD)**.† This coincidence led some scientists to suggest that the two genes might be one and the same. Furthermore, one research group discovered three (rather than two) copies of the amyloid gene in three people with Alzheimer disease, as well as in two people with nontrisomy Down syndrome (Chapter 11). This finding suggested that people with Alzheimer disease might have a small duplicated region on one chromosome 21. If

---

* But when present in excess in a pregnant mother's serum, it may signal the presence of spinal cord abnormalities in the fetus.

† Symptoms appear during one's 40s or 50s rather than after age 65.

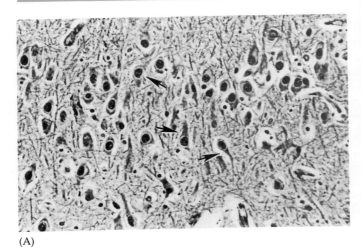

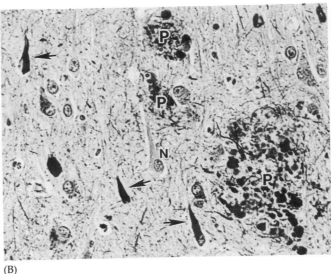

Figure 10.6
Plaques and tangles. (A) Normal brain tis-sue. Arrows point to neurons. (B) Brain tissue from an adult with Down syndrome. P, plaques; N, neurons. Arrows point to tangles.

this were true, perhaps Alzheimer disease is caused by overproduction of the β-amyloid protein or some mutant form of it.

But recent research shows that the situation is not nearly so simple. What we call Alzheimer disease may actually be several clinically similar conditions involving several independent genes, and these genes may also interact differentially with unknown environmental factors. For example, combining family studies with molecular studies, scientists discovered that the genes for the amyloid protein and FAD could actually be separated from each other. Yet a few families with hereditary Alzheimer disease do have rare mutations in their APP gene. But in other families AD has recently been traced to entirely different genes on chromosome 14 and on chromosome 19. Perhaps there are additional, undiscovered AD loci as well.

Over a decade of intensive AD research has yielded many tantalizing clues, but still not answered some of the most basic questions about this tragic disease. Are all cases of AD inherited? In particular, is the (much more common) late-onset AD inherited? Do early-onset and late-onset AD have different causes? Are the β-amyloid plaques and tangles the cause or the result of senile dementia? If (as it now appears) healthy cells also generate β-amyloid as a breakdown product of APP, then what causes the accumulation of β-amyloid in the brains of people with Alzheimer disease? Recent evidence suggests that APP can be broken down in *lysosomes*, where the β-amyloid protein resists further degradation and tends to pile up. Does this lysosomal pathway become more predominant as AD develops? Clearly, there are many more questions than answers about Alzheimer disease.

Animal Models for Down Syndrome and Alzheimer Disease

Any disorder found only in humans is extremely difficult to study. If a condition occurs naturally or can be experimentally induced in other animals (preferably in small laboratory mammals whose physiology and genetics are well understood), then scientists can much more readily study its development and do preliminary tests of experimental treatments.

The creation of an animal model for human trisomy 21 has opened up another line of research on Down syndrome. Charles Epstein and colleagues at the University of California in San Francisco have produced mice that are trisomic for mouse chromosome 16, which carries several genes that are homologous to human chromosome 21 genes (Figure 10.7). Although the mouse fetuses do not survive quite until birth, their multiple defects can be studied in great detail. Many of these trisomic mice exhibit the same type of heart defect that is found in some people with Down syndrome; they also show defects in the nervous system and immunological abnormalities.

By combining trisomic mouse cells with normal diploid mouse cells, Epstein and his co-workers have also produced *mosaic mice* that survive beyond birth, and these, too, are being intensively studied. But researchers stress that because mouse chromosome 16 is a much larger chromosome than human chromosome 21, the two trisomies cannot be entirely homologous. Thus, a *partial trisomy* of mouse chromosome 16, but carrying genes similar to those found on the distal region of human chromosome 21q, would be much better for comparative studies.

There is good evidence that differences in gene dosage lead to proportional differences in the amount of enzyme produced. But the effects of aneuploidy must also involve dosage differences for cell receptors, regulatory genes, homeotic and other developmental genes, and genes that produce subunits for important structural proteins such as collagen. Although no links between the imbalance of any specific genes and specific phenotypes have been absolutely proved, scientists are starting to make some headway in such studies.

Researchers have also been trying to produce animal models for Alzheimer disease. One method is to inject the human *APP* gene into mouse zygotes and then recover a few so-called *transgenic* mice (Chapter 13) that have incorporated the human locus into their genome. Initial reports in 1991 suggested that the brains of several *APP* transgenic mice contained some deposits resembling the β-amyloid plaques and tangles seen in people with AD. But some of these results could not be verified when the experiments were repeated, so the excitement over a good mouse model for AD was very short-lived.

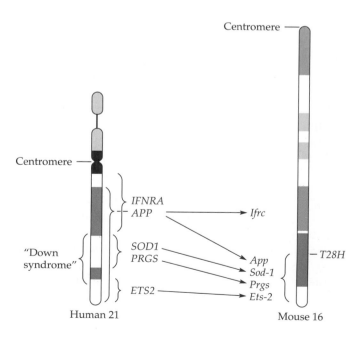

Figure 10.7
Comparison of the ends of human chromosome 21q and mouse chromosome 16q, showing the relative (and approximate) positions of a few gene loci: interferon receptor alpha (IFNRA), amyloid precursor protein (APP), superoxide dismutase 1 (SOD1), phosphoribosylglycineamide synthetase (PRGS), and oncogene ets-2 (ETS2). Note: Several genes at the very tip of human 21q have their homologues on mouse chromosomes 3, 10, or 17, rather than on mouse 16. (From Epstein 1989.)

Search for Factors Altering Nondisjunction Rates

The well-documented maternal age effect in the incidence of Down syndrome has focused the greatest attention on the mother, particularly the older mother. This is because oocytes may spend several decades in prophase of meiosis I, whereas sperm are continually formed anew. But nobody knows for sure what causes nondisjunction in women of any age. Many possible factors (thyroid disorders, viral infections, radiation, caffeine, drugs, hormones, contraceptives, reduced frequency of coitus, etc.) have been suggested and then usually rejected for lack of convincing evidence. For example, clusters of Down syndrome cases sometimes occur following epidemics of infectious hepatitis or German measles. Another possibility is that X-radiation increases the rate of nondisjunction in aging females. This outcome is known to happen in fruit flies and may also occur in mice, but the data on humans are not clear-cut.*

Down syndrome is occasionally seen together with sex chromosome aneuploidy. Although the data are scanty, it appears that in some families these *double trisomics* (for sex chromosomes and chromosome 21) may occur more often than would be expected by chance alone. This finding has led some geneticists to postulate the existence of genes that increase the probability of nondisjunction. Such mutant genes have been identified in fruit flies and corn. Additional evidence for their presence in humans are reports of a few families in which several members have different single aneuploidies—say, one trisomy 21 and one 47,XXY—again, more frequently than expected by chance alone. But even if a genetic predisposition to nondisjunction were proved in some families, it could not account for more than a fraction of all the cases that occur. Ann Chandley (1989) sounds a note of despair:

> For it is clear that after decades of aneuploidy research embracing plants and animal species from yeasts to humans, and of studies into tubulin, microtubules, centromeres, kinetochores, centrioles and spindles, the question of what causes nondisjunction has still not been answered. As one of the most prevalent problems in human genetic disease, aneuploidy remains one of its least understood phenomena.

Down Syndrome Mosaics

Roughly 2% of all live-born people with Down syndrome are mosaics, possessing both normal and trisomic cells. These individuals show great phenotypic variability in all respects. The degree of abnormality varies, depending on what proportion of tissues end up being trisomic and which specific tissues these are. If cells of the nervous system turn out to be euploid, for example, the individual may have normal intelligence. Those individuals with only a tiny proportion of aneuploid cells may have a completely normal phenotype and escape detection entirely.

In some families, multiple cases of trisomy 21 can be traced to aneuploidy in a mosaic parent's gonadal tissue. These instances of *germinal mosaicism* are important to detect early, because unlike the usual sporadic pattern, the occurrence of nondisjunction in the mosaic parent's trisomic germ cells can lead to multiple Down syndrome births in a family. Indeed, about half of the gametes from trisomic oocytes or spermatocytes may contain two rather than one chromosome 21.

Some small fraction of Down syndrome cases may be due to mitotic nondisjunction in the *embryo* rather than to meiotic nondisjunction in a

---

* But even if X-rays have no effect on the incidence of human aneuploidy, they are known to cause developmental defects in fetuses. Thus, prospective mothers should avoid exposure to X-rays, especially during early pregnancy.

parent. Recall that nondisjunction during the *first* cleavage division (Figure 10.3A) would result in one trisomic daughter cell and one monosomic daughter cell. The latter usually dies, leaving the trisomy 21 and its descendant cells to make up the entire embryo. But nondisjunction at the *second* cleavage division (Figure 10.3B) produces an embryo with two normal cells, one trisomic cell, and one monosomic cell. This individual will be a mosaic and will likely show some characteristics of Down syndrome.

Other Autosomal Aneuploids

The only other autosomal trisomies that occur with any significant frequency in newborns, **trisomy 18** and **trisomy 13**, were first reported in 1960. Estimates for frequency of either syndrome range from 1 in 4,000 to 1 in 10,000. There is a maternal age effect, but it is not so striking as that found for Down syndrome. Both syndromes are associated with gross abnormalities of nearly all systems, leading to early death. In fact, they share many characteristics, with considerable overlap: Profound mental and developmental retardation, heart defects, kidney defects, low-set and malformed ears, small eyes, and a small, receding jaw are almost always observed. Prominent heel bones and flexion (bending) deformities of various joints are often present. Other traits also are associated with these two syndromes.

The most distinctive additional features of *trisomy 18 syndrome* are the occurrence of simple arch patterns on three or more fingertips (rare in normal individuals) and a long skull that bulges in the back (Figure 10.8). Because of abnormally increased muscle tone, the fists are tightly clenched, with the index finger bent sideways across the third finger, and the limbs and the hip joints are so stiff that they can hardly be moved. About 75% of babies with trisomy 18 have "rocker bottom" feet with big toes that are short and bent upward. Approximately half have extra folds of skin on the neck. Distinctive facial features—including a round face, small and wide-spaced eyes, and a small mouth—also help with diagnosis. For unknown reasons, about 80% of all trisomy 18 births are female. Birth weight is low, and the mean survival time is about 2.5 months, 90% of all cases dying within a year.

For *trisomy 13 syndrome*, the most distinctive additional traits, occurring in about 75% of all cases, are a small head, cleft palate or lip (or both), port-wine-colored birth marks, and extra fingers and toes (Figure 10.9). Unusual dermatoglyphics are the rule too. Almost all babies with trisomy 13 appear to be deaf, and some are blind. Extreme "jitteriness" and seizures are not uncommon. Males may have undescended testes, and females may have a uterus divided into two parts. The fingers may be abnormally bent. The nose is often large and triangular, and the eyes are frequently wide-spaced and defective. The mean survival time is about 3 months, 80% of all cases dying within a year.

Over two dozen cases of *trisomy 22* have been reported. Their phenotypes are variable but include mental and growth retardation and many serious deformities. Only a handful of *trisomy 8* cases are known, but several dozen mosaics for trisomy 8 have been described. The latter show characteristic facial features, as well as bone and joint defects. But mental retardation is moderate and life span is normal.

Although monosomy of autosomes is generally lethal in humans, a few cases of apparent *monosomy 21* have survived beyond birth, with severe mental retardation and multiple abnormalities of other systems. Several mosaics for monosomy 21 have been described, as well as a few cases in which only part of chromosome 21 is missing.

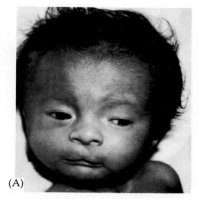
(A)

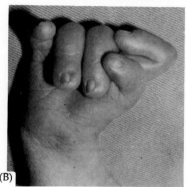
(B)

Figure 10.8
Trisomy 18. (A) Child with trisomy 18. (B) The hand of an infant with trisomy 18 demonstrates typical flexion deformities. (From Summitt 1973.)

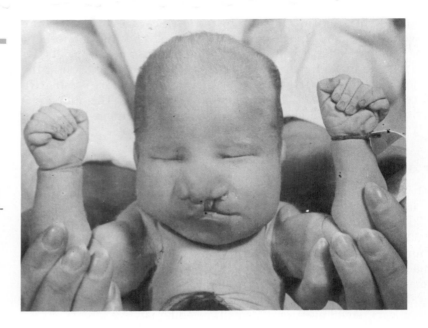

Figure 10.9
Trisomy 13. This infant was not photographed as a living patient but was found in the Department of Anatomy at the University of Western Ontario, where it had been preserved for 30 years. The features typical of trisomy 13 leave no doubt as to the diagnosis. Note the cleft lip, extra fingers, large triangular nose, and wide-spaced eyes. (Photograph by Murray J. Barr; from Valentine 1986.)

In conclusion, we emphasize that these disorders result from chromosome imbalance rather than from gene mutation. That so many systems of the body are deranged when one small chromosome is present in excess suggests that normal metabolism and development are finely attuned not just to the nature of the gene products but to their concentration as well. This finding is also true of other species that have been studied. Furthermore, the relative rarity of autosomal aneuploidies among live-born children indicates that most imbalances of even small chromosomes interfere so seriously with metabolism that embryonic development is aborted in early pregnancy. Sex chromosome aneuploidy presents a somewhat different story.

NONDISJUNCTION OF SEX CHROMOSOMES

Recall that in the late 1950s scientists discovered that the human Y chromosome plays a major role in sex determination. Until then, it had been thought that the critical factor was (as in the fruit fly) the number of X chromosomes—two X's leading to femaleness and one to maleness, with the Y chromosome needed only for fertility in males.

But soon after the human chromosome number was established, the role of the human Y chromosome in sex determination was clarified by two previously known syndromes. People with **Klinefelter syndrome** have one too many sex chromosomes: Their karyotype is 47,XXY, and their phenotype is male. Conversely, people with **Turner syndrome** have one too few sex chromosomes: Their karyotype is 45,X, and their phenotype is female. Thus, two X's do not always make a female, and one X does not always make a male. Findings on other karyotypes rounded out the picture: In humans, the presence or absence of a Y chromosome is much more important to male development than is the number of X chromosomes.

Later, some **47,XYY males** were found among the inmates of penal institutions, raising the question of whether an extra Y chromosome may somehow predispose its host to criminal behavior. In the rest of this chapter, we describe the major sex chromosome anomalies and some aspects of their influence on behavior.

Klinefelter Syndrome: 47,XXY Males

In the 1940s Harry Klinefelter and his colleagues at Massachusetts General Hospital described a syndrome occurring in males and usually not detected until after puberty. Some years later, in 1959, the Scottish cytogeneticists Patricia Jacobs and J. A. Strong showed that males with Klinefelter syndrome are usually, but not always, 47,XXY.

The signs almost always observed in these males (Figure 10.10) are very small testes (about one-third normal size), absence of sperm, and androgen (male sex hormone) deficiency. The penis and scrotum are usually of normal size, however. Additional but much more variable symptoms include poorly developed male secondary sexual characteristics (such as scanty facial hair) and the presence of some feminine traits (such as breast development). Sexual behavior is fairly normal among young males with Klinefelter syndrome, who experience spontaneous erections and (usually spermless) ejaculations. A few have fathered children, however. Many marry and maintain sexual relations, but impotence is common among older men.

Males with Klinefelter syndrome often have unusually long limbs and average about 2–4.5 inches taller than unaffected males; their hands and feet may be large too. But many males with Klinefelter syndrome show normal height and intelligence and function well in society. Indeed, some 47,XXY males exhibit no symptoms of Klinefelter syndrome except infertility. These individuals may live their entire lives without the slightest inkling that they are in any way unusual. But others exhibit mild retardation, which—together with an increased tendency to emotional and social problems—may account for their clear overrepresentation in mental and penal institutions.

For the more serious cases, surgical removal of breast tissue can relieve some psychological stress. Although the sterility is unalterable, treatment with testosterone does promote development of sex organs, body hair, musculature, deeper voice, and so on; it may also improve social adjustment, behavior, and learning ability of some males with Klinefelter syndrome. A warm and stimulating family and school environment early in life also seems to help mental and emotional development.

Klinefelter syndrome is not rare in the general population. Its overall frequency among newborn males is 1 in 500 to 1 in 2,000. In subpopulations of tall men (over 183 cm, or 6 feet), the frequency of males with Klinefelter syndrome may be as high as 1 in 260. Among residents of mental institutions, it is even greater—about 1 in 100; and roughly the same percentage is found in penal institutions. Perhaps 1 in 20 patients seen in fertility clinics has Klinefelter syndrome. Because of underdevelopment of secondary sexual characteristics, they also appear relatively frequently among hospital patients. These differences in incidence among various groups of males point up the importance of carefully defining each test group and of not trying to project the results from any one group onto the general population.

Within families, Klinefelter syndrome seems to occur randomly. The extra chromosome is thought to arise almost entirely from nondisjunction in a parent—about 67% occurring in mothers and about 33% in fathers. As with Down syndrome, it appears that the chromosomes in older eggs are more prone to nondisjunction, but the reasons remain unknown. No paternal age effect has been observed for Klinefelter syndrome.

Klinefelter Variants. About 10% of males with Klinefelter syndrome exhibit karyotypes that are different from the usual 47,XXY. These include 48,XXXY, 48,XXYY, 49,XXXXY, and 49,XXXYY. Klinefelter syndrome involving single cell lines of 48 or 49 chromosomes seem to have more

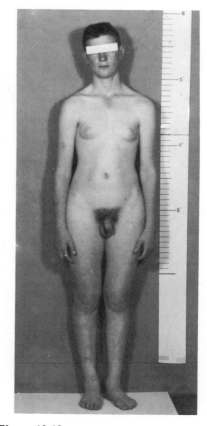

Figure 10.10
A 47,XXY male with Klinefelter syndrome. Note the breast development and female pattern of pubic hair growth. (Photograph by Earl Plunkett; from Valentine 1986.)

223

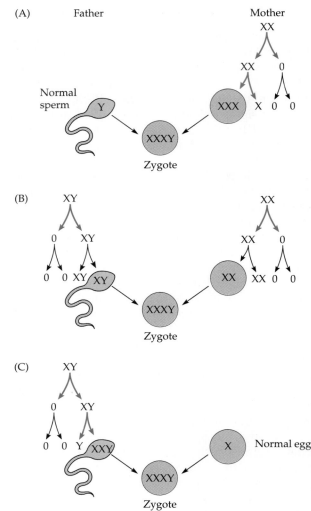

Figure 10.11

Three ways, each requiring a total of two nondisjunctional events, by which a 48,XXXY zygote could be formed. Nondisjunction is indicated by heavy arrows at (A) first and second meiotic divisions in the mother, (B) first divisions in both mother and father, and (C) first and second meiotic divisions in the father.

extreme problems of all sorts, including severe mental retardation. Indeed, most cases are detected through chromosome surveys in mental and penal institutions.

How do all these unusual karyotypes arise? To account for the karyotypes with 48 or 49 chromosomes, two nondisjunctional events are required. For example, a 48,XXXY zygote can arise in several ways, as shown in Figure 10.11: a Y-bearing sperm fertilizing an XXX egg; an XY sperm fertilizing an XX egg; or an XXY sperm fertilizing an X-bearing egg. Sometimes the pattern of inheritance of X-linked genes, chromosomal banding patterns, or special DNA sequences called RFLPs (Chapter 13) can distinguish among the various possibilities.

Several types of mosaicism collectively account for about 10% of males with Klinefelter syndrome. As with any mosaic, the phenotypes of the individuals differ, depending on what proportions of the cells are aberrant and also on how the aberrant cells are distributed among the various tissues. Mitotic nondisjunction or chromosome loss, or both, must be invoked to explain the origin of these different mosaics. Perhaps some began as aneuploid zygotes after meiotic nondisjunction in the parents.

The 47,XYY Karyotype

In 1961 Avery Sandberg and some colleagues at the Roswell Park Memorial Institute in Buffalo, New York, discovered the first 47,XYY male by chance —as the father of a child with Down syndrome. He was tall and of average intelligence, and he had no serious physical problems. His chromosomal constitution resulted from paternal nondisjunction during meiosis II, producing a YY sperm. Several other 47,XYY karyotypes were reported later, but they did not begin to attract widespread interest until 1965. Studying the chromosomes of 197 mentally subnormal males with dangerous, violent, or criminal tendencies in an institution in Scotland, Jacobs and colleagues found seven males with the 47,XYY karyotype—a surprisingly high frequency of 1 in 28 that was not observed among other groups of males they tested. Their preliminary report also noted that the average height of the seven 47,XYY inmates was 186 cm (6 feet, 1 inch), compared with a mean height of 170 cm (5 feet, 7 inches) among the 46,XY males in the same institutional sample. Jacobs, Brunton, and Melville (1965) commented:

> The finding that 3.5 per cent of the population we studied were XYY males must represent a marked increase in frequency by comparison with the frequency of such males at birth. . . . At present it is not clear whether the increased frequency of XYY males found in this institution is related to their aggressive behavior or to their mental deficiency or to a combination of these factors.

Other geneticists began testing tall institutionalized males and finding additional 47,XYY males. Unfortunately, not all investigators were careful to define their samples or to study and describe control males from the same populations.* Based sometimes on individual cases found in preselected environments and without much knowledge of how 47,XYY males in the general population behaved, some very extravagant claims were made about the phenotypic effects of an extra Y chromosome. Soon articles linking Y chromosomes to violent or criminal behavior began appearing in the popular press as well as in journals read by psychologists, sociologists, and lawyers. The so-called XYY syndrome was also sensationalized in a few murder trials, where it was suggested that the defendants' 47,XYY karyotypes rendered them less responsible for their actions. Have post-1965 reports altered the early impressions about the phenotype associated with this karyotype?

One would think that reliable estimates of 47,XYY frequencies among various groups of males could be estimated by pooling the results of several comparable studies. But even a simple trait like tallness is not standardized, and criteria for studying other subpopulations (such as mental patients, inmates in prisons, and inmates in mental-penal institutions) are even more difficult to sort out. Thus, it is not surprising that estimated frequencies vary widely from study to study. And coupled with the tallying problem is the possibility that different geographical or ethnic groups may exhibit different frequencies of the trait in question. Thus, estimates of a given trait in a certain population category can vary greatly.

With these warnings, we present some estimates of the 47,XYY frequencies. Among males at birth or in the *general population*, the frequency of 47,XYY is on the order of 1 in 1,000 males; among *tall males* in the general population it is perhaps 1 in 325. As samples narrow down to *tall mental patients* or *tall penal inmates* and then to *tall mental-penal inmates*, frequency estimates increase to about 1 in 30.

Why are XYY males overrepresented in institutionalized populations?

---

* Of course, one cannot help but find criminals when testing is restricted to prisons.

Many of the early claims about XYY males were based on small numbers of cases found in highly selected populations without adequate control studies of 46,XY males in the same selected populations. Or controls, when used, were not always well matched. Most early studies had another serious defect: The researchers who worked up psychological profiles on these men often knew what their karyotypes were, perhaps leading to an unconscious bias in the evaluations. The only way to eliminate such bias is to conduct **double-blind studies**, in which those who do the karyotyping know nothing of the subjects' histories, and those who compile the histories know nothing about their subjects' karyotypes. Only after both sets of data are independently and fully completed should they be matched up to see how the behavioral profiles of 47,XYY males compare with those of 46,XY males.

One large-scale retrospective study that avoided many of these pitfalls made use of extensive records kept by the Danish government on every 26-year-old male reporting to his draft board. These data included the results of a physical exam and intelligence test as well as educational history, social class, and criminal convictions. To save money, investigators restricted their search to 4,139 tall males. Among this group, they found 12 XYY karyotypes, a frequency of 1 in 345.* The remaining tall, karyotypically normal 46,XY males made up the control group.

How did the 12 tall 47,XYY males compare with their tall 46,XY counterparts? The mean height of the XYY males was 4 cm (1.5 inches) greater than that of the controls. Five of the XYY males (42%) had criminal records, compared with only 9% of the controls, but their crimes were not violent, and their sentences were light. The mean educational level of the XYY males and their test scores on the army intelligence test were significantly lower than those of tall control males. But even among the control males, those with criminal records had a significantly lower mean educational level and lower test scores than those without criminal records. This finding suggests that tall males with lower intellectual functioning are more likely to be convicted of crimes, regardless of their karyotypes.

> If the syndrome is as common as recent surveys suggest, there are probably a large number of undiagnosed XYY males who are neither in prisons nor in frequent barroom brawls. A particularly acceptable social adjustment . . . would be as a pro-football linebacker, but it is also possible that the majority of individuals with XYY chromosomes are not sufficiently large or aggressive for this vocation. (Nora and Fraser 1989)

Turner Syndrome: 45,X Females

In 1938 an American endocrinologist, Henry Turner, described seven grown females who lacked breasts and other secondary sexual characteristics, failed to menstruate, and were sterile. Although the body form was reasonably well proportioned, they were very short (adult height under 150 cm, or 4 feet, 11 inches). Their necks were short and webbed, and their forearms showed greater than normal angling away from the body when the palms faced forward (Figure 10.12). Most females with Turner syndrome have (instead of ovaries) primitive streak gonads that lack both germ cells and hormone-producing tissue. Their oviducts, uterus, and vagina remain small and immature, and the external genitals are also infantile. Up to 10% menstruate and ovulate, however, and in rare cases have given birth.

Other defects might include widely spaced nipples on a broad chest,

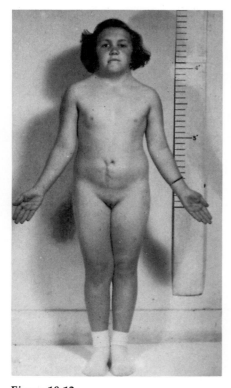

Figure 10.12
A 45,X female with Turner syndrome. This 14-year-old girl is 142 cm (4 feet, 8 inches) tall. Note the lack of sexual development, the webbed neck, the broad chest with wide-spaced nipples, the old-looking face, and the bent forearms. (Photograph by Earl Plunkett; from Valentine 1986.)

---

* Also identified were 16 47,XXY males and a few 46,XY males with other chromosomal abnormalities.

short ring fingers, poorly developed nails, many small pigmented moles on the neck and torso, a narrowed aorta (large vessel that carries blood away from the heart), horseshoe-shaped kidneys, and double ureters. Additional skeletal deformities are sometimes seen: a high, arched palate, receding chin, mismatching of the upper and lower teeth, low-set ears, and abnormally low bone density. The hairline at the back of the neck may be low, and there may be an increased tendency to scarring after injury. In contrast to the childlike body form, the face often looks old. Indeed, premature aging sometimes occurs, and life expectancy may be reduced owing to defects of the heart and other organs. Females with Turner syndrome are often identified at birth by characteristic skin folds on the back of the neck, by swelling of the hands and feet, by abnormally large fingerprint patterns, and by low birth weight. But many females with Turner syndrome show no phenotypic abnormalities even as adults, except for infertility and below-average height.

Although some females with Turner syndrome may show a slight to moderate decrease in IQ (as low as 70), most are completely normal. Some seem to have trouble with a certain type of space perception, however. Their general behavior is normal during childhood. But failure to undergo puberty, coupled with small size, makes it difficult for them to keep up with their peers socially as teenagers. It is often reported that these females tend to be timid, unambitious, impressionable, and immature—but this could be due to their smallness and lack of sexual development more than to their specific karyotype. Hormone treatments may enhance skeletal growth and promote the development of breasts and other secondary sex characteristics. This, plus corrective surgery if necessary, may alleviate some of their social problems and also make normal sexual relations possible.

In 1959 it was reported that females with Turner syndrome have a 45,X karyotype. Nondisjunction in the father accounts for about 75% of all live-born cases, and there is no maternal or paternal age effect. Chromosome loss during early cleavage of the zygote may account for another 10% of females with Turner syndrome, and many types of mosaics have been found. Phenotypes of mosaics range from fully affected to normal, depending on the proportion and distribution of the different cell lines. In a few cases, only part of an X chromosome is missing: Those lacking one short arm (Xp) express the usual Turner phenotypes, but those missing one long arm (Xq) are taller and may appear normal except for the absence of sexual development.

Turner syndrome has an overall frequency of about 1 in 2,500 to 10,000 live-born females. But it is highly lethal in embryos, being the most common karyotype among spontaneous abortions and accounting for about 20% of all chromosomally abnormal aborted embryos. Indeed, it appears that at least 98% of all 45,X zygotes are lost during the first three months of pregnancy. Yet for unknown reasons, the few that survive are not so severely deformed, especially with regard to mental capacities, as are the other live-born chromosomal aneuploidies.

Poly-X Females

Females with the **47,XXX karyotype**, first reported in 1959, present no distinctive phenotype aside from a tendency to be tall and thin, and many of them seem to be completely normal. Their frequency among newborns in the general population is roughly 1 in 1,000 to 2,000. Some triple-X females seem to show a tendency to mental retardation and/or psychosis, however, because they are detected more frequently among institutionalized females (1 in 225 to 425) than among newborn females. Prospective studies have also detected some educational and behavioral problems among 47,XXX subjects. As with other chromosomal anomalies, mosaics are also observed.

Although some XXX females have menstrual difficulties, many menstruate regularly and are fertile. At least a dozen have delivered a total of over 30 children, most of whom were karyotypically and phenotypically normal. This outcome is surprising, because we would expect that half the eggs from a 47,XXX female would be XX, resulting in progeny that are 47,XXX or 47,XXY. Perhaps, during meiosis in XXX females, the XX nuclei are more likely to be shunted into polar bodies; aberrant segregation of this type is known to occur in other organisms.* Studies of the inheritance and expression of X-linked genes suggest that the origin of the extra X chromosome in 47,XXX females is usually nondisjunction in meiosis I of the mother; a maternal age effect has also been noted.

Females with more than three X chromosomes are extremely rare, perhaps fewer than 1 in 10,000 live births. Only about 30 cases of the 48,XXXX karyotype are known; although normal in sexual development, these females are severely retarded and also exhibit a wide range of physical abnormalities. Even fewer 49,XXXXX females have been found in populations of mentally retarded individuals, and none in the general population. Keep in mind, however, that only the abnormal cases are likely to be discovered. Individuals with unusual karyotypes but normal phenotypes are unlikely to be ascertained.

SUMMARY

1. Nondisjunction is either the failure of homologous chromosomes to separate (disjoin) during a meiotic division or the failure of sister chromatids to separate during mitosis or meiosis. In either case, the two daughter cells will have unbalanced chromosome sets; that is, they will be aneuploid.

2. Meiotic nondisjunction in a parent leads to aneuploidy of gametes, which can lead to aneuploidy in all cells of an offspring.

3. Mitotic nondisjunction leads to mosaicism within an individual. The degree of abnormality expressed depends on when nondisjunction occurs, which tissues are affected, and the viability of the aneuploid cells relative to the normal cells.

4. Aneuploidy often causes gross abnormalities or death of cells or organisms as a result of generalized gene imbalance. Because of lyonization, sex chromosome aneuploidy is better tolerated than autosomal aneuploidy; aneuploidy of large autosomes is usually lethal. Trisomies are better tolerated than monosomies.

5. In humans, only three well-defined autosomal aneuploidies are observed among newborns: trisomy 21, trisomy 18, and trisomy 13. All affected individuals are mentally retarded and exhibit deformities of most organ systems.

6. Trisomy 21, or Down syndrome, accounts for about 10% of the retarded population. Sporadic Down syndrome has a frequency of about 1 in 800 live births. Advanced maternal age is a strong determining factor, and about 91% of all trisomy 21 cases are due to nondisjunction in the mother, usually in meiosis I. Rare multiple cases in a family may be due to nondisjunction in a phenotypically normal parent with germinal mosaicism for trisomy 21.

---

* A similar deficiency of aneuploid offspring has also been noted from 47,XYY fathers, whose offspring are not known to include any 47,XXY or 47,XYY sons.

7. Many older people with Down syndrome develop Alzheimer disease, which is characterized by dementia due to degenerative changes in certain areas of the brain. β-amyloid protein, which accumulates in and around abnormal nerve cells, is derived from a larger molecule called amyloid precursor protein. The latter is produced by a gene on chromosome 21. Nearby on the same chromosome is a gene that is associated with a few (but not all) cases of familial Alzheimer disease.

8. In humans the presence of a Y chromosome determines maleness, irrespective of how many X chromosomes are present. The absence of a Y chromosome determines femaleness.

9. Males with a 47,XXY karyotype usually have Klinefelter syndrome, characterized by small testes, infertility, and underdeveloped secondary sexual characteristics. They also tend to be tall and may exhibit mild mental retardation and/or some breast development.

10. Males with the XYY karyotype tend to be tall, and some may show below-average intelligence; but they have no other distinctive traits. Misconceptions about 47,XYY males arose partly because data from small and selective studies were prematurely generalized. Very little is known about 47,XYY males in the general population.

11. Females with the 45,X karyotype have Turner syndrome. They are short, have primitive streak gonads, and lack secondary sexual characteristics. They are usually of normal intelligence.

12. Females with the 47,XXX karyotype show no distinctive phenotype, but some may have slightly lowered intelligence.

KEY TERMS

47,XXX karyotype	chromosome loss	maternal age effect	retrospective study
47,XYY karyotype	disomic	meiotic nondisjunction	senile plaque
α-fetoprotein (AFP)	double-blind study	mitotic nondisjunction	trisomic
amyloid precursor protein (APP)	Down syndrome	monosomic	trisomy 13
	euploid	neurofibrillary tangle	trisomy 18
aneuploid	germinal mosaicism	nullisomic	trisomy 21
β-amyloid	Klinefelter syndrome	prospective study	Turner syndrome

QUESTIONS

1. Diagram the loss of one member of a chromosome pair during (a) meiosis I and (b) meiosis II of spermatogenesis. Compare the results with those in Figure 10.2.

2. Describe the zygote formed by the fertilization of a 21-disomic egg by a 21-nullisomic sperm.

3. With reference to chromosome 21, what kinds of eggs could be produced by a female with Down syndrome? Include a diagram as part of your answer.

4. At least one case has been reported of a pair of otherwise identical twins (same blood types, etc.) of which one member has Down syndrome and the other is unaffected. How might this situation have come about?

5. In one published study, 4 out of 20 females with Turner syndrome showed (X-linked recessive) red-green color vision defects, whereas none of the 55 tested males with Klinefelter syndrome did. Color blindness is usually expressed much more frequently in males than in females. What is the genetic explanation for these observations?

6. In other studies, green-shifted color perception has been noted in both females with Turner syndrome and males with Klinefelter syndrome having both mothers and fathers with normal color vision. Does this give any information on the nondisjunctional events that occurred in the parents? (Assume no mutation.) Explain.

7. A pair of twins had identical blood groups and could successfully exchange skin grafts (another criterion for genotypic identity). One twin was a normal male, but the other was a female with Turner syndrome. How can this be explained?

8. Diagram all types of nondisjunction of the sex chromosomes that can occur in a 46,XX female and a 46,XY male during meiosis I alone, meiosis II alone, and both meiosis I and II. Assume that no more than one pair of centromeres fails to disjoin in any one division.

9. Referring to the diagrams for question 8, list the ways by which a 47,XXY zygote could be formed from karyotypically normal parents. Which are most likely to occur?

229

10. List all types of sperm that can be formed by a 47,XYY male. To distinguish between the two Y chromosomes, call them Y_1 and Y_2. Assume that the three chromosome pair as a trio and then separate as $1 + 2$ at the first meiotic division.

11. The most common type of mosaicism found in males with Klinefelter syndrome is XXY/XY. What two types of postzygotic events could account for this?

12. The phenotypes of males with Klinefelter syndrome can vary widely. Data from one study showed that those who were identified at fertility clinics were generally more masculine than those found at an endocrinology (hormone study) clinic:

Sample	Poor growth of facial hair	Breast development	Feminine pattern of pubic hair
Subfertile males	32%	26%	32%
Endocrinology patients	88%	63%	54%

How can the relationship between phenotype of these males with Klinefelter syndrome and their method of ascertainment be explained?

13. A few prospective studies have followed the development of 47,XYY males from birth through adolescence, but these investigations have been very controversial. List all the advantages and disadvantages that you can think of for such 47,XYY studies.

FURTHER READING

A number of books and articles cover the material in Chapters 10 and 11, that is, aneuploidy in general. These include Resnick and Vig (1989), Valentine (1986), and Epstein (1986, 1988). Reviews of chromosomal disorders include Gardner and Sutherland (1989), De Grouchy and Turleau (1990), and Obe and Natarajan (1990). On sex chromosome anomalies, see Evans et al. (1991) and de la Chapelle (1990).

Many articles and books are available on Down syndrome alone. Some written for the lay public include Pueschel (1990), Selikowitz (1990), Trainer (1991), Cunningham (1988), and Patterson (1987). More technically oriented are Epstein (1989), Patterson and Epstein (1990), and Stewart, Hassold, and Kurnit (1988). Alzheimer disease is discussed by Selkoe (1991), Marx (1992a,b), Ezzell (1992), Hardy, Chartier-Harlan, and Mullan (1992), Yankner and Mesulam (1991), and St George-Hyslop et al. (1990).

Other Chromosomal Abnormalities

11

Bruce Bryer suffered from four rare and severe inherited conditions: Duchenne muscular dystrophy (a muscle-wasting disease), retinitis pigmentosa (an eye disorder), McLeod red cell phenotype (an abnormal blood type), and chronic granulomatous disease (an immune disorder). From the time of his birth in 1966, in Spokane, Washington, he gamely battled the effects of these illnesses, attending school whenever possible and even becoming an accomplished organist. He died at age 17 after a car accident.

Bruce Bryer also earned a permanent place in the annals of medical genetic history. When his case was discovered by human geneticists, only muscular dystrophy was known to be X-linked, and its gene location was still a mystery. Could it be possible that Bruce was missing a section from his one X chromosome that contained the gene loci responsible for *all four* conditions? Because the total loss of several important genes is usually lethal, no such males had ever been reported. But when Dr. Uta Francke (then at Yale University) first looked at Bruce's X chromosome from a culture of his blood cells, it seemed as though a tiny piece might be deleted from the short arm (Figure 11.1).

It took her and many other researchers almost three years of molecular tests to prove that this was indeed the case and that all four genes occupy neighboring loci in normal X chromosomes. This discovery made the front pages of newspapers around the world. It also led to the development of molecular tests for predicting whether or not a male fetus carries a defective allele of the muscular dystrophy gene.

In the preceding chapter we discussed abnormalities involving too many or too few whole chromosomes. Here we concentrate on what happens when *parts* of chromosomes are lost, gained, or moved to new positions in the genome. Some of these *structural abnormalities* occur when chromosomes break and fail to reattach properly. Instead of rejoining to each other, the two ends from one break may attach to the ends from other breaks. Depending on the number and positions of breaks, many kinds of rearrangements can be formed.

Such changes have been extensively analyzed in fruit flies, corn, mice, and other experimental organisms, where chromosome breaks can be induced by radiation, by a wide variety of chemicals, and by certain viruses. But detailed analyses of structural aberrations in humans have become possible only recently as technical advances have allowed investigators to analyze the fine banding and molecular structure of human chromosomes. These developments have led to the discovery of many new chromosomal syndromes. Also, through some detailed analyses of rare aberrations (such as that of Bruce Bryer), geneticists have been able to find the locations of several important genes that had eluded them for years.

However they arise and whatever they may contribute to genetic analyses, such defects inflict tremendous pain and suffering on human

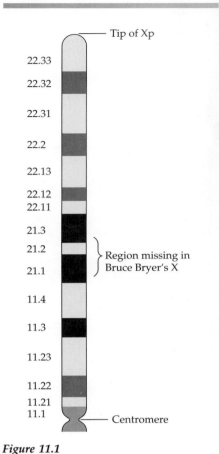

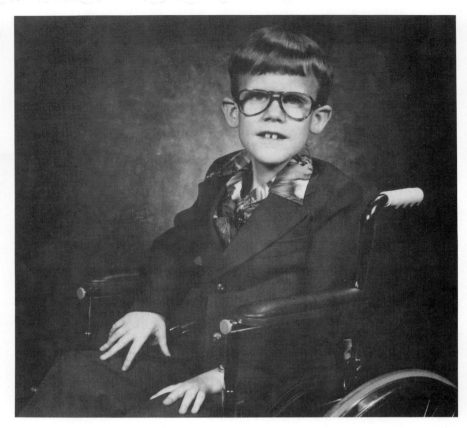

Figure 11.1
Bruce Bryer and a diagram of his X chromosome. Shown here is the standard Q-banding pattern of the X chromosome short arm, as presented by Francke et al. in 1985. The brace indicates the missing section in Bruce Bryer's X chromosome. This deletion included the loci for Duchenne muscular dystrophy as well as chronic granulomatous disease, retinitis pigmentosa, and McLeod's red cell phenotype. Figure 11.14 shows the fine structure of the region and the order of some of these genes. (Photograph © Eastern Washington State Historical Society; courtesy Muscular Dystrophy Association.)

carriers and their families. Virtually all of the chromosomal syndromes involve mental retardation and growth defects; seizures are very common, as are unusually low birth weight and failure to thrive. Heart and other vascular defects, abnormalities in the sex organs, and unusual arrays of fingerprint patterns are frequently observed. Different syndromes are recognized by different combinations of these abnormalities. Among the 2,000+ chromosomal abnormalities that have been cataloged, about 100 are classified as syndromes, roughly half of which show some uniformity in phenotypic traits.

DELETIONS AND DUPLICATIONS

Pieces of chromosomes can be lost or gained. The loss of a chromosome piece, that is, a **deletion**, can take place in several ways. As shown in Figure 11.2A, a single break can lead to the loss of a chromosome tip, which is what usually happens in humans. Less frequent in humans is the occurrence of two breaks with loss of the intervening segment (Figure 11.2B). If breaks occur at both chromosome ends, the acentric fragments will get lost, and a ring may be formed by union of the broken ends (Figure 11 2C). Deletions may also occur as a by-product of rearrangements between different chromosomes.

Except in the X chromosome, deletions of more than a few genes tend to be lethal even when heterozygous (i.e., when the altered chromosome is paired with a normal, nondeleted chromosome). Those deletions that are not lethal are often associated with severe phenotypic defects. Deletions of the short arms of chromosomes 4, 5, 9, 11, and 18, as

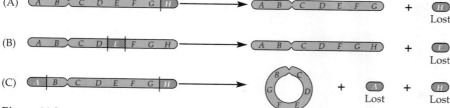

Figure 11.2
Origin of deletions. (A) One break near a chromosome tip. A small piece is lost. (B) Two breaks followed by the loss of a small internal piece. (C) Two breaks followed by the loss of both tips and formation of a ring chromosome. Chromosome fragments that are acentric (i.e., lacking a centromere) will be lost from the nucleus in the subsequent cell division.

well as the long arms of 11, 13, and 18, are the ones most often seen in infants. But deletions for every chromosome tip and for many interstitial (internal) segments are known as well. Chromosomes 4 and 5 seem to be particularly susceptible to breakage. Or perhaps breaks in these two chromosomes are simply less lethal than those occurring in some other chromosome and thus lead to a type of sampling bias.

The most common deletion found in humans (about 1 in 50,000 live births) is loss of much of the short arm of chromosome 5. Because newborns with 5p− usually have a high-pitched mewing cry like that of a kitten, this aberration is called the *cat-cry syndrome* (Figure 11.3).* Other features are variable, but they usually include a small head with a round face, wide-set eyes with epicanthal folds on the upper lids (covering the inner corners of the eyes), low-set ears, and slow growth. Most affected individuals survive beyond childhood, but with severe mental retardation. Deletion sizes differ from case to case, but the crucial region absent in all children with the cat-cry syndrome is a tiny segment near the middle of 5p15.

* Recall that p refers to the short arm of a chromosome and q to the long arm. A minus sign following the p or q means that the arm is too short; a plus sign means that it is too long.

Figure 11.3
A patient with the cat-cry syndrome, associated with a 5p− (or partial 5p−) deletion. (A) Affected newborn, showing moon face and wide-set eyes. (B) Same child at four years; the cat-cry and moon face have disappeared, but the epicanthal skin fold has persisted. The ears are somewhat misshapen and low set. (Photos from Valentine 1986.) (C) Chromosomes 4 and 5 from an affected individual, showing a deletion of part of the short arm of one chromosome 5. (Chromosome photographs courtesy of Irene Uchida, Oshawa General Hospital.)

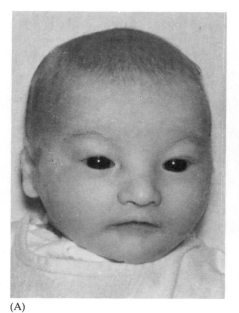

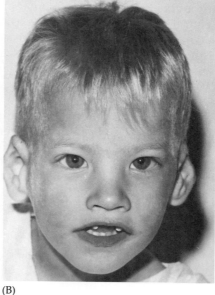

(A)   (B)

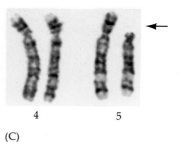

4   5

(C)

233

(A)

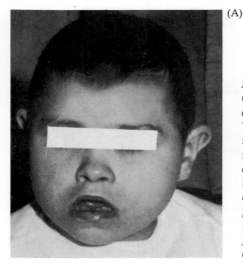

(B)

18

Figure 11.4
(A) A child with deletion of one-fourth to one-half of the long arm of chromosome 18. The downturned mouth, absence of the midline indentation that runs between the nose and the mouth, and the peculiar shape of the ears are characteristic of these patients. The carp mouth trait also occurs in certain other chromosomal syndromes. (A, photograph by F. Sergovich, from Valentine 1986. (B) Chromosomes from an affected individual. (B, photographs courtesy of Irene Uchida, Oshawa General Hospital.)

The next most common deletion involves the short arm of chromosome 4 (i.e., 4p−) and is often associated with improper fusion of the midline of the body. Affected infants usually have wide-set eyes, a broad nose, low-set and malformed ears, a very small lower jaw, and a cleft palate. Heart, lung, and skeletal abnormalities are common, and in males the penis may be improperly fused. Birth weight is very low. These babies generally fail to thrive, and most of them die young.

Large deletions of chromosome 18 are also known. Loss of the short arm (18p−) results in growth retardation and mental retardation. Loss of part of the long arm (18q−) causes these anomalies and also defects of the ears, eyes, mouth, and facial structure (Figure 11.4). Heart and minor skeletal defects are also frequent. In addition, both deletions are often associated with the lack of a blood serum protein called immunoglobulin A (IgA).

Pieces of chromosomes can be gained as well as lost. These **duplications** may result when, following three chromosome breaks, a segment of one chromosome is inserted elsewhere in the homologous chromosome or into a different chromosome (Figure 11.5A). Duplications can also arise through errors in chromosome replication, through errors in crossing over (Figure 11.5B), or by normal crossing over between chromosomes that are heterozygous for certain other structural abnormalities. When the resultant **unbalanced** gamete (i.e., one having extra or missing pieces of chromosomes) combines with a normal gamete, the zygote will possess an extra batch of genes, ranging from just a few genes to an entire chromosome arm.

The phenotypic effects of duplications vary according to their size and position. In general, however, small duplications are less harmful (even when homozygous) than are deletions of comparable size. Thus, there seem to be fewer well-defined human syndromes associated with duplications alone. One example is *trisomy 9p*, which invariably involves mental retardation. Other features vary, but may include a small head, a bulbous nose, wide-set, squinting eyes, a lopsided grin, and a "worried" expression.

Geneticists believe that duplications have played an important role in the evolution of many species by providing additional pieces of DNA that are then free to mutate to different genes. For example, it appears that the independent but very similar genes controlling the production of certain protein chains in human hemoglobin all arose from duplications of a single ancestral gene (Chapter 12).

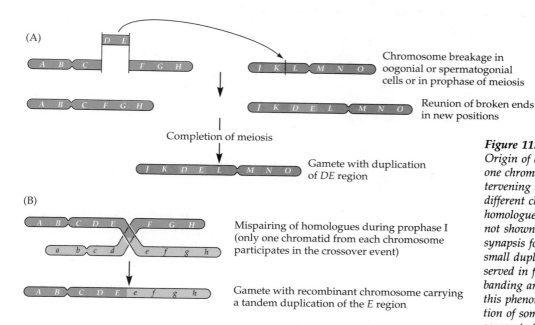

(A)

Chromosome breakage in oogonial or spermatogonial cells or in prophase of meiosis

Reunion of broken ends in new positions

Completion of meiosis

Gamete with duplication of *DE* region

(B)

Mispairing of homologues during prophase I (only one chromatid from each chromosome participates in the crossover event)

Gamete with recombinant chromosome carrying a tandem duplication of the *E* region

Figure 11.5
Origin of duplications. (A) Two breaks in one chromosome, with insertion of the intervening segment into a third break in a different chromosome. **Note:** *The normal homologues of these two chromosomes are not shown here. (B) Rare out-of-register synapsis followed by crossing over. Such small duplicated segments have been observed in fruit flies, in which very precise banding analysis is possible. In humans, this phenomenon accounts for the formation of some regions with many duplicated segments (Figure 13.13). The gametes with duplicated regions, when combined with normal gametes, produce zygotes with extra genetic material. Gametes with deletions can also be produced in these processes, but these have not been indicated on the diagrams.*

Microdeletions and Microduplications

In the past few years, a new class of chromosomal syndromes has come to light. These disorders were thought to be transmitted as simple Mendelian traits, often showing autosomal dominant inheritance. But high-resolution banding analyses have revealed that in a few cases they are associated with extremely tiny deletions or duplications. Sometimes such **microdeletions** or **microduplications** are so small that they can be detected only by molecular techniques. These syndromes are often associated with additional, unrelated features that may result from the loss or addition of genes located next to the locus in question. The half dozen known microdeletion syndromes, for example, usually involve mental retardation. The great phenotypic variability seen in these syndromes probably reflects the great variability in the size of the microdeletions and microduplications.

Tiny deletions in 15q, near the centromere, are associated with the *Prader-Willi syndrome (PWS)*—a disorder characterized in infants by poor muscle tone, poor reflexes, poor feeding, and underdevelopment of the gonads and external genitals.* Children with PWS are mentally retarded and extremely obese; they are also very short, with small hands and feet. Tiny internal deletions in 13q are associated with *retinoblastoma* (tumor of the retina of the eye); their incidence is about 1 in 100,000 live births. A deletion in 11p causes the *WAGR syndrome*, a complex that includes <u>W</u>ilms tumor (a cancer of the kidney), <u>a</u>niridia (absence of the iris of the eye), ambiguous <u>g</u>enitals and gonadoblastoma (tumor of the gonad), and mental <u>r</u>etardation. Such cancer-related deletions are discussed in Chapter 15.

The first-known example of a microduplication syndrome in humans was reported in 1991. *Charcot-Marie-Tooth (CMT) disease*, named after the physicians who first described it in 1886, is a fairly common neuromuscular condition with a frequency of about 1 in 2,500. Type 1 CMT (CMT1) is usually inherited as an autosomal dominant, fully penetrant disorder with variable phenotype and variable age of onset. Often beginning in their early teens, affected individuals show weakness and wasting of certain lower leg muscles, which then spread to other muscles of the legs, arms, feet, and hands. Deformities and loss of tendon reflexes and feeling in the feet are common, and loss of muscle coordination of the arms may occur

---

* But about 33% of individuals with PWS show no detectable chromosomal defects. Recall also that the PWS locus is subject to imprinting.

Table 11.1

Some important deletion and duplication syndromes. Many of the autosomal syndromes involve mental retardation and growth retardation, in addition to some characteristic features.

DELETIONS

Whole Arm	Partial Arm	Microdeletion
4p (Wolf-Hirshhorn)	partial 4q	8q24 (Langer-Giedion)
5p (cat-cry)	7qter	11p13 (Wilms tumor)
9p	interstitial 11p	13q14 (retinoblastoma)
12p	partial 11q	15q11 (Prader-Willi)
18p	distal 13	17p11 (Smith-Magenis)
Xp (Turner)	interstitial 13q	17p13 (Miller-Dieker)
Xq (Turner)	partial 18q	22q11 (DiGeorge)
	ring 18	Xp21 (Duchenne muscular dystrophy)
	ring 21	
	proximal 21	
	ring 22	
	ring X	
	partial Xp	
	partial Xq	

DUPLICATIONS

Whole Arm	Partial Arm	Microduplication
4p	partial 4q	4p16.3 (Huntington disease)[a]
5p	partial 6p	11p15 (Beckwith-Wiedemann)
9p	6qter	17p11–12 (Charcot-Marie-Tooth)
10p	partial 7q	19q13 (Myotonic dystrophy)[a]
12p	partial 9q	Xq21–22 (Kennedy disease)[a]
20p	partial 10q	Xq27.3 (Fragile X syndrome)[a]
Xq isochromosome	partial 11q	
	11;22 translocation	
	partial 13q: proximal	
	partial 13q: distal	
	proximal 14	
	proximal 15	

Source: Adapted from de Grouchy and Turleau 1990, and from de la Chapelle 1990.

[a]The extra genetic material consists of a repeated triplet.

too. The cause is gradual degeneration of the myelin sheath of affected nerves, followed by nerve thickening and blockage of electrical signals. In fact, abnormally low nerve conduction speeds are detectable at birth, long before the other symptoms appear.

Autosomal recessive and X-linked forms occur too, and there are at least three different loci associated with CMT1. In the early 1980s a CMT1 gene was mapped to chromosome 1 in several large pedigrees, but later investigations found that this linkage did not hold for all cases. For example, a chromosome 17 linkage has been reported for a number of large pedigrees from widely varied ethnic groups. Molecular analyses reveal, quite surprisingly, that the defective allele is not a point mutation or a deletion. Rather, it is a submicroscopic duplication near the centromere in 17p. How the duplication arose and how it causes the CMT phenotype are not known. But because the CMT1 phenotype can be confused with other neuromuscular conditions, the new discovery has immediate practical value as a diagnostic tool. It has also led geneticists to wonder if microduplications may be associated with other human diseases.

Table 11.1 summarizes the most important deletion and duplication syndromes of various types: whole arm, partial arm, and submicroscopic.

TRANSLOCATIONS

Breaks in two or more nonhomologous chromosomes, followed by reattachments in new combinations, can lead to the formation of **translocations** (Figure 11.6). This process is fairly common, and in some cases the rearranged chromosomes may be transmitted through many successive generations. If the rearrangement of chromosome parts is complete, with no leftover pieces, the translocation is **reciprocal**. Because all of the genetic material is still present but in a different arrangement, a heterozygote for such a translocation is said to be **balanced**, and the phenotype is usually normal.

Translocations between and among acrocentric chromosomes (i.e., numbers 13, 14, 15, 21, and 22) represent a special case known as a **Robertsonian translocation**, in honor of an investigator who studied such translocations in other organisms. Robertsonian translocations are the most common structural abnormality in humans. Figure 11.7A shows a reciprocal exchange: The long arms of two acrocentric chromosomes are joined together at the centromere, as are the two heterochromatic tips. The tiny heterochromatic chromosome so formed, which apparently carries no important genes, is usually lost. But the longer translocated chromosome contains the full complement of essential genes from two nonhomologous chromosomes. A gamete with this balanced chromosome, when combined with a normal gamete, gives rise to a balanced heterozygote. The resulting individual has a total of only 45 chromosomes, yet is phenotypically normal.

Another special type of aberration is a metacentric chromosome consisting of two identical arms (Figure 11.7B); this aberration may or may not represent an actual translocation. Such chromosomes are called **isochromosomes** (Greek *iso*, "alike," "equal") and are thought to be formed by misdivision of centromeres. Instead of the usual lengthwise separation

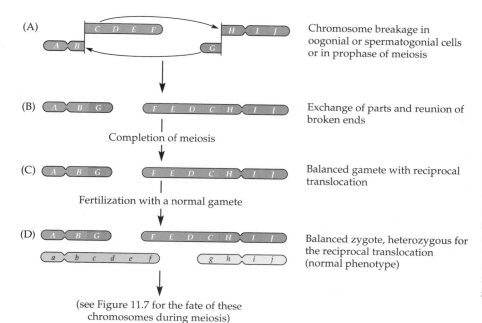

(A) Chromosome breakage in oogonial or spermatogonial cells or in prophase of meiosis

Completion of meiosis

(B) Exchange of parts and reunion of broken ends

(C) Balanced gamete with reciprocal translocation

Fertilization with a normal gamete

(D) Balanced zygote, heterozygous for the reciprocal translocation (normal phenotype)

(see Figure 11.7 for the fate of these chromosomes during meiosis)

Figure 11.6
Origin of a reciprocal translocation. (A) Breaks occur in each of two nonhomologous chromosomes. **Note:** The normal homologues of these two chromosomes are not shown here. (B) The pieces without centromeres exchange places. (C) One possible pattern of segregation puts the two translocated chromosomes into one gamete. This condition is balanced because all chromosomal segments are present; they are simply rearranged. (D) Combination of this gamete with a normal one gives a zygote that is heterozygous for the translocation; that is, two chromosomes carry translocations and their homologues do not. Notice that two pairs of chromosomes are involved.

237

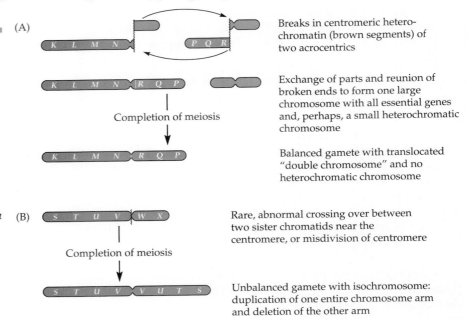

(A)

Breaks in centromeric hetero-chromatin (brown segments) of two acrocentrics

Exchange of parts and reunion of broken ends to form one large chromosome with all essential genes and, perhaps, a small heterochromatic chromosome

Completion of meiosis

Balanced gamete with translocated "double chromosome" and no heterochromatic chromosome

(B)

Rare, abnormal crossing over between two sister chromatids near the centromere, or misdivision of centromere

Completion of meiosis

Unbalanced gamete with isochromosome: duplication of one entire chromosome arm and deletion of the other arm

Figure 11.7
Special types of translocations. (A) Robertsonian translocation. Two breaks occur very near the centromeres of acrocentric chromosomes, whose short heterochromatic arms carry no essential genes. Thus, a gamete that carries only the chromosome with the joined long arms is essentially balanced. **Note:** *The normal homologues of these two chromosomes are not shown here. (B) Isochromosome. An isochromosome is formed by the union of two identical chromosome arms, usually by misdivision of the centromere.*

between duplicate chromatids, the centromere seems to divide crosswise, thereby joining two identical chromatid arms. Union of broken sister chromatid arms also can produce an isochromosome. During the course of meiosis, the two isochromosomes separate; 50% of the gametes contain a duplication of one arm and a deletion of the other.

Isochromosomes for the long arm of group D or G chromosomes, when heterozygous with normal chromosomes, lead to effective trisomy for that chromosome, even though the total number of chromosomes is only 46. Indeed, a few cases of Down syndrome are known to have a long-arm isochromosome of 21 and a normal chromosome 21.

Behavior of Translocations in Meiosis

The meiotic behavior of reciprocal translocations, when heterozygous with structurally normal chromosomes, is rather complicated. Because prophase I synapsis involves gene-by-gene pairing, the translocated chromosomes must form some unusual patterns to match up with their partners. The simplest translocations will arrange themselves into a distinctive crosslike shape (Figure 11.8), from which pairs of chromosomes segregate in three possible ways during anaphase I. If segregation is alternate (i.e., if chromosomes that are diagonally opposite move to the same pole), then the resultant gametes and zygotes are balanced. Note that among these balanced gametes, half will carry the two translocated chromosomes and half will carry the two normal chromosomes. If separation is not alternate (i.e., if it occurs in the vertical plane or in the horizontal plane shown in Figure 11.8), then the resultant gametes will be unbalanced. Any zygote formed by combination with a normal gamete will be unbalanced and abnormal.

The three kinds of segregation are not equally frequent, but enough unbalanced gametes are produced to reduce the fertility of translocation heterozygotes. In experimental plants and animals, such reduced fertility is passed on to about half the offspring (i.e., to translocation carriers). Such a dramatic decline in reproductive fitness may not be so obvious in humans because they have few offspring and are likely to "replace" mis-

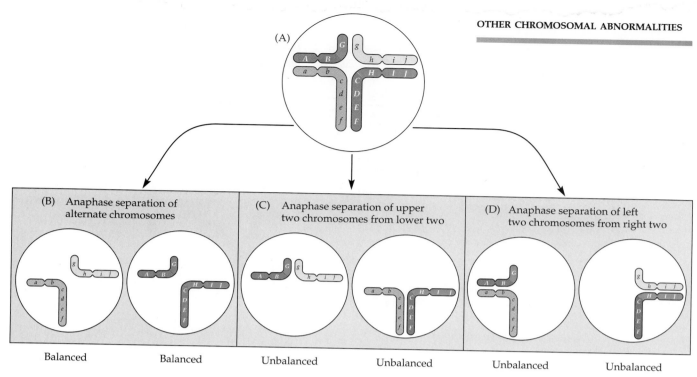

(A)

(B) Anaphase separation of alternate chromosomes

Balanced   Balanced

(C) Anaphase separation of upper two chromosomes from lower two

Unbalanced   Unbalanced

(D) Anaphase separation of left two chromosomes from right two

Unbalanced   Unbalanced

Figure 11.8
Meiotic synapsis and segregation from the reciprocal translocation heterozygote shown in Figure 11.6. (A) The only configuration that allows gene-by-gene pairing between two translocated and two nontranslocated chromosomes is a crosslike configuration. (For simplicity, we have not shown the chromosomes in their actual doubled state.) From this, the chromosomes usually segregate two-by-two at anaphase 1. (B) Of the two balanced combinations, one has two normal chromosomes and the other, two translocated chromosomes. The latter, if combined with gametes carrying normal chromosomes, will again give rise to translocation heterozygotes. (C, D) Four unbalanced combinations, each with the duplication of one segment and deletion of another.

carriages, stillbirths, or early infant deaths with subsequent pregnancies. Yet some translocation heterozygotes will show a clear reduction in fertility, and most are likely to exhibit a higher-than-normal rate of loss of progeny.

Translocation Down Syndrome

Recall that nearly all cases of Down syndrome occur sporadically in families whose other members are unaffected. The affected individuals have the karyotype 47,XX,+21 or 47,XY,+21, the extra chromosome coming from nondisjunction in one of the karyotypically normal parents, usually the mother. In contrast to this situation, about 4% of people with Down syndrome have 46 rather than 47 chromosomes, one of which is abnormally long. This long chromosome is a Robertsonian translocation, usually involving chromosomes 14 and 21. Because nondisjunction is not the cause, no maternal age effect is seen here. Although the percentage of such cases is very small, they are important to identify because this form of Down syndrome may recur in the same family.

A carrier parent of an individual with translocation Down syndrome may have only 45 chromosomes, one of which is 14q21q. As shown in Figure 11.9A, three chromosomes in these translocation heterozygotes contain nearly all of the material present in the original four chromosomes. The tiny tips of chromosomes 14 and 21 are lost, but the fact that the translocation carrier is unaffected means no essential genes are missing.

Now consider the kinds of gametes that a carrier might produce (Figure 11.9B). During meiosis, two chromosomes will usually proceed to one pole and one to the other; thus, six types of gametes are possible. Of these, two types are balanced: one bearing the normal chromosomes 14 and 21 and another carrying the single large 14q21q translocation. When combined with structurally normal chromosomes during fertilization (Figure 11.9C), the former will yield a normal zygote, and the latter will give

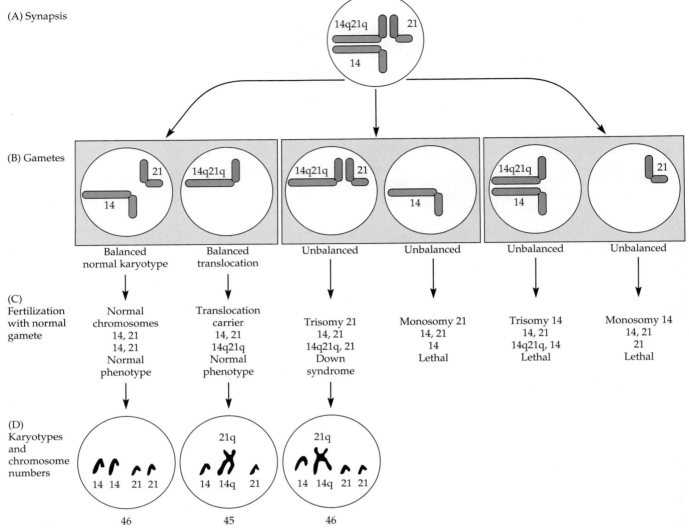

(A) Synapsis

(B) Gametes

| Balanced normal karyotype | Balanced translocation | Unbalanced | Unbalanced | Unbalanced | Unbalanced |

(C) Fertilization with normal gamete

| Normal chromosomes 14, 21 14, 21 Normal phenotype | Translocation carrier 14, 21 14q21q Normal phenotype | Trisomy 21 14, 21 14q21q, 21 Down syndrome | Monosomy 21 14, 21 14 Lethal | Trisomy 14 14, 21 14q21q, 14 Lethal | Monosomy 14 14, 21 21 Lethal |

(D) Karyotypes and chromosome numbers

14 14 21 21   46

21q 14 14q 21   45

21q 14 14q 21 21   46

Figure 11.9

Consequences for a phenotypically normal person heterozygous for a 14q21q Robertsonian translocation. (A) Synapsis of the 14q21q translocation with the normal 14 and normal 21. Missing is the small reciprocal chromosome containing the short arms of 14 and 21. Compare with Figures 11.7A and 11.8. (B) Segregation (diagonal, up-down, left-right) gives rise to six types of gametes, of which four are unbalanced and two are balanced. (C) Fertilization with normal gametes results in three lethal zygotes, one zygote leading to Down syndrome, and two producing normal phenotypes. Of the two normal phenotypes, one is karyotypically normal and one is again heterozygous for the translocation, like the parent. (D) Karyotypes.

rise to a translocation heterozygote with a normal phenotype. Thus, the translocation may pass unnoticed through a number of generations. Among the types of unbalanced gametes produced by a carrier, only one will lead to a viable zygote after combining with a gamete with normal chromosomes. This zygote will have an extra dose of chromosome 21 and will develop Down syndrome. A carrier parent may therefore produce up to three kinds of offspring: phenotypically and karyotypically normal, phenotypically unaffected translocation heterozygote, and translocation Down syndrome (Figure 11.10). The theoretical risk of having a child with Down syndrome is thus 1/3, but the actual observed frequencies show considerable variation.

Although the multiple occurrence of Down syndrome in a family may suggest the presence of a (14q21q) translocation, it is not a certainty. In fact, most cases of multiple Down syndrome are due to multiple instances of sporadic nondisjunction in karyotypically normal parents.

Other Translocations

Translocations are not uncommon in humans, occurring with a frequency of about 1 in 500 among newborns and a few percent among people who are mentally retarded. Collectively they involve all chromosome arms,

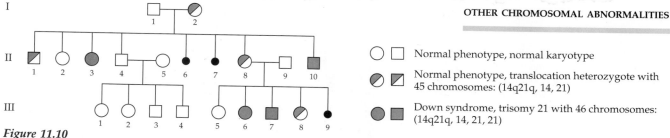

Figure 11.10

A hypothetical pedigree of translocation Down syndrome. Note that several affected persons may appear in the same sibship and be related to other affected persons through translocation carriers with normal phenotypes. This pattern contrasts with that of nondisjunction Down syndrome, in which the occurrence of affected persons is usually sporadic.

more or less at random—with the exception of a few high-affinity combinations such as 11q with 22q, 9 with 22, and 9 with 15. An unbalanced translocation involving chromosomes 11 and 22, for example, has given rise to several disorders associated with partial trisomies of 11q. People with balanced translocations are usually phenotypically normal, although their fertility may be reduced by repeated spontaneous abortions. X/autosome translocations have played a key role in mapping the genes for a number of X-linked recessive diseases.

Autosomal translocations are sometimes involved in specific types of cancer; balanced reciprocal translocations, for example, occur in leukemias and lymphomas. They will be discussed further in Chapter 15.

OTHER ABNORMALITIES

Four additional types of chromosomal aberrations are seen humans. Three are compatible with survival beyond the fetal stage, and one is not.

Inversions

When two breaks occur in one chromosome and the intervening segment gets inverted before the broken ends rejoin, the resultant aberration is known as an **inversion** (Figure 11.11A). The reversed segment may or may not include the centromere. An inversion does not ordinarily produce a distinctive phenotype. Instead, its presence is detected by unusual meiotic events or unusual segregation patterns in organisms heterozygous for the inversion. Although estimates differ widely, the incidence of inversions in humans might be as high as 1–2%. The particular chromosomes involved and the locations of breakpoints appear to be highly nonrandom. Chromosome 9, for example, accounts for over 10% of all known inversions, and certain breakpoints recur with great frequency.

Gene-by-gene pairing during synapsis of an inversion heterozygote requires an inverted chromosome and its normal homologue to form a loop during prophase I (Figure 11.11B). What happens then depends on (1) whether or not crossing over occurs within the inversion loop and (2) whether or not the centromere is included within the inversion. When no crossovers occur within the inversion loop, no unbalanced chromosomes are formed and there is no effect on fertility.

But when crossing over occurs within the inversion loop, the two crossover chromatids end up with both duplications and deletions. In addition, crossover chromatids may be either **dicentric** (have two centromeres) or **acentric** (have no centromeres), resulting in highly unbalanced gametes and zygote death. (For details, see Mange and Mange 1990 or

241

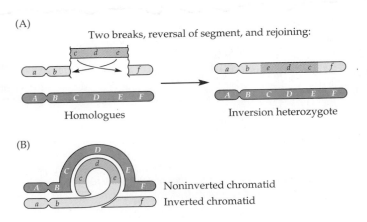

any general genetics textbook.) The overall effect on fertility varies with the frequency of crossing over within the inversion. Thus, larger inversions may give rise to more defective gametes and to greater infertility.

Ring Chromosomes

Recall that when two breaks occur in one chromosome, **ring chromosomes** can be formed (Figure 11.2C). *Acentric rings*, lacking a centromere, usually get lost in subsequent cell divisions; but *centric rings*, because they have a centromere, can be passed on to daughter cells. If there is no crossing over during meiosis of a centric ring chromosome, it will be transmitted intact. But with certain crossovers, a double-sized *dicentric ring* (or other abnormal structures) will be formed. During anaphase the dicentric ring will break, and both daughter cells may receive centric rings of different sizes. This process can give rise to many different cell lines in one individual.

Rings have been reported for every human chromosome, but there is great variation in phenotypes, ranging from normal to highly abnormal. Although syndromes have been described for many different ring chromosomes, only those involving ring chromosomes 13, 14, 18, 21, and 22 present any uniformity of phenotypic features.

Polyploidy

Entire chromosome sets may also be present in excess, a situation called **polyploidy** (Greek *poly*, "many"; English *ploid*, "set"). An organism may be **triploid**, with three representatives of each chromosome, or **tetraploid**, with four of each chromosome, and so on. Triploids may arise when two sperm fertilize one egg. Tetraploids may arise when the failure of a meiotic division produces diploid rather than haploid gametes. Two diploid gametes could then unite to form a tetraploid zygote.

Polyploidy is quite common in plants and often has great commercial value. Many garden flowers, ornamental plants, fruit and forest trees, grasses, and crop plants fall into this category. Because of irregularities in synapsis during meiosis, many polyploids are infertile. (This inability to set seeds can give rise to seedless varieties of fruits.) Although during synapsis no more than two chromosomes can be paired for any given segment of a chromosome, different segments of the same chromosome may be paired with different partners. The resultant gametes will contain one representative of certain chromosomes and multiple representatives of others, leading to zygote lethality.

Polyploidy is rare among animals and lethal in humans. It seems that when sex determination depends on a certain balance between sex chro-

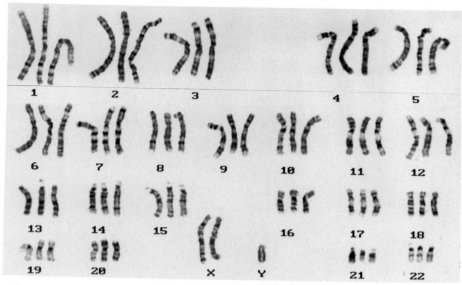

Figure 11.12
The triploid (Q-banded) karyotype 69,XXY of a stillborn infant with congenital malfor-
mations. Trypsin-Giesma banding allows each set of three homologous chromosomes to be
distinguished from others. (Courtesy of Irene Uchida, Oshawa General Hospital.)

mosomes and autosomes or on the XY karyotype, changes in ploidy generally lead to the formation of many intersexes. In humans, triploidy is very common among embryos that are spontaneously aborted during the first three months of pregnancy (Figure 11.12). Many are incompletely formed, but others show few or no gross abnormalities, and there is no distinctive phenotype. The most common defect seen is an enlarged placenta, sometimes with grape-shaped cysts. Growth of the embryos is stunted, resulting in low birth weight of the few live-born triploids. Intersexuality is also the rule. Webbing of the fingers and toes is common, and various abnormalities of the head, skeleton, heart, kidneys, and brain have also been reported. The few babies with triploidy who have survived beyond birth are mosaics with some diploid tissues. They almost always show severe mental and motor retardation.

Triploid embryos generally have one parent's genome in double dose and the other parent's genome in single dose. Does it matter which parent contributes the double dose? Yes, it does. Recall from our discussion of imprinting in Chapter 6 that human triploids with two sets of paternal chromosomes have a large placenta, whereas those with two sets of maternal chromosomes have an underdeveloped placenta. Indeed, the former develop the phenotype, we have just described, and the latter develop so poorly that they usually abort in very early pregnancy.

Uniparental Disomy

The basic rules of inheritance assume that an individual gets one chromosome of each homologous pair from the mother and the other one from the father. So when unaffected parents have a child affected with a recessive disorder, we know that (barring nonpaternity or a new mutation) both parents must be heterozygous for the mutant allele. Recently, however, a few cases have been reported in which a child with a recessive disorder has only *one* carrier parent, the other parent being homozygous normal at that locus. The child's karyotype looks normal, but molecular analyses show that the child has received *both* mutant alleles from the

Normal situation; no uniparental disomy	(A1) MII nondisjunction in one parent & MI or MII nondisjunction in other parent	(A2) MI nondisjunction in one parent & MI or MII nondisjunction in other parent	(B) Meiotic nondisjunction in one parent & duplication of monosomic chromosome in zygote
Gametes	Disomic Nullisomic	Disomic Nullisomic	
Zygote			Duplication
Embryo			

carrier parent and *no* member of that particular allele or chromosome from the other parent. This odd phenomenon is called **uniparental disomy** (*uni* meaning "one", and *disomy* referring to the presence of exactly two chromosomes of a homologous pair).*

As shown in Figure 11.13, there are several ways by which uniparental disomy can arise. In all cases it requires at least two events of abnormal cell division. Nondisjunction in both parents, followed by the complementary union of a disomic and a nullisomic gamete, is one possible cause (Figure 11.13A). Note that if the offspring's two disomic chromosomes are *identical*, nondisjunction must have occurred during meiosis II in the parent who contributed them. But if the offspring's two disomic chromosomes are *different*, nondisjunction had to occur during meiosis I in the contributing parent. Additional scenarios, involving chromosome loss or duplication in zygotes or in embryonic cells, can also be invoked (Figure 11.13B–D).

Uniparental disomy has been reported in a few cases of cystic fibrosis (chromosome 7), Prader-Willi and Angelman syndromes (chromosome 15), and homozygotes for chromosomal aberrations. The actual frequency of this phenomenon in the human population is unknown. But some geneticists suggest that it may be more common than previously realized and should be considered in the following cases:

Rare homozygous recessives having only one heterozygous parent
Rare recessive disorders accompanied by unexpected features such as mental retardation or growth retardation
Females affected with rare X-linked recessive disorders
Father-to-son transmission of X-linked recessive disorders
Transmission of a rare recessive disorder from parent to child
Transmission of an apparently balanced chromosomal aberration from a healthy parent to a child with developmental defects

FREQUENCIES OF CHROMOSOMAL ABNORMALITIES

Karyotype data are available for several categories of cells and phenotypes in humans, ranging from gametes to newborns to adults who are mentally

* Note that *bi*parental disomy is the normal state of affairs.

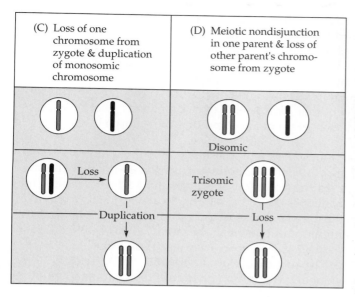

Figure 11.13
Some mechanisms (each involving two errors of cell division) that can give rise to uniparental disomy, whereby a child with a recessive condition has just one heterozygous parent. (For the sake of simplicity, we consider an autosomal pair of chromosomes and ignore crossing over.) Each dark red chromosome inherited from one parent carries the same recessive allele. Black chromosomes are from the other parent, as shown for the normal situation on the far left. (Adapted from Spence et al. 1988.)
(A) Nondisjunction during meiosis in both parents, followed by the complementary union of a disomic and a nullisomic gamete. (A1) If the disomic gamete resulted from nondisjunction during the second meiotic division (MII), the two uniparental chromosomes in the zygote will be identical. (A2) If it resulted from nondisjunction during the first meiotic division (MI), the two uniparental chromosomes will be nonidentical. (B) Nondisjunction in one parent, followed by union of a nullisomic and a normal gamete. The monosomic chromosome in the zygote undergoes mitotic duplication (without anaphase separation). (C) Loss of one chromosome from a normal zygote, followed by mitotic duplication of the remaining chromosome in the zygote. (D) Nondisjunction during meiosis in one parent, followed by the union of a disomic and a normal gamete. The trisomic zygote loses the other parent's chromosome.

retarded. Some of the frequency estimates are fairly crude stemming from the different methods of sampling a given group, the different karyotyping techniques used over a period of years, and the small sample size for some groups.

Gametes and Prenatal Tests

The chromosomes in human gametes are very difficult to study, but researchers are improving their techniques. Quite surprisingly, even normal fertile males produce about 10% karyotypically abnormal sperm, with chromosomal abnormalities increasing as sperm count decreases.

Chorionic villus sampling (CVS), described in Chapter 19, provides a way to detect certain chromosomal and metabolic disorders in fetal tissue during early pregnancy. Studies indicate that about 3% of 10-week-old embryos have chromosomal abnormalities. This frequency is a little higher than the roughly 2% of chromosomal abnormalities found with *amniocentesis*, a prenatal test that is done at approximately the fifteenth or sixteenth week of pregnancy—and probably accounts for the estimated rate of loss (from spontaneous abortion) of embryos and fetuses between weeks 10 and 16+ of pregnancy.

Abortions, Stillbirths, and Neonatal Deaths

Getting reliable data on spontaneous abortions (commonly called miscarriages) is very difficult. For one thing, it is not always clear whether an abortion (defined as termination of pregnancy before 20 to 22 weeks of development or when embryonic weight is less than 400–500 g) is spontaneous or induced. Yet the distinction is critical, because the former show a much higher frequency of abnormalities of all types than do the latter.

At least 15–20% of all conceptions are known to end up as *detectable* spontaneous abortions, and the actual frequency of early embryonic loss may be as high as 50–60%.* But researchers can now detect tiny amounts of a hormone (human chorionic gonadotropin) secreted by a week-old implanted embryo, which allows them to measure very early pregnancy loss. In one study of healthy women who were trying to conceive, 22%

---

* Some investigators think that over 75% of all human conceptions are spontaneously aborted, most before the first missed menstrual period.

of pregnancies ended before the pregnancy was clinically noticeable—that is, after implantation but before the sixth week of development. An additional 9% were lost following clinical confirmation, totaling 31% of pregnancies ended *after* implantation. What is still not known is how much loss occurs *before* implantation, that is, during the first week of development. Estimates range widely, from about 15% to 55%. Researchers basically agree, however, that about 90% of spontaneous abortions occur during the first trimester of pregnancy and that the embryo is usually less than 8 weeks old.

Fetal loss occurs for a variety of reasons. Overall, scientists estimate that at least one-half of all known spontaneous losses involve chromosomal abnormalities. In the youngest abortuses (up to 4 weeks old), the frequency of abnormalities is 90–100%; among those aged 5–8 weeks, it is about 60%; and in those 9–12 weeks of age, it ranges from 12% to 32%. The single most common chromosomal aberration is 45,X; indeed, it is thought that only 1% of all 45,X zygotes ever reach full term. As a group, trisomics are the most frequent class of abnormality, with trisomy 16 by far the most common among these. Polyploids are the next most frequent class of abnormality. Virtually all autosomal monosomies seem to be lost very early in development, probably before implantation.

In contrast to spontaneous abortions, only about 5% of induced abortions exhibit chromosomal abnormalities. But here, too, the earlier the gestational age, the more frequent the abnormalities. About half of the aberrations are autosomal trisomies. Triploidies and 45,X are also fairly frequent.

Roughly 1% of all recognized pregnancies end in *stillbirths*, whose observed rate of chromosomal abnormalities is about one-tenth of that found in spontaneous abortuses. Trisomy 18 is by far the most common aberration. A striking maternal age effect has been noted in these studies: About 35% of stillbirths and neonatal deaths from mothers over age 40 had karyotypic aberrations, whereas only 6% of those born to mothers under 40 showed chromosomal abnormalities.

Newborns and Some Adults

There is a further tenfold decrease in the frequency of chromosomal abnormalities found in live births (below that found in stillbirths). The reported rate of abnormal karyotypes is 1 per 160 to 200 births. Among the various abnormal karyotypes, about 35% are considered of clinical importance; these include the autosomal trisomies, the Klinefelter and Turner syndromes, and the unbalanced translocations that are accompanied by congenital malformations.

No comprehensive karyotypic studies of randomly chosen adults have been reported. But small surveys have detected a variety of chromosomal abnormalities, as well as an increase in mosaicism of sex chromosome aneuploidy (e.g., the number of 45,X cells), in both aging males and females. Among couples who have had two or more spontaneous abortions, the rate of chromosomal abnormality is greater than in randomly selected adults.

Studies of people who are mentally retarded reveal many types of abnormalities. Down syndrome (trisomy 21) accounts for about 10% of all retarded children, and the fragile X syndrome comes in second. Also, among institutionalized males who are mentally subnormal, about 3% are either 47,XXY or 47,XYY. Balanced translocations and inversions also appear more frequently among people who are mentally subnormal than among newborns.

It is clear that chromosomal abnormalities constitute a significant medical problem. Yet the grief and the burden to families and to society would be enormously increased if a large proportion of karyotypically abnormal conceptions were not spontaneously aborted early in pregnancy. In fact, live-born humans still show a much higher rate of chromosomal abnormalities than that found in any other species—even after spontaneous abortion eliminates over 95% of defective conceptions. The reasons for this situation are not clear. Perhaps it is the price we pay for human evolution. Or perhaps a high rate of mutation, chromosomal abnormalities, and spontaneous abortions leads to the spacing out of births in a species that requires an extremely long period for child raising. In any event, as Dorothy Warburton (1987) points out:

> In these days of efficient contraception, postponement of childbearing, and a small number of carefully planned pregnancies, low fertility and embryonic or fetal loss often cause disappointment and grief when they occur. Few couples realize how common such problems are, and many think that they, or their physicians, are somehow inadequate. . . . Except for maternal age, almost no genetic or environmental factors are known to alter the rates of human chromosomal abnormalities. Thus, physicians and the public should be taught that some degree of reproductive loss is normal and that the best insurance against being childless may be not to allot too short a period for childbearing.

CHROMOSOMAL ABNORMALITIES AND GENE MAPPING

Chromosomal abnormalities provide useful tools for narrowing down the locations of specific genes. This fact is especially true for human geneticists, who for decades had to rely entirely on pedigree studies and whatever turned up in the way of chromosomal variants. But since the early 1980s, the precision of all mapping studies has been greatly refined and enhanced by molecular genetic technologies (Chapter 13).

Deletion mapping hinges on the principle that the absence of a chromosome segment should be correlated with reduced expression, non-expression, or abnormal expression of genes on that segment. Males with X chromosome deletions, for example, usually have some kind of defect or disorder. Duplications have also been used to assign genes to specific regions of chromosomes. From analyses of enzyme levels in several individuals who carry different duplications or deletions in the same chromosome, it is sometimes possible to determine the location of the gene responsible for that enzyme. For example, two loci on chromosome 21—the *superoxide dismutase (SOD) enzyme* and antiviral protein—show dosage effects: Trisomics produce excess product, and monosomics have a reduced amount.

Recall that isolation of the Y-linked locus for TDF (testis-determining factor) was made possible by DNA analyses of several dozen "sex-reversed" individuals. These individuals were XY females who had a deletion in their Y chromosome and XX males who had a tiny piece of a Y chromosome attached to one X (Figure 6.7). By relating the sex of the carrier to the original location of the deleted or added piece of Y, and also by looking for the smallest "common denominator," the researchers were able to pinpoint the critical male-determining region.

Studies of pedigrees with translocations (sometimes accompanied by

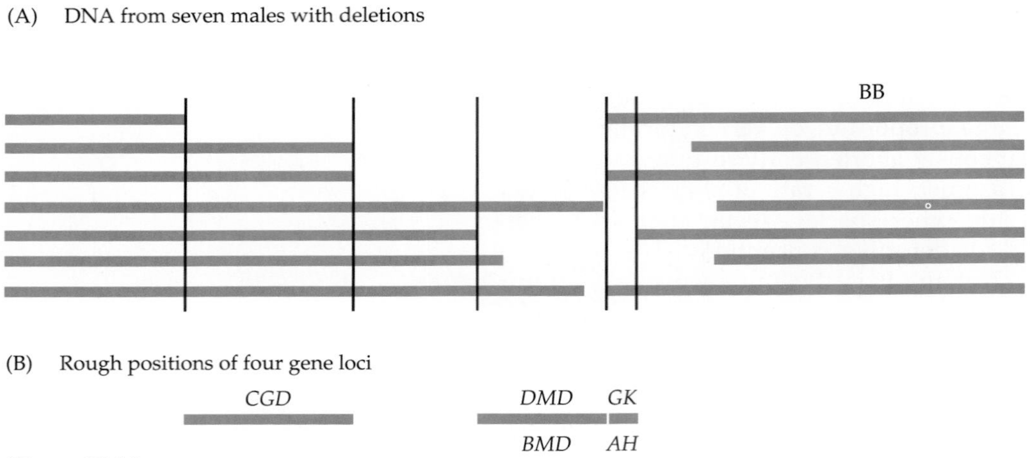

(A)   DNA from seven males with deletions

BB

Patient does (+) or does not (−) have the following disorder(s):

CGD	DMD/BMD	GK	AH
+	+	−	−
−	+	+	+
−	+	−	−
−	−	+	+
−	+	+	+
−	+	+	+
−	+	−	−

(B)   Rough positions of four gene loci

CGD   DMD   GK

             BMD   AH

Figure 11.14
Deletion mapping of the Duchenne muscular dystrophy (DMD) locus, based on analyses of DNA from individuals with deletions that overlap in region 21 of the X chromosome short arm. The centromere lies to the left and the telomere to the right. (A) The open spaces in the seven lines represent deleted segments of DNA from seven patients with one or more of the following disorders: chronic granulomatous disease (CGD), Duchenne/Becker muscular dystrophy (DMD/BMD), glycerol kinase (GK) deficiency, and adrenal hypoplasia (AH). The top line is Bruce Bryer's DNA. (B) Approximate positions of four linked gene loci, as derived from molecular data. Note: AH is now known to lie just to the right of GK. Not included in this study were the genes for Bruce Bryer's retinitis pigmentosa (which probably lies just to the left of CGD) and McLeod red cell phenotype (which lies between CGD and DMD). (Adapted from van Ommen et al. 1986.)

deletions or duplications) have also been important in gene mapping. Translocation and deletion mapping, combined with molecular techniques, finally allowed geneticists to localize the X-linked recessive Duchenne muscular dystrophy (DMD) gene in 1986. The first clue to its general position came in the 1970s, with reports of a few rare DMD females who had X/autosome translocations. Although the autosomal breakpoints varied from case to case, every DMD female had a breakpoint at Xp21—the large band in the middle of the short arm. Several years later, researchers began to analyze the DNA of a few rare DMD males who suffered from additional X-linked recessive disorders. One of these patients, Bruce Bryer (described at the beginning of this chapter), also had chronic granulomatous disease (CGD), retinitis pigmentosa (RP), McLeod syndrome —and a tiny deletion in Xp21. Another DMD male also had glycerol kinase deficiency (GK) and adrenal hypoplasia (AH)—and a similar-sized, partially overlapping deletion in the same region. By fitting together all the various puzzle pieces of cytogenetic, phenotypic, and molecular data, a team of Dutch, American, and Canadian researchers was able to come up with a concise description of the chromosomal region involved in these diseases (Figure 11.14). They determined the relative positions of the various loci. They also constructed a map of the molecular landmarks in this region, covering about 4 million DNA base pairs.

Comparable studies of other important disease loci and any closely linked chromosomal abnormalities will continue to add greatly to our understanding. Meanwhile, the discovery of so many deletions associated with "monogenic" disorders such as Duchenne muscular dystrophy has muddied the distinction between simple point mutations (such as a base substitution, deletion, or addition) and tiny chromosomal aberrations. As Epstein (1988) points out, aneuploidy may be a continuum affecting single genes as well as chromosome segments and whole chromosomes.

SUMMARY

1. Chromosomal rearrangements arise when segments of broken chromosomes reattach in different ways. Detached acentric pieces of chromosomes are lost.

2. Deletions of more than a few genes are usually lethal, even when heterozygous with a normal chromosome.

3. Deletions and duplications are often the by-products of meiotic segregation from other rearrangements. Extra chromosomal pieces (duplications, trisomies) are phenotypically less harmful than the loss of chromosomal material (deletions, monosomies). Microdeletions and microduplications are associated with several severe disorders.

4. Meiotic segregation from heterozygous translocations leads to aberrations in some gametes. Translocation Down syndrome is associated with specific types of rearrangements.

5. Crossing over within the pairing loop of inversion heterozygotes usually leads to imbalance in some gametes. In some cases, dicentrics and acentrics are also formed.

6. Ring chromosomes are unstable during cell division, giving rise to several different cell lines in one individual.

7. Polyploidy is rare in live-born animals, but triploidy accounts for about 6% of early human spontaneous abortions.

8. Uniparental disomy occurs when an individual receives two chromosomes of a homologous pair from one parent and none from the other. The two chromosomes may include both parental homologues or two copies of exactly the same chromosome. Thus, a child with only one carrier parent may be affected with a recessive disorder.

9. The frequency of chromosomal abnormalities (mostly trisomies, polyploids, and 45,X) among early spontaneous abortuses is at least 50%. The observed spontaneous abortion rate of 15–20% would be much higher if very early (preimplantation) losses could be detected.

10. About 5% of stillbirths and roughly 0.5% of live births exhibit chromosomal abnormalities. About half of the latter consist of structural aberrations (mostly translocations), and half are trisomies.

11. Many persons who are mentally subnormal are also karyotypically abnormal, having either Down syndrome or the fragile X syndrome.

12. Traditional pedigree studies involving chromosome aberrations have proved very useful, when combined with molecular methods, in localizing genes to specific regions of the chromosomes.

KEY TERMS

acentric
deletion
deletion mapping
dicentric
duplication

inversion
isochromosome
microdeletion
microduplication
polyploidy

reciprocal translocation
ring chromosome
Robertsonian translocation
tetraploid
triploid

uniparental disomy
variant chromosome

QUESTIONS

1. Diagram the synaptic configurations (with gene-by-gene pairing) of a pair of chromosomes *heterozygous* for (a) an interstitial deletion, (b) a tandem duplication, and (c) a simple (i.e., nonreciprocal) translocation with one chromosome arm being attached to the end of a different chromosome.

2. Diagram the synaptic configurations that would be seen in chromosome pairs *homozygous* (a) for an inversion and (b) for a reciprocal translocation.

3. Among the parents of people with Down syndrome, a few have 45 chromosomes, including a Robertsonian translocation between the two chromosomes 21. (a) What types of gametes would they form? (b) Could they produce unaffected progeny?

4. A published case history describes a woman with 45 chromosomes, one of which was a Robertsonian translocation between the two chromosomes 15. She was childless, having had 12 spontaneous abortions. What was the reason for this?

5. How can a metacentric chromosome be changed to an acrocentric one?

6. List two ways by which a triploid could arise.

7. Is there any way by which fertilization between two aneuploid gametes produced by individuals with the same type of translocation heterozygote could produce a euploid zygote? (*Hint:* See Figure 11.8.)

8. 46,XX individuals with one X chromosome that is an iso-chromosome of the long arm are often mosaics, possessing a 45,X cell line as well. How could this come about? (Assume that after replication in the zygote, the two duplicates of the isolated Xq form an isochromosome.)

9. Vidaud et al. (1989) report a case of father-to-son transmission of hemophilia A, with nonpaternity ruled out. How could this come about? Be specific.

10. In Figure 11.13A1 and 2, describing mechanisms for uniparental disomy, we assume no crossing over between homologues in either parent. How would the outcomes be changed with the occurrence of crossing over?

11. How can uniparental disomy be useful in chromosome mapping?

12. Lupski et al. (1991) report a mating between two people affected with Charcot-Marie-Tooth syndrome. Both of their children were affected, the second one very severely and with onset before the age of one year. What is the most likely explanation for the second child's phenotype?

FURTHER READING

For details about chromosome structure and abnormalities, consult Gardner and Sutherland (1989), de Grouchy and Turleau (1984, 1990), and Valentine (1986). Dellarco, Voytek, and Hollaender (1985), Bond and Chandley (1983), and Epstein (1986, 1988) concentrate on particular aspects of aneuploidy. Weiss (1990) and Engel (1980) discuss the concept and potential effects of uniparental disomy. Ledbetter and Cavenee (1989) describe microdeletion and microduplication syndromes. For reviews of frequencies of chromosomal abnormalities in abortions, live births, and other categories, see Chandley (1990), Hassold (1986), Boué, Boué, and Gropp (1985), and Hook (1985).

DNA *in the* Organism

12

Our genes, segments of long DNA molecules, are one of the hottest subjects for modern biological research. Because genes provide the blueprints for development and growth, they affect the entire range of human differences from the obvious to the subtle, from the trivial to the life-threatening. Thus, unraveling the mysteries of DNA structure and function promises not only to refine the practice of medicine, but also to profoundly change how we see ourselves and others.

The more we learn about our genes, and the more deftly we are able to manipulate them, the less clearly defined become the boundaries between normal and abnormal conditions. For example, because human growth hormone can be produced by culturing genetically engineered cells, it is now available in substantial amounts. The question then arises as to who should receive growth hormone treatment to alter their growth patterns. In other words, "How short is too short?" (Werth 1991). As public debate heats up on more and more genetic fronts, many people worry about unwarranted uses of biotechnology and fear the possibility of unfair discrimination based on information about an individual's DNA.

But some situations are not subject to doubt. Scientists know, for example, that extraordinarily small alterations in DNA may lead to grave genetic diseases. In 1957 Vernon Ingram, a biochemist then working in England, reported his findings about the new concept of molecular disease. He showed that the abnormal hemoglobin in the red blood cells of patients with *sickle-cell anemia* differs from normal hemoglobin by just one amino acid, the basic building block of proteins. A few years later, researchers learned that this amino acid change results from the alteration of just one base among the many thousands making up the genes that code for hemoglobin. Unfortunately, however, new scientific understanding does not always lead to quick cures, and we still have no effective long-term treatment for what one patient described as the "pain, strokes, seizures, and leg ulcers, the ridicule from peers, low self-esteem, desire to die, and diminishing hope for the future associated with sickle cell anemia." (Huntley 1984)

In Chapter 2, we presented a short introduction to **DNA, deoxyribonucleic acid**, pointing out that its basic structure is a double helix. This discovery, made in 1953 by Nobel prize winners James Watson and Francis Crick, was perhaps the most important biological advance of the century. In this chapter we describe in greater detail the activity of DNA, showing that genes are like the three-by-five cards in a recipe box. Just as a recipe tells a cook how to put together a simple sauce or an elaborate soufflé, the DNA segment that forms a particular gene tells the cells of the body how to assemble a particular protein. The ingredients for this synthesis are there in the cells, derived from the things we eat.

In Chapter 13, we will look at the many clever manipulations of DNA that have provided both basic knowledge and practical applications. The

THE DOUBLE HELIX
Complementary Bases
Replication
Bubbles and Forks

GENE EXPRESSION: DNA →
RNA → PROTEIN
Types of RNA
Proteins
Translation

THREE-LETTER WORDS
Cracking the Code
Properties of the Code

GENE ORGANIZATION
Exons and Introns
Gene Families
Repetitive DNA
Mobile Genes

MUTATIONS
Substitutions and Frameshifts
Chemical Mutagenesis
Radiation Mutagenesis

first breakthrough in DNA technology occurred in 1973, when Paul Berg of Stanford University and his colleagues invented methods for hooking together pieces of DNA from different species. The product of such splicing operations is called *recombinant DNA*. Other scientists have come up with additional techniques for handling the genetic material. Biochemical expertise has now been so sharpened that DNA molecules can be copied, tagged, sequenced, sized, sliced, spliced, and hybridized in precise ways.

Chapters 14 through 16 discuss details of some inherited genetic disorders, the genetic aspects of cancer, and the intricacies of the immune system. Together, Chapters 12 through 16 illustrate the power of genetic research and point up the need for an informed public to help guide and regulate the biological revolution.

THE DOUBLE HELIX

In his best-seller *The Double Helix*, James Watson (1968) recounts how early one morning he was playing with cardboard cutouts of the individual parts of DNA. Suddenly he became aware of possible regularities of structure that made a lot of sense—and "no fudging was required" to make the parts fit together. His colleague, Francis Crick, did not get more than halfway through the door before Watson let loose that the answer to everything was at hand. They excitedly worked and reworked a wire model of the DNA double helix. Less than two months later, their historic one-page report announcing the structure of the genetic material appeared in print (Watson and Crick 1953).

Watson, a virus geneticist, and Crick, a physicist, had begun their collaboration at Cambridge University in 1951 (Figure 12.1). They realized

Figure 12.1
James Watson (left) and Francis Crick at Cambridge University in 1953 with their original wire model of the DNA double helix. Watson was 25 years old at the time and Crick 37. (From Crick 1988.)

that determining the structure of DNA would be a grand coup, because evidence was accumulating that DNA was the genetic material. For example, the amount of DNA was known to be constant for all cells of a species (except for gametes, which of course contained half as much), a pattern that was not observed for other cell components. In 1944, it had been shown that free DNA, but not other cell components, could enter and change the genotype of bacteria. These changes were then passed on to subsequent, untreated generations. By 1952, even skeptics were convinced by these and other findings that heredity was controlled by the nucleic acids (DNA or, in a few cases, RNA).

We present here enough about DNA so that you can appreciate the nature of the genetic material and how genes replicate, mutate, store information, and have this information direct the structure of proteins. (For more depth, consult a detailed textbook of genetics, molecular biology, or biochemistry.)

Watson and Crick knew that DNA contained: (1) the sugar (**S**) deoxyribose (Figure 12.2); (2) phosphoric acid (**P**), and (3) four bases: adenine (**A**), thymine (**T**), guanine (**G**), and cytosine (**C**) (Figure 12.3). Furthermore, Erwin Chargaff at Columbia University had discovered that in DNA the amount of A equals the amount of T, and the amount of G equals the amount of C.

Watson and Crick also knew about the X-ray studies of DNA fibers done by Rosalind Franklin and Maurice Wilkins at King's College of the University of London.* They deduced that DNA had a helical structure, but it was unclear how many strands were present and whether the bases stuck out from a narrow helical core or in from a wider helix. The fit of the DNA parts had to satisfy the laws of structural chemistry, as well as Chargaff's rules and some dimensions of the helix revealed by the X-ray photographs. The structure also had to make some biological sense. Before they assembled the components in the right way (amply confirmed later), they invented and discarded a long series of wrong choices.

The predominant form of DNA in living cells is called *B-DNA*, a *right-handed* helix with certain physical dimensions. (Right-handed helices are like screws that advance into wood when twisted to the right.) B-DNA is the form modeled by Watson and Crick and sketched in Chapter 2: two

---

* With Watson and Crick in 1962, Wilkins also won a Nobel prize for this work. Unfortunately, Franklin died of cancer before the prizes (awarded only to living persons) were announced.

DNA IN THE ORGANISM

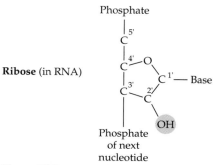

Figure 12.2
Partial structures of the five-carbon sugars, ribose and deoxyribose, present in nucleic acids. Each sugar includes a ring of four carbon atoms and one oxygen. The fifth carbon atom is outside the ring, and the carbons are numbered 1' to 5'. As indicated, the 3' and 5' carbons are bonded to phosphate groups, making a long alternating sugar and phosphate backbone structure. Each sugar binds one of four possible bases, whose structures are shown in Figure 12.3. Note that the "deoxy" sugar in DNA has one less oxygen than the sugar in RNA (see colored circles).

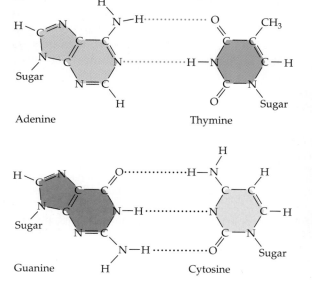

Figure 12.3
Structures of the four bases present in DNA. The bases are shown in their base-pairing configurations, that is, as if one base were on one strand of DNA and its pairing partner were on the other. Each pair consists of a larger purine (A or G) and a smaller pyrimidine (T or C), and could be flipped left for right. The hydrogen bonds that connect the purine-pyrimidine pair are indicated by dotted lines — two hydrogen bonds for an AT pair and three for a GC pair.

253

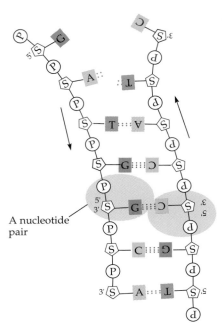

Figure 12.4
A portion of a double-stranded molecule of DNA showing AT and GC base pairing joining the two strands. (The helical form of the molecule is not shown.) The order of bases on one chain is irregular, but given the bases on one strand, those on the other are determined. Note that the two strands have opposite directionality, represented here by arrows that point from the 5' to the 3' end of a given molecule of deoxyribose sugar. In this diagram, the two chains are separating (at the top) as they might during DNA replication.

smoothly spiraling sugar-phosphate backbones with bases extending toward each other like the rungs of a twisted ladder, albeit a flexible, thread-like ladder. The width of the helix is constant at about 2 nm, and its length may be thousands to millions of times more than its width, depending on the source. (See the labyrinthine traces of DNA in Figure 2.7A.)

Complementary Bases

The point we emphasize here is not the three-dimensional measurements of the molecule but *the pairing of the bases*. The uniform width throughout the B-form helix means that each rung of the ladder is the same size, a larger purine (A or G) being paired with a smaller pyrimidine (T or C). Chemical restraints require, further, that purine A pair only with pyrimidine T, and purine G only with pyrimidine C (Figure 12.4). Thus, the sequence of bases in one strand is always **complementary** to the bases on the other strand according to the pairing rules: A opposite T (or vice versa) and G opposite C (or vice versa). The percentage of pairs that are AT or GC depends on the sequence along the length of a strand, however, and this varies from region to region, from molecule to molecule, and from organism to organism.

The basic building block of DNA is called a **nucleotide,*** and it consists of a deoxyribose sugar, a phosphate attached to the sugar's 5' carbon atom, and a base attached to the sugar's 1' carbon. (The sugar's 3' carbon is attached to the *next* nucleotide.) The nucleotides that are incorporated into DNA are synthesized from simpler compounds in the cell. Because the four nucleotides present in DNA differ from one another only in which base they carry, the terms *nucleotide pairing* and *base pairing* express the same complementarity.

Another interesting point of DNA structure is that the two sugar-phosphate backbones of the double helix run in "opposite" directions, like two-way traffic. Specifically, the orientations of the sugars, marked by labeling the 3' and 5' carbons, are reversed on the two strands. Note also that at the tip of a DNA molecule (refer to Figure 12.4), the sugar on one strand has a 5' carbon unattached to any more nucleotides, whereas the sugar on the other strand has its 3' carbon so exposed.

Replication

Complementary base pairing suggested to Watson and Crick a mechanism for the replication of DNA. Replication could occur by separation of the two strands, with each single strand acting as a **template** (i.e., a mold or pattern) for the formation of a new strand. Individual nucleotides, present in the cell nucleus, are captured and strung together successively, one at a time, opposite their complements in the old template strands, a process resulting in two identical double helices where there was once just one.

That DNA actually replicates in the way Watson and Crick suggested was shown by ingenious experiments with *Escherichia coli* by Matthew Meselson and Franklin Stahl (1958) at the California Institute of Technology. The bacterial cells were grown for several generations in medium with nutrients containing the heavy isotope nitrogen-15, rather than the common, lighter isotope nitrogen-14.[†] The purines and pyrimidines syn-

---

* More precisely, the nucleotides in RNA are called *ribo*nucleotides because the sugar component is *ribose*. Those in DNA are called *deoxyribo*nucleotides because the sugar is *deoxyribose*.

† Isotopes are variants of the same chemical element that may differ in some physical properties. For example, because it possesses an additional proton, nitrogen-15 (a rare isotope) is slightly heavier than nitrogen-14 (the common form). Some other isotopes are radioactive; for example, phosphorus-32 is sometimes used to label molecules of DNA, and sulfur-35 is sometimes used to label protein molecules.

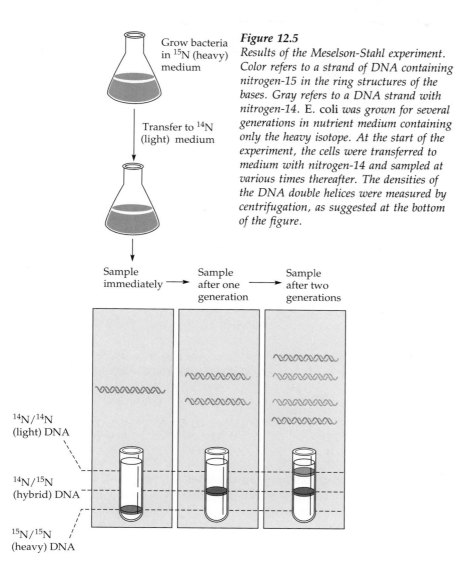

Figure 12.5
Results of the Meselson-Stahl experiment. Color refers to a strand of DNA containing nitrogen-15 in the ring structures of the bases. Gray refers to a DNA strand with nitrogen-14. E. coli was grown for several generations in nutrient medium containing only the heavy isotope. At the start of the experiment, the cells were transferred to medium with nitrogen-14 and sampled at various times thereafter. The densities of the DNA double helices were measured by centrifugation, as suggested at the bottom of the figure.

Grow bacteria in ^{15}N (heavy) medium

Transfer to ^{14}N (light) medium

Sample immediately → Sample after one generation → Sample after two generations

$^{14}N/^{14}N$ (light) DNA

$^{14}N/^{15}N$ (hybrid) DNA

$^{15}N/^{15}N$ (heavy) DNA

thesized by the bacteria incorporated nitrogen-15 into their structure, and therefore the bacterial DNA became slightly heavier than if the cells had been grown in the common isotope of nitrogen. At the start of the experiment, then, both strands of DNA in the double helix contained nitrogen-15, and the DNA was considered *heavy* (Figure 12.5A).

Next, the *E. coli* cells were grown for one generation in a medium containing only nutrients with nitrogen-14. If DNA replicates as outlined here (Figure 12.5B), then each new double helix should contain one heavy strand and one light strand, thus being *intermediate* in density. This was indeed the observed result. As further predicted, when the *E. coli* cells were allowed to undergo *two* generations of growth in nitrogen-14 (Figure 12.5C), half the DNA double helices were of *intermediate* density and half were *light*. (See also question 3.)

Bubbles and Forks

In higher organisms, the single DNA molecule that makes up each chromosome begins to replicate at thousands of sites, rather than at just one site as in *E. coli*. Local untwistings of the two strands at many places along the length of the helix expose sequences of bases to which enzymes can

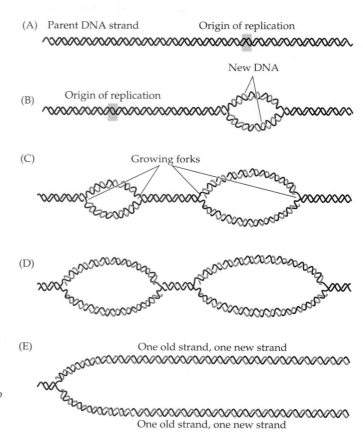

(A) Parent DNA strand   Origin of replication

New DNA

(B)   Origin of replication

(C)   Growing forks

(D)

(E)   One old strand, one new strand

One old strand, one new strand

Figure 12.6
DNA replication arising through local separations of the double helix and progressive synthesis. Newly synthesized DNA strands are shown in color. (B, C) Two bubbles are initiated, and new complementary strands are synthesized opposite each of the old strands. (D) The bubbles increase in size at both ends. (E) The bubbles have joined with each other (and with a third bubble off to the right). During a subsequent cell division, the newly formed double helices segregate to the two daughter cells.

bind and attach complementary nucleotides (Figure 12.6). Under an electron microscope, each local separation appears as a bubble in the otherwise still closely paired double helix.

The replication of DNA by complementary base pairing requires the presence of an enzyme called **DNA polymerase**, which travels along the old DNA strand and aids in the selection, placement, and bonding together of the new nucleotides. One of the interesting facts of biochemical life is that the new nucleotides are always added to the 3' end of the sugar of the nucleotide already in place in the growing new strand of DNA. The two strands of the original double helix run in opposite directions; consequently, the two new strands forming at one of the ends of a replicating bubble cannot both be synthesized in the same direction. Consider one of the Y-shaped replication forks at the right side of a bubble (Figure 12.7). The upper branch of the replication fork can continuously add nucleotides to the 3' end, but replication on the lower branch must wait until a length of the template strand has been exposed and then proceed "backward." The new fragments thus formed—called *Okazaki fragments* after their discoverer—are about 150 nucleotides long in mammalian cells. They are eventually joined together by another enzyme called **DNA ligase**, thus making this new strand continuous too.

In higher organisms, DNA is replicated in preparation for each mitotic cell division and prior to the first meiotic division of gametogenesis. Accurate replication is needed for transmission of exact genetic information from cell to daughter cells and from one generation to the next. High-fidelity reproduction is accomplished by several mechanisms, including what is called the *proofreading mechanism* of DNA polymerase. The last-attached nucleotide at the growing tip of the new strand must "settle in," becoming firmly base-paired to its complementary nucleotide on the template strand before another nucleotide can be added. DNA polymerase will remove any improperly matched nucleotide and try again for an exact AT or GC fit before continuing down the line. It is estimated that only one replication mistake remains for every 10^{10} base pairings—that's an average of roughly one mistake for each replication of a human cell with about 6×10^9 nucleotide pairs. For comparison, just one typographical error in this book of about 3×10^6 characters would constitute an error rate 2,000 times greater than that occurring in DNA replication.

GENE EXPRESSION: DNA → RNA → PROTEIN

Within cells, information encoded in DNA helps to direct the development and maintenance of the organism. Decoding of the information proceeds in two steps. First, as shown in Figure 12.8, base sequences in DNA are **transcribed** into complementary sequences of **RNA, ribonucleic acid**. Note that DNA is not turned into RNA; rather, DNA is like a recipe: It provides information that the cell uses to put together RNA. These molecules of RNA are made at specific chromosomal sites and then migrate across the nuclear membrane into the cytoplasm. In the second step, occurring at ribosomes in the cytoplasm, the information now encoded in the base sequence of RNA is **translated** into a string of amino acids, that is, a polypeptide. Again we emphasize that RNA is not turned into polypeptides; rather, RNA provides information that the cell uses to synthesize a polypeptide. Many polypeptides are the enzymes (or parts of enzymes) required for cell metabolism; other polypeptides are hormones, antibodies, structural elements, or other vital components of our bodies.

The terms *transcription* and *translation*, the two steps in **gene expression**, are well chosen. The former term means a transformation within the *same* language, such as from oral to written English, or from a string of deoxyribonucleotides to one of ribonucleotides (but still a language of nucleotides). The latter term describes a more profound change: from one language to another, say from English to French, or from a string of nucleotides to a string of amino acids.

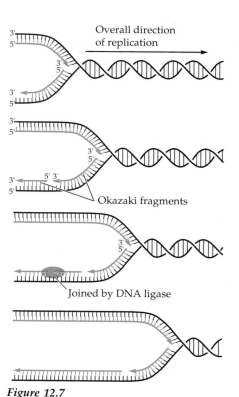

Figure 12.7
Events at a replicating fork. Newly synthesized DNA strands are shown in color. DNA polymerase adds new nucleotides on only the 3' end of the sugar already in place. Because the two strands run in opposite directions (see Figure 12.4), replication of one strand—the lower one here—occurs in discontinuous steps. The double helix must open for a length before synthesis can start at the point of the fork and proceed "backward" in the proper 5' to 3' direction. These segments butt up against each other, and their sugar-phosphate backbones are joined into a continuous strand by DNA ligase.

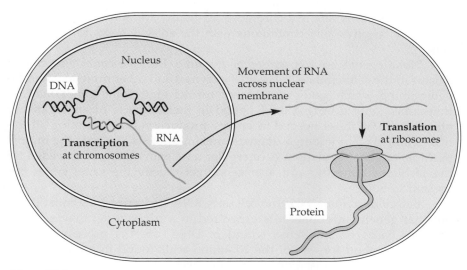

Figure 12.8
Decoding a gene. In a two-step process, a particular string of nucleotides in DNA directs a cell to assemble a particular string of amino acids into a polypeptide. The RNA molecules are intermediaries, carrying copies of genetic information from the nucleus to the cytoplasm.

257

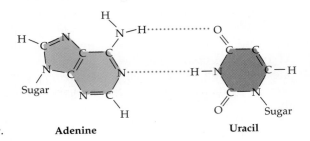

Figure 12.9
The pyrimidine base uracil, present in RNA. During transcription and translation, uracil pairs with adenine, as indicated. Thus, uracil plays the same role in RNA as the chemically similar thymine does in DNA. Comparison with Figure 12.3 shows that uracil lacks only the methyl group (CH3) present in thymine.

The process of transcription is similar to DNA replication, except that the complementary molecule that is formed is RNA, containing the sugar ribose (rather than deoxyribose—see Figure 12.2) and the base uracil (**U**) instead of thymine (Figure 12.9). Transcription is controlled by the enzyme **RNA polymerase**, which links together ribonucleotides one after another into a sequence complementary to the sequence of bases along one of the strands of DNA.

The process is illustrated in Figure 12.10. The place where RNA polymerase first attaches is a specific succession of bases called the **promoter**. Although promoters vary from gene to gene and from organism to organism, they all contain certain sequences in common: In eukaryotes, a seven-base sequence of virtually all T's and A's (the so-called TATA box) is part of all promoters. Transcription begins at a spot after the promoter, with new ribonucleotides being added in turn to the 3' end of the preceding ribonucleotides (following the same direction of synthesis used in DNA replication). The RNA polymerase moves off the DNA molecule at a certain **termination sequence**, completing the transcription of a gene into a complementary RNA copy. Each human chromosome includes a single, very long molecule of DNA partitioned into thousands of genes, each (usually) with its own promoter and terminator.

The question of which strand of a double helix acts as the template for the formation of complementary RNA is crucial, since one strand is obviously different from the other. For a particular gene, the strand that is actually transcribed is the one with the promoter sequence and is called the **sense strand**; the other, the **antisense strand**, is not used in transcription (but is used during DNA replication). The strand of the DNA double helix that makes sense at one place, however, may not be the one that makes sense elsewhere in a chromosome; that is, the sense strand for one gene may be physically continuous with the antisense strand of another gene.

Not all genes are transcribed all the time. Intricate controlling mechanisms in the cell regulate the flow of enzymes and other proteins, thereby determining which genes should be "turned on." Some genes may be transcribed only during a brief period in the life cycle of an organism; others, such as those needed in energy production, are operating all the time. Animals that undergo dramatic metamorphosis illustrate the important role of gene regulation: A caterpillar uses many genes not needed by the resultant butterfly, and vice versa; yet exactly the same genes are present in both.

Before describing how the cell uses RNA transcripts, we must point out that genetic information is expressed in a *triplet* code; that is, three adjacent bases on the sense strand of DNA, via the process of gene expression, stand for one of the 20 amino acids found in proteins. Consequently, during translation, three ribonucleotides in sequence along an RNA molecule are needed to position one amino acid in a polypeptide chain. But neither DNA nor the intermediary RNA are actually marked

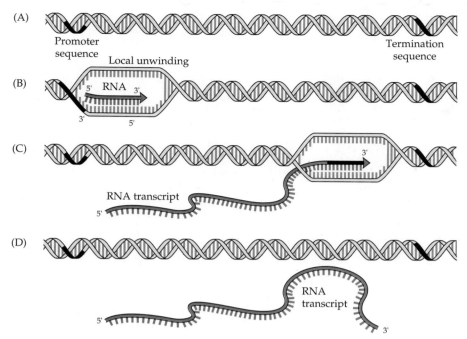

Figure 12.10
The process of transcription. (A) The gene to be transcribed lies between the promoter and terminator sequences (indicated by thick lines) on the sense strand of DNA. (B) Local unwinding of the DNA double helix occurs as the RNA polymerase (not shown) attaches to the promoter. One after another, ribonucleotides complementary to the DNA bases are added to the RNA transcript (shown in color) at its 3' end. (C) The growing RNA transcript peels off as it is formed. (D) When the polymerase reaches the termination site, both the enzyme and the new RNA transcript separate from the DNA, which assumes its original hydrogen-bonded state.

off into threes by any special chemical structure. Rather, *successive groups of three bases* are interpreted by the cellular chemistry as units of information. We present some details of the genetic code later in this chapter.

Types of RNA

There are three main types of RNA: *messenger* RNA (mRNA), *transfer* RNA (tRNA), and *ribosomal* RNA (rRNA), each transcribed from different sets of genes. All of the RNAs enter the cytoplasm, where they play specific roles during translation (Figure 12.11).

Messenger RNA (mRNA) plays the key informational role in protein synthesis. Only mRNA is actually translated into polypeptide chains, so mRNA is what is meant in the shorthand scheme DNA → RNA → protein. Indeed, a gene is often defined as the sequence of nucleotides that dictates the order of amino acids in a polypeptide via an mRNA transcript. The lengths of different mRNA molecules are not the same. Because most polypeptides are at least 100 amino acids long, most mRNA molecules are at least 300 nucleotides long.

Each successive group of three nucleotides in mRNA represents the code for one amino acid and is called a **codon**. There are as many codons strung together in a particular mRNA as there are amino acids in the corresponding polypeptide. (Actually, as we will see later, this statement is an oversimplification.) Because there are tens of thousands of different polypeptides, there are tens of thousands of different mRNA molecules. Any one mRNA molecule may be translated many times, although it is eventually degraded.

Transfer RNA (tRNA) molecules act as adapters, binding to specific amino acids in the cytoplasm and fitting them to appropriate codons of mRNA. They accomplish this task through a double specificity: One end of a particular tRNA molecule is able to bind one and only one of the 20 amino acids (Mlot 1989). The other end can recognize a specific codon of mRNA by means of a sequence of three bases called the **anticodon**. For example, the tRNA anticodon CCC recognizes and binds to the mRNA

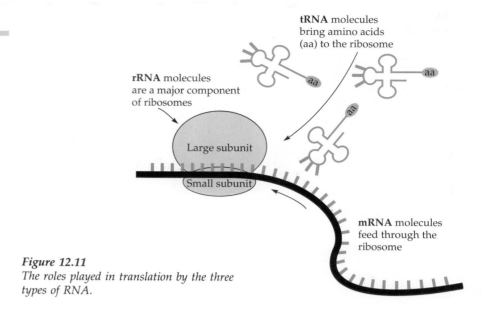

rRNA molecules
are a major component
of ribosomes

tRNA molecules
bring amino acids
(aa) to the ribosome

Large subunit

Small subunit

mRNA molecules
feed through the
ribosome

Figure 12.11
The roles played in translation by the three
types of RNA.

codon GGG by the usual rules of base pairing. The anticodon of a given tRNA molecule and the amino acid that it picks up are "matched" to each other; for example, the structure of the tRNA molecule with the anticodon CCC allows the tRNA to enzymatically bind to only one specific amino acid (proline, in this case).

There are several dozen different tRNA molecules, one or a few different types for each amino acid. Each tRNA molecule is about 80 nucleotides long and is coiled and folded into an intricate three-dimensional structure. The folding is brought about in part by complementary base pairing between different parts of the single-stranded molecule (Rich and Kim 1978).

Ribosomal RNA (rRNA), the third type of RNA, helps form the ribosomes, the "workbenches" on which protein synthesis occurs. The ribosomes in eukaryotes are composed of four types of rRNA and about 80 types of protein molecules arranged in a complex three-dimensional structure (Lake 1981). Recall that some of the rRNA molecules are transcribed off special chromosomal regions called *nucleolus organizers*. There are ten of these regions in the human genome, one on each short arm of the five pairs of acrocentric chromosomes. These regions have multiple copies—perhaps hundreds—of rRNA genes. Like the tRNA genes, the rRNA genes are transcribed *but not translated*.

The ribosomes come apart into two subunits—one small and one large—enabling them to clamp onto a molecule of mRNA, which then feeds through the ribosome during protein synthesis. The amino acids needed for incorporation into the polypeptide chains are transported to the ribosomes by the tRNA molecules. Thus, all three types of RNA are brought together at the ribosomes, as shown in Figure 12.11.

Proteins

During the process of translation, anywhere from a dozen to a thousand or more **amino acids** are strung together to form unbranched **polypeptide chains**. One or more of these chains—coiled, folded, intertwined, cross-linked—constitutes a **protein**. The marvelous diversity of proteins is suggested by the following rather random list: muscles, spider webs, tiger

Table 12.1
Representative proteins.

Name of Protein	Type of Protein[a]	Specific Function	Polypeptide Chains
DNA polymerase	Intracellular enzyme	Replicates and repairs DNA	One chain of about 1,000 amino acids
Trypsin	Digestive enzyme	Cleaves proteins into smaller polypeptides in small intestine	One chain of 223 amino acids (bovine)
Insulin	Hormone	Aids in moving glucose from the bloodstream into cells	Two chains of 21 and 30 amino acids
Hemoglobin	Transport protein	Carries oxygen in red blood cells from lungs to tissues	Four chains: two with 141, two with 146 amino acids; plus heme (nonprotein part)
Gamma globulin	Protective protein	Antibody molecules, which react with and incapacitate foreign substances	Four chains: two with 214 and two with 446 amino acids
Myosin	Contractile protein	Sliding arrays of myosin and actin molecules cause muscles to contract	Six chains: two fiberlike chains of about 1,800 amino acids; four small globular chains
Collagens	Structural element	Connective tissue in cartilage, tendons, bones, arteries, etc. (one-third of body protein)	Three chains of about 1,000 amino acids each woven into various triple helices
Keratin	Structural element	Main fibrous component of hair, skin, nails (and wool, silk, horns, hooves, feathers, turtle shells)	Complex helical or zigzag structures of thousands of amino acids each

[a] The first type, intracellular enzymes, is the largest class of proteins. The first five types are largely globular in shape and water-soluble. The last three types are largely fiberlike and water insoluble.

claws, snake venoms, eye lenses, blood clots, and catalysts for DNA replication (Table 12.1). The properties of a cell are often determined by the proteins it makes. For example, red blood cells manufacture large amounts of hemoglobin; certain white blood cells make antibodies; liver cells make an array of enzymes; and hair follicle cells make keratin. The human body contains more than 10,000 different proteins, each with a characteristic structure and function and each assembled from the same set of 20 amino acids.

Amino acids all have the same general structure: a central carbon atom (C) bonded to four groupings, two of which—the basic or *amino group (NH₂)* and the acid or *carboxyl group (COOH)*—give the class its name. A third group is a single hydrogen atom (H). The fourth group, the *side chain* (Figure 12.12), accounts for the differences in the properties of the various amino acids. The simplest amino acid is *glycine*, in which the side chain is another hydrogen atom. In *alanine*, the side chain is a methyl group (CH₃). In *phenylalanine*, a six-membered ring of carbons (a phenyl or benzene group) substitutes for one of the three hydrogens of the alanine methyl group. A few more structures are shown in Figure 12.12, and all 20 amino acids are listed along with their common abbreviations in Table 12.2.

Amino acids in polypeptide chains are joined to one another by the so-called *peptide bond* between the carboxyl group of one amino acid and the amino group of the next. The successive bonding of any number of amino acids into polypeptide chains leaves a free amino group on one end of the molecule and a free carboxyl group on the other. The complete

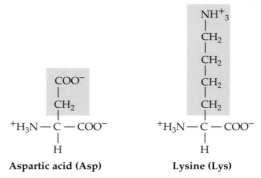

(A) Uncharged side groups

Glycine (Gly)

Amino group

Carboxyl group

Alanine (Ala)

Valine (Val)

Phenylalanine (Phe)

Tyrosine (Tyr)

Cysteine (Cys)

(B) Charged side groups (shown in charged state)

Aspartic acid (Asp)

Lysine (Lys)

Figure 12.12
Representative amino acids found in proteins. The unshaded portions are common to all the amino acids.

Table 12.2
The common abbreviations for the amino acids found in proteins.

Amino Acid	Abbreviation
Alanine	Ala
Arginine*	Arg
Asparagine	Asn
Aspartic acid	Asp
Cysteine	Cys
Glutamine	Gln
Glutamic acid	Glu
Glycine	Gly
Histidine*	His
Isoleucine	Ile
Leucine*	Leu
Lysine*	Lys
Methionine*	Met
Phenylalanine*	Phe
Proline	Pro
Serine	Ser
Threonine*	Thr
Tryptophan*	Trp
Tyrosine	Tyr
Valine*	Val

* The so-called essential amino acids. We cannot synthesize these particular amino acids for ourselves, but must have them premade in the proteins of the foods we eat.

sequence of amino acids in a small two-chain protein, human insulin, is illustrated in Figure 12.13. Note that covalent bonding between the sulfur atoms of some of the cysteine side chains (called disulfide bonds) holds the two chains together. (In addition, there is a disulfide bond between the cysteines at positions 6 and 11 of the A chain.) Not all proteins with multiple chains are held together this way, however.

The three-dimensional shape of a protein is crucial to its biological activity. Most proteins are globular, in part or in whole, because the polypeptide chains are compactly folded and looped, often in complex ways. In enzyme molecules, some groups of amino acids fold in from the surface to form a precisely shaped cavity called the *active site*, into which fit the substances that the enzyme acts on. Critical sites of function are also present in other proteins, such as the oxygen-binding sites of hemoglobin or the antigen-binding sites of antibody molecules.

Translation

The way a protein works ultimately comes down to its amino acid sequence, which is determined by the sequence of codons in mRNA (which in turn is determined by the string of corresponding DNA triplets). As noted earlier, translation occurs on the ribosomes with the aid of tRNAs that have been loaded with specific amino acids. In Figure 12.14, we see how the sixth amino acid of a growing polypeptide chain becomes bonded to the first five, which are already linked together. The large subunit of the ribosome has pockets for *two* tRNA molecules with their attached amino acids. The right pocket will fit only the tRNA with a *particular anticodon*, because a *particular codon* of mRNA forms the lower boundary of the pocket (Figure 12.14A). For example, if the sixth mRNA codon is CUC, then by the rules of base pairing, only that tRNA with the anticodon GAG will fit the pocket. The tRNA molecule that has the anticodon GAG is specific for the amino acid leucine (more about this in the next section). In this way, whenever the codon CUC passes through the ribosome, leucine is brought into a nearby position (with its amino group to the left and its carboxyl to the right).

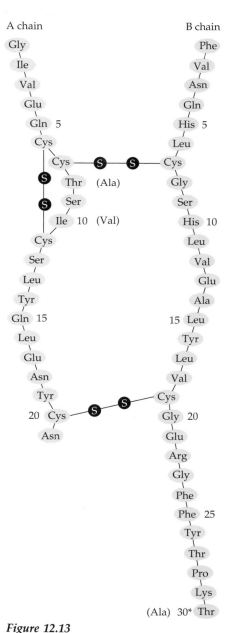

Figure 12.13
The two polypeptide chains of human insulin, but not showing the three-dimensional structure of the molecule. The amino acids are numbered from the free amino ends of the chains (top). Bovine insulin, which differs in the three positions indicated, was the first protein to be sequenced; this was accomplished by Cambridge University chemist Frederick Sanger, a Nobel laureate in 1958.

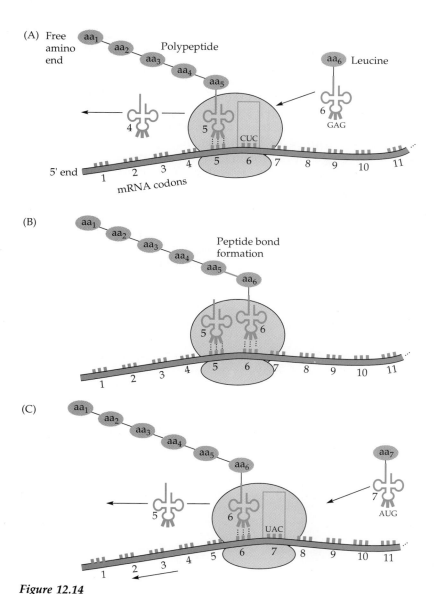

Figure 12.14
The process of translation. For clarity, the bases in mRNA are crowded a little into three-base codons; in reality, the bases are equally spaced along the molecule. The 5' end of the messenger feeds through the ribosome first, and the first mRNA codon leads to the placement of the amino acid that ends up with a free amino group. See text for further explanation.

263

In Figure 12.14B, the tRNA-leucine unit is stationed in the right pocket. In the left pocket is the fifth amino acid attached to its tRNA. As the fifth amino acid is released from its tRNA and becomes attached instead to leucine by a peptide bond, the length of the polypeptide is increased by one amino acid.

In Figure 12.14C, the mRNA molecule has moved one codon to the left; at the same time, the leucine-tRNA has skipped to the left pocket, since the two RNAs are joined by base pairing, codon to anticodon. We are back to a diagram like that in Figure 12.14A, but now one amino acid has been added to the polypeptide. The right pocket is free to accept the seventh tRNA and its associated amino acid. The fifth tRNA molecule, having contributed its amino acid to the polypeptide, is free to bind another molecule of the same amino acid to be incorporated later (if called for) as the mRNA feeds through the ribosome.

The mRNA molecule contains not only the codons for the polypeptide but also certain head and tail sequences that are not translated. In particular, the first translated codon (some distance in from the 5' tip of the molecule) always calls for the amino acid methionine, containing a modified amino group. (After protein synthesis is completed, this first amino acid may be cut off from the polypeptide.) Following the last translated codon is a so-called *stop codon*, or *termination signal*. At any given time, many ribosomes may be "working" the same mRNA molecule in single file, looking like beads on a string.

Note that the processes of transcription and translation both depend on base pairing of T with A (or U with A) and G with C (Figure 12.15). To say that a gene "codes for" a protein means that successive DNA triplets provide the information for the assembly of successive amino acids through two rounds of base pairing. In bacteria, in which there is no separation of nucleus from cytoplasm, one end of the mRNA molecule (the 3' end) is still being transcribed while the other end (5') is already being translated! Some remarkable electron microscope photographs of a bacterial gene being multiply transcribed as each of the mRNA transcripts is being multiply translated can be seen in Miller (1973).

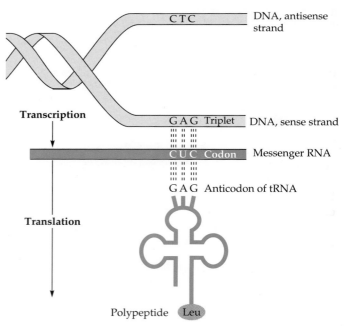

Figure 12.15
Base pairing during transcription and translation. The DNA triplet GAG in a polynucleotide provides the information that finally leads to the incorporation of the amino acid leucine into a polypeptide.

THREE-LETTER WORDS

By 1960 molecular biologists had established the importance of base pairing in the reactions of DNA and RNA. But they were pessimistic about quickly cracking the code—that is, discovering *which* specific sequences of nucleotides caused the assembly of *which* specific amino acids. Yet by 1965, several groups of investigators had ingeniously worked out the complete coding dictionary.

Because proteins are assembled from 20 amino acids, a string of three nucleotides was considered the most reasonable size for a coding unit: *One* nucleotide can specify only 4 items (Table 12.3A); *two* nucleotides in sequence can specify only 16 items (Table 12.3B); *three* nucleotides, however, carry more than enough information, 64 items in all (Table 12.3C), to code for all 20 amino acids. (Note that these numbers are successive powers of 4: 4^1, 4^2, and 4^3.) Using a type of mutation in which a base is added to (or deleted from) a gene, Crick and his associates were able to show that indeed *three* was the correct coding unit. (These gene alterations are called *frameshift* mutations and are discussed later in the chapter.)

Cracking the Code

Early experimenters trying to determine which specific group of three bases coded for which amino acid made their own *synthetic messenger RNA*. These informational molecules were added to a test tube containing all the ingredients that a cell uses to make proteins:

- All the individual amino acids
- All the transfer RNA molecules
- Ribosomes extracted from *E. coli*
- ATP and other energy-rich compounds
- A batch of appropriate enzymes

Remarkably, these in vitro systems spontaneously synthesized polypeptides. For example, when the artificial mRNA was polyuracil, which has a monotonous sequence of one base,

. . . UUUUUUUUUUUU . . .

the test tube was found after a time to contain a polypeptide containing only phenylalanine,

. . . Phe−Phe−Phe−Phe . . .

The conclusion was clear: UUU in messenger mRNA (or, by inference, AAA in DNA) was the genetic word for phenylalanine.

As biochemists developed more sophisticated means for hooking together different ribonucleotides in specific orders, the nucleic acid vocabulary was extended. Here are two examples. (See also question 11.)

Example 1. When synthetic mRNA with a repeating set of two nucleotides,

. . . UGUGUGUGUGUG . . .

was added to the protein-synthesizing system, the polypeptide

. . . Val−Cys−Val−Cys . . .

was made. Thus, valine is coded for by UGU and cysteine by GUG, or

Table 12.3
One-, two-, and three-letter code words that can be formed from an alphabet with the four letters A, G, T, and C.

(A) THE 4 DIFFERENT ONE-LETTER WORDS

A	G	T	C

(B) THE 16 DIFFERENT TWO-LETTER WORDS:

AA	GA	TA	CA
AG	GG	TG	CG
AT	GT	TT	CT
AC	GC	TC	CC

(C) THE 64 DIFFERENT THREE-LETTER WORDS:

AAA	AGA	ATA	ACA
AAG	AGG	ATG	ACG
AAT	AGT	ATT	ACT
AAC	AGC	ATC	ACC
GAA	GGA	GTA	GCA
GAG	GGG	GTG	GCG
GAT	GGT	GTT	GCT
GAC	GGC	GTC	GCC
TAA	TGA	TTA	TCA
TAG	TGG	TTG	TCG
TAT	TGT	TTT	TCT
TAC	TGC	TTC	TCC
CAA	CGA	CTA	CCA
CAG	CGG	CTG	CCG
CAT	CGT	CTT	CCT
CAC	CGC	CTC	CCC

265

vice versa. Because the in vitro system lacks a proper signal for initiating the translation process, it could start at the first or second or third base. These different starting places are said to produce different **reading frames** for the mRNA. In this particular example, however, the different reading frames give the same polypeptide product, an alternating string of valines and cysteines (although the first amino acid in the string does depend on the reading frame).

Example 2. The synthetic mRNA

. . . UUCUUCUUCUUC . . .

(a repeating trinucleotide) made *three different* polypeptides:

. . . Phe−Phe−Phe−Phe . . .
. . . Ser−Ser−Ser−Ser . . .
. . . Leu−Leu−Leu−Leu . . .

The three reading frames for the mRNA lead to repeated codons UUC or UCU or CUU; thus, these three codons stand for the three amino acids, although which stands for which is not clear from this experiment by itself. (You can confirm that a coding unit of two bases or four bases would have generated molecules consisting of three different amino acids in one polypeptide.)

On the basis of similar experiments, especially those using RNA molecules only three nucleotides long, researchers gave all 64 possible codons in mRNA the unambiguous assignments shown in Table 12.4. Crick (1988) provides an engaging account of this exciting period of biochemical sleuthing.

Table 12.4
The genetic code displayed as DNA triplets (boldface) and as mRNA codons (in parentheses).

FIRST BASE[a]	SECOND BASE			
	A (U)	**G** (C)	**T** (A)	**C** (G)
A (U)	**AAA** (UUU) **AAG** (UUC) } Phe **AAT** (UUA) **AAC** (UUG) } Leu	**AGA** (UCU) **AGG** (UCC) **AGT** (UCA) **AGC** (UCG) } Ser	**ATA** (UAU) **ATG** (UAC) } Tyr **ATT** (UAA) **ATC** (UAG) } Stop	**ACA** (UGU) **ACG** (UGC) } Cys **ACT** (UGA) Stop **ACC** (UGG) Trp
G (C)	**GAA** (CUU) **GAG** (CUC) **GAT** (CUA) **GAC** (CUG) } Leu	**GGA** (CCU) **GGG** (CCC) **GGT** (CCA) **GGC** (CCG) } Pro	**GTA** (CAU) **GTG** (CAC) } His **GTT** (CAA) **GTC** (CAG) } Gln	**GCA** (CGU) **GCG** (CGC) **GCT** (CGA) **GCC** (CGG) } Arg
T (A)	**TAA** (AUU) **TAG** (AUC) **TAT** (AUA) } Ile **TAC** (AUG) Met	**TGA** (ACU) **TGG** (ACC) **TGT** (ACA) **TGC** (ACG) } Thr	**TTA** (AAU) **TTG** (AAC) } Asn **TTT** (AAA) **TTC** (AAG) } Lys	**TCA** (AGU) **TCG** (AGC) } Ser **TCT** (AGA) **TCC** (AGG) } Arg
C (G)	**CAA** (GUU) **CAG** (GUC) **CAT** (GUA) **CAC** (GUG) } Val	**CGA** (GCU) **CGG** (GCC) **CGT** (GCA) **CGC** (GCG) } Ala	**CTA** (GAU) **CTG** (GAC) } Asp **CTT** (GAA) **CTC** (GAG) } Glu	**CCA** (GGU) **CCG** (GGC) **CCT** (GGA) **CCC** (GGG) } Gly

[a]By convention, the first base is the one on the 3′ side of the DNA triplet and on the 5′ side of the mRNA codon.

Properties of the Code

Examination of Table 12.4 reveals some interesting properties of the genetic code.

1. *Several codons often stand for the same amino acid*. This so-called **code degeneracy** might be expected when 64 codons are available to specify just 20 amino acids. The most striking examples are the six codons for leucine, the six for serine, and the six for arginine. Fourfold degeneracy is seen for several other amino acids (e.g., valine); here the third base of the codons can be ignored, effectively reducing the genetic dictionary to two-letter words. Twofold degeneracy is seen for many others (e.g., phenylalanine); here the third RNA base is either of the two pyrimidines (U or C) or either of the two purines (A or G). Thus, code degeneracy usually involves the third base. The number of synonymous codons for an amino acid (1, 2, 3, 4, or 6) is roughly proportional to the actual frequency of that amino acid in proteins. For example, tryptophan, the least frequent amino acid, has a single codon; serine, the most frequent, has six codons.

Although the code is degenerate, the genetic language is not imprecise: A specific DNA triplet, through its mRNA codon, will direct the assembly of a specific amino acid.

2. *Three of the 64 codons do not specify any amino acid*. While deciphering the code, investigators found that the RNA codons UAA, UAG, and UGA failed to incorporate any amino acid. These sequences are sometimes called *nonsense codons* because they do not code for amino acids; the term **chain termination (stop) codons**, however, is more accurate, because UAA, UAG, and UGA stop the translation (but not transcription) process, releasing the now-completed polypeptide from the ribosome. The evidence for this action came from mutant strains of microorganisms in which the synthesis of a polypeptide stopped prematurely. This could happen, for example, if the third base in a tyrosine mRNA codon was changed to A or G (see Table 12.4).

3. *The code is nearly universal*. Although the code was first deciphered using synthetic RNA molecules and translation components from the bacterium *E. coli*, further tests showed that almost all organisms use the same array of codons. Biochemists can obtain specific genes or messenger RNA molecules in large enough quantities to determine the ordering of the bases; matching the codons with the amino acids in the resultant proteins confirms that the genes of viruses, bacteria, yeast, maize, worms, flies, mice, humans, and most other organisms rely on the same genetic blueprint.

Furthermore, protein synthesis can run smoothly even with mixed components. Viral proteins, for example, are normally translated on the ribosomes of the host species. Toad oocytes that are injected with the messenger RNA for rabbit globin chains synthesize rabbit globins; when injected with bee venom mRNA, they synthesize bee venom. Researchers have even made a remarkable strain of tobacco that glows in the dark, accomplishing this feat by transferring to the tobacco the firefly gene that encodes the luminescence enzyme *luciferase* (Root 1988). Pigs injected as embryos with human genes can sometimes make human hemoglobin and transmit this property to their offspring (Hilts 1991). As described in the next chapter, bacterial cells can also replicate, transcribe, and translate human genes that are incorporated into their DNA.

Scientists have found several exceptions to the codon assignments in Table 12.4. For example, in all mitochondrial DNAs that have been studied, the mRNA codon UGA encodes tryptophan rather than chain termination. A few other exceptions are also known. The near universality of the code, however, suggests that it evolved just once in ancient life forms, perhaps 3 billion years ago. Possibly the code began as a sequence of two nucleotides, because some proteins in very primitive bacterial species have fewer than 16 amino acids. But how each of the 20 amino acids in the current coding dictionary came to be specified by a particular nucleotide triplet is not known.

GENE ORGANIZATION

The preceding sketch gave a short account of DNA construction, replication, and expression. In this section we add a few details about the way base sequences are arranged along the DNA molecule, including some mysteries about the nature and operation of the genetic material. (The way DNA is packaged into chromosomes was presented in Chapter 2.)

Exons and Introns

In the years following the 1953 discovery of DNA structure by Watson and Crick, geneticists developed the view that the DNA from a chromosome consists of a succession of thousands of genes. Each gene was seen as an unbroken string of nucleotides, separated perhaps by "spacer" DNA or control regions that allowed polymerases to act. For most of the genes of higher organisms, however, this portrait turns out *not* to be entirely true.

Observations leading to a more complex view included electron micrographs of the sense strand of a gene (isolated from the nucleus) joined with its complementary messenger RNA (isolated from the cytoplasm). These threads are hybrid molecules in which one strand is DNA and the other RNA. The traditional picture would have been a simple double-stranded structure throughout its length (Figure 12.16A), because DNA, base after base, would have paired exactly with the complementary mRNA, base after base.

In fact, what is often seen are short double-stranded DNA-RNA regions separated from one another by single-stranded loops of unpaired DNA (Figure 12.16B). As expected, the paired regions consist of DNA segments hybridized with their mRNA product. These are the parts of the gene, called **exons** or **expressed sequences**, that actually code for a polypeptide. The unpaired loops of DNA, called **introns** or **intervening sequences**, consist of segments of a gene that have no corresponding mRNA product. The introns are often many and long: In the gene seen in Figure 12.16B, eight exons (labeled 1–8) coding for successive segments of the polypeptide are separated from each other by seven introns.

What an extraordinary surprise to discover that some nucleotide sections within the boundaries of a gene have no mRNA complement and therefore code for no amino acids in the polypeptide product. An analogy would be for readers of this book to confront scattered discourses on music or sports or perhaps just pages of nonsense. Investigators have discovered that most genes of vertebrates are interrupted this way by one or more introns—segments that may be much longer than the coding regions of the gene they interrupt. Among human genes, the one that codes for clotting factor VIII (and is mutated in persons with hemophilia) is an extreme (but not the most extreme) example: There are 26 exons coding for about 2,000 amino acids but constituting only 4% of the total length of the gene.

The genes that code for the polypeptides of hemoglobin provide

(A) Continuous gene

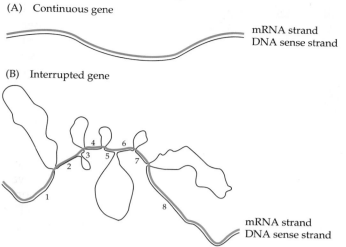

mRNA strand
DNA sense strand

(B) Interrupted gene

mRNA strand
DNA sense strand

Figure 12.16
The results of hybridizing the DNA sense strand of a gene with its finalized messenger RNA. Double-stranded segments result from complementary base pairing between DNA and RNA. Single-stranded segments result whenever one strand has no complement. (A) A continuous gene transcribed into mRNA, all of whose nucleotides code for amino acids. Genes of this type are found in bacteria and viruses. (B) Interrupted genes of higher organisms consist of exons *(coding segments) separated by* introns *(noncoding segments). The regions of the mRNA that are transcribed from the introns are cut out of the molecule before it is used in translation. Hybridization of the gene with mature mRNA, therefore, results in the looping out of those regions of the gene that have no corresponding sequences in the mRNA. This drawing, based on the chicken gene coding for the egg-white protein ovalbumin (Chambon 1981), shows eight exons (numbered) separated by seven looping introns.*

other well-studied examples. The protein (globin) part of the molecule includes two α polypeptide chains and two β polypeptide chains. Each type of chain is encoded by a different gene, and each gene has three exons separated by two introns. The gene that codes for the β polypeptide chain, for example, contains 1,606 bases. The initial RNA transcript of the human β-globin gene (Figure 12.17) contains 1,606 nucleotides, an exact complement of the entire gene. But the 130 bases complementary to intron 1 and the 850 bases complementary to intron 2 are then *cut out* of this RNA molecule before it leaves the nucleus. The splicing together of the cut ends of successive *exons*, called **RNA splicing**, must be done with pinpoint precision to preserve the step-by-step progression of three-base

Figure 12.17
The structure of the gene encoding the β polypeptide chain of human hemoglobin. The crosshatched leader segment of exon 1 is an untranslated 50-base sequence plus a 3-base sequence encoding an initial methionine that is cut off the final β polypeptide. The crosshatched trailing segment of exon 3 is an untranslated sequence of 135 bases. The remaining 90 + 222 + 126 = 438 bases of the three exons specify the 146 amino acids of the β chain. Note that intron 2 is much longer than all three coding regions together.

3' end of
sense
strand

5' end of
sense
strand

Transcribed β-globin gene

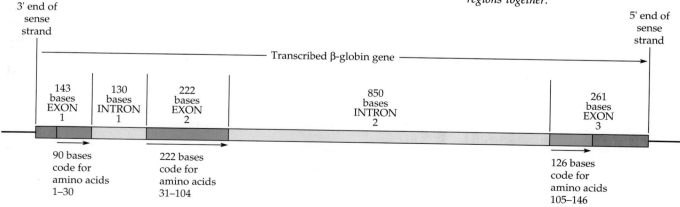

| 143 bases EXON 1 | 130 bases INTRON 1 | 222 bases EXON 2 | 850 bases INTRON 2 | 261 bases EXON 3 |

90 bases code for amino acids 1–30

222 bases code for amino acids 31–104

126 bases code for amino acids 105–146

269

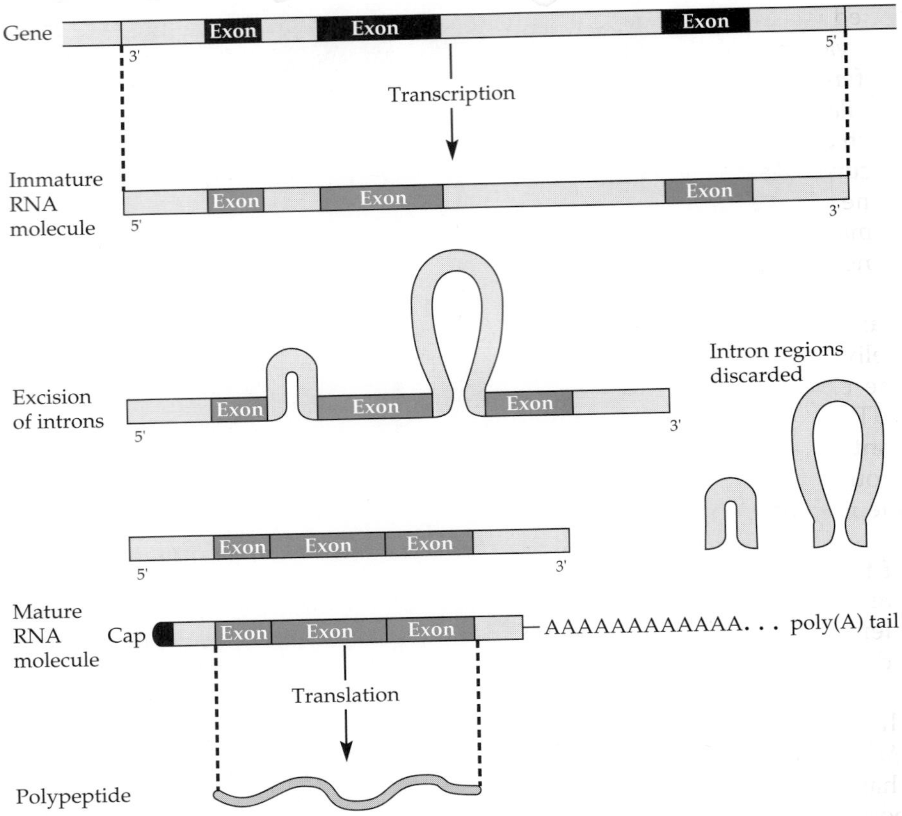

Figure 12.18
Messenger RNA processing. Between transcription and translation, mRNA is modified in three ways: The regions corresponding to introns are cut out, and all the exons are joined into an uninterrupted sequence; a special nucleotide is put on the 5' end (the head cap); and a string of 50 to 200 nucleotides, all with adenine, is put on the 3' end, forming the poly(A) tail.

codons in the different segments. Specific base sequences at the two ends of an intron mark the region to be cut out.

In some genes, individual exons seem to code for key functional regions in the resultant protein. Such regions, called domains, might be useful to the organism as parts of other proteins with similar functions. The same exon might therefore be expected to be found in different genes. This work on "mix-and-match" exons, summarized by Angier (1990), suggests a mechanism for gene evolution, an idea we return to later.

Alterations to RNA molecules after transcription but before translation are collectively called **mRNA processing**. The removal of introns and the joining together of adjacent exons constitute just one part of the process, however. Prospective mRNA is modified in two other ways before it leaves the nucleus (Figure 12.18). After transcription, some nucleotides are cut off from the 3' end of the molecule, and a string of 50 to 200 adenine nucleotides is added in their place; this *poly(A) tail* aids in transporting the RNA molecule into the cytoplasm. At the 5' end of the molecule, a *cap* consisting of one special nucleotide (a modified G nucleotide in reverse orientation) is added; its function is to attach the mRNA to the ribosome. The term *messenger RNA* is usually reserved for the finished product that attaches to ribosomes, that is, the sequence of codons specifying amino acids and leader and trailing sequences.

Gene Families

Within an individual, there are several different but similar genes that code for various hemoglobin chains. Apparently, a gene present in a human ancestor millions of years ago was duplicated, so that it was present twice rather than once. Over successive generations, the duplicated loci evolved independently of each other to produce genes that code for variants of a single polypeptide (or in other cases, polypeptides so different that they developed different functions). Further replications of similar loci can lead to a family of many related genes. The members of such a **gene family** may be clustered on a chromosome, or if translocations occurred nearby, they may be scattered on different chromosomes. Known gene families include those that code for histones, antibodies, and ribosomal RNA components, but one of the best understood is the family of genes that code for the β (and β-like) polypeptides of hemoglobin.

Recall that hemoglobin (the red pigment of blood) binds oxygen and transports it from the lungs to all parts of the body. The oxygen that is delivered to tissues is essential for cell respiration, the process by which the chemical energy of food is converted to the high-energy compound ATP (Appendix 2). Each oxygen molecule is associated with an iron atom contained within a nonprotein *heme* group of hemoglobin. Four heme groups are pocketed in folds of the four polypeptides, the *globin* part of the molecule (Figure 12.19).

Humans produce several kinds of hemoglobin. In adults, about 98% of the hemoglobin is characterized as $\alpha_2\beta_2$: Two of the four polypeptides are identical *alpha (α) chains* with 141 amino acids each, and two are identical *beta (β) chains* with 146 amino acids each. The α and β polypeptides are coded for by the two unlinked genes.

About 2% of hemoglobin in adults consists of two α chains and two chains similar to, but not the same as, the β chains. These are the *delta (δ) chains*, in which just 10 of the 146 amino acids differ from those in β chains. This hemoglobin is therefore characterized as $\alpha_2\delta_2$. A very small portion of adult hemoglobin (under 1%) has two β chains and two *gamma (γ) chains*, in which 39 of 146 amino acids differ from those of β. There are two different types of γ chains, one with the amino acid glycine in position 136 (called $^G\gamma$ or G-gamma chains) and one with alanine (called $^A\gamma$ or A-gamma chains). During most of fetal life, hemoglobin with two γ chains is the predominant form of hemoglobin. *Early* embryos, however, possess yet another β-like polypeptide, the *epsilon (ε) chain*, which differs from the β chain in 35 amino acids.

To summarize, the non-α chains of hemoglobin are of *five* types—ε, γ of two types, δ, and β—depending on the stage of development. Each polypeptide is coded for by a different gene, and the five genes form a cluster on the short arm of chromosome 11. Each gene has three exons and two introns similar to those of the β-globin gene depicted in Figure 12.17. Although no other genes are in this region, the five β-like genes occupy only a small fraction of the total length of this stretch of DNA. The nucleotides that constitute the "space" between the genes are not transcribed and translated, but they have many intriguing aspects, including the following two types of sequences.

One is a gene-sized length of DNA called *psi beta (ψβ)*, which in structure and nucleotide sequence is similar to the β-like genes. Because it has no polypeptide product, however, it is called a **pseudogene**. Pseudogenes represent evolutionary dead ends, having accumulated one or more critical mutations that inactivate the gene. Usually the gene cannot be transcribed, or if transcribed, its mRNA cannot be translated. The human ψβ pseudogene, for example, has small insertions that muddle the reading of the triplet code and the intron-exon junctions. Globinlike pseudogenes have been found elsewhere; several are present in a clustered family of α-globin genes.

A second curious aspect of the intergenic DNA in the β-globin cluster are four very similar sequences of about 310 bases that code for nothing

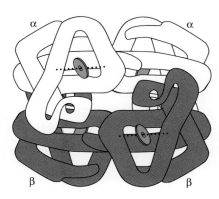

Figure 12.19
The folding of the four polypeptide chains in normal adult hemoglobin, $\alpha_2\beta_2$. Each chain envelops a heme group (dark colored disk) containing an iron atom (small sphere) that reversibly attaches a molecule of oxygen.

271

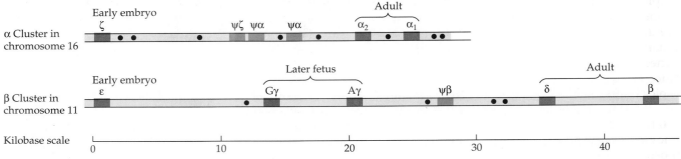

Figure 12.20
The human α-globin and β-globin gene clusters. Active genes are shown by colored bars, along with the stages of development when they are expressed. Note that the order of expression in development corresponds to the order along the chromosomes. Pseudogenes are shown by dark gray bars, and each is prefixed by the Greek letter psi (ψ). Dots represent Alu sequences, tracts of about 310 bases that are repeated a half million times in the whole human genome.

in particular. These repetitious sequences are called ***Alu* repeats**, and there are other *Alu* repeats nearby. In fact, *Alu* sequences are a strikingly common feature of our genome, and they are discussed further in the next section. Figure 12.20 shows how the various elements of the α-globin and β-globin clusters are arranged.

Repetitive DNA

Geneticists have traditionally studied genes that are responsible for the production of enzymes and other protein molecules. These genes are called the *structural genes*, and each is present twice in every diploid cell of the body. Taken collectively, the structural genes are known as *unique sequence DNA*, or **single-copy DNA**, because the bases exhibit no particular repeating patterns. This state of affairs is expected, since the structural genes are coding (largely) for proteins with different amino acid sequences.

Only about 3% of human DNA, however, is transcribed into mRNA molecules or into tRNA or rRNA. The question of what the remaining 97% of DNA does is largely unanswered, and much of it may simply be nonfunctional "junk." In the 1960s researchers found that about half of this DNA exists as multiples of relatively short sequences of nucleotides called **repetitive DNA**. Because most repetitive DNA never gives rise to protein products, repeated sequences are generally not associated with any detectable phenotypes. Repetitive DNA is classified as highly repetitive or moderately repetitive.

The *highly repetitive* portion consists of short lengths of DNA, up to a few hundred bases, that are repeated hundreds of thousands of times. Highly repetitive sequences include the *Alu* repeats, such as those in the globin region noted previously. There are about half a million copies of *Alu* generously spread throughout the haploid complement of 23 chromosomes, between genes and within introns. This amount of DNA represents about 5% of human DNA, the rough equivalent of a whole average-sized chromosome! Alu repeats have been identified in DNA extracted from the ancient archaeological specimens (see Box B, in Chapter 18), but their origin and function are matters of speculation.

Other types of highly repetitive DNA, also constituting about 5% of human DNA, are very short sequences of about half a dozen to a dozen bases repeated in tandem many times. Rather than being interspersed among the structural genes, this highly repetitive DNA is often concentrated in the centromere regions of chromosomes; it also accounts for over half the DNA in the human Y chromosome.

The *moderately repetitive* portion consists of varied lengths of DNA that are repeated tens to thousands of times in the genome. Some genes with well-known functions are moderately repetitive, a redundancy perhaps reflecting the need for large amounts of the gene product. For example, the genes that code for ribosomal and transfer RNA molecules are present in multiple copies. Other sequences that have repetitive elements are the gene families coding for histones and for antibodies. (The extraordinary nature of the antibody genes is discussed in Chapter 16.) Also, the genes encoded in mitochondrial DNA represent a type of repetitive DNA, since a mitochondrion has about ten identical molecules of DNA and a typical human cell has hundreds of these organelles.

Not transcribed into any product, the sequence TTAGGG is repeated 250 to 1,500 times at the tips of all chromosomes. These special structures, known as **telomeres** (Greek *telos,* "end"), are vital to the stability of chromosomes. Conserved through many millions of years of evolutionary time, exactly the same sequence acts as telomeres in the chromosomes of organisms as widely separated as yeast and humans.

Mobile Genes

Similar genetic structures in related species suggest that the organization of DNA is *stable* over long periods of time. Yet there are now known to be genetic elements that actually move around within the genome. This phenomenon was first demonstrated in the 1940s by Barbara McClintock of Cold Spring Harbor Laboratory, in New York. By examining color patterns on kernels of corn, she was able to show that certain genetic elements within the maize genome modified the expression of other genes at adjacent sites. From time to time these *control elements* seemed to disappear from their old locations and reappear at new locations. McClintock's interpretations were considered innovative and unusual at the time, but did not lead to studies on other organisms until the 1960s, when somewhat similar DNA segments called **transposable elements**, or **transposons**, were discovered in bacteria. Various types of "jumping genes" with a variety of effects have since been found in many species, including yeast, fruit flies, and mice.

Typically, a piece of DNA that can move from one position to another consists of one or several genes in the middle, flanked by end sequences that are identical to each other. It is believed that these *terminal repeats* help in the insertion and excision of the element when it moves.

MUTATIONS

A mutation is a change in DNA that may adversely affect the phenotype of an organism. Here we will be primarily concerned with **germinal mutations**, which occur in cells that are destined to become eggs or sperm. A germinal mutation does not affect in any obvious way the person in whom it occurs, but it can be transmitted to and harm future generations. **Somatic mutations**, on the other hand, occur in body cells (liver, lung, intestine, etc.) not ancestral to gametes. A somatic mutation can affect the phenotype of its carrier, but it will not be transmitted to offspring. Somatic mutations are involved in one or more steps that lead to the development of cancer (Chapter 15).

The occurrence of an altered phenotype may be far removed in time from the germinal mutation itself, especially if the mutated allele behaves in development as a recessive, as many do. In these cases, it must be combined in a zygote with another allelic recessive mutation for its effect to be revealed during the course of development. Thus, the mutational event itself and the mutant phenotype that it produces may be separated by generations.

The effect of a mutated gene on the phenotype of an organism ranges from trivial to lethal. It all depends on the properties of the protein encoded by the altered gene and how that protein affects development. Many mutations produce inconsequential phenotypic effects. Examples include clockwise versus counterclockwise cowlick patterns on top of the head, cleft versus noncleft chin, and soft versus brittle ear wax. In contrast are the genes that cause death during infancy (Tay-Sachs disease), young adulthood (Duchenne muscular dystrophy or cystic fibrosis), or later (Huntington disease). In between are genetic conditions that are inconvenient or burdensome to lesser or greater degrees (e.g., brachydactyly, pattern baldness, ichthyosis, hemophilia, sickle-cell anemia, albinism, dwarfism, and diabetes).

Mutations that occur haphazardly, irrespective of known environmental conditions, are called **spontaneous mutations**. To a large extent, they result from random thermal motions of atoms and molecules in and near DNA, often at the time of DNA replication, but at other times as well. This motion is characteristic of all matter. In addition, it is likely that some spontaneous mutations are due to foreign chemicals or radiation that get inside cells and near DNA. Because we cannot observe the mutational events, it is not possible to pinpoint the precise cause (heat, chemicals, radiation) of any spontaneous mutation, especially since the observable phenotypic effects may not be evident for a long time.

Induced mutations refer to gene changes following exposure of an organism to an agent, called a mutagen, known to produce mutations. Mutagens of concern include components of cigarette smoke, some industrial chemicals, pesticides, various substances in waste landfills, food additives, drugs, medical and dental X-rays, ultraviolet light, and radiation from past atomic weapons tests, atomic warfare, and the nuclear power industry—a list that reflects our technological society, attitudes toward the environment, and modern life-styles. On the other hand, some largely inescapable *natural* products are known to be mutagens, including the radioactivity found in soil, rocks, and our bodies. Many common raw or cooked foods also contain substantial amounts of mutagenic chemicals; some of these are manufactured by plants as natural pesticides, toxic to the species that would otherwise eat them (Ames 1983).

Substitutions and Frameshifts

At the molecular level, the actual genetic change may be a *replacement* of one base by another, called a **base substitution mutation**. The most innocuous substitution mutation would change a base in the *noncoding* spacer DNA between genes. Also innocuous would be the change of a *coding* DNA triplet to a synonym for the same amino acid. For example, a DNA base substitution in an exon that changes AAA to AAG would still lead to the incorporation of phenylalanine (see Table 12.4); such a mutation would probably be undetectable except by sequence analysis of DNA or mRNA molecules. Because one base can change to any of the other three bases, a DNA triplet can mutate in nine different ways by a single base substitution.

Sickle-Cell Anemia. The classic example of a base substitution is the one that causes **sickle-cell anemia**. This illness was the first to be dubbed a *molecular disease* (by the American Nobel chemist Linus Pauling), one brought on by an altered protein encoded by a mutated gene. The mutation occurs in the gene coding for the β chain of hemoglobin. The middle base of a DNA triplet coding for the amino acid *glutamic acid* is mutated

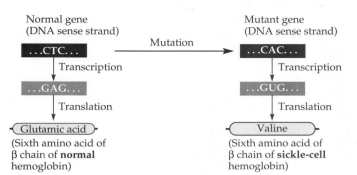

Normal gene
(DNA sense strand)

...CTC...

→ Mutation →

Mutant gene
(DNA sense strand)

...CAC...

Transcription

...GAG...

Transcription

...GUG...

Translation

Glutamic acid

(Sixth amino acid of
β chain of **normal**
hemoglobin)

Translation

Valine

(Sixth amino acid of
β chain of **sickle-cell**
hemoglobin)

Figure 12.21

The sickle-cell mutation. The normal gene for the a polypeptide chain of hemoglobin incorporates glutamic acid as the sixth amino acid from the free amino end (left). The mutation is a base substitution (A for T) in one of the two possible codons for glutamic acid. (Verify the coding by referring to Table 12.4).

so that the triplet now codes for *valine* instead (Figure 12.21). The resultant amino acid switch changes the shape of hemoglobin so that the altered molecules stack into narrow crystals, distorting the smooth, rounded surface of red blood cells (Figure 12.22). Distorted cells have a life span of only a few weeks rather than the normal life span of about four months, leading to anemia (i.e., too few red blood cells). In addition, the sickled cells become trapped in small vessels throughout the body, leading to local oxygen depletion, tissue damage, infections, and periodic episodes of excruciating pain. These pleiotropic effects of the sickle-cell mutation markedly reduce the quality of life, reproductive ability, and life span.

Affected persons are homozygous (Hb^S/Hb^S, or more simply S/S) for the sickling allele; heterozygotes (A/S, A being the normal allele) are said to have the sickle-cell *trait*; they are essentially normal and rarely come to medical attention. The sickle-cell allele is prevalent in African populations and in their descendants worldwide because of a curious interaction with malaria (Chapter 18). Despite much clinical work, there is no good basic treatment for the disease. Pain can be relieved, and the frequency and duration of crises can be lessened by rest, transfusions, and antibiotics. Drugs can reduce sickling, but they have major side effects and remain experimental. Prenatal diagnosis of sickle-cell anemia is possible, however, so heterozygous couples have the option of aborting affected fetuses.

Thalassemias. When the α or β polypeptides of hemoglobin are synthesized in reduced amounts, the resultant disease is **thalassemia** (Greek *Thalass-emia*, "sea-blood"), a seriously debilitating or lethal condition com-

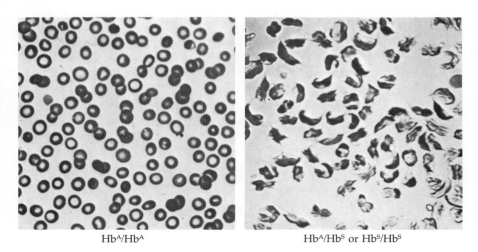

Hb^A/Hb^A

Hb^A/Hb^S or Hb^S/Hb^S

Figure 12.22

Red blood cells from normal homozygotes, Hb^A/Hb^A, and individuals with sickle-cell anemia, Hb^S/Hb^S. Although the red blood cells within heterozygotes, Hb^A/Hb^S, generally look rounded and function normally (as at left), they can be made to sickle (as at right) in a sealed film of blood on a microscope slide, where the amount of oxygen is reduced. (Micrographs courtesy of C. L. Conley.)

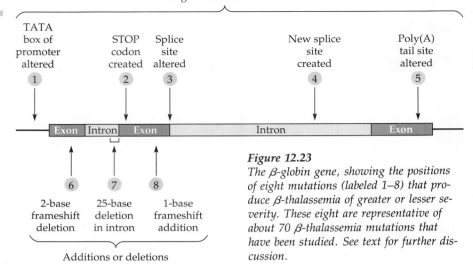

Single-base substitutions

Figure 12.23

The β-globin gene, showing the positions of eight mutations (labeled 1–8) that produce β-thalassemia of greater or lesser severity. These eight are representative of about 70 β-thalassemia mutations that have been studied. See text for further discussion.

mon in some Mediterranean peoples and in their worldwide descendants. Like sickle-cell anemia, affected children cannot usually be treated very satisfactorily, but the disease can be diagnosed prenatally.

Different forms are designated by the globin chain that is deficient. The **β-thalassemias** constitute one of the most serious health problems worldwide, accounting for hundreds of thousands of childhood deaths per year. Heterozygotes have *β-thalassemia minor* and usually only a mild anemia, but homozygotes—afflicted with *β-thalassemia major*—often die before ten years of age. Many kinds of mutations produce β-thalassemia. They affect not only the actual amino acid coding regions of the β-globin exons, but also sites surrounding the gene and even within the noncoding introns. Some of these mutations are shown in Figure 12.23.

Mutation 1 in Figure 12.23 alters the promoter region of the gene, thereby markedly reducing the efficiency of transcription. Mutation 2 changes codon 39 from a triplet specifying glutamine to a stop codon. (What was the base substitution? See Table 12.4.) Mutations of this type —from a codon specifying an amino acid to a stop codon—are sometimes called *nonsense mutations*. They have the effect, as you might suspect, of stopping translation prematurely, producing a too-short polypeptide. In this case, the polypeptide is only 38 amino acids long, no functional β chains are present, and homozygotes are severely anemic. This particular mutation accounts for about 30% of β-thalassemia alleles in the Mediterranean region and for almost all of the disease on the island of Sardinia.

Mutations 3 and 4 perturb RNA splicing so that the intron regions are not cut out of the molecule properly. Mutation 5 causes the poly(A) tail on the messenger RNA to be attached in the wrong place. As a consequence, the mRNA is unstable and a mild thalassemia results.

Mutations 6, 7, and 8 are **frameshift mutations**, which are defined as the addition or deletion of one or more bases. Frameshift mutations not only alter the codon in which they occur, but may shift the *reading frame*, the division of the DNA message into *successive* three-letter words. If an extra base is inserted (as in mutation 8), all the coding units after the insertion are read incorrectly (Figure 12.24). The term *point mutation* is often used for both base substitutions or small frameshift mutations, since they are not microscopically visible.

The **α-thalassemias** usually result from large deletions that eliminate

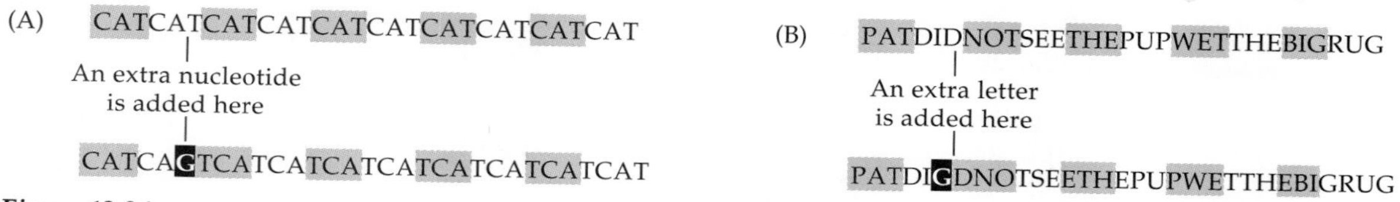

(A) CATCATCATCATCATCATCATCATCATCAT

An extra nucleotide
is added here

CATCA**G**TCATCATCATCATCATCATCATCAT

(B) PATDIDNOTSEETHEPUPWETTHEBIGRUG

An extra letter
is added here

PATDI**G**DNOTSEETHEPUPWETTHEBIGRUG

Figure 12.24

The consequences of a frameshift mutation resulting from a single base addition. (A) A hypothetical gene in which all triplets are CAT. If a base (here G) is added, all triplets beyond the insertion are TCA because of a one-base shift in the reading frame. (B) An English language analogy in which a sentence with only three-letter words becomes gibberish after a one-letter addition. It is assumed that the sentence—like DNA—continues to be read in successive three-letter words.

one or both of the two duplicate α-globin genes present on chromosome 16. The severity of the disease is proportional to how many of the α-globin genes (one, two, three, or four) are deleted in a given individual. Some hemoglobin is made out of four γ chains (during embryonic life) or out of four β chains (postnatal). But these γ_4 or β_4 hemoglobins are poor oxygen carriers. When all four α-globin genes are deleted, chain synthesis is totally suppressed; the fetus dies in utero and is miscarried late in pregnancy.

Chemical Mutagenesis

The agents that induce mutations are called **mutagens**. They fall into two broad categories: chemicals and radiation. Direct information on how these agents affect human populations is scanty, but geneticists who work with experimental organisms are able to obtain some fairly precise data that may or may not apply to humans. They can, for example, deliberately subject laboratory animals to known mutagens to produce phenotypic variations in the descendants of the treated animals. Three chemical mutagens often used in a laboratory setting are:

1. *Proflavin*, a brownish dye used as an antiseptic in veterinary medicine, causes the addition or deletion of single bases during the synthesis of new DNA strands. Proflavin's mutagenic action is caused by its shape: It is a very flat molecule that wedges itself between the base pair rungs of the DNA helix.
2. *Nitrous acid* is a powerful mutagen that replaces amino groups (— NH_2) with keto groups (= O). By this action, the base cytosine is converted to uracil, and adenine is converted to a base called hypoxanthine. Because the pairing properties of changed bases are modified, DNA treated with nitrous acid produces errors during subsequent replications.
3. *5-bromouracil* is only one of many *base analogues*, substances that so closely resemble the natural bases that they can be mistakenly incorporated into DNA at the time of replication. A molecule of 5-bromouracil is like thymine except that the methyl group of thymine is replaced by a bromine atom. Whereas thymine base-pairs with adenine, 5-bromouracil can sometimes (but not always) pair with guanine and lead to mistakes during subsequent replications.

Although these three chemicals are not part of the general environment, the human population is exposed daily to a variety of substances that may be mutagenic. It is possible that substantial genetic damage is occurring without our ever being able to link a particular mutagen with

IS CAFFEINE A MUTAGEN?

Caffeine is present naturally or artificially in coffee, tea, chocolate, many soft drinks, and a wide range of common medicines. The plants that synthesize caffeine in their tissues apparently use it as a natural pesticide. Caffeine is a purine, very similar in structure to adenine and guanine, but different enough (with three methyl groups) that it does not substitute for the normal purines in DNA (see figure). When tested on most bacteria and fungi, caffeine produces mutations. It is thought that caffeine may act not by being a mutagen itself, but by interfering with the ability of cells to repair damage to DNA from other causes. In plant, hamster, and human cell cultures, high concentrations of caffeine have been shown to cause chromosome breaks, but the relationship between breakage and point mutations is unclear. With fruit flies, the evidence is contradictory. Extensive experiments with mice yield no indication of caffeine-induced mutations. In humans, ingested caffeine is rapidly demethylated and excreted, although it does reach the gonads and does cross the placental barrier. Fetuses appear to be unaffected by even heavy maternal coffee consumption, however. Thus, while pondering all the data over a cup of coffee, we probably

Caffeine

Adenine

Guanine

need not worry too much about the caffeine in it. On the other hand, risk taking is personal, and you may wish to investigate whether caffeine has other detrimental or beneficial effects (Dews 1984; Lecos 1988).

its eventual deleterious phenotype. We are unable to match a mutagen to its effect because induced mutations are generally not different in their phenotypic effects from spontaneous mutations, and their effects may not appear for generations.

Because it is neither ethical nor feasible to deliberately expose humans to a chemical in order to detect possible mutations, no reliable data have been obtained directly on human populations. Instead, test systems have been developed that use bacteria, fungi, plants, insects, mammals, or mammalian cells in tissue culture. In principle, because DNA is DNA, mutagenicity in one organism should mean mutagenicity in another. But in practice, differences exist between species, both in the degree of protection afforded germinal tissue and in the type of chemistry developed by an organism over evolutionary time. For example, in a bacterium, a single cell membrane and cell wall separate its DNA from its surroundings. In humans, an ingested potential mutagen is first exposed to the harsh environments of the mouth, stomach, and intestines before it is absorbed into the blood via the capillaries of the digestive organs. This blood is then delivered to the liver, where absorbed materials may be processed by myriad enzyme systems. Only then are ingested substances or their by-products transported to other sites throughout the body (including the germinal cells). Different mammals may metabolize particular chemicals in different ways. Even within a species, certain members may be more susceptible than others to the action of a mutagen. Thus, mutagenicity is not always straightforward, as illustrated by the case of caffeine (Box A).

Because it is costly and time-consuming to test small mammals directly with compounds that may be mutagenic or carcinogenic (cancer causing), assays on lower organisms are used to prescreen a wide variety of chemicals. Then those chemicals that give positive results can be investigated further by using more difficult, but more pertinent, test systems. Of the many screening tests for chemical mutagenicity, perhaps the most rapid, simple, sensitive, and economical test was developed by Bruce Ames, of the University of California at Berkeley. The **Ames test**, as it is called, can be used to suggest whether a compound is mutagenic or possibly carcinogenic in animals by determining whether it is mutagenic in bacteria.

The Ames test starts with strains of *Salmonella* bacteria that are unable to make the amino acid histidine because they carry a mutant allele his^- (coding for a defective enzyme). Histidine is a vital amino acid and the bacteria need it to grow. Bacterial cells carrying his^- are spread over the surface of nutrient medium lacking histidine. Therefore, at the start of the test, the cells cannot grow; but they *will* grow if a random mutation changes the allele coding for the defective enzyme to an allele coding for a functioning enzyme (a so-called back mutation, symbolized $his^- \rightarrow his^+$). Individual bacteria with back mutations will eventually be visible as discrete bacterial colonies (Figure 12.25).

Several different histidine-requiring strains of *Salmonella* can be used in different tests of the same potential mutagen. Some tester strains are unable to synthesize their own histidine because of a base substitution mutation, and others because of a frameshift mutation. Thus, a mutagen with a particular mode of action that might be missed using one tester strain may be detected using another. For example, a frameshift mutation in a gene due to a single-base *deletion* can be most simply "corrected" by a nearby single-base *addition*. No base substitution can correct a frameshift mutation.

The tester strains of *Salmonella* also carry other heritable traits that make the test very sensitive. A modification of the cell wall allows easy entry of the test chemical into the bacteria. In addition, the bacterial enzymes that are normally used to repair DNA damage are inactivated. Sometimes added to the growth medium are rat liver enzymes that are capable of modifying some test chemicals from nonmutagenic to mutagenic form, or vice versa. In this way the bacterial test is made to resemble, to some extent, a mammalian system in which nutrients absorbed into the bloodstream first travel to the liver.

Radiation Mutagenesis

All living things are exposed to natural, and largely unavoidable, radiation emanating from the sun, from cosmic rays, and from the radioactivity of uranium, radium, and other unstable elements in rocks, soils, food, and air. Very soon after the discovery of X-rays and radioactivity in the 1890s, it became clear that the body could be damaged by these agents that could not be seen, felt, or smelled. In the 1920s, Herman J. Muller, then at the University of Texas, reported that X-rays induce mutations in fruit flies; at about the same time, Lewis J. Stadler, of the University of Missouri,

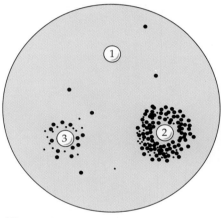

Figure 12.25

An Ames test. In a petri dish, a thin, nearly invisible lawn of bacteria is spread on a nutrient medium that lacks the vital amino acid histidine. Because the bacteria are his^- and thus unable to synthesize histidine, they fail to grow. A mutation from his^- to his^+, however, would permit a cell to make its own histidine and thus to reproduce. Each mutated cell, dividing and redividing, forms a visible colony—the small, dark spots. Three drops (open circles) were placed on the lawn. Drop 1 contained no mutagen. Drop 2 contained a strong mutagen that diffused into the growth medium; mutagenicity is indicated by the halo of colonies surrounding the test drop. Drop 3 contained a weaker mutagen. The several randomly spaced colonies represent spontaneous mutations from his^- to his^+. (Many other types of mutations not revealed by this test presumably also occurred in the bacteria.)

got the same effects in barley plants. Public awareness of the biological consequences of radiation began in the 1950s as a result of the atomic bombing of two Japanese cities at the end of World War II and the continued atmospheric testing of nuclear weapons. Current medical uses and the presence of about 90 nuclear power plants in the United States (about 400 worldwide) continue to focus public attention on the implications of radiation. Interest has been heightened by aging nuclear power facilities and accidents at Three Mile Island in 1979 and Chernobyl in 1986.

To be mutagenic, radiant energy must usually reach DNA. When it does, radiation that is called *ionizing* may knock an electron out of an atom in DNA; this event leaves the atom as a charged ion—a very reactive state subject to further chemical changes. A direct hit on DNA is not necessary, however; nearby ionized chemicals can subsequently affect DNA. With our emphasis on germinal mutations, we are not concerned here with *ultraviolet* light, since it does not penetrate below the skin (although ultraviolet radiation is implicated in the induction of skin cancer). Nor are we concerned here with the *electromagnetic fields* (EMFs) associated with power lines. This radiation is nonionizing (not able to produce ions); no evidence connects EMF with germinal mutations, and the evidence that it can cause cancer is inconclusive and contradictory (Pool 1990).

The amount of radiation absorbed is called the *radiation dose*. It can be calculated from physical principles and measured by various devices (e.g., Geiger counters or badges containing sensitive film) that determine the ions produced or the energy absorbed. Radiation can be received at different rates. A dose received from a high-intensity source is called *acute* radiation, whereas a dose received at low intensity is called *chronic* or low-level radiation. The public is concerned primarily with chronic doses, because these characterize many natural and artificial sources to which the general population is exposed.

A typical American receives 10–11 rem (a unit of radiation dose) of low-level ionizing radiation from all sources over the average reproductive cycle of 30 years (Box B). Roughly 80% of this is from natural sources, including radioactive isotopes in our environment and within our bodies. The major artificial source of radiation is from medical procedures. Only a fraction of this radiation reaches the gonads, where it could induce germinal mutations. For abdominal or pelvic X-rays, a greater fraction is genetically significant because the gonads (especially the ovaries) cannot always be shielded effectively.

Radiation has many different kinds of effects. These include chromosome breaks and point mutations in germinal and somatic tissues, the induction of cancer (particularly leukemia), localized radiation burns and tissue scarring, and at even higher doses, generalized radiation sickness (nausea, vomiting, diarrhea, internal bleeding, general weakness) and death. One important and much-debated question is whether or not there is a "safe dose" or **threshold**, an amount of radiation below which there is no risk. To answer this question, investigators irradiate experimental animals with smaller and smaller doses (necessitating greater costs and the use of more and more test organisms) and then measure a particular effect, for example, point mutations. The charted result is a *dose-response curve*, which unfortunately will have no experimental points below a certain dose. The form of the curve, however, gives theoretical information on the way radiation induces damage, as well as practical information to help formulate medical and public health policies.

The *somatic* effect of major concern with respect to low-level radiation is the possible induction of cancer. Using Japanese data on atomic bomb survivors and patients treated with radiation, as well as other data, re-

EXPOSURE TO LOW-LEVEL IONIZING RADIATION

Every day we are bombarded with radiation from both natural and artificial sources. In the pie diagram shown here, we give rough estimates of the amount of low-level ionizing radiation to which an average American is exposed (BEIR V 1990). The total amount is 10–11 rem per reproductive generation (30 years). For each source, the numbers give the 30-year dose in rem and the corresponding percentage of the total radiation exposure.

Natural radiation, which is largely unavoidable, accounts collectively for about 82% of our total exposure. Only recently has it been recognized that a sizable proportion (55% of the total) comes from radioactive radon gas escaping from the earth. Some of this radon seeps into houses from soil and groundwater; the tighter and more energy-efficient the house, the higher the concentration of radon. Although the matter is still uncertain, radon appears to pose a risk much greater than that from other radiation sources, especially in certain geographical areas (Abelson 1990). Here are some other sources of natural radiation:

Internal emitters (11%) refer to radioactive elements that become part of our body structure through eating and breathing. A major source is radioactive potassium in many common foods.

Cosmic rays (8%) reach the earth from the sun and interstellar space. Progressively filtered by layers of the atmosphere, the dose is several times greater in mountains than at sea level.

Terrestrial radiation (8%) is due largely to the radioactive elements uranium and thorium in rocks and soil. Individual doses vary considerably from one geographical area to another.

The sources of artificial ionizing radiation are the following:

X-rays for medical diagnosis (11%) and other health uses of radiation differ from those previously listed in that they are deliberate,

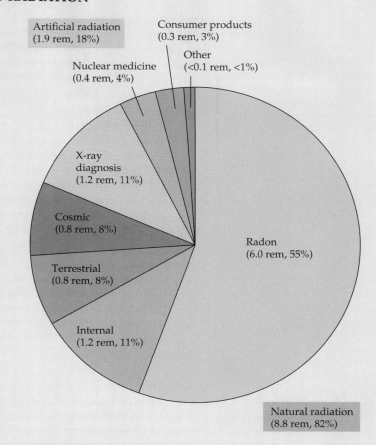

rather than incidental to the activities of living. Exposures include dental and chest X-rays, CAT scans, and mammography. Fluoroscopic examinations produce much higher doses than most other types of X-ray diagnoses. Pregnant women should be irradiated only in unusual circumstances, because the developing fetus is particularly sensitive to radiation damage.

Nuclear medicine (4%) includes many diagnostic and therapeutic techniques. Radioactive isotopes are often used to trace the path of chemicals as they are metabolized in the body or to make an image of an organ or tissue, such as the heart, liver, lung, bone, or thyroid gland.

Consumer products (3%) include a variety of sources, such as radioactive materials in tobacco smoke (polonium), building materials (uranium), well water (radon), and lawn and garden fertilizers. Smaller exposures come from color televisions, computer screens, fire detectors, and luminous dials.

The *other* category (less than 1%) includes occupational risks to miners, medical personnel such as X-ray technicians, and flight crews (because they work at high altitudes). (In the diagram, these exposures are assumed to be spread over the total U.S. population.) Also included here are sources of considerable political interest: nuclear and coal-fired power production and fallout from testing of nuclear weapons.

searchers have estimated that 1 rem of X-radiation per person (above that normally received) leads very roughly to 8 to 27 cancer deaths per 10,000 people (Davis and Bruwer 1991). For comparison, note that roughly 1,500 of 10,000 births (15%) currently end in a cancer death. No single cancer fatality can be unequivocally identified as radiation induced, because (1) the time elapsing between irradiation and the detection of the malignancy is long—years or even decades, and (2) the cancers induced by excess radiation are not different in kind from those occurring from other causes (although leukemia seems to be induced at a greater rate by radiation than by other agents). Thus, it is unclear whether there is a radiation threshold for the induction of cancer.

The *germinal* effect of major concern is the possible induction of point mutations or chromosomal aberrations that may cause harm to the immediate offspring or later descendants. Reliable estimates of the germinal effect of radiation are especially difficult to obtain directly from humans because many new mutations are recessive; therefore the appearance of the mutant phenotype may be several generations removed from the mutational event (if it is ever expressed at all). The direct studies from Hiroshima and Nagasaki of the germinal effects of radiation were also plagued by great uncertainties about the doses received by the survivors. On the basis of animal (especially mouse) and human data, researchers have estimated that 1 rem per generation (over successive generations) produces roughly one case of genetic disease (Mendelian or chromosomal) per 7,000 births. For comparison, note that approximately 3% of all newborns are currently affected with a moderate to severe Mendelian or chromosomal disorder. On theoretical grounds and from the experiments with flies and mice, most geneticists think that any amount of radiation carries with it a proportional risk of mutation. *There is probably no threshold, no safe dose, for germinal effects.*

On the basis of all information available, we conclude that the genetic hazards of radiation for the general human population are small. Different groups of researchers have estimated that 100–400 rem of chronic radiation would be required to double the rate at which mutations occur spontaneously (Neel 1991). This amount is called the **doubling dose**. Over a reproductive lifetime, however, a typical person receives only 10–11 rem of radiation (Box B); so it appears that the radiation we normally receive cannot account for most of the mutations that occur spontaneously.

Studies of risk taking have shown that people's perceptions of the hazards of life often correlate poorly with real dangers. For example, actuarial estimates of the causes of death in the United States show that cigarettes, alcohol, motor vehicles, and handguns are far more dangerous than police work, plane travel, nuclear power, and pesticides (Upton 1982). Yet we often tolerate the more serious risks while concentrating on the lesser ones, perhaps because some risks are voluntary and others are imposed on us. With regard to radiation, we also tend to ignore medical and dental X-rays as a significant source of radiation over which we have some personal control. Although the benefits from diagnostic and therapeutic X-rays are often substantial, their risks should be evaluated and minimized when possible, and alternative methods should be considered.

SUMMARY

1. DNA, the genetic material, consists of two nucleotide chains wound in a helix. In the backbone of each chain, the sugar deoxyribose alternates with phosphate groups. Attached to the sugars of both strands are the paired bases: A opposite T, and G opposite C.

2. DNA is replicated by separation of the helix, each old strand becoming a template for the enzymatic synthesis of a new complementary strand. DNA polymerase, the replicating enzyme, proceeds in only one direction along the helix.

3. DNA is transcribed by the enzyme RNA polymerase. The process forms complementary single-stranded copies of RNA of three types: (a) Messenger RNA molecules are the transcripts of genes that will be translated to produce polypeptides. Successive groups of three nucleotides are the coding units, or codons, for amino acids. (b) Transfer RNA molecules act like adapters during translation. One end binds a particular amino acid; the other end has a three-base sequence, the anticodon, that pairs with a codon on messenger RNA. (c) Ribosomal RNA molecules are structural elements of ribosomes. Neither ribosomal nor transfer RNA is translated into a polypeptide.

4. As it feeds through a ribosome, the mRNA is translated into a polypeptide. The complex process links amino acids together into a chain. The particular order of amino acids causes the polypeptide molecule to fold into a unique three-dimensional shape that determines its characteristics. Proteins consist of one or several polypeptide chains with a total of dozens to thousands of amino acids.

5. Specific codons were deciphered by adding artificial RNA molecules of known composition to an in vitro protein-synthesizing system. One genetic code, which includes several codons for most amino acids and signals to stop translation, applies to almost all forms of life.

6. Genes are often interrupted by noncoding regions called introns. Although the introns are transcribed, they are cut out of RNA molecules before translation.

7. The β-globin genes represent a gene family. The cluster consists of five genes coding for β-type chains of hemoglobin and a pseudogene that has no polypeptide product.

8. (a) Moderately repetitive DNA sequences are present up to thousands of times per genome. Examples include the genes that encode ribosomal and transfer RNA molecules. (b) Highly repetitive DNA sequences may occur as many as a million times per genome. Some short tandem sequences are concentrated at certain chromosomal locations, but longer sequences, like the *Alu* repeats, may be interspersed throughout the genome.

9. Some segments of DNA with terminal repeats can move around in the genome.

10. Mutations are usually base substitutions or frameshift mutations occurring in somatic or germinal tissue. They may occur spontaneously or be induced by a mutagen.

11. The serious symptoms of sickle-cell anemia stem from a single base substitution. The thalassemias result from the underproduction of functional α or β chains of hemoglobin. Studies of the mutational sites in the globin genes have shown many different ways by which transcription and translation can be modified.

12. Chemical mutagens in our environment can be identified through screening tests using microorganisms, cells grown in tissue culture, or small mammals. The often-used Ames test is a simple system for determining mutagenesis in bacteria.

13. The human population is exposed to both natural and artificial sources of radiation, which can cause cancer via somatic mutations and hereditary defects via germinal mutations.

14. Although there appear to be only small effects of low-level radiation, no dose is absolutely safe with regard to inducing germinal mutations.

KEY TERMS

Alu repeat
Ames test
amino acid
anticodon
base substitution mutation
chain termination (stop) codon
code degeneracy
codon
complementary base
DNA ligase
DNA polymerase
doubling dose

exon
frameshift mutation
gene expression
gene family
germinal mutation
induced mutation
intron
ionizing radiation
messenger RNA (mRNA)
mRNA processing
mutagen
nucleotide
polypeptide

promoter
protein
pseudogene
purine (A, G)
pyrimidine (C, T, U)
reading frame
repetitive DNA
ribosomal RNA (rRNA)
RNA
RNA polymerase
RNA splicing
sense strand
somatic mutation

spontaneous mutation
telomere
template
termination sequence
threshold
transcription
transfer RNA (tRNA)
translation
transposon

QUESTIONS

1. The elements C, H, N, O, P, and S are the common constituents of organic matter. Which is present in DNA but not in protein? Which is present in protein but not in DNA?

2. Chargaff's rules (A = T; G = C) do *not* hold for the DNA in the virus φX174 and in some other small viruses, nor do comparable rules (A = U; G = C) hold for the RNA of retroviruses and most other molecules of RNA. Explain these exceptions.

3. If the cells in a Meselson-Stahl experiment (Figure 12.5) were grown for three generations in nitrogen-14, what fraction of the DNA double helices would be intermediate in density? What fraction would be light?

4. A messenger RNA molecule contains the following base composition: 21% A, 33% U, 28% G, and 18% C. What is the base composition of (a) the sense strand of DNA from which it was transcribed, (b) the antisense strand, and (c) both strands of DNA considered together? Assume there are no introns in this gene.

5. A tRNA molecule that binds the amino acid glycine has the anticodon CCC; that binding phenylalanine has AAA. Give the sequences of bases in mRNA and in the sense strand of DNA that code for the dipeptide glycine-phenylalanine.

6. The digestive enzyme trypsin specifically cleaves a protein molecule on the carboxyl side of lysines and arginines. What peptide pieces result from the trypsin digestion of the B chain of human insulin (Figure 12.13)?

7. What fraction of the bases transcribed from the β-globin gene actually codes for amino acids in the β polypeptide chain of hemoglobin?

8. The β-globin cluster of genes extends over about 50 kilobases. It includes the five genes that code for polypeptide products, the β, δ, ^{A}γ, ^{G}γ, and ε chains of hemoglobin. Each of these chains is 146 amino acids long. What fraction of that segment of DNA actually codes for amino acids in polypeptides?

9. Let + stand for a one-nucleotide addition and − for a one-nucleotide deletion. Would the following mutant genes (with *combinations* of nearby additions and deletions) produce a frameshift beyond the last mutational site? (a) +− (b) +−+ (c) +++ (d) +−−+ (e) +++−

10. The human brain hormone somatostatin was the first product of recombinant DNA technology. Bacteria synthesized the hormone after biochemists made and inserted into the cells DNA with the following base sequence (sense strand):
CGA CCA ACA TTC TTG AAG AAA ACC TTC TGA AAG
TGA AGC ACA
What is the amino acid sequence of somatostatin?

11. What polypeptides could be made by each of the following artificial mRNA molecules?
(a) polyadenine, or poly(A)
(b) . . . UCAUCAUCAUCA . . .
(c) . . . UUCCUUCCUUCC . . .

12. List the nine possible DNA triplets and corresponding amino acids that would be formed by a single base substitution mutation in the CCC triplet for glycine. Which do not produce an amino acid change?

13. In what way is the Ames test for chemical mutagenicity in bacteria made to resemble a mammalian test system?

14. A man who worked in a radiation laboratory sued his company because his son was born with hemophilia, which is due to a recessive X-linked allele. He claimed that his working environment was responsible for his son's disease, inasmuch as hemophilia was not present in him, his wife, or any other relative. You are asked to testify in a pretrial hearing as an expert witness in genetics. What would you say?

15. What is the argument for the statement that most spontaneous mutations in humans are not caused by radiation?

16. Newborns homozygous for β-thalassemia are often healthy, the severe disease symptoms developing during the first several months *after* birth. On the other hand, newborns for α-thalassemia are often severely affected *at* birth or are spontaneously aborted. Explain.

FURTHER READING

For more detailed information on all aspects of gene structure and function, we recommend the beautiful books by Darnell, Lodish, and Baltimore (1990), Lewin (1990), Alberts et al. (1989), and Watson et al. (1987). Scriver et al. (1989) provide individual chapters on the molecular basis of human diseases as well as several general chapters. Bunn and Forget (1986) give readable details on the molecular, genetic, and clinical aspects of hemoglobin. Among the best of several histories of the events leading to the Watson-Crick double helix is Judson (1979). See Watson (1968) and Crick (1988) for lively personal accounts, and Sayre (1975) for another view on the role of Franklin in the discovery of the DNA double helix.

Among the wealth of *Scientific American* articles on genes and proteins, we recommend the older articles by Crick (1966) on the genetic code, Kornberg (1968) on the test tube synthesis of DNA, Stent (1972) on "premature" discoveries, Temin (1972) on the transcription of DNA from RNA, Miller (1973) for beautiful micrographs of transcription and translation, Lane (1976) on toad eggs that synthesize rabbit proteins, Rich and Kim (1978) on transfer RNA, Perutz (1978) on hemoglobin structure, Chambon (1981) on split genes, Howard-Flanders (1981) on DNA repair, and Lake (1981) on ribosomes.

More recent *Scientific American* articles include Upton (1982) on low-level radiation, Dickerson (1983) on Z-DNA, Darnell (1983) on RNA processing, Nomura (1984) on ribosome structure and function, Gallo and Montagnier (1988) on the AIDS virus, Radman and Wagner (1988) on DNA replication, Steitz (1988) on intron splicing, and Moyzis (1991) on telomeres. In addition, the October 1985 issue is devoted to the molecules of life and includes articles on DNA by Felsenfeld, RNA by Darnell, and proteins by Doolittle.

13 DNA *in the* Laboratory

Starting with a bit of minced tissue from a cow, students in introductory biology courses can isolate nearly pure DNA during one laboratory period. The students first add a detergent to dissolve the cell membranes, letting the insides spill out. Then the proteins in the resultant cellular soup are removed by gentle shaking with an organic solvent. This treatment produces a separate water layer containing the DNA, which is very gooey because the long, thin, stretchy threads are intertwined with each other. Adding alcohol precipitates the DNA molecules, which can be twirled around the tip of a glass rod and lifted from the liquid as a glistening globule.

Obtained this way or by other methods, DNA is used for the ever-expanding areas of research and applied science called *molecular biology* and *biotechnology* (also called *bioengineering* or *genetic engineering*). These fields require techniques for multiplying specific stretches of DNA, for investigating their structure and properties, and for joining together DNA molecules from different species. A major feature of DNA research is the speed with which new discoveries are currently reported. In his forward to the 125 papers in *Molecular Biology of Homo sapiens* (Cold Spring Harbor Symposia 1986), James Watson writes, "The scientific advances . . . amaze, stimulate, and increasingly often overwhelm us. Facts that until recently were virtually unobtainable now flow forth almost effortlessly." The practical benefits—as well as ethical and legal problems created by the new information—will eventually permeate medicine, agriculture, and other human endeavors in important ways that cannot be fully foreseen.

FOUR INGENIOUS IDEAS

Here we describe a few of the many techniques for manipulating DNA, as well as a few of their applications to human genetics. Rather than taking a chronological approach, we start with one of the newest, simplest, and cleverest techniques, a remarkably simple chemical method (discovered in 1985) for **cloning** selected pieces of DNA. (The term *cloning* refers in general to making multiple copies of something. Identical twins, for example, are "whole-body" clones of each other. In this chapter, however, we are dealing only with molecular clones.) We continue with three other innovative and well-established processes: the cloning of segments of DNA in microorganisms (from 1973), electrophoresis (an "ancient" technique from 1937), and the sequencing of bases in DNA (from 1977).

The Polymerase Chain Reaction

To study any chemical process, we need adequate amounts of the starting materials, often in relatively pure form. But recent technical advances allow researchers to analyze extraordinarily tiny samples of DNA that may be impure, fragmented, or ancient. Indeed, scientists in diverse fields are now learning how to deal with DNA extracted from pathology collections, old and dusty museum exhibits, archaeological specimens such as Egyptian mummies and extinct animals—"all sorts of dead things" (Sykes 1991). They may also start with the DNA present in the cells at the base of a single human hair—perhaps the only evidence found at the scene of a crime.

The process begins with the **polymerase chain reaction (PCR)**. Essentially, PCR singles out a stretch of about 50 to 2,000 nucleotides from the total DNA present in a sample and multiplies this *target DNA* a millionfold or more in just a few hours! Thus, starting with just one or a few copies of the target DNA, investigators can rapidly amplify it enough to do whatever analyses are called for. This technique was conceived and developed in 1985 by Kary Mullis and other researchers of the Cetus Corporation, a California biotechnology firm (Mullis 1990). PCR quickly became one of the most convenient tools of molecular biology, with widespread and growing applications.

To use it, however, researchers must already know the order of at least 20 nucleotides at each end of the target region. Lack of this sequence information usually prevents the use of PCR, requiring that other methods of cloning DNA be used instead. (These alternatives, technically more difficult but perhaps more useful in certain situations, predate the discovery of PCR. They are described in the next section.)

The first step in PCR is to synthesize a pair of **oligonucleotide primers** (Greek *oligo*, "few"). These are single-stranded DNA molecules whose base sequences are complementary to the known sequences flanking the target DNA. Each primer contains about 20 nucleotides. (Oligonucleotides of any sequence up to about 50 bases long are routinely synthesized using automated devices and ingredients such as purines, pyrimidines, and deoxyribose.)

The next step in PCR is to separate into its single strands the double-stranded DNA that includes the region of interest, a process called **denaturation**. Then the key reagents, the two oligonucleotide primers, are added to the denatured DNA, along with the ingredients for DNA replication: DNA polymerase and the four nucleotides. Because of their complementary sequences, the primers will anneal to one or the other end of the short target region present in the long, original strands of denatured DNA. Figure 13.1 shows in detail how the primers bind to *opposite strands* at *opposite ends* of the region to be duplicated.

After binding to the denatured DNA, each primer acts as a starting point for replication. Using the original DNA strands as templates, replication proceeds under the influence of the enzyme *DNA polymerase* in the usual 5' to 3' direction. That is, DNA polymerase adds new complementary nucleotides only to the 3' end of DNA segments already in place (in this case, the primers). In the first cycle of replication shown in Figure 13.2, note that synthesis of new strands proceeds from each primer in the direction of the other primer. The joining of nucleotides into a new strand of DNA encompasses the region of interest and continues past the place where the other primer can anneal. Thus, in the second cycle of copying, both old and new strands provide sites for the binding of primers.

After the third cycle of replication, some of the double-stranded DNA is in short segments, containing only the target region with its flanking primer sites (Figure 13.3). After many cycles of replication, the short segments, doubling (approximately) in each cycle, will outnumber by far the longer segments that contain extraneous sequences to the left and

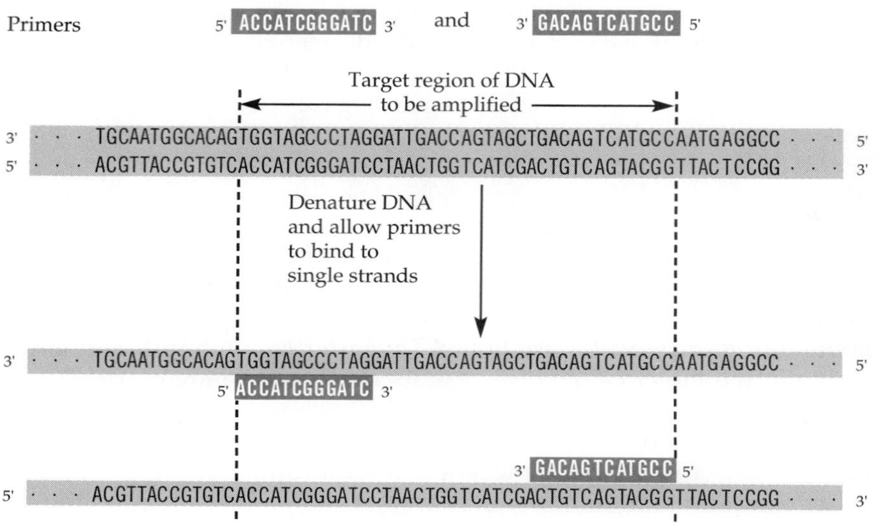

Primers   5′ ACCATCGGGATC 3′   and   3′ GACAGTCATGCC 5′

Target region of DNA
to be amplified

3′ · · · TGCAATGGCACAGTGGTAGCCCTAGGATTGACCAGTAGCTGACAGTCATGCCAATGAGGCC · · · 5′
5′ · · · ACGTTACCGTGTCACCATCGGGATCCTAACTGGTCATCGACTGTCAGTACGGTTACTCCGG · · · 3′

Denature DNA
and allow primers
to bind to
single strands

3′ · · · TGCAATGGCACAGTGGTAGCCCTAGGATTGACCAGTAGCTGACAGTCATGCCAATGAGGCC · · · 5′
5′ ACCATCGGGATC 3′

3′ GACAGTCATGCC 5′
5′ · · · ACGTTACCGTGTCACCATCGGGATCCTAACTGGTCATCGACTGTCAGTACGGTTACTCCGG · · · 3′

Figure 13.1
The primers used in the polymerase chain reaction are chosen so that they bind to opposite ends of opposite strands of the DNA section to be amplified. In this oversimplified diagram, the primers are only 12 bases long, rather than about 20. In the next step, the primers will be elongated at their 3′ ends (see Figure 13.2).

right of the target region. The numbers of copies are shown in Table 13.1 (see also question 1). The entire PCR process occurs automatically and continuously in a single small reaction tube with the initial set of reagents. No further additions to the tube are required, although the starting reagents must include a great deal of the primers and four nucleotides, since more and more of these materials are needed in successive cycles of replication.

Changes in the temperature of the reaction mixture control the phases of each cycle: 95°C to denature, 40°C to anneal the primers, and 70°C to extend the complementary copies with polymerase. Each cycle takes about 4 to 5 minutes. The temperatures are programmed into and controlled by a small machine called a *thermal cycler*. After about 2 hours, the target DNA region—that spanned by the two primers—is amplified a million times or more in the original test tube.

The process makes use of a special replicating enzyme called **Taq DNA polymerase**, isolated from *Thermus aquaticus*, bacteria that live in the hot springs and geysers of Yellowstone National Park.* Having adapted to its environment, the enzyme can survive extended incubation at 95°C (just under the boiling point of water) during the denaturation phases. Virtually all ordinary enzymes are inactivated by such high temperatures.

* Because of its importance, thermally stable *Taq* DNA polymerase was chosen by *Science* magazine as its 1989 "Molecule of the Year" (Koshland 1989).

Table 13.1
The amplification of DNA segments in the polymerase chain reaction, starting with one double-stranded molecule.

	NUMBER OF SEGMENTS AFTER CYCLE NUMBER (n)										
Segment[a]	1	2	3	4	5	6	7	8	9	10	11
Short	0	0	2	8	22	52	114	240	494	1,004	2,026
Intermediate	0	2	4	6	8	10	12	14	16	18	20
Long	2	2	2	2	2	2	2	2	2	2	2
Total (2^n)	2	4	8	16	32	64	128	256	512	1,024	2,048

[a] The short, intermediate, and long segments are identified in Figure 13.3.

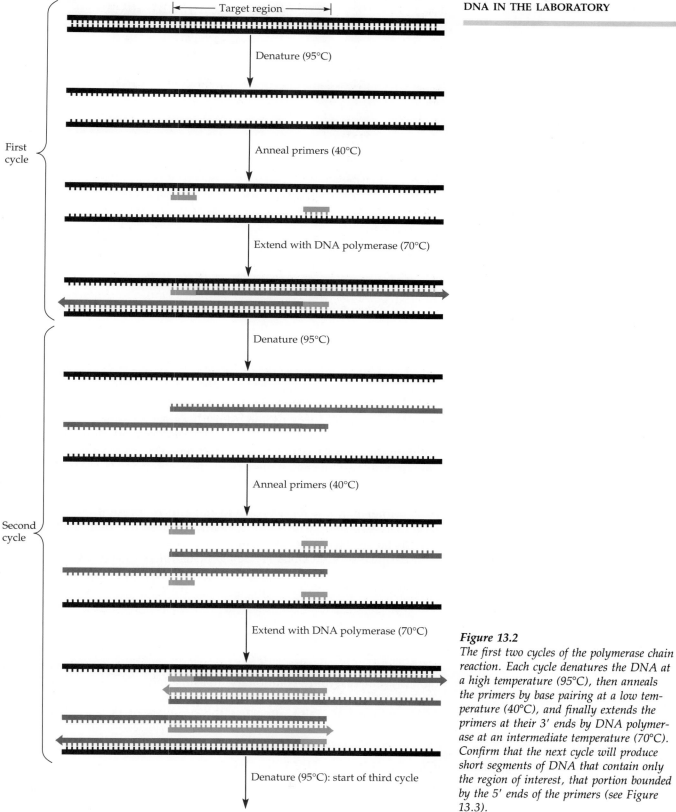

Figure 13.2
The first two cycles of the polymerase chain reaction. Each cycle denatures the DNA at a high temperature (95°C), then anneals the primers by base pairing at a low temperature (40°C), and finally extends the primers at their 3' ends by DNA polymerase at an intermediate temperature (70°C). Confirm that the next cycle will produce short segments of DNA that contain only the region of interest, that portion bounded by the 5' ends of the primers (see Figure 13.3).

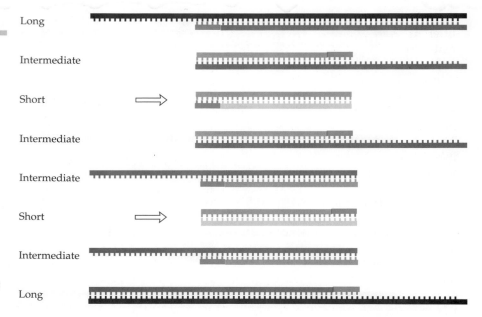

Figure 13.3
The products of the third cycle of PCR. The arrows indicate short segments that increase exponentially in subsequent cycles (see Table 13.1).

You may wonder why the primers do not anneal to complementary 20-base sequences elsewhere in the DNA sample, leading to the amplification of unwanted segments. We can answer with a probability calculation: What is the likelihood of finding a particular sequence of 20 bases, assuming a random order of nucleotides? Because each base can be any one of four items (A, T, G, or C), the answer is 1 in 4^{20}. This number is about 1 in 10^{12}, or 1 in a trillion. Since the total bases in a person's DNA is less than 1% of a trillion, a specific sequence of 20 nucleotides (of nonrepetitive DNA) is not likely to be repeated by chance. Although a one- or two-base mismatch might still allow some annealing of the primer, the PCR conditions tend to minimize such spurious binding. Furthermore, for the PCR amplification to be spoiled to a significant degree, *both* 20-base primer binding regions would have to be present outside the real target DNA, on opposite strands, relatively close together, and with their 3' ends facing each other (see question 2). These conditions *could* happen. And because so little DNA is needed as template to start PCR, laboratory workers take unusual care to guard against the amplification of a minute amount of a DNA contaminant, perhaps from prior PCR runs.

The generous amounts of target DNA taken from the thermal cycler can be used in an ever-increasing number of applications. These include basic research, prenatal diagnosis of genetic disease, detection and identification of infectious microorganisms (including the AIDS virus), crime investigation where cells are left behind (e.g., rape), in analysis of archaeological samples, and evolutionary investigations. Some of the techniques explained in the following sections—especially cloning in bacteria and DNA sequencing—also often use the products of PCR as their starting materials.

Cloning Human DNA in Other Species

To a large extent, the same basic chemicals and metabolic reactions, especially those needed for energy production, occur in all cells, from bacterial to human. This biochemical unity of life includes the genetic code, as well as the processes of DNA replication, transcription, translation, and mutation. Given the high degree of biochemical similarity, it was inevitable that scientists would try to mix and match the genes themselves.

By 1973 the technology for splicing together DNA segments from different sources had been developed by researchers at two neighboring institutions, Stanford University and the University of California at San Francisco. Using complementary base pairing as a kind of glue, they (and others) were able, for example, to insert DNA from the primitive frog *Xenopus* into the DNA of the bacterium *E. coli*, where the frog DNA subsequently reproduced itself as if it were "at home." Such a hybrid molecule (in this case, a frog-bacterial molecule) is called **recombinant DNA**. Its formation does not involve reassortment of independent or linked genes of a species, as described in Chapter 9. Rather, the making of recombinant DNA, which is also called **gene splicing**, refers to the attachment of a piece of DNA from one species to that of a second species, followed by insertion of the hybrid molecule into a host organism (often a bacterium).

The high-precision biological tool for making recombinant DNA is a special class of enzymes that cut across the double helix in an interesting way. Called **restriction enzymes**, these special cutting enzymes are widespread among bacteria. Their properties were first investigated by Werner Arber, a Swiss biochemist at the University of Geneva who later received a Nobel prize for this work. Acting like scissors that cut the DNA of invading viruses, the restriction enzymes in bacteria protect their owners from viral infection. Thus, they *restrict* the range of viruses that can successfully attack. Hundreds of different restriction enzymes have been isolated from many species of bacteria; each recognizes a different target sequence of usually four to seven bases.

For example, one such enzyme called *Eco***RI*** recognizes the following sequence of six base pairs wherever it occurs in long DNA molecules:

```
...G A A T T C...
...C T T A A G...
```

The enzyme causes breaks in the backbones between the G and A nucleotides on *both* strands. Then the weak hydrogen bonds joining the four intervening base pairs also break, and a staggered cut results:

$$...G A A T T C... \rightarrow ...G \qquad A A T T C...$$
$$...C T T A A G... \qquad ...C T T A A \quad + \quad G...$$

Double-stranded DNA molecules with protruding single-stranded tips like these are said to have **sticky** or cohesive ends. This is because each broken end has four unpaired bases, AATT (A being the terminal base), which tend to spontaneously pair with any complementary end that is available —with the single-stranded tip from which it has just been separated or with the tip of any other DNA molecule cut with the *same* restriction enzyme. A key point is that the DNA molecules that join together need not be from the same organism or even from the same species.

The hosts for much recombinant DNA work are special strains of *E. coli* carrying deleterious mutant genes that prevent them from surviving outside a laboratory. Each *E. coli* cell has a circular chromosome, a DNA double helix coding for about 4,000 polypeptides. In addition to chromosomal genes, some strains of *E. coli* possess extra genes that confer attributes that the cells can usually do without. These dispensable genes are present not on the large bacterial chromosome, but on small, circular pieces of DNA called **plasmids**, which often contain just a few genes

---

* *Eco*RI stands for *E. coli*, strain R, first (Roman numeral I) restriction enzyme isolated.

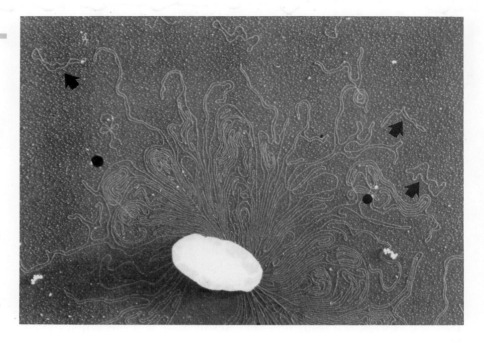

Figure 13.4
An electron micrograph showing DNA extruded from a ruptured E. coli *cell (the white area). The long, tortuous path of chromosomal DNA is easily seen, as well as at least three small plasmids (arrows). (Courtesy of David Dressler and Huntington Potter, Harvard Medical School.)*

(Figure 13.4). For example, some plasmids possess genes that make the *E. coli* resistant to common antibiotics.* Anywhere from one to a few hundred copies of a plasmid may be present in a cell, and they replicate and are inherited by daughter cells the same as chromosomal DNA. Because researchers can easily purify the small plasmids, operate on them, and cause them to be taken up again by other strains of *E. coli*, the plasmids are good tools for manipulating DNA.

One particular plasmid, dubbed pBR322, is of special interest to recombinant DNA workers. Along its length of about 4,000 base pairs, it has exactly one GAATTC sequence recognized by the restriction enzyme *Eco*RI. A researcher can isolate pBR322 plasmids in a test tube and treat them with *Eco*RI to open up each DNA circle at that one point, exposing two sticky, single-stranded ends. *The crux of gene splicing is to mix the opened plasmids with foreign DNA that has been treated with the same restriction enzyme* (Figure 13.5).

When the plasmid DNA is enlarged by the addition of a piece of foreign DNA, it becomes a vehicle, or **vector**, for transporting the foreign DNA into other *E. coli* cells. The recipient bacteria are treated with calcium chloride, which alters the cell wall and permits an occasional plasmid to enter. Because any bacterium carrying the new plasmid has thousands of its own genes but only a few genes from the foreign source (equivalent, say, to adding just one word to six pages of this book), we cannot really call it a new species. The bacterium is essentially its old self—except that it harbors a bit of foreign DNA that replicates along with the plasmid DNA.

Manufacturing Useful Proteins. Because the bacteria can be grown in prodigious quantities (a 10,000-gallon industrial fermenter has roughly the volume of a 12 × 14-foot living room), the foreign gene can be gotten in substantial amounts for investigative or practical purposes—all this from as little as one copy of one gene inserted into one plasmid that was taken

---

* Such plasmids, called R or resistance plasmids, are often found in hospital populations of pathogenic bacteria. Because these germs may possess resistance to more than one antibiotic, they are difficult to control.

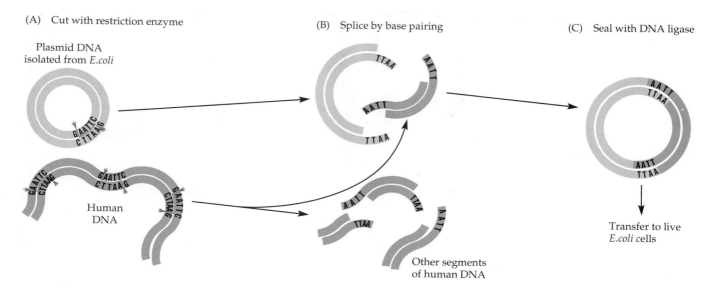

(A) Cut with restriction enzyme

Plasmid DNA
isolated from E.coli

(B) Splice by base pairing

(C) Seal with DNA ligase

Human
DNA

Other segments
of human DNA

Transfer to live
E.coli cells

up by one bacterial cell! The foreign DNA has been cloned, just as in the polymerase chain reaction. But this type of molecular cloning goes beyond the capabilities of PCR. Because bacteria are living cells, they not only replicate their genetic material, but also transcribe it into messenger RNA and translate it into polypeptide products. The bacteria can do the same with foreign genes.

A California research team was the first to report on a human polypeptide "engineered" this way, having tricked E. coli cells into making a small hormone called **somatostatin**. Just 14 amino acids long, somatostatin is normally made in the brain and acts to limit the production of other hormones. Although no inherited disease is known to involve this polypeptide directly, a deficiency of somatostatin can lead to the overproduction of growth hormone and a form of gigantism called *acromegaly*.

Knowing the amino acid sequence of somatostatin and the genetic code, these researchers first deduced a possible base sequence for the gene. Then they constructed a double-stranded helix, both sense and antisense strands, by synthetically linking together the appropriate nucleotides. They also tacked the DNA triplet for methionine onto the beginning and *two* stop triplets (just to be sure) onto the end (Figure 13.6A). Affixing appropriate single-stranded tips to this double-stranded structure of 17 coding units, they used two different restriction enzymes to splice it into a section of an E. coli plasmid. When the plasmid was incorporated into E. coli, the live cells transcribed and translated the foreign insert into somatostatin as if it were a host gene.

Actually, the somatostatin molecule that the researchers isolated from the E. coli cells was attached to the end of another polypeptide (Figure 13.6B), because the somatostatin gene had been deliberately spliced into the middle of a bacterial gene. Thus, the E. coli could use the promoter of the bacterial gene to start transcription of this section of the plasmid DNA. The resultant polypeptide began as a bacterial protein and ended as a human protein! But investigators were able to detach somatostatin from the rest by treating the hybrid protein molecule with a reagent (called cyanogen bromide) that specifically cleaves a polypeptide after any methionine residue (Figure 13.6C). This clever biochemical trick worked only because somatostatin itself contains *no* methionine. Thus, the cyanogen bromide stratagem cannot be used to isolate the more typical proteins of a hundred or more amino acids that usually contain methionine. Alto-

Figure 13.5
Preparation of bacterial-human recombinant DNA. (A) In one test tube, plasmids isolated from E. coli are treated with the restriction enzyme EcoRI. In another test tube, human DNA is treated with EcoRI. In both cases, the enzyme makes staggered cuts as shown, leaving single-stranded AATT tips. (B) When the cut-up DNAs are combined, some human DNA is spliced into the E. coli plasmids by complementary base pairing. (C) The chemical bonds in the sugar-phosphate backbones are rejoined by another enzyme, DNA ligase. The enlarged plasmids can then be transferred into live E. coli cells.

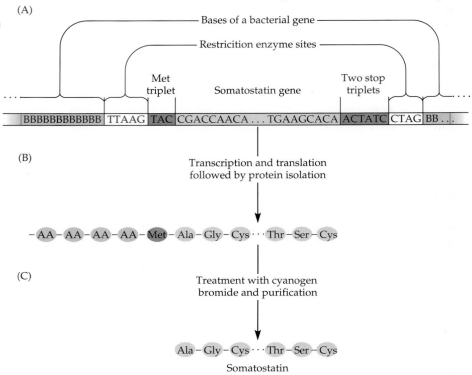

(A)

Bases of a bacterial gene

Restricition enzyme sites

Met triplet   Somatostatin gene   Two stop triplets

... |BBBBBBBBBBBB|TTAAG|TAC|CGACCAACA...TGAAGCACA|ACTATC|CTAG|BB...|...

(B)

Transcription and translation
followed by protein isolation

–AA–AA–AA–AA–Met–Ala–Gly–Cys···Thr–Ser–Cys

(C)

Treatment with cyanogen
bromide and purification

Ala–Gly–Cys···Thr–Ser–Cys

Somatostatin

Figure 13.6
Making the human hormone somatostatin in E. coli *cells. (For simplicity, the diagrams omit eight triplets in the middle of the somatostatin gene and eight amino acids in soma-tostatin.) (A) The bases coding for somatostatin, with some leading and following se-quences, are spliced into a bacterial gene (shown as ... BBBBB ...). Only the sense strand of the double helix is shown. (B) Transcription and translation of this region of DNA produces the first part of the bacterial protein (... −AA−AA− ...) attached to the amino acid methionine, which is attached to the 14 amino acids of somatostatin. This peculiar hybrid protein is isolated from the* E. coli *culture by standard chemical techniques. (C) Treatment with cyanogen bromide frees somatostatin from methionine and the frag-ment of the bacterial protein.*

gether, the in vivo bacterial synthesis of human somatostatin was a bio-chemical tour de force.

Many valuable pharmaceuticals have been produced by recombinant DNA technology. Human gene products include three hormones—*insulin* for diabetics, *growth hormone* for pituitary dwarfs, and *erythropoietin* for people with severe anemia—and a clot-dissolving enzyme called *tissue plasminogen activator* for heart attack patients. At the time of this writing, eleven different recombinant DNA products, having gone through years of clinical testing, had been approved for commercial use by the U.S. Food and Drug Administration (Table 13.2). Hundreds of other recombinant DNA products for medicine (and also for agriculture and industry) are in various stages of development. Whether the basic discoveries, testing, and development are carried out in universities or in private industries, the general aim is the same: to have a constant and reliable source of pure, concentrated polypeptides (which may or may not be available from other sources). Biotechnology companies—about 400 of them in the United States—are, of course, interested in turning a profit too.

The recombinant DNA production of human insulin in 1982 (the first recombinant product approved for commercial use) was very similar to that of somatostatin. Researchers synthesized the gene for the A polypep-

Table 13.2

Eleven human recombinant DNA proteins approved by the FDA.

Protein	Year Approved	Approved for	Cloned in
Insulin	1982	Children with insulin-dependent diabetes, to replace missing insulin	*E. coli*
Growth hormone	1985	Children with pituitary dwarfism, to replace missing growth hormone	*E. coli*
α-Interferon	1986	Cancer patients with hairy cell leukemia, AIDS-related Kaposi sarcoma; also for genital warts and hepatitis C	*E. coli*
Hepatitis B vaccine	1986	Health care workers exposed to blood; also for drug users and homosexuals; to ward off liver infection	Yeast
Tissue plasminogen activator	1987	Heart attack patients, to dissolve blood clots blocking coronary arteries	Hamster cells in culture
Erythropoietin	1989	Patients with kidney failure, to treat severe anemia due to deficiency of erythropoietin (made in kidney); also approved for AIDS-related anemia	Hamster cells in culture
γ-Interferon	1990	Patients with chronic granulomatous disease, a rare genetic defect of the immune system, to prevent severe infections	*E. coli*
Granulocyte colony-stimulating factor	1991	Cancer patients undergoing chemotherapy, to prevent infections by stimulating growth of white blood cells	*E. coli*
Granulocyte/macrophage colony-stimulating factor	1991	Patients with leukemia or Hodgkin disease, to speed the growth of white blood cells in transplanted bone marrow	Yeast
Interleukin 2	1992	Kidney cancer patients, although side effects may be serious	*E. coli*
Clotting factor VIII	1992	Patients with hemophilia A (classical hemophilia)	Hamster cells in culture

tide chain from scratch (see Figure 12.13) and inserted it into a plasmid in one strain of *E. coli*. Then they made the gene for the B chain from scratch and inserted it into a second strain of *E. coli*. Subsequently, they combined the two chains. The strictly chemical formation of the disulfide bonds between and within the A and B chains was inefficient, however, so researchers tried an alternative method of making insulin, which allowed the disulfide bonds to form more naturally. To explain what "naturally" means, we point out that a *single* human insulin gene codes for a *single* polypeptide about twice the length of insulin. The longer molecule, called *proinsulin*, folds up spontaneously; then the disulfide bonds are formed, and a section between the A and B chains is removed, producing active insulin. In the alternative method for the bacterial production of human insulin, investigators inserted into a single *E. coli* plasmid the nucleotides for the *proinsulin* polypeptide.

The recombinant DNA production of human growth hormone (HGH) differed from that of somatostatin and insulin in that the gene was not primarily synthesized from individual nucleotides. Instead, most of the gene was copied (by the enzyme reverse transcriptase)* from the HGH messenger RNA that had been isolated from pituitary glands. Recombinant HGH was approved by the FDA just a few months after they had

---

* This enzyme is isolated from retroviruses (such as HIV, the human immunodeficiency virus that causes AIDS), which use it to make a DNA copy of their RNA genetic material.

banned the only other source of HGH—pituitary glands from human cadavers (Norman 1985). The ban on cadaver-derived HGH was instituted because several people so treated had died of Creutzfeldt-Jakob disease, a very rare and mysterious viral disease. If the bacterially derived HGH had not existed, several thousand American children and adolescents with HGH deficiencies would have been without effective treatment.

Blood clotting *factor VIII* is a very large protein (2,332 amino acids) that is missing in patients with hemophilia. Its synthesis by recombinant DNA techniques required unprecedented technical creativity (Lawn and Vehar 1986). After screening a collection of thousands of randomly cloned human DNA segments, investigators found the factor VIII gene; it turned out to be 186,000 bases long, with 26 separate exons that totaled about 9,000 base pairs. Eventually, they managed to clone a gene sequence containing just the exons. In December 1992, recombinant factor VIII was approved by the FDA for general clinical use. It will be a boon to hemophiliacs. Since the 1960s, the only treatment for the disease has been frequent injections of a protein concentrate from pooled donor blood, a program costing about $8,000 per year per patient. Furthermore, treated hemophiliacs have been chronically infected with hepatitis viruses (and were also at risk to acquired immune deficiency syndrome until appropriate assays became available).

Other Ways to Clone. Vectors other than plasmids can be inserted into *E. coli* cells. Viruses that infect bacteria, called *bacteriophage* (literally "bacteria-eaters") or simply *phage*, consist of a molecule of DNA inside a protein coat. *Lambda* (λ) *phage* that attacks *E. coli* can have a 20-kilobase length of its DNA replaced with foreign DNA without altering its infectivity. Another type of vector, called a *cosmid*, consists of a plasmid joined to very short end pieces (so-called *cos* sites) of λ DNA and housed in a λ protein coat. Both phage and cosmids are convenient vectors in many situations, in part because they accommodate somewhat larger pieces of foreign DNA than plasmids.

Organisms other than *E. coli* can also be used for cloning. An increasing favorite (used also for many other types of genetic studies) is baker's yeast, *Saccharomyces cerevisiae*. This one-celled organism is easily grown and manipulated in laboratory glassware and can be maintained in either a haploid or a diploid state. Yeast is, moreover, a *eukaryotic* organism. Although we have stressed the biochemical unity of all organisms, there are clear distinctions in cell architecture and function between prokaryotes, like bacteria (with no distinct nucleus enclosing the chromosomes), and eukaryotes like yeast and humans (with a distinct nucleus). With only about four times the DNA content of *E. coli*, the 16 discrete chromosomes of yeast contain hundreds of mapped genes. Researchers have developed some very useful vectors that they can shuttle back and forth between yeast and *E. coli*. For example, a **yeast artificial chromosome (YAC)** has been made to order by combining DNA from several different sources. Specific restriction enzyme sites are positioned along the artificial chromosome for inserting foreign DNA. A major advantage of this vector is its ability to incorporate very large inserts—200 to 2,000 kilobases (kb), as opposed to inserts of only 10 to 40 kb of foreign DNA in plasmids, phage, and cosmids.

YACs, as well as phage and cosmids, are useful vectors for making *libraries* of human DNA. Such libraries are assembled by cloning random pieces of DNA into the vectors and selecting so many of them that the probability is high that every piece of human DNA is present in one or another clone. For a **genomic library**, the starting material is total human

Figure 13.7
A transgenic goat that secretes the human enzyme tissue plasminogen activator (tPA) in her milk. This goat developed from an egg that was injected with human DNA encoding tPA. She has passed the human gene on to two of her offspring. During the process of normal wound healing, tPA converts the serum protein plasminogen to an active form that dissolves blood clots. For this reason, tPA is also useful in treating heart attack patients and is available now as a recombinant DNA product (see Table 13.2). The drug is, however, very expensive —over $2,000 per treatment. Opinion is divided over the value of tPA versus other clot-dissolving products that cost far less. (Courtesy Karl M. Ebert, Tufts University School of Veterinary Medicine, and Genzyme Corporation.)

DNA isolated from any cell type, for example, sperm or a bit of liver tissue. A genomic library contains all types of DNA, including coding regions, introns, control sequences, pseudogenes, and repetitive sequences.

For a so-called **cDNA library**, the starting material is the DNA that is obtained by copying from the messenger RNA present in a particular cell type. The "c" in cDNA stands for *copy* or *complementary*, and it is obtained from RNA via the enzyme reverse transcriptase. A cDNA library contains the coding regions of genes, as well as the leading and trailing sequences present in mRNA, but no introns or intergenic DNA. Not all genes are represented, however—just those genes that are expressed in the cells from which the messenger RNA was obtained. For example, cDNA derived from immature red blood cells consists almost entirely of the α-globin and β-globin genes that encode the polypeptides of hemoglobin, since these genes are transcribed at high levels.

There are other cloning systems too. The human factor VIII gene, for example, was cloned in hamster cells grown in laboratory culture dishes. Many different methods have been developed for getting foreign genes into cultured mammalian cells, where they may become stably incorporated into the host's DNA. One method is to treat the foreign DNA with calcium phosphate, making tiny clumps that a cell readily engulfs with its cell membrane. Still another method is to insert the foreign gene into a virus that naturally infects the cells. Still another method is to microinject the foreign DNA into the nucleus of the cells with a very fine needle. Each method has advantages and disadvantages.

It is also possible to inject DNA into the *fertilized eggs* of mammals. When such eggs are then implanted in surrogate mothers, they can lead with low frequency to the development of newborns with the foreign gene in every cell, so-called **transgenic** animals. The injected foreign gene seems to be incorporated into the host DNA at random places (if it is incorporated at all), with consequences that have not been fully explored. Research teams in the United States, Scotland, and the Netherlands have produced some unusual transgenic goats, sheep, and cows that secrete small to moderate amounts (grams per liter) of the protein product of a *human* gene into the female animals' milk (Figure 13.7)! To accomplish this feat, researchers manipulate the DNA segment to be injected into the eggs so that the human gene is joined to the promoter of a host gene that encodes a milk protein. By this maneuver, the human gene is expressed, and the resultant protein is secreted in the mammary gland during lactation. Although the embryonic development of such a transgenic animal is unpredictable, it is possible that just *one* useful transgenic animal could transmit its new capability to many descendants. The secretion of a useful pharmaceutical in milk gives a convenient and possibly generous source of supply, since an average cow makes 10,000 liters of milk per year. Purifying the drug from the complex mix of components in milk may be difficult, however, and the issue of safety still needs to be addressed. In any case, research in such barnyard biotechnology—called "pharming" in news reports—continues (Bialy 1991), although it will be many years before drugs produced this way will be available for general human use.

Public Concerns. Many people believe that the techniques just outlined promise great benefits to human welfare; they are confident that accompanying risks can be evaluated and managed. But others claim that the risks to public health and to the environment are unacceptably large or that it is morally wrong to tinker with nature in this manner.

In fact, some molecular biologists were the first to voice safety concerns. In the summer of 1973, a small group attending a conference on nucleic acid research requested that the National Academy of Sciences (NAS) study the potential hazards of recombinant DNA research. One year later, a distinguished NAS committee headed by Paul Berg (a 1980 Nobel laureate in chemistry) recommended that scientists throughout the world refrain from certain types of recombinant DNA experiments pending further evaluation. This call for a moratorium on research by the researchers themselves was unprecedented in the history of science.

As one consequence, an international panel of scientists and lay people met in 1975 at the Asilomar Conference Center in California. After days of spirited discussion, the conferees decided that research with recombinant DNA should be regulated by a set of safety rules. By the time the National Institutes of Health (NIH) set forth these rules in July 1976, considerable public interest had been generated by eminent scientists who spoke out for or against such government regulation. (For a lively documentary history of the recombinant DNA debate, see Watson and Tooze 1981.)

The original NIH guidelines for recombinant DNA research provided successively more rigorous standards for experiments that were perceived to be increasingly dangerous to public health and safety. (Strictly speaking, the guidelines applied only to research funded by NIH, but it was hoped that other groups—including pharmaceutical companies—would also observe them.) At the lowest level (P_1), commonsense laboratory methods for handling nonpathogenic microorganisms are sufficient. At intermediate levels (P_2 and P_3), laboratory access is limited, experimental methodology is much more meticulous, and special construction and ventilation features must be built in. For handling microorganisms that are extremely hazardous, a P_4 laboratory is needed.

In addition to the successive levels of *physical containment*, increasingly risky experiments require increasingly enfeebled strains of *E. coli*. This type of safeguard is referred to as *biological containment* and seeks to ensure that the engineered organisms do not survive if released into the general environment. The guidelines also called for an outright ban on some experiments, such as cloning in *E. coli* any DNA derived from known pathogens or cancer-causing viruses. Experience to date suggests that the original NIH guidelines for laboratory research were too stringent; indeed, they have been relaxed several times since 1976 without untoward effects. In thousands of laboratories around the world, no known harm has yet been documented or even suspected to have occurred as a consequence of recombinant DNA research. Most biologists feel that in complying with the NIH guidelines, the hazards of handling recombinant organisms are no greater than the hazards of handling the same species as nonrecombinants.

But controversy continues, with recent debate focusing on the environmental safety of recombinant agricultural products that are tested in open fields. Examples of such products include engineered crop plants that resist pests and herbicides, tolerate drought or soil contaminants, provide better nutrition, or can be more easily transported and preserved. Microorganisms are being created that attack weeds and insects, clean up toxic chemicals, enhance oil recovery, and leach minerals from their ore beds. For some people, it would be morally wrong *not* to vigorously pursue these technologies in the face of environmental degradation from a burgeoning human population. For others, genetic engineering is the route to ecological disaster, spearheaded by exotic blights such as mutated pathogens or uncontrollable weeds. Small farmers are concerned about their jobs, because biotechnological advances are seen as benefiting only large-scale operations. Still others worry that wrong decisions will be made because the public is not objective in balancing benefits and risks and lacks a basic understanding of probability. Academic and industrial personnel, environmental activists, biotechnology critics and proponents, the Environmental Protection Agency, the Department of Agriculture, and residents near proposed test fields may all become involved in spirited give-and-take of fact and opinion (Hoy 1991; Maranto 1986).

Electrophoresis

We now back up half a century to present one of the most widely used methods in chemistry. **Electrophoresis** (Greek *phoresis*, "to be carried") is the movement of electrically charged molecules, such as proteins or nucleic acids, in a solution to which an electric field is applied. The use of electrophoresis to separate one protein from another was first accomplished in 1937 by the Swedish Nobel-prize-winning chemist Arne Tiselius. (He was a graduate student at the time of his invention.) Although Tiselius carried out electrophoresis with proteins free in solution, more modern methods utilize a supporting material through which the charged molecules move. The support can be simply a sheet of filter paper or a gel-like substance such as starch, agar, or often a polymer called *polyacrylamide*.

Figure 13.8 shows a basic setup for paper electrophoresis to separate proteins in solution. An electric current causes the proteins to migrate

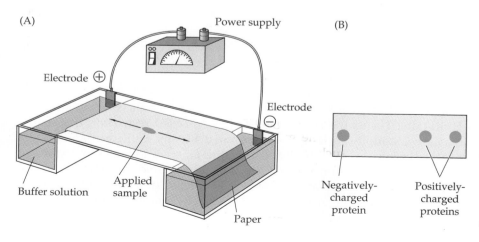

(A) Power supply (B)

Electrode ⊕

Electrode ⊖

Buffer solution

Applied sample

Paper

Negatively-charged protein

Positively-charged proteins

Figure 13.8

Paper electrophoresis of a solution of different proteins. (A) A small drop of the protein sample is placed at the middle of a wet sheet of filter paper whose ends dangle in buffer solutions connected to positive (+) and negative (−) electrodes. The power supply is turned on, so that a direct electric current passes through the filter paper. The current carries positively charged proteins toward the negative electrode and negatively charged proteins toward the positive electrode. (B) After a time (often an hour or more), the filter paper is removed from the apparatus and stained for protein. In this particular experiment, the original mixed solution (applied sample) is resolved into two different proteins that are positively charged (one more than the other) and a protein that is negatively charged.

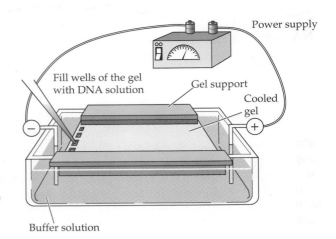

Figure 13.9

An electrophoresis apparatus for separating lengths of DNA by size. Hot liquid agarose is poured into the plastic gel support, which is taped shut at both ends and which holds a comb with rectangular teeth to make the wells. After the gel cools and sets, the tape and the comb are removed. Test solutions are pipetted into the wells of the slightly submerged gel, and voltage (20–200 volts) is applied to the two ends. The gel contains a fluorescent dye that binds to DNA, making the DNA visible.

Figure 13.10

A hypothetical gel as it would appear when DNA 20 kb long is cut with restrictions enzymes and electrophoresed, as noted in the text. When the electric current is applied, the negatively charged DNA molecules move out of the well toward the positive pole — the smaller, the faster. Each size of molecule moves out of the well and forms a band of migrating DNA. Because of the contained dye, bands appear bright orange under ultraviolet light. In a real situation, the size standards (shown in lane 5) would not necessarily exactly match the bands in the unknowns in lanes 1–4.

according to their charge: The greater the charge, the farther the molecule will travel in a set time. Although the proteins are invisible in the solutions, they can be made evident after electrophoresis by treating the filter paper with a protein stain. Radioactive marking can also be used, but special care must be exercised by laboratory personnel in handling radioactive materials.

In gel electrophoresis of DNA molecules, the DNA molecules migrate through a thin slab of gel material placed in an electric field. Because their regularly spaced phosphate groups are negatively charged, DNA molecules will always move toward the positive pole. And because the supporting medium impedes the migration of bigger DNA molecules, the smaller pieces of DNA will migrate faster through the pores of the gel. Under good conditions a DNA segment with n bases will migrate noticeably farther than a piece with $n + 1$ bases.

An apparatus for gel electrophoresis of DNA is illustrated in Figure 13.9. The wells at the left of the gel hold the solutions in the starting gates. For example, let us assume that we have a pure solution of DNA molecules 20 kilobases long. We cut different samples with restriction enzymes— *Eco*RI (which cuts DNA at the sequence GAATTC), or *Bam*HI (which cuts DNA at the sequence GGATCC), or both enzymes. We then place each sample in a separate well of the gel.

    lane 1: DNA cut with *Eco*RI
    lane 2: DNA cut with *Bam*HI
    lane 3: DNA cut with both *Eco*RI and *Bam*HI
    lane 4: uncut DNA
    lane 5: a standard solution that possesses different species of DNA
          of known lengths

How many recognition sites does the original DNA have for the restriction enzymes, and how far apart are they? The results of gel electrophoresis, shown in Figure 13.10, provide an answer to this problem of mapping the relative positions of recognition sequences. We note first that *Eco*RI (lane 1) cuts the DNA into two pieces, called **restriction fragments**, that are 2 and 18 kb long, so there must be one *Eco*RI recognition site relatively close to, say, the left end of the diagram below. The results of lane 2 show that the one *Bam*HI site is nearer the middle, since it cuts the DNA into two pieces 7 and 13 kb long. Whether the 7-kb or 13-kb *Bam*HI segment contains the *Eco*RI site is answered by lane 3, where the DNA is cut with both enzymes: The 7-kb piece is replaced by pieces 2 and 5 kb long, but the 13-kb piece is still intact. We conclude that the map of the DNA molecule is

Such mapping of the recognition sites of restriction enzymes, called **restriction mapping,** is a common practice in molecular biology. It provides signposts along a length of DNA and points to where genetic manipulations may be possible. In addition, investigators can cut out the piece of gel that contains a single band, chemically remove the gel material, and thereby obtain a single, purified segment of DNA. This DNA could, for example, be the starting material for cloning or sequence determination.

DNA Sequencing

DNA sequencing means determining the exact order of bases along a length of (single-stranded) DNA. Practically all extended manipulations of DNA involve sequence determination somewhere along the way. One increasingly used application is the study of evolutionary relationships among a group of related species. Another major application is the fine-grained analysis of mutations that cause genetic diseases. Knowledge of the nucleotide sequence of a specific human gene may also be used to infer the amino acid sequence of the encoded protein—often a crucial step in the diagnosis and management of an inherited disorder.

There are two widely used sequencing methods, the details of which are largely beyond the scope of this book (but see Mange and Mange 1990). The inventors were Nobel prize winners Frederick Sanger,* of the Medical Research Council in Cambridge, England, and Walter Gilbert, of Harvard University. Both methods rely on four reaction tubes (labeled G, C, A, and T in the Sanger method). Each tube determines the positions of a particular base along a segment of DNA. Within the G tube, for example, a chemical reaction produces many fragments of DNA of different lengths, but all ending in G. The fragments are separated from each other in one lane of an electrophoretic gel, parallel to lanes for the other three reactions. The resulting bands in the G lane represent the positions of all the G's in the original DNA segment. Combining the information from the four lanes gives the positions of every G, C, A, and T.

The DNA fragments in the gel are made visible by incorporation of radioactive phosphorus into the DNA. After researchers complete electrophoresis, they lay a sheet of X-ray film over the gel. The high-energy particles striking the film from the radioactive DNA develop the silver grains in the film and produce a high-resolution image, called an **autoradiograph** (self-image). The fastest-migrating band of DNA represents position 1, the next fastest position 2, and so on (Figure 13.11). Thus, the sequence of bases is read directly off an X-ray film exposed to the gel, from the bottom up. A length of several hundred nucleotides can be read off one gel. Longer sequences can be obtained by combining the information of several cloned DNA segments that overlap one another.

Automated DNA sequencing devices are being perfected. The prototype for these expensive machines was developed by Leroy Hood, now at the University of Washington in Seattle. Although capable of sequencing

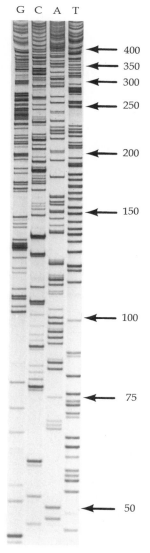

Figure 13.11
Autoradiograph of a sequencing gel. The bands in the four parallel, vertical lanes represent the positions of different-sized segments of DNA ending in either G, C, A, or T. (The bands appear dark because the radioactivity in the DNA sensitized the X-ray film.) To determine the sequence of bases starting at, say, position 100, note that the band at this level is in the A lane. Moving slightly upward in the gel, note that the next higher band is in the C lane. Continuing upward and switching back and forth between lanes as required, successive bands give the sequence of bases. Confirm that the sequence of bases from position 100 to position 110 is AC-GAGCCGGAA. (From Williams et al. 1986.)

---

* Having invented first a method for determining the sequence of amino acids in proteins and then a method for determining the sequence of bases in DNA, Sanger is one of a very few individuals to win *two* Nobel prizes (in 1958 and 1980). The first person to win two Nobel prizes was Marie Curie, in 1903 and 1911.

tens of thousands of bases a day, such a machine does not avoid the need for gel electrophoresis to separate a series of DNA fragments differing in length by single bases.

The DNA sequencing procedure is now so good that in more and more cases, the amino acid order of a *protein* is determined by sequencing the *gene* that encodes the protein. Researchers then infer the amino acids of the protein from knowledge of the genetic code. This approach was taken in 1987 for X-linked Duchenne and Becker muscular dystrophies. In a large cooperative research effort led by Louis Kunkel, of Boston's Children's Hospital, the massive gene was cloned and shown to extend over 2.3 million bases, by far the largest gene known. The expressed portion of the gene is divided into about 75 exons that account for only about 1% of the total sequence. That is, the gene is about 99% noncoding introns. After excision of the corresponding intron regions of the RNA transcript, the final messenger RNA encodes a protein with 3,685 amino acids. Dubbed *dystrophin*, the previously unrecognized protein is present in muscle tissue, but in barely detectable amounts. It is absent in boys with Duchenne muscular dystrophy and modified in boys with Becker muscular dystrophy. Studies of dystrophin are in progress, and some results were presented in Chapter 5.

Many extended sections of human DNA have been sequenced—for example, 73,000 bases in the neighborhood of the β-globin cluster of genes on chromosome 16 and 66,000 bases encompassing the growth hormone gene on chromosome 17. Included in these segments are start and stop signals, exons, introns, repetitive DNA, and other genetic features. Many such sequences, from a wide variety of organisms, have been brought together in computerized databases. One of these, GenBank, sponsored by the U.S. National Institutes of Health (NIH), now contains information on over 10,000 segments of human DNA—some important genes and some "junk"—totaling over 10 million bases of sequence information. Thus, a small start has been made in knowing the finest details of the human genome. We say *small*, because 10 million is only 0.3% of 3 billion, the approximate number of total bases in a haploid set of human chromosomes (Figure 13.12).

The Human Genome Project. Instead of accumulating sequence data unsystematically, several national and international organizations have been set up for coordinating efforts to map and sequence all 3 billion bases of the human genome. It is expected that the endeavor—known as the Human Genome Project—will take more than a decade of work by many technical teams and will cost billions of dollars. Despite early doubts about the goals of the colossal project and worry about draining funds from other desirable research, most investigators now agree that the known and possible benefits are worth the costs. In any event, the enterprise is moving slowly forward. The American Society of Human Genetics (representing about 4,000 physicians, scientists, and genetic counselors) strongly supports the project:

> The mapping and sequencing of the human genome will provide extensive new knowledge about the genes involved in inherited diseases, birth defects, later-onset diseases with a genetic component—including heart disease, neurological and behavioral disorders, and many forms of cancer—and even susceptibility to infection. This knowledge will have profound impact on the understanding, detection, prevention, and treatment of these disorders. (Caskey, Worton, and Watson, 1991)

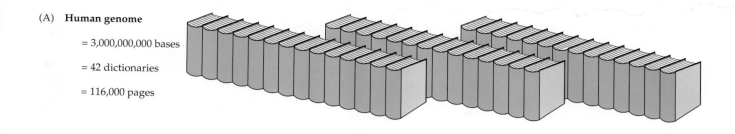

(A) **Human genome**

= 3,000,000,000 bases

= 42 dictionaries

= 116,000 pages

(B) **Fruit fly genome**

= 160,000,000 bases

= 2 dictionaries

= 5,600 pages

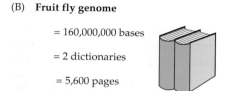

(C) **E. coli genome**

= 5,000,000 bases

= 200 pages

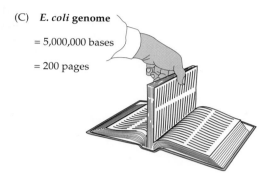

(D) **Large insert cloned in a YAC**

= 1,000,000 bases

= 40 pages

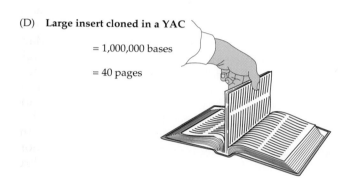

(E) **Average insert cloned in a plasmid, phage, or cosmid**

= 30,000 bases

= 1 page

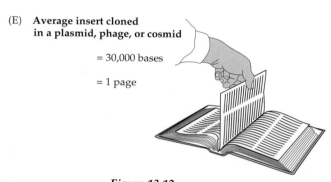

Figure 13.12
The approximate number of bases in three different genomes and in some cloning vectors, compared to the number of characters in Webster's Unabridged Dictionary *(Merriam-Webster, Springfield, Massachusetts, 1986). This dictionary has about 26,000 characters per page and 2,770 pages, for a total of 72 million characters.*

There are likely to be economic spin-offs, as well, when the methods are applied to farm crops and animals. And as is usual in research, unexpected results and insights are anticipated.

Because most human DNA (about 97%) does *not* code for proteins, tRNA, or rRNA, some criticism of the project has centered on its potential wastefulness. Why bother with all those repetitive sequences—the 500,000 copies of the *Alu* repeat, for example? The criticism loses its force, however, if researchers use *cDNA* rather than "raw" genomic DNA. Recall that cDNA is transcribed from messenger RNA and therefore contains essentially the coding triplets for the amino acids of cell proteins. In cDNA there are no introns or regulatory sequences or any of the DNA sequences between genes. Researchers at the National Institutes of Health, for example, have begun to analyze the cDNA obtained from the RNA found in human brain tissue. (It is thought that 30–60% of all genes are expressed in the cells of the brain.) By sequencing a few hundred bases in each of about 600 different molecules of cDNA, these researchers found several dozen genes that were previously sequenced, plus 230 genes that appear to be newly revealed.

In the absence of any other information, how do researchers know that a newly sequenced length of DNA is possibly a gene that codes for a protein? First they scan the mystery sequence for stop triplets. If a stop

303

triplet is *not* found for a considerable distance—say, 300 nucleotides—the sequence is called an **open reading frame (ORF)**, and any ORF could be a gene. Then the inferred amino acid order can be examined to see if it is similar to other polypeptides or if it is in any way, "proteinlike." Note that for every nucleotide sequence, these considerations must be checked for *each* of the three possible reading frames.

Improvements in the technologies of mapping and sequencing have been forthcoming, but more are needed to better automate the process. Advances in computer-based methods for collecting, storing, distributing, and analyzing the data are also anticipated. As noted in Figure 13.12, just to print 3 billion nucleotides would fill 42 unabridged dictionaries. For this reason, much sequence information will only exist as an (unprinted) electronic database. Analysis of variation in nucleotides among people, as well as between the two haploid genomes of a single person, will necessitate still greater computer-related capabilities.

Computerization of sequence information will also allow researchers in different laboratories to investigate the same region of DNA without actually exchanging biological samples. Researchers will simply send to a centralized database the sequence of, say, 500 nucleotides of the DNA segment they are working with. This short 500-base segment is called a **sequence tagged site** or **STS**. Researchers in other laboratories can recreate the DNA segment of interest by PCR and their own DNA libraries.

Interest in the Human Genome Project is centered in Europe, Japan, and the United States under the umbrella of the Human Genome Mapping Organization (HUGO). In this country, organizational and financial support comes from the National Institutes of Health and the Department of Energy. James Watson was the first to head the National Center for Human Genome Research. In addition to coordinating research efforts, encouraging the development of new technology, and overseeing the massive job of data analyses, these organizations and others are also concerned with the possible misuse of the information. For example, maintaining confidentiality of medical data and ensuring freedom in personal decisions are already familiar problems in health care, insurance, and employment. But they will become much more difficult as the increased precision of personal health data becomes a reality.

OTHER APPLICATIONS

The molecular techniques presented so far have proved to be versatile, with ever-increasing uses in science, medicine, industry, and other human activities. We present here a few applications that seem interesting and important.

DNA Fingerprinting

Fingerprinting with ink and paper, based on fingertip ridge patterns, is the standard method for personal identification. Constant throughout life, fingerprints can be used to distinguish one person from any other—even one identical twin from the other, although their patterns are very similar. The eugenicist Francis Galton (see Box A in Chapter 8) began fingerprint classification in the 1890s, making practical the filing and retrieval of records. **DNA fingerprinting** has a similar aim, but it is based on certain nucleotide sequences that differ greatly among people. (Actually, the method should be called simply *DNA printing* or *DNA typing*, since it has nothing to do with fingers.) The protocol begins by using a restriction enzyme to cut a person's DNA into fragments and then using electrophoresis of the fragments to separate different lengths. Pinpointing specific length differences between people involves some additional techniques.

Recall from Chapter 12 that there are many places in human DNA where a particular sequence of nucleotides is repeated over and over. For example, the half million *Alu* repeats (each several hundred bases long) are dispersed throughout the genome and make up about 5% of the total DNA in a human cell. In other cases a much shorter sequence is repeated in tandem many times within a defined DNA region, and the same pattern with perhaps a different number of repeats occurs at other places. For example, a particular sequence that we all have in our DNA is 16 nucleotides long:

AGAGGTGGGCAGGTGG

This unit is repeated in tandem hundreds of times at dozens of different regions scattered over most, if not all, of the human chromosomes. The repetitive regions are not in genes and do not code for polypeptides. They are just there.

DNA fingerprinting—that is, individual-specific DNA identification—is made possible by the finding that no two people are likely to have the same number of copies of repetitive DNA units at all of the places where the sequences occur. Such regions are called **VNTRs**, which stands for **variable number of tandem repeats**. It is not known why people differ so much in the number of repetitive DNA units anymore than it is known why people have different fingerprints. But the variation may result from a process called **unequal crossing over** occurring in past generations (Figure 13.13).

The variation in copy number leads directly to variation in the lengths of the repetitive regions—lengths that can be measured. For example, let us assume that two blood samples are at hand, one from a blood stain found at the scene of a crime and another from a suspect. Since the samples, especially the crime evidence, may yield only tiny amounts of

(A)

(B)

(C)

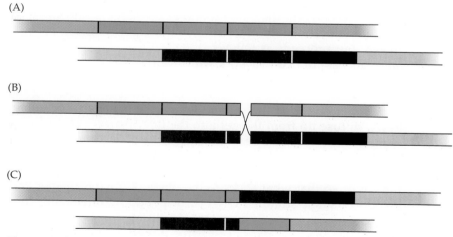

Figure 13.13

Unequal crossing over occurring along a segment of chromosome where the same nucleotide sequence is repeated many times. Here, the long rectangles represent the same unit repeated three times on each chromosome, one shown in color, one not. (A) Since all units are the same, synapsis may be misaligned—in this case offset by one unit. (B) Crossing over at a point in the misaligned region. The particular site of the crossover does not matter. (C) The products of crossing over, either one of which could be included in a gamete and transmitted to offspring, consist of one chromosome with four repeats and another with two repeats.

DNA (isolated from the white blood cells), the polymerase chain reaction can be used as a preliminary step to make adequate amounts of DNA. The analysis then proceeds with the following steps:

1. We separately treat the two DNA samples with a restriction enzyme such as *Eco*RI, which cuts DNA at all GAATTC sites. It is important that this recognition site *not* occur in the repetitive unit to be analyzed, say, the 16-base sequence already noted. *Eco*RI cuts human DNA into about 100,000 pieces of varying sizes, a few of which will include tandem repetitions of the 16-base sequence in question. For example, letting r represent one of the 16-base units and **x** represent the sequence GAATTC where *Eco*RI cuts, we see that one person's DNA might have the following pattern at a particular place on one chromosome (12 repeats of r):

—— x —— rrrrrrrrrrrr ———————— x ——

At the corresponding place, another person's DNA might have 25 repeats of r and the same *Eco*RI sites to the left and right:

—— x —— rrrrrrrrrrrrrrrrrrrrrrrrr ———————— x ——

The distance between cutting sites is greater in the second case, thus yielding a bigger fragment.

2. We use electrophoresis to spread out the 100,000 fragments of DNA according to size. The two samples are run in side-by-side lanes along with control lanes containing DNA segments of known size. Although there are so many fragments that they smear into each other, somewhere along the length of the gel, invisible at this time, are the several bands that contain the repetitive DNA. We want to know their exact positions, which are indicative of their lengths. If the lengths are different, then the suspect's blood and the blood found at the scene of the crime are different. The method for revealing the positions is a bit involved, but proceeds as follows.

3. Since gels are difficult to work with, we transfer the pattern of DNA bands in the gel to a piece of nitrocellulose filter paper. As shown in Figure 13.14, this is accomplished by simply laying the filter paper on top of the gel and letting the materials, including DNA, diffuse up from the gel into the filter paper. This technique was invented by Scottish biochemist E. M. Southern in 1975 and is appropriately called **Southern blotting**. The liquid that moves up through the gel and filter paper also denatures the DNA, so that what sticks to the fibers of the filter paper are lengths of single-stranded DNA corresponding to the bands in the gel.

4. We remove the filter paper and bathe it with a so-called **radioactive probe**, which is single-stranded DNA complementary to the repetitive DNA. In this case, we can use a probe developed by English geneticist Alec Jeffreys, who was instrumental in introducing DNA fingerprinting (and who coined the phrase). This probe, consisting of 29 repetitions of our 16-base sequence, will hybridize by base pairing and therefore stick to the corresponding repetitive DNA sequences wherever they occur on the filter paper.

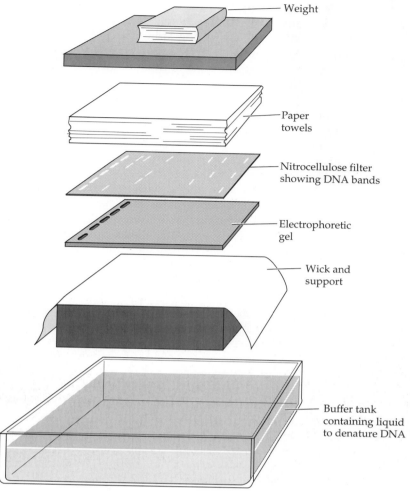

Weight

Paper towels

Nitrocellulose filter showing DNA bands

Electrophoretic gel

Wick and support

Buffer tank containing liquid to denature DNA

Figure 13.14
The setup for Southern blotting, one of the most frequently used procedures in molecular biology. A stack of absorbent paper towels draws up the liquid from the buffer tank through the wick, through the gel, and through the nitrocellulose filter paper. DNA molecules from the gel are denatured in the upwardly moving liquid and stick to the fibers of the filter paper (where they can be easily treated one way or another). Although the diagram shows bands of DNA, they are actually invisible at this stage. For clarity the drawing includes extra space between the several layers.

5. The final step is *autoradiography*, which was described earlier with regard to DNA sequencing. If we lay a sheet of film on the filter paper, bands will appear wherever the radioactivity of the probe develops the film. That is, the film will show dark bands representing the various lengths of DNA "discovered" by the probe.

Jeffreys's probe detects about 15 distinct bands that are between 4 and 20 kb long. (Some shorter repetitive DNA segments migrate off the gel, and other bands are faint or smeared.) Some of the detected bands represent DNA fragments that are "alleles" of each other; that is, they represent the same region on the two homologous chromosomes, but they have different lengths because of different numbers of copies of the repeat unit. Other bands are "nonalleles," deriving from different places in the genome. A probe, like Jeffreys's, that detects many different regions is called a **multilocus probe** (Figure 13.15A). The degree of variation in bands from one person to another is so large that the theoretical probability that the bands seen in one individual are present in another unrelated individual may be one in many millions. Essentially, each person has a unique DNA fingerprint. But problems in doing the test and uncertainty over how to calculate the probabilities have cast some doubt on the legal uses of DNA typing.

(A)

Multi-locus probe

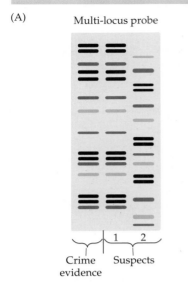

Crime
evidence

Suspects

1   2

(B)

One single-
locus probe

Another single-
locus probe

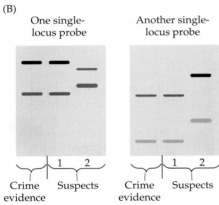

Crime
evidence

Suspects

1   2

Crime
evidence

Suspects

1   2

Figure 13.15
Typical DNA fingerprinting autoradio-
graphs, which implicate suspect 1 and ex-
onerate suspect 2. (A) Results using a
multilocus probe that "discovers" about
15 bands. Each band represents a DNA
fragment of a particular length, dependent
on the number of tandem repeats of the
short nucleotide that binds the probe. In
part the bands represent alleles, and in
part nonalleles, but it is not possible to say
which are which. (B) Results using two
different single-locus probes, each of
which binds to just two bands representing
allelic DNA fragments.

A **single-locus probe** will hybridize to just one region in an individual's DNA and is easier to analyze mathematically. Only one or two bands can be present in such an autoradiograph, corresponding to the two alleles. But because there is so much variation in DNA fragment sizes (owing to the variation in copy number of the repetitive unit), these alleles will usually be different and lead to distinguishable bands. More critically, the *two* bands from one person will almost always be different in some way from the *two* bands in another unrelated individual (Figure 13.15B).

Calculating the exact probability that unrelated people will match depends on how **polymorphic** the VNTR locus is, that is, on how many different alleles there are in the population from which the samples are drawn and on the frequencies of those alleles. (*Polymorphic* means "many forms." A gene with two or more alleles with appreciable frequencies is said to be polymorphic.) Some repeat sites may have more than 10, or even more than 100, alleles distinguishable from each other by the fingerprinting process. Because a gene with 100 alleles can produce 5,050 different genotypes,* even a single-locus probe can do a good job of distinguishing unrelated persons. Combining the information in several different single-locus probes can provide each person with a unique DNA fingerprint.

DNA fingerprinting was first used in England during a rape-murder trial in which DNA isolated from the rapist's sperm was compared with DNA from the white blood cells of possible suspects (Box A). It has also been used by the English immigration service to positively match children with their putative parents. More recently, the procedure has been used in an increasing number of U.S. courts—in several hundred trials in about 40 different states. It is estimated that thousands of cases involving DNA testing never get to trial because of the strength of the evidence.

Controversy over the use of DNA fingerprinting continues, however, (Epstein 1991; Lander 1991; Lewis 1989; Neufeld and Colman 1990). One of several practical problems is called *band shifting*. Two DNA samples prepared separately for electrophoresis may have slightly different concentrations of DNA, salt, acid-base chemicals, or contaminants. Also, the DNA in the samples may be degraded to varying degrees. For any of these reasons, two samples that contain the same DNA alleles may have their DNA bands shifted in the electrophoretic gel relative to the other. By artificially including in each sample a known DNA, technicians can, however, usually control for band shifting. But there is the problem of interpretation of faint, fuzzy, or overlapping bands, and what constitutes a match. How close must two bands in different lanes be before calling it a match?

Another difficulty deals with probability calculations. Although accuracy depends on knowing the nature of the VNTR polymorphism—how many alleles and how frequent each allele is—this knowledge is currently incomplete. The specifics of the polymorphism may differ from one ethnic group to another. Thus, it may not be enough to break down frequency data as representative of European or African or Central American populations. We might have to compile basic information about much narrower groups of people—for example, Scots, or Danes, or Poles—if their frequencies differ significantly from each other.

In short, DNA fingerprinting detects variation in the lengths of DNA fragments based on restriction digests followed by electrophoresis, Southern blotting, and autoradiography. As a method for the unique identification of human beings, it is very good in theory—much more discriminating than blood typing and more useful than traditional fingerprinting.

* The formula for the number of different genotypes is $n(n + 1)/2$, where n is the number of alleles. Try out the formula for $n = 1, 2,$ or 3. Mathematically minded readers should try to derive it.

THE FIRST USE OF DNA FINGERPRINTING IN CRIMINAL INVESTIGATION

In 1983, on a narrow footpath that passed by a psychiatric hospital in the central English village of Narborough, 15-year-old Lynda Mann was raped and strangled. Investigating detectives found few substantive clues. Conventional forensic tests on recovered semen, for example, could only narrow the identity of the killer to about 10% of the population. Despite a year of wide-ranging police work, the case was eventually dropped when hundreds of leads all proved futile.

Three years later, 15-year-old Dawn Ashworth was similarly sexually attacked and slain within a few hundred yards of the first crime, and semen was again recovered from a vaginal swab. Richard Buckland, a 17-year-old kitchen worker in the psychiatric hospital, had known the victim and had a history of sexual misbehavior. He was arrested, and after extensive interrogation confessed to Ashworth's rape and murder.

The unsolved Lynda Mann case was then reopened by sending the semen samples taken from Mann, Ashworth, and Buckland to Professor Alec Jeffreys at nearby Leicester University. Jeffreys had used his newly developed technique of DNA fingerprinting in a highly publicized immigration dispute. On the basis of DNA band matching using a multilocus probe, Jeffreys concluded that both girls had been raped by the same man, but the man was not Buckland! In fact, Buckland — mentally a bit dull and easily confused — had not even fit into the 10% class of possible suspects.

Police persevered in their investigation of the two frightful crimes. Men and teenage boys in Narborough and nearby villages were asked to voluntarily submit to blood drawing, and any blood that met the right criteria by traditional tests was sent to Jeffreys for DNA analysis. The police hoped that this massive undertaking would somehow flush out the killer. After thousands of samples, however, representing about 98% of possible males, nothing turned up.

Colin Pitchfork, a 28-year-old bakery worker who was well known for creative cake decoration, had not given blood when sent his notice. Worried about prior convictions for flashing (not in the Narborough community), he had used a stand-in with falsified identification: Pitchfork's passport with the stand-in's picture. The break in the case came in a local pub over a few pints of ale, when the stand-in casually spoke of his trickery to fellow workers, a disclosure that was reported to the police. Pitchfork, who had managed to conceal his sociopathic personality, was eventually arrested and convicted of the crimes against both girls, as well as two earlier unreported sexual assaults. Pitchfork's blood was the last of 4,583 to be drawn and tested. His DNA proved a perfect match to the genetic fingerprint obtained from the semen samples (Wambaugh 1989).

It requires further standardization and quality control, however, to be universally accepted as a tool to convict the guilty and exonerate the innocent.

Mapping with RFLPs

The use of restriction enzymes has revolutionized the mapping of human genes, that is, locating the position of genes along a chromosome with respect to each other. (You may wish to review the principles of mapping in Chapter 9.) Mapping is often the first step in a long research journey that leads to better diagnosis and treatment of genetic diseases. Like DNA fingerprinting, the new mapping process relies on variation in the lengths of DNA fragments. Unlike DNA fingerprinting, the variation arises not because of variable numbers of tandem repeats but because of "trivial" base substitution mutations.

Extensive DNA sequencing has revealed that the genomes of two unrelated, normal people differ from each other in roughly one base in several hundred. Yet there is usually no obvious corresponding phenotypic differences, because the variable nucleotides usually lie in the spacer regions between genes or in introns. We have noted before that these noncoding sequences constitute most of a person's DNA. Nonetheless, these mutated sites represent allelic differences at the DNA level, just like the variability in the number of repeat units.

But short of sequencing everybody's DNA, how do we identify and visualize such inconsequential differences? Consider what would happen if some of the phenotypically neutral base changes in DNA altered—just by chance—the recognition sites of restriction enzymes. Then, at those particular spots, the enzyme would cut one person's DNA but not another's. For example, suppose that a certain section of John's DNA has four *Eco*RI sites (shown by **x**'s) spaced as follows:

7 kb   **1 kb**   **4 kb**

—x————————————x——x————x—

whereas the same section of Jan's DNA has only three *Eco*RI sites:

8 kb   **4 kb**

—x——————————————x————x—

The difference between John and Jan could have resulted from a substitution mutation in any one of the bases, GAATTC, that make up John's second *Eco*RI recognition site. This base change, occurring perhaps many generations back, produced two alleles—one allowing the restriction enzyme to cut (GAATTC in John) and one preventing the cut (say, AAATTC in Jan). Actually, the mutation might have been the reverse: from the sequence in Jan to the sequence in John. In any case, such allelic differences can be detected by the Southern blot and autoradiography techniques if a radioactive probe is available whose sequence is complementary to this region. The experimenter would take the *Eco*RI digests of John's DNA and Jan's DNA, run them in an electrophoretic gel, denature and blot the migrated DNA onto a nitrocellulose filter, hybridize with the probe, and autoradiograph. The result would look like this:

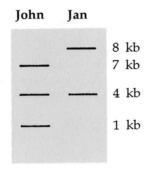

Both John and Jan have the 4-kb segment, but their other DNA restriction fragments have different lengths.

Assume that some people resemble John and some resemble Jan (and some may have yet different band patterns, based on mutations of the other *Eco*RI sites). Recall that we introduced the term *polymorphic* to describe a locus at which there are several common genetic variants. Here we are studying a polymorphic gene that leads to different patterns of restriction fragments—thus the phrase **restriction fragment length polymorphism**, or **RFLP**. Actually, we could also have used this term to describe the variability underlying DNA fingerprinting.*

A mutant site producing an RFLP represents a genetic locus with two alleles, a situation similar in principle to blood groups. As with blood groups, the resulting phenotypes are not readily apparent but can be detected by a laboratory worker. In our example of John and Jan, we implicitly assumed that each was homozygous at the restriction sites. But people can be heterozygous, one allele like John's and one like Jan's. (You

---

* If you like acronyms, call the DNA fingerprinting fragment variation the VNTR type of RFLP.

should be able to predict the banding pattern for such a heterozygote: bands at the 8-, 7-, 4-, and 1-kb levels, with the 4-kb band extra thick.) Recall from Chapter 9 that heterozygotes are needed to map loci with respect to each other. Because the DNA sites that result in RFLPs occur at many random places throughout the genome, and because the heterozygous state is common and easy to detect, RFLPs are very useful for mapping (White and Lalouel 1988).

Huntington Disease. The pioneering work of several teams of scientists led by James Gusella, of Harvard Medical School, established in 1983 that the dominant autosomal gene for **Huntington disease (HD)** is linked to an RFLP site located near the tip of the short arm of chromosome 4. This was the first use of molecular markers to map a genetic disease gene starting with virtually no information about its location in the genome. Recall from Chapter 5 that the symptoms of HD begin in middle age; they include the gradual loss of muscle control and mental function and lead to death after some years. Many persons with HD produce their children before any symptoms occur, so these children live with the knowledge that they have a 50% chance of being similarly affected.

Gusella and his colleagues obtained white blood cells from the members of several extended families in which HD occurred. Among these was a large kindred of affected individuals in some isolated villages in Venezuela. DNA was extracted from each cell sample and digested with a restriction enzyme. In a process we have now described several times, the resulting fragments were electrophoresed, transferred by the Southern blot method, and hybridized with a radioactive probe. In this case, the probe was a random segment from a library of small pieces of human genomic DNA. The plan was to systematically try one random probe after another, until one of them produced a pattern of DNA bands that correlated with the inheritance of HD. That is, the researchers looked for a distinctive set of electrophoretic bands when the DNA sample was from a person with Huntington disease, and a different pattern when the DNA came from a normal individual. The interpretation of such a result would be this: The HD gene locus is next to an RFLP site, and the mutant HD allele is on the chromosome associated with one distinctive band pattern (and the normal HD allele is on the other).

Although the investigators thought that they would have to try hundreds of random probes (as well as dozens of different restriction enzymes), they were very lucky: Within the first few runs, a random probe (called G8) detected an RFLP site linked to HD. As shown in Figure 13.16, this RFLP region actually involves *two* nearby variable restriction sites (labeled 1 and 2). As the figure shows, this double variation produces four

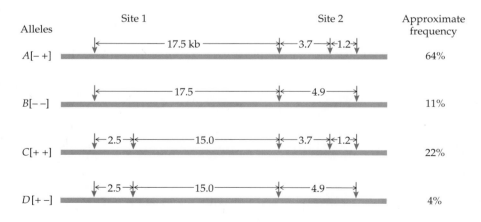

Figure 13.16
The RFLP sites in the region of chromosome 4 that hybridizes to the G8 probe. Four alleles (labeled A, B, C, and D) are produced by variants at two nearby recognition sites (labeled 1 and 2). Along each chromosome, the red arrows symbolize the positions of restriction enzyme cuts, allele C having five such positions. The other alleles are mutated at site 1 or 2 or both, so that the restriction enzyme does not cut. Within brackets at the left, + means cut *and − means* no cut *at sites 1 and 2, respectively. The lengths of the resulting DNA fragments are shown in kilobases. These four alleles are present in the general population in the frequencies shown at the right.*

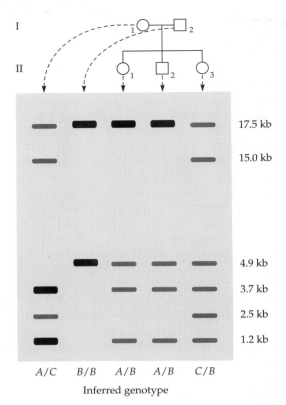

Figure 13.17
Banding patterns from five family members. Each person's DNA was digested with the restriction enzyme (used in Figure 13.16), then electrophoresed, Southern blotted, probed with G8, and autoradiographed. A control lane (not shown) allowed investigators to identify the lengths of DNA in each band, as indicated at the right. You should be able to deduce the genotypes by noting that the bands in each vertical lane correspond to two alleles depicted in Figure 13.16. A person might, however, be either homozygous (thick band) or heterozygous.

different chromosomes or alleles (sometimes called *haplotypes*). Each occurs in appreciable frequencies (4–64%), so that the heterozygotes, such as *A/B* or *B/C*, needed for mapping studies are also frequently encountered. The banding patterns from members of a hypothetical family, as well as the deduced genotypes, are shown in Figure 13.17.

In the large Venezuelan pedigree of over 3,000 people, the HD gene was almost always associated with the *C* allele. Crossovers were seen infrequently, corresponding to about 4 map units. For the younger symptom-free members of HD pedigrees who choose to be tested with the G8 probe, the 4% recombination leads to serious uncertainties. For example, let *A*, *B*, and *C* represent RFLP alleles, *H* the HD allele, and *h* the normal allele, and assume 4% recombination. Then one such mating would be

HD parent × non-HD parent

$$\frac{A\ h}{C\ H} \qquad\qquad \frac{B\ h}{B\ h}$$

which yields four genotypes among the children:

$\underline{C\ H}\ /\ \underline{B\ h}$ 48%

$\underline{C\ h}\ /\ \underline{B\ h}$ 2% (a recombinant type)

$\underline{A\ h}\ /\ \underline{B\ h}$ 48%

$\underline{A\ H}\ /\ \underline{B\ h}$ 2% (a recombinant type)

Among the children who possess the *C* allele (first two lines), 48/50, or 96%, will also carry the *H* allele; but 2/50, or 4%, of the time a prediction of Huntington disease will be wrong, because the *H* allele was exchanged by crossing over. Conversely, anyone with the *A* allele will most likely carry the *h* allele as well, but 4% of the time a comforting prediction of freedom from Huntington disease will be wrong. Although not huge, a 4% error rate in such a fateful situation is not satisfactory.

Fortunately, the exact location and sequence of the Huntington gene has now been established. It is expected that direct testing on DNA with virually no errors in diagnosis will soon be available. (Recall from Chapter 7 that the abnormal allele has more than 40 repeats of the trinucleotide CAG.) For family members at risk, taking even a 100% predictive test, however, will be a highly personal decision, fraught with difficulties, because there is currently no treatment to ameliorate the disease symptoms. Skilled counseling and support are needed to help deal with possibly serious emotional and psychological problems in people who are told they will succumb to an agonizing illness. And if one family member does not want to know what the future holds, will closely related individuals be denied information to which they are rightfully entitled?

Somatic Cell Techniques

Twenty-five years ago, not a single autosomal gene could be assigned to a specific chromosome; but now about 2,000 autosomal genes have been located. With relative ease, researchers make hundreds of such assignments every year (and with more difficulty they often determine the relative positions of the genes on a given chromosome). This stunning turnabout in putting human genes in the right pigeonhole started in 1967, when Mary Weiss and Howard Green, at New York University School of Medicine, located on chromosome 17 the gene for thymidine kinase, an enzyme needed in the synthesis of DNA. They used a curious technique called **somatic cell fusion**, in which laboratory cultures of human cells and mouse cells are joined together to form hybrid cells (Ephrussi and Weiss 1969). (We hasten to add that these hybrid cells remain in their petri dishes and never develop into any kind of critter.) Used in conjunction with traditional family studies, somatic cell fusion revitalized mapping based on pedigrees. Now also used in conjunction with the molecular techniques described in this chapter, it is a powerful tool for human genetic analysis.

Somatic cells of different species (say, human and mouse) are placed together in a culture dish with nutrient medium and left for hours or days to grow and divide. In the mixed population, several human-mouse hybrid cells will form spontaneously, but the frequency of fused cells can be enhanced up to a thousandfold by adding to the culture medium either a chemical (polyethylene glycol) or an inactivated virus (the Sendai virus). These agents cause cells to clump together, aiding in their intimate fusion. The chromosomes of both species eventually collect into one nucleus in the hybrid cells, which divide and redivide by mitosis. Furthermore, the genes of both parental types may be expressed in the hybrid cells.

The cardinal property of fused cells that enables researchers to localize genes is the loss of chromosomes over the course of cell generations (Figure 13.18). Interestingly, it turns out that in human-mouse hybrid cell lines, only *human* chromosomes are lost. The reason for a preferential loss is not clear, but it may hinge on the slower replication rate of human chromosomes. Which particular human chromosomes are lost and which ones are retained seems to be fairly random. The process of elimination continues until only one or a few human chromosomes are left in a hybrid nucleus. At this point, the various subclones (shown at the bottom of Figure 13.20) become stabilized.

This loss of chromosomes is accompanied by a loss of the proteins encoded by the genes present on those chromosomes. By monitoring the

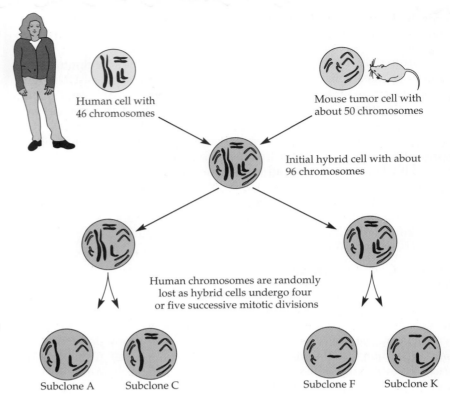

Figure 13.18
The derivation of stable cells containing one or a few human chromosomes. All the chromosomes of the two somatic cell parents are present in the initial hybrid cell. Human chromosomes (shown in black), however, are successively lost during divisions of the hybrid cells. Because the loss is random, each eventual subclone usually ends up with a different array of human chromosomes (but a full complement of mouse chromosomes). No further loss occurs when the subclones are reduced to one or a few human chromosomes.

Human cell with 46 chromosomes

Mouse tumor cell with about 50 chromosomes

Initial hybrid cell with about 96 chromosomes

Human chromosomes are randomly lost as hybrid cells undergo four or five successive mitotic divisions

Subclone A   Subclone C   Subclone F   Subclone K

loss of particular chromosomes by standard karyotyping and also the loss of proteins, researchers gather information on which functions are associated with which chromosomes (Ruddle and Kucherlapati 1974). Although all mammalian cells contain mostly the same proteins to regulate their metabolism, the amino acids usually differ somewhat from species to species. Thus, two species may possess different molecular forms of an enzyme, which can often be detected by gel electrophoresis. Like the separation of DNA segments by migration in an electric field, the separation of protein molecules depends on charge and size variations. In the case of similarly sized proteins, their differential migration rates come about primarily because of charge differences in their constituent amino acids.

For example, in the early 1970s, Frank Ruddle and his colleagues at Yale University analyzed 26 hybrid clones for the presence or absence of a human enzyme called peptidase C, which migrates a little faster than the corresponding mouse peptidase C. In 14 hybrid cell lines that retained a chromosome 1, they found the human enzyme; but in 12 lines that had lost both chromosomes 1, they failed to find the human enzyme. Thus, the human gene for peptidase C must reside on human chromosome 1.

There is one difficulty that we need to address. Fused cells are not necessarily any different in microscopic appearance from the parental mouse and human cells, so that the isolation of the hybrids becomes a problem. For this task, a number of biochemical mutants of mammalian cells have been collected and maintained. As an example, suppose that the human cells lack an enzyme H that is necessary for the production of a vital substance, say the purine nucleotides that go into DNA (call these cells H^-). Assume that the mouse cells lack a different enzyme, T, needed to produce the purine nucleotides (call them T^-). To grow, a cell normally needs both enzymes. To keep H⁻ and T⁻ cells going, special culture conditions can provide an alternative way to obtain the purine nucleotides, bypassing the need for enzymes H and T.

What happens in a laboratory dish containing H⁻ cells, T⁻ cells, and the hybrid cells derived from their fusion, *without* any special culture conditions (Figure 13.19)? Since H⁻ cells carry the normal allele for the synthesis of T (they are really H⁻T⁺), and T⁻ cells carry the normal allele for the synthesis of H (H⁺T⁻), a combination of the two types will be mutually complementary, that is, phenotypically H⁺T⁺. The fused cells can therefore survive and form colonies where neither parental line can. Thus, geneticists can detect and isolate even extremely rare hybrid cells from a background of mostly parental cells. A selective method similar to this scenario is called the *HAT* technique, where H and T stand for enzymes needed in the synthesis of DNA.

Other Methods for Gene Assignment. The strategy we have outlined for assigning a locus to a specific chromosome is limited to genes that are *expressed*, that is, transcribed and translated into detectable amounts of the protein product. Hence, this approach is suitable for only some of the genes present in the cell. A newer approach can locate *any* gene, provided researchers have a corresponding DNA probe.

The experimental procedure starts, as before, with the isolation of a human-mouse hybrid cell line containing a small number of human chromosomes whose identities are established by standard cytogenetic techniques. A researcher will likely have handy dozens of hybrid cell lines, each with a different array of human chromosomes. From each hybrid cell line, DNA is extracted, cut with a restriction enzyme, spread out by size on an electrophoretic gel, denatured, and transferred to a nitrocellulose filter by Southern blotting. The filter is hybridized with a radioactive probe complementary to the gene in question and autoradiographed. A band will appear in a particular lane of the gel only if the original hybrid cell contained a chromosome with the gene in question.

Gusella and his colleagues localized the Huntington disease gene to chromosome 4 by this method. Recall that they found a probe called G8 that bonded to an RFLP site located about 4 map units from the disease gene. They examined 18 different human-mouse hybrid lines for the presence or absence of DNA sequences complementary to the G8 probe. The data for six of the lines are given in Table 13.3. The radioactive G8 probe bound only to those lanes on the Southern blot filter containing DNA from human-mouse hybrid cells containing chromosome 4.

A rather different method for gene localization is called **in situ hybridization (ISH)**. Using this technique, researchers prepare human metaphase chromosomes on a microscope slide in the standard way (Figure

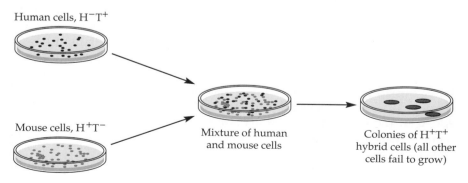

Human cells, H⁻T⁺

Mouse cells, H⁺T⁻

Mixture of human and mouse cells

Colonies of H⁺T⁺ hybrid cells (all other cells fail to grow)

Figure 13.19

Selection of human-mouse hybrid cells in the presence of a background of large numbers of parental cells. The human cells (in black) and mouse cells (in color) are mixed under conditions where neither can grow. The human cells are mutant for gene H (but not T), and the mouse cells are mutant for T (but not H). Only a fused hybrid cell (H⁺T⁺) can grow and form visible colonies.

Table 13.3

Assignment of the Huntington disease gene to chromosome 4 by analysis of human–mouse hybrid cells.

Hybrid Cell Line	Presence (+) or Absence (−) of G8 Sequences	Human Chromosomes Present							
W-5	+	**4** 17 18 21 X							
J-22	+	**4** 6 10 11 14 17 18 20 21							
N-16	+	3 **4** 5 7 12 17 18 21 X							
W-2	−	8 12 17 21 X							
N-5	−	12 14 15 16 18 20							
R-11	−	11 13 16 20 21 X							

Source: Gusella et al. 1983.

13.20). First the chromosomes are stained and photographed to reveal the banding patterns. Next the DNA that remains in place (i.e., "in situ") is denatured to form single strands. Researchers bathe the slide with a radioactive DNA probe specific for a given gene, wash off excess probe, and layer a photographic emulsion on top of the slide to make an autoradiograph. Finally, they take photomicrographs of the exposed and developed film on top of the same chromosome spreads that had been previously stained and photographed. Thus, the developed silver grains in the film can be juxtaposed to the banded metaphase chromosomes to locate the position of the gene complementary to the probe.

In a variant on this technique, the DNA probe is coupled with a fluorescent tag, so that the hybridized probe is revealed by fluorescence under ultraviolet light rather than by radioactivity. Called **FISH**, for **fluorescence in situ hybridization**, these nonradioactive techniques have become increasingly sensitive and have been used in recent years to diagnose many chromosomal aberrations prenatally, including translocations and trisomies (Figure 13.21). The technique requires a probe that binds to, say, the DNA at the breakpoints of a translocation or the centromeric DNA of a specific chromosome (as in Figure 13.21).

ROAD MAPS OF THE HUMAN GENOME

Geneticists now deal with the geography of chromosomes at several levels, as reflected in the different kinds of maps they make. The more traditional *genetic linkage* map and *cytological* map go back to the early decades of this century. On the linkage map, distances between two genes are measured by the percent recombination that is observed in the phenotypes of children from doubly heterozygous parents (Chapter 9). The biological phenomenon underlying recombination of linked genes is crossing over during meiosis. The recombinational length of all 23 pairs of chromosomes taken together is about 3,300 map units.

On the cytological map—seen through a light microscope—deletions, duplications, breakpoints of translocations, and other chromosomal changes provide visible signposts among the banding patterns. Recently added to the cytological maps are the sites of genes and DNA markers made evident by in situ hybridization using radioactive or fluorescent labels.

At the finer *molecular* level, geneticists use the technologies described

Figure 13.20
Basic methodology for in situ hybridization using a radioactive marker.

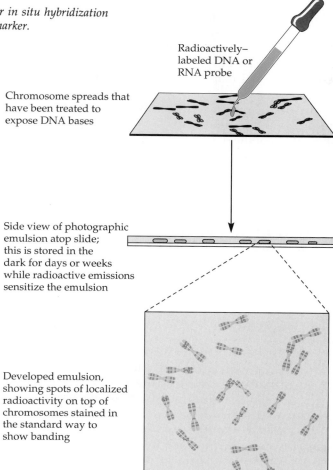

Radioactively–labeled DNA or RNA probe

Chromosome spreads that have been treated to expose DNA bases

Side view of photographic emulsion atop slide; this is stored in the dark for days or weeks while radioactive emissions sensitize the emulsion

Developed emulsion, showing spots of localized radioactivity on top of chromosomes stained in the standard way to show banding

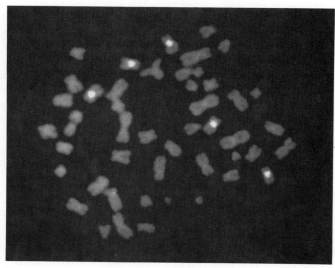

Figure 13.21
Fluorescence in situ hybridization. In this photomicrograph, metaphase chromosomes had been hybridized with a yellow fluorescent DNA probe specific for sequences in the centromere region of the X chromosome. The five yellow regions indicate an aneuploid 49,XXXXX karyotype. A red fluorescent counterstain (called propidium iodide) was used to color all chromosomes about equally. (Courtesy Oncor Inc., Gaithersburg, Maryland.)

in this chapter to isolate, dissect, and analyze stretches of nucleotides that represent genes, parts of genes, or various types of intergenic DNA. Road distances are measured in bases or kilobases, and the total amount of DNA in the haploid human genome is about 3 billion nucleotide pairs. Therefore, for the entire genome, about a million bases correspond to one recombinational map unit (3×10^9 total bases divided by 3.3×10^3 map units).

Molecular studies themselves involve two levels of mapping. The finest type of information is provided by nucleotide sequencing, but only about 0.3% of the 3×10^9 bases have been sequenced to date. Less detailed information along the nucleotide highway is provided by DNA markers. We presented several types in this chapter: restriction enzyme sites (RFLPs), variations in the number of tandem repeats (VNTRs), and sequence tagged sites (STSs). Another DNA marker consists of sites where very short sequences (a doublet like CA, or a triplet, or a quadruplet of bases) are repeated dozens or hundreds of times. These are called *microsatellite repeats*. Still other DNA signposts consist of cloned segments that are shown to overlap one another. Such a region of ordered segments, called a *contig* (for contiguity), forms a long section of DNA that can eventually be sequenced or searched for a disease-causing gene. Chromosome 21 and the functional (nonrepetitive) portion of the Y chromosome are nearly 100% covered by contigs cloned in yeast artificial chromosomes.

The various kinds of maps are roughly comparable to one another. Although the genes are always found in the same order along a chromosome, the relative distances are not always exactly corresponding. For example, map distances measured through genetic recombination in doubly heterozygous human females are routinely 40–50% greater than distances based on doubly heterozygous males, although molecular distances along the chromosomes are the same in the two sexes. This discrepancy arises because crossing over occurs more frequently in females than in males.

In the 1980s, RFLPs and other molecular mile markers revitalized the traditional genetic linkage technique by removing a major impediment. Recall that the accurate measurement of recombination distances requires a lot of doubly heterozygous parents. Except for the blood groups and a few other genes, however, the scarcity of common polymorphisms had made those double heterozygotes hard to come by. Molecular markers, however, now fill the "polymorphism gap." The approximately 10,000 DNA markers that have been positioned throughout the human genome are increasingly relied on to find linkage distances to the 2,500 human structural genes whose locations are known. Within a matter of a few years, all intrinsically interesting genes should be within 2 or 3 map units of a DNA marker or from each other. Current approaches to studying the genetic or physical features of the human genome thus combine the traditional and molecular methodologies. This fruitful combination will continue to yield more sophisticated understanding, diagnosis, and treatment of human genetic disease.

SUMMARY

1. The polymerase chain reaction (PCR) can quickly replicate a section of DNA a millionfold or more in a thermal cycler, provided that one has available short primer DNAs complementary to the two ends of the target sequence. The polymerase chain reaction has many applications, including the prenatal diagnosis of genetic disease.

2. DNA segments from different species can be spliced together to make recombinant DNA molecules. The first step in this procedure uses restriction enzymes, precision cutting tools of molecular biologists. Once inserted into a plasmid of an *E. coli* host cell, a foreign gene can be multiplied to any desired amount.

3. Genes for insulin, growth hormone, interferon, and other products can be cloned and expressed in bacterial or cell culture systems, permitting commercial-scale synthesis of these valuable medical products.

4. Transgenic animals possess a foreign gene. Investigators are hoping that transgenic farm animals can be produced that secrete useful amounts of a human protein in their milk.

5. DNA molecules can be separated from one another by electrophoresis, which sorts them according to size. By electrophoresing DNA that is cut with different restriction enzymes and comparing the resultant banding patterns, scientists can map the various restriction sites along the molecule.

6. The sequence of bases along a segment of DNA can be readily determined. Electrophoresis of four preparations allows researchers to determine the positions of the four bases. The order of bases is read sequentially from an autoradiograph.

7. The methods for DNA sequencing are now so good that more and more often a human protein is identified and characterized by first cloning and sequencing the gene that encodes it.

8. The Human Genome Project seeks to sequence all 3 billion bases of our DNA in a coordinated international effort. To avoid sequencing long tracts of possibly uninteresting repetitive or intronic DNA (about 97% of the total), some investigators may make use of cDNA (DNA copied from messenger RNA) rather than raw genomic DNA.

9. DNA fingerprinting is based on the individual-specific variable number of tandem repeats (VNTR). Even tiny amounts of DNA (such as that in a sperm sample or a few hairs) can sometimes be used for forensic analysis, because parts of the DNA can be amplified by the polymerase chain reaction.

10. The fingerprinting process makes use of restriction enzyme digests, electrophoresis, Southern blotting, and autoradiography using a radioactive probe. In Southern blotting, the pattern of DNA bands in a gel is transferred to an overlying sheet of filter paper, where the DNA is more easily handled.

11. Restriction fragment length polymorphism (RFLP) occurs whenever the DNA from individuals differs in one of the bases included in a recognition site of a restriction enzyme. Thus the cutting or not cutting of that restriction site identifies alternative alleles that can be detected by Southern blotting.

12. An RFLP site found *near* the Huntington disease (HD) locus provides an imperfect aid for early diagnosis and genetic counseling. Researchers have now identified the mutational site within the HD gene.

13. Human and mouse somatic cells, when put together in tissue culture, can fuse to make hybrid cells. In further divisions, human chromosomes are randomly lost until only one or several are left. Correlating the loss of chromosomes with the loss of specific proteins can identify which chromosome the responsible gene is on.

14. Other methods for gene location include in situ hybridization, in which a gene probe is labeled with a radioactive or fluorescent tag.

15. The construction of road maps of the human genome involves both genetic linkage studies and molecular and cytological techniques.

KEY TERMS

QUESTIONS

1. Using the last line of Table 13.1, indicate how many PCR cycles are required to produce a million copies of a target sequence, starting with one copy. How long will this take at, say, 5 minutes per cycle?

2. A *palindrome* is a sequence of letters that reads the same forward and backward: WASITACATISAW, for example. In what way is the sequence recognized by *Eco*RI (GAATTC on one strand) palindromic?

3. Assume that you have cloned a 1-kb human gene in a plasmid using restriction enzyme *Eco*RI as outlined in Figure 13.5. You have subsequently isolated in a test tube a high concentration of the recombinant plasmid shown at the bottom of Figure 13.5. What is the first step in getting the human gene separated from the plasmid DNA?

4. Assume that YAC vectors hold, on the average, 300-kb inserts of foreign DNA. Ideally, how many YACs would be needed in a library to accommodate all 3 billion bases of human DNA?

5. Assume restriction sites as follows, letting **E** = *Eco*RI site (GAATTC) and **H** = *Hin*dIII site (AAGCTT):

    |—3 kb—**E**——6 kb——**H**——6 kb——**E**——7 kb——|

What bands would be present in an electrophoretic gel when this DNA is digested with *Eco*RI alone, *Hin*dIII alone, and with both enzymes?

6. A researcher sets up a gel with three experimental lanes in which the same DNA sample has been cut with either *Eco*RI, *Bam*HI, or both enzymes. Comparing the resultant bands with a standard, she finds that the sizes of DNA in the various bands are as follows:

    *Eco*RI lane: 3 bands with lengths of 3, 5, and 14 kb
    *Bam*HI lane: 2 bands with lengths of 7 and 15 kb
    "Both" lane: 4 bands with lengths of 1, 2, 5, and
        14 kb

Draw the restriction map; that is, indicate on a segment of DNA where the restriction recognition sites are.

7. From Figure 13.11, determine the sequence of the ten bases from position 50 to position 59.

8. When 30 million bases of human DNA have been sequenced, what percentage of the human genome will remain to be sequenced?

9. A researcher investigates a section of DNA of unknown function and finds the following sequence. Using Table 12.4, determine if this sequence, in any frame, could represent a structural gene. How do you know?

    AAGGGTCCGTTGGCTTCCTCTCGAGATCATCGGATTCAACCAATG

10. From Figure 13.16, predict the pattern of bands on a Southern blot from individuals *heterozygous* for the *A* and *C* alleles.

11. Variation in the lengths of DNA fragments underlies both DNA fingerprinting and the RFLPs used to map the HD gene. The variation, however, comes about for different reasons. Explain.

12. Owerbach et al. (1980) examined 15 human-mouse hybrid cell lines by Southern blotting for the presence or absence of DNA sequences complementary to the human insulin gene. The data for six of the lines are given below. Which chromosome carries the gene?

Hybrid cell line	Insulin sequences detected(+) or not(−)	Human chromosomes present
W-8	+	6 7 10 11 14 17 18 20 21 X
T-5	+	4 5 10 11 12 17 18 21
D-5	+	3 5 11 14 15 17 18 21
W-2	−	8 10 12 15 17 21 X
T-2	−	2 5 6 10 12 18 20 21 X
T-8	−	17 18 20

13. Would it be possible to assign human genes by means of human-mouse cell hybrids if mouse chromosomes were randomly lost and human chromosomes retained?

14. Would you expect that genes found on the same chromosome by cell fusion methods will always be found linked by traditional pedigree linkage studies? Explain.

Designed for general college students, the nicely written books by Drlica (1992) and Micklos and Freyer (1990) cover much of the material in this chapter. Journalists Bishop and Waldholz (1990) have written an interesting popular account of the motivations and work of the scientists who pioneered the molecular biology revolution.

Much has been written on the polymerase chain reaction. Good elementary articles include Baskin (1990), Arnheim, White, and Rainey (1990), and Eisenstein (1990). Thayer (1991) examines the production of valuable proteins by biotechnology companies, and Watson (1990), Maddox (1991), and Kevles and Hood (1992) discuss the Human Genome Project. Green and Waterston (1991) relate the Genome Project to clinical medicine. Good sources on DNA fingerprinting are Lewis (1989), Annas (1990), Office of Technology Assessment (1990), and National Research Council (1992).

Some of the many *Scientific American* articles on biotechnology are Neufeld and Colman (1990) on DNA fingerprinting, Mullis (1990) on the discovery of the polymerase chain reaction, White and Lalouel (1988) on RFLPs, Lawn and Vehar (1986) on cloning the hemophilia gene, Anderson and Diacumakos (1981) on gene transfer into mammalian cells, and Ruddle and Kucherlapati (1974) on human-mouse hybrid cells.

14 Inborn Errors of Metabolism

In 1887 an American neurologist, Bernard Sachs, described the case history of an infant who had a strange condition and died at the age of two years:

> The little girl . . . was born at full term, and appeared to be a healthy child in every respect; its body and head were well proportioned, its features beautifully regular. Nothing abnormal was noticed until the age of two to three months, when the parents observed that the child was much more listless than children of that age are apt to be . . . and that its eyes rolled about curiously. . . . The child would ordinarily lie upon its back, and was never able to change its position; muscles of head, neck, and back were so weak that it was not able either to hold its head straight or to sit upright. . . . It could not be made to play with any toy, did not recognize people's voices, and showed no preference for persons around it. During the first year of its life, the child was attracted by the light, and would move its eyes, following objects drawn across its field of vision; but later on absolute blindness set in. . . . Hearing seemed to be very acute . . . the slightest touch and every sound were apt to startle the child. . . . The child never learned to utter a single sound. . . . [T]he child grew steadily weaker, it ceased to take its food properly, its bronchial troubles increased, and finally, pneumonia setting in, it died.

Later in the chapter we discuss the genetics and biochemistry of this devastating and untreatable illness, which is now known as Tay-Sachs disease. Its cause, like that of phenylketonuria, described in Chapter 1, was ultimately traced to a missing or defective enzyme. Thus, both conditions are examples of inherited metabolic disease, or inborn errors of metabolism, which were first described around the turn of the century.

THE CONCEPT OF INHERITED METABOLIC DISEASE

At the time that Mendel's work was being discovered, English physician and biochemist Archibald Garrod was studying an obscure condition known as *alkaptonuria*. Otherwise healthy newborns with this disorder usually had one striking feature: Their urine, upon standing in contact with air, turned black. (In their later years, most people with alkaptonuria develop arthritis.) The dark substance in the urine was the oxidation product of alkapton, now called *homogentisic acid*, which contains a six-carbon ring structure. Other researchers had shown that the excretion of homogentisic acid in urine was increased when people with alkaptonuria were fed excess protein or the amino acids tyrosine and phenylalanine, both of which contain benzene rings.

Garrod suggested that homogentisic acid is an ordinary product of metabolism that is broken down in normal people but is not degraded in people with alkaptonuria. This **metabolic block** causes homogentisic acid

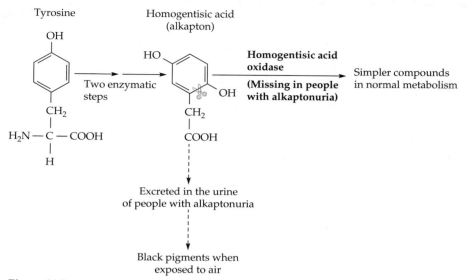

Figure 14.1

The metabolic block in persons with alkaptonuria. The ring of six carbon atoms—depicted ⬡—is referred to as a benzene ring or phenyl group. The splitting of the ring by homogentisic acid oxidase—indicated by ✁—during normal metabolism was proposed in 1902 by Garrod. His idea that people with alkaptonuria lacked this enzyme because of a gene defect was far ahead of its time.

to be excreted intact in the urine. This idea was supported by the fact that *unaffected* individuals who ingest excessive amounts of homogentisic acid never excrete any of it in their urine. More specifically, Garrod thought that people with alkaptonuria do not properly metabolize ring-containing fractions of proteins because they lack a special enzyme that is present in unaffected individuals. He proposed that during the normal breakdown of proteins, phenylalanine is converted to tyrosine, which in turn is changed to homogentisic acid and then to simpler products (Figure 14.1).

Garrod's analysis did not stop with the concept of metabolic blocks, however. The brilliance of his proposal lay in the genetic connection that he and population geneticist William Bateson were able to make. Garrod noted that alkaptonuria tended to occur in several sibs of a family whose parents were nevertheless unaffected and also that many affected children were the offspring of first-cousin marriages. This pattern of inheritance fit the requirements for a rare recessive trait, the first human application of Mendel's newly discovered laws. Garrod classified alkaptonuria and three other conditions that he had studied as **inborn errors of metabolism**.

He suggested that these four genetic conditions must represent only a tiny fraction of the errors of metabolism that exist in human populations, pointing out that many of them may produce no obvious phenotypic effects. Citing known variations in hemoglobins and in muscle proteins as examples of chemical diversity within and between species, Garrod correctly prophesied in 1909 that

> it is among the highly complex proteins that such specific differences are to be looked for. . . . [We] should expect the differences between individuals to be . . . subtle and difficult of detection. . . . Even those idiosyncrasies with regard to drugs and articles of food which are summed up in the proverbial saying that what is one man's meat is another man's poison presumably have a chemical basis.

But Garrod's ideas, like Mendel's, were largely ignored for over 30

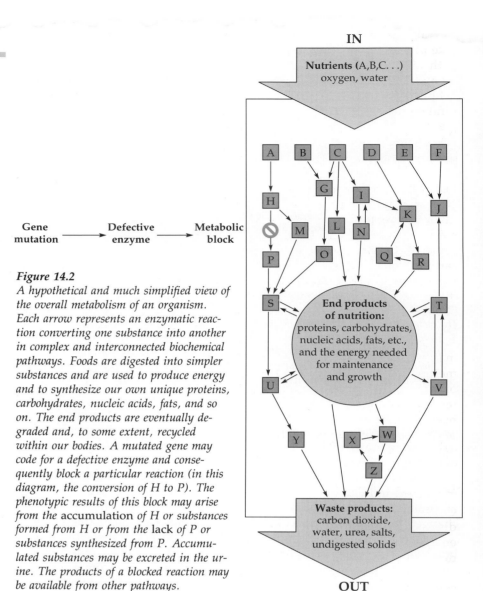

IN

Nutrients (A,B,C . . .)
oxygen, water

Gene mutation → Defective enzyme → Metabolic block

Figure 14.2

A hypothetical and much simplified view of the overall metabolism of an organism. Each arrow represents an enzymatic reaction converting one substance into another in complex and interconnected biochemical pathways. Foods are digested into simpler substances and are used to produce energy and to synthesize our own unique proteins, carbohydrates, nucleic acids, fats, and so on. The end products are eventually degraded and, to some extent, recycled within our bodies. A mutated gene may code for a defective enzyme and consequently block a particular reaction (in this diagram, the conversion of H to P). The phenotypic results of this block may arise from the accumulation of H or substances formed from H or from the lack of P or substances synthesized from P. Accumulated substances may be excreted in the urine. The products of a blocked reaction may be available from other pathways.

End products of nutrition: proteins, carbohydrates, nucleic acids, fats, etc., and the energy needed for maintenance and growth

Waste products: carbon dioxide, water, urea, salts, undigested solids

OUT

years. During that period, studies of pigment formation in flowers, in the fur of animals, and in the eyes of fruit flies laid the groundwork for more extensive research on the nutritional requirements of the pink bread mold *Neurospora*. In a series of experiments done at Stanford University and reported in 1941, George Beadle and Edward Tatum irradiated wildtype *Neurospora* to produce a class of mutants that could survive only if specific chemicals were added to the nutrient medium. By collecting and analyzing dozens of these strains, they were able to show that each one had a single metabolic block—presumably due to an enzymatic defect—that could be traced to the mutation of a single gene (Figure 14.2). Although Beadle and Tatum were not the first to conceive of the one gene–one enzyme hypothesis, their relatively simple testing system greatly accelerated research in this developing field of biochemical genetics and earned them a Nobel prize as well. (Later work refined their proposal to a **one gene–one polypeptide** concept—which we now know to be an oversimplification too, given the recent discoveries of some overlapping, nested, split, and shuffled genes.)

By the mid-1940s it was clear that (1) the biochemical processes that take place in an organism are genetically controlled; (2) every biochemical pathway can be resolved into a series of individual steps, each mediated by a different enzyme; and (3) each enzyme is usually encoded by one or a few genes. In fact, by analyzing groups of mutant strains that have different blocks in the same metabolic pathway, it is possible to determine the exact order in which these metabolic steps normally occur (see question 10). Clear-cut examples of biochemical mutants have now been detected in a wide variety of organisms.

Indeed, we can generalize to say that if a condition is simply inherited, it is usually caused by an abnormality in a single protein molecule. Not all single-gene disorders involve enzyme defects, however. Many involve receptors, hormones, transport proteins, immunoglobulins, and collagens. (Immunoglobulins include antibodies and antibody-forming proteins. Collagen is the main supportive protein of skin, tendon, bone, cartilage, and connective tissue.) Although nearly all are individually rare, together the known single-gene disorders account for over 5% of all pediatric hospital admissions. Their frequency in the general population is about 1%: roughly 0.65% autosomal dominants, 0.28% autosomal recessives, and 0.06% X-linked.

Since 1950, the discovery of metabolic diseases has increased greatly, in large measure owing to the development of new methods for identifying metabolites of various kinds. At least 250 different human metabolic disorders are known, distributed among many categories (Table 14.1). In recent years our understanding of these diseases and their genetic bases has also been vastly enriched by molecular techniques. From the blending of old and new knowledge, some useful generalizations about metabolic disorders have emerged.

Detection of Metabolic Disease in Newborns

Certain symptoms tend to be common to a wide variety of metabolic disorders. Because some syndromes can be treated if recognized early enough, it is important that physicians and parents be alert to these diagnostic signals.

Many affected babies appear normal at birth but very soon develop severe problems. Some of the most telling signs of a metabolic disorder are (1) severe vomiting, beginning in the first few days of life; (2) abnormal odor to the urine, skin, or breath; (3) abnormal concentrations of ketones or acids in the body fluids; (4) abnormalities in pigment development; (5) seizures, especially when accompanied by muscle spasms in a particular area of the body; (6) failure of muscle coordination; (7) ambiguous genitals; (8) failure to thrive; (9) overwhelming, unexplained illness; and (10) a family history characterized by the death of sibs in early life from unknown causes. Another important sign, which may not be detected for several months, is severe mental retardation. By this time, however, the damage is usually irreversible. Nevertheless, once recognized in a family, a metabolic disorder can be anticipated in subsequent pregnancies or births, carefully monitored, and perhaps treated. Most of these illnesses are rare, but as Leon Rosenberg (1972) has commented:

> The rarity of any one of these disorders is no longer an acceptable explanation for failure to consider them, particularly when a simple battery of laboratory studies and a safe form of therapeutic trial should be available in almost every hospital. I have yet to see bereaved parents comforted by the knowledge that their child died of a rare disease.

Table 14.1
Some additional examples of metabolic disorders.

Disorder	Inheritance[a]	Defective/Deficient Protein	Clinical Features
Amino Acid Metabolism			
Maple sugar urine disease	AR	Branched chain keto-acid dehydrogenase	Sugary-smelling urine; breathing/feeding problems, mental retardation, death.
Ornithine transcarbamylase deficiency	XD?	Ornithine transcarbamylase	Irritability, severe mental and physical retardation, coma. Fatal if untreated.
Carbohydrate Metabolism			
Hereditary fructose intolerance	AR	Fructose-1-phosphate aldolase	Hypoglycemia and vomiting after fructose intake; liver and kidney damage. Fatal if untreated.
Galactosemia	AR	Galactose-1-phosphate uridyl transferase	Inability to digest milk and milk products; vomiting, enlarged liver, jaundice, cataracts; mental retardation and death if untreated.
Nucleic Acid Metabolism			
ADA deficiency	AR	Adenosine deaminase	Toxicity to lymphocytes, causing severe combined immunodeficiency disease. Fatal if untreated.
Gout (one of many types, with various causes)	XR	Phosphoribosyl pyrophosphate synthetase	Superactive enzyme leading to excess of purines and uric acid; possibly deafness.
Organic Acid Metabolism			
Glutaric acidemia type I	AR	Glutaryl-CoA dehydrogenase	Brain damage, causing seizures and involuntary movements of head, trunk and arms. Fatal.
MCAD deficiency	AR	Medium chain acyl-CoA dehydrogenase	Variable; may resemble Reye syndrome; hypoglycemia, vomiting, lethargy, liver damage; sudden death may occur.
Lipoprotein and Lipid Metabolism			
Familial LCAT deficiency	AR	Lecithin: cholesterol acyltransferase	Corneal opacities, anemia, kidney problems.
Metal Metabolism			
Hemochromatosis	AR	? (Defect in iron absorption)	Excess iron damages liver, heart, pancreas, skin, joints. Fatal if untreated.
Menkes (steely hair) disease	XR	Possibly copper-transporting ATPase (Defect in copper absorption)	Copper deficiency; grayish, broken hair; abnormal facial features, cerebral degeneration, bone damage, arterial rupture. Fatal.
Wilson disease	AR	? (Defect in copper transportation)	Excess copper; liver disease, ring around iris, tremor, emotional and behavioral effects.

(continued)

Table 14.1 (continued)

Disorder	Inheri-tance[a]	Defective/Deficient Protein	Clinical Features
Lysosomal Enzymes			
Hurler syndrome (Mucopolysaccharidosis I)	AR	α-L-iduronidase	Enlarged liver, spleen; skeletal deformities, coarse facial features, large tongue, hearing loss, corneal clouding, heart disease, mental retardation, death.
Hunter syndrome (Mucopolysaccharidosis II)	XR	Iduronate sulphate sulfatase	Severe form: coarse facial features, short stature, skeletal deformities, joint stiffness, mental retardation, death.
Gaucher disease, type I (adult, or chronic, form)	AR	Glucocerebrosidase	Variable; enlarged spleen, bone defects and fractures, arthritis.
Peroxisomal Enzymes			
Adrenoleukodystrophy[b]	XR	Possibly ALD membrane transport protein	Variable; excess of very long-chain fatty acids; adrenal insufficiency. Childhood form includes dementia, seizures, paralysis, loss of speech, deafness, and blindness. Fatal.
Zellweger syndrome	AR	? (No peroxisomes)	High forehead and other facial features; weakness; defects of eyes, brain, liver, kidneys, heart. Fatal.
Hormones			
X-linked ichthyosis	XR	Steroid sulfatase	Dry, scaly "fish-skin"; mild corneal opacity.
Goiterous cretinism (one of many types of hypothyroidism)	AR	Iodotyrosine dehalogenase	Dwarfism, mental retardation, goiter, coarse skin and facial features.
Blood Proteins			
von Willebrand disease, type I	AD	von Willebrand factor	Varies greatly; bruising, bleeding from gums, cuts, and gastrointestinal tract; skin hematomas; heavy menstrual bleeding.
Hemophilia A	XR	Factor VIII, subunit a	Spontaneous bleeding, especially into large joints and muscles; hematomas, chronic arthritis; can be fatal if untreated.
Connective Proteins			
Osteogenesis imperfecta, type I	AD	Procollagen type I	Osteoporosis, bone fractures, blue sclera (outer layer of eyeball), hearing loss.
Ehlers-Danlos syndrome, type IV	AD	Collagen, type III	Fragile, thin skin; veins visible; bruising; bowel, arterial, and uterine rupture; no stretchy skin or loose joints.
Marfan syndrome	AD	Fibrillin	Long, thin fingers and limbs; chest deformity; loose joints; curved spine; displaced lens, cardiovascular problems, including tendency to aortic rupture.

[a] AR = autosomal recessive; AD = autosomal dominant; XR = X-linked recessive; XD = X-linked dominant.
[b] Featured in the 1992 movie *Lorenzo's Oil*.

The mode of inheritance is a good clue to the basic defect in genetic disorders. In those cases where the basic defect is known, recessive traits usually involve enzymes (especially enzymes that break down substances to simpler compounds) or peptide hormones. They also tend to show a fairly uniform phenotype and an early age of onset. Dominant traits, on the other hand, usually involve nonenzymatic or structural proteins or enzymes that are involved in complex control systems. They also tend to be much more variable in phenotype and often have a later (i.e., adult) age of onset. Thus, biochemical analyses of dominant disorders are more difficult, and relatively fewer of them are known.

Here we present a few examples of inborn errors of metabolism, starting with the case described at the beginning of this chapter.

TAY-SACHS DISEASE

Dr. Sachs had referred his sick little patient to an ophthalmologist, who discovered a peculiar *cherry-red spot* on the retina of each eye (Figure 14.3A). In the next few years, Sachs encountered several more children with the same malady, including a sib of the first case and four sibs in another family. A search of the medical literature for previous descriptions of this condition uncovered two reports published in the 1880s by an English ophthalmologist named Warren Tay. He had described a family in which the cherry spots occurred in several members, all of whom died by the age of three years. Within a decade, about two dozen cases were recorded, almost exclusively among Jewish families.

This still untreatable illness is called infantile **Tay-Sachs disease (TSD)**. Its progression is very much as we have described. Usually the first symptom noticed is the "startle reaction" to sharp noises. After one year of age, the child's condition degenerates rapidly; generalized paral-

Figure 14.3
Manifestations of Tay-Sachs disease in the retina of the eye and in nerve cells of the brain. (A) The cherry-red spot on the retina can be seen through the pupil of the eye. The spot is caused by the absence of cells that usually overlay the fovea centralis, the region of sharpest vision. Neither the red spot nor the pronounced whitish halo is seen in normal retinas. (B) Darkly staining membranous cytoplasmic bodies in a brain cell of a deceased patient contain excess lipid. The structures are absent in normal nerve cells. (A from Sloan and Fredrickson 1972; B from Terry and Weiss 1963.)

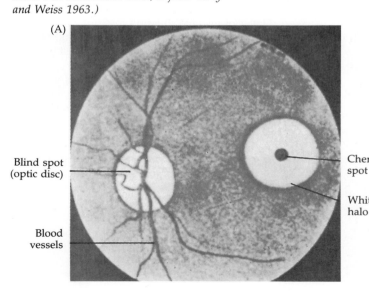

(A)

Blind spot (optic disc)

Blood vessels

Cherry-red spot

Whitish halo

(B)

Nucleolus

Nucleus

Membranous cytoplasmic bodies

ysis, blindness, gradual loss of hearing, severe feeding difficulties, and some enlargement of the head are the rule by 18 months. By two years the child is completely immobile and may need to be institutionalized. Most patients die from respiratory infections by the age of three or four.

TSD occurs predominantly among Ashkenazi Jews (the descendants of Jews who settled in eastern and central Europe), although carrier screening in high-risk groups has drastically reduced the frequency in recent years (Chapter 19). Because of unusual founder effects, high rates of TSD are also found among French Canadians in Quebec, among some non-Amish Pennsylvania Dutch, and among southwest Louisiana Cajuns (descendants of French Acadians expelled from Canada in the 1700s). Among Jewish American and Canadian populations, the incidence (before screening programs) was roughly 1 in 6,000 births—much higher than the incidence of about 1 in 550,000 births found among the non-Jewish American and Canadian populations. Roughly 1 in 30 to 40 Jews is heterozygous for the gene, whereas only 1 in about 300 to 400 non-Jews is heterozygous. Why this harmful allele has been maintained at such a high frequency is not known.

Structural and Biochemical Abnormalities

The cerebrum of affected individuals becomes greatly enlarged by the swelling of individual nerve cells. This abnormality is due to the accumulation of a specialized type of lipid known as a **ganglioside**. Within individual nerve cells, this material collects in the form of *membranous cytoplasmic bodies* (Figure 14.3B), which fill up and expand the cell.

Gangliosides, found mostly in the brain, are important components of normal membranes. These large molecules are synthesized by the addition of several simple sugars to a long-chain lipid and degraded in the lysosomes by the removal of these sugars—one by one—from the lipid part of the molecule. This latter process is defective in Tay-Sachs disease, resulting in the massive accumulation of a ganglioside called G_{M2}. The basic defect is a missing *hexosaminidase* enzyme that normally splits off the terminal sugar group (hexosamine) from the rest of the G_{M2} molecule. There are *two* forms of the enzyme in the tissues of unaffected children, but only *one* form in the tissues of Tay-Sachs patients. The form missing in Tay-Sachs children is called **hexosaminidase A**, or **Hex-A** (Figure 14.4). The form present in both unaffected children and children with Tay-Sachs disease is *hexosaminidase B*, or *Hex-B*.

Molecular Biology and Genetic Variants

The Hex-A enzyme actually consists of two subunits, α and β chains, which are encoded by genes on chromosome 15 and chromosome 5, respectively (Figure 14.5). TSD results from a deficiency of α chains due to mutations in the α locus; about a dozen variants are known. Affected Ashkenazi Jews usually make no complete messenger RNA and no α chains for this locus. Two mutations account for nearly all cases in this population. The major defect, occurring in about 75% of carriers, is a four base-pair *insertion* that causes a frameshift and a premature termination of the polypeptide. (This mutation is also the most frequent *TSD* allele in the Louisiana Cajun population.) Another mutant, a *base substitution*, probably results in defective splicing of the messenger RNA. It is found in about 17% of carriers. These two mutations occur infrequently (less than 5% of carriers) elsewhere.

Most affected French Canadians also make no α chains, but they

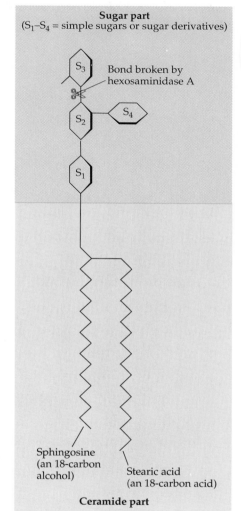

Sugar part
(S_1–S_4 = simple sugars or sugar derivatives)

S_3

Bond broken by hexosaminidase A

S_4

S_2

S_1

Sphingosine (an 18-carbon alcohol)

Stearic acid (an 18-carbon acid)

Ceramide part

Figure 14.4

Ganglioside G_{M2}, the lipid that accumulates in the cells of persons with Tay-Sachs disease. This large, complex molecule is a component of the membranes of brain cells. In unaffected persons, the enzyme hexosaminidase A cuts off the terminal sugar group (S_3) in one step of a pathway that degrades the molecule. Infants with Tay-Sachs disease are homozygous for an autosomal recessive allele and lack a functional enzyme.

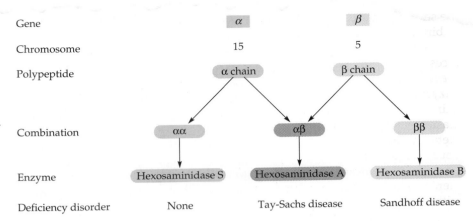

Figure 14.5
The α- and β-hexosaminidase system. Independent genes produce the α chains and β chains needed for the breakdown of ganglioside G_{M2}. The hexosaminidase A enzyme consists of both α and β chains. Hex-A deficiency causes infantile Tay-Sachs disease. Deficiency of the hexosaminidase B enzyme, which consists of β subunits only, causes infantile Sandhoff disease. Hexosaminidase S is a minor enzyme, apparently unassociated with a particular disorder. (Adapted from Sandhoff et al. 1989.)

have a large *deletion* at the 5' end of the gene. Some affected persons who belong to neither of these populations make altered α chains that form defective Hex-A with low activity or low stability. Other variants have also been described in populations around the world. In all groups, some affected individuals are **genetic compounds** (i.e., heterozygous for two different mutant alleles) rather than being homozygous for a single allele.

Other types of TSD (juvenile, adult, chronic) are known. In these conditions the symptoms have a later onset or are less severe, and their enzymes show weak activity. Affected persons may make α chain precursors that do not associate with β chains and are not converted to the mature form. Or they may carry a classic *TSD* allele plus a milder mutant variant.

The Hex-B enzyme, consisting of β subunits only, is produced by the chromosome 5 *β* chain locus. Mutation at this locus causes absence of both Hex-A and Hex-B enzyme activity and is associated with another fatal but very rare lipid storage disease that is quite similar to Tay-Sachs disease. *Sandhoff disease* is characterized by the massive accumulation of a different form of G_{M2} ganglioside in the cerebrum and in other organs.

Carrier Detection and Prenatal Diagnosis

Heterozygous carriers of the Tay-Sachs gene have Hex-A levels that are intermediate between those of unaffected controls and affected individuals. Thus, they can be detected by measuring the level of this enzyme in a small sample of their blood. In Chapter 19 we describe the screening programs set up to identify carriers in a number of Jewish populations.

The hexosaminidase assay also allows for the diagnosis of *TSD* homozygotes: Cells from affected individuals contain Hex-B but no Hex-A. The first prenatal identification of a Tay-Sachs fetus occurred in 1970. Since then, TSD has been almost eliminated from U.S. Jewish populations—by detecting carrier couples *before* they produce an affected offspring and by diagnosing Tay-Sachs disease during early pregnancy. Current techniques also make use of DNA analyses to detect the mutations directly.

PHENYLKETONURIA AND SOME RELATED CONDITIONS

Phenylketonuria (PKU) is one of the most common defects of amino acid metabolism and is also among the most common inherited metabolic diseases affecting brain development. It once caused hopeless mental and physical degeneration; until the early 1960s it accounted for about 1% of the severely retarded people in institutions. If detected within a few weeks of birth, however, it is treatable by dietary therapy.

Recall from Chapter 1 that this condition is inherited as an autosomal recessive. The metabolic defect is an inability to convert the amino acid *phenylalanine* to *tyrosine* (Figure 14.6A), owing to the absence of a liver enzyme called **phenylalanine hydroxylase (PAH)**. Some of the phenylalanine that accumulates in the body fluids (cerebrospinal fluid, blood plasma, and sweat) is converted to phenylpyruvic acid, which in turn is metabolized to several other derivatives. It is the excess of these compounds that appears to cause the brain damage. Because tyrosine is deficient in these individuals, so, too, are its derivatives, including the pigment melanin. Thus, untreated affected persons tend to be fair skinned, with blond hair and blue eyes.

Phenylketonuria affects about 1 in 10,000 newborns in the United States and Europe. It is more frequent in certain northern European populations and their descendants (1 in 4,000 among the Irish, and 1 in 8,000 among the Scotch and Scandinavians), but rare among Finns, southern and eastern Europeans, and Asians. The frequency of carriers likewise varies from group to group, but among Caucasians it averages about 1 in 50.

Figure 14.6
The metabolism of phenylalanine and tyrosine, showing the metabolic blocks that lead to several genetic disorders. (A) In phenylketonuria, phenylalanine cannot be converted to tyrosine; as a consequence, phenylalanine and its breakdown products accumulate and are excreted in the urine. The missing enzyme is phenylalanine hydroxylase. (B) In classic albinism, tyrosine cannot be converted to DOPA and dopaquinone, thereby preventing the formation of melanin pigments. The missing enzyme is tyrosinase. (C) In alkaptonuria, homogentisic acid cannot be broken down further and appears in the urine (see Figure 14.1). The missing enzyme is homogentisic acid oxidase.

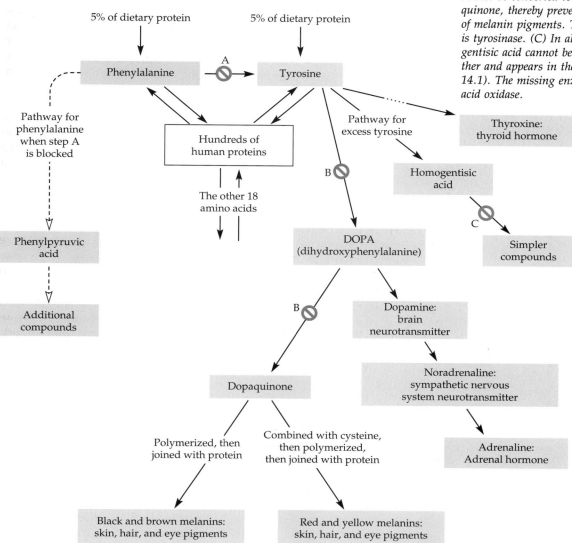

The only clear-cut trait shown by newborn babies with PKU is the accumulation of excess phenylalanine in the blood plasma, starting several days after birth. No consistent clinical signs occur during the first month or two of life. After two or three months, however, rapid and progressive deterioration of central nervous system development sets in. The brains of untreated persons with PKU tend to be smaller than normal, but show no obvious structural defects. At the cellular level there is inadequate development of *myelin*, a fatty substance that normally forms an insulating sheath around certain nerve fibers and thereby speeds up the transmission of nerve impulses.

Babies and children with untreated PKU are hyperactive and un-coordinated; some never sit, walk, or develop bowel and bladder control. The majority never learn to talk and do not progress beyond a mental age of two years. They are extremely agitated, with awkward and jerky movements caused by abnormally increased muscle tone. Behavioral problems range from fearfulness and irritability to violent and destructive temper tantrums. In addition to nervous system disorders, anomalies of the skin (eczema), teeth (defective enamel), and bones (small skull, growth retardation) may occur. About 75% of untreated people with PKU die before the age of 30.

Unfortunately, the precise cellular actions of this biochemical defect are still unknown, and there are no naturally occurring animal models to study. Some researchers have tried to produce animal models by injecting rats and mice with excess phenylalanine, but the resulting phenotypes do not closely mimic human phenylketonuria. Another approach is to insert human PKU alleles into laboratory animals.

Dietary Treatment

In the 1950s, physicians began to treat PKU by restricting the dietary intake of phenylalanine. Synthetic mixes of amino acids are supplemented with vitamins, minerals, fats, carbohydrates, fruits, vegetables, and certain other foods. The diet is nutritious but not so delicious. Patients need to be continually monitored, and diets are adjusted to maintain the optimal concentration of serum phenylalanine: either too much or too little can lead to brain damage.

Therapy begun after the age of three to six months shows little or no benefit, apparently because the brain has already been irreparably damaged. But when started within the first two months after birth, and properly controlled, the low-phenylalanine diet is very effective. Hyperactive and difficult babies become alert and responsive and usually develop normal or near-normal intelligence. The practice now is to encourage continuation of the strict diet for 12 to 15 years, or even indefinitely. Compliance is especially important for females of childbearing age who have PKU (discussed shortly).

Neonatal Screening and PKU Variants

To get affected infants on the diet before brain damage occurs, they must be identified within a few weeks after birth. This goal became possible in 1963 with the development of a simple, reliable, and inexpensive test for excess serum phenylalanine (Chapter 20). Most infants are now tested for this disorder before they leave the hospital in which they were born.

Even before early screening for PKU became possible, it was known that some individuals homozygous for mutant PKU alleles developed normal or near-normal intelligence. Until the early 1960s, virtually *all* cases

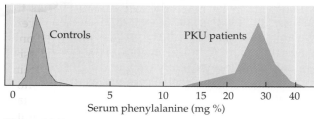
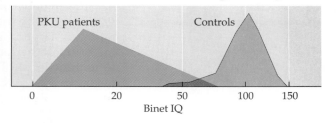

Figure 14.7

The distribution of phenylalanine in blood serum and the distribution of IQ scores in persons with PKU and in unaffected individuals. (The nonlinear measurement scales reflect a certain kind of statistical analysis.) Note a complete absence of overlap in serum phenylalanine levels and only a slight overlap in IQ scores. This study was done before screening programs had begun, so that most affected individuals were in institutions. Following the adoption of PKU screening programs in the 1960s, a few newborns with intermediate levels of serum phenylalanine were discovered, leading to problems of classification and treatment. (Redrawn from Penrose 1951.)

of PKU were ascertained because of their mental retardation, and all had high levels of phenylalanine in their blood—a condition called **hyperphenylalaninemia** (Figure 14.7). But did *all* individuals with hyperphenylalaninemia become mentally retarded? Physicians suspected that some people with hyperphenylalaninemia actually developed normally. It soon became clear that this was the case, and screening programs began turning up new classes of individuals with *intermediate* levels of serum phenylalanine. Clinicians were then faced with the vexing problems of deciding which ones should undergo dietary restriction and what concentrations of phenylalanine are safe at what ages.

Further studies revealed that the metabolism of phenylalanine involves a pathway of several enzymes, with each complete or partial deficiency leading to a different type of hyperphenylalaninemia. Even classic PKU itself is not a single genetic entity. The five main types of hyperphenylalaninemia (Figure 14.8) are all inherited as autosomal recessives. *Classic phenylketonuria* is due to a virtually complete absence of activity of the enzyme *phenylalanine hydroxylase (PAH)* (number 1 in Figure 14.8). It accounts for roughly 60% of all hyperphenylalaninemia cases. Partial PAH deficiencies lead to *mild hyperphenylalaninemia* of two types (numbers 2 and

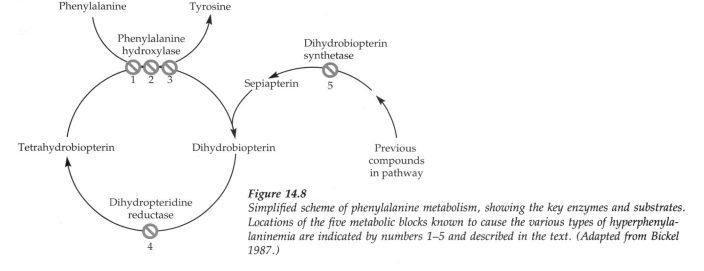

Figure 14.8

Simplified scheme of phenylalanine metabolism, showing the key enzymes and substrates. Locations of the five metabolic blocks known to cause the various types of hyperphenylalaninemia are indicated by numbers 1–5 and described in the text. (Adapted from Bickel 1987.)

333

3). They account for about 35% of all cases of hyperphenylalaninemia and require either no treatment or only early dietary restriction. In the remaining 5% of hyperphenylalaninemia cases the PAH enzyme is normal, but one of two other enzymes (numbers 4 and 5) in that metabolic pathway is nonfunctional. These two conditions involve abnormal responses to dietary treatment: Even after phenylalanine levels are under control, brain function continues to deteriorate.

Molecular Biology and Prenatal Screening

The *PAH* locus resides near the tip of the long arm of chromosome 12. It is a large gene and highly conserved, with the human and rat *PAH* genes showing about 98% homology. Over 30 different *PAH* mutations have been identified. The various PAH-deficient phenotypes may be homozygotes, or they may be compound heterozygotes for different *PAH* mutant alleles. Different alleles tend to occur in different populations, and specific combinations of alleles have been correlated with certain clinical phenotypes.

Molecular analyses of DNA from affected Danish individuals indicate that some of the more common *PKU* alleles may be more deleterious as well. The most common *PKU* allele (38%) has a base substitution in intron 12 that leads to defective processing of PAH mRNA and loss of PAH enzyme activity. The second most prevalent *PKU* allele (20% in Denmark) has a base substitution in exon 12. A totally different *PKU* mutation — deletion of an entire exon — has been found among Yemenite Jews in Israel.

Despite its effect on the brain, PAH is expressed only in the liver. Thus, its activity or inactivity cannot be detected prenatally in fetal amniotic cells. But molecular techniques make it possible to detect the mutant DNA itself. Because RFLP sites are linked to the *PAH* gene, most families already known to be at risk for PKU can be effectively screened and diagnosed. Yet for those families with uninformative segregation patterns, and for carriers without a family history of PKU, this kind of analysis alone will not work. Direct detection of the mutant alleles in the DNA of some carriers or homozygotes is also necessary. Such analyses — using oligonucleotide probes specific for each mutation — are relatively simple, quick, and cheap. They can also be automated, an advantage that opens the door to carrier screening in the general population.

Maternal Phenylketonuria

Before the advent of mass screening for PKU, very few affected females had babies. These women usually had profound mental retardation and were often institutionalized. But as dietary treatment rescued more and more of them from retardation and allowed them to lead normal lives that included motherhood, they began to produce non-PKU offspring who were retarded. When mothers with PKU go untreated during pregnancy, over 92% of their babies are mentally retarded and about 75% have a small head (microcephaly). In addition, these babies often have low birth weight and heart defects. Despite the fact that they are heterozygotes for *recessively* inherited PKU, many also exhibit hyperphenylalaninemia.

The origin of the newly observed **maternal PKU** soon became clear. Most of these mothers had been allowed to abandon the diet therapy by their early teens, after which their serum phenylalanine levels increased greatly. This dietary change seemed to have limited ill effects on them, but it caused irreparable harm to their developing offspring by transferring high phenylalanine levels across the placenta. Indeed, if these women reproduce at average rates, the incidence of PKU-related mental retardation could return to its original level within just one generation!

The problem can be prevented if women with PKU remain on diet therapy through adulthood, or return to it before (or perhaps within a few weeks after) they become pregnant. But many females of childbearing age are unaware of their PKU until after they have produced an abnormal baby. Oddly enough, many have no recollection of ever being on a special diet, and their parents often see no reason to tell them. As mentioned in Chapter 1, the treatment of maternal PKU raises many medical, legal, and ethical questions.

ALBINISM

Although not too common in humans, the absence of coloration is such a striking trait that most of us have seen it at some time. **Albinism** has been noted by writers as far back as the first century A.D. Indeed, from the description of his birth in ancient Old Testament writings, it appears that Noah might have been an albino. European explorers were fascinated by the "white Negroes" they saw in Africa and by the "moon-eyed people" found in some indigenous tribes of Central and North America.

Albinism occurs in a wide variety of animals, from insects to mammals (Figure 14.9). The basic defect is usually the absence of melanin pigments, inherited as an autosomal recessive. Garrod (1909) regarded albinism as an inborn error of metabolism caused by the lack of an end product rather than by the excess of some biochemical intermediate. He suggested that an intracellular enzyme was missing. Again he was correct.

Figure 14.9
An albino child in a Liberian village. (Photograph © Richard Dranitzke/Science Source/Photo Researchers.)

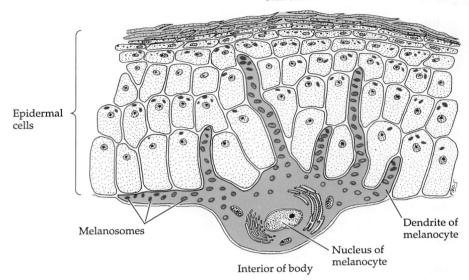

Skin surface

Epidermal
cells

Melanosomes

Dendrite of
melanocyte

Nucleus of
melanocyte

Interior of body

Figure 14.10
Association of one melanocyte with several dozen overlying epidermal cells. Pigment-containing melanosomes migrate up through the dendrites and are transferred to the surrounding epidermal cells.

Complete, or classic, albinism is now known to be caused by the absence of the functional enzyme **tyrosinase** (Figure 14.6B). But the biochemistry of melanin formation is very complex, and other types of albinism are also known, including a few X-linked or dominant types. About 60 different genes affect the skin and fur colors of mice, which suggests that genetic influences on human pigmentation may be equally complex.

Tyrosinase is active only in specialized cells known as **melanocytes**. These arise in the neural crest—blocks of tissue on either side of the neural fold (Chapter 6)—during very early embryonic development. Later they migrate to their final destinations. Melanocytes occur in the lower layer of the skin epidermis and the hair bulbs as well as in certain parts of the eye, ear, nervous system, and mucous membranes. Inside the melanocytes, tyrosinase is restricted to small specialized structures known as *melanosomes* (Figure 14.10). During normal pigment formation, melanin is deposited in these melanosomes, which are transferred to the adjacent epidermal cells and come to lie over the nucleus of each cell. By capping the epidermal nuclei and absorbing much of the ultraviolet radiation that enters the cell, melanin shields the DNA from the mutagenic effects of sunlight and other sources of ultraviolet radiation.

People with albinism have normal melanocytes with structurally normal melanosomes that lack only pigment. Among unaffected persons, skin color is determined not by the number of melanocytes—which appears to be about the same in all races—but rather by the number and distribution of melanosomes produced by these cells, as well as by the type of melanin. In all races, the rate of production of melanosomes varies with amount of sunlight, hormonal changes, and genetic constitution.

Biochemistry

As shown in Figure 14.6B, tyrosine is normally converted to *dihydroxyphenylalanine (DOPA)*, a key biological compound. In melanocytes, DOPA is converted to *dopaquinone*, which finally is polymerized with proteins in the melanosomes to form *black and brown melanins*. Alternatively, dopaquinone may combine with cysteine before being conjugated with melanosome protein to form *red and yellow melanins*. The first two steps in this scheme (tyrosine → DOPA → dopaquinone) require the enzyme tyrosinase, whose gene resides on the short arm of chromosome 11.

In humans, several types of albinism are known, including even a few with normally pigmented skin. Although the severity of the clinical features varies, all types of albinism involve reduced or absent eye pigmentation and structural abnormalities of the eye. Severe squinting, poor vision, and *nystagmus* (rapid, jerky, involuntary movements of the eyeballs) also occur.

All affected individuals lack binocular vision as a result of misrouting of optic nerve fiber pathways in the brain. Studies on albino cats suggest that this impairment is due to a lack of pigment in certain nerve cells during early development. Albinism affects hearing as well as sight. In the inner ear, nerve fibers are normally routed to one or the other side of the brain, and melanin is apparently necessary for proper development of these neural pathways. The amount of melanin found in the inner ear correlates closely with the amount present in the iris of the eye, and affected people lack pigment in both.

About a dozen types of albinism are known, most of which are poorly understood. They fall into two main groups. **Oculocutaneous albinism** (*oculo*, "eye"; *cutaneous*, "skin"), or **OCA**, involves the absence or partial deficiency of melanin in the eyes, skin, and hair, as well as the visual defects previously described. But with **ocular albinism (OA)**, defects are limited to the eyes; the skin is normally pigmented or only slightly underpigmented.

The most extreme clinical form, the classic one studied by Garrod, is **tyrosinase-negative OCA**. It is the only type for which the metabolic defect is known. Absence of the enzyme tyrosinase leads to the total lack of melanin in the eyes, skin, and hair. Extreme sensitivity to sunlight causes excessive roughness, wrinkling, and folding of the skin as well as a high frequency of skin cancer. (This is especially true in tropical regions. In Nigeria, half of all Ibos with albinism develop cancer by the age of 26 and die by age 40.) The irises of the eye are gray to blue, containing no visible pigment. Affected individuals are usually cross-eyed and very nearsighted, often to the point of legal blindness.

The overall frequency of this disorder varies greatly among ethnic groups. Among the Irish, for example, the frequency is about 1 in 10,000 to 15,000; in the United States the frequency is about 1 in 28,000 among blacks and about 1 in 39,000 among whites; but in British Columbia, Canada, it is about 1 in 68,000. Heterozygotes for the various types of albinism are fairly common, however, totaling 1–2% of the general population.

The other major type of oculocutaneous albinism is called **tyrosinase-positive OCA** because tyrosinase is present. The nature of the metabolic error is unknown, but it might involve a defect in the transport of tyrosine into the melanosomes rather than in its subsequent conversion to melanin. Tyrosinase-positive OCA is the most common type, although less frequent in Caucasians (about 1 in 37,000 to 60,000) than in blacks (1 in 14,000 to 15,000). Among Nigerian Ibos, the frequency is about 1 in 1,000. It is also quite prevalent among certain Native American tribes (about 1 in 140 to 240 for the Tele Cuna, Hopi, Jemez, and Zuni), perhaps owing to the special place of albinos in these cultures. The gene for tyrosinase-positive OCA has recently been mapped to chromosome 15q. It may be homologous to the pink-eyed dilute locus (*p*) on mouse chromosome 7—which, as in humans, causes decreased production of black and brown melanins.

Because both types of OCA are autosomal recessives, matings between two tyrosinase-negative or two tyrosinase-positive albinos are expected to produce all albino progeny. But because the two autosomal loci are independent, matings between tyrosinase-positive and tyrosinase-negative albinos produce all unaffected offspring.

LESCH-NYHAN SYNDROME

In 1962 a seriously ill boy was referred to Johns Hopkins School of Medicine. There he was examined by a physician, William Nyhan, and by a medical student, Michael Lesch. The child suffered from various problems: blood in the urine, extremely high concentrations of uric acid in the urine and blood, mental retardation, spastic cerebral palsy (a motor disorder that includes uncontrollable spasms of the legs and arms), and self-mutilation. After ruling out other conditions known to be associated with excess production of uric acid, Lesch and Nyhan concluded that they were seeing a new metabolic disease.

Babies with **Lesch-Nyhan syndrome** look healthy at birth and seem to develop normally for several months. Nevertheless, early signs of anomaly are sometimes noted during the first weeks of life, such as the presence of orange "sand" (uric acid crystals) in the diapers or the occurrence of colic or uncontrolled vomiting. The latter may cause dehydration, which is followed by further concentration of uric acid, kidney stones, obstruction of the urinary tract, and other problems.

Within a few months, delays in motor development set in. Within a year, the muscles develop excess tone and the child begins to exhibit uncontrollable writhing movements of the hands and feet. Involuntary spasms of the arms, legs, ankles, neck, and trunk prevent the child from walking unassisted and often lead to dislocated hips or club feet. Feeding problems contribute to the patient's small size and frail condition. Poor muscle control also prohibits the child from speaking clearly or from being toilet trained. Seizures occur in about half of these children.

After two or three years of age, some children begin to exhibit the most unusual trait associated with this disorder: compulsive biting of the fingers, lips, tongue, and inside of the mouth. Also typical is aggressive behavior toward others (hitting, pinching, swearing, and spitting). Patients with Lesch-Nyhan disease were sometimes classed as severely retarded, but some researchers suspect that their poor scores are partly due to difficulties in communication. These children often seem alert, cheerful, responsive, and genuinely anxious to control their destructive tendencies. The degree of self-mutilation varies considerably among patients and also in the same patient under different circumstances. Although survival into the 20s or 30s is known, most affected individuals die before then.

Biochemistry

Lesch and Nyhan determined that the basic error was a block in purine metabolism. As shown in Figure 14.11, the purine nucleotides needed for DNA and RNA production can either be synthesized from scratch or be rescued from degraded nucleic acids and recycled by means of a salvage pathway. In unaffected humans this latter pathway works very efficiently, recycling 90% of free purines; but in affected persons it fails to function, and excess purines are converted to uric acid.

Affected individuals show less than 1% of the normal activity of an enzyme, **hypoxanthine-guanine phosphoribosyl transferase (HPRT)**. In unaffected individuals this enzyme functions in all cells, but is most active in the brain, especially in *basal ganglia*—paired masses of gray matter that are embedded in the middle of the cerebrum and that control certain

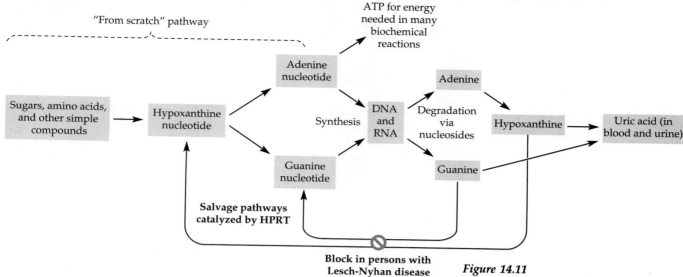

Figure 14.11
The metabolism of purines, showing the
block in persons with Lesch-Nyhan disease.
The purine nucleotides that are incorporated into RNA and DNA are mostly synthesized from simple compounds (far left).
The free purines (adenine and guanine) resulting from the degradation of old nucleic
acids can be recycled, however. These salvage pathways use the enzyme hypoxanthine-guanine phosphoribosyl transferase
(HPRT), which transfers a phosphate-plus-
ribose group to either hypoxanthine or to
guanine. Lesch-Nyhan infants lack HPRT
and excrete large amounts of uric acid. In
this metabolic scheme, most arrows represent several individual chemical reactions.

movements of skeletal muscles. How a deficiency of HPRT can bring about
such highly stereotyped behavior is still a mystery, but all evidence suggests that the neurological defect is biochemical rather than anatomical.

Partial deficiency of HPRT (1–20% activity) leads to a severe form of
gout, with high uric acid production. Affected persons (about 1 in 200
males with gout), have a normal life span and can reproduce. Although
about 20% show some neurological symptoms (e.g., retardation, spasticity,
seizures), none of them self-mutilate.

Genetics and Molecular Biology

Total HPRT deficiency is inherited as an X-linked recessive. No females
with Lesch-Nyhan syndrome are known, and affected males do not reproduce. The trait is relatively rare (1 in 100,000 to 380,000 births) and is
not associated with any particular ethnic groups. Over 15 X-linked recessive variants with partial deficiencies of HPRT enzyme activity have also
been identified.

In humans the *HPRT* gene is located near the tip of Xq, closely linked
to the fragile X and deutan loci. Although large, its actual coding region
is of modest size and gives rise to a 218-amino-acid protein product.
Sequence comparisons with mouse and hamster DNAs reveal that homology between the three species is over 95% in the coding regions of
these genes.

Armed with the complete sequences for both the *HPRT* gene and the
protein, researchers can analyze specific mutations in persons with Lesch-
Nyhan disease or gout. The alterations turn out to be very heterogeneous.
In one study the cell lines from at least 16 out of 24 HPRT-deficient persons
contained unique mutations. So far, about 85% of analyzed mutations are
point mutations or very tiny chromosomal aberrations. Only 15% are major
deletions or rearrangements.

Carrier Detection and Prenatal Diagnosis

Heterozygotes, although clinically normal, may show some subtle defects
in purine metabolism, including increased levels of uric acid in their urine.
Because the HPRT enzyme is detectable in all cells of the body, including

amnion and chorion cells, prenatal diagnosis by amniocentesis or chorionic villus sampling is an option for women known to carry the defective allele. The occurrence of mild forms of HPRT deficiency complicates the process of prenatal diagnosis, however. Great care must be taken to distinguish fetuses affected with Lesch-Nyhan disease from those with *partial* HPRT deficiency, because the abortion of the latter type is unwarranted.

Work by some British researchers on HPRT-deficient ($HPRT^-$) mice has raised the possibility of detecting $HPRT^-$ embryos within a few days of conception, that is, before implantation. They were able to detect the absence of HPRT activity in single mouse cells removed from eight-cell preimplantation embryos derived from carrier female mice. Experimental procedures like these have been extended to humans for the purpose of sex determination (Chapter 19).

Animal Models and Possible Gene Therapy

No animals with phenotypic symptoms corresponding to the Lesch-Nyhan syndrome are known, but researchers can now study genetically engineered HPRT-deficient mice. The production of mutant $HPRT^-$ genes can be accomplished by chemical or retroviral means or by the selection of preexisting spontaneous $HPRT^-$ mutants among XY embryonic cells. Because these methods allow researchers to produce and isolate almost *any* biochemical mutant and incorporate it into laboratory mice, it should now become possible to study in the laboratory many other inborn errors of metabolism known to afflict humans.

The devastating nature of Lesch-Nyhan syndrome, the lack of effective treatment for its neurological symptoms, and the special characteristics of the *HPRT* gene—that is, being expressed in all cells, being a reasonable size, and being readily selectable—make this disorder a prime candidate for gene therapy. The technical and ethical problems associated with this issue will be discussed in Chapter 20. Here we will only point out that some important first steps have been taken by scientists who have successfully substituted normal *HPRT* alleles for mutant $HPRT^-$ alleles (or vice versa) in mouse embryonic stem cells. The results of this work suggest that it may be a practical method for producing specific, predetermined germline changes in experimental animals. Researchers hope that these various methods will allow them to generate mice of any desired genotype.

PHARMACOGENETICS AND ECOGENETICS

Box A presents one dramatic example of how undiagnosed metabolic disease can be mistaken for other disorders—or even for foul play. But not every inborn error of metabolism causes death, mental retardation, or even serious illness. In fact, an individual can be born with a metabolic defect and live for years, perhaps a lifetime, without any health problems. If the affected biochemical pathway is a relatively unimportant one or needed only under special circumstances, no ill effects may ever be noticed.

As technology introduces new substances into our environment, however, it is inevitable that some people will be unable to metabolize some of them because of the presence of previously unrecognized enzyme variants. In recent years, the discovery of more and more cases of this type has spawned new fields of study called **pharmacogenetics** and **ecogenetics**. These new areas of research deal with the genetic variation leading to differences in the ways that individuals metabolize drugs, foods,

METABOLIC ILLNESS AND CRIMINAL JUSTICE

In 1990 Patricia Stallings was convicted of first-degree murder for poisoning her baby, and sentenced to life imprisonment. Her ordeal began in 1989, when she brought her critically ill three-month-old son to a hospital in St. Louis. The attending physician decided that the baby's symptoms (severe vomiting, gastric distress, and labored breathing) were caused by poisoning with ethylene glycol, a substance found in antifreeze. This diagnosis was confirmed by laboratory analysis of the infant's blood. The baby recovered from the crisis and was placed in a foster home. But after one parental visit, during which Stallings had fed her child a bottle of milk, he became acutely ill again and died. The hospital laboratory and a commercial laboratory then analyzed his blood and the empty milk bottle. After both labs reported the presence of ethylene glycol, Stallings was charged with murder and jailed to await trial.

Meanwhile, she had become pregnant again. While in custody she gave birth to another son, who was immediately placed in foster care. But about two weeks later he began to exhibit the same severe symptoms that killed his brother. This time, however, the diagnosis was *methylmalonic acidemia (MMA)*—a rare, autosomal recessive disorder of amino acid metabolism. Thus, the possibility existed that her other child had died from MMA rather than from poisoning. Yet the crucial evidence was withheld during her trial, and she was convicted and sentenced to life imprisonment.

After hearing about the case, two St. Louis University experts in metabolic disorders decided to do their own analyses of the dead baby's blood and urine. (The incriminating bottle had disappeared.) All their tests for ethylene glycol were negative, but the samples gave clear evidence (i.e., large amounts of accumulated methylmalonic acid and other substances) that the child did have MMA. When these results were presented to the prosecuting attorney, he withdrew the charges and Patricia Stallings was released. But what about the original lab reports? When reviewed, all were found to be in error for one reason or another.

and other substances in their environment. The results of such studies will play an increasingly important role in medicine.

Glucose-6-Phosphate Dehydrogenase Deficiency

One example is a common inborn error of sugar metabolism involving abnormalities of an enzyme called **glucose-6-phosphate dehydrogenase (G6PD)**. The *G6PD* locus lies near the tip of Xq, next to the *HPRT* gene. About 325 variants of this X-linked recessive gene are known, many of which are rare. A few kinds of G6PD deficiency are of great medical interest—not only because of their striking effects, such as *acute hemolytic anemia* caused by breakdown of red blood cells, but also because they are surprisingly common, affecting about 100 million people (mostly males) worldwide. But virtually all the affected people live in (or come from) tropical or subtropical countries where malaria is endemic. Here the mutant genes have some selective advantage because heterozygous females are more resistant to malaria than are $G6PD^+$ (unaffected) homozygotes.

The G6PD enzyme functions only under special circumstances in a minor biochemical pathway of red blood cells, where it acts on glucose-6-phosphate. Coupled to this reaction is a second step (catalyzed by a different enzyme) that generates *reduced glutathione (GSH)*. The two steps are crudely summarized as follows:

$$\text{glucose-6-phosphate} \xrightarrow{\text{G6PD}} \text{(intermediate)} \xrightarrow[\text{enzymes}]{\text{other}} \text{reduced glutathione}$$

Reduced glutathione is necessary for maintaining the integrity of the cell membrane. But because this pathway is scarcely utilized, under usual conditions even a low activity of G6PD is sufficient to keep red blood cell membranes intact. Only in "emergency" situations, when GSH is rapidly depleted, does the need arise for a normal full-strength, quick-acting G6PD enzyme to reverse this reaction.

The rapid depletion of GSH can occur when certain foods or drugs, such as *fava beans* or the antimalarial drug *primaquine*, are ingested. In persons with certain mutant *G6PD* alleles, the low activity of the G6PD enzyme cannot meet the challenge: The concentration of GSH falls below required levels and the red blood cells rupture. Depending on genetic and environmental circumstances, these attacks of hemolytic anemia vary from mild to acute. In some cases they can be fatal if blood transfusions are not given.

There are two major variants of *normal* G6PD: B (100% activity), which is common throughout the world, and A (88% activity), which is confined to black Africans and their descendants. The amino acid sequences of these two variants are identical except at one position. **Favism** is associated with a mutation of the *B* allele known as the *Mediterranean* or B^- allele. Despite its severe enzyme deficiency and clinical effects,* this mutant is extraordinarily common in Greece, Sardinia, northwest India, and southern Italy and among the Sephardic Jews of Israel. **Primaquine sensitivity** among blacks of African descent is associated with a mutation of the *A* allele known as A^-. Partly because the enzyme's activity is somewhat higher than in the case of the B^- mutant, it can maintain GSH in young blood cells but not in old ones. The frequency of the mutant allele is over 20% among males in many parts of Africa and about 10–15% among black American males. Two other common *G6PD* alleles are the *Canton variant* among southern Chinese and the *Constantine variant* among Arabs.

Porphyria

Sometimes a drug reaction unmasks the presence of a previously unsuspected metabolic disease. This happens in people who carry the dominant allele for **acute intermittent porphyria**. Such individuals may live for decades without any inkling of the potentially fatal condition they harbor. When given barbiturates and certain other sedatives or anesthetics, however, they suffer attacks characterized by severe pains in the abdomen and limbs, muscular weakness, and mental instability. Sometimes abdominal surgery is performed, but when nothing is found amiss, the physician may suggest that the problem is entirely psychosomatic, especially in view of the patient's seemingly abnormal behavior. Ironically, barbiturate-type anesthetics may be used during the operation and more sedatives prescribed thereafter; life-threatening symptoms that include delirium, excruciating pain, and paralysis may follow. Other factors known to trigger these episodes include infections, increased levels of steroid hormones (especially estrogens), and severe dieting or starvation.

Because its symptoms mimic so many other known conditions, porphyria may be misdiagnosed. The best clue to this disorder is the striking change in the color of the urine—to a brilliant red hue like that of port wine—when it is exposed to light and air. This is due to the excessive excretion of two metabolic precursors of *heme*, the red-colored part of a hemoglobin molecule. (Heme belongs to the *porphyrin* class of molecules, which forms the base of all respiratory pigments in animals and plants.) The basic metabolic error is a deficiency of the enzyme *porphobilinogen deaminase*. About six variants are known, most of which exhibit only about half the enzymatic activity shown by normal controls, as would be expected for heterozygotes. Porphyria is an exception to the general finding that enzyme deficiencies are inherited as recessives. The gene has been mapped to the long arm of chromosome 11. The vastly increased use of barbiturates in recent years has led to its detection in many populations

---

* In people with favism, reactions may also be triggered by certain infections as well as by sulfa drugs, chloramphenical, vitamin K, and aspirin. Other oxidizing substances, such as mothballs, can cause severe reactions too.

—notably in Sweden and Lapland, where the frequency is about 1 in 1,000. Elsewhere in the world its frequency may be about 1 in 50,000.

Another autosomal dominant disorder, **porphyria variegata**, is especially common among whites in South Africa and in Finland. Affected adults, in addition to suffering acute drug-induced attacks like those already described, have a variety of skin problems. These include blistering and fragility of those areas exposed to sunlight, with frequent infection and scarring of the lesions, as well as excess hairiness and overpigmentation. Here the deficient enzyme is *protoporphyrinogen oxidase*.

Two cases of genetic sleuthing on porphyria variegata are particularly interesting. Dean (1957) traced its inheritance in a huge South African kindred back to one very prolific early Dutch settler, and Macalpine and Hunter (1969) reviewed the medical history of King George III, the "mad" British monarch who reigned in the late 1700s, and suggest that he had porphyria. In addition to the mental and physical symptoms, they note references to the king's dark, discolored urine. Other descendants of Mary, Queen of Scots, also suffered intermittent "colics," progressive paralysis, and madness. Among the sufferers was her son, King James VI, who claimed to have urine the color of wine.

Other Drug Reactions

Dozens of examples are now known in which a drug that benefits most people can cause severe reactions or even death in some unlucky patients. Consider the muscle relaxant *suxamethonium* (succinylcholine chloride), which is used before the administration of general anesthesia to allow the easy insertion of a tube into the windpipe. It is also used by psychiatrists in patients undergoing electric shock treatment. Normally, the relaxing effect lasts only 2 to 3 minutes, because an enzyme known as *pseudocholinesterase* quickly inactivates the drug by splitting it apart. But about 1 in 2,000 whites of European extraction and about 1 in 50 Alaskan Eskimos is homozygous for autosomal recessive mutations of the pseudocholinesterase gene. Because the variant enzymes are much less effective at inactivating the drug, paralysis lasts much longer—hours or even days—and may be fatal if the patient's breathing is not maintained by a respirator during that time. The gene has been mapped to the long arm of chromosome 3.

One of the most common enzyme deficiencies of humans occurs in so-called **slow metabolizers**. These people are the 5–10% of North American and European whites and 30% of Hong Kong Chinese who show adverse side effects to ordinary doses of about 25 commonly prescribed medications—including some β-blocker blood pressure drugs (such as *debrisoquine*), drugs for angina and abnormal heart rhythms, asthma medicines, cough medicines, and antidepressants. The deficiency is in a *cytochrome enzyme*. (The cytochromes are important liver enzymes that play a major role in the detoxification of drugs and environmental pollutants; their genetics is very complex.) By cloning and sequencing the key human gene from both normal and poor metabolizers, researchers discovered several variants. All these mutations occurred within introns, giving rise to mRNAs with splicing errors and to defective enzyme products. This great variability and high total frequency (35–43%) of mutant alleles promise to create major problems for drug companies and regulatory agencies as well as for patients and physicians. But the good news is that poor metabolizers also seem to show a much reduced risk to cancers of the liver, gastrointestinal tract, and lung. The mutations, which seem be to inherited as recessives, have been mapped to chromosome 22.

In the 1950s an extremely common enzyme variant was discovered among tuberculosis patients being treated with the drug *isoniazid*. Many of them inactivated this drug so slowly that they required only a fraction of the normal dose. In fact, the standard dose produced toxic side effects. Here the difference in response is caused by a variant of the *N-acetyltransferase* enzyme, which is inherited as an autosomal recessive trait. Homozygous recessive slow inactivators are widespread (40–70%) in populations of European, African, and Jewish origin. They are much less common (10–20%) among Far Eastern populations and are least often found (5%) among Eskimos. In cases like this, where two forms of a gene are well established, the mutant allele is presumed to have some selective value in a population—such as the ability to better metabolize certain types of food or other substances in the environment.

Not all drug reactions are so dramatic as the ones just discussed, but there is no longer any doubt that metabolic variability must be taken into account in the dispensation of drugs. Any serious side effects should be investigated for possible genetic bases, and if the situation warrants it, other members of the family should be tested for similar metabolic responses. Garrod's prophesy of a chemical basis for "idiosyncrasies with regard to drugs and articles of food" is amply borne out as large numbers of people are being hospitalized—not for their original medical problems, but rather for unexpected reactions to their treatment.

Lung Disease and Ecogenetics

Emphysema means, literally, "inflated tissues." In pulmonary emphysema the lungs become permanently inflated owing to loss of elasticity in the walls of their tiny air sacs, the alveoli. Affected people cannot exhale properly, and the exchange of oxygen and carbon dioxide across alveolar walls is reduced. As the disease progresses, damaged alveoli combine to form larger (and even less efficient) air sacs and also become infiltrated with fibrous tissue. Because of damage to alveolar capillaries, the heart must work much harder to pump blood through the lungs. Infections also occur more frequently. The patient becomes more and more incapacited and dependent on extra oxygen.

What causes the alveoli to lose their elasticity? Some environmental factors (especially smoking) have been implicated, and the inherited deficiency of an enzyme called **α₁-antitrypsin (AAT)** may also be a key element. AAT is a small protein produced in liver cells and secreted into the blood. From there it diffuses into body tissues, mainly the lungs, and functions as a *protease inhibitor*. It counterbalances the action of certain lysosomal *proteases* (intracellular enzymes that digest proteins), which in this case are released by infection-fighting white blood cells in the lungs. Proteases are useful in digesting bacteria and other dead cells. But they can also attack elastin fibers in the walls of the alveoli if they are not bound and inactivated by α₁-antitrypsin.

The AAT gene, called *Pi* (for protease inhibitor), maps to chromosome 14q. It has been cloned, and its mutant DNA sequences have been identified. The locus is highly polymorphic, with over 70 known alleles whose frequencies vary among different populations. The most common of these codominant variants is the normal one, Pi^M (82–96% in whites, 98–100% in nonwhites). The most clinically significant alleles are Pi^Z, Pi^S, and Pi^{QO}. The first two are point mutations and fairly common (1–2%) in whites (especially those of Scandinavian, Spanish, and Portuguese origin), but rare or absent in black or Asian populations. They produce defective polypeptides that tend to get hung up in the liver cells, rather than being secreted into the bloodstream. The third, a rare "null" mutant, makes no polypeptide at all.

Homozygotes for the Pi^Z and Pi^{QO} alleles suffer from severe lung disease, with smokers showing an earlier onset. Indeed, all AAT disease phenotypes develop from an interplay between genotype and environment. Thus, AAT deficiency is an *ecogenetic* disorder, with cigarette smoking and other chemical irritants being major hazards for people having this inherited tendency to lung disease. As studies of disease susceptibilities become more extensive and sophisticated, scientists expect to find many more examples of ecogenetic traits.

SUMMARY

1. All metabolic events are mediated by enzymes under genetic control. Thus, a mutation in an enzyme-producing gene is likely to induce a metabolic block in one step of a biochemical pathway. The consequences of an excess of precursors or lack of end products depend on the pathway involved.

2. Archibald Garrod first proposed the occurrence of gene-determined inborn errors of metabolism and described several examples in humans.

3. Enzymatic disorders are usually inherited as recessive traits, although many so-called homozygotes are actually compound heterozygotes, carrying two different mutant alleles of the same gene.

4. Tay-Sachs disease is an autosomal recessive error of lipid metabolism caused by deficiency of the Hex-A enzyme. Resulting accumulation of a certain lipid in nerve cells leads to progressive mental deterioration, blindness, paralysis, and death in early childhood. Tay-Sachs disease occurs in all ethnic groups, but mostly among Ashkenazi Jews and French Canadians.

5. Classic phenylketonuria (PKU), a defect of amino acid metabolism, leads to profound mental retardation unless it is detected near the time of birth and treated with a low-phenylalanine diet. Properly treated, these children have intelligence in the normal range. Many alleles are known; they differ among ethnic groups and in clinical effects.

6. Women with PKU who become pregnant while not on the restrictive diet give birth to babies who, although heterozygous and expected to be normal, suffer brain damage from the high levels of phenylalanine in the mother's blood. Unless women with PKU return to diet therapy before pregnancy, the success in treating homozygotes will be totally offset by maternal PKU occurring in their children.

7. Albinism is the inherited absence of melanin pigments. All albinos have visual problems as a result of melanin deficiencies in the eyes; most have pigment deficiencies in the skin and inner ear too. The classic (tyrosine-negative oculocutaneous) albinism is caused by the deficiency of an enzyme that normally converts tyrosine to DOPA. About a dozen other types of albinism exist, but their genetic defects are poorly understood.

8. Boys with Lesch-Nyhan syndrome (an X-linked recessive disorder of purine metabolism) are retarded, lack motor control, and exhibit stereotyped biting behavior. Partial HPRT deficiency leads to a milder condition characterized by severe gout but no retardation or behavioral problems. Mutations of the *HPRT* gene are very heterogeneous.

9. Pharmacogenetics and ecogenetics deal with inherited enzyme variants that lead to individual differences in the metabolism of drugs,

diet, or other environmental substances. In particular, poor inactivation of a drug can lead to serious illness.

10. G6PD deficiency is an X-linked recessive defect of sugar metabolism that causes hemolytic anemia when affected individuals eat fava beans or take certain drugs, such as primaquine. *G6PD* mutant alleles are most common in populations where malaria occurs, because heterozygotes are more resistant to malaria than are unaffected homozygotes.

11. The porphyrias, errors of porphyrin metabolism, involve sensitivity to barbiturates and certain other sedatives. Porphyrias are inherited as autosomal dominants—an exception to the rule that the mutations causing enzyme disorders are usually recessive.

12. Some kinds of lung disease are ecogenetic disorders, involving an interplay between environmental factors and inherited deficiencies of the enzyme α_1-antitrypsin.

KEY TERMS

α_1-antitrypsin (AAT)	hyperphenylalaninemia	metabolic block	pharmacogenetics
ecogenetics	inborn error of metabolism	one gene–one polypeptide	primaquine sensitivity
ganglioside			

QUESTIONS

1. In humans and other organisms, what is the difference between essential and nonessential amino acids? See also Chapter 12.

2. Sometimes the intermediate substance just preceding a metabolic block is *not* excreted in excess. Suggest an explanation.

3. In a progress report on an early newborn screening program in Massachusetts, researchers state that 27 cases of phenylketonuria were detected out of 217,752 babies tested. This frequency of roughly 1 in 8,000 is considerably higher than that estimated for the country as a whole. Suggest an explanation.

4. A yellow variant of albinism phenotypically resembles tyrosinase-positive albinism, but affected individuals have a distinctly yellowish cast to the hair and the ability to tan slightly when exposed to sunlight. These individuals give equivocal results with the hair bulb test, however. After incubation with tyrosine, some hair bulbs show a small amount of pigmentation, but others do not. After incubation with cysteine (in the presence of tyrosine and DOPA), some yellow pigment is formed, but no black pigment develops. Refer to Figure 14.6 and suggest where the biochemical block may occur in this mutant.

5. Females who are heterozygous at the *G6PD* locus nevertheless show levels of enzyme activity that range from near-deficiency to within the normal range. Why?

6. With rare exceptions, all children up to the age of two years are able to digest the milk sugar lactose; they can do so because they have large amounts of the enzyme *lactase*, which splits lactose into glucose and galactose. After the age of two, however, most of the world's people no longer produce much of this enzyme and thus become lactose in-

tolerant. Drinking more than a moderate amount of milk causes bloatedness, stomach cramps, gassiness, and severe diarrhea. Lactose-intolerant children who continue to drink milk may become malnourished and perhaps even die. (Eating cheese and yogurt presents no problem, however.) The vast majority of northern Europeans, Americans of European descent, and a few dairying tribes in Africa possess a dominant mutation that causes the continued production of lactase through adulthood. But 70% of U.S. blacks and nearly all Africans and Asians are lactose intolerant. Comment on the usefulness of aid programs that involve the shipping of large amounts of dried milk to Africa and Asia.

7. Milk is not the only source of lactose in industrial countries. Either dried milk or lactose is added as a filler to most processed and prepackaged foods. How does this affect the ability of lactose-intolerant individuals and people with galactosemia to restrict their dietary intake of lactose?

8. Hunter syndrome and Hurler syndrome are recessive lysosomal storage disorders of mucopolysaccharide metabolism. When researchers put cultured Hunter syndrome cells in medium from Hurler syndrome cell cultures, or Hurler syndrome cells in medium from the Hunter syndrome cell cultures, both metabolic defects are corrected. What does this suggest about possible allelism of the two defective mutants?

9. Fibroblast cells and hair follicles taken from females who are heterozygous for HPRT deficiency show mosaicism in the expression of this enzyme's activity, being either deficient or normal. When red or white blood cells are sampled from the same females, however, all show normal HPRT activity. Suggest one or more reasons why the Lyon effect may not be randomly expressed in all tissues.

10. The occurrence of several biochemical mutants involving the same metabolic pathway allows researchers to determine the relative positions of individual steps in a pathway. The general principles are: (1) the substance just prior to a block accumulates, and (2) the mutant organisms cannot grow on any substances prior to the block but will grow if provided with substances just after the block.

For example, the following table summarizes the growth responses of several different mutant strains of the bacterium *Salmonella typhimurium* in several different media. Wildtype *Salmonella* can grow on minimal medium, but none of the four mutant strains can. In this table, + means growth and 0 means no growth.

Diagram the correct order of all the steps in this biosynthetic pathway, indicating the location of the metabolic block imposed by each mutation. *Hint:* The fewer substances a mutant can grow on, the further along is its metabolic block.

			GROWTH RESPONSE			
			Minimal Medium Plus			
Mutant strain	Accumulated product	Minimal medium	Indole	Anthranilic acid (Ant)	Tryptophan (Trp)	Indole glycerol phosphatase (IGP)
2	Ant	0	+	0	+	+
6	Indole	0	0	0	+	0
3	IGP	0	+	0	+	0
8	None of the 4 compounds	0	+	+	+	+

FURTHER READING

Harris (1980) writes very lucidly on the principles of human biochemical genetics. Emery and Rimoin (1990), Scriver et al. (1989), King, Rotter, and Motulsky (1992), Schaub, Van Hoof, and Vis (1991), and Fernandes, Saudubray, and Tada (1990) are gold mines of information, containing dozens of detailed chapters written by experts on all types of metabolic disorders. Motulsky (1960) provides an interesting review of metabolic variants and the role of infectious diseases in human evolution. On ecogenetics and pharmacogenetics, see Grandjean (1991) and Kalow, Goedde, and Agarwal (1986).

15

Cancer

Although most forms of cancer show little tendency to run in families, the diseases are nonetheless primarily genetic. Rather than being transmitted in germ cells, however, the genetic damage is usually confined to a person's *somatic* (body) cells. The somatic mutations that lead to cancer, like germinal mutations that lead to typical inherited disorders, may be induced by environmental agents such as radiation and chemicals. The resulting damage to the DNA of somatic cells causes the orderly succession of mitotic divisions to go awry. Researchers have begun to unravel which particular genes are involved and how their encoded proteins regulate—or fail to regulate—cell growth.

Through new insights into the genetic control of the cell division cycle, researchers hope to provide much better methods for preventing, diagnosing, and treating cancer. The official United States War on Cancer began in 1971 (Box A). But many of the returns on the research investment—especially in prevention—still lie in the future, for the toll from all types of cancer combined has remained high and substantially unchanged for several decades. True, the mortality rates from some cancers have been decreasing owing to earlier detection (uterine cancer), changing diets (stomach and cervical cancer), and more effective radiation and chemical therapies (childhood leukemias and lymphomas). Yet the small increase in the incidence in breast cancer (Marshall 1991) and the large increase in lung cancer (first noticed among men in the 1930s and among women in the 1960s) have overshadowed the decreases. The tally of new cases of all forms of cancer among Americans is now about 1 million annually (Table 15.1). Deaths from cancer number about 500,000 per year and account for 16% of all deaths—more than any other cause of death among women, and second to heart disease among men. As mortality from heart disease continues to decline (a result of healthier life-styles, earlier detection, and better treatments), the presence of cancer in our lives may actually increase.

CANCEROUS CELLS

Unlike normal cells injured by infectious diseases, cancerous cells are quite healthy by the usual standards—so much so that they interfere with the well-being of neighboring tissue. A mass of cells that multiplies when it should not is called a *tumor* or *neoplasm* (a medical term meaning "new form"). A **benign tumor** remains localized in the place of origin, often separated from the surrounding cells by a layer of connective tissue. (Common warts, for example, are benign skin tumors occurring primarily in children and young adults.) Although compact and slow growing, benign tumors may eventually contain billions of cells, become quite large, and cause damage by pressing against a vital organ. They do not, however,

THE BEGINNINGS OF THE WAR ON CANCER

The National Cancer Act of 1971 was ceremoniously signed into law during the Christmas season by President Richard M. Nixon before 100 guests in the State Dining Room. Popularly called the War on Cancer, the legislation built on people's fears of the dreadful disease. Several months before, columnist Ann Landers had urged her readers to write their Washington representatives to support an attack on cancer, and they had responded with an unprecedented avalanche of mail on Capitol Hill. With the 1972 presidential election coming, it did no harm for Nixon to preempt the cancer issue, however humanitarian his motives may have been.

Expectations for the conquest of cancer within the decade ran high (Greenberg 1986, 1991). After all, well-funded national projects had allowed Americans to build an atomic bomb in the 1940s and to walk on the moon in the 1960s. Congress thus resolved that a similar undertaking should eliminate cancer by 1976, America's two-hundredth anniversary. But cancer proved much too powerful an opponent. The major obstacle to victory was a lack of understanding of the biology of the disease and a lack of emphasis on the requisite basic research. Instead, early priorities were on treatment by the standard techniques of surgery, radiation therapy, and chemotherapy. Although there have been some successful campaigns in dealing with childhood cancers, these afflictions account for but a small part of the total cancer burden. The war will eventually turn decisively against cancer when greater knowledge of biological processes and environmental factors help provide clearer targets and more effective weapons. This chapter is a general survey of the battleground.

Table 15.1
New cancer cases in the United States in 1990.

Cancer	Number of Cases	Percent of Total	Some Major Causes
Lung	157,000	15	Tobacco
Colon and rectum	155,000	15	Animal fat, low fiber, possibly alcohol
Breast	150,900	15	Ovarian hormones
Prostate	106,000	10	Testosterone, possibly estrogen
Bladder	49,000	5	Tobacco
Non-Hodgkin's lymphoma	35,600	3	Unknown
Uterus	33,000	3	Estrogen
Mouth and throat	30,500	3	Tobacco, alcohol
Pancreas	28,100	3	Tobacco
Leukemia	27,800	3	X-rays
Melanoma	27,600	3	Ultraviolet light (sunburning)
Kidney	24,000	2	Tobacco, possibly pain relievers and diuretics
Stomach	23,200	2	Salt, tobacco
Ovary	20,500	2	Ovulation
Brain and nerves	15,600	2	Trauma, X-rays
Cervix	13,500	1	Possibly papilloma virus
Liver	13,100	1	Hepatitis virus, alcohol, possibly tobacco
Larynx	12,300	1	Tobacco, alcohol
Thyroid	12,100	1	Possibly iodine excess
Multiple myeloma	11,800	1	Unknown
Esophagus	10,600	1	Alcohol, tobacco
Hodgkin's disease	7,400	<1	Unknown
Testis	5,900	<1	Unknown
All others	69,500	7	
Total	1,040,000	100	

Source: Adapted from Henderson et al. (1991).

spread to other sites or recur following surgical removal. In benign tumor cells, the chromosomal makeup is usually completely normal.

The cells of a **malignant tumor**, or **cancer**,* however, have the following characteristics:

1. They do not respond to the mechanisms that normally restrict cell numbers to those required for growth or replacement. Like the growth of benign tumors, the proliferation of cancer cells may be slow, but it continues unabated.
2. Cancer cells become disorganized and may revert to a more primitive form, losing their specialized structures and functions. Pathologists look for these characteristic changes under the microscope to distinguish benign from malignant tumors.
3. Cancer cells are invasive, damaging adjacent tissues and often producing internal bleeding. They may also be shed from the primary site and circulate through the blood and lymph to other locations in the body, where they initiate secondary cancers. The spreading to distant sites, called **metastasis**, poses a severe problem in treatment. Deaths of cancer patients often result from the destruction of vital organs at secondary sites or from overwhelming infections due to general debilitation and damage to the immune system.
4. Cancer cells almost always have abnormal karyotypes. These include translocations, inversions, deletions, isochromosomes, monosomies, and extra chromosomes—the latter sometimes far beyond the normal diploid number. Specific chromosomal aberrations are sometimes associated with particular cancers.

Tumor cells pass through stages as they go from relatively benign to highly malignant. This process of **tumor progression** may follow the order of characteristics given in the preceding list; that is, cells first divide without control, then become disordered and less distinctive, and then invade surrounding tissue and metastasize. These steps may be accompanied by increasing chromosomal aneuploidy or by changes in genes that code for proteins that regulate growth and division. It is possible that each of us has some of the early mutational changes in some of our body cells that could progress to cancer. But being based on chance mutational and chromosomal events, tumor progression is not inevitable. Yet as variant cell populations compete with one another, some cell lines may emerge with greater growth potentials. Then a cell may arise with a critical load of accumulated genetic damage. The *clone* of cells resulting from successive divisions of this one highly aberrant cell constitutes the unmistakable cancerous growth (Figure 15.1).

Of the more than 200 different kinds of human cancer, about 85% are **carcinomas**, solid tumors derived from epithelial tissues—skin, linings of the respiratory, digestive, urinary, and genital systems, plus various glands, the breasts, and nervous tissue. Carcinomas of the lung, large intestine, and breast account for about half of all human cancers in Western societies (top three lines of Table 15.1). Also classified as solid tumors, **sarcomas** (2% of all cancers) are composed of closely packed cells derived from bone, cartilage, muscle, and fat. Two types of malignancies affect white blood cells: **Leukemias** (3% of all malignancies) are cancers of the white blood cells manufactured in the bone marrow; **lymphomas** (5% of

---

* *Cancer* in Latin means "crab." The ancient physician Galen thought that the distended veins in a breast cancer looked like the extended legs of a crab. The crab reference had been used even earlier by the "father of medicine," Hippocrates (460–370 b.c.). He thought that invasive growths of a cancer appeared to grasp surrounding tissues.

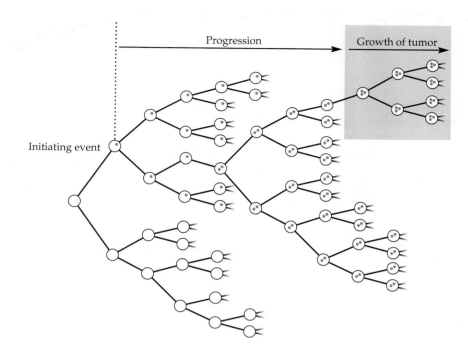

Progression

Growth of tumor

Initiating event

Figure 15.1
The cellular pathway by which the accumulation of genetic damage leads to cancerous growth. Each colored dot represents some type of genetic change: mutation, loss or amplification of genetic material, translocation, inversion, and so on. Disturbing orderly cell divisions in some way, each change acts in either a dominant or recessive fashion. In the figure, three changes are assumed to produce cancerous growth, but the actual number of such accumulated changes may vary.

all cancers) affect the white blood cells present in the spleen and lymph nodes. These and some other cancers and the tissues they affect are listed in Table 15.2. Although cancers are primarily associated with processes of aging, they can occur in younger persons as well. Leukemia, for example, is a major cause of death in children.

DOMINANT SOMATIC MUTATIONS PREDISPOSING TO CANCER

A major advance in understanding cancer came with the recognition in the 1970s of a class of genes whose normal alleles regulate cell division. A mutant allele, when present in a single dose, led to a cancerous line of cells. These genes were first found in chickens and mice, although many identical or similar genes were later shown to be present in human cells too. In chickens and mice, but not in humans, these genes were picked up by *viruses*, altered to become abnormal alleles, and carried to other cells.

Table 15.2
Some cancers and the tissues that are affected.

Cancer	Affected Tissue
Carcinoma	Epithelial tissues: skin and linings of organs
Adenocarcinoma	Glandular tissues
Hepatoma	Liver
Leukemia	White blood cells in bone marrow
Lymphoma	White blood cells in spleen and lymph nodes
Melanoma	Melanin-forming cells of skin
Myeloma	Bone marrow
Neuroblastoma	Embryonic cells of nervous tissue
Sarcoma	Connective tissues: bone, cartilage, muscle
Osteosarcoma	Bone (often the femur)

Animal Viruses

Because viruses have been key agents in understanding the general genetic basis of malignancy, we discuss them briefly, although they have been implicated in only a few forms of human cancer. Viruses are, however, responsible for a wide range of human diseases other than cancer: AIDS, yellow fever, mumps, chicken pox, measles, rubella, polio, hepatitis, shingles, warts, mononucleosis, influenza, and common colds.

Many viruses consist of little more than a molecule of nucleic acid—DNA or RNA—surrounded by a protein coat. The surface or coat proteins are able to recognize receptors on the cells of their hosts. The specificity of this recognition determines what species, and what cell types, a virus can attach to and invade. Then by reproducing themselves inside the cells of a host organism, the viruses may disrupt the cell's normal metabolism and lead in some cases to cell death. Although not seen until the invention of the electron microscope, the presence of viruses was suggested early in this century as so-called *filterable agents*, particles so small that they passed right through the pores of filters developed by Pasteur to stop the passage of bacteria.

That some viruses can also be *carcinogenic*—that is, cancer causing—was shown early in this century by Peyton Rous while he was a young research assistant at Rockefeller Institute. Rous found that a cell-free filtrate from a ground-up chicken sarcoma produced the same kind of tumor when injected into other chickens. Despite careful, well-designed experiments, his work was generally disregarded because it contradicted prevailing views about the causation of cancer. Not until the 1950s was it generally recognized that viruses could transmit some forms of cancer. In 1966 Rous, at age 86, was awarded a Nobel prize for his pioneering work (Dulbecco 1976).

The *Rous sarcoma virus (RSV)* is one of a family of viruses called **retroviruses**, which have been implicated in many types of animal cancers. A retroviral genome consists of single-stranded RNA, about 10 kb long, that codes for only about a half dozen proteins. One of the proteins is the enzyme **reverse transcriptase**. Immediately after entry of the retrovirus into a cell, this enzyme is used to reverse transcribe the RNA genome into a DNA copy (Figure 15.2). In double-stranded form, the DNA is inserted as a stable **provirus** at a random site of a host chromosome. During successive cell divisions of the host cells, the provirus is replicated and segregated along with the host DNA. The proviral DNA can also be transcribed (in the usual forward direction) into viral RNA and translated into viral proteins, which are packaged into complete viruses. These can escape and infect other cells.

Oncogenes

In addition to, or in place of, their usual genes, some retroviruses carry a gene that is unneeded by the virus but is able to convert normal host cells growing in culture into cancer cells. Such a gene is called an **oncogene** (Greek *onco*, "mass").* The insertion of a single provirus of the Rous sarcoma virus is sufficient to accomplish this **transformation**, the change of cultured cells from normal to malignant. The cancer-inducing gene of RSV is called *src* (for sarcoma); when present in the virus, it is further designated **v-src**.

Where do viral oncogenes come from? And how do they induce cancer? On the question of origin, the oncogene in the Rous sarcoma virus came from a normal chicken cell some time in the past. One line of

* The medical specialty dealing with cancer is called *oncology*.

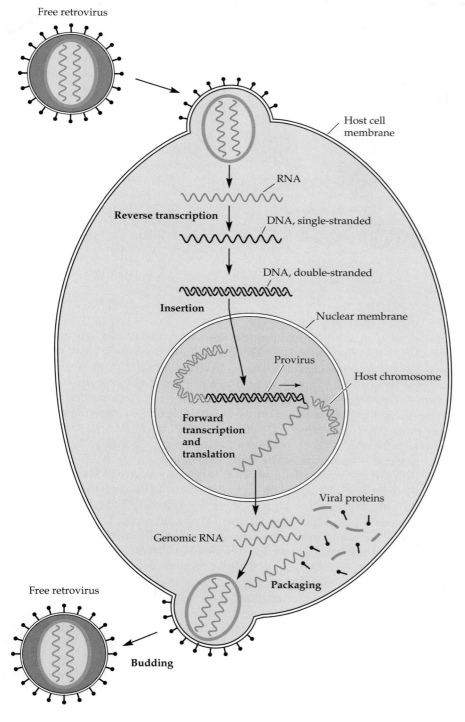

Free retrovirus

Host cell membrane

RNA

Reverse transcription

DNA, single-stranded

DNA, double-stranded

Insertion

Nuclear membrane

Provirus

Host chromosome

Forward transcription and translation

Viral proteins

Genomic RNA

Packaging

Free retrovirus

Budding

Figure 15.2
Schematic outline of the life cycle of a retrovirus. During the provirus stage (center), the cell may divide by mitosis in the usual way, replicating the proviral DNA as if it were host DNA. The expression of the provirus into viral components and the subsequent release of free retroviral particles (bottom) can result in the disruption and death of the cell.

supporting evidence was the demonstration that DNA sequences very similar to v-*src* are present in normal uninfected chicken cells, as well as in the normal uninfected cells of other birds. The bird gene that is similar to v-*src* is called **c-*src***, the **cellular** oncogene; usually, the normal c-*src* gene is called a **proto-oncogene** (Greek *proto*, "first" or "earliest form of "). The presence of the proto-oncogene in many bird species suggests that it is essential to the regular metabolism of all avian cells.

Although the *normal* chicken gene c-*src* is the source of the *abnormal*

Table 15.3

Some oncogenes carried by retroviruses and the homologous human genes.

Oncogene	Animal Source of Virus	Human Chromosome with Homologous Proto-Oncogene/ Oncogene	Human Cancer Implicated
abl	Mouse	9	Chronic myelogenous leukemia
*erb*B	Chicken	7	Unknown
fos	Mouse	14	Unknown
H-*ras*	Rat	11	Bladder carcinoma
K-*ras*	Rat	12	Lung and colon carcinomas
N-*ras*	Unknown	1	Neuroblastoma and leukemia
jun	Chicken	1	Unknown
myc	Chicken	8	Burkitt lymphoma
L-*myc*	Chicken	1	Small-cell lung carcinoma
N-*myc*	Chicken	2	Neuroblastoma
sis	Monkey	22	Unknown
src	Chicken	20	Unknown

viral gene v-*src*, the two cannot be the same. Otherwise, every chicken cell would become cancerous. The general explanation is this: When the provirus (of an invading retrovirus) is inserted into the DNA of an infected cell (see Figure 15.2), it may sit next to a proto-oncogene of the host. Then when the provirus is transcribed to produce new virus particles, part or all of the neighboring proto-oncogene is included in the transcript. But the process is imperfect. Deletions, point mutations, recombination events, or several of these changes combined produce an altered allele—an oncogene—now part of the new retroviruses.

Researchers have found several dozen proto-oncogenes in birds and rodents that were picked up by some retrovirus to become viral oncogenes (Table 15.3). Remarkably, almost all of the animal proto-oncogenes have homologous DNA sequences in the *human* genome. The chromosomal sites of the human genes have been determined by somatic cell fusion or by in situ hybridization, methods described in Chapter 13. Note, however, that the human proto-oncogenes are in no way associated with viruses.

Investigators have now shown directly that specific, dominantly acting mutant genes (i.e., oncogenes) can indeed influence the course of human cancer. One line of experimentation was carried out by Robert Weinberg (1983) and colleagues at the Massachusetts Institute of Technology. These researchers began with DNA extracted from cells of a human bladder carcinoma. Using this DNA, they were able to transform a special line of noncancerous mouse fibroblast cells growing in tissue culture. During the transformation process, short sections of the human DNA entered the mouse cells and were inserted into the mouse DNA, although with low frequency (Figure 15.3). The transformed mouse cells were identified by their abnormal growth behavior: Rather than remaining in a flat layer—one cell thick—on the surface of the nutrient medium, they grew into a mound of cells during a two- to three-week period.

Weinberg was able to identify and clone the piece of human DNA—the oncogene—that was responsible for the transformation. This oncogene hybridized strongly with a DNA sequence in *normal* human cells—the presumed proto-oncogene. Further research showed that the nucleotide sequence of the bladder carcinoma oncogene is similar to that of a rodent oncogene called v-*ras* (for rat sarcoma) carried by a retrovirus. Thus,

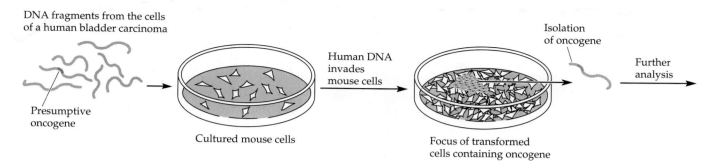

DNA fragments from the cells of a human bladder carcinoma

Presumptive oncogene

Cultured mouse cells

Human DNA invades mouse cells

Focus of transformed cells containing oncogene

Isolation of oncogene

Further analysis

Figure 15.3
The initial steps in experiments that demonstrated the existence of a dominant human oncogene. DNA extracted from a human bladder carcinoma was used to transform mouse cells growing in culture. In cancer research, to transform means to produce in vitro a small focus of tumorous cells growing abnormally into a small mound. The researchers were able to isolate, clone, and sequence the cancer-causing gene from this focus.

it appears that the same proto-oncogene can be changed to oncogenic forms either by some sort of mutational event not associated with viruses or by passage through a retrovirus. In humans, oncogenes from a number of different tumors, including carcinomas of the lung, colon, bladder, and pancreas, are in the *ras* family.

By sequencing both the bladder carcinoma oncogene and its corresponding proto-oncogene, Weinberg and his colleagues (and other investigative teams) were able to pinpoint the mutational event that converted the one into the other. The DNA molecules differed by a single base: G in the proto-oncogene was substituted by T in the oncogene. The change is summarized as follows:

$$\text{proto-oncogene} \xrightarrow[\text{mutation}]{\text{base substitution}} \text{oncogene}$$
$$\text{(recessive allele)} \qquad\qquad\qquad \text{(dominant allele)}$$

The DNA triplet with this mutation specifies glycine in the proto-oncogene protein and valine in the oncogenic product. Studies on the consequences of this change are discussed in the next section.

About 60 oncogenes and their corresponding proto-oncogenes have been located in the genomes of humans and other animals by procedures involving DNA transformation, by studies of oncogenic retroviruses, and by other means. It has been found that a single change by itself is probably never sufficient to cause a cell to become cancerous. Several mutational events—some dominant, some recessive—are required for malignancy. Each misstep in the progression may foster additional abnormalities characteristic of cancer: cell proliferation, invasion of adjacent tissue, metastasis, and establishment of new tumorous sites. Although not all of these events may be directly precipitated by *genetic* damage, recent knowledge of oncogenes has provided a basic unifying mechanism to view the myriad aspects of carcinogenesis.

The Proteins Encoded by Oncogenes

Researchers are beginning to understand the physiological functions of the protein products of proto-oncogenes and oncogenes. Although the genes and proteins that have been studied appear to act in diverse ways, a common thread is emerging (Hunter 1984). Many of the proto-oncogenes code for proteins in information pathways that extend from the cell surface to nuclear genes—specifically those genes that control growth and division. This line of chemical communication from the external environment, across the cell membrane, to the nucleus of a cell is called **signal transduction** (Berridge 1985). When one or more of these chemical signals goes awry because of an oncogenic mutation, the coordination that is necessary for orderly cell division is disrupted, and uncontrolled growth of cells may result.

Beginning the process of signal transduction are proteins called *growth factors* that bind to receptor molecules on the surface of cells. Growth factors encoded by normal proto-oncogenes are delivered to receptors at cell surfaces in a regulated manner, stimulating cell growth only at appropriate times. The oncogene-encoded protein may be produced continually or at inappropriate times or places. Proto-oncogenes also code for the *receptor proteins*. The oncogenic form of a receptor may mislead the cell into thinking that the growth factor is bound to the receptor when it is not—similar to a switch being stuck in the "ON" position. Other membrane molecules involved in signal transduction are called *G proteins*. The *ras* proto-oncogenes referred to earlier code for G proteins (Linder and Gilman 1992).

Many products of oncogenes appear to be **protein kinases** (Greek *kinase*, "to move"), enzymes that modify other proteins. Some receptor molecules or G proteins have kinase activity. The kinases act by *phosphorylation*, that is, by adding phosphate groups (from ATP) to other proteins. The transfer alters the three-dimensional shape of the phosphorylated protein, either increasing or decreasing its activity. One molecule that is phosphorylated, called *vinculin*, is involved in binding cells to surfaces and to each other. Researchers speculate that increased phosphorylation of vinculin in cancer cells may produce their rounded shape and decreased adhesion.

The cascade of events continues into the cell cytoplasm and into the nucleus. A number of oncogenic proteins have been shown to act in cell nuclei as **transcription factors** that turn genes on or off (noted also in the events of early development in Chapter 6). An example is the protein encoded by c-*jun*, which interacts with specific DNA promoter sequences to regulate gene expression. The product of another proto-oncogene, c-*fos*, binds with the c-*jun* product to enhance its activity.

In summary, carcinogenesis is at least partly precipitated by stepwise genetic damage occurring in somatic cells. The mutations from recessive proto-oncogenes to dominant oncogenes alter the encoded proteins, which in turn disturb elaborately balanced signaling systems extending from the cell surface to the nucleus. The general action of a single dominant oncogene (even when heterozygous with its standard allele) is to abnormally accelerate cell division.

Transgenic Mouse Models

Laboratory strains of mice that have a foreign gene stably incorporated into their germline DNA are called **transgenic mice**. One example is the strain of mice carrying the human *HPRT* gene described in Chapter 14. An important new approach to the study of cancer derives from the corresponding ability to insert an oncogene into the germinal cells of a mouse, allowing copies of the oncogene to be transmitted from generation to generation. The study of such mice has been useful in exploring how a particular oncogene affects the multistep process leading to cancer.

To produce a transgenic mouse carrying an oncogene, experimenters may first clone in a bacterial plasmid the DNA sequences constituting the oncogene. A solution with this DNA is then injected via a thin glass needle (and skillful hands) into the nucleus of a fertilized mouse egg (Figure 15.4). The eggs that survive the microinjection are then surgically implanted into a foster mother. Some of the embryos that come to term will have incorporated the oncogene into their own DNA (at random sites) in some or all of their cells. In about 10% of births, the oncogene is also incorporated into the germline and can thus be propagated by appropriate matings with

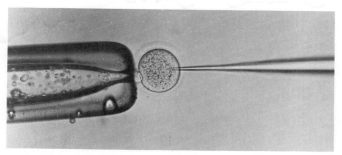

Figure 15.4
Injection of foreign DNA into a fertilized mouse egg. The egg is held on the blunted tip of a glass micropipet by slight suction (left). The injecting microneedle (right) is inserted into the nucleus of the egg, delivering about 1 picoliter (1 pL = 10^{-12} L) of DNA solution. The eggs are in a petri dish placed on the stage of a low-power microscope, and the movement of the needle is controlled precisely through the knobs of a micromanipulator. Successfully injected eggs are then surgically transferred to the oviducts of a foster mother for development. (From Gordon and Ruddle 1983.)

standard strains. The numerous mice descended from a single transgenic progenitor will all carry the same oncogene and will be genetically similar.

The "transplanted" oncogene will often elicit tumors in a reproducible fashion in all members of a transgenic mouse strain. Different tissues appear to be particularly susceptible to transformation by specific oncogenes. For example, the *fos* oncogene expresses itself in transgenic mice by pathological changes in bone and thymus tissues that are similar in nature to those that arise via infection of mice by retroviruses carrying *fos*. Several dozen strains of mice, each transgenic for a different oncogene, have been made and studied.

The development of cancers in the transgenic mice is usually not direct, but is preceded in almost every case by the appearance of proliferating cell lines that are not themselves cancerous. Thus the study of transgenic mice confirms that the presence of an oncogene is a necessary but not sufficient condition for manifestation of malignancy. The primary action of the oncogenes seems to be the loss of restraints on cell division; secondarily, one or more of these cells may become cancerous as the result of further cellular changes.

RECESSIVE SOMATIC MUTATIONS PREDISPOSING TO CANCER

The changes that prod cells along the road toward uncontrolled growth are not all dominantly acting like the mutant oncogenes that we have previously described. Another group of cancer-predisposing genes act in a recessive fashion within somatic cells. That is, the mutant allele must be homozygous to contribute to the disease state. If the mutant allele, when homozygous, allows unrestrained cell growth, then the normal dominant allele must hold growth back (Miller 1990). For this reason, the normal dominant alleles of such genes are called **tumor suppressor genes.** In Table 15.4, we summarize the somewhat cumbersome terminology applied to dominant and recessive cancer genes. About a dozen specific human tumor suppressor genes have now been identified. The first of these genes to be studied in detail was the one that, when mutated, predisposes its carriers to the disease retinoblastoma.

Table 15.4
Terminology and characteristics of the normal and abnormal alleles of cancer genes.

Normal Allele		Allele Predisposing to Cancer
Proto-oncogene (recessive)	Mutation (activation) ⟶	Oncogene (dominant)
Tumor suppressor gene (dominant)	Mutation or loss (inactivation) ⟶	Mutant or missing allele (recessive)

Retinoblastoma and Wilms Tumor

A cancer of the retina of one or both eyes, **retinoblastoma** occurs with a frequency of about 1 in 20,000 persons. The malignant tumors develop primarily in young children, because cancerous changes almost always start in *dividing* cells, and retinal cells stop dividing early in life. (The term *blast* means "immature cells.") Studies of cultured retinoblastoma cells suggest that the cancers develop specifically in the color-perceiving cone cells.

A key symptom of the disease is the glassy appearance of the pupil of the eye in which the tumor is growing (Figure 15.5). If untreated, the cancer expands along the optic nerve to the brain; it may also metastasize to the skeletal system, liver, or other organs, thereby causing death. If the eye tumors are detected early enough, however, and the affected eye or eyes are removed by surgery or treated by cryotherapy, chemotherapy, and radiation, patients may live to adulthood and parenthood (Angier 1987).

The disease has a very unusual pattern of occurrence: About 40% of cases run in families (owing to an inherited gene called *Rb*), while 60% of cases occur sporadically and are not inherited. Researchers usually find that inherited cases involve tumors in both eyes, whereas noninherited cases involve just one eye. We will now discuss why this might be expected.

Some of the *heritable* cases are accompanied by a small visible deletion in the long arm of one chromosome 13. These deletions vary in size; but they all include the band q14 (Figure 15.6A). Persons who inherit either the q14 deletion or a mutant *Rb* allele (that is not cytologically visible) have it on just one chromosome 13 in all cells. These cells are not cancerous, although the normal function of one *Rb* locus is presumed lost. Alfred Knudson, of the Fox Chase Cancer Center in Philadelphia, suggested in the 1970s that a second mutation occurring at the corresponding locus on the other chromosome 13 in any retinal cell is the precipitating factor in the retinoblastoma malignancies. Thus, the cancers develop only when a particular retinal cell becomes doubly abnormal for the mutant *Rb* allele, the deletion, or one of each (Figure 15.7A). Thus, the expression of retinoblastoma at the cellular level is a *recessive* trait, requiring abnormal states at both of the relevant q14 sites on chromosome 13.

The likelihood of the second somatic mutation in inherited retinoblastoma is sufficiently high that several tumors develop, thus explaining why these cases usually affect both eyes. Note that loss of function of both q14 sites on chromosome 13 can occur either via two mutations at the nucleotide level, two visible deletions, or one of each. In addition,

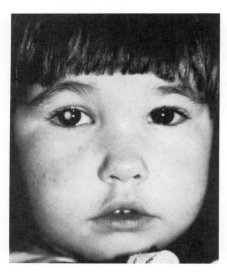

Figure 15.5
Light reflected off the surface of an eye tumor in a young girl with retinoblastoma. (Courtesy of B. L. Gallie, The Hospital for Sick Children, Toronto.)

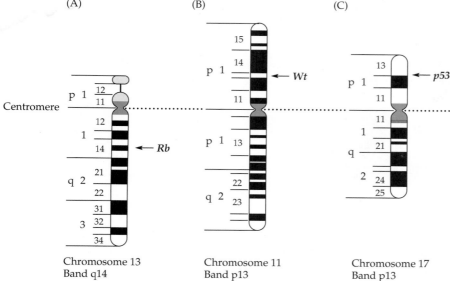

(A)

Centromere

Chromosome 13
Band q14

(B)

Chromosome 11
Band p13

(C)

Chromosome 17
Band p13

Figure 15.6
The chromosomal locations (arrows) of three tumor suppressor genes. (A) Retinoblastoma gene, Rb. (B) Wilms tumor gene, Wt. (C) p53 gene. Deletion or mutation of these sites is often involved in the development of cancer—of the colon, breast, lung, and other organs in the case of p53.

somatic nondisjunction resulting in the loss of an entire chromosome 13, or other cytological events, can produce the second "mutation" needed for malignancy.

Although the expression of the retinoblastoma gene is *recessive* at the level of somatic cells, the disease is transmitted between generations in a *dominant* fashion. This oddity is because the germline deletion (or mutation) of the gene at band q14 is inherited by half the offspring of a carrier parent. The error is then reproduced in all the cells of the child's body, both somatic and germline. The child develops retinoblastoma through a second mutation—a somatic mutation—with high probability (about 90%).

In retinoblastoma that is not inherited, both chromosomes 13 in the zygote are normal at the relevant q14 locus (Figure 15.7B). The mutational or cytological changes of the *Rb* loci must occur on *both* chromosomes 13 in the *same* retinal cell for retinoblastomas to develop. Such persons typically have a tumor in only one eye—a tumor that develops later in childhood, as expected by the very low probability of two independent rare events occurring in a single cell. The idea that several steps are needed to produce the tumors of retinoblastoma has been a model for explaining the general multistage development of cancer.

The DNA region including the *Rb* gene has been identified, cloned, and sequenced. The Rb protein product is phosphorylated; that is, it has phosphate groups attached to the side chains of some of the amino acids. The number of phosphate groups varies in a set pattern during the cell cycle. The Rb phosphoprotein is found in cell nuclei, where it associates with the protein products of several proto-oncogenes to regulate the transcription of cell cycle genes. Scientists are now studying the protein kinases that phosphorylate the Rb protein.

The *Rb* gene appears to be transcribed in cells throughout the body, and it is not clear why the mutant allele affects the retinal cone cells in particular. That *other* cells can also be targeted is suggested by the increased risk of bone cancers among the survivors of retinoblastoma. In addition, some persons who do not get retinoblastoma but develop certain forms of breast, lung, and other cancers have structural abnormalities of the retinoblastoma gene. Thus, the *Rb* gene may play a key role in susceptibility to several forms of cancer.

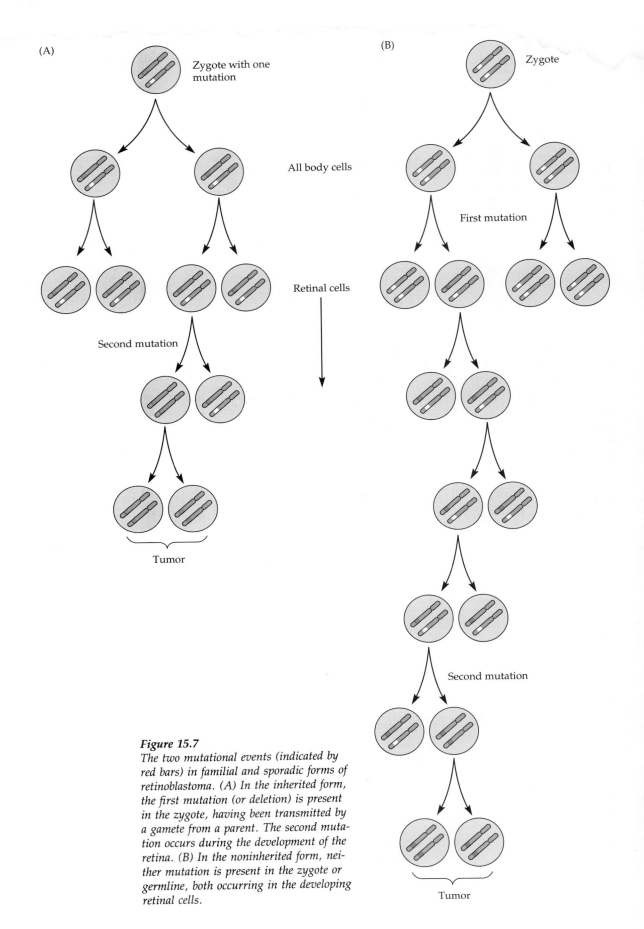

(A) Zygote with one mutation

All body cells

Retinal cells

Second mutation

Tumor

(B) Zygote

First mutation

Second mutation

Tumor

Figure 15.7
The two mutational events (indicated by red bars) in familial and sporadic forms of retinoblastoma. (A) In the inherited form, the first mutation (or deletion) is present in the zygote, having been transmitted by a gamete from a parent. The second mutation occurs during the development of the retina. (B) In the noninherited form, neither mutation is present in the zygote or germline, both occurring in the developing retinal cells.

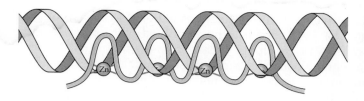

Figure 15.8
The product of the Wilms tumor gene is a zinc-finger protein with four loops (color). The zinc atoms bend the protein into the fingerlike projections by binding to specific amino acids along the length of the chain. Zinc-finger proteins may interdigitate with specific sequences of the DNA double helix to regulate transcription.

Another cancer, called **Wilms tumor**, develops in the embryonic cells of one or both kidneys. Occurring in about 1 in 10,000 children under the age of seven, Wilms tumor is often curable by surgery, radiotherapy, and chemotherapy. Its pattern of occurrence and the types of genetic changes closely resemble those in patients with retinoblastoma. For example, a noninherited form usually affects one kidney; a less common, inherited form due to a dominant germline gene *Wt* may affect both kidneys. In a few patients with inherited Wilms tumor, a visible deletion occurs in all somatic cells in band p13 of one chromosome 11 (Figure 15.6B).* As with retinoblastoma, the development of tumors is usually contingent on the occurrence of the abnormality (deletion or mutation) at both *Wt* loci on homologous chromosomes.

The Wilms tumor gene has been cloned, and its protein product is also a transcription factor. It belongs to a class called **zinc-finger proteins**, in which zinc atoms bind to the protein and bend the molecule sharply into a finger shape (Rhodes and Klug 1993). The *Wt* protein has four such fingers that interlock with the turns of DNA molecules (Figure 15.8), binding specifically to the sequence CGCCCCCGC.

For both retinoblastomas and Wilms tumors, scientists have noted a peculiar nonrandom inheritance of the mutant allele. In both cases, the germline mutation (or deletion) is usually on the chromosome inherited from the father, and the subsequent somatic damage that leads to homozygosity and cancer is on the chromosome inherited from the mother. Similarly, in noninherited cases, the first somatic event is usually on the paternal chromosome and the second event on the maternal chromosome. Such apparent nonrandom events are confusing, since there seems to be no reasonable explanation. One possibility is *imprinting*, which was initially discussed in Chapter 6. Genes of maternal or paternal origin may be differentially methylated, producing perhaps different responses to the cell environments.

Li-Fraumeni Syndrome and Gene p53

In 1969 two researchers at the National Cancer Institute, Frederick Li and Joseph Fraumeni, described a syndrome in which individuals are extremely susceptible to not just one type of cancer but to a great many: cancers of the breast, brain, bone, white blood cells, as well as other sarcomas and carcinomas. The cancers usually develop in children and young adults, and if the victims survive a first bout with one type of cancer they often are afflicted with a second type. The devastating **Li-Fraumeni syndrome (LFS)** runs in families in an unmistakable autosomal dominant fashion (Figure 15.9). The disease is very rare, only about 100 families being known worldwide.

In a different line of investigation during the 1980s, cancer investigators discovered another tumor suppressor gene located in band p13 of chromosome 17 (Figure 15.6C). It was called *p53* because it coded for a

---

* When the deletion is present, additional effects are often seen: absence of the irises of the eyes (aniridia), abnormalities of the genital and urinary systems, and mental retardation. It is presumed that the p13 deletion on chromosome 11 includes genetic material affecting these other aspects of the phenotype.

Family 1

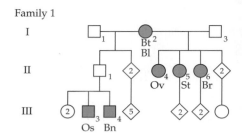

Family 2

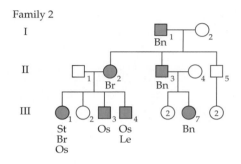

Bl = Bladder cancer
Bn = Brain cancer
Br = Breast cancer
Bt = Carcinoma of the bile duct
Le = Leukemia
Os = Osteosarcoma
Ov = Ovarian cancer
St = Soft tissue sarcoma

Figure 15.9
Two families with the Li-Fraumeni syn-
drome, showing autosomal dominant in-
heritance (Malkin et al. 1990). As noted
on the pedigrees, affected individuals had a
great variety of cancers. In family 1, the
age of onset of the cancers was from 16 to
53; in family 2, from age 2 to 38.

protein of molecular weight 53,000. Mutations of the *p53* gene (when they became homozygous) were shown to be associated with many sporadic (noninherited) cancers of the lung, breast, colon, esophagus, bladder, brain, bone, and white blood cells. Since these sporadic cancers matched some of the common ones in Li-Fraumeni patients, *p53* became a so-called *candidate gene*, one that could possibly account for the cancers in LFS pedigrees.

With this clue, researchers showed that patients with the Li-Fraumeni syndrome indeed have germline mutations of the *p53* gene. To do this, they extracted DNA from normal and cancerous cells of patients with LFS, amplified the *p53* gene by the polymerase chain reaction, and sequenced the amplified product to identify any mutations. Affected individuals in family 1 (Figure 15.9), for example, were shown to have a particular germline base substitution mutation in one copy of the *p53* gene; the other *p53* allele was missing altogether in the tumorous tissues. In family 2, the germline mutation was a different base substitution. The story of mutations and deletions knocking out the function of both copies of a tumor suppressor gene was similar to that in retinoblastoma and Wilms tumor.

The *p53* mutations are present not only in cancer-prone families with the Li-Fraumeni syndrome, but also (as already noted) in many sporadic cancers. In fact, a *p53* alteration is the most common genetic change seen in human cancers, occurring (as one of several mutational steps) in perhaps half of all human cancers, especially those of the colon, breast, and lung. But the particular *p53* mutation varies with particular cancers.

In line with the general notion that several genetic changes are needed for the development of cancer, the *p53* mutation is not the only one required for the development of cancer. The stepwise progression in noninherited colon cancers has been studied in detail by Bert Vogelstein and colleagues at the Johns Hopkins Oncology Center in Baltimore and by other groups (Radetsky 1991). As illustrated in Figure 15.10, mutations of proto-oncogenes and tumor suppressor genes can be correlated with the clinical progression of tissue changes leading to malignant colon tumors. A half dozen genetic mutations have been discovered that are needed to produce colon cancer.

The first such change leading to noninherited colon cancer is a small deletion in chromosome 5. This site also appears to harbor a dominant mutation that leads to an inherited form of colon cancer called *familial adenomatous polyposis*. In the inherited disease, which strikes about 1 in 5,000 persons, thousands of tiny blebs grow in the lining of the colon. These benign polyps may progress to malignant growths unless the colon is removed. Thus, like other types of cancer, colon cancer exists in more common sporadic forms (accounting for perhaps 80% of colon cancers) as well as less common inherited forms. A similar situation exists for breast cancer; at least 90% occur sporadically, while the remainder (usually earlier-onset breast cancer) is inherited, with some cases involving mutations of the tumor suppressor gene *p53* (Wright 1990).

In summary, tumor suppressor genes such as *Rb*, *Wt*, and *p53* in their unmutated dominant forms act as brakes on cell division. Mutations of these genes to recessive alleles promote cancer in somatic cells only when they become homozygous and release the brake. Two complications are seen: (1) The mutations may sometimes occur in the germline, predisposing several family members to a particular cancer; more usually, however, the mutations occur only in somatic tissue, producing noninherited, sporadic cases of the cancer. (2) In the familial cases, the inheritance from one generation to the next follows a dominant pattern; at the cell level, however, these cancers develop in a recessive fashion, that is, only when the cancer-promoting allele is homozygous.

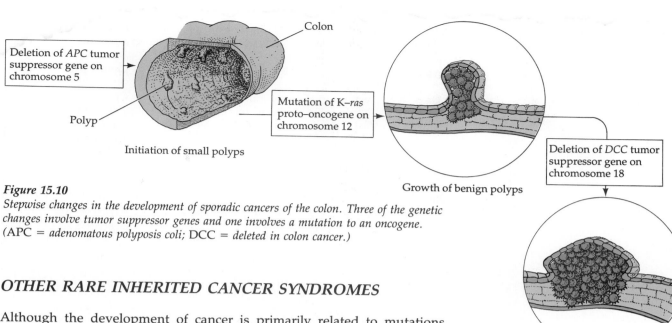

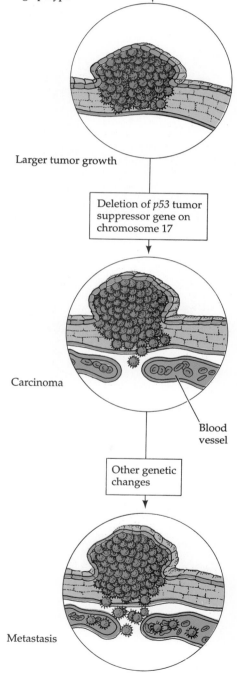

Figure 15.10
Stepwise changes in the development of sporadic cancers of the colon. Three of the genetic changes involve tumor suppressor genes and one involves a mutation to an oncogene. (APC = adenomatous polyposis coli; DCC = deleted in colon cancer.)

Labels in figure: Colon; Deletion of *APC* tumor suppressor gene on chromosome 5; Polyp; Initiation of small polyps; Mutation of K–*ras* proto–oncogene on chromosome 12; Growth of benign polyps; Deletion of *DCC* tumor suppressor gene on chromosome 18; Larger tumor growth; Deletion of *p53* tumor suppressor gene on chromosome 17; Carcinoma; Blood vessel; Other genetic changes; Metastasis

OTHER RARE INHERITED CANCER SYNDROMES

Although the development of cancer is primarily related to mutations occurring in somatic cells, our previous examples included cancer-causing mutations that are carried in eggs or sperm and copied to all cells of the developing child. What is inherited in these "inherited cancers" are genetic defects that increase the likelihood of cancerous changes in at least one of a person's trillions of somatic cells. Some additional inherited cancer syndromes show either simple Mendelian ratios or less distinct transmission patterns that exhibit only a small increase in risk above the general population. Each of the cancers with a familial component is uncommon, and collectively they constitute only about 20% of all malignancies.

Among the cancers that run in families are disorders in which cells of the body are defective in DNA replication, repair, transcription, or recombination. These routine processes require enzyme systems that recognize and correct base-pairing errors in DNA, or breaks in the sugar-phosphate backbone, or abnormal bonding of the bases. When these processes go awry, the likelihood of cancer may rise. About a half dozen autosomal recessive disorders are referred to as *chromosome breakage syndromes* or *DNA repair syndromes*. Four of them are presented in Table 15.5, and one is discussed in greater detail here. The causation of the variable phenotypes is only partly understood because many of the genes and gene products are not precisely characterized.

Ataxia-Telangiectasia (AT)

Occurring with a frequency of about 1 in 40,000 children, **ataxia-telangiectasia (AT)** is an autosomal recessive disease that is pleiotropic and variable in expression. Although normal at birth, affected infants soon begin to display the diverse symptoms. *Ataxia* refers to loss of muscle control (often beginning with an unsteady gait) due to progressive damage in the cerebellum, the portion of the brain concerned with posture, balance, and coordination. Patients may later be confined to wheelchairs and have difficulty speaking. *Telangiectasia* refers to skin redness (especially on the eyes, ears, and neck) resulting from tiny dilations of capillaries. Immunological deficiencies are common, contributing to a high frequency of lung infections, a major cause of early death. About 15% of patients develop malignancies, especially lymphomas and leukemias, but also carcinomas of the stomach, ovary, breast, mouth, and other organs.

A distinctive feature of AT is extreme sensitivity to ionizing radiation

Table 15.5
Chromosome breakage syndromes inherited as autosomal recessive disorders.

Syndrome	Clinical Appearance	Chromosome Effects
Ataxia-telangiectasia	Loss of muscle control; skin redness; immunological problems; predisposition to cancers, especially leukemias; extreme sensitivity to X-radiation	Chromatid breaks; formation of dicentric and acentric chromosomes
Xeroderma pigmentosum	Sensitivity to sunlight, leading to multiple metastasizing skin cancers; mental deterioration	Defective DNA repair, producing abnormal crossovers between sister chromatids
Bloom syndrome	Dwarfism, disfiguring facial rashes made worse by sunlight; immunological problems; predisposition to cancers, especially leukemias and lymphomas	Chromosome breaks; abnormal crossovers between sister chromatids
Fanconi anemia	Decreased blood cells; short stature; arm malformations; patchy dark skin pigmentation; predisposition to cancer, especially leukemia	Chromatid breaks, gaps; exchange of segments between chromosomes

(but not to ultraviolet light). For example, X-rays used for therapeutic treatment of the cancers in AT patients often produce devastating death of normal tissues. X-radiation also easily kills cultured cells from AT patients. In the cells that survive, X-ray-induced DNA breaks in one or both strands of the double helix may go unrepaired. Resulting chromosomal breaks, acentrics, dicentrics, and rearrangements are common (Figure 15.11). Anomalies are also seen in the replication of DNA. Observations such as these point to abnormalities in DNA processing or repair systems among AT patients, although the precise biochemical defects are not yet known.

Different families with ataxia-telangiectasia may be variable in their responses to ionizing radiation. Part of this variation apparently stems from different genetic mutations. Genetic heterogeneity can be demonstrated in somatic cell hybrids formed from different AT patients. Restoration of normal function following X-radiation in such cell fusions indicates that whatever is wrong in the one cell line is corrected by the other, and vice versa. At least four groups have been delineated, a finding suggesting that at least four different genes can produce somewhat similar AT phenotypes.

Although heterozygotes for AT are generally healthy, they are more susceptible to cancer, especially breast cancer, than people in the general population. In a recent study (summarized by Ezzell 1992), researchers at the University of North Carolina at Chapel Hill examined women who were very likely heterozygous for AT; that is, they were closely related to affected AT homozygotes.* These females developed breast cancer about

---

* Since the recessive AT gene has not been mapped or cloned or otherwise characterized, there is at present no direct method to test for heterozygosity.

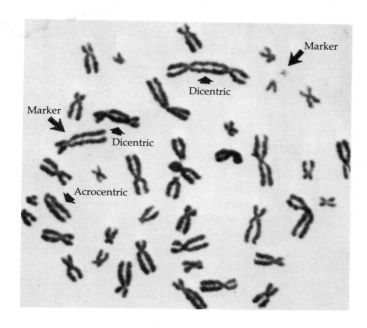

Figure 15.11

Abnormal chromosomes in a white blood cell from a patient with ataxia-telangiectasia. The karyotype includes two dicentric chromosomes, one abnormal acrocentric, and two unusual "marker" chromosomes (that have a subtle characteristic feature apparent to a cytologist). (From Harnden 1974.)

five times more frequently than women in a matched control group consisting of normal homozygotes. Knowing that about 1% of randomly chosen persons is likely to be heterozygous for AT, the researchers calculate that about 10% of all patients with breast cancer in the United States would be expected to be heterozygous for AT.

Heterozygotes for AT are particularly vulnerable to the induction of cancer by radiation. This finding raises questions about the advisability of routine mammography for all women. Although the X-ray dose during mammography is very low and AT carriers have been proportionately represented in numerous studies that have concluded that mammography, properly performed, is safe, the risk-benefit ratio specifically for AT heterozygotes is uncertain.

CANCER CYTOGENETICS

For many years, researchers have seen in cancerous tissues abnormalities of chromosome structure or chromosome number. The karyotypes of the cancer cells may become even more abnormal over time, a trend that signals a poorer prognosis for patients. But from the time of inception of the malignancy, some *specific* aberrant chromosomes are consistently present in some *specific* cancers. Thus, a structural change in a chromosome can itself contribute to cancer induction. This discovery was first made in 1960 at the University of Pennsylvania in Philadelphia and focussed on the role of the so-called *Philadelphia chromosome* in leukemia. More recently, this aberration has been shown to affect the activity of a proto-oncogene at one of the breakpoints of a reciprocal translocation.

Proto-Oncogenes at Translocation Breakpoints

The various leukemias are cancers of immature white blood cells. Induced in part by ionizing radiation, leukemia syndromes occur with increased incidence among radiologists exposed to excessive amounts of X-radiation, among patients receiving large doses of therapeutic radiation, and among

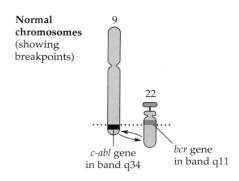

Normal
chromosomes
(showing
breakpoints)

9

22

c-abl gene
in band q34

bcr gene
in band q11

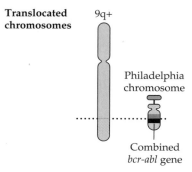

Translocated
chromosomes

9q+

Philadelphia
chromosome

Combined
bcr-abl gene

Figure 15.12
The reciprocal translocation between chromosomes 9 and 22 that produces the Philadelphia chromosome characteristic of chronic myelogenous leukemia. Breakage and reunion join together the c-abl proto-oncogene on chromosome 9 with a so-called breakage cluster region (bcr) on chromosome 22.

the people exposed to the atomic bomb explosions. In *myelogenous leukemia*, the cancer proliferates from stem cells (myeloblasts) in bone marrow that mature into so-called granular white blood cells. In *lymphocytic leukemia*, the cancer proliferates from stem cells (lymphoblasts) that mature into so-called B and T immune cells (Chapter 16). Additional leukemias arise as a result of the abnormal progression of other types of white blood cells.

In the bone marrow, the massive buildup of leukemic cells crowds out mature red blood cells (thereby producing anemia) and platelets (thereby producing internal bleeding). In addition, the liver, spleen, and other vital organs are damaged by large numbers of infiltrating cancer cells. Advance of the disease can either be slow (*chronic*) or rapid (*acute*), the latter type leading to death in a few weeks or months if untreated. A frequent cause of death in acute leukemias is uncontrolled infection due to the deficiency of mature white blood cells.

Chronic myelogenous leukemia (CML) affects mainly people of middle and older ages and is invariably fatal. It is easily recognized because the leukemic cells usually have a distinctive chromosomal marker, the so-called **Philadelphia chromosome**, which is a shortened chromosome 22 originally thought to result from a simple deletion. The introduction of banding techniques, however, revealed that end pieces of the long arms of chromosomes 9 and 22—unequal in size—had switched places through a reciprocal translocation (Figure 15.12). Blood examinations of the atomic bomb survivors who developed CML show that about eight years pass between the original translocation that produces the Philadelphia chromosome and the development of clinical symptoms.

The regions around the breakpoints of chromosomes 9 and 22 have been cloned and sequenced. The breakpoint in chromosome 9 occurs in the cellular proto-oncogene c-*abl*. The breakpoint in chromosome 22 occurs in a gene that is called the *breakage cluster region (bcr)*, but whose specific function is not known. The fused gene in the Philadelphia chromosome is transcribed and translated into an abnormally long protein whose immediate effect in cells is a matter of speculation.

Another translocation with a breakpoint near another proto-oncogene, c-*myc* on chromosome 8, is a consistent feature of **Burkitt lymphoma (BL)**. Occurring commonly in parts of central Africa (about 1 case per 10,000 children per year) and rarely in other regions of the world, BL is a malignancy of B lymphocytes, which become antibody-producing cells. In a wide belt of equatorial Africa, it is the most frequent cancer among children, the average age at onset being about seven years. First described in 1958 by Denis Burkitt, a British surgeon working at Makerere University in Uganda, the solid tumors typically affect two areas: the bones of the jaws and the organs in the abdomen. (Unlike leukemia, the bone marrow is not usually involved.) Extremely fast growing and aggressive, the tumors can reach prodigious size in a matter of days and obstruct neighboring organs (Figure 15.13). The kidneys may also be damaged by the excessive elimination of uric acid, a by-product of the breakdown of DNA, which results from the rapid growth and death of cancer cells. Fortunately, treatment with anticancer drugs often gives dramatic improvement and leads to long-term survival in about half the cases.

The malignant immune cells of BL patients have a translocation involving the c-*myc* proto-oncogene at band q24 on chromosome 8. The other participant in the translocation in most cases is chromosome 14, or less commonly chromosome 22 or 2 (Figure 15.14). The breakpoints in the latter three chromosomes are all within genes that code for parts of antibody molecules (Croce and Klein 1985).

As with the *bcr-abl* fusion gene, the juxtaposed genes of c-*myc* on

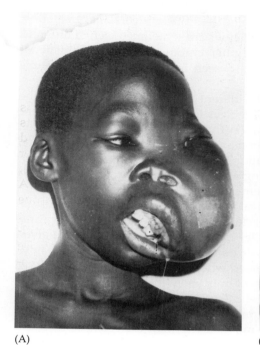 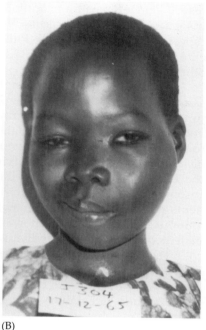

(A)   (B)

Figure 15.13
(A) A massive jaw tumor characteristic of Burkitt lymphoma in a nine-year-old Ugandan girl. (B) The same child 3½ weeks after two injections of an anticancer drug, cyclophosphamide. (From Owor and Olweny 1978.)

chromosome 8 and one of the antibody genes have been analyzed in detail at the nucleotide level. Unlike the *bcr-abl* protein product, however, the normal c-*myc* protein remains intact and unattached to any other polypeptide. As a result of its new neighbor, however, c-*myc* becomes *deregulated*, being transcribed and translated at inappropriate times or at increased rates. This deregulation is thought to be one of the steps required to produce the malignant transformation, but it is probably not the only step.

Other Specific Chromosomal Abnormalities

In addition to chronic myelogenous leukemia and Burkitt lymphoma, many other cancers of blood cells are associated with chromosomal abnormalities, although usually with less consistency. The most common

Chromosome:	8	14	22	2
Breakpoint band:	q24	q32	q11	p11
Breakpoint gene:	c-*myc*	*IgH*	*Igλ*	*Igκ*
Approximate percentage of cases:	100	89–90	5–15	5

Figure 15.14
The breakpoints and genes (solid rectangles) involved in the translocations producing Burkitt lymphoma (shown prior to the rearrangements). In all cases, a breakpoint occurs in chromosome 8 at the c-myc proto-oncogene. The other breakpoint (to the right of the dotted line) occurs in either chromosome 14, 22, or 2 at the genes, IgH, Igλ, or Igκ. These three genes code for the polypeptides of antibody molecules (see Chapter 16).

367

rearrangements are translocations, and less often inversions, deletions, trisomies, and so on. The same aberrations are sometimes seen in different cancers. For example, the Philadelphia chromosome characteristic of chronic myelogenous leukemia is also seen in the cancer cells of some other leukemias.

On the other hand, several different aberrations may lead to the same cancer. In these cases, the prognosis for patients and their response to treatment may vary, depending on the specific chromosomal defects. For example, Jorge Yunis, of Hahnemann Medical School in Philadelphia, found 17 different types of aberrations among 99 patients with *acute non-lymphocytic leukemia*. Those carrying an inversion of chromosome 16 had the best prognosis, many of them surviving many years. Those with trisomy 8 had intermediate survival—about one year. A number of patients with complex chromosomal defects usually lived only two or three months. The hope is that knowledge of the specific aberrations in a given patient can lead to more individualized therapy.

The leukemias and lymphomas whose chromosomal aspects have been so intensively studied represent only about 8% of all human cancers. Less is known about the cytogenetic aspects of the other 92%, including the common solid tumors of the breast, lung, colon, prostate, and pancreas. Because they divide much more slowly than the cells of blood cell malignancies, the cells of most solid tumors are more difficult to grow in culture and less likely to be seen during the mitotic divisions needed to obtain karyotypes. Another problem is that solid tumors of the same type have a range of chromosomal defects, sometimes even within the same tumor (Figure 15.15). Dozens of different secondary changes, often characterized by a great many extra chromosomes, may obscure a primary karyotypic event that led directly to the cancerous transformation.

The idea that secondary chromosomal changes in cancer cells can lead to greater malignancy has been confirmed by the discovery of repetitive elements that contain an oncogene. In some lung cancers and in some cancers of neural tissue (*neuroblastomas*), researchers have found the *myc* proto-oncogene repeated many times—a phenomenon sometimes called *gene amplification*. The increase in copy number of the *myc* proto-oncogene often occurs in later stages of the cancers, probably contributing to the progression to more highly malignant forms.

We note finally that the chromosomal changes discussed here occur in the *cancerous* cells but *not* in the normal cells of the body and *not* in the germline, where they might be included in eggs or sperm.

ENVIRONMENTAL FACTORS

Although this chapter has been devoted primarily to the *genetics* of cancer, we emphasize again that the disease is *not* usually transmitted from parents to offspring in the manner of typical heritable traits due to either single genes or the combined effects of several genes. All cancer is probably genetic, however, in the sense of something being transmitted from one *cell* generation to the next during the development and maintenance of a person's organs and tissues. *Thus, the genetics of cancer is played out primarily in somatic rather than germinal cells.* The environmental agents that initiate cancer in somatic cells are called **carcinogens**, and most of these apparently act by causing mutations—perhaps mutations in proto-oncogenes or tumor suppressor genes. In fact, about 90% of carcinogens (revealed by animal testing) are mutagens (demonstrated by the Ames test; Chapter 12).

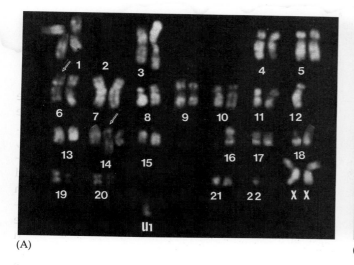

(A)

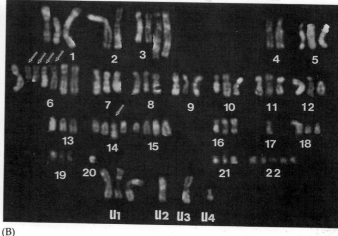

(B)

Figure 15.15
Two cells that are derived from the same ovarian cancer but show different karyotypes. (A) A cell showing a 6;14 translocation (arrows) and a small chromosome of unknown origin (U1), with a total of 45 chromosomes. (B) A cell showing the same 6;14 translocation, several chromosomes of unknown origin (U1–U4), and a total of 77 chromosomes. The translocation is thought to be the primary cancer-causing cytogenetic change. (From Sandberg 1988.)

Identifying environmental causes of cancer in humans is usually difficult, because years or even decades may pass between the action of carcinogens and the clinical detection of tumors. A clear illustration of this time lag involves the sex hormone diethylstilbestrol (DES). First administered to women during the 1940s, when it was thought to help maintain pregnancy, DES was later found to be ineffective, even detrimental, for this purpose. Among the teenage daughters resulting from these pregnancies—perhaps a million of them—about 1 in 1,000 has developed vaginal cancer, a type of malignancy that is otherwise extremely rare in young women. The clinical effect was clearly delayed for nearly two decades, because these cancers were initiated prior to birth. Note that if DES had instead induced a more common type of cancer, such as lung or breast cancer, its carcinogenic effect might never have been detected.

Extensive testing of substances for their carcinogenic potential stems in part from the 1958 Delaney amendment to the Food, Drug, and Cosmetic Act, which bars from the market any additive found to cause cancer in experimental animals, *whatever* the dosages tested.* Because (1) the animal tests must be conducted within a reasonable time on a finite budget and (2) the cancerous transformation of any given cell is a rare event, the experimental animals must be exposed to very high concentrations of the test agent. Only in this way is the testing procedure practical. Although the extrapolation from rodent bioassays based on high doses to humans exposed to low doses is far from direct, some reliance on animal tests is unavoidable.

Most investigators—epidemiologists and others—think that about 80% of all cancers are caused by various environmental factors, including life-style considerations (Table 15.6). At the top of the list are tobacco and diet. Cigarette smoke contains about 50 known carcinogens and by itself accounts for perhaps 30% of all cancer deaths. Smoking is the greatest single *preventable* cause of cancer.

Dietary factors are thought to be responsible for another 30% of cancer deaths, although the uncertainty in this figure is greater than that for tobacco smoke. Many food plants manufacture their own toxic chemicals to ward off insects and other attackers. A number of these substances

---

* Some critics claim that trying to legislate a 100% risk-free environment is simply unrealistic. Furthermore, the Delaney amendment targets only a small part of the carcinogenic burden, ignoring, for example, naturally occurring carcinogens and toxins.

Table 15.6
Percent of cancer deaths attributable to different environmental factors.

Factor or Class of Factors	Best Estimate of Percent of Cancer Deaths	Range of Acceptable Estimates
Tobacco	30	25–35
Diet	30	10–70
Reproductive and sexual behavior	7	1–13
Occupation	4	2–8
Alcohol	3	2–4
Pollution		
Atmospheric	1	<1–5
Water	<1	
Ionizing radiation		
Background	1	
Medical procedures	0.5	1–2
Industry	<0.1	
Industrial products	<1	1–2
Medical drugs	<1	<1–2
Ultraviolet light	0.5	<0.2–1
Food additives	<1	−5–2
Other and unknown	>10(?)	?

Source: After Doll and Peto (1987).

are animal carcinogens, or they are converted to carcinogens by storage or preparation of foods or by metabolic events in the body. Examples include *hydrazines* in mushrooms, *piperines* in black pepper, and *pyrrolizidines* in herbal teas. Ames (1984) states that we "are eating more than 10,000 times more of nature's pesticides than of man-made pesticides." On the other hand, some dietary components (including *vitamin E*, *β-carotene*, and *selenium*) have been proposed as *protective* agents against cancer. The health consequences of eating this or that type of food—for example, the fats and fiber that one can hardly escape hearing about from food companies and government agencies—are continuing areas of research (Cohen 1987).

Chemicals and radiation in the workplace and in the general environment are responsible for just a few percent of cancers. We hear a lot about some of the carcinogenic agents in these categories: metals, asbestos, and compounds used in the manufacture of pesticides, plastics, dyes, and paints. Pollution, food additives, alcohol, and medical drugs also appear to be responsible for only a few percent of cancers.

Certainly, we need to know more about both the genetic and environmental factors that contribute to the genesis of cancer cells. Although the common cancers—those of the colon, breast, and lung—have no discernable heritable basis generally, there are subsets of each type of cancer that appear in increased frequency within families. The relationships between substantially environmental types of cancer and substantially heritable types, as well as the roles of proto-oncogenes, tumor suppressor genes, and their protein products, need to be worked out. Many initially unconnected lines of investigation are now converging toward a fuller understanding. It is hoped that the new knowledge—especially at the molecular level—will eventually open up ways to alleviate the great suffering that accompanies cancer.

1. Cancerous cells reproduce without end. They lose their normal form and function, invade nearby tissue, and spread to remote sites. The cells of benign tumors also reproduce without end, but they remain localized. Tumors usually stem from single cells and progress through several stages as they become more malignant.

2. Dominantly acting oncogenes were discovered in retroviruses, which use reverse transcriptase to copy their genetic information from RNA to DNA. The DNA copy, the provirus, is inserted into the DNA of their host. Oncogenes in retroviruses were derived from normal proto-oncogenes in host species (often birds or rodents) sometime in the past.

3. Human oncogenes have been detected directly by using DNA taken from human cancer cells to transform mouse cells. Some dominant human oncogenes differ from their normal recessive alleles (the proto-oncogenes) by single base substitutions.

4. The protein products of proto-oncogenes are involved in cell growth and division—especially in the chemical pathway from the cellular surface to the nuclear genes. A number of proto-oncogenic proteins are kinases that alter the activity of other proteins.

5. Transgenic mice that carry oncogenes in all their cells are being used to study the development of cancer.

6. Retinoblastoma and Wilms tumor are cancers that can be either dominantly inherited from a parent or not inherited at all. In the inherited form, all somatic cells receive (through successive mitoses) one DNA lesion, a mutation or deletion of a tumor suppressor gene. A second somatic lesion at the homologous locus is required to initiate the cancer. Although the diseases are transmitted through the germline as dominant traits, they are expressed at the cellular level in a recessive manner.

7. In the noninherited form of retinoblastoma and Wilms tumor, two somatic mutations or deletions (at homologous loci in the same somatic cell) are required to initiate cancer.

8. A dominantly inherited mutation of the tumor suppressor gene *p53* is present in families with the Li-Fraumeni syndrome. Many non-inherited cancers, especially of the colon, breast, and lung, also involve a *p53* mutation that becomes homozygous in a somatic cell.

9. Ataxia-telangiectasia is an example of a recessively inherited disease that shows increased cancer susceptibility due to defects in DNA repair. Ataxia-telangiectasia involves enzymes that repair X-ray-induced damage.

10. In chronic myelogenous leukemia, a translocation with one breakpoint at the *abl* proto-oncogene produces the Philadelphia chromosome. The protein product of *abl* is thereby altered.

11. In Burkitt lymphoma, a translocation has one breakpoint at the *myc* proto-oncogene and the other breakpoint at one of three different antibody genes in three different chromosomes. The protein product of *myc* is not altered, but it is expressed at the wrong times or in the wrong amounts.

12. Researchers sometimes find the same chromosomal aberrations in different cancers and different aberrations in the same cancer. The progression of some solid cancers is accompanied by secondary chromosomal aberrations or by the occurrence of gene amplification.

13. Carcinogens are environmental agents that initiate and promote malignant changes, and they often act by causing a series of mutations. Agents in tobacco smoke and diet are major factors in cancer causation, but the clinical effects of a carcinogen may be delayed for years or decades.

ataxia-telangiectasia
benign tumor
Burkitt lymphoma
cancer
carcinogen
carcinoma
leukemia

Li-Fraumeni syndrome
lymphoma
malignant tumor
metastasis
oncogene
p53
Philadelphia chromosome

protein kinase
proto-oncogene
retinoblastoma
retrovirus
sarcoma
signal transduction
transcription factor

transformation
transgenic mouse
tumor progression
tumor suppressor gene
Wilms tumor
zinc-finger protein

QUESTIONS

1. A tumor the size of a marble, about 1 cubic centimeter (cc) in volume, may contain 10^9 cells. How many cell generations (starting from a single cell) are required to produce this tumor? (*Note:* Such a lump is about the size of a breast tumor that a woman can find by self-examination. Mammography can detect smaller tumors.)

2. Some uterine tumors consist of as many as 10^{11} (100 billion) cells. In women heterozygous for a particular X-linked gene, researchers have discovered that *every* cell of such a tumor has the *same* active X-linked allele. Explain this observation in terms of the Lyon hypothesis.

3. Although cancer researchers generally agree that the path to malignancy is a *multistep* process, Weinberg and his colleagues were able to transform mouse cells in culture in *one* step, as outlined in Figure 15.3. Can you suggest an explanation for the apparent discrepancy?

4. What experiment would show that transformation of mouse cells by the DNA taken from a bladder carcinoma was not just a property of normal DNA?

5. The proto-oncogene *erb*B encodes the cell surface receptor for a growth factor. When the growth factor outside the cell joins to the receptor, it signals the cell to divide. Speculate on how a mutation in the *erb*B proto-oncogene might lead to malignancy.

6. In recent years, oncogenes that are detected in retroviruses or by transformation experiments are often the same as oncogenes previously known. What does this suggest about the total number of oncogenes?

7. What are transcription factors? Which figures in this chapter have relevance to transcription factors?

8. Although often seen in leukemias and lymphomas, particular chromosomal defects are not usually identifiable in the common solid tumors of the lung, breast, and colon. Yet investigators suggest that this lack is just an artifact of working with solid tumors. What are the difficulties in finding the *specific* chromosomal defect in a solid tumor?

9. Distinguish between dominant inheritance and recessive expression for retinoblastoma.

10. Mutant alleles of the tumor suppressor genes discussed in this chapter can sometimes be transmitted in germ cells, leading to hereditary patterns of cancer occurrence. In other cases, the mutant alleles may be induced only in the somatic cells of individuals and not be passed on in eggs or sperm. Discuss with regard to *p53*.

11. Researchers have found that breast cancer is not common among *homozygotes* affected with ataxia-telangiectasia, but breast cancer is the most frequent type of cancer among *heterozygotes* for ataxia-telangiectasia. The researchers think that this oddity may be a consequence of the ages of the people within the two different groups. Can you give a reasonable explanation?

FURTHER READING

Excellent general discussions of cancer for nonscience students and lay persons include those by Roberts (1984), Prescott and Flexer (1986), LaFond (1988), and especially Varmus and Weinberg (1993). More detailed accounts but also at an introductory level are given by Franks and Teich (1986) and by Micklos and Freyer (1990).

Scientific American articles about cancer have been collected by Friedberg (1986). Subsequent *Scientific American* articles (not in the collection) are by Croce and Klein (1985), Cairns (1985), Cohen (1987), Weinberg (1988), Rosenberg (1990), and Liotta (1992). For browsing on the topic of cancer, see Library of Congress classifications RC 254–282, where a variety of books, journals, and review series can be found.

Immunogenetics

16

In 1971, a Texas family anxiously awaited the birth of their third child. They had already lost one son to an X-linked recessive disorder called *severe combined immunodeficiency disease (SCID)*, and prenatal diagnosis showed that this baby was also a male with a 50% chance of inheriting the lethal condition. SCID patients have no defense against microbes, so they easily fall victim to pneumonia, influenza, bronchitis, meningitis, and a host of fungal diseases. But this time, the family and its physicians had some advance warning. Abortion was offered and declined. After delivery by cesarean section, baby David was quickly placed in a sterile chamber. Attending physicians believed that a cure for David's SCID (for he had inherited the abnormal allele) was at most a few years away, so the germ-free isolation was considered a temporary situation. This view turned out to be overly optimistic.

As David grew, so did his marvelously devised chambers and life-support machinery at Texas Children's Hospital in Houston (Figure 16.1). He was surrounded and loved by his parents, sister, friends, physicians, and dozens of devoted hospital staff, whom he loved in return. He had a positive attitude, a sharp mind, and a fine sense of humor. Although his days were filled with activities made as normal as possible, his outlook on the world for a dozen years originated almost entirely from two multi-chambered plastic bubbles, one in the hospital and one at home. Tests showed that David's perceptions of space were distorted; he was unable, for example, to imagine a four-sided building. We will never fully understand how he felt about the tight limits on his freedom, although he seemed more comfortable with his confinement than did those around him. In any event, the case of the "bubble boy" and the increasingly technological approaches to his treatment received tremendous attention in the media (Carol Ann with Demaret 1984; Rennie 1985). At age 12, he received a bone marrow transplant from his sister, but it failed, and he died four months later.

Immune systems help to protect us against threats to our health from harmful microbes. *Immunology* is the study of these protective mechanisms, and *immunogenetics* deals with the genes that are involved. Beginning with Louis Pasteur and others in the late 1800s, a golden age of immunology brought under control many infectious diseases (smallpox, yellow fever, diphtheria, whooping cough, tetanus) that had ravaged Western countries. By the 1970s, the deadly smallpox virus, which left disfiguring pockmarks on its surviving victims, was wiped out completely. Now the only existing smallpox viruses are locked away at research facilities in Moscow and Atlanta, and these stocks are soon to be destroyed (Caldwell 1992a). All that will remain is a list of the 175,000 bases of the genome of several different strains. Researchers hope to arrange a similar fate for a new threat to our health, the human immunodeficiency virus (HIV), but the time scale for the conquest of acquired immune deficiency syndrome (AIDS) is unclear.

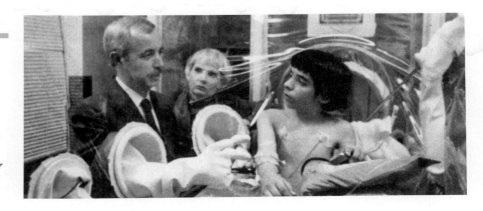

Figure 16.1
SCID patient David in his bubble chamber
at Texas Children's Hospital in Houston.
(Courtesy Baylor College of Medicine.)

The growth of molecular biology ushered in a second golden age of immunology that is providing a deeper understanding of the immune system. Researchers have shown that the mechanisms of immunity consist of marvelously intricate webs of interactions among and between cells and molecules. These processes are of practical importance in the prevention, diagnosis, and treatment of both infectious diseases (such as AIDS) and noninfectious diseases (such as cancer). But the immune system also causes serious problems in blood transfusions and organ transplants. In trying to understand such wide-ranging situations, immunologists have discovered unusual elements of gene structure and novel patterns of gene expression. These several aspects of immune processes are discussed in this chapter.

THE IMMUNE RESPONSE

The immune system can (1) recognize substances as either native to one's own body or foreign (i.e., self or nonself); (2) react specifically against any foreign substance; and (3) remember a particular foreign substance and respond more strongly to it upon a later exposure. **Antigens**, the substances that provoke the immune system, are often protein molecules embedded in the surface of bacteria, viruses, or other small particles such as pollen grains. Some antigens, however, are combination protein molecules: nucleoproteins (nucleic acid and protein), lipoproteins (fat and protein), or glycoproteins (carbohydrate and protein). Antigens may also differ among the various cell types in the body and among various members of a species—the latter accounting for different blood and tissue types.

Main Components of the System

The vital immunological properties—recognition, reaction, and memory—depend on several types of **white blood cells** that develop in the bone marrow and enter the bloodstream. From there they slip through capillary walls and wander the spaces between tissue cells. These spaces are filled with a fluid called **lymph**, which is derived largely from **blood plasma**, the liquid, somewhat yellowish portion of blood. The lymph drains into inconspicuous networks of lymphatic vessels that run more or less parallel to, but separate from, the network of veins. The largest lymph vessel empties into a main vein near the heart, completing the joint blood-lymph circulation.

Associated with this *lymphatic system* are the following small organs: several dozen *lymph nodes* spaced along major lymph vessels; a *spleen* lying behind the stomach; and a *thymus gland* lying in the chest cavity between the lungs (Figure 16.2). These organs are rich in small, roundish white blood cells called **lymphocytes**, which respond specifically to foreign antigens. In addition, as lymph seeps through the finely divided spaces within the nodes, large, irregularly shaped white blood cells called **macrophages** engulf and digest foreign substances. (Macrophages are said to be *phagocytic*—literally, "to eat cells.") Lymph nodes in the neck may become swollen and painful during colds and other infections. The tonsils and adenoids in the throat are somewhat similar to lymph nodes in structure and function.

The thymus, which is particularly important in the immune response, consists of a spongy mass of lymphocytes. It is fully developed at birth but gradually regresses during adolescence, remaining very small throughout adult life. Experiments have shown that newborn mice whose thymus glands are removed grow fairly normally but cannot respond immunologically to some types of antigens. Mice that are homozygous for a certain autosomal recessive allele develop no thymus at all. Called *nude* because they are hairless, these mice will die unless they are protected from infection in a germ-free environment.

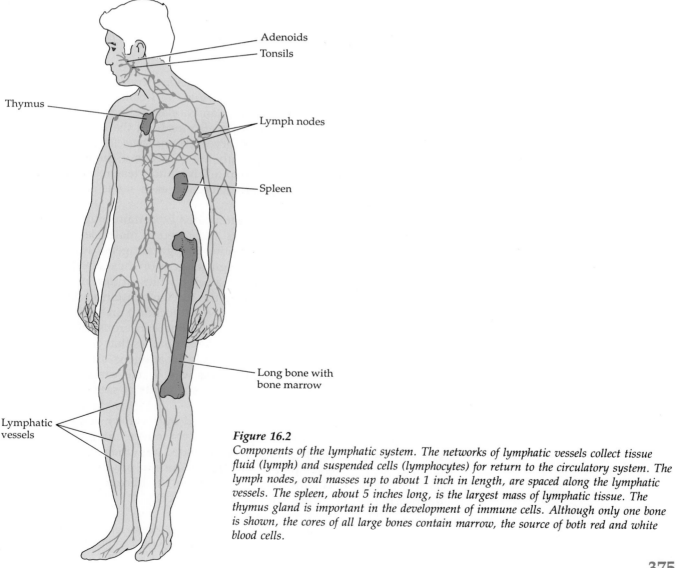

Figure 16.2
Components of the lymphatic system. The networks of lymphatic vessels collect tissue fluid (lymph) and suspended cells (lymphocytes) for return to the circulatory system. The lymph nodes, oval masses up to about 1 inch in length, are spaced along the lymphatic vessels. The spleen, about 5 inches long, is the largest mass of lymphatic tissue. The thymus gland is important in the development of immune cells. Although only one bone is shown, the cores of all large bones contain marrow, the source of both red and white blood cells.

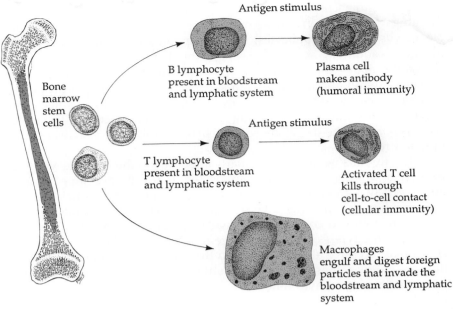

Figure 16.3

Differentiation of the white blood cells involved in the immune response. Although the plasma cell actually secretes antibody molecules, the T cells and macrophages are also required for initiation of antibody production. The T cells and macrophages have additional immune functions (see text). The cells are magnified about 4,000 times.

Two major types of small lymphocytes are involved in the immune response: the **B cells** (or **B lymphocytes**) and the **T cells** (or **T lymphocytes**) (Figure 16.3). The precursors of human B cells are made and mature in the bone marrow. Human T cells are also made in the bone marrow but migrate to the thymus in early embryos. There they "learn" what is self and what is not by associating with the cells about them before being released to the circulation. Both B and T lymphocytes are found throughout the blood, lymph, and lymphatic organs; each site is rich in one type or the other.

The B and T cells can be distinguished by certain properties of their membranes (e.g., different antigens and receptor molecules) and by the strengths of their responses to various foreign antigens. Human B cells respond particularly strongly to foreign antigens on the surface of invading bacteria by differentiating into **plasma cells**, which in turn make **antibodies**, protein molecules that recognize the foreign substances.* (Antibodies will be described more fully in the next section.) In contrast, T cells respond particularly strongly to viral infections, cancerous cells, and tissue grafts. Unlike B cells, T cells do not make antibodies that circulate in blood plasma. But they do respond specifically to antigens and interact with them by direct cell-to-cell contact. In this way, T cells can bring about the death of, say, foreign cells present in tissue grafts of skin, bone marrow, or other organs.

Both B cells and T cells are regulated by a variety of chemical messengers called **lymphokines** (for "lymphocyte movement"), and the whole system is finely tuned to produce an integrated immune response throughout the body. For example, B cells that mature to produce antibodies do so only after contact with T cells. In addition, the phagocytic macrophages interact in complex ways with both B and T cells.

---

* *Plasma cells* should not be confused with *blood plasma*. Plasma cells are white blood cells that manufacture and secrete antibodies. They are suspended in blood plasma, and the antibodies are dissolved in blood plasma. Nor should *antibody* be confused with *antibiotic*, a drug (e.g., penicillin) that nonspecifically attacks invading bacteria.

The B Cell Response

By the 1950s it was known that plasma cells congregate in large numbers at the sites of bacterial infection. These cells, rich in the organelles of protein manufacture, are the source of antibodies directed against the many different antigenic sites on the surface of the bacteria. By culturing individual plasma cells, scientists showed that *one* plasma cell makes antibody against only *one* specific antigen. This result was itself surprising, but the most intriguing question about the total immune response was how it could manage to identify so *many* different foreign antigens. Over the course of a lifetime, a human being must be exposed (via food, air, skin contact, bites, etc.) to hundreds of thousands or even millions of different nonself materials, and the immune system can respond specifically to each one. Indeed, researchers have shown that plasma cells can make antibodies to *synthetic* chemicals that no species has ever encountered over the course of evolution! The seemingly unlimited individual challenges that the immune system meets and overcomes is its most extraordinary property.

One model of the immune response that has been supported by much evidence was suggested by two Nobel prize winners: Niels Jerne, director of the Basel Institute of Immunology, and F. Macfarlane Burnet, director of Melbourne's Institute of Medical Research. According to Burnet's **clonal selection theory**, each lymphocyte is *preprogrammed* to recognize and respond to just one antigen, but there is no advance "knowledge" of whether it will ever be called on to perform (Ada and Nossal 1987). The idea is illustrated in Figure 16.4, which shows a small sample of B cells. The preprogrammed antigen specificity of each cell is indicated by a number. When antigen 17 is present, for example, it interacts with the surface molecules on one or a few lymphocytes specific for 17, stimulating them to divide and differentiate over a period of several days to form a clone of several hundred identical plasma cells. These mature plasma cells secrete large amounts of antibodies, Y-shaped protein molecules with two pockets at the upper tips of the Y. The pockets, called the **combining sites**, have a shape complementary to the shape of the initiating antigen.

Recognition of antigen 17 by the B cells that are preprogrammed to manufacture anti-17 antibody begins when antigen 17 joins to receptor molecules embedded in the surface of the B cells (step 1 of Figure 16.5). On any given B cell, all the receptors are the same and virtually identical to the Y-shaped antibody molecules that the mature plasma cell will eventually secrete in large amounts. At this stage, however, the B cell is not secreting any antibody into the extracellular space; it is just displaying its name tag so that it can be greeted accurately.

Another event is needed before the B cell can proliferate and mature: Antigen 17 must be processed inside the cell and a portion of it displayed on the surface together with another protein (step 2 in Figure 16.5). This other membrane protein is synthesized within the B cell from genes called *MHC*, an abbreviation for major histocompatibility complex. (MHC proteins are discussed more fully in later sections.) This association of foreign antigen plus MHC protein is then recognized by a specific type of T cell called a *helper T cell* (which had been previously stimulated by the same antigen). The helper T cell then secretes a hormonelike chemical (a lymphokine called interleukin 2) that stimulates the B cell to multiply and differentiate (step 3). The daunting complexity of this recognition system apparently helps the immune system discriminate between foreign substances and the body's own cells.

A week or so after initial contact with bacterial antigen 17, the re-

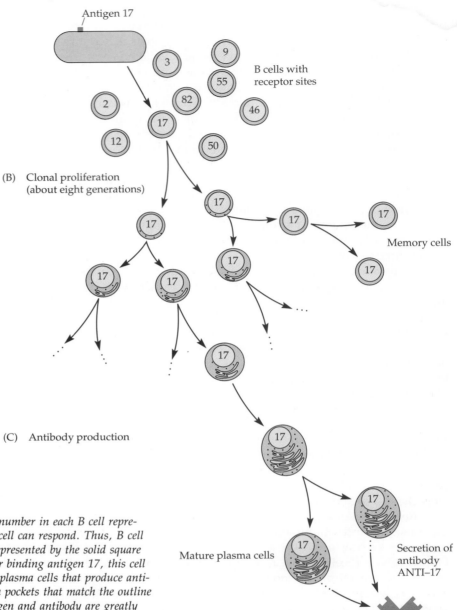

(A) Antigen recognition

Antigen 17

B cells with receptor sites

(B) Clonal proliferation (about eight generations)

Memory cells

(C) Antibody production

Mature plasma cells

Secretion of antibody ANTI–17

Figure 16.4

Burnet's clonal selection hypothesis. (A) The number in each B cell represents the one particular antigen to which the cell can respond. Thus, B cell 17 recognizes only foreign antigen 17 (here represented by the solid square shape on the surface of a bacterium). (B) After binding antigen 17, this cell divides repeatedly, and (C) differentiates into plasma cells that produce antibody anti-17 (represented by the Y shape with pockets that match the outline of antigen 17). Note that the sizes of the antigen and antibody are greatly exaggerated compared to the cell sizes. Details of part (A) are shown in Figure 16.5.

sponding plasma cells begin to secrete antibodies into the fluid portion of blood and lymph (Figure 16.4C). The combining sites of the antibodies recognize and join with antigen 17 on the surface of an invading bacterium. When the tips of antibody 17 are bound to antigen 17, the stem of the Y-shaped antibody molecule sticks out from the surface and can react with receptor sites on the surface of macrophages. This reaction stimulates the phagocytic action of the macrophages, and the bacterium is engulfed and destroyed.

Alternatively, the antigen-antibody complex that forms on some cell surfaces invites other plasma proteins, called *complement*, to join up. Complement proteins injure the cell membranes of the bacteria, allowing fluid to enter. The bacteria swell and break open, a process called *lysis*. Thus, we see that antibody does not itself destroy the invading bacteria. Rather, antibody molecules mark cells for destruction by other agents.

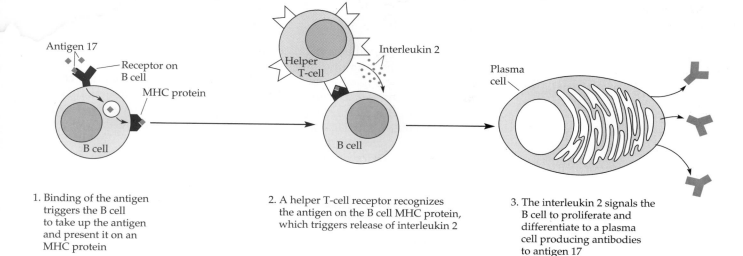

1. Binding of the antigen triggers the B cell to take up the antigen and present it on an MHC protein

2. A helper T-cell receptor recognizes the antigen on the B cell MHC protein, which triggers release of interleukin 2

3. The interleukin 2 signals the B cell to proliferate and differentiate to a plasma cell producing antibodies to antigen 17

Figure 16.5
The recognition steps that are needed to select and activate a specific B cell for proliferation. Steps 1 and 2. An antigen (say, antigen 17) binds to a specific B cell receptor that is bound to the cell surface. (Note that this receptor is shaped exactly like the antibody molecule later secreted by the cell.) The antigen is then drawn into the cell, processed, and passed to another membrane protein (MHC). Step 3. The receptor of a helper T cell binds to the antigen-MHC complex. The helper T cell then secretes a lymphokine (interleukin 2) that stimulates the B cell to divide and differentiate into a plasma cell.

The antibody-producing plasma cells of a selected clone do not live long after the removal of the stimulating antigen. Other members of the same clone that have not fully differentiated, however—called **memory cells**—can survive for years. Once primed by a given antigen, they will initiate the so-called **secondary immune response** when they meet the same antigen again. This response is characterized by a quicker, stronger production of antibody, primarily because more plasma cells are produced. The booster shots given in an immunization program have the effect of renewing the supply of memory cells.

The T Cell Response

Like B cells, T cells also respond by clonal selection, multiplying and differentiating after they are specifically stimulated. Showing the same high degree of specificity, T lymphocytes have on their surfaces recognition molecules, called **T cell receptors**, able to make the necessary fine distinctions between foreign antigens (Marrack and Kappler 1986). Although the T cell receptors differ from the B cell receptor molecules, the two systems share some amino acid sequences and are presumed to be evolutionary products of a common primordial molecule. As we have seen, the B cell receptors are membrane-bound forms of the antibody molecules that the B cells will eventually secrete when they are mature. The T cell type of receptor molecules, by contrast, are never secreted. Instead, these antigen-specific molecules perform their functions by cell-to-cell contact.

The antigens that stimulate specific T cells to proliferate into a clone do not join directly to the T cell receptors. Rather, a foreign protein antigen is first phagocytized by macrophages and processed internally. A portion of the antigen is displayed on the macrophage surface along with an MHC protein. Only then is the antigen in a form that the T cell receptor can recognize. This complicated system of recognition is similar in some ways to that already described for the stimulation of B cells.

An additional complexity of the T cell system is the existence of different types of T cells: **cytotoxic T cells**, which do the actual killing achieved by the T cell system, and **helper T cells**, which modulate the activities of both B cells and other T cells. Because helper T cells occupy such a central regulatory role in the immune system, their preferential destruction by the AIDS virus leaves AIDS victims susceptible to a variety of infections and malignancies (Laurence 1985); we return to this topic later in the chapter).

379

ANTIBODIES

Collectively, antibodies are called **immunoglobulins (Ig)**, a general term describing many immune molecules that are dissolved in blood plasma. (*Globular* proteins like the immunoglobulins have their polypeptide chains folded into more or less rounded shapes, rather than being long and *fibrous* like collagen.) Because antibodies secreted by B cells are so varied in their antigen-combining abilities, protein chemists long wondered how one differed from another. But the chemists were hampered by the large size of the molecules and by their inability to purify enough of any one antibody to analyze it chemically. This problem was solved by the discovery that persons suffering from *multiple myeloma*, a cancer of bone marrow, had in their plasma large amounts of a *single* antibody species. But each patient had a different antibody. Accounting for as much as 95% of the immunoglobulins in a myeloma patient, the unusual antibody presence resulted from one antibody-producing plasma cell becoming malignant.

Structure

The researchers were able to show that the major form of antibody, **IgG (Immunoglobulin G,** or **gamma globulin),** consists of four polypeptide chains connected to each other at several places (Figure 16.6). Each of the two identical **heavy (H) chains** consists of four regions of about 110 amino acids each. The regions, called **domains**, have similar amino acids. They are depicted by the elliptical outlines in Figure 16.6. Domains similar to those in the heavy chains also occur twice in each of the two identical **light (L) chains** that are present in the two branches of the Y-shaped molecule. The attachment of the branches to the stem of the Y is somewhat flexible.

Comparison of the amino acid sequences of light chains from different antibodies reveals a remarkable pattern: Within the domains at the top of the Y, about 25% of the amino acids in one antibody differ from those in another; thus, this region is called the **variable (V) domain**. In the lower half of the light chain, the amino acid sequences are almost exactly the

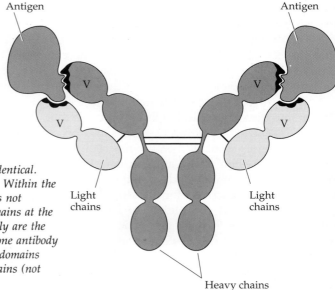

Figure 16.6

The Y-shaped IgG antibody molecule. The left and right halves are identical. Each heavy chain is 446 amino acids long and each light chain, 214. Within the outline of each chain, the complex folding of the polypeptide chains is not shown. The colored regions with V's symbolize the four variable domains at the upper end of the two heavy chains and two light chains. Shown boldly are the amino acid regions of the variable domains that vary the most from one antibody molecule to the next and make close contact with the antigen. These domains have been likened to the claws of a lobster. The remaining eight domains (not colored) do not vary much.

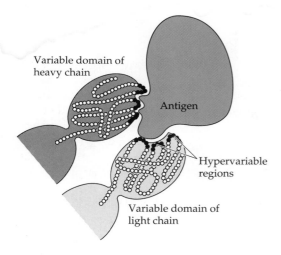

Variable domain of
heavy chain

Antigen

Hypervariable
regions

Variable domain of
light chain

Figure 16.7

One combining site of an antibody. The specific looping pattern of the polypeptide chain within the variable domains of the heavy and light chains causes certain amino acids (indicated by color) to line the pocket binding the antigen. These amino acids are in hypervariable regions, which contain most of the differences between antibodies of different specificities. The actual folding of the polypeptides is more tortuous than indicated here.

same from one antibody to another; thus, this region is called the **constant (C) domain**. When the heavy chains from different antibodies are examined, the domains at the top of the Y are also found to be variable; the remaining three domains of a heavy chain are constant. Interestingly, most of the variation in amino acids within variable domains is restricted to a few subregions totaling 20 to 30 amino acids. These so-called **hypervariable regions**—three in each light chain and three in each heavy chain—form the lining of the antigen-binding cavity (Figure 16.7). Antibody specificity arises from the three-dimensional shape of the cavity and from the particular chemistry of the hypervariable amino acids. The surfaces of the antigen and the heavy and light chains conform like parts of a jigsaw puzzle, but the fit may be either snug or loose.

IgG, whose structure we have been describing, constitutes about 80% of the immunoglobulin molecules. Different immunoglobulin classes (G, A, M, E, in the order of their concentration in plasma) have been analyzed: Each has different amino acids in the *stem* region of the heavy chains, giving the four classes somewhat different functions. For example, IgG readily crosses the placenta and provides protection to newborns before they develop their own immunological competency. The IgA class is found not only in plasma but also in body secretions, such as saliva, nasal mucus, sweat, and breast milk. IgM molecules have somewhat larger heavy chains than the other classes do, and five of the Y-shaped molecules may be joined together in a circle. Because each IgM antibody then has ten combining sites, IgM molecules bind more firmly to antigens. IgE is involved in allergic reactions.

It is important to note that antibody of any of the classes can have the same combining site, directed against, say, antigen 17 (in our previous example). That is, the four functional classes of antibodies, differing in the *stems* of the Y's, can have exactly the same antigen specificity in the *branches* of the Y's. A given B cell can synthesize antibodies of any of the classes, although not at the same time.

Antibody Diversity

As a group, the B cells of a person can make a million or more different IgG antibodies, each characterized by the distinctive array of amino acids lining the pockets where the antigens bind. Like any protein, antibodies are encoded by genes. Yet we could not possibly have a million antibody genes, each one coding for the amino acid sequence of a different antibody,

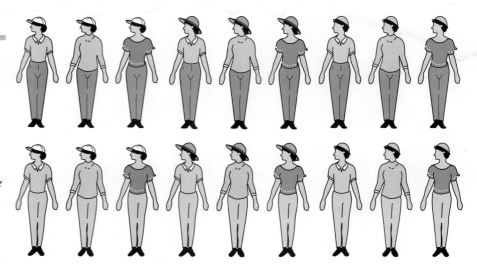

Figure 16.8
A person with a set of three different hats, three different shirts, and two different slacks can be dressed with the (3 × 3 × 2) = 18 different outfits depicted here. The large number of antibody "outfits" are assembled in a similar manner—from several sets of DNA "clothing elements" assembled in numerous combinations (see Figure 16.9).

because that is more genes than we possess. The geometry of each combining site is formed, however, by *two* polypeptide chains, one heavy and one light. Thus, if there were 1,000 different H chains (from 1,000 genes) and 1,000 different L chains (from another 1,000 genes), then all possible combinations of the two proteins would generate 1,000 × 1,000 = 1 million different spatial patterns. Thus, 2,000 antibody genes might suffice.

Actually, research has shown that the enormous repertoire of antibody molecules is encoded by just a few hundred genetic elements! The primary solution to this number puzzle is simply stated: We inherit and pass on in germ cells several separate DNA sequences that, when put together, code for the combining sites. To generate diversity, the pieces are shuffled around and combined in different ways during the formation of individual B cells. By analogy, imagine outfitting somebody with a hat, shirt, and slacks (Figure 16.8). If the person has 10 different hats, 10 different shirts, and 10 different slacks (30 elements), you can arrange 10 × 10 × 10 = 1,000 different costumes. With a *second* person and an *additional* 10 hats, 10 shirts, and 10 slacks, you can generate another 1,000 outfits. All together, from just 60 unique elements, you will be able to assemble a million different *pairs* of outfits! In a somewhat similar multiple-choice situation, the various DNA segments constituting the parts of antibody genes are shuffled together as B cells develop from undifferentiated stem cells.

The shuffling comes about by a process called **somatic recombination**, which has been studied by several researchers, including Susumu Tonegawa, then at the Basel Institute of Immunology and a Nobel prize winner in 1987. Tonegawa identified, mapped, and cloned the DNA elements that undergo somatic recombination (Leder 1982; Tonegawa 1985). Most of his work was done with mouse DNA, and only later have the details of human antibody genes become known. Researchers have found, for example, that an extended region near the tip of the long arm of human chromosome 14 codes for the heavy chain. The top line of Figure 16.9) shows the general arrangement of heavy chain DNA elements as they are inherited. Note that there are four families of elements: *V* (variable), *D* (diversity), *J* (joining), and *C* (constant). During the maturation of a particular B cell, intervening DNA regions are cut out, promoting the splicing together, by somatic recombination, of *one* member of *each* family to yield a unique *V-D-J-C* **gene** (second line of Figure 16.9). The gene is then transcribed and translated into a heavy chain.

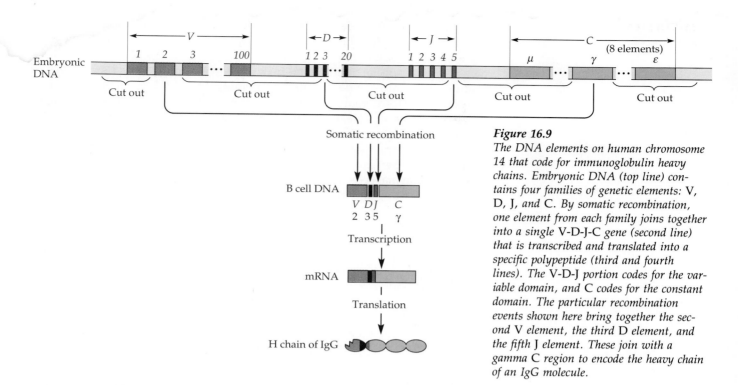

Figure 16.9

The DNA elements on human chromosome 14 that code for immunoglobulin heavy chains. Embryonic DNA (top line) contains four families of genetic elements: V, D, J, and C. By somatic recombination, one element from each family joins together into a single V-D-J-C gene (second line) that is transcribed and translated into a specific polypeptide (third and fourth lines). The V-D-J portion codes for the variable domain, and C codes for the constant domain. The particular recombination events shown here bring together the second V element, the third D element, and the fifth J element. These join with a gamma C region to encode the heavy chain of an IgG molecule.

The combined V-D-J portion of the heavy chain gene codes for the variable domain (including the hypervariable regions that characterize the antigen specificity); the *C* portion codes for all of the constant domains, which determine the class of the antibody: IgG, IgM, IgA, or IgE. The number of different V-D-J regions that a cell can construct is the product of the number of *V*, *D*, and *J* elements. For humans, the numbers are not known precisely, but there are approximately 100 *V*, 20 *D*, and 5 *J* elements. On the basis of these estimates, $100 \times 20 \times 5 = 10,000$ different heavy chain variable domains can be made.

Actually, the number of different heavy chain variable domain polypeptides is more than the 10,000 predicted by the combinatorial shuffling described here. The greater possibilities result from other unusual phenomena, including **somatic mutation**. In this process, base substitution mutations occur at a substantial rate in the immunoglobulin DNA during the differentiation of a B cell into a mature plasma cell. It is thought that some of these base substitutions lead to variant antibodies that fit even better with the antigen.

Note that in the case of immunoglobulins, several genes ("minigenes" or "genelets") cooperate to encode one polypeptide, leading to a novel *two (or more) genes–one polypeptide* idea. Note also that we have described the generation of variability only for the H chains. To make an antibody, the H chain joins with an L chain from a somewhat similarly processed gene on chromosome 2 or 22 (two different L chain genes). The degree of light chain variation is less, because a *V* element joins directly with a *J* element—no family of *D* elements exists.

Antibody Engineering

A single antigen injected into a mouse will generally raise a mixture of antibodies, each recognizing a slightly different part of the antigenic surface. If the antigen is complex, like a whole virus or a cell with myriad proteins, lipoproteins, or glycoproteins on its surface, then dozens or hundreds of antibodies will be made, each by a different clone of plasma cells. All the antibodies mix together in the blood plasma, where they are chemically inseparable.

But a special method of making and isolating very homogeneous antibodies with predetermined combining abilities has generated much interest in recent years because of its demonstrated and potential value in the diagnosis and treatment of disease. Each pure and uniform antibody, called a **monoclonal antibody**, is secreted by a clone of cells grown in culture in virtually unlimited numbers. The special clone is derived by fusing together an antibody-secreting *plasma* cell with a *cancer* cell. Although plasma cells by themselves do not continue to live in culture, cancer cells do, and the hybridization renders the plasma cell immortal. The plasma cell–cancer cell combination is called a **hybridoma**. The process of producing hybridomas is a variant of the somatic cell fusion technique described in Chapter 13.

Monoclonal antibodies were first developed at Cambridge University by German researcher Georges Köhler and Argentina's Cesar Milstein (1980), who were awarded Nobel prizes for this work. The basic procedure for producing monoclonal antibodies is outlined in Figure 16.10. First, a mouse is injected with the antigen of interest in order to stimulate the proliferation of plasma cells that make antibodies against that antigen (Figure 16.10A). The mouse's spleen, containing a mixture of lymphocytes, is later removed. A suspension of spleen cells, including (it is hoped) some activated antibody-producing plasma cells, is mixed with cells derived from a mouse myeloma (Figure 16.10B). As in the procedure described in Chapter 13, cell mixture occurs in a medium that slows or stops the growth of unfused cells. Therefore, any hybrid cells that form will rapidly outgrow the parental cell lines. Several hundred fused cells may be formed from the spleen of one mouse. The resultant clones of cells from each hybrid are isolated and tested for the presence of antibody against the original immunizing antigen (Figure 16.10C). Any positive lines—there may be

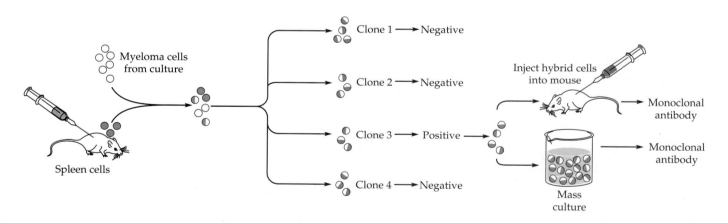

(A) Inject antigen into mouse, then remove spleen

(B) Fuse two kinds of cells

(C) Clone hybrid cells and test for antibodies

(D) Propagate hybridoma

Figure 16.10

The production of monoclonal antibodies. (A) Inject and reinject a mouse with an antigen against which an antibody is desired; later, remove the mouse's spleen. (B) Fuse the spleen cells (including, it is hoped, specific antibody-producing plasma cells) with mouse myeloma cells in a culture medium that allows only hybrid cells to grow. (C) Clone the hybrid cells individually, testing the medium in which each clone is growing for the presence of antibody directed against the original immunizing antigen. (D) Grow the hybridomas (antibody-producing hybrid cells) in tissue culture, or inject them into mice to form myelomas that continue to produce pure monoclonals.

none or at best a small number—can be propagated in culture medium or inside living mice as myeloma tumors (Figure 16.10D). Each isolated hybridoma clone arises from a single plasma cell that secretes a single antibody. Thus, no matter how complex the stimulating antigenic substance is, this protocol yields a homogeneous crop of antibodies.

Dozens of pharmaceutical and biotechnology companies are developing monoclonal antibodies, which are being put to a growing array of uses. For example, researchers routinely use monoclonal antibodies to type blood and tissue (see next sections) and to distinguish between cytotoxic and helper T cells. Clinical uses of monoclonal antibodies for diagnosis and therapy of infectious and other diseases are still largely experimental, but increasing. For example, a monoclonal antibody that attacks specific kinds of T cells has been used to help prevent rejection of transplanted kidneys. Cancer cells are also vulnerable to immunological attacks, because such cells acquire tumor-associated (newly "foreign") antigens in the process of becoming malignant. Thus, monoclonal antibodies directed against lymphomas, melanomas, and cancers of the lung, colon, ovary, and breast are in various phases of clinical trials. Researchers have been trying to couple the monoclonal antibodies directed against a tumor antigen with a toxic chemical to make a so-called magic bullet, an *immunotoxin* that will kill malignant cells but no others (Collier and Kaplan 1984). Such cancer therapies would be much better than current chemotherapies or radiation treatments, which kill all rapidly dividing cells, including normal ones.

A major problem with monoclonal antibodies, however, is that the best ones are made by *mouse* cells growing in culture, not human cells. As such, they are foreign to humans and can cause a patient receiving a monoclonal antibody to make antibodies against it, producing serious side effects. For this reason, investigators have been trying to make monoclonal antibodies derived from a *human* plasma cell fused with a *human* myeloma cell. But this approach has not been very successful.

More promising ways to produce useful monoclonal antibodies involve joining together polypeptide elements from mouse and human sources. Figure 16.11 illustrates two possibilities based on cloning the antibody genes from hybridomas. In a *chimeric antibody*, the whole V domains derived from mouse cells are joined with the constant domains from human cells. In a *reshaped antibody* (sometimes called a *humanized antibody*), just the hypervariable regions from mouse cells (the amino acids lining the antigen-combining cavity) are inserted into an otherwise human-derived antibody. Eventually, it may be possible to clone the raw elements (the *V*, *D*, *J*, and *C* minigenes) of human antibody genes in a virus or bacterium. If the host microorganism can be induced to join the minigenes together and express the result, the combinatorial library might yield all possible antibodies, even those that have never been made by real people (Lewis 1990).

RED CELL ANTIGENS AND BLOOD TRANSFUSION

Blood transfusions were first attempted centuries ago, but interest in this form of treatment waxed and waned because patients exhibited an unfortunate tendency to die from it. The following graphic account by a French physician (Denis 1667) describes one patient's reaction to a second transfusion with calf's blood:

> We observ'd a plentiful sweat over all his face. His pulse varied extreamly at this instant, and he complain'd of great pains in his Kidneys, and that he was not well in his Stomach, and that he was ready to choak unless they gave him his liberty.

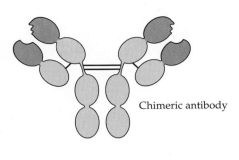

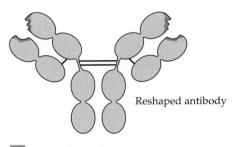

Chimeric antibody

Reshaped antibody

■ Mouse–derived regions
▨ Human–derived regions

Figure 16.11
Two types of monoclonal antibodies of mixed origin. Both are mostly of human origin to avoid an immunological problem in treated patients, but partly of mouse origin to make it easier to produce the molecule. (A) The chimeric molecule has the entire variable region from a mouse, while (B) the so-called reshaped or humanized molecule uses only the hypervariable regions from mouse cells. (After Co and Queen 1991.)

Presently the Pipe was taken out that convey'd the blood into his Veins, and whilst we were closing the wound, he vomited. . . . He made a great glass full of Urine, of a colour as black as if it had been mixed with the soot of Chimneys.

The consequences of this second transfusion are typical of the hemolytic shock that may accompany an incompatible blood transfusion.

The discovery in 1900 of the ABO blood group system led again to the use of transfusions to treat wounded soldiers in World War I. Even then, the procedure sometimes produced unexpectedly disastrous results. Only after the discovery in 1940 of the Rh blood group system did transfusions become generally safe. Although ABO and Rh are the most important blood groups from a medical point of view, about 20 additional systems (not discussed here) make blood groups very useful in chromosome mapping, paternity testing, forensic medicine, and anthropology.

Blood makes up about 8% of body weight (5 quarts in a 125-pound person). The plasma transports nutrients, hormones, enzymes, metabolites, immunoglobulins, clotting factors, and waste materials to and from various parts of the body. Suspended in the plasma are white blood cells (which have the immune functions described earlier), platelets (which initiate the clotting process), and red blood cells (the focus of this section).

Red blood cells, or **erythrocytes,**, make up about 42% of blood volume in females and 47% in males. Shaped like doughnuts with partly filled holes (Figure 16.12), they can pass in single file through the finest capillaries. The average red blood cell circulates for about four months before it is eventually engulfed by macrophages in the spleen. Red blood cells contain no nuclei or other organelles, although they are derived from precursor cells (*erythroblasts*) that do. The primary function of red blood cells is to carry oxygen to the tissues (where it participates in energy production) and to return carbon dioxide (the waste product of these reactions) to the lungs. The transport of O_2 and CO_2 involves hemoglobin, which is packed inside red blood cells.

Figure 16.12
Red blood cells seen in a scanning electron microscope, which produces a three-dimensional appearance (×5,000). The doughnut shape, with the hole partly filled, is clearly evident. (From Morel, Baker, and Wayland 1971.)

10 μm

Embedded in the membrane of a red blood cell are many types of proteins: structural elements, enzymes, receptors, transmembrane transport proteins, and *cell recognition proteins*. The latter stick out from the cell membrane and may be combined with glucose, galactose, or other sugars into molecules called *glycoproteins*. Similar sugar groupings may be combined with lipid molecules in the membrane to form *glycolipids*. Cell surface molecules mediate cell-to-cell recognition and communication, regulate tissue growth and differentiation, and may be altered as cells become cancerous. Each biological species has some cell surface molecules that act as antigens because they are foreign to other species. Even within a given species, individuals may differ from one another in the red cell antigens they possess. These differences provide the basis for human blood grouping.

The membranes of red blood cells are loaded with antigens. One set of antigens determines the ABO blood group, another set determines Rh, and so on. Each set of antigens is controlled by a different gene, which often has multiple alleles. When an antigen is present on red blood cells, a corresponding allele is present in a person's genotype. The alleles of one set, with their corresponding antigens, are referred to as a **blood group system** or simply a **blood group**.

Some blood group genes have one very common allele and one very rare allele. Because almost all people would be homozygous for the common allele, these genes have limited general interest. Some blood group genes, however, have sizable proportions of *two or more* alleles or phenotypes. Such genes are said to be **polymorphic** (*poly*, "many"; *morph*, "form"). For example, among Americans, the ABO blood types are distributed roughly as follows:

Blood type	Percent
A	40
B	7
AB	3
O	50

For a polymorphic gene, two people picked at random are likely to have different genotypes, and a relatively high proportion of them will be heterozygous. The *ABO* and *Rh* genes are highly polymorphic.*

The antigens present on a person's red blood cells determine his or her blood types. The presence of any antigen is revealed by a chemical reaction with a corresponding antibody. We have noted that the antigen-antibody reaction is a very specific one, so that antigen Z, say, will react only with the antibody anti-Z. The reaction is often done on a microscope slide or in a test tube. A drop of red blood cells is mixed with a drop of **serum** (plural, sera) containing a known antibody, say, anti-Z. (Blood serum is blood plasma whose clotting elements have been removed or inactivated.) Either the red blood cells clump, a process called **agglutination**, or nothing happens. Agglutination with anti-Z means that the corresponding antigen, Z, is present on the red blood cells; no agglutination means that the antigen Z is absent.

A microscope is rarely needed to see the clumps of red blood cells resulting from agglutination. The phenomenon is illustrated in Figure 16.13, showing a sample of red blood cells with antigen A. Two tests are shown, one with anti-A and another with anti-B. Because each antibody has *two* combining sites that can react with the antigen, adjacent red blood

---

* About a dozen blood group genes are polymorphic. One group, called Xg, is due to an X-linked gene. Several (Duffy, Kell, Kidd, Lewis, Lutheran) are named after the family in which they were first discovered.

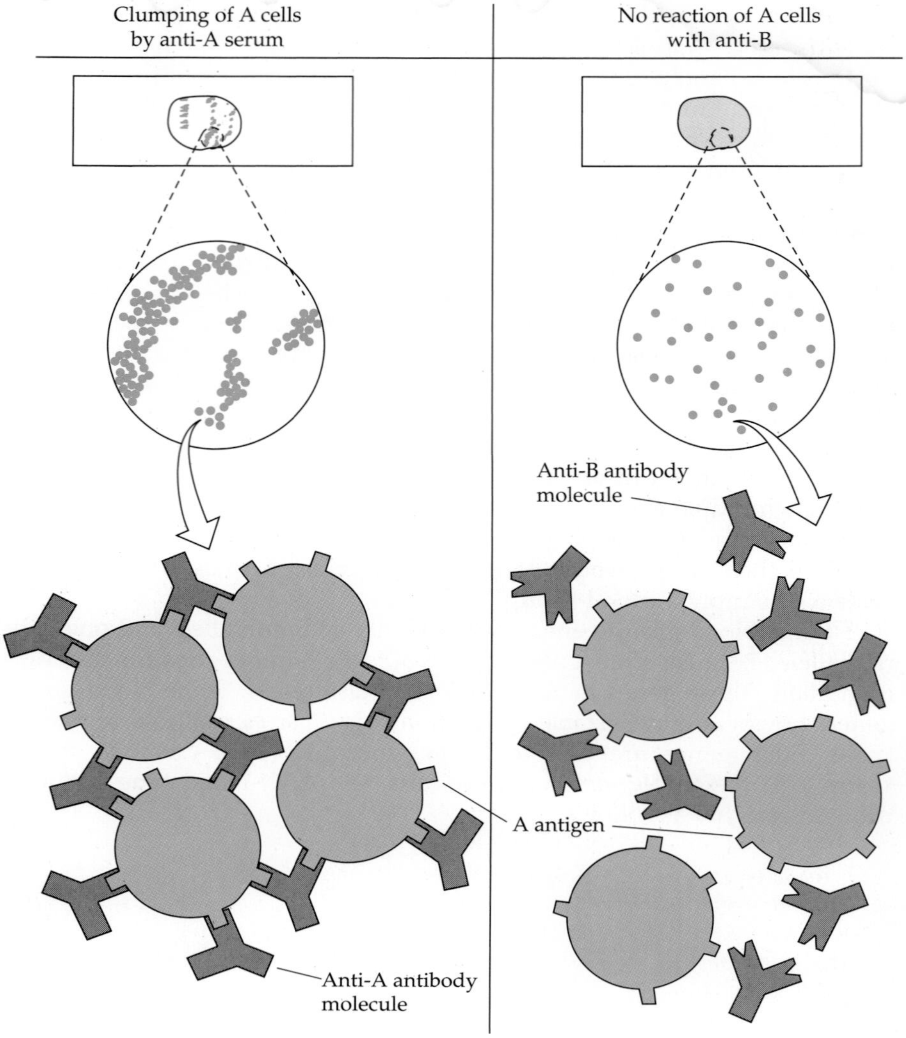

Clumping of A cells
by anti-A serum

No reaction of A cells
with anti-B

Anti-B antibody
molecule

A antigen

Anti-A antibody
molecule

Figure 16.13
The appearance of the red blood cells of a person with antigen A when mixed with serum containing anti-A or anti-B. Three levels of observation are shown: the unaided eye (top), a medium-power microscope about ×500 (middle), and the molecular level (bottom). In the molecular view, the shapes of the A antigen and antibodies are conjectural. The size of these molecules is much exaggerated compared to the size of the red blood cells.

cells can be linked into a network of clustered cells large enough to be seen easily. Anti-B is unable to recognize and react with the A antigen because its combining sites are not complementary to the antigen.

Some of the antibodies that laboratories use for blood typing are obtained by injecting rabbits with human red blood cells. The animals respond to the foreign antigens by manufacturing antibodies, which can then be retrieved simply by drawing blood from the animals. The serum, when separated from its suspended particles, is called immune serum; it will usually contain, in significant concentration, many kinds of antibodies against the many human red cell antigens. The immune serum sample with multiple antibodies can often be purified so that it contains just one antibody type. Several dozen such typing sera, each with an antibody specific for one antigen, may be used when determining human blood groups.

In addition to immunizing animals with human red blood cells to obtain typing antibodies, blood workers rely on several other sources. Commonly used are *normal* human serum (for ABO typing only) or the serum of people who have had multiple transfusions or pregnancies. In addition, monoclonal antibodies now provide very pure, very consistent typing reagents.

Table 16.1
The ABO blood group system.

REACTION OF RED BLOOD CELLS WITH[a]		Antigens on Red Blood Cells	Blood Type (Phenotype)	Genotype
Anti-A	Anti-B			
+	0	A only	A	*A/A* or *A/O*
0	+	B only	B	*B/B* or *B/O*
+	+	Both A and B	AB	*A/B*
0	0	Neither A nor B	O	*O/O*

[a] + = agglutination, 0 = no reaction.

In summary, the blood groups are defined in terms of *antigens on the surface of red blood cells*. An antigen (1) elicits the production of the corresponding antibody when injected into an animal to which the antigen is foreign and (2) combines with the specific antibody to cause red cell agglutination on a microscope slide or in a test tube. It is important to recognize that the blood group antigens and the resultant phenotypes are generally neutral attributes, not directly causing disease.

The ABO Blood Group System

Already outlined in Chapter 4 to illustrate Mendel's second law, the **ABO blood group system** is shown again in Table 16.1. The antibodies anti-A and anti-B are used to determine the presence or absence of the corresponding antigens on red blood cells* (left two columns). In the last column, we note that these antigens are controlled by the three alleles *A*, *B*, and *O* of the *ABO* gene (using shorthand notations for genotype). Note that the symbol A can mean an antibody, an antigen, a blood type (phenotype), or an allele (genotype). The context in which the symbol is used makes clear which is meant.

Note that blood types AB and O correspond to genotypes *A/B* and *O/O*, respectively (Table 16.1). Blood types A and B can result either from the corresponding homozygote or from the heterozygous combination with the *O* allele. Thus, the *O* allele is recessive to *A* and to *B*, whereas the *A* and *B* alleles are *codominant* to each other. The specific genotype for phenotypes A and B can be inferred from the blood types of related individuals. For example, consider the mother, father, and daughter with the ABO blood types shown in the following diagram.

The mother cannot be genotypically *A/A*, for she would then have transmitted an *A* allele to her daughter (but her daughter does not have an *A* allele). The daughter cannot be *B/B*, because her mother does not have a *B* allele to transmit. The father, however, can be either *B/B* or *B/O*.

It turns out that the typing antibodies, anti-A and anti-B, can be isolated from the *normal* serum of some persons, namely, those who *lack* the corresponding antigens. Thus, anti-A is found in the serum of persons of blood type B or O, and anti-B is found in the serum of persons of blood

* Although found primarily on red blood cells, the ABO antigens also occur on many other cell types in the body.

THE DISCOVERY OF THE ABO BLOOD TYPES

Karl Landsteiner of the University of Vienna knew that mixing bloods of *different* species produced severe reactions. This knowledge made him wonder whether lesser reactions might occur between the bloods of individuals of the *same* species. In 1900, Landsteiner took blood from himself and from five of his colleagues and by centrifugation separated the sera from the red cells. He then recombined the sera and cells in all possible pairings. He noted, for example, that the serum of Dr. Pletschnig strongly agglutinated the cells of Dr. Sturli, but did not affect his own. Thus, the serum of Pletschnig seemed to have an antibody directed against an antigen present on one person's red blood cells but not on another's.

The following data are culled from Landsteiner's original paper (1901). (Note that + = agglutination and 0 = no reaction.)

Serum from	Red blood cells from		
	Ple	Stu	Lan
1. Ple	0	+	0
2. Stu	+	0	0
3. Lan	+	+	0

Interestingly, it is possible to conclude from this table that Landsteinder was blood type O, his serum thus containing both anti-A and anti-B. The reason is that the sera of Ple and Stu agglutinated each other's red cells (lines 1 and 2), so one was type A and the other B. Landsteiner's serum, however, agglutinated the red cells of both his colleagues (line 3).

Only the ABO blood groups could be discovered in the way that Landsteiner did his simple experiment, because only ABO antibodies occur in *normal* serum, that is, in the serum of persons who have not been exposed to foreign antigens. Also for this reason, the ABO antigens and antibodies of persons who are involved in transfusions must be very carefully checked in order to prevent a serious antigen-antibody reaction within the patient's bloodstream.

For his investigations of antigen-antibody reactions over a period of several decades, Landsteiner received a Nobel prize in 1930. He was the first of many subsequent winners whose research was related to genetics (Dixon 1984).

type A or O (Box A). It is not known why these antibodies are present. For no other red cell antigens are any antibodies normally present in human plasma.

As you may have figured out, the presence of anti-A and anti-B in some normal human plasma was the main stumbling block to successful transfusion. For example, type A red blood cells inadvertently given to a type B person will be attacked by the patient's anti-A antibodies. This antigen-antibody reaction within the body results in the *hemolysis* (breaking open) of the red blood cells, the spilling of hemoglobin into plasma, and *anemia* (too few red blood cells to carry oxygen). Such sudden hemolytic anemia results in fatigue, breathlessness, heart fluttering, jaundice and kidney dysfunction (from breakdown of hemoglobin), low blood pressure—and perhaps death, since the transfusion reaction occurs in addition to the loss of blood that necessitated the transfusion.

The reactions involving the *donor's red cell antigens* (A or B or both) and the *patient's plasma antibodies* (anti-A or anti-B or both) are the serious ones when whole blood is transfused. In the reverse situation—that is, the donor has antibodies against patient antigens—the transfusion is usually considered safe. Here the donor antibodies usually do no harm because they are diluted out when one or a few pints of blood are added to the several quarts of blood in the patient. The prohibited combinations are indicated by X's in Table 16.2. The absence of an X, however, does not guarantee a safe combination. There are two additional precautions in the choice of a donor:

1. The donor must be matched for Rh type, positive for positive, negative for negative (discussed shortly). It is especially important that Rh-negative females not be given Rh-positive blood.
2. After selecting donor blood by the ABO and Rh label, the donor and patient bloods should be tested for compatibility by **cross**

Table 16.2
Transfusions prohibited (X) because donor's red blood cells will be attacked by patient's ABO antibodies.

Donor's Blood Type	PATIENT'S BLOOD TYPE			
	A	B	AB	O
A		X		X
B	X			X
AB	X	X		X
O				

matching. The patient's serum is added to the donor cells to test for any significant antibody that is present for any reason. If positive reactions occur, another donor is chosen. Positive reactions may result because of a previous transfusion or pregnancy.

Clearly, the use of the term *universal donor* for persons of type O (because they have neither the A nor the B antigen on their red blood cells) is inappropriate. There are many reasons not related to ABO why a type O person may be an unsafe donor. Similarly, the term *universal recipient* for persons of type AB—because they have neither anti-A nor anti-B—may be a dangerous label, for they may have other antibodies that could react with red blood cells transfused to them.

The safest transfusions of all, called **autologous blood transfusions**, occur when people donate blood to themselves. About two-thirds of all transfusions accompany surgery. When an operation is planned well in advance, some persons can have blood withdrawn ahead of time and get it back when needed. Autologous blood transfusions are particularly valuable for patients with very rare blood types or for those who may have built up many antibodies as the result of prior transfusions or pregnancies. Furthermore, such transfusions carry no risk of blood-borne infections (hepatitis or AIDS), and they reduce the demand on blood banks.

The *ABO* gene has now been isolated and cloned, and the chemical structures of the A and B antigens are known. They are short carbohydrate chains of six sugars that are attached to a carrier molecule, either protein or lipid (Figure 16.14). The carrier is embedded in the red blood cell membrane, and the carbohydrate chain sticks out to form the business end of the molecule. That is, the antigenicity—what makes A react differently from B—is a property of the carbohydrate chain. Whereas the terminal sugar in antigen B is galactose, the sugar in antigen A is *N*-acetylgalactosamine. The small difference in these chemicals so alters the immunological specificity of the antigens that severe consequences ensue from a wrong transfusion. Such fine tuning of antigen-antibody reactions was first studied by Karl Landsteiner, who showed that small chemical groups attached to larger molecules were often the fundamental antigenic units.

The A and B antigens are both derived from the same chemical, the **H substance**, by the addition of the appropriate terminal sugars (see also Figure 16.14). That is, the *ABO* gene codes for an enzyme that adds to the H substance either galactose (by the enzyme encoded by the *B* allele) or *N*-acetylgalactosamine (by the enzyme encoded by the *A* allele). Nucleotide sequencing has revealed that the *A* and *B* alleles differ by four bases, and the resultant enzymes differ by four amino acids. The *O* allele has a frameshift mutation (a single base deletion); its enzyme product is inactive, having no effect on the H substance. Although everyone has some H substance on their red blood cells, blood type O people have the most. Since the H substance can be an antigen in its own right, it is not quite correct to say that blood type O people lack antigens on their red blood cells.

The Rh Blood Group System

Like the antigens of the ABO system, the antigens of the **Rh blood group system** are found on the surface of human red blood cells. They were discovered through an experiment by Landsteiner and colleagues that began by injecting rabbits with the red blood cells of *rhesus* monkeys. The complete Rh system has a dozen or so antigens controlled either by mul-

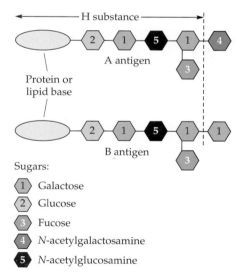

Figure 16.14

Chemical outlines of the A and B blood group antigens. The only difference between the two is the terminal sugar (in color) of the six-unit carbohydrate portion of the molecules. Enzymes encoded by the ABO *gene add these terminal sugars to a common precursor called the H substance.*

tiple alleles of a gene or by a set of closely linked genes (no one is certain which). Despite the complexity, all of the antigens can be classified as Rh positive or Rh negative on the basis of a laboratory test with the single typing antibody anti-Rh:

Reaction of red cells with anti-Rh	Phenotype	Genotype	Frequency (%) in U.S.		
			Whites	Blacks	Natives
Agglutination	Rh positive	*R/R* or *R/r*	60	70	100
No reaction	Rh negative	*r/r*	40	30	0

This classification of people as Rh positive or Rh negative on the basis of whether or not they possess the Rh antigen on their red blood cells is the medically relevant fact. Because red blood cells from both *R/R* and *R/r* carry the Rh antigen and are agglutinated equally well by anti-Rh, the *R* allele is dominant to *r*.

Upon its discovery in 1940, the Rh antigen was immediately recognized as a strong antigen, one that could be responsible for severe transfusion reactions. If blood with the Rh antigen is transfused to a person who lacks it (Rh positive into Rh negative), the recipient responds by making the anti-Rh antibody. Although this antibody will not usually produce problems after a first transfusion—and it disappears after a time—the patient's immune system, through its memory cells, is now sensitized to the Rh antigen. Upon additional transfusions of blood containing the Rh antigen, increasingly severe reactions can be expected as more and more anti-Rh antibodies are produced by the Rh negative patient.

The Rh antigen is also implicated in a disease called **hemolytic disease of the newborn (HDN)**, whose occurrence has been drastically reduced since the 1970s. HDN results from an antigen-antibody reaction that occurs within the fetal blood circulation prior to and at the time of birth, but is restricted to *Rh positive fetuses* carried by *Rh negative women*. The sequence of events leading to HDN is outlined in Figure 16.15.

During one or more Rh-positive pregnancies, red blood cells with the Rh antigen leak into a mother's circulation (Figure 16.15A). Although the bloodstreams of mother and fetus do not generally mix, parts of the placental structure separating the two circulatory systems are very thin. Tiny breaks may allow some fetal red blood cells to enter the mother's system. Alternatively, at the time of birth, when the placenta separates from the uterine wall, it is possible that larger amounts of fetal blood enter the maternal circulation. In response to the Rh antigen on these fetal red blood cells, the woman is sensitized. But by the time the woman has built up any significant amount of anti-Rh, the child has already been born.

When the woman carries a later Rh positive child, the leakage of just a small amount of fetal blood during pregnancy (Figure 16.15B) is sufficient to stimulate a lot of anti-Rh from memory cells. Then anti-Rh passes across the placenta and damages red blood cells of the fetus, who at the time of birth is observed to have HDN. The probability of HDN increases with subsequent Rh positive pregnancies. Jaundice is one of the distinguishing symptoms. Another consequence is indicated by the medical term for HDN: *erythroblastosis fetalis*. When mature red blood cells are destroyed in large number, the body tries to compensate by releasing into the circulation immature ones, erythroblasts.

HDN is a potential problem whenever a fetus possesses any strong red cell antigen that is not present in the mother; this situation is called **maternal-fetal incompatibility.** In these cases, the allele for the antigen is absent in the mother but present in the father, who transmits the relevant

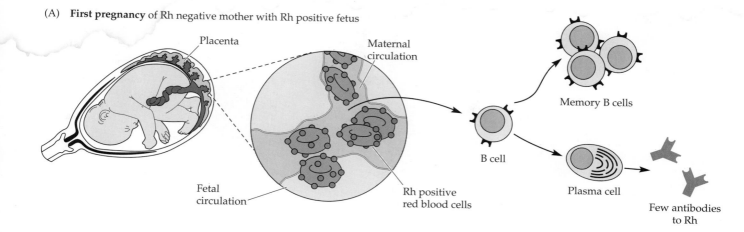

(A) **First pregnancy** of Rh negative mother with Rh positive fetus

Placenta

Maternal circulation

Fetal circulation

Rh positive red blood cells

B cell

Memory B cells

Plasma cell

Few antibodies to Rh

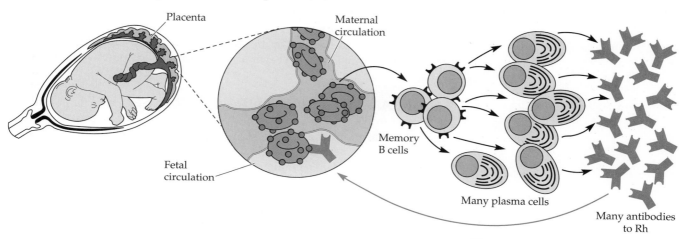

(B) **Later pregnancy** of Rh negative mother with Rh positive fetus

Placenta

Maternal circulation

Fetal circulation

Memory B cells

Many plasma cells

Many antibodies to Rh

allele to the fetus. (If the father is heterozygous, however, not every fetus will be incompatible with its mother.) Mothers who are incompatible with their fetuses for the Rh antigen account for most cases of HDN, although it can happen because of other red blood cell antigens too. Rh-caused cases, however, are generally more severe, over 10% of affected fetuses being stillborn prior to full term.

Since the 1970s, couples starting their families have not needed to worry much about Rh-caused HDN, because the incidence of the disease has been reduced about 90%. The action required to prevent the disease is very simple: Within a few days of the birth of *every* Rh positive offspring to an Rh negative mother, the mother is given a dose of the anti-Rh antibody. It is important that the Rh negative woman be injected starting with the *first* Rh positive fetus—whether it is born alive, stillborn, aborted, or miscarried.

It might seem strange to administer the very antibody that is *un*-wanted. But by injecting anti-Rh, the mother is prevented from making her own anti-Rh. Bear in mind that Rh-caused HDN occurs after the mother has been sensitized by the Rh antigen as a consequence of a prior pregnancy. Upon each subsequent exposure to the Rh antigen, her own immune mechanism responds more strongly to it through more and more memory cells. Most of the leakage of fetal red blood cells seems to occur at the time of birth, when the placenta separates from the uterine wall. Thus, the preventive shot of anti-Rh soon after destroys the invading fetal cells before their Rh antigen has an opportunity to stimulate the mother's

Figure 16.15
The development of hemolytic disease of the newborn. (A) In a first pregnancy, some fetal red blood cells with the Rh antigen leak into the Rh negative mother. In response, the mother makes some anti-Rh and memory cells, but the baby is born before any significant damage is done. (B) In a later pregnancy, leakage of fetal red blood cells from an Rh positive fetus stimulates the mother's memory cells to quickly make much anti-Rh. Entering the fetus, the anti-Rh destroys the Rh positive fetal red blood cells.

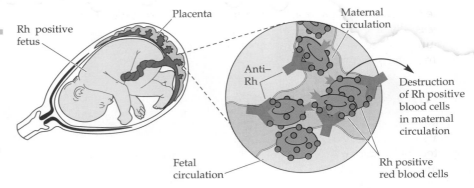

Placenta

Rh positive fetus

Maternal circulation

Anti–Rh

Destruction of Rh positive blood cells in maternal circulation

Fetal circulation

Rh positive red blood cells

Figure 16.16
Protection against hemolytic disease of the newborn due to Rh. Following each Rh positive birth, an injected dose of anti-Rh destroys any fetal red blood cells that leak into the maternal circulation. The mother is therefore never sensitized to the Rh antigen. The injected anti-Rh disappears after a few months.

immune system (Figure 16.16). The Rh antigen is never "registered" in the mother's immune memory, and the injected anti-Rh disappears after a few months (Clarke 1968). Immunologically speaking, the treated mothers enter their second pregnancy, and any succeeding pregnancy, as if it were their first. In other words, whenever the possibility of Rh sensitization occurs, it is prevented by the anti-Rh treatment. Of course, Rh negative mothers must never receive a transfusion of Rh positive blood.

TISSUE ANTIGENS AND ORGAN TRANSPLANTATION

The Australian immunologist G. J. V. Nossal (1978) has written,

> There is perhaps something ghoulish in the thought that immediately after a person's death, a whole tribe of surgeons may descend on the body ready to remove kidneys, heart, liver, lungs, pancreas, and any other bits and pieces that they can get hold of. This block is largely in our minds. Most people in good health would agree that their own organs would be better employed, after their death, in helping to keep another person alive than in being burnt in a crematorium or buried in a coffin.

It is, of course, not the role of medicine to confer immortality by continually replacing worn-out parts. Rather, each of us hopes to live with a minimum of ill health and, after some scores of years, die with dignity. In achieving these goals, the transplantation of organs has a limited but increasing role. For an individual who might lead a productive life except for damage to one vital organ, a transplant operation is sometimes a last but logical recourse. The most common organ to be transplanted, and the one with the most hopeful prognosis over a long period of time, is a kidney. But this operation requires more than a skilled surgeon working in a modern medical setting. The body's "search and destroy" immune system must also be taken into account. If the new kidney's degree of foreignness is too great, reaction by the recipient's T cells will lead to the destruction of the intrusive tissue.

To improve the chances of a "take," two avenues are open. One is to repress the normal activity of the recipient's immune system. Drugs are available that interfere with the normal functions of T cells,* but side effects may be serious. Mainly, the patient's ability to ward off infection is greatly reduced. This deficiency may lead to a fatal infection shortly after a transplant operation or to cancer later in the recovery period. The second method for thwarting the immune system is to choose a donor organ with a minimum degree of foreignness. This matching is done by

---

* The immunosupressant *cyclosporin*, approved by the Food and Drug Administration in 1983, is generally credited with making organ transplantation much less risky.

tissue typing the recipient and potential donors for so-called *transplantation antigens*. By a combination of antigen matching, surgical expertise, and immune suppression, a transplanted kidney today has a very good chance of functioning for many years.

The HLA System and Tissue Typing

Knowledge of transplantation antigens was first obtained in the 1940s from grafts of skin between different inbred strains of mice. The fate of grafted skin depended on so-called **histocompatibility antigens** (*histo*, "tissue") on cell surfaces. Rejection occurred if tissue from the donor strain had histocompatibility antigens foreign to the recipient. Although many different genes and antigens were involved, researchers found that one series of closely linked genes coded for particularly strong cell surface antigens that led to quick rejection if not matched.

Beginning in the 1950s, work on human transplantation genes and antigens was done by several researchers, including Nobel prize winner Jean Dausset at the University of Paris. He found that the immune sera of a few patients who had received multiple transfusions reacted with the *white* blood cells of some, but not all, French people. His methodology, similar to that used to detect red cell blood groups, led to the discovery of white cell blood groups. Further worldwide research has revealed dozens of antigens on white blood cells, many of which also occur on all nucleated cells of the body. (For the latter antigens, white blood cells are simply a convenient cell source.) The typing tests are more complicated than the simple agglutinations used to classify red cell antigens, and initially they gave inconsistent results. But now, pure preparations of monoclonal antibodies for typing white cell and tissue antigens have produced uniform results worldwide.

The strongest tissue antigens are encoded by six or more genes on human chromosome 6, spanning a region of over 3 million bases (Figure 16.17). The genes and the cell surface antigens are known collectively as either the **MHC system** (for major histocompatibility complex, a general name applying to any species) or the **HLA system**, (for human leukocyte

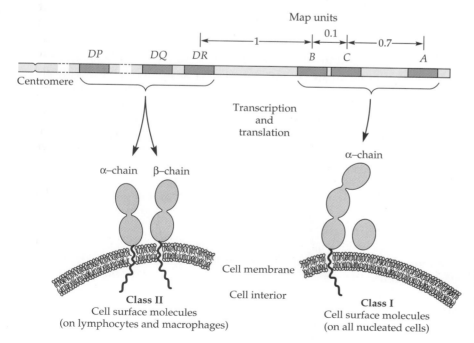

Figure 16.17
The genes and polypeptides of the HLA system. The genes on chromosome 6 are separated by map distances shown at the top. The polypeptide chains of class I and class II molecules have domain structures (depicted by the colored ovals) similar to one another and to the domains of the immunoglobulins. Such homologies reflect a common evolutionary ancestry. The polypeptides also have about 50 amino acids that continue the chains through the cell membrane into the cell interior. (The small uncolored chain of the class I molecules is not encoded by the HLA system.)

395

antigens). (Leukocyte is another term for white blood cell.) The gene products are polypeptides that combine with carbohydrates to form cell surface *glycoproteins*. (The carbohydrate portions are not shown in Figure 16.17.) Two classes of cell surface molecules are encoded by the HLA genes. Class I antigens are encoded by genes *B*, *C*, and *A*; class II antigens are encoded by the several genes within the *D* region, called *DP*, *DQ*, and *DR*. Class I molecules are found on the surface of *all nucleated cells* in the body, whereas class II molecules are found only on the cells of the immune system. In addition to the genes depicted in Figure 16.17, genes coding for some components of serum complement have been mapped to the DNA segment between the *D* and *B* genes. Thus, this region of chromosome 6 is rich in immunological functions.

The histocompatibility antigens were briefly noted earlier in this chapter (Figure 16.5). Recall that to be identified by the cells of one's immune system, a foreign antigen must be presented on cell surfaces in conjunction with one's own MHC protein. This recognition step of the immune process is said to be *MHC-restricted*, fine-tuning immune cells to perceive foreignness by "seeing" both self and nonself at once. Helping to discriminate between self and nonself is the *real* function of the protein molecules encoded by the MHC genes, rather than tissue rejection (Caldwell 1992b).

With large numbers of fairly common alleles, the HLA genes are the most polymorphic known in humans. (True, the VNTRs used in DNA fingerprinting, discussed in Chapter 13, are more polymorphic; but they are not really genes, because they are not transcribed into RNA and translated into a polypeptide product.) About 50 different alleles of the *B* gene, about 10 of the *C* gene, and about 25 of the *A* gene control the presence of roughly 85 different white cell antigens (detected by 85 corresponding HLA antibodies). In any given group of people, the frequencies of many of the alleles are substantial (Table 16.3). Differences between populations have been well documented by anthropologists and others. For example, in Table 16.3, note that the most frequent *B* allele among American whites is the one designated *44* (at 14%); among American blacks, it is allele *58* (at 10%); and among Mexicans, it is allele *35* (at 24%).

A single chromosome 6 will, of course, have all three genes, *HLA-B*, *-C*, and *-A*. The alleles of these genes will almost always be inherited together, because crossing over is uncommon within an interval of less than 1 map unit. The number of possible combinations of *B*, *C*, and *A* alleles on one chromosome is about $50 \times 10 \times 25 = 12,500$, since (in theory) any *B* allele can be associated with any *C* allele and with any *A* allele. A particular combination is called a **haplotype**; for example, a chromosome 6 carrying *B7*, *C4*, and *A3* has the haplotype *7 4 3*. (It is understood that the alleles of the three genes are written in the left-to-right order *B*, *C*, *A*, the way they occur on the chromosome.) For simplicity, we have omitted discussion of the *D* loci, although polymorphisms for these genes are also very important in transplants.

The HLA genotype of a person consists of two haplotypes—one for each chromosome 6. Considering just the *B*, *C*, and *A* genes, we calculate that there are more than 78 million—(12,500)(12,501)/2—different HLA genotypes, and much, much more if we were to consider the dozens of alleles at the *D* loci as well. Because no one genotype is particularly common, no two unrelated people are likely to have exactly the same set of white cell antigens. This great diversity of haplotypes is one reason why organ transplants may fail when the donor tissue has not been specifically chosen to match the recipient.

Figure 16.18 illustrates how the HLA haplotypes for the genes *B*, *C*, and *A* can be inferred from the tissue-typing reactions. Note that any

Table 16.3
The more common alleles of the HLA-B, HLA-C, *and* HLA-A *genes.*

	PERCENT AMONG		
Allele	American Whites	American Blacks	Mexicans
B7	10	9	3
B8	9	3	3
B35	8	6	24
B44	14	7	19
B58	1	10	1
C2	5	12	4
C3	12	9	10
C4	10	16	22
A1	14	3	10
A2	27	15	25
A3	14	7	3
A23	3	10	1
A24	7	3	14
A25	2	0	11
A30	3	15	2

Source: Tiwari and Terasaki (1985).

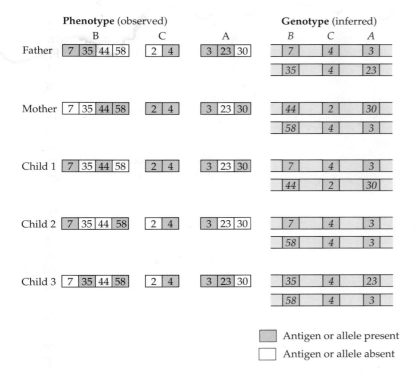

Figure 16.18
The inheritance of histocompatibility haplotypes in a hypothetical family, assuming no crossing over. The antigens that are present (+ reactions) are determined by using corresponding HLA antibodies. We can then infer the genotypes, each consisting of two B C A haplotypes. Note that child 2 is homozygous at both the C and A loci, which establishes the phases for the parental haplotypes. The father, for example, cannot have the B alleles interchanged with respect to the A alleles and still be the father of child 2. Note that the fewer the +'s for an individual, the more easily we can infer the genotype.

individual will type positive for either one or two antigens encoded by each of the three genes—three to six antigens altogether. The complete specification of a person's genotype often requires additional information from close relatives. Once the haplotypes are inferred, however, the inheritance patterns are exactly like that for codominant alleles. In other words, a given *B C A* haplotype is inherited as a unit, that is, as if it were a single allele of a simple gene, because crossing over occurs rarely within this region. Note that each parent can have, at most, a total of two haplotypes. Calling them, for simplicity, 1/2 in the father and 3/4 in the mother, only four different genotypes are possible among their children (excluding the possibility of recombination): 1/3, 1/4, 2/3, and 2/4.

Transplantation

The fewer foreign HLA-encoded antigens present on a donor organ, the more likely it is to escape rejection by the host's T cells. By foreign, we mean HLA antigens on the donor tissue that are different from the host's own HLA antigens. In addition to the HLA system, transplants need to be matched for the ABO blood group system, because the strong A and B antigens appear on many cell types besides red blood cells. There are also a number of *minor histocompatibility systems*.

Kidney Transplantation. The best donor for a kidney transplant, after an identical twin, is a sib with the same HLA and ABO antigens as the recipient. (Rejection episodes due to the minor histocompatibility antigens can usually be suppressed by drug therapy.) In more than 90% of these cases, the transplanted kidney is still functioning a year later; in more than 80% of cases, the kidney survives five years. The survival time of a grafted kidney is reduced when a sib donor has one or more HLA antigens foreign to the recipient. In any event, the donor must be exceptionally well motivated, since the removal of an organ and subsequent existence with only one kidney entail some risk.

Kidneys from unrelated people can also be successfully transplanted, and tens of thousands of such operations have been performed since the 1960s. The donor is often an auto accident victim who is certified brain dead (the usual criterion for death), but whose other body systems can continue to function with external life-support machinery until organs are removed. Although not all transplant surgeons believe that the advantages of tissue typing outweigh possible technical and administrative delays, many studies show that the long-term survival of grafts is related to the closeness of the HLA matching. For example, in one British study (Strachan and Harris 1990), increasing the number of matching HLA haplotypes increased the five-year survival of transplanted cadaver kidneys in the following way:

Number of HLA haplotype matches	Survival of graft at 5 years (%)
0 or 1	63
2	71
3	76
4	90

Another problem with any kidney transplant, however, is that the disease that destroyed the patient's own kidneys may affect the grafted one as well.

Bone Marrow Transplantation. Unlike solid organs, the cells of the bone marrow continually renew themselves from stem cells, and a portion may be removed from a living donor without long-term loss of function. The physician repeatedly inserts a long needle into the upper part of the hip bone of the donor (who is under general anesthesia) to remove about 500 milliliters (1 pint) of the spongy marrow contained within. The cell suspension is filtered to remove chips of bone and clumps of fat and then slowly injected into a vein of the recipient. Such transplants are increasingly successful in treating patients with marrow defects, such as leukemia, immunodeficiency diseases, and some types of anemia. Marrow transplants have also been tried on a number of patients suffering from inborn errors of metabolism (including thalassemia and sickle-cell anemia) that are clinically expressed in the cells of bone marrow. Even more generalized genetic diseases may be treatable if the critical enzyme is transported from grafted marrow to tissues where it is needed.

The **leukemias** are a group of diseases caused by the malignant proliferation of one or another type of white blood cell (Chapter 15). In the marrow, the massive buildup of cancer cells crowds out both red blood cells (thereby producing anemia) and platelets (producing internal bleeding). Also, vital organs are damaged by huge numbers of infiltrating leukemic cells. Advance of the disease can be either slow (chronic) or rapid (acute), the latter leading to death in a few months if untreated. *Acute lymphoblastic leukemia* is the most frequent type of childhood cancer. (A lymphoblast is an immature lymphocyte.)

Depending on the course of the disease and a patient's response to other forms of treatment, bone marrow transplantation is now an accepted procedure for some types of leukemia. Before the transplant, physicians try to kill 100% of the leukemic cells with drugs and radiation, forms of treatment that destroy all rapidly dividing cells (and also cause nausea and hair loss). Then they inject donor marrow cells, taken from a sib with the same HLA type, that may seed the patient's bone marrow cavities and eventually restore immune functions.

But these procedures may have several adverse consequences: (1) recurrence of leukemia if all malignant cells were not destroyed, (2) serious microbial, viral, or fungal infections while the patient's immune system is knocked out, or (3) **graft-versus-host disease**. The latter reaction occurs when the lymphocytes normally present in the *donor* blood marrow mount a pervasive and painful immunological attack against antigens on the cells of the recipient—the reverse of the usual transplant complications. It is fatal in about 25% of transplant patients with acute leukemia.

When a person with leukemia does not have an HLA-identical sib, the physician may instead remove bone marrow from the patient. The marrow is later returned, but first technicians try to purge it of the leukemic cells by monoclonal antibodies directed against cancer-specific surface antigens. In the United States, where a person averages little more than one sib, about 70% of the population do not have an HLA-identical sib; consequently, this form of therapy is being actively investigated.

Other Considerations. Because fewer transplant operations have involved other organs (e.g., heart and liver), the importance of the HLA antigens on these grafts has generally not been fully evaluated. A liver transplant in particular is technically very difficult and expensive ($300,000 or more), requiring great surgical skill and complex support systems.

The problem of sharply limited supply hampers all organ transplants, and many patients die while waiting for their last chance to live. Important questions remain unanswered: Who gets and who pays for an organ? How much effort should be directed toward the use of animal organs (chimpanzees and baboons) or the development of artificial parts? Is it acceptable to consider the organs of anencephalic newborns, who will die because they are missing most of their brains? Should medical resources be directed to entirely different endeavors? These questions invite political, social, and moral commentary (e.g., Engelhardt 1984).

Finally, we note the following transplantation-related puzzle: Why does a pregnant female not reject a fetus "grafted" to her uterus? The developing fetus is likely to have several foreign histocompatibility alleles inherited from the father. Presented to the mother as an organ transplant, the encoded antigens would elicit antibody responses and graft rejection. The significant role of HLA antigens in pregnancies is underscored by some studies that show, surprisingly, that HLA *mismatches* between mother and fetus are advantageous for a successful pregnancy (Rodger and Drake 1987). Thus, the maternal-fetal union seems to have developed interesting, but poorly understood, modifications of immune mechanisms.

THE IMMUNE SYSTEM AND DISEASE

Here we briefly look at several types of disorders that involve the immune system.

HLA Associations and Autoimmune Disorders

An intriguing aspect of the HLA genes is the association of particular alleles with particular diseases (Table 16.4). The prime example is *ankylosing spondylitis*, an inflammation of the areas where tendons and ligaments attach to bones, especially the bones of the hip and spine. In studies of Caucasian groups, those who possessed the HLA antigen B27 increased their risk of the disease 90 times over those who did not have the B27 antigen. In the absence of B27, the disease hardly ever develops. In fact,

Table 16.4
Some associations between disease and HLA antigens.

Disease	HLA Antigen	Relative Risk[a]
Ankylosing spondylitis (joints)	B27	90
Narcolepsy (nerves: sleep disorder)	DR2	34
Juvenile diabetes mellitus (many organs)	DR4, DR3	33
	DR4 only	6
	DR3 only	3
Celiac disease (intestine)	DQ2, DR3	17
21-hydroxylase deficiency (sex development)	B47	15
Rheumatoid arthritis (joints)	DR4	5
Hemochromatosis (many organs)	A3	4
Psoriasis vulgaris (skin)	C6, DR7	4
Multiple sclerosis (nerves)	DQ1, DR2	4

Source: Kostyu and Amos (1989).

[a] Relative risk is the incidence of the disease among persons *possessing* the antigen divided by the incidence among persons *lacking* the antigen. For example, people with the antigen DR2 are 34 times more likely to have narcolepsy than are people without DR2.

ankylosing spondylitis is fairly uncommon (the overall frequency is about 1 in 2,000 persons), and most people with B27 remain healthy.

Different theories have been advanced to explain this correlation. It could be that susceptibility to ankylosing spondylitis is due to an allele of a gene that is closely linked to the *HLA-B* locus. This hypothesis further supposes that chromosomes that contain both the ankylosing spondylitis allele and the *HLA-B* allele *B27* are more common than would be expected on the basis of random combination. However, it could be that the polypeptide encoded by the *B27* allele simply increases susceptibility to the disease. Perhaps the B27 antigen is a cell surface receptor for a virus or other pathogen. Alternatively, the B27 molecule may mimic antigens on a pathogen, so that antibodies directed against the pathogen also attack host tissues carrying B27. At this time no one knows which hypothesis is correct.

One certain physical linkage in this region of chromosome 6 is the gene for *hemochromatosis*, which maps very close to the *HLA-A* locus and is often associated with allele *A3*. Inherited as an autosomal recessive, hemochromatosis (literally, "abnormally colored blood") is characterized by excessive absorption of iron from the intestine into the bloodstream—so much iron that some patients set off metal detectors at security gates (Monmaney 1989). Many more men are affected than women. The skin can turn gray-bronze, and iron deposits can damage the pancreas, liver, heart, and joints. If detected early, the symptoms may be prevented by the simple treatment of bloodletting. (Regular blood donations will do.) If untreated, the symptoms of disease set in gradually, often appearing in men in their 40s and women in their 50s, the latter gaining some protection from menstruation. Although easy to diagnose by blood analysis, the disease is unfortunately often misdiagnosed or not diagnosed at all until irreversible damage occurs.

Although they affect many different body systems, the HLA-associated diseases share some general features. Typically, the disorders are chronic, have late onset, can often be treated to some extent, and involve immune processes in one way or another. They also have a genetic component, but do not show simple Mendelian inheritance, a finding suggesting multiple causative genes and substantial environmental influences. A particular group of such diseases are called **autoimmune disorders**, because so-called *autoantibodies* are present that attack one's own tissues (Table 16.5). Autoimmune diseases affect 5–7% of the population, but it is not clear how they come about. T cells that could damage the body by binding to self-antigens are normally removed by the thymus gland. Perhaps when this mechanism is imperfect, immune cells that damage one's own tissues remain in the body (von Boehmer and Kisielow 1991).

In *systemic lupus erythematosus*, for example, autoantibodies are directed against cell surface components, nuclear substances (DNA, RNA, histones), and cytoplasmic elements (mitochondria, lysosomes, ribosomes) throughout the body. With periods of greater or lesser suffering, the illness affects many organs, most frequently the brain, joints, kidneys, and skin (Rosenthal 1989). Serious kidney damage results from deposits of antigen-antibody complexes that block fluid filtration. Lupus erythematosus is chronic and is weakly associated with the antigens DR2 or DR3 for reasons that are not clear.

Juvenile diabetes (type I, insulin-dependent diabetes mellitus, or IDDM) affects at least half a million people in the United States. Although the onset of clinical symptoms may be sudden, the disease develops only after about 80% of the insulin-producing cells in the pancreas have been destroyed over a course of time by autoantibodies and autoimmune T cells. Histocompatibility antigens DR3 and DR4 are important contributors to the onset of disease in ways that are not understood. In fact, 95% of diabetic individuals—but only 40% of the general population—have one or the other antigen or both. Since about 1 in 20 children of a diabetic parent gets diabetes (versus 1 in 500 for the general population), several genes are probably involved. Furthermore, no one knows what genetic or environmental factors trigger the autoimmunity. Some researchers suggest the involvement of viruses that display antigenic proteins that happen to mimic proteins in insulin-producing cells of the pancreas (Atkinson and Maclaren 1990). The standard therapy for juvenile diabetes includes careful control of diet and daily injections of insulin.

Immune Overreactions

In addition to abnormal responses that characterize autoimmune disease, overstimulation of the immune system occurs in a variety of ways. For example, recent research suggests that the toxins produced by *staphylococcal* bacteria (in "staph" infections) act as **superantigens** (Johnson 1992). Most antigens are ingested, processed, and displayed by macrophages before they are specifically recognized by a small number of T cells. But

Table 16.5
HLA-associated autoimmune disorders.

Disorder	Target of Autoantibodies
Graves disease	Thyroid gland
Juvenile diabetes	Insulin-making cells of the pancreas
Multiple sclerosis	Fatty sheaths around nerve cells
Myasthenia gravis	Receptors in nerve-muscle junctions
Rheumatoid arthritis	Joints
Systemic lupus erythematosus	Brain, joints, skin, kidneys, blood vessels, and other organs

sidestepping this normal route of immune identification, superantigens combine directly with MHC molecules and with an unusually large number of T cells. The increased number of stimulated T cells respond with an increased amount of interleukin 2 (see Figure 16.5), which makes people feel ill with nausea, diarrhea, and fever. Staph infections that lead to abnormally high levels of interleukin 2 occur in some types of *food poisoning* and in the dangerous *toxic shock syndrome* that can be caused by the use of high-absorbency tampons.

The discomforts associated with **allergies** also appear to be unfortunate by-products of certain antigen-antibody reactions. In some people, the antigens, or *allergens* (in recognition of their effects), cause immunological reactions when *inhaled* (e.g., ragweed pollen and animal danders), or *eaten* (chocolate, nuts, bananas), or *touched* (wool, poison ivy), or *injected* (bee venom, penicillin)—often in very small amounts.

Allergic reactions that follow quickly upon exposure to the allergen are provoked by a B cell response leading to circulating antibodies of a special class, IgE (Buisseret 1982). The fast antigen-antibody reaction stimulates certain tissue cells to release *histamine*. The physiological effects of histamine include contraction of the muscles of the lungs and dilation of capillaries. The resulting leakage of fluid into the surrounding tissues is responsible for the discomforts of allergy. *Anaphylaxis*, a severe and rapidly developing form of allergy (e.g., to a bee sting), can sometimes be fatal if not quickly treated with antihistamines.

Immune Deficiencies

In another group of about a dozen diseases, the immune system functions poorly or not at all. These immune deficiency disorders may involve just the B cells, just the T cells, or both.

Agammaglobulinemia, for example, is a rare, X-linked, recessive disease (Figure 16.19). Affected male infants have no detectable plasma cells and virtually no immunoglobulin, except for their initial charge of maternally transmitted antibodies. (Recall that IgG, the major antibody class, readily crosses the placenta. A normal baby's ability to manufacture its own antibodies develops gradually after birth.) Infants remain well for about a year, but thereafter suffer from recurrent bacterial infections. Survival depends on periodic injections of gamma globulin that contain antibodies to common bacterial diseases. Because they possess normal T cells, however, the patients are no more susceptible to common viral diseases than other infants.

Severe combined immunodeficiency disease (SCID) involves the absence of both B and T cells. Infants fail to thrive because of devastating bacterial, fungal, and viral infections, and they usually die in their first years. A few SCID patients have been treated with a bone marrow transplant. Such was the case with David, the "bubble boy," described at the beginning of this chapter. He received bone marrow from his older sister, who was the most closely (but still not perfectly) HLA-matched donor available. Prior to infusion into David, her marrow cells were treated with a monoclonal antibody against an antigen present on mature T cells. The object was to kill donor cells that might otherwise initiate graft-versus-host disease. Although most likely due to transplant complications, the exact cause of David's death is not clear (Simmons 1984).

A distinctive variant of SCID called *adenosine deaminase deficiency* is inherited as an autosomal recessive. This disease, in which DNA synthesis is abnormal, is discussed in Chapter 20, because two young girls with adenosine deaminase deficiency have become the first patients to be managed by the new procedure of *gene therapy*.

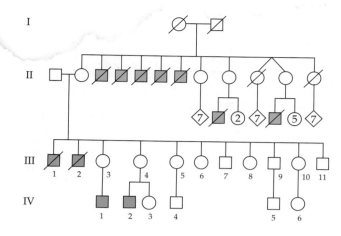

∅ ⊘ Dead at the time of investigation

▨ Died of infection before 2 years of age

▨ Clinical and laboratory evidence of agammaglobulinemia

Figure 16.19
*Pedigree illustrating X-linked inheritance of agammaglobuline-
mia. Affected males are hemizygous for a recessive mutant allele, and their mothers are heterozygous. (From Conley et al.
1986.)*

Acquired immune deficiency syndrome (AIDS) is due to the human immunodeficiency virus (HIV), which invades several cell types, especially
helper T cells (Gallo and Montagnier 1988). (Two variants of the virus,
HIV-1 and HIV-2, are recognized.) Immune functions related to both T
cells and B cells gradually deteriorate. Usually, no overt symptoms occur
for several months to a year after infection with HIV, although laboratories
can detect anti-HIV antibodies against several different proteins in the
viral coat long before clinical onset. For most patients, swollen lymph
nodes are the first sign of disease, as B cells are initially overactive in
making antibodies and helper T cells begin to die. Slowly, over a period
of years, other symptoms develop. *Thrush,* a fungal growth in the mouth,
is only one of several diseases that afflict the mucus membranes and the
skin. So-called **opportunistic infections**—those caused by microbes that
are commonly present but usually effectively controlled—take increasing
advantage of a weakening immune system (Mills and Masur 1990). Implicated microorganisms include the parasite *Pneumocystis carinii,* which
causes pneumonia. (In 1981, five surprising cases of this previously very
rare type of pneumonia were the first indication of AIDS in the United
States.) Fever, diarrhea, and weight loss weaken the body. Cancer, typically the otherwise rare Kaposi sarcoma, and brain damage are not uncommon. Although some of the disease symptoms can be prevented or
eased with medical care, the hard reality is that most patients die five to
ten years after being infected with HIV.

Containing RNA rather than DNA, HIV is a retrovirus with about
10,000 bases. It uses its own reverse transcriptase enzyme to copy the viral
RNA into DNA, which is then inserted into the genome of a helper T cell
(Chapter 15). Once integrated, the provirus replicates along with the cell's
own DNA. Later, the proviral DNA can be transcribed and translated,
leading to the synthesis and release of large numbers of intact virus
particles and the death of the cell. The drugs AZT (azidothymidine, or
zidovudine) and DDI (dideoxyinosine, or didanosine) work by inhibiting
the reverse transcription step. Unfortunately, currently available AIDS
drugs are neither completely effective nor free from serious side effects.

AIDS is transmitted almost entirely by sexual acts or blood exchange
and from mother to fetus or nursing baby. It is prevalent among, but not
limited to, homosexual men, intravenous drug users and their sexual
partners, and recipients of blood products. Because the virus is very
mutable (leading to variation in the coat proteins), and because the disease
involves vital immune cells in complex ways, the development of protective vaccines and effective treatments has proved difficult. Education,
testing, and modification of transmission-related habits appear to be the
best current strategy for containment (Francis and Chin 1987).

1. The immune system identifies an antigenic molecule as foreign, responds specifically to its shape and chemistry, and reacts more vigorously to it upon subsequent exposures.

2. Immune responsiveness depends on various types of white blood cells. B cells manufacture antibodies; T cells kill by cell-to-cell contact or regulate other elements of the immune system; and macrophages present antigen to T cells and engulf foreign particles.

3. As an immune cell matures, it develops receptors that respond to one specific antigen. When stimulated by that antigen, it grows into a clone of immunologically active cells and a reserve of memory cells.

4. Antibodies are Y-shaped molecules composed of two heavy and two light polypeptide chains. The specificity of the antigen-binding pockets at the tips of the Y is determined by the particular amino acids of the hypervariable regions.

5. The great diversity of antibodies that plasma cells can make results largely from the multitude of ways that DNA elements recombine with each other during somatic differentiation of lymphocytes.

6. Although the many antibodies that are present in any person's plasma are a mixed lot, a monoclonal antibody preparation is pure and homogeneous. Monoclonals are produced by hybridomas, each made by fusing a specifically activated plasma cell with a cancer cell.

7. Red blood cells have antigens that determine the various blood group systems. Each set of antigens is controlled by a different—often polymorphic—blood group gene. Red blood cells carrying a given antigen are detected by agglutination with the corresponding antibody.

8. The ABO antigens are the most important in transfusions because the antibodies (anti-A and anti-B) occur naturally in the plasma of persons lacking the corresponding antigen on their red blood cells.

9. The A and B antigens are similar molecules differing by a single sugar residue. They are synthesized from the H substance via enzymes coded for by the *A* and *B* alleles, respectively. The O enzyme is inactive, thereby leaving a lot of H unaltered on the red cell surface.

10. Hemolytic disease of the newborn can be due to maternal-fetal incompatibility for any strong antigen, but most cases are due to the Rh (rhesus) antigen. Nearly complete protection against Rh-caused hemolytic disease is afforded by injections of anti-Rh following every Rh positive pregnancy of Rh negative women.

11. The HLA antigens on the surface of cells are encoded by several highly polymorphic, closely linked genes—the major histocompatibility complex (MHC). The particular alleles of the linked HLA genes are inherited as a group called a haplotype.

12. Matching the HLA antigens of donor and host helps make transplants successful by avoiding rejection of grafted cells by the patient's T cells.

13. Bone marrow transplantation is useful for some types of leukemia and other diseases. Besides the usual complications of transplants, patients are subject to graft-versus-host disease.

14. Particular HLA alleles are associated with some chronic, partly genetic diseases. Often the body makes autoantibodies. Discomfort and disease can also result from overstimulation of the immune system.

15. Another group of usually severe diseases results from poor functioning of the immune response. These may involve B cells, T cells, or both. In AIDS, the immunodeficiency virus kills T cells, leaving the victim vulnerable to opportunistic infections.

ABO blood group system
agglutination
antibody
antigen
autoimmune disorder
autologous blood transfusion
B cell
blood plasma
clonal selection theory
combining site
constant (C) domain
cross matching

cytotoxic T cell
erythrocyte
graft-versus-host disease
haplotype
heavy (H) chain
helper T cell
hemolytic disease of the
   newborn
histocompatibility antigen
HLA system
H substance
hybridoma

hypervariable region
immunoglobulin G
light (L) chain
lymphocyte
lymphokine
macrophage
major histocompatibility
   complex (MHC)
maternal-fetal
   incompatibility
memory cell
monoclonal antibody

opportunistic infection
plasma cell
polymorphic
Rh blood group system
secondary immune response
somatic recombination
superantigen
T cell
variable (V) domain

QUESTIONS

1. What is the cellular basis of immunological memory, that is, the stronger antibody response to a second exposure to an antigen?

2. Explain how patients with multiple myeloma contributed to the determination of antibody structure.

3. A protein-splitting enzyme called papain (from the papaya tree) cuts the antibody molecule into exactly three pieces. Two are identical to each other and can still bind antigen well; the third piece does not bind antigen at all. Where does papain split the antibody molecule (see Figure 16.6)?

4. In 1945 a California woman accused Charlie Chaplin of fathering her child. The ABO blood types were: woman, A; Chaplin, O; child, B. When a jury decided that Chaplin was the father, an editorial writer commented that California had declared that black is white and up is down (*Boston Herald*, April 19, 1945). What do you think?

5. Rare baby mix-ups have sometimes been resolved by blood groups. For example, assume that Mrs. W returned home from the hospital with the name label Y on her newborn child, while Mrs. Y's newborn had the name label W. Assume that the ABO blood types of the individuals were:

Mrs. W: O   Mr. W: AB   Newborn with W label: A
Mrs. Y: A   Mr. Y: B   Newborn with Y label: AB

Can you decide whether the hospital switched the labels or the babies?

6. Substances that resemble the A, B, and H antigens are widespread in nature, and we undoubtedly eat some. Does this suggest where normally occurring anti-A and anti-B in human plasma might come from?

7. Considering the *ABO* and *Rh* loci together, what *phenotypes* would you expect among the offspring of the mating *A/B R/r* × *A/O R/r*? What is the expected proportion of each phenotype?

8. A baby is born with hemolytic disease of the newborn, but both the baby and its mother are Rh positive. Give a reasonable explanation for this occurrence.

9. Within the pedigree symbols is recorded the reaction of each person's red blood cells with anti-A, anti-B, and anti-Rh, in that order. Give the ABO and Rh *genotypes* of each person. (*Note:* + = agglutination, 0 = no reaction.)

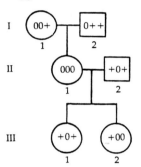

10. In the mouse, there are approximately 150 *V* elements, 12 *D* elements, and 4 *J* elements coding the variable domain of the H chain of IgG. How many different heavy chain variable domains can a mouse make?

11. What are the HLA genotypes (for just the *B* and *A* genes) of the family members below? Indicate properly as haplotypes. (*Note:* + = antigen present, blank = antigen absent.)

	B antigens			A antigens			
	5	8	40	2	3	9	11
Father	+		+	+		+	
Mother	+	+			+		+
First child	+			+	+		
Second child	+	+		+			+
Third child		+	+			+	+

12. For the family in question 11, would the first or third child make a better kidney donor to the second child?

13. For the family in Figure 16.19, assuming no crossing over, what additional genotype is possible among children of the given parents?

14. For the family in Figure 16.19, what four genotypes could the children have if crossing over occurred in the mother between the *C* and *A* loci?

15. Why are the HLA white cell antigens more useful than the ABO or Rh red cell antigens in cases of disputed paternity?

16. The cornea (the transparent front surface of the eyeball) has no blood or lymphatic vessels. Does this suggest why corneal transplants from unrelated donors are usually successful without using immunosuppressants?

17. In bone marrow transplants, patients are usually irradiated to destroy their immune cells, making way for sites to be populated by the donor's immune cells. Yet when he received a bone marrow transplant from his sister, David, the "bubble boy," was not irradiated. Why?

18. Since transplants do not occur in nature, antigens that are important in tissue transplantation did not arise over evolutionary time to counter transplants. So how did people develop the ability to routinely and strongly reject foreign tissue?

19. In conjunction with other modern medical practices, organ transplants have led to a reexamination of the definition of the time of death. Discuss.

FURTHER READING

The immune system is a favorite subject for articles in *Scientific American*, *Discover*, and other basic science magazines. Collections have been edited by Burnet (1976) and Paul (1991). Additional recent articles are by Caldwell (1991) and Johnson, Russell, and Pontzer (1992) on superantigens; von Boehmer and Kisielow (1991), Rennie (1992), and Caldwell (1992b) on the self-versus-nonself learning process; and Golde (1991) on the body's production of blood cells. Mills and Masur (1990) describe treatments for AIDS infections, and Wachter (1992) discusses political activism by AIDS advocates. Lax (1984) gives a reporter's account of a hospital unit devoted to bone marrow transplantation. More details about blood groups can be found in Mange and Mange (1990).

Population Concepts
17

Through newborn testing programs, Sarah and John M. learned that their firstborn had the phenylketonuria (PKU) genotype. This autosomal recessive disorder (genotype p/p), if untreated, leads to profound physical and mental abnormalities (Chapter 1). A diet with carefully controlled amounts of phenylalanine maintained for a decade or more can, however, lead to normal or near-normal phenotypes. Although fairly effective, the diet is expensive, not very tasty, and therefore difficult to adhere to. If the diet is not maintained long enough, additional problems may be in store for the children of treated females (Chapter 14).

Knowing that they were both heterozygous for PKU (P/p) but encouraged by treatment possibilities, Sarah and John had additional children. The second was a heterozygote, the third homozygous normal (P/P), and the fourth had again the phenylketonuria genotype. The p/p children were, of course, put on low-phenylalanine diets and grew to enjoy full and happy lives, including marriage and children.

Although this kind of scenario has been played out many times, much to the joy of parents and children, the story has a seemingly unfavorable side effect. Sarah and John and their children who carry the PKU allele (either as homozygotes or heterozygotes) knowingly contributed harmful genes to subsequent generations. This situation prompts important questions about genetics and society: How much responsibility should any generation accept for the genetic material that it transmits to the next generation? And how much harm to future generations is produced when medical science allows individuals with genetic disorders—who formerly would have had no children—to live well enough and long enough to reproduce?

These questions deal with populations and thus require a kind of thinking that we have not previously considered. This chapter and the next present some basic concepts in the field of **population genetics**, which is based partly on logical extensions of Mendel's ideas. Human population genetics seeks to understand how inherited traits are distributed in diverse groups of people and why the statistics that apply to genetic conditions may change over time. You may be surprised to learn that for rare recessively inherited diseases like PKU, only very slow frequency changes over time are expected when treated individuals live long enough to reproduce.

POPULATION FREQUENCIES

A **population** is simply a collection of individuals. To describe the proportion of a certain type within a population, we use the term *frequency*, which is similar in meaning to the term *probability*. Both words express a group concept; a single individual is not usually described by a frequency or probability.

Suppose that all members of a population—a group of parents—have their genotypes tested, giving these results:

Genotype	Number	Frequency
B/B	114	57%, or 0.57
B/b	56	28%, or 0.28
b/b	30	15%, or 0.15
Total	200	100%, or 1.00

Each **genotype frequency** is the proportion that that particular genotype is of the whole, expressed as a percentage or, more usually, as a decimal fraction. For example, freq(B/B), the frequency of the B/B genotype, is 114/200 = 0.57. Note that the sum of all genotype frequencies must necessarily be 1, a rule that can provide a check on arithmetic.

If each genotype in a population corresponds to a different phenotype, then genotype frequencies and **phenotype frequencies** are the same. If B is dominant to b, however, then the genotypes B/B and B/b are phenotypically indistinguishable; thus we have:

Phenotype	Number	Frequency
Dominant	170	0.85
Recessive	30	0.15
Total	200	1.00

A little more arithmetic is needed to obtain what are called the **allele frequencies**, the proportions of the total number of alleles represented by B or b. We note that each individual possesses two alleles per locus, so that among the 200 parents, there are 400 alleles to be tallied as either B or b. Assuming that B/B's and B/b's are distinguishable from each other (no dominance), the 114 B/B homozygotes together possess 228 B alleles. The 56 heterozygotes add to the tally of both allelic forms, 56 of B and 56 of b. The total number of B alleles is therefore 228 + 56, and the total number of b alleles is 60 + 56. Allele frequencies can then be calculated:

For the B allele: freq(B) = (228 + 56)/400 = 284/400 = 0.71
For the b allele: freq(b) = (60 + 56)/400 = 116/400 = 0.29
Sum of allele freq = 1

To make this calculation, homozygous and heterozygous genotypes must be distinguishable from each other. We will see later in this chapter what can be done to estimate allele frequencies when this condition is not met. The allele frequencies are often symbolized more briefly as p = **freq(B)** and q = **freq(b)**. Using this notation, $p + q = 1$.

An easily remembered relationship exists between allele frequencies and genotype frequencies:

freq(B) = freq(B/B) + (1/2)freq(B/b)
freq(b) = freq(b/b) + (1/2)freq(B/b)

In words, the top equation says that the frequency of the B allele (left side of equation) equals the frequency of the homozygous genotype plus half the frequency of the heterozygous genotype. This is a commonsense expression, a weighted average. That is, the alleles in the homozygote are *all B*, while the alleles in the heterozygote are *one-half B*. (If you like algebra, you might want to derive the formulas.)

Assuming that the frequencies given earlier apply to a *parent* generation, can we now predict the proportions of the various genotypes among their *children*? The answer is yes, but we must first know what matings occur among the parents. In particular, we first seek the proportions of the six possible matings:

1. $B/B \times B/B$
2. $B/b \times B/b$
3. $b/b \times b/b$
4. $B/B \times B/b$
5. $b/b \times B/b$
6. $B/B \times b/b$

The frequency of mating 1, for example, is the proportion that this cross is of all crosses in the population. But to find its value, we need to know the rules of the mating game. For example, if "like attracts like," it may be that the only matings that occur are 1, 2, 3, and 4, in which both mates have the same phenotype. But many other mating schemes can be envisioned. In short, we need to know the **mating frequencies**.

Random Mating

One mathematically convenient and sometimes realistic assumption, called **random mating**, is that mating occurs without regard to a person's genotype or phenotype. This assumption is true (or very nearly true) for some human traits—blood groups or HLA types, for example. It means that an individual is equally likely to mate with any member of the opposite sex. Random mating frequencies depend only on the frequencies of the two pertinent genotypes (which are multiplied together). For example, for mating 1, we calculate

$$\text{freq}(B/B \times B/B) = \text{freq}(B/B\ \male) \times \text{freq}(B/B\ \female)$$
$$= (0.57)(0.57) = (0.57)^2$$

The expression $\text{freq}(B/B\ \male)$ means the frequency of B/B among males, and $\text{freq}(B/B\ \female)$ means the frequency of B/B among females. We are assuming that these are equal, that is, that genotypes are distributed similarly in the sexes.

Matings 4, 5, and 6 are a bit different from 1, 2, and 3 in that the two mates are genotypically different from each other. As a consequence, one must add together the frequencies of *reciprocal* crosses. For example, the mating $B/B \times b/b$ (mating 6) means either $B/B\ \male \times b/b\ \female$ or the reverse, $B/B\ \female \times b/b\ \male$. Thus,

$$\text{freq}(B/B \times b/b) = \text{freq}(B/B\ \male) \times \text{freq}(b/b\ \female)$$
$$+ \text{freq}(B/B\ \female) \times \text{freq}(b/b\ \male)$$
$$= (0.57)(0.15) + (0.57)(0.15)$$
$$= 2(0.57)(0.15)$$

The factor 2 is present in the random mating frequencies of matings 4, 5, and 6. It is not present in matings 1, 2, and 3 because the spouses have the same genotype; reversing the sex is not a different situation.

For the specific numbers assumed, the random mating frequencies for the three genotypes are summarized in Table 17.1. Note that just as the two possible *alleles* in a population have frequencies that add to 1, and the three possible *genotypes* have frequencies that add to 1, so, too, do the six *matings* have frequencies that add to 1.

Using the mating frequencies in Table 17.1, we can find the frequency of any particular genotype among the offspring. Note, for example, that

Table 17.1
Random mating frequencies, assuming that genotypes B/B, B/b,
and b/b have frequencies 0.57, 0.28, and 0.15, respectively, and
that individuals mate randomly with respect to each other.

Mating	Frequency under Random Mating
1. freq(*B/B* × *B/B*)	$(0.57)^2 = 0.3249$
2. freq(*B/b* × *B/b*)	$(0.28)^2 = 0.0784$
3. freq(*b/b* × *b/b*)	$(0.15)^2 = 0.0225$
4. freq(*B/B* × *B/b*)	$2(0.57)(0.28) = 0.3192$
5. freq(*b/b* × *B/b*)	$2(0.15)(0.28) = 0.0840$
6. freq(*B/B* × *b/b*)	$2(0.57)(0.15) = \underline{0.1710}$
	Total = 1.000

the genotype *b/b* is expected among the children of matings 2, 3, and 5. From these parents, the expected fractions of *b/b* children based on Mendel's first law, are

Mating	Probability of *b/b* offspring
2. *B/b* × *B/b*	1/4
3. *b/b* × *b/b*	1
5. *b/b* × *B/b*	1/2

Based on this and similar tables for other offspring genotypes, we could then predict the overall frequencies of the three genotypes among the offspring. These calculations are not conceptually difficult, but they are a bit tedious. There is an easier way to get at the answer.

THE HARDY-WEINBERG LAW

G. H. Hardy was an English mathematician and Wilhelm Weinberg was a German clinical physician working in the early part of this century. In 1908, Hardy and Weinberg applied the recently rediscovered rules of Mendel to randomly mating populations in a manner somewhat similar to that used in our earlier example. They expressed their results in what has become known as the **Hardy-Weinberg law**. For two alleles of an autosomal gene, *B* and *b*, the Hardy-Weinberg law can be stated in two parts:

1. Under so-called Hardy-Weinberg conditions, the genotype frequencies in a population become, after *one* generation,

freq(*B/B*) $= p^2$
freq(*B/b*) $= 2pq$
freq(*b/b*) $= q^2$

where *p* and *q* are allele frequencies, that is, $p = $ freq(*B*) and $q = $ freq(*b*). These equations mean that the genotype frequencies can be predicted from the allele frequencies in any generation. Here are the conditions under which the Hardy-Weinberg law is applicable:

A. Approximately random mating
B. A fairly large population (several hundred or more)
C. A negligible amount of mutation between the *B* and *b* alleles

D. A negligible amount of migration into and out of the population
E. A negligible amount of selection, with all genotypes about equally viable and equally fertile

2. As long as the Hardy-Weinberg conditions prevail, the allele and genotype frequencies do not change. That is, generation after generation, the frequencies of the three genotypes remain constant at p^2, $2pq$, and q^2. The constancy over time is called an **equilibrium**, meaning no net change in frequencies. Of course, the individuals in succeeding generations are not the same; it is the expected frequencies of their genotypes that stay the same.

Strictly speaking, the Hardy-Weinberg conditions are more stringent than we have indicated. They include an *infinitely* large population, *no* mutation, *no* migration, and *no* selection. But for practical purposes, the law can be applied where these conditions are only approximated. There is no living population that exactly meets the mathematical model. But there are many genes in many populations that meet the conditions sufficiently well that the Hardy-Weinberg law becomes usable and useful.

A proof of the Hardy-Weinberg law is based on the equivalence of random mating of genotypes and random combination of gametes at the time of fertilization. If it is a matter of chance which genotypes mate, it must also be true that it is a matter of chance which allele (*B* or *b*) contained in an egg combines with which allele (*B* or *b*) contained in a sperm. We set up a checkerboard for the entire population at once, rather than for a particular mating (as we have done before). As in any other checkerboard, we label the rows and columns with the parental gametes—in this case, two sorts of eggs and two sorts of sperm with frequencies p and q:

The offspring genotypes are written in the four squares of the checkerboard by combining the alleles and multiplying the frequencies heading the rows and columns. Thus, among the children,

freq(B/B) = p^2 [from top left square]
freq(b/b) = q^2 [from bottom right square]
freq(B/b) = $pq + pq = 2pq$ [from other squares]

The second part of the Hardy-Weinberg law, that which asserts constant frequencies over successive generations, is shown by constructing a checkerboard for the *next* generation. To label its rows and columns, we need to first calculate the allele frequencies (equal to the gamete frequencies) among the children from the previous checkerboard. Recall that an allele frequency is equal to all the homozygotes and half the heterozygotes. Thus,

freq(B) = $p^2 + (1/2)(2pq) = p^2 + pq = p(p + q) = p(1) = p$
freq(b) = $q^2 + (1/2)(2pq) = q^2 + pq = q(p + q) = q(1) = q$

We see that the allele frequencies are unchanged; thus, the headings and body of the next checkerboard are unchanged from the last. Simply stated, the same little checkerboard is repeated over and over, showing the constancy of genotype frequencies at p^2, $2pq$, and q^2.

The Meaning of the Hardy-Weinberg Law

The Hardy-Weinberg equilibrium is a statement of no change—of no evolution. Over long periods of time, it is likely to be unrealistic, because we know that populations do evolve. But over short periods, the Hardy-Weinberg stability may apply. The maintenance of alleles at constant frequencies also means that genetic variability within a population is not blended away over successive generations. This Mendelian notion of discrete hereditary units was an important contribution to evolutionary thought.

A population evolves when the conditions of the Hardy-Weinberg law are not met: by departures from random mating, by unpredictable changes that occur because a population is small, by mutation, by migration, and especially (in many cases) by natural selection. These mechanisms lead to allele frequency changes, with corresponding changes in the average phenotype of a population. The accumulation of modifications over time brings about the evolutionary patterns that have been documented by paleontologists and others. Although the grand sweep of phylogenetic changes over millions of bygone years cannot be correlated with specific genes, more recent evolutionary events can sometimes be studied by analyzing allelic changes. The emergence of high frequencies of melanistic (darkly pigmented) forms of moths in industrially polluted regions is a well-known example. Such analyses use the Hardy-Weinberg law, which is modified as needed.

It is important to note that the frequency of an allele remains constant under Hardy-Weinberg conditions whether it is dominant or recessive. Dominance is *not* an evolutionary force; it is merely a label for how alleles express themselves in a heterozygote. An allele does not automatically come to have a frequency of 1/2 (giving the dominant phenotype to three-quarters of the population). The equilibrium frequency of an allele depends primarily on how it affects reproductive success or failure. Some dominant alleles impair the ability of persons to have offspring, as do some recessive alleles. The frequency of an allele then changes because it affects reproduction, not because it is dominant or recessive.

The actual frequency of a hereditary disease in human populations, far from being 75% (for a dominant allele) or 25% (for a recessive), is (in either case) 1 in thousands of people to 1 in millions. The low frequency persists whether it is inherited in a dominant or a recessive pattern. As Hardy says in his 1908 paper, "There is not the slightest foundation for the idea that a dominant character should show a tendency to spread over a whole population, or that a recessive should tend to die out."

The frequency of an allele can theoretically be any number from 0 to 1 inclusive. As noted, its actual frequency will be the result of a long evolutionary history. The genotype frequencies predicted from the Hardy-Weinberg law will vary accordingly. A graphical display of some representative values is presented in Figure 17.1. As freq(A) varies from 1 to 0 (reading from top to bottom), the frequency of the A/A homozygote does likewise; but since the genotypic proportion is a squared function, the decrease in freq(A/A) is rapid. For example, as freq(A) = p decreases from 1 to 0.5, freq(A/A) = p^2 decreases from 1 to 0.25. If the frequency of an allele is quite low (say, 1 in 100), then the corresponding homozygote is

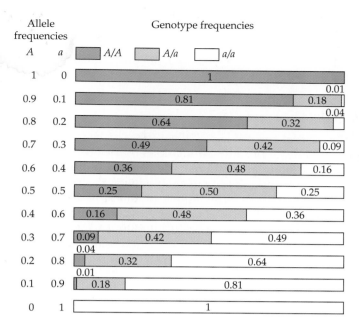

Figure 17.1
The expected genotype frequencies in a Hardy-Weinberg population for different values of the allele frequencies. A fuller geometrical interpretation for three of the lines is given in Figure 17.2.

rare indeed (1 in 10,000). The frequency of heterozygotes is largest when the two alleles are equal in frequency and decreases to 0 at both extremes. In the middle third of the range of allele frequencies, heterozygotes are the largest genotypic class. The main features of the Hardy-Weinberg equilibrium are also nicely shown by a geometrical scheme based on checkerboards (Figure 17.2).

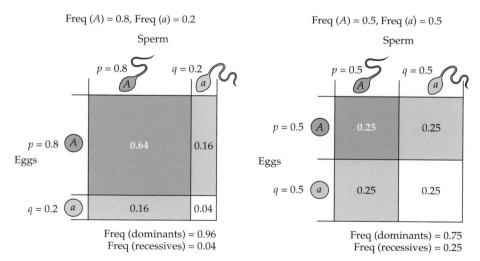
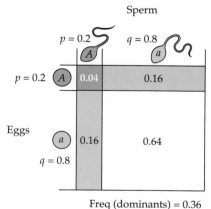

Figure 17.2
Checkerboards used to represent three Hardy-Weinberg equilibriums. The frequencies of the parental gamete types, A or a, correspond to allele frequencies, since each gamete carries one allele. The linear dimensions of the row and column headings are proportional to the allele frequencies in the parents, and the areas within the checkerboards are proportional to the genotype frequencies in the offspring. Assume that A is dominant to a; then the shaded areas correspond to the dominant phenotype, and the unshaded to the recessive phenotype.

SOME APPLICATIONS

Now let us see how Hardy-Weinberg calculations can give some interesting results.

The Sickle-Cell Allele

We noted in Chapter 12 that the sickle-cell allele (Hb^S, or more simply S) has a relatively high frequency in some populations that were exposed to the malarial parasite over many generations. The black population of the United States, for example, although not now generally at risk for malaria, derives in large part from African populations of the seventeenth and eighteenth centuries that were at risk. In some African groups the frequency of the sickle-cell allele is as high as 20% today, and it may have had a similar frequency in centuries past. A random sample of 2,000 newborn American blacks (successive births in a hospital, say) would typically consist of the following genotypes:

Shorthand genotype	Number	Frequency
A/A	1,804	0.902
A/S	192	0.096
S/S	4	0.002
Total	2,000	1.000

Thus, about 1 in 500 newborns (4/2,000) has sickle-cell anemia (S/S), and almost 1 in 10 (192/2,000) is a carrier (A/S) with a normal phenotype. We can calculate the frequency of the sickle-cell allele, S, in this sample as

$$\text{freq}(S) = q = 0.002 + (1/2)(0.096) = 0.050$$

In words, about 1 in every 20 hemoglobin alleles present in U.S. blacks codes for sickle-cell hemoglobin.

The random frequency of matings between two carriers is

$$\text{freq}(A/S \times A/S) = (0.096)^2 = 0.0092$$

or a little less than 1% of matings among U.S. blacks. It is primarily these matings that produce offspring at risk to sickle-cell anemia, since most S/S individuals do not become parents: They either die before reproductive age or, as adults, do not have children. To be exact about the expected fraction of heterozygote × heterozygote matings, we should modify the numbers in our table to reflect the *mating* population rather than the *newborn* population. If we assume, for simplicity, that no S/S genotypes reproduce but that all others are equally viable and fertile, the effective *mating* population becomes:

Genotype	Number	Frequency
A/A	1,804	0.9038
A/S	192	0.0962
S/S	0	0
Total	1,906	1.000

We do not wish to belabor the tiny difference between these two sets of frequencies, except to note that the nonreproduction of most S/S persons over the many generations since the seventeenth century accounts for part of the decline in the frequency of the S allele from approximately 20% to about 5%. As will be explained in the next chapter, the introduction of alleles from white populations into black populations through interracial marriages also accounts for some of the reduction in freq(S).

The Problem of Dominance

Among 250,000 randomly chosen infants, perhaps just one will be affected with the genetic disorder alkaptonuria, whose genotype is k/k. The diagnosis is simple—the darkening of urine in the diapers—but there is currently no test to tell which of the remaining 249,999 infants are normal carriers, K/k, and which are normal homozygotes, K/K. These observations are summarized as follows:

Genotype	Number	Frequency
$K/-$	249,999	$0.999996 \approx 1$
k/k	1	$0.000004 = 4 \times 10^{-6}$
Total	250,000	1.000

When we try to calculate the frequency of the alkaptonuria allele, $q = \text{freq}(k)$, a problem arises: We do not know the number of heterozygotes. Therefore, the previous method, which requires this information, cannot be used. Perhaps some assumptions about the population might aid in calculating q. Because the disease is very rare and relatively benign, a reasonable supposition is that the distribution of genotypes approximates the Hardy-Weinberg formulation. Making this assumption, we would then expect

$$q^2 = \text{freq}(k/k) = 4 \times 10^{-6}$$

Taking the square root of both sides, we get an estimate of the frequency of the recessive allele for alkaptonuria:

$$q = \sqrt{4 \times 10^{-6}} = 2 \times 10^{-3} = 0.002$$

Thus, among all the alleles of this gene, 1 in 500 gives rise to a defective enzyme if the population in question meets the Hardy-Weinberg conditions. What we are doing is equating an *observed* frequency (4×10^{-6}) with an *expected* frequency (q^2), given some hypothesis about the population. If the hypothesis does not apply, we err in proceeding with the calculation.

Although the Hardy-Weinberg law seems to be acceptable, the frequency estimate of 0.002 is not very good for another reason. If, by chance, the sample of 250,000 includes just a few more affected individuals, our estimate might be twice as big. On the other hand, if the sample does not happen to include even one alkaptonuric, then we have an allele frequency of 0. The wide range of possible estimates arises because we are looking at very small numbers of affected individuals.* If we could identify and tally all of the normal *heterozygotes* (of which there are many more), our potential error would be much less. But we have no choice in this case, and we accept $q = 0.002$ as the best possible estimate of the allele frequency under the circumstances.

---

* Some political, social, economic, and medical decisions are based on estimates of this type; lacking extensive knowledge about uncommon events, we rely on whatever imperfect information is at hand.

Having estimated the allele frequency, however roughly, we can now estimate the proportion of heterozygotes in the population:

$$\text{freq}(K/k) = 2pq = 2(0.998)(0.002) \approx 0.004$$

Thus, about 1 person in 250 is a carrier, a somewhat surprising result when we started with the information that only 1 in 250,000 has the metabolic disorder. For every person with alkaptonuria, there are about 1,000 carriers.

X-Linked Genes

Genes on the X chromosome, but not on the Y, will be present twice in females, but only once in males. For this reason, the equilibrium frequencies under Hardy-Weinberg assumptions must be stated separately for the two sexes:

	Genotype	Hardy-Weinberg equilibrium frequency
Among females:	H/H	p^2
	H/h	$2pq$
	h/h	q^2
Among males:	H/(Y)	p
	h/(Y)	q

H and h are used here to represent the two alleles of any X-linked gene (e.g., hemophilia); p and q represent the frequencies of H and h, respectively, in either sex. Among females, the Hardy-Weinberg frequencies are just as they would be for an autosomal gene. A male's *one* allele, however, either H or h, determines his X-linked phenotype. Therefore, the frequency of male genotypes or phenotypes are the same as the allele frequencies. For example, approximately 1 in 10,000 males is afflicted with classic hemophilia and therefore has the genotype h/(Y). This proportion tells us that $q = \text{freq}(h) = 1/10,000 = 10^{-4}$. The Hardy-Weinberg formulation suggests why the trait is exceedingly rare (but not absent) among females. An affected female must have the genotype h/h and the expected frequency $q^2 = (10^{-4})^2 = 10^{-8}$, or 1 in 100 million. If the total population of the United States (more than 250 million) were one big randomly mating Hardy-Weinberg population, we would expect to find just a few female hemophiliacs. Indeed, a few cases of homozygous female bleeders, with bleeder fathers and carrier mothers, have been reported.

INBREEDING

Populations do not always mate randomly as assumed in our previous discussions. For some characteristics, people tend to mate **assortatively**, that is, like with like. Human groups mate assortatively with regard to stature (e.g., tall with tall), skin color, intelligence, and some other quantitative traits. Parallel to matings that occur between couples with *similar phenotypes* are matings that occur between couples who are *related* to each other. Such couples may possess copies of precisely the same gene inherited from a common ancestor. A mating between relatives is called **consanguineous** ("with blood"),* and the offspring are said to be **inbred**.

---

* Although consanguinity derives from shared *alleles* rather than shared *blood,* the use of the latter term persists in some everyday expressions: "bloodline," "full-blooded," "blue-blood," and so on.

Inbreeding is maximized in self-fertilizing species. Mendel himself theorized about successive generations of self-fertilization. He assumed that an initial generation contained known numbers of garden peas of the three genotypes A/A, A/a, and a/a. Then each plant—mating with itself—contributed equally to the next generation. He showed that continued self-fertilization led to a regular *decrease in the fraction of heterozygotes* and a corresponding *increase in the fraction of homozygotes*. This change is the basic result for all forms of inbreeding, but the rate of change depends on the strength of the inbreeding.

In humans and other species with separate sexes, the closest form of inbreeding comes from parent-child or brother-sister mating. Such matings, called **incestuous**, are usually prohibited by law and by religion. The incest taboo probably evolved very early in prehistory to minimize adverse effects on offspring and to help maintain the stability of family units. Because incest involves only family members and because those involved almost always feel guilt and shame, the matter is not often resolved in open court, and its actual frequency is unknown.* In some cultures and during some periods of history, however, brother-sister marriages were encouraged among royalty. For example, Cleopatra may have been a child of a brother-sister mating (Figure 17.3).

Plant and animal geneticists may use continued programs of brother-sister matings to produce so-called *inbred strains* of corn, flies, mice, and so on. The likelihood that the organisms are homozygous increases quite rapidly in successive generations of sib matings, but not as rapidly as with self-fertilization. In either case, however, the likelihood of any individual being homozygous approaches certainty after a moderate number of generations.

Uncle-niece and aunt-nephew marriages are very rare in human groups except where social custom dictates this form of consanguinity. In southern India, for example, Hindu men favor marriage with a sister's daughter, but may not marry a brother's daughter. In one study of several caste groups representing very different socioeconomic levels (from well-to-do to so-called untouchables), about 19% of all marriages were between uncle and niece, and another 22% were between first cousins (Reddy 1987).

Marriage between first cousins is the most frequent type of human consanguineous mating, in part because cousins may be close in age and their families are often known to each other. Usually, the fraction of first-cousin marriages is well below 1%, but it may be higher in some cultures, as noted for southern India. In small Japanese towns, first-cousin marriages are currently 1–3%, but they were much more frequent (5–12%) a generation ago (Imaizumi 1986).

Note that two people who are related to each other share one or more ancestors. Sibs have the same mother and father, and first cousins share a set of grandparents (Box A). Two individuals in the same close ethnic group may have many identifiable common ancestors within a half dozen or so generations back. Extending even further back in time, we are all related one way or another. Consider that the maximum number of ancestors n generations back for one person is 2^n. Thus, you have 2 parents, 4 grandparents, 8 great-grandparents, 16 great-great-grandparents, and so on. This number rapidly exceeds the entire population of the world at some time in the not-too-distant past and shows that all of us must have many ancestors in common.

---

* Here we define incest *narrowly* as sexual intercourse between sibs, half sibs, parent–child, grandparent–grandchild, or uncle–niece (aunt–nephew). Social scientists usually define incest more *broadly* to include abuse of young children within the home by older persons, especially adults, including step-relatives. These investigators are concerned with psychological harm to the victim and approaches to effective therapy (Gelinas 1983). Recent investigations suggest that incest in the broad sense is more common than once thought.

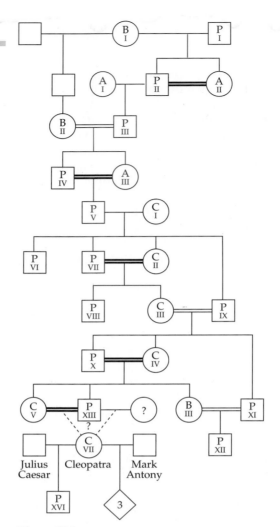

Figure 17.3
A pedigree of the Ptolemies, kings of Egypt from 323 to 30 B.C. The dynasty was often turbulent, marked by intrigue, intrafamily violence, and war. The famous Cleopatra is C-VII. It is unclear whether she was a child of a brother-sister marriage or whether she was not inbred at all. She was to marry her younger brother (half brother?), but he met an untimely death in war. In the manner of her forebears, Cleopatra dispatched another brother (or half brother) by poison. She and her son by Caesar, P-XVI, were the last of the Ptolemaic line. (Pedigree mostly after Weigall 1924.)

Measuring Inbreeding

Whether as a preferred social custom, regular system of mating, or an irregular happening, consanguinity increases the likelihood of homozygosity among the inbred progeny. A person can be homozygous without being inbred, of course, as is clear from the Hardy-Weinberg formulation: The probability of homozygosity in a randomly mating population is equal to p^2 (for A/A) plus q^2 (for a/a). Inbreeding pushes the proportion of homozygotes above this value. The *excess homozygosity* (above $p^2 + q^2$) is what produces the greater likelihood of adverse effects in inbred persons. The increase in homozygosity comes about because of shared ancestry; a specific deleterious allele from a joint forebear can be successively replicated and transmitted through *both* lines of descent to mates who are related, and thence to their offspring. Two alleles that have a common history are said to be **identical by descent**. They not only code for the

COUSINS—CLOSE AND FAR-FLUNG

Two people are related to each other if they have one or more shared ancestors. Sibs, of course, have the same mother and father, but most other relatives have common ancestry more remote than the parental generation. In Western civilizations, the system of naming related persons, beyond close relatives, is based on the term *cousin* and a few modifiers. In the pedigree here, first cousins (G and H) have a common set of grandparents (A and B), and second cousins (I and J) have a common set of great-grandparents (A and B again). For first cousins, one each of their parents are sibs; for second cousins, one each of their parents are first cousins. The term *removed* pertains to a difference in generations: thus G and J are first cousins once removed, and G and L are first cousins twice removed.

If D and E had the same father but different mothers (or vice versa), they would be *half sibs*, and their children would then be *half* first cousins, and so on. If G and H had two common sets of grandparents, they would then be *double* first cousins; this relationship would arise, for example, if their fathers were brothers and their mothers were sisters. Note that the diamond symbol is used when the sex of the individual is not relevant.

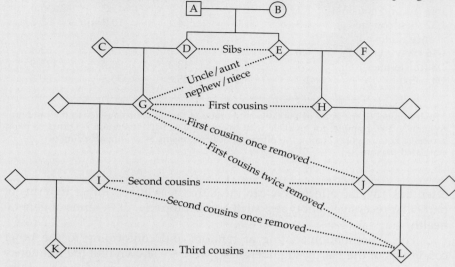

same polypeptide but also trace back to exactly the same segment of DNA in a common ancestor.

The usual measure of inbreeding, called the **coefficient of inbreeding**, was invented by the much honored and long-lived American geneticist Sewall Wright (Figure 17.4). The coefficient measures the probability that alleles are identical by descent. Its value for the children of parents related in various ways is as follows:

Relationship between parents	Coefficient of inbreeding of their children = f
Siblings	1/4
Uncle-niece or aunt-nephew	1/8
First cousins	1/16
First cousins, once removed	1/32
Second cousins	1/64

Note that the inbreeding coefficient, usually designated by f, decreases by a factor of 1/2 for each looser degree of relationship. The value $f = 1/16$ for a person whose parents were first cousins means that one-sixteenth of the time a pair of alleles—for any gene being considered—is expected to be identical by descent.

The Effects of Inbreeding

Whether inbreeding is good or bad depends on whether increased homozygosity is good or bad. For the human species, in which matings are usually between unrelated or only remotely related persons, inbreeding

Figure 17.4
Sewall Wright (1889–1988) was a leading population geneticist. He was still publishing papers at the age of 92 when this picture was taken. (Courtesy of William B. Provine, Cornell University.)

419

Table 17.2

The increase in a/a children from consanguineous matings compared to random matings.

Coefficient of Inbreeding, f, of Offspring (Relationship between Parents)	RATIO OF INBRED FREQ(*A/A*) TO RANDOM FREQ(*A/A*) FOR:[a]				
	q = 1/10	*q* = 1/30	*q* = 1/100	*q* = 1/300	*q* = 1/1,000
1/64 (second cousins)	1.1	1.5	2.5	5.7	16.6
1/32 (first cousins once removed)	1.3	1.9	4.1	10.3	32.2
1/16 (first cousins)	1.6	2.8	7.2	19.7	63.4
1/8 (uncle-niece)	2.1	4.6	13.4	38.4	125.9
1/4 (sibs)	3.2	8.2	25.8	75.8	250.8

[a] Each number in the body of the table is the ratio of the inbred frequency to the random frequency. For example, for an allele with frequency $q = 1/100$, there are 7.2 times as many recessive homozygotes among the offspring of first cousins as among offspring of unrelated persons.

often has adverse effects. We can measure these effects either in terms of particular recessive genotypes, such as alkaptonuria, or in terms of general health.

For a population mating at random and conforming to the other conditions of the Hardy-Weinberg law, we have shown that the frequency of a recessive phenotype is q^2, where q is the frequency of the recessive allele. For alkaptonuria, where freq(*k*) = 1/500, the *random* frequency of alkaptonuria is freq(*k/k*) = $(1/500)^2 = 4.0 \times 10^{-6}$.

When a person is *inbred*, we need to know two items to calculate probability of homozygosity for a particular gene: the person's coefficient of inbreeding, f, and the frequency of the allele, q. The general formula for a recessive allele, *a*, is

$$\text{freq}(a/a) = qf + q^2(1 - f)$$

Notice that this expression for freq(*a/a*) reduces to q^2 when $f = 0$ (as it must because $f = 0$ corresponds to random mating).

Using the formula, we can list the increase in the likelihood of homozygosis among the offspring of various consanguineous matings and for various allele frequencies, as seen in Table 17.2. Reading down a column, note that the effect of inbreeding increases as the parental consanguinity increases. Reading across a row, note that the rarer the allele, the greater the likelihood of homozygosity from consanguineous mating relative to random mating. This is because it is extremely unlikely that very rare alleles would come together in an individual independently of each other. But related persons can inherit the identical allele from a common ancestor. Thus, inbreeding is usually bad for people because it tends to reveal detrimental recessive alleles (for alkaptonuria, phenylketonuria, albinism, cystic fibrosis, Tay-Sachs, etc.) that are usually hidden in heterozygotes.

Recessive mutations that do not result in specific, recognizable diseases may also cause harm by their small cumulative effects when made homozygous. Many studies have compared aspects of health and survival of individuals with different coefficients of inbreeding. The traits that have been tallied include congenital malformations, mental illnesses, IQ, fertility, and mortality within given time periods. Examples of the latter include miscarriages, stillbirths, neonatal deaths (i.e., within a month of birth),

Table 17.3
Effect of parental consanguinity on infant mortality in Japan.

Coefficient of Inbreeding (*f*) of Offspring (Relationship between Parents)	INFANT DEATHS/LIVE BIRTHS (= % DEATHS)[a]		
	Hiroshima	Nagasaki	Total
0 (unrelated)	145/4,089 = 3.5%	273/7,988 = 3.4%	418/12,077 = 3.46%
1/64 (second cousins)	32/722 = 4.4%	50/1,312 = 3.8%	82/2,034 = 4.03%
1/32 (first cousins once removed)	42/585 = 7.2%	53/1,073 = 4.9%	95/1,658 = 5.73%
1/16 (first cousins)	101/1,651 = 6.1%	173/3,296 = 5.2%	274/4,947 = 5.54%

[a] These inbreeding results by Schull and Neel (1965) were separate from studies by the same investigators on the effects of radiation from the atomic bombs.

and deaths during childhood and the teens. All of these traits are influenced in many ways by both genetic and environmental factors—some large and some small.

The most dramatic data deal with incest, although such information is difficult to collect. One investigation involved 161 Czechoslovakian children from 88 father-daughter matings, 72 brother-sister matings, and 1 mother-son mating (Seemanová 1971). About 48% either died before one year of age or had significant mental or physical defects, a rate five times greater than in a control group of noninbred children. Another study from Canada included 21 mothers who, at the time of contact, were still pregnant from the incestuous relationship (Baird and McGillivray 1982). Of the ensuing 21 children (7 from father-daughter matings and 14 from brother-sister matings), 43% had severe malformations or retardation without known cause or (in one case) a specific autosomal recessive disorder. Even though the parents of these children may have represented a biased sample (as was also true in the Czechoslovakian study), it was evident that the risk for abnormalities in children of incest is high.

An extensive and detailed study in postwar Japan of the effects of less severe (and more usual) inbreeding is that of William J. Schull and James V. Neel (1965), of the University of Michigan. Table 17.3 gives a small portion of the results. Reading down a column, infant mortality is generally seen to increase a small (but significant) amount among the more inbred Japanese infants. Data like those in Table 17.3 can be used to estimate how many harmful recessive alleles an average person carries in a *heterozygous* state (thereby causing no damage). This estimation is done by extrapolating the amount of mortality or disease among children with known *f* values to what it would be if *f* = 1 (completely homozygous). Although this calculation is not very precise, the results suggest that the average person carries several recessive genes that would lead to a premature death if homozygous.

The Prognosis for a First-Cousin Marriage. What should be said to first cousins who contemplate marriage and are anxious about possible harm to their offspring? Popular opinion holds that such children are likely to suffer malformations or be less intelligent than their peers. But the data presented here suggest that first cousins carry only a small additional risk of having children with genetic defects of greater or lesser severity. All couples, whether related or not, face the likelihood (perhaps 2–3%) of having a child with a serious defect, depending on the criteria used to define "serious." Another few percent of children may fall outside a commonly accepted definition of "normal" but not have a serious defect.

A genetic counselor will certainly obtain a detailed family history from both of the cousins. If there is a suggestion of a deleterious recessive allele in one of them or in an ancestor (common or otherwise), a specific probability calculation can usually be made. The couple can then consider both the risk, severity, and burden of the disorder. More often than not, however, there will be nothing in the family history on which to base a specific calculation, because, as we have seen, rare recessive detrimental alleles will usually remain hidden in heterozygous condition generation after generation. Only vague statements of the small but real increased risk from first-cousin consanguinity can then be made. The couple will have to evaluate this not very satisfactory information and come to a decision primarily on the basis of their own feelings.

Inbreeding in Isolates

An **isolate** is a community whose members marry within the group for generations, thereby separating themselves genetically from the general population. The fewer the immigrants such a group receives, the more clannish it becomes. Isolates, intriguing in themselves, have several properties that also make them genetically useful: (1) Uniform living conditions, which often prevail in isolates, can aid in the study of the relative effects of genetic variation versus environmental differences. (2) Small, closed populations allow investigation of some aspects of human evolution, especially *genetic drift* (random changes in allele frequencies); (3) Excellent genealogical records, kept because the members of isolates are often proud of their culture and ancestry, provide very thorough analyses of inbreeding (Box B).

The most isolated groups are characterized by geographical remoteness or strongly held religious beliefs. Examples that have been studied include the island populations of Pitcairn in the Pacific and Tristan da Cunha in the Atlantic, the practically inaccessible villages of the Jicaque Indians in Honduras, and the Xavante Indians in the interior of Brazil (Jacquard 1974). Religious isolates include the Old Order Amish of Pennsylvania, Ohio, and Indiana and the Hutterites of the western prairies of the United States and Canada.

The Amish sect originated in the seventeenth century in Switzerland as an offshoot of the Mennonites. Both the Amish and the Hutterites were forced to migrate when they incurred the wrath of European Catholic and Protestant churches for deviations from accepted religious practices. About 200 Amish people moved to Pennsylvania between 1720 and 1770, and their descendants in Lancaster County currently number about 14,000. Other Amish people settled in Ohio and Indiana. The Amish follow a Bible-centered life in which religion and economics are strongly interdependent. Remaining aloof from the outside world within highly ordered farming communities, they are characterized by high morals, devout pacifism, adult baptism, hard work, thrift, and mutual aid. The familiar horse and buggy seen on country roads in Pennsylvania symbolize Amish conservatism, which rejects most modern technology as well as formal education beyond the legal minimum.

In a survey of the Amish by Victor McKusick and his colleagues at Johns Hopkins University (1964), several ordinarily rare diseases were found in increased frequency, each one in a different subgroup of the population. One of these diseases, the Ellis-van Creveld syndrome (Figure 17.5), is a form of dwarfism in which the forearms and lower legs are

Figure 17.5
An Amish child with the Ellis-van Creveld syndrome. Note the shortened arms and the extra finger on the left hand. (Courtesy of Victor A. McKusick, Johns Hopkins University.)

COMMON LAST NAMES AS A MEASURE OF INBREEDING IN SMALL GROUPS

In many cultures a woman assumes a man's last name at the time of marriage. A person's surname is then like a genetic trait, but one with a peculiar pattern of inheritance: It is transmitted from a father to all his children. The sons continue to transmit the trait, but the daughters do not, since they take their husband's name. Under this system, ¼ of first cousins will have the same surname—that is, whenever their fathers are brothers, but not when their mothers are sisters or when they are related through both sexes. Second cousins will have the same surname $\frac{1}{16}$ of the time; third cousins, $\frac{1}{64}$; and so on.

A husband and wife who, before marriage, happen to have the same surname have a certain probability of being related to each other; if they are related, their children will be inbred. These facts suggest that the occurrence of like surnames among marriage partners, called **parental isonymy**, can be used to measure the average inbreeding in a small, closely knit group (Crow and Mange 1965). The relationship turns out to be very simple: *The average inbreeding coefficient of children is ¼ the frequency of parental isonymy.*

The explanation goes like this: The children of first cousins have $f = \frac{1}{16}$, that is, ¼ the likelihood of parental isonymy. The children of second cousins have $f = \frac{1}{64}$, again 1/4 the likelihood of parental isonymy. The children of third cousins have $f = \frac{1}{128}$, again ¼ the likelihood of parental isonymy. For most common relationships, the formula holds. It just doesn't matter how parents are re-

lated; in any situation, the inbreeding coefficient of children, on the average, is ¼ the likelihood that their parents have the same surname.

Based on the isonymy method, the average inbreeding coefficient calculated for a population is often greater than that based on a large compilation of pedigree information. The reason is that the isonymy method takes *all* remote ancestry into account, whether it is explicitly known from pedigrees or not. Several types of errors are inherent in the method, however, and it may overestimate or underestimate inbreeding levels in some populations. Nevertheless, isonymy is now much used for the analysis of the structure of human populations (Lasker 1985).

disproportionately short. Many patients die soon after birth, but a few milder cases reach adulthood and may rarely have children. Among the approximately 8,000 Amish then living in Lancaster County, the investigators found 43 persons with the syndrome (genotypically e/e). The frequency of the recessive allele (e) in this subgroup was thus estimated as the square root of 43/8,000, or about 1/14. Elsewhere, even among other Amish groups, the allele frequency is very low, less than 1/1,000. Two factors, both related to unpredictable changes in allele frequencies that may occur when a population is small, account for the allele's exceptionally high frequency in the Lancaster County Amish: the founder effect and random genetic drift.

The Founder Effect. In general terms, **founder effect** refers to a newly formed population in which an allele has a frequency different from its frequency in the general population from which the founders came. The smaller the number of founders, the more likely they are to be unrepresentative of the larger population. This concept is illustrated by random sampling from a bag containing, say, 900 black marbles and 100 yellow ones. If you draw a large number of marbles, close to 10% are likely to be yellow. But if you randomly pick just 10 marbles, it would not be at all surprising to find that as little as 0% ($P = 0.35$) or as great as 20% ($P = 0.19$) are yellow.

The ancestry of all the Lancaster County families with the Ellis-van Creveld syndrome traces back to a Mr. and Mrs. Samuel King who immigrated in 1744. The recessive abnormal allele was undoubtedly present in the one or the other in heterozygous form. Thus, the frequency of the allele among the founders of the Lancaster County Amish was perhaps 1/400 if present only in Mr. or Mrs. King. But its frequency was not likely to have been as high as 1/14, its value in 1964. Another factor must have operated to increase its frequency further.

Random Genetic Drift. A generation of persons derived from just a small number of parents may not possess genotypes representative of the parental generation. One parental couple with an unusual genotype may be prolific while another remains childless, or Mendelian segregation of alleles within families may depart significantly from expected ratios. Thus, allele frequencies may change capriciously, irrespective of whether they are "good" alleles or "bad" ones. Such changes, which are increasingly likely the smaller the population, are called **random genetic drift**, or simply *drift*. The founder effect is really a special case of drift occurring in an initial generation. Another special case is the *bottleneck effect*, in which a community's sample of alleles is changed when the population is reduced in numbers by natural catastrophe, infectious disease, or other factors.* Broadly speaking, drift refers to haphazard changes in allele frequencies due to sampling from small numbers.

In the case of the Lancaster County Amish, the Kings and their descendants had larger families than others in their community, despite the presence of the semilethal recessive allele for a form of dwarfism. As a consequence, the frequency of the deleterious allele "drifted" higher in succeeding generations, especially in the early generations when the population was smaller.

SUMMARY

1. Population genetics extends the Mendelian rules of inheritance from families to larger groups, that is, to mixed populations that can be described by the frequencies of the various genotypes, phenotypes, alleles, and matings.

2. Allele frequencies can be calculated from genotype frequencies: The frequency of an allele is equal to the frequency of homozygotes plus half the frequency of heterozygotes.

3. The Hardy-Weinberg law states that population statistics remain constant when evolutionary forces are negligible. For p = freq(A) and q = freq(a), the equilibrium genotype frequencies of A/A, A/a, and a/a become p^2, $2pq$, and q^2, respectively, in one generation.

4. Some aspects of evolution begin with modifications of the Hardy-Weinberg law. The frequency of a particular allele today depends primarily on evolutionary forces in the past and not on whether the allele acts in a dominant or recessive fashion.

5. Hardy-Weinberg arithmetic can be applied to specific human traits to obtain statistical information on allele and genotype frequencies. The Hardy-Weinberg law can be extended to situations involving X-linked genes.

6. A mating between relatives is said to be consanguineous, and the resulting children are inbred. People who are related to each other have one or more ancestors in common.

7. An inbred person may be homozygous not only for independently inherited alleles (as in a noninbred person), but also for alleles identical by descent. The basic effect of inbreeding is to increase homozygosity for alleles identical by descent.

8. Whether inbreeding is beneficial or detrimental depends on the consequences of increased homozygosity. On the average, inbreeding in humans has small, adverse effects on the offspring of first cousins, the most common consanguineous mating.

---

* For example, the world's population of cheetahs (totaling 20,000 animals in two African locations) is now threatened by a lack of genetic variability because of severe and prolonged bottlenecks about 10,000 years ago. O'Brien et al. (1986) write that the populations "must have dropped to a very few individuals, escaping extinction by a whisker."

9. The rarer a recessive allele, the greater is the risk of its appearing homozygously in inbred persons as compared to noninbred persons. Mortality statistics for increasingly inbred persons suggest that each of us possesses several genes that, if made homozygous, would kill us before maturity.

10. Isolates may have allele frequencies different from the those of the general population. The high frequency of certain alleles in isolated groups may be due to the founder effect and random genetic drift.

KEY TERMS

allele frequency	founder effect	isolate	random genetic drift
alleles identical by descent	genotype frequency	mating frequency	random mating
coefficient of inbreeding (f)	Hardy-Weinberg law	p and q (allele frequency)	sib mating
consanguineous mating	inbred offspring	phenotype frequency	
equilibrium	incestuous mating	population	

QUESTIONS

1. In a certain population, there were 100 people of genotype A/A, 400 of genotype A/a, and 500 of genotype a/a. Calculate the frequencies of the alleles A and a.

2. For the allele frequencies that you calculated in question 1, what would be the genotype frequencies in the next generation if the population now meets the conditions of the Hardy-Weinberg law? (Note that these values turn out to be different from the genotype frequencies in question 1. Evidently, the parental population in question 1 had not previously met one or more of the Hardy-Weinberg conditions.)

3. For each of the populations below, calculate the expected frequencies of the phenotypes for the MN blood group, assuming that the conditions of the Hardy-Weinberg law are met. Phenotype M and phenotype N are homozygotes, and phenotype MN is heterozygous. (The data in this and some other problems are taken from Mourant et al. 1976.)

	Allele frequency	
Population	M	N
Pima Indians of Arizona	0.70	0.30
Bushmen of Botswana	0.60	0.40
Amish of Indiana	0.50	0.50
Polynesians of Easter Island	0.40	0.60

4. Assume that the frequency of the Rh negative allele, r, in a particular city is 0.1. (a) What genotypic frequencies for R/R, R/r, and r/r are predicted by the Hardy-Weinberg law? (b) There are 40 children in one of the city's kindergartens. What is the probability that all 40 are Rh positive, $R/-$? (Just set up the proper expression.)

5. Among native Hawaiians, about 90% of alleles at the Rh locus are R; thus, freq(R) = 0.9 and freq(r) = 0.1. (a) In what percentage of matings is one person Rh positive and the other Rh negative? (b) In what percentage of matings is the woman Rh positive and the man Rh negative?

6. A group of 100 people splits away from a larger population and establishes a separate society. With respect to the MN blood types, the emigrants number: type M = 41, type MN = 38, type N = 21. (a) What are the allele frequencies? Do you need to assume the Hardy-Weinberg law to do these calculations? (b) If this group and their descendants now meet the conditions of the Hardy-Weinberg law, what are the expected frequencies of the MN phenotypes in subsequent generations? (Although the group is fairly small, assume that genetic drift is negligible.)

7. Consider the sickle-cell gene, S, and its normal allele, A. In an adult African population, an investigator finds the following numbers of the three genotypes:

Genotype	Number
A/A	605
A/S	390
S/S	5
Total	1,000

What numbers (out of 1,000) should the investigator have expected if this population had met the conditions of the Hardy-Weinberg law?

8. In a certain population, 1 person in 10,000 is albino (c/c). What fraction of this population is expected to be heterozygous? Do you need to assume the Hardy-Weinberg law?

9. Among human males, about 10% are color-blind because they carry an allele of an X-linked gene. What percentage of women are expected to be color-blind? Assume the Hardy-Weinberg law.

10. If 60% of men showed a *dominant* X-linked trait caused by genotype $A/(Y)$, what percentage of women would be expected to show the trait, genotype $A/-$? Assume the Hardy-Weinberg law.

11. Among French Canadians in Quebec, the frequencies of the ABO alleles are roughly freq(A) = 0.3, freq(B) = 0.1, freq(O) = 0.6. What are the expected Hardy-Weinberg frequencies of the ABO blood types? Note that this question extends the Hardy-Weinberg law to a case with *three* alleles of a gene. As with two alleles, the expected frequency of any homozygote is the square of its allele frequency; the expected frequency of any heterozygote is twice the product of the corresponding allele frequencies.

12. Draw the pedigree of John and Mary, who are double first cousins.

13. In Figure 17.3, what is the relationship (a) between wife B-II and husband P-III and (b) between wife C-III and husband P-IX?

14. What are the coefficients of inbreeding for the children of incest listed in the footnote on p. 417? Use the in-text table on p. 419, noting that each closer step in relationship increases the f value by a factor of 1/2, and each looser step decreases the f value by a factor of 1/2.

15. We noted that the frequency of the allele for the Ellis-van Creveld syndrome among the Lancaster County Amish is freq(e) = 0.07. (a) What is the expected frequency of the syndrome among the offspring of unrelated parents, that is, under random mating? (b) What is the expected frequency of the syndrome among the offspring of first cousins in this population? (c) How many times more frequent is the e/e genotype in case (b) than in case (a)?

16. Assume that outside Lancaster County the frequency of the e allele is 1/1,000 (although it is probably less than this). (a) What is the expected frequency of the syndrome in a randomly mating population? (b) How many times more frequent would it be among the offspring of first cousins? Find the answer in Table 17.2. Contrast this value with the corresponding factor for the Lancaster County Amish (question 15).

FURTHER READING

Introductions to population genetics are clearly presented in the short books by Ayala (1982) and Hartl (1987) and in the general genetics books by Strickberger (1985) and Crow (1983). Interesting historical insights into population genetics are provided by Crow (1987, 1988) and at greater length by Provine (1971). Jameson (1977) reprints important papers, including ones by Hardy and Weinberg. Accounts of human population genetics include Bodmer and Cavalli-Sforza (1976) and Cavalli-Sforza and Bodmer (1971). Hostetler (1980) has given an interesting account of Amish life, and McKusick (1978) has edited a large collection of medical genetic studies of the Amish.

Evolutionary Concepts

18

A wealth of evidence indicates that we and the creatures around us have evolved from other creatures over the course of millions of years. The idea that this gradual **evolution** came about in large part by the mechanism of **natural selection** was the most important scientific concept of the 1800s. It occurred independently to two Victorian naturalists, one much better known than the other, although each of them set forth equally convincing documentation to back up his claims.

DARWIN AND WALLACE

The better-known man, Charles Darwin (1809–1882), was the son of a wealthy country doctor (Figure 18.1A); he did not have to work to earn a living. As a young man he took a five-year voyage of discovery on the surveying ship, the *Beagle*. Based on his findings, Darwin then investigated and wrote about the mysteries of the origin of species by natural selection, or, as he first called his concept, "descent with modification." The lesser-known and younger man, Alfred Russel Wallace (1823–1913), was the son of a financially insecure businessman (Figure 18.1B). Wallace, too, traveled widely as a young man, exploring the Amazon region for four years and the islands of Indonesia for eight (Figure 18.2). During the latter trip he collected and preserved over 125,000 plants and animals. Indeed, he earned his living by selling beautiful and exotic specimens to museums and by lecturing and writing about his many tropical adventures.

The interaction between Darwin and Wallace is a favorite story of biologists. Both men were avid readers, Wallace having examined (in his early 20s) Darwin's 1839 account of the voyage of the *Beagle* as well as Thomas Malthus's 1798 *Essay on the Principle of Population*. Malthus outlined how human populations tended to grow faster than the resources on which they depended and how human numbers were held in check by disease and famine, misery and vice. His essay greatly influenced the thinking of both investigators.

While in Indonesia, Wallace published a paper in which he wrote, "Every species has come into existence coincident both in space and time with a pre-existing closely allied species" (Wallace 1855). After reading this article, Darwin wrote Wallace that he agreed with almost every word and that he was preparing a major work on the subject. Although he gave no details to Wallace, Darwin had already arrived at a reasonable theory to explain evolution. In fact, he had been struggling with his manuscript, *On the Origin of Species by Means of Natural Selection*, for over a dozen years—ever since *he* had read the essay by Malthus—but he was reluctant to publish anything.

At first a mechanism to explain speciation eluded Wallace, who continued his collecting in Indonesia. While suffering from a bout of fever

Figure 18.1
(A) Charles Darwin at age 60. (B) Alfred Russel Wallace at age 55.

and chills, however, Wallace recalled the population essay of Malthus, recognized that it applied to any species, and asked himself why *particular* individuals lived and others died.

> And the answer was clearly, that on the whole the best fitted live. From the effects of disease the most healthy escaped; from enemies, the strongest, the swiftest, or the most cunning; from famine, the best hunters or those with the best digestion; and so on. Then it suddenly flashed upon me that this self-acting process would necessarily *improve the race*. . . . There would be ample time for the change to be effected by the survival of the best fitted in every generation. In this way each part of an animal's organization could be modified . . . and the clear *isolation* of each new species would be explained. The more I thought it over the more I became convinced that I had at length found the long-sought-for law of nature that solved the problem of the origin of species. (Wallace 1908)

Wallace wrote these ideas to Darwin. His letter was a shocker to Darwin, who had hit upon precisely the same idea a score of years earlier, but whose detailed documentation on the subject lay in his study (with £400 and instructions to his wife to have it published should he suddenly die). Darwin declared, "I never saw a more striking coincidence; if Wallace had my MS. sketch written out in 1842, he could not have made a better short abstract!" (F. Darwin 1887). Darwin's hand was forced; through the intercession of friends, the views of both Wallace and Darwin were read at the next meeting of the Linnean Society of London, a group of eminent natural historians (Darwin and Wallace 1858). Wallace, meanwhile, continued to work in the Indonesian islands, gathering evidence for his own important additional works on biogeography of island populations. In 1859, Darwin finally published *On the Origin of Species*, whose first printing of 1,250 copies sold out in one day.

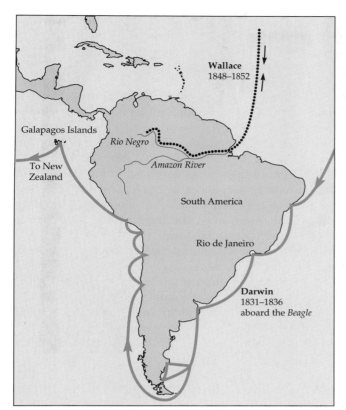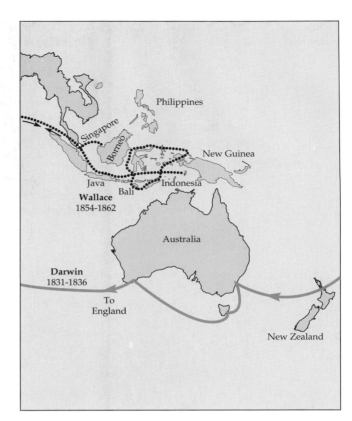

Figure 18.2
The travels of Darwin (dashed lines) and Wallace (dotted lines).

The two men were honorable, generous, and cordial, but Wallace was usually content to bask in the light of England's better-established scientists. Dubbed a moon to Darwin's sun, Wallace was often deferential; for example, his popular 1889 book summarizing the principles of natural selection was entitled *Darwinism*.

The *idea* of evolution actually predated the work of Darwin and Wallace by many years. In the early 1880s, for example, the Frenchman Jean-Baptiste de Lamarck argued strongly for evolution and reasoned quite logically that it occurred by the inheritance of so-called acquired characteristics. But convincing scientific support for this particular mechanism for evolution was not brought forward by Lamarck, nor by anyone else. On the other hand, Darwin and Wallace marshaled considerable evidence that evolution occurred by natural selection. This mechanism for evolution, coupled with the workings of heredity, can account for the step-by-step, nonrandom, adaptive changes characteristic of the evolution of populations over space and time.

After Darwin and Wallace

In the early decades of the 1900s, the ideas of Darwin and Wallace were integrated with the discoveries of Mendel into a mathematics of evolutionary processes. This work was accomplished in large part by two Englishmen, J. B. S. Haldane and Sir Ronald Fisher, and an American, Sewall Wright. Along with most other scientists (since about 1800), they agreed that populations of organisms have changed and diverged over eons to produce the untold number of current species. Each man, however, put his special stamp on the theories behind the facts; that is, they differed among themselves, sometimes vigorously, on the mechanisms that might explain how evolution occurred—and is still occurring. The evolutionary mechanisms that they argued about—in addition to natural selection—were **mutation, migration**, and **drift**, factors that by themselves lack the "management skills" of selection. That is, these factors act in a more random way to bring about change. The mathematical models that they constructed were abstruse and became better understood through the writings of others. An especially important work was the 1937 book *Genetics and the Origin of Species*, by Theodosius Dobzhansky, of Columbia University.

Recent work in evolutionary theory has emphasized variation at the molecular level, in the nucleotide sequences of DNA. Some base substitution mutations, for example, are likely to have no phenotypic manifestation, because they are in an intron or in the third position of a codon that does not change the resultant amino acid. Such mutations might therefore be expected to be selectively neutral and perhaps be useful to researchers in tracing evolutionary lineages.

We will not broach here the question of biological evolution versus special creation.* Rather, we start with the knowledge that gradual, cumulative, and nonmiraculous changes over many millions of generations account for the wonderful complexity of adaptations and the astonishing diversity of species, including our own. Evolution—the fundamental thread tying together all fields of biology—has been repeatedly investigated and thoroughly confirmed. Even in 1932, Haldane could write that it is "quite as well proven as most other historical facts." The documentation is presented especially well by Dawkins (1986) and more fully in textbooks of evolutionary biology (e.g., Strickberger 1990). Here are some of the many interlocking lines of evidence that support evolution:

1. The fossil record of ages past, showing temporal sequences of more or less continuous forms
2. The geographical distributions of past and present species over wide or narrow ranges (intensely studied by Wallace)
3. The use of artificial selection by plant and animal breeders to generate a prodigious variety of pets, flowers, livestock, crops, and so forth (intensely studied by Darwin)
4. The retention of developmental stages of remote ancestors during the embryology of current species (Figure 18.3)
5. The similar underlying structures of some functionally dissimilar body parts as revealed by studies of comparative anatomy
6. The greater similarity of cellular and molecular structures (chromosomes, nucleotide sequences, amino acid sequences, etc.) among more closely related species
7. The nearly universal genetic code and other similarities and subtle differences at the biochemical level among all living creatures

MUTATION

Mutations are the ultimate source of the variation on which evolution depends. Recombination of chromosome segments or migration of indi-

---

* Regarding this issue, the United States Supreme Court has ruled that it is unconstitutional to require public school teachers to discuss special creation whenever they discuss evolution (Edwards v. Aguillard 1987). The justices argued that special creation reflects a particular religious doctrine (the Book of Genesis); teaching it would therefore violate the separate status of church and state mandated by the First Amendment. A prior court case emphasized that so-called "creation science" is not science in any sense of the word (McLean v. Arkansas Board of Education 1982). The myth of creation is also examined in Berra (1990).

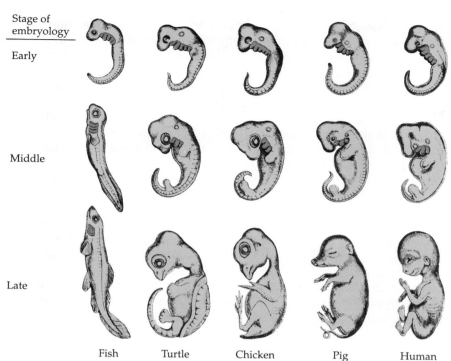

Stage of embryology

Early

Middle

Late

Fish   Turtle   Chicken   Pig   Human

Figure 18.3
Similar embryonic development of diverse vertebrate species, reflecting a common evolutionary origin. All possess gill slits (color) in early embryology, although only those in fishes go on to develop into organs for extracting oxygen from water. The fish-like embryonic period of all vertebrates is consistent with their possession of common genes, modified over time to adapt the adults to other life-styles.

viduals can bring about reshuffling of alleles, yielding perhaps beneficial combinations, but all current genetically based differences between organisms derive from random mutations at some time in the recent or remote past. In Chapter 12 we noted that mutations can occur in any cell at any time, but many of them probably happen at the time of DNA replication. We noted that mutations may be due to chemicals or radiation, but there is usually no way of knowing the source of a particular spontaneous mutation. Mutations are often *recessive*, and one good, unmutated copy of a gene may be sufficient to produce a normal phenotype. Mutations can affect any aspect of phenotype: physical, biochemical, or behavioral. Most mutations result in *detrimental* alleles, because the change alters a system already attuned to its internal and external environments. (By analogy, how likely is it that a *random* change in the innards of an adequate TV set will improve the picture?) Occasionally, a mutation might be beneficial initially (especially if it produces only a small change in the phenotype), or it might become beneficial in a later generation as a result of an environmental change. For example, mutations resulting in penicillin resistance are of no use to bacteria unless they encounter penicillin in their environment.

The time required for a mutation occurring in an egg or sperm to be expressed in an individual's phenotype varies. For example, an autosomal mutation to a *dominant* allele will be immediately evident if transmitted to an offspring. On the other hand, the number of generations that elapse between an autosomal *recessive* mutation and its eventual expression in a homozygote depends on the rate at which the particular mutation recurs, on the pattern of matings, and on chance events. A high mutation rate and inbreeding, for example, will hasten the formation of the homozygotes in which the recessive allele will be expressed.

Measuring Spontaneous Mutation Rates

Determining a spontaneous mutation rate is inherently difficult because the events to be counted are very rare and, in the case of recessive mutations, generations removed from their ultimate expression in an animal or plant. These problems can often be overcome by using experimental organisms, especially *haploid* microorganisms, in which the problem of dominance and recessiveness vanishes. Bacteria and viruses can be grown in prodigious numbers, and their environments can be manipulated to allow mutants, no matter how rare, to stand out from a background of millions of unmutated cells. For example, if a drug-sensitive population is exposed to that drug, only the preexisting drug-resistant mutants will survive. But mutation can be studied in diploid organisms too. In the 1920s, The American Nobel prize winner H. J. Muller invented methods to measure collectively the rate of mutation of the few thousand genes on one fruit fly chromosome, although the genes were not individually identified.

In mice, researchers have developed tester stocks that are homozygous recessive for several different genes affecting fur color, eye color, or other easily scored phenotypes. For one locus, the test goes as follows: Wildtype mice, homozygous for the dominant allele (B/B = black fur), are mated to the tester stock (b/b = brown fur). In the absence of mutation in the wildtype animals, all the offspring are black (B/b). But a mutation to a recessive in the germline of a wildtype parent ($B \rightarrow b$) leads to an offspring that is brown (b/b). By using a tester stock that is homozygous recessive for a half dozen different genes, it becomes feasible—barely—to carry out mutation rate research in mice (studying, for example, the effects of a mutagen such as X-rays).

The time, space, and money needed to determine mutation rates increase dramatically when we move from lower to higher organisms. In a few days at a small lab bench on a modest budget, one geneticist can determine fairly accurately some specific mutation rates in a virus or bacterium. To collect mutation rate data in fruit flies takes many months and requires much technical help to raise, examine, and count offspring. With mice, years of work, extensive animal-rearing facilities needing careful supervision and care, and a sizable investment of money can produce the approximate value of the mutation rate for relatively few genes.

For humans, experimental methods are not possible; instead, investigators must ferret out informative families from among worldwide populations. Even though mutations are infrequent, there are a lot of us to use as a base population, and large collections of hospital records may be available. It is possible to systematically tally at least small numbers of mutant phenotypes and thereby obtain approximate values for some mutation rates. We can do this fairly directly for dominant mutations, but only indirectly for recessive mutations.

For example, we can identify new dominant mutations that arise from recessive alleles by noting persons with the dominant (usually abnormal) trait who have normal parents. The genotypes involved are:

Genotype of normal parents		Gamete	Genotype of affected offspring
Father: *b/b*	germinal mutation →	B	
			B/b
Mother: *b/b*	⟶	*b*	

In this example, the sperm is assumed to contain the mutated allele ($b \rightarrow B$) appearing sometime in the germline. In practice, an investigator examines all the births in a given region during a given time period. Then the mutation rate is one-half the number of affected newborns with normal

Figure 18.4
An apparent achondroplastic dwarf, Giacomo Favorchi, painted by the Dutch artist Karel van Mander around 1600. (Courtesy of Statens Museum for Kunst, Copenhagen.)

parents, divided by the total number of newborns. The factor 1/2 occurs because an affected newborn arises from two gametes, only one of which carries the new mutant allele.

In an early study using this method, Mørch (1941) tallied eight achondroplastic dwarfs (Greek *a*, "lack of"; *chondros*, "cartilage"; *plasia*, "molding") with normal parents among 94,075 births in a Copenhagen hospital over a 20-year period. In a more recent study, two achondroplastic dwarfs with normal parents were seen among 69,277 infants born at a Boston hospital over a 10-year period (Nelson and Holmes 1989). This dominant disease produces abnormal bone development, but the specific biochemical pathway is unknown. The result is abnormally short extremities set on a normal-sized but swayback trunk and an overlarge head (Figure 18.4).* Severe cases are stillborn or die in infancy, but patients who survive the first year of life often become healthy adults of normal intelligence and may reproduce.

For the two studies, the mutation rates are computed as follows:

For Copenhagen: mutation rate $= (1/2)(8/94,075) = 4.3 \times 10^{-5}$
For Boston: mutation rate $= (1/2)(2/69,227) = 1.4 \times 10^{-5}$

These results indicate about 1 to 4 mutations per 100,000 genes per generation. Clearly, the rate cannot be known very precisely because of the small number of mutations on which the calculations are based. In addition, achondroplasia may result from more than one cause (genetic heterogeneity), so the calculated rate may apply to several genes together.

---

* Achondroplastic dwarfs are quite different from normally proportioned pituitary dwarfs, who are deficient in growth hormone (see Figure 4.5). An organization called The Little People of America was formed in 1960 to help meet some of the social difficulties faced by all types of dwarfs.

For rare *recessive* traits, the preceding method of calculation will not work because affected individuals with normal parents almost never signify newly occurring mutations. Rather, the affected person is a homozygote arising by Mendelian segregation from heterozygous parents: $B/b \times B/b \to b/b$. The mutational event ($B \to b$) occurred some generations in the past. Nevertheless, a mutation rate calculation can sometimes be made by assuming that the gain in detrimental alleles by mutation is equal to the loss of detrimental alleles through death or nonreproduction. This method is also not very precise, because many so-called recessive alleles actually have a small effect on the phenotype when heterozygous. That is, some recessive alleles are not completely recessive; they have an adverse consequence when they are present in just single dose. This *heterozygous effect*, even if very small, has a large consequence on the mathematical calculation, because so many more individuals in a population are heterozygous than homozygous recessive.

The Average Mutation Rate for Human Genes

Table 18.1 gives mutation rate data for a few genes. The rates are heterogeneous, ranging from a high of about 73 to a low of 2 mutations per million gametes per generation. These values are likely to be overestimates of the true mutability of human genes, simply because the traits have been documented in several to many families (for a calculation to be made). In contrast, some hereditary phenotypes are known in just one or a few families. These latter, truly rare, diseases undoubtedly have much lower mutation rates. Thus, an average rate of 1 mutation per million gametes per generation (for genes that lead to visible phenotypic effects) may be nearer the truth than the higher rates in Table 18.1.

The tabulated traits are also atypical in that they lead to easily identifiable, usually severe phenotypes. An "ordinary" mutation may be more benign, leading to an amino acid substitution that has little, if any, effect on the functioning of its protein. Some researchers, especially the Japanese mathematical geneticist Motoo Kimura (1979), believe that most base substitution mutations occurring over evolutionary time may be neutral or nearly neutral in their effect on phenotype.

An interesting question is whether mutation rates differ between the sexes. A sex difference is predicted if, as is thought, many mutations occur at the time of DNA replication and fewer at other times in the cell cycle. Vogel and Motulsky (1986) estimate that about 24 cell divisions ensue between an XX zygote and the eggs that a woman produces. On the other hand, sperm cells from a 25-year-old male have passed through about 300 divisions starting from the XY zygote. Thus, division-dependent mutation rates would be expected to be greater in males—perhaps much greater—but experimental data on this point are scanty and contradictory.

Because spermatogonial stem cells continue to proliferate at the rate of about 23 divisions per year, the mature sperm from, say, a 38-year-old man may have passed through 300 more divisions than the mature sperm of a 25-year-old man (13 more years × 23 divisions per year). Thus, a further prediction of the division-dependent hypothesis is that older fathers should have a greater probability of passing a mutated gene on to children than younger fathers do. This outcome does seem to occur for some diseases (e.g., achondroplasia), but not for all diseases for which data are available. This age effect would not be expected to occur in females.

All in all, a gene is a remarkably stable chemical entity, because a probability of 10^{-6} can be interpreted to mean that a single gene followed

Table 18.1
Fairly reliable mutation rates for some human genes.

Trait and Description[a]	Country	Mutations per Million Gametes per Generation[b]	
Neurofibromatosis (AD): multiple fibrous tumors of the skin and nervous tissue; scattered areas of brownish pigment in skin (known as café-au-lait spots)	Russia USA	46 100	avg. = 73
Duchenne-type muscular dystrophy (XR): progressive degeneration of muscular tissue from infancy; usually death before adulthood; due to absence of dystrophin in muscle cells (see Chapter 5); the many reliable estimates derive from the relatively easy diagnosis and zero fitness of affected persons	England Poland England Germany Northern Ireland Japan Switzerland USA USA England	43 46 47 48 60 65 73 92 95 105	avg. = 67
Hemophilia A (classical) (XR): internal bleeding, especially at joints, sometimes leading to crippling deformities; delayed clotting results from lack of factor VIII in plasma	Finland Germany	32 57	avg. = 44
Achondroplasia (AD): dwarfism with disproportionate shortening of arms and legs; nasal bridge is set back, and forehead protrudes; a defect of cartilage growth	Germany Denmark Northern Ireland	8 10 13	avg. = 10
Retinoblastoma (AD): a cancer of the retina of one or both eyes, appearing in children who may live to adulthood if the affected eyes are removed; otherwise fatal in childhood (see Chapter 15)	France Hungary Japan Netherlands	5 6 8 12	avg. = 8
Marfan syndrome (AD): unusually long limbs and digits, accompanied often by defects of the heart and eyes; an abnormality of a connective tissue protein called fibrillin	Northern Ireland	5	avg. = 5
Hemophilia B (Christmas disease) (XR): symptoms similar to classical hemophilia but generally milder; defect results from lack of factor IX in plasma	Finland Germany	2 3	avg. = 2

Source: The data are selected from a more extensive table in Vogel and Rathenberg (1975).
[a]AD = autosomal dominant; XR = X-linked recessive.
[b]Multiple determinations for the same gene represent independent studies.

from gamete to gamete will change only once in a million generations—about 30 million years. Nevertheless, the total number of mutations among all of the 50,000 or so genes per individual is substantial and quite adequate to provide the raw material for evolution.

SELECTION

With regard to selection, the important variable to consider is called **Darwinian fitness**, or simply **fitness**, a measure of the average number of offspring left by an individual or class of individuals. Differences in fitness depend on differences in both *survival* and *fertility*. For example, Tay-Sachs disease represents zero fitness due to infant death, whereas Turner syndrome represents zero fitness due to infertility. Zero Darwinian fitness

BOX A

WHY SHOULD I TAKE ALL MY PILLS?*

When your doctor prescribes an antibiotic to treat an infection, you are told to take the entire supply—down to the last pill in the bottle. Even if you feel better after taking just some of the pills, the doctor insists that you continue to the end. This directive is based on good evolutionary biology. After a short time the drug may have killed many of the bacteria, but the few survivors are those that are resistant to the drug. If the pills are stopped before *all* the bacteria are killed, the resistant survivors will contribute their drug-fighting genes to succeeding generations. The result could be unfortunate for your well-being: the same adverse symptoms as before, but an illness that requires a more potent dose of the antibiotic or a different antibiotic. Either course of action may have undesirable side effects.

* After Berra 1990.

would also apply to members of religious orders who practice celibacy. But consider persons with Huntington disease, a crippling condition characterized by progressive brain destruction; they may have near-average Darwinian fitness because the devastating dominant allele often does not manifest itself until after they have produced children.

The greater the overall fitness of individuals, the more their genes tend to be represented in future generations. This idea is the salient point of evolution by natural selection. Fitness is not usually an all-or-none condition, however; a difference of a few percentage points in health or fertility may distinguish different categories of people. The catchword *survival of the fittest* usually means just somewhat greater procreation by a particular group of individuals. In addition, fitness—a phenotypic trait—relates to prevailing environments. If surroundings change, the fitness of individuals may change. This relation is particularly striking when severe genetic diseases are treatable: The phenylketonuria genotype is nearly lethal in one dietary situation but nearly normal in another.

That natural selection occurs in nonhuman species is well documented. When species are exposed to new environmental conditions, the differential reproduction of the more fit types has led to new genetic strains: poison-resistant rats, DDT-resistant mosquitoes, penicillin-resistant bacteria (Box A). In industrial areas with soot-covered tree trunks, pale-colored moths have evolved to dark strains that are apparently better able to survive predation by birds. And more recently, in areas where tree trunks have lightened as a result of pollution control, the dark moths are evolving back to the pale forms from which they came (Bishop and Cook 1975).

New strains of animals and plants have been bred by **artificial selection**. Through this process, breeders have enhanced certain economical or esthetic traits for the benefit of people. The target species itself may or may not benefit, but if overall fitness is reduced, as often happens, the breeding stock can be coddled. Chickens laying more eggs? Mink with unusual colors? Leaner pork? Higher-protein corn? Fancier goldfish? Some progress toward any goal is almost always possible—in either direction: Cow's milk can be selected for either higher or lower butterfat; dogs, for long legs or short, for short fur or long. The implication of these examples is that mutation has supplied plenty of genetic variation to be exploited by artificial (or natural) selection (Figure 18.5).

It has been suggested that natural selection in our species is no longer

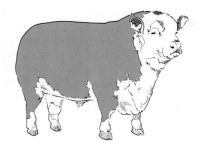

A. Hereford bull

B. Angus bull

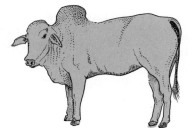

C. Brahman bull

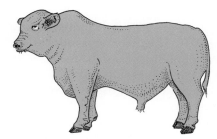

D. Santa Gertrudis bull

E. Holstein cow

F. Jersey cow

Figure 18.5

A few of the over 250 varieties of cattle achieved by selection and interbreeding among and between two closely related originally wild species: a tropical type with a hump above the shoulders and a temperate type without a hump. A–D are primarily beef cattle, E–F dairy cattle. (A) Herefords, developed in England, are reddish with white faces. They are sturdy and adaptable to extreme range conditions. (B) Angus cattle, developed in Scotland, are black, compact, and short-legged. They are noted for the rich marbling of their meat. (C) The gray Brahmans were developed in the United States by interbreeding four types of cattle from India. They are resistant to heat and humidity. (D) The reddish Santa Gertrudis cattle line was developed from crosses between Brahman and Texas Shorthorn; it is the best-known American breed, being well adapted to semiarid ranch conditions. (E) Black and white Holstein dairy cows, familiar to travelers of American roads, originated in the Netherlands. Holsteins are relatively large and produce correspondingly large quantities of milk. (F) Fawn colored Jersey cattle originated on the island of Jersey in the English Channel. Relatively small, they are known for the high butterfat content of their milk.

operating—that technology preserves the more fit and the less fit alike. According to this view, the unrelenting forces of nature still operate in other species, but selection has now been *relaxed* for humans. Medical science has surely been successful in preserving a few once-harmful genotypes. Patients with phenylketonuria eat foods low in phenylalanine; hemophiliacs receive clotting factor VIII; diabetics take insulin. Surgery can correct some faults of anatomy and physiology (whether heritable or not). Myopic individuals wear eyeglasses, increasing their fitness above what it might have been in a hunting culture. To the extent that the treatments preserve previously (but no longer) detrimental alleles, selection has indeed been relaxed.

However, technological civilizations may discriminate against some genetic constitutions. How much do individuals differ in genes (and in the resultant enzymes, receptors, or other proteins) that affect sensitivities and allergies to an increasing number, and more widespread use, of newly synthesized chemicals? Is heritable resistance to pulmonary, nervous, and circulatory diseases a significant variable when individuals (including fetuses) are challenged by contaminants—albeit at a low level—in air, food, and water? Examples of pervasive pollutants that regularly make the news

Table 18.2
Simple models of different types of natural selection.

Model	A/A	A/a	a/a	Type of Selection
Case I	x	x		Against a dominant phenotype
Case II			x	Against a recessive phenotype
Case III	x		x	Against both homozygotes; heterozygote has highest fitness

Note: An "x" indicates a genotype that leaves, on the average, fewer offspring than the other genotypes; that is, the x'ed genotypes have lower Darwinian fitness.

include car exhaust, industrial fumes, second-hand cigarette smoke, vapors from consumer products and construction materials (especially in tight buildings), asbestos fibers, pesticide and herbicide residues on fruit and vegetables, food additives, and runoff into water supplies of chemicals from agricultural use and waste disposal. Are psychological disorders that may have some genetic basis intensified in crowded urban environments and complex social situations? The answers are few, but there are enough questions to indicate that a suspension of natural selection cannot yet be proclaimed.

We should also note that medical science treats relatively few genetic diseases. People still die childless from a long list of formidable hereditary conditions for which medicine can currently do little. In addition, natural selection evidently still operates quite unchanged in that part of our life cycle from gametogenesis through the beginning of fetal life. It is estimated, for example, that perhaps 50% of fertilizations do not come to term. Thus, some weeding out of unfit alleles continues in the relatively unseen early stages of our lives.

Here we will examine only the simplest aspects of selection: changes in allele frequencies of a *single* gene in a *constant* environment. In reality, the reproductive fitnesses of various organisms depend on a complex interaction of their total genotypes with varied and changing environments. Three specific types of simple selection are outlined in Table 18.2. In case I, the dominant allele A is deleterious or perhaps lethal. If A is lethal, it will be reduced to near-zero frequency in one generation. Note that "lethal" need not mean that the allele kills its carrier—only that it prevents reproduction. Each generation, all new cases of the dominant trait must arise from recurrent mutation ($a \rightarrow A$), since only a/a's have progeny.

Some Long-Term Consequences

Case II considers a recessive detrimental allele. We will examine the extreme case, a **recessive lethal**, in which the fitness values of the genotypes can be assigned as follows:

	Genotype		
	A/A	A/a	a/a
Fitness	1	1	0

The fitness values are relative measures of reproduction, spanning the range from 1 (assigned to the most productive genotype) to 0 (assigned to a genotype that leaves no offspring). In this case, A/A and A/a are equally fit. Notice that the situation is very different from that of a dominant lethal, because the potentially harmful recessive a alleles that are present in normal heterozygotes are temporarily protected from selection.

How much is the frequency of the a allele reduced in one generation? To make things concrete, we assume that a/a individuals are live-born but die prior to reproduction. We also assume that the population meets all the Hardy-Weinberg conditions except for the indicated selection. Under these circumstances, the frequency of a recessive lethal, freq(a) or simply q, is reduced in one generation from q to q', where $q' = q/(1 + q)$.*

The change from q to $q/(1 + q)$ is a small one when freq(a) is small to begin with (as would be the case for any lethal allele). For example, if $q = 0.02$, then the frequency of a recessive lethal will be reduced in one generation to $q' = 0.0196$. The difference, $q' - q$, is called Δq (delta q). Here

$$\Delta q = 0.0196 - 0.0200 = -0.0004$$

That Δq is negative means that freq(a) is decreasing. The changes (expressed as percentages) for a recessive lethal over a range of frequencies are as follows:

q = freq(a)	$q' = q/(1 + q)$	Change = $\Delta q = q' - q$	Percentage change = $100(\Delta q/q)$
0.200 = 1/5	0.167 = 1/6	−0.033	−16.7
0.0200 = 1/50	0.0196 = 1/51	−0.0004	−1.96
0.002000 = 1/500	0.001996 = 1/501	−0.000004	−0.200

By contrast, for a dominant lethal allele, the change would have been 100% regardless of its initial frequency. Furthermore, *the rarer the recessive lethal becomes, the less effective selection is in reducing its frequency further* (reading down the last column). For example, it is possible to show that it takes 45 generations to reduce the frequency of a recessive lethal from 0.2 to 0.02, but 450 generations to go the next factor-of-ten decrement, from 0.02 to 0.002. In human terms, 450 generations is a long time: $450 \times 30 = 13,500$ years. This many years ago, humankind was just learning primitive agriculture.

The reason why selection against a recessive lethal is progressively less effective becomes clear when we examine the ratio of heterozygotes to homozygous recessives. In a population of 1 million, these ratios are as follows (using Hardy-Weinberg arithmetic):

q = freq(a)	Number of A/a	Number of a/a	Ratio A/a:a/a
0.2	320,000	40,000	8:1
0.02	39,200	400	98:1
0.002	3,992	4	998:1

Thus, as the allele gets rarer (reading from top to bottom), heterozygotes

---

* In arriving at this change, we multiplied the frequencies of the newborn genotypes (p^2, $2pq$, and q^2) by the fitness of each (1, 1, and 0) to get the *relative* proportions in the mating population (p^2, $2pq$, 0). The *absolute* genotype frequencies are obtained by dividing each of these terms by their sum. Freq(a) in the mating population—call it q', since it will have changed in value from q—is obtained by adding the frequency of the a/a homozygotes, 0, to half the frequency of A/a heterozygotes, $2pq/(1 - q^2)$. The mating population under Hardy-Weinberg conditions then produces the genotypes of the next newborn generation without any further change in the frequency of the alleles.

are not reduced in numbers nearly as quickly as homozygotes are. In the last line there are nearly 1,000 times more phenotypically normal heterozygotes, which *mask* the recessive lethal from selection, than homozygous recessives, which *expose* the lethal to selection.

Eugenic proposals in the early part of this century suggested that our genetic endowment could be much improved if we got rid of heritable diseases by sterilization of affected persons. As far as recessive traits are concerned, we can see that such a plan could not succeed. For example, if albinos were considered undesirable, their nonreproduction would yield allele frequency changes over three generations of

$$\overset{1}{0.0100} \rightarrow \overset{2}{0.0099} \rightarrow \overset{3}{0.0098} \rightarrow 0.0097$$

This *tiny* reduction would represent the progress of 100 years of sterilization of all albinos!

As a corollary, we can show that treating rare recessive lethal genotypes such as phenylketonuria, so that these individuals can lead productive and fertile lives, will increase the frequency of the causative alleles

Figure 18.6

The geographical distribution of malaria in the Old World before about 1930, compared with the distributions of three hereditary diseases: sickle-cell anemia, thalassemia, and glucose-6-phosphate dehydrogenase deficiency. The nature of the interaction of malaria with the last two diseases is not very well understood. (After Motulsky 1960.)

only very slowly. Thousands of years may be involved for appreciable frequency changes. The snail's pace of evolutionary change characteristic of recessive *lethals* is slower yet for recessive *detrimental* alleles, which have fitness values greater than 0 but less than 1. Substantial reductions in the frequency of severe recessive diseases might be achieved by identifying heterozygous persons (e.g., by biochemical tests) and counseling them as to risks. Voluntary modification of reproduction by heterozygotes could achieve what sterilization of affected persons could not.

For case III in Table 18.2, there is a well worked out human example—that of sickle-cell anemia in a particular environmental setting. In regions of the world where malaria is prevalent, there is a **superior heterozygote** whose fitness is higher than that of either homozygote. The reasons for this odd circumstance are as follows: The malarial parasite, which spends part of its life cycle inside red blood cells, flourishes less well when these cells possess some sickle-cell hemoglobin. Therefore, heterozygotes, Hb^A/Hb^S, are less susceptible to malaria than normal homozygotes, Hb^A/Hb^A. Although persons with sickle-cell anemia, Hb^S/Hb^S, are also resistant to malaria (as you would expect), they often die young of sickle-cell complications, so their malarial resistance is of little benefit to them. About 15% of Hb^A/Hb^A's and 90% of Hb^S/Hb^S's die between birth and reproduction.

As a result of losing a certain proportion of *both* homozygotes, a balance is struck, leading to stable intermediate frequencies. This so-called **genetic equilibrium** occurs in a population when the frequency of the sickle-cell allele is about 15%. Because of the equilibrium, sickle-cell anemia, despite being a serious disease, is not eliminated in regions where malaria occurs, but recurs in about 1 in 45 newborns (Figure 18.6).

In regions of the world where malaria is not a problem, the fitness of *A/A* homozygotes is as high as that of *A/S* heterozygotes, because the absence of malaria erases the selective advantage of being heterozygous. Thus, only the *S* alleles in persons with sickle-cell anemia (*S/S*) are lost through selection. In the United States, for example, sickle-cell anemia acts like a simple recessive detrimental, and freq(*S*) continues to decrease at a slow rate in successive generations. It remains a serious health problem among U.S. black populations, however.

OTHER FORCES OF EVOLUTION

We have described how mutation provides the raw material for evolution and how selection can progressively modify allele frequencies over successive generations. Now we consider allele frequency changes due to migration between populations and due to haphazard events that occur just because population numbers are small.

Migration and Gene Flow

Migration is the movement of people. Prehistoric migrations (e.g., Asians into the New World) often involved movement into previously uninhabited regions, where new environments may have imposed new selective pressures. More recent migrations have been into already occupied territory (e.g., Europeans and Africans into the New World). Although the immigrants may remain genetically isolated from their neighbors, more often some interbreeding occurs, leading in subsequent generations to mixed populations. When the original groups differ in certain allele frequencies, it may be possible to document the amount of **gene flow**, that is, the introduction of new alleles into a population. To the extent that a population receives new alleles, the genetic variability *within* it is in-

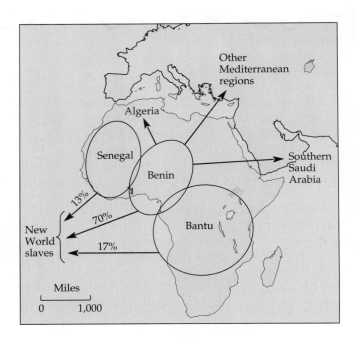

Figure 18.7
Areas and migration of three African groups—from Senegal, Benin, and the Bantu-speaking region—carrying different Hb^S chromosomes. Each chromosome is identified by the pattern of cuts (and noncuts) by different restriction enzymes in the region around the β-globin gene. All three regions were raided for slaves for the New World. Blacks in Mediterranean regions and in Saudi Arabia, however, are exclusively of the Benin type, having followed established caravan trade routes across the Sahara. (Map after Labie et al. 1986; New World data from Nagel 1984.)

creased. But gene flow between populations causes the populations to be more similar to each other than they would otherwise be.

Analysis of gene flow has been used to complement historical knowledge of migration for various groups:

American blacks: a two-way mix of Africans (of various cultures) and Europeans (of various cultures, but primarily English)

Brazilian subgroups: a three-way mix of native American Indians, Sudanese and Bantu Africans, and Europeans from Portugal, Italy, and Spain

Hawaiian subgroups: multiple mixtures of native Polynesians, whites, Japanese, Chinese, and Filipinos

To obtain information on the extent of the mixing, we concentrate on one allele of a gene, for example, the R^0 allele of the Rh system.* (Some selection also occurs at the Rh locus because of incompatibility, but the resulting changes in allele frequencies are small.) We illustrate the method by considering American blacks.

About 400,000 slaves were brought to the United States, mostly during the eighteenth century. They came from western regions of Africa, extending from Senegal in the north, through central (Benin) areas, to Bantu-speaking regions in the south. As noted in Figure 18.7, the percentages from each area are approximately known from molecular data based on the β-globin gene. In the United States, the offspring of slaves and Europeans, and their subsequent descendants, were and are almost always considered black. Only rarely is a person with mixed ancestry thought of as white. Thus, by definition, the mixing has been virtually all in one direction.

For contemporary U.S. blacks, we let M_1 be the fraction of their ancestry that is African and M_2 be the fraction of their ancestry that is European (where $M_1 + M_2 = 1$). The fractions can be related this way:

$$P = M_1P_1 + M_2P_2$$

* R^0 is one of several Rh positive alleles.

where the P's are the frequencies of the R^0 alleles:

P = frequency of R^0 among contemporary U.S. blacks
P_1 = frequency of R^0 among the ancestral African black population
P_2 = frequency of R^0 among the ancestral European white population

An analogy to this expression is the mixing of two paints: The final color (P) results from combining so much (M_1) of one paint color (P_1) and so much (M_2) of a second color (P_2). The value of P will be intermediate between the values of the ancestral frequencies, or paint colors. Solving this equation for M_1 (after noting that $M_2 = 1 - M_1$) gives

$$M_1 = \frac{P - P_2}{P_1 - P_2}$$

Thus, if we know the P values, we can calculate the M values.

Although we can measure the current P value directly, it is unclear what values to use for P_1 and P_2, the frequencies of R^0 among the ancestral populations that were intermixed many years ago with no geneticists around to test their Rh types. The best that can be done is to test contemporary samples of West Africans and Europeans and assume that the current population samples are representative of the ancestral populations. Another difficulty is that U.S. blacks are not homogeneous. For example, $freq(R^0)$ varies as follows:

$freq(R^0)$	Sampled in
0.56	Charleston, South Carolina
0.49	Oakland, California
0.36	New York, New York

Recognizing the uncertainties, typical figures might be as follows:

P = $freq(R^0)$ among U.S. blacks $= 0.5$
P_1 = $freq(R^0)$ among African ancestors $= 0.6$
P_2 = $freq(R^0)$ among European ancestors $= 0.025$

from which

$$M_1 = \frac{P - P_2}{P_1 - P_2} = \frac{0.5 - 0.025}{0.6 - 0.025} = 475/575 = 0.83$$

$$M_2 = 1 - M_1 = 0.17$$

Although this representative calculation indicates that about 17% of the ancestry of U.S. blacks is European in origin, specific calculations for black populations in the southern United States typically yield lower figures (4–10%) and in northern populations higher figures (20–30%). A recent study in Pittsburgh estimated the value at 25.2% (Chakraborty et al. 1992). These investigators averaged the results of 52 alleles of 15 different genes coding for serum proteins. Because their calculations included several dozen alleles whose frequency was 0 in one or the other source population, the 25.2% estimate is fairly precise—assuming that the source populations they used were properly representative. Volunteer blood donors in Pittsburgh provided values for P (contemporary U.S. blacks) and P_2 (European ancestors). Donated blood of volunteer students, teachers, and civil service workers in Benin City, Nigeria, provided a value for P_1 (African ancestors).

Table 18.3
Genetic drift illustrated by picking four jelly beans at random from a large bowl with 70% red and 30% black.

Jelly Beans Picked				Probability Factors	Probability Calculation	Overall Distribution
1st	2nd	3rd	4th			
●	●	●	●	.7 .7 .7 .7	$(.7)^4 = .2401$	4 red, 0 black
●	●	●	●	.7 .7 .7 .3		
●	●	●	●	.7 .7 .3 .7	$4(.7)^3(.3)^1 = .4116$	3 red, 1 black
●	●	●	●	.7 .3 .7 .7		
●	●	●	●	.3 .7 .7 .7		
●	●	●	●	.7 .7 .3 .3		
●	●	●	●	.7 .3 .7 .3		
●	●	●	●	.3 .7 .7 .3	$6(.7)^2(.3)^2 = .2646$	2 red, 2 black
●	●	●	●	.7 .3 .3 .7		
●	●	●	●	.3 .7 .3 .7		
●	●	●	●	.3 .3 .7 .7		
●	●	●	●	.7 .3 .3 .3		
●	●	●	●	.3 .7 .3 .3	$4(.7)^1(.3)^3 = .0756$	1 red, 3 black
●	●	●	●	.3 .3 .7 .3		
●	●	●	●	.3 .3 .3 .7		
●	●	●	●	.3 .3 .3 .3	$(.3)^4 = .0081$	0 red, 4 black

Genetic Drift

The concept of **drift** was introduced in the last chapter to explain the relatively high frequency of the Ellis-van Creveld syndrome among the Pennsylvania Amish. By chance, the initial frequency of the causative allele among the group of founders was higher than among the source population. This founder effect is one aspect of drift and is illustrated further in Table 18.3 for a very small population of jelly beans. Here a generation of just four jelly beans (representing, say, the alleles in one male and one female) is picked from a base population with allele frequencies of 0.7 (red) and 0.3 (black). Such a sample of four jelly beans is likely to be unrepresentative of the preceding generation. It is not even possible, of course, to get exactly 0.7 and 0.3 with four jelly beans. We can get close, that is, 0.75 (3 red) and 0.25 (1 black), but that result is expected to happen only about 41% of the time (as calculated in the table). Note that it would not be unusual with such a small sample to completely eliminate one allele or the other: About 24% of the time, the four jelly beans are expected to be all red, and about 1% of the time, they would be all black. Under these circumstances, only mutation (or perhaps migration) would bring back the other allele to the group of founders and their descendants.

In general, genetic drift refers to unrepresentative sampling in successive generations of a population. Drift occurs in part because different matings may not leave equal numbers of offspring. For example, the Amish families that harbored the allele for the Ellis-van Creveld syndrome appeared to have been more prolific over time than average Amish families. Drift can also occur because the successful gametes from heterozygotes may differ from 50:50 expectations.

The importance of drift in evolutionary history is disputed. Fisher, who downplayed its role, thought that most of evolution occurred in large populations by the kinds of selective systems we talked about earlier. Such evolutionary changes are very slow, to be sure, but long periods of time have been available. Wright, an advocate of drift, emphasized that evolution can occur more rapidly when a species is broken up into groups of moderate size between which there is a limited amount of migration. When populations are too small, however, drift leads inevitably to the loss of alleles, to the reduction of variability, and to the consequent inability of populations to respond to environmental changes. Wright (1930) wrote that at intermediate sizes "there will be continuous kaleidoscopic shifting of the prevailing gene combinations, not adaptive itself, but providing an opportunity for the occasional appearance of new adaptive combinations of types which would never be reached by a direct selection process." With regard to human evolution, drift was possibly more important in prehistory when our species was broken up into small hunting camps, tribes, and clans.

We emphasize that the argument is not *whether* evolution occurred, but *how* evolution occurred. To explain the marvelous complexities of the living world, virtually all professional biologists see the many lines of evidence as favoring stepwise evolutionary changes over many millions of years. Some others may try to misuse the sometimes spirited arguments among biologists about the *mechanisms* of evolution to discredit evolution itself. But the arguments only attest to the vigor of the science and, to some extent, the clash of personalities.

HUMAN RACES

The terms *species* and *races* are familiar; both suggest the idea that members of a group have more in common with one another than with outsiders. It is often difficult, however, to delimit the groups and to objectively measure the differences between them.

We begin with the easier term. A **species** is a group whose members are capable of interbreeding with one another but incapable of producing viable and fertile offspring with members of other species. The divergence of two or more separate species from a common ancestral stock is one of the main themes of evolution. During the branching process extending over long periods of time, geographically isolated groups come to have different allele frequencies and perhaps different amounts of chromosomal material, but one way or another, they evolve the intersterility that distinguishes them as separate species.

Tracing the evolution of the several species that were ancestral to our own is unfinished business because the reconstruction is based largely on a limited number of often isolated fossil discoveries. The broad outline of current knowledge points to the separation of a humanlike (hominid) lineage from apelike ancestors about 5 to 6 million years ago (Figure 18.8). A number of species that were intermediate between the apelike ancestors and the more recognizably human types have been grouped in a genus called *Australopithecus*; these creatures were small-brained but, unlike fossil or modern apes, they walked upright on two legs. Although some Australopithecine branches seem to have been evolutionary dead ends, one branch with a more lightly built skeleton is thought to have evolved about 2 million years ago into populations of the genus *Homo* that are our direct ancestors. A succession of species—*H. habilis*, *H. erectus*, *H. sapiens*, whose points of separation are somewhat arbitrary—represent increasing brain size and greater use of the hands for making tools, brandishing weapons, and carrying food and babies.

Just as separate species can differentiate from each other given iso-

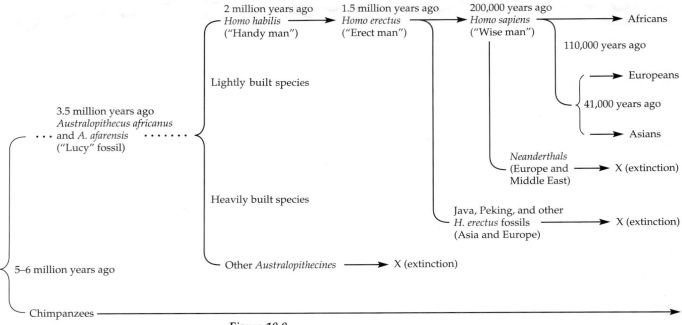

Figure 18.8
Some generally, but not universally, accepted landmarks in the evolution of human beings and closely related species. The discovery of new fossils and the compilation of new molecular data may well lead to modifications of this evolutionary tree. Migrations between human groups, especially in the last 50,000 years, complicate this simple presentation.

lation and enough time, so, too, can subgroups within a species. Extensive wanderings of *Homo sapiens* led to dispersed settlements worldwide starting about 100,000 years ago. Group differences developed through natural selection operating in widely differing environments or through other evolutionary means. Some differences relate to easily visible traits, such as skin color and facial features. Other differences, such as blood groups, enzymes, and RFLPs, are usually observed only by laboratory workers.

Races can be defined as the subdivisions of a species that have come to differ, to a greater or lesser degree, in the frequencies of the alleles they possess. But human races have not achieved anything like separate species status because (1) the time available has been relatively short and (2) the continual migrations of people throughout history have prevented the degree of isolation necessary for speciation. It is likely that few, if any, human groups have remained completely isolated genetically for more than a few generations. Races depend on finer distinctions than species, but the differences are sufficient to call attention to themselves and to invite some sort of classification.

Although there is theoretical agreement on what constitutes separate species, there is no single handy criterion by which to define races. Consequently, there are many opinions as to how the job should be done. Among anthropologists, the "lumpers" recognize relatively few different races—commonly just Africans, Caucasians, and Asians—whereas the "splitters" come up with many more divisions. The problems of classification relate to the number and placement of lines between more or less merging categories. Perhaps no practical purpose is served by drawing lines at all. But the groups are of scientific interest, primarily for investigating how they evolved. That is, the measurement of variability within and between divisions of humankind provides insight into the processes of biological and cultural evolution.

Human Variability

The traditional measurements of racial differences relate to the size, shape, and color of various parts of the body. Cultural characteristics, especially language and behavior, have also been studied. These traits are substantially influenced by environmental variables and depend on many genes whose precise functions are unknown and that interact during development in unknown ways. Attempts to delineate racial differences by overall morphologies are also made difficult because the traits are measured along a continuous scale; the uninterrupted sequence of data makes division arbitrary.

To avoid some, but not all, of these problems, anthropologists have increasingly turned to simply inherited traits that may be present or absent in one or more populations. As early as 1918, even before ABO inheritance was understood, anti-A and anti-B were used by army physicians on the Balkan front to survey different groups of soldiers. Since then, blood samples have been channeled into anthropologists' and geneticists' test tubes from literally millions of people living in virtually every habitat around the globe. Using the Hardy-Weinberg law to calculate allele frequencies, masses of blood group data have pointed up simple differences between human races.

For example, current knowledge of the worldwide distribution of the B allele of the ABO gene is presented in Figure 18.9. The maximum frequencies of about 30% appear in central Asia. Frequencies decrease westward into Europe, diminishing to about 5% and even less in the Basque populations of France and Spain. The gradual changes in the distribution of the B allele across Europe seem to be due to the Mongolian invasions of the fifth through fifteenth centuries. The allele is absent among Indian populations of the New World and among the aboriginal populations of Australia.

Other blood groups, especially MN, Rh, and Duffy, exhibit even larger allele frequency differences. (The specifics of variation differ from gene to gene.) Table 18.4 shows some typical Rh allele frequencies. The Rh negative allele (r) has its highest frequency among Caucasians and is

Figure 18.9
The distribution of the B allele of the ABO gene among indigenous populations of the world. (From Mourant et al. 1976.)

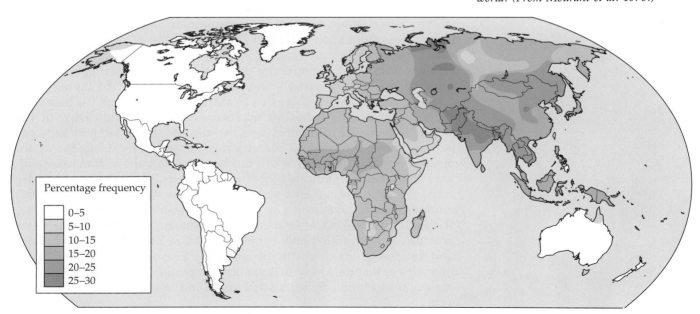

Percentage frequency

	0–5
	5–10
	10–15
	15–20
	20–25
	25–30

Table 18.4

Typical Rh allele frequencies in different world populations.

Population	ALLELE FREQUENCIES (%)	
	R	r
Caucasians		
Spanish Basques	50	50
American whites		
Australian whites	60	40
Spaniards		
Africans		
American blacks	70	30
Angolan Bantus	75	25
Nigerians		
Asians		
Native Americans		
Australian aborigines	95+	near 0
Japanese		
Chinese		

Source: Selected from the extensive compilation of Mourant et al. (1976).

least common in Asians. As you would expect, racial origin is more important than current geographical location: Compare American whites, American blacks, and Native Americans.

More recently, other kinds of simply inherited traits have been added to the repertoire of tools. These include various plasma proteins, red blood cell enzymes, hemoglobin types, histocompatibility types, and restriction enzyme sites and DNA sequences of both nuclear and mitochondrial DNA.

The Message from Mitochondrial DNA

The use of **mitochondrial DNA (mtDNA)** has yielded especially interesting results. Recall that mitochondria are small organelles that generate most of a cell's chemical energy. Each mitochondrion has a small amount of DNA—circular molecules of 16,569 nucleotides each. The base sequences encode transfer RNAs, ribosomal RNAs, and a few dozen enzymes that help synthesize the energy-carrying molecules of ATP. Mitochondrial DNA is relatively easy to purify, and its simple structure (no introns or repetitive sequences) makes it ideal for some types of analysis (Box B). Although each individual has over a trillion cells, and each cell has several hundred mitochondria, and each mitochondrion has about ten copies of the DNA circle, virtually all of the over 10^{15} molecules of mtDNA within a person seem to be identical. Recall further that all mitochondria in a fertilized egg come from the egg only; the mitochondria that are present in the tail section of a sperm are excluded during the formation of the zygote. Thus, the genes in mtDNA follow an exclusively maternal inheritance pattern generation after generation.

The mutation rate of mtDNA is much higher than that of nuclear DNA. Mitochondrial DNA is thus a sensitive indicator of evolutionary modification that can lead to racial differences. Some recent sequencing studies have examined base changes that occur along a particular 600-nucleotide segment. In these studies, total DNA was extracted (in many cases, from a single hair plucked from the scalp of individuals). Researchers then used the polymerase chain reaction to amplify the target DNA by using primers that hybridized to unique sequences at the left and right ends of the region.

Figure 18.10 shows data on mitochondrial DNA differences from a study of 189 individuals selected from diverse populations worldwide. The branching tree was constructed by a computer program that minimized the number of mutations that had to be invoked to account for the variation. A time scale was deduced by comparing an average human mtDNA sequence with that of an average chimpanzee. Because the human-chimpanzee divergence occurred about 5 million years ago, the investigators were able to calculate that about 14 substitutions occur per 100 bases per million years in this region of mitochondrial DNA. Knowledge of the nucleotide sequence in chimpanzees also allowed the researchers to "root" the tree, that is, to trace all present-day human mtDNA sequences to a common origin. The patterns of mitochondrial mutations, translated into twigs and branches, indicated that this origin was 200,000 years ago, more or less. Statistical arguments suggested that the source of all human mitochondria must have been a *single* individual—nicknamed Eve by some. (The mitochondrial ancestor had to be a woman, of course, since mitochondria are transmitted only by females.) Many other females and males lived contemporaneously with this unknown Eve, and many other females and males of that time have had modern descendants, but this one woman was the source of all mitochondrial genes in the present human population.

RAISING THE DEAD

The curious half-zebra-like creature posing here at the London Zoo in 1870 is a *quagga*, an animal that has now been extinct for over 100 years. From a salt-preserved quagga skin, researchers have cloned and sequenced a length of mitochondrial DNA and documented the base substitution mutations that distance it from some present-day relatives.

DNA molecules, even in degraded form, are absent in virtually all fossilized material because fossils are completely mineralized. But useful DNA exists in many specimens preserved for museum displays, as well as in tombs, frozen soil, peat bogs, and amber. Because there are hundreds of mitochondria per cell, each with multiple copies of one DNA molecule, the likelihood that archaeological specimens contain clonable fragments of mtDNA is thousands of times greater than for single-copy genes in nuclear DNA.

Researchers have, however, cloned and sequenced *repetitive nuclear* DNA from long-dead specimens. For example, Pääbo (1985) extracted DNA from the leg skin of a 2,400-year-old Egyptian mummy and discovered the base sequences of *Alu* repeats. Doran et al. (1986) identified both *Alu* segments and mitochondrial segments in 8,000-year-old human brains preserved in a peat bog in Florida. Thus, both wet and dry human remains have provided DNA for analysis. A remarkable endurance record belongs to magnolia leaves. Having fallen to the bottom of

an Idaho lake 17 to 20 million years ago, the leaves were still green when researchers cleaved open the slate in which they had become embedded. Using the polymerase chain reaction to multiply part of a gene that codes for one of the enzymes used in photosynthesis, Golenberg and colleagues (1990) were able to demonstrate that the ancient gene differed by 17 base substitution mutations from the corresponding present-day magnolia gene.

The current record holder for DNA preservation is a plant-eating weevil from Lebanon embalmed in

amber 130 million years ago. Amber is a form of fossilized tree sap. Sealed in amber, small creatures dry out and are protected from decay-causing bacteria. From several such insects, researchers have been able to extract and amplify (by PCR) both nuclear and mitochondrial genes (Hoppe 1992). It is considered highly improbable that 200-million-year-old dinosaur DNA could be obtained from amber-preserved blood-sucking insects, but the thought has crossed the minds of scientists, novelists, and moviemakers.

The genealogy depicted in Figure 18.10 also suggests that Eve lived in Africa, because the major branches from the 200,000-year-old trunk have only African descendants (top right quadrant). In general, African populations are much more variable in mitochondrial DNA than others, and the branches reach back further in time. Under this view, our species *Homo sapiens* spread relatively recently over the world from an African origin and *replaced* preexisting groups of *H. erectus* in various geographical locations. This view supposes that *H. erectus* in Europe and Asia died out without contributing any genetic material to the *H. sapiens* who arrived there from Africa perhaps 50,000 years ago. Racial diversity within our species thus stems from about this time.

The details of the interactions between residents in Europe and Asia

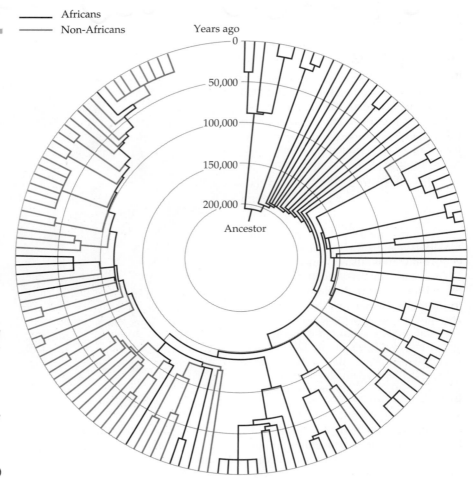

Figure 18.10
A genealogical tree of different mitochondrial DNA types found in a survey of 189 people from around the globe. The tree is derived from mutational differences in a 600-base segment of the control region of mitochondrial DNA. Around the periphery are the 135 different mitochondrial DNA types that were found. The branchings arise from a computer program that seeks the minimum number of mutations necessary to account for the differences between the individuals. The "years ago" time scale is based on a mutation rate in this region of mitochondrial DNA of about 14 base substitutions per 100 bases per million years. (Adapted from Vigilant et al. 1991.)

and migrants from Africa are perplexing. Some anthropologists and archaeologists, in fact, vigorously dispute the relatively recent **out-of-Africa theory** of human races (Angier 1991). They claim that the mitochondrial dating technique seriously underestimates the time of emergence of our species. More critical yet are recent arguments that the computer program that generated the branching tree was flawed. Many other equally likely trees could come out of the computer analysis—trees that suggest an origin other than 200,000 years ago and a place other than Africa (Wilford 1992). These critics suggest instead that human races evolved over perhaps a period of 1 million years, each race from a separate lineage of *H. erectus* in Africa, Europe, and Asia. Calling their view the **multiregional model** of racial differentiation, these researchers suggest greater separateness between the races because of the greater time available for racial divergence (Thorne and Wolpoff 1992).

Mitochondrial DNA is now being used for many studies of migrations and origins. For example, mtDNA analysis of American Indians gives a longer than customary span for their existence in the New World. Contrary to a long-standing belief that people walked over a land bridge from Asia into Alaska about 12,000 years ago, the mtDNA evidence suggests that the earliest band of pioneers arrived 20,000 or more years ago (Horgan 1992). But disagreements are spirited among and between archaeologists, anthropologists, linguists, molecular biologists, and others.

Individuality

The existence of so much variation, however measured, means that the idea of a pure race is fallacious. Purity implies constancy of type, which implies homozygosity, as found in the pure-breeding strains of Mendel's peas, alike in their pods. The absence of pure races is also due to the population intermixtures that result from migration. Gene flow is not just a recent phenomenon. Fossils increasingly indicate that several different forms of *Homo sapiens* coexisted tens of thousands of years ago. Successive separations and fusions of human groups suggest that racial classifications are not static.

It is easy to see traits such as skin color or eyelid shape, which are determined by just a few of the tens of thousands of human genes. Even though these particular characteristics allow pigeonholing, we know that the degree of variability within any one grouping is very large. The genetic variability among the individual members of any race, however defined, is greater than the average difference between races. As a fairly obvious example, the difference in height between the average American male and the average Japanese male is about 4 inches, whereas it would be usual to have differences much greater than that within even a small group of American males or within a small group of Japanese males.

Further evidence of human variability has come from the English geneticist Harry Harris (1980), who summarized data on human enzymes, tabulating the results on 104 different genes. Of these, 33 genes exhibit two or more common variants in at least one major ethnic group. Using these data, Harris calculated that the average person is heterozygous for about 6% of the genes that code for enzymes with electrophoretic variants. But because electrophoresis detects only a minority of enzyme variants, the actual number of heterozygous genes per person is much higher.

However we view our biological endowments, the chance of finding two persons with the same total genotype (discounting identical twins) is effectively zero. Genetically, each person is unique.

SUMMARY

1. Independently of each other in the mid-nineteenth century, Darwin and Wallace developed the idea that gradual evolution by natural selection can account for the vast array of complex living forms. Many diverse lines of evidence support the concept of evolution.

2. Evolution proceeds largely by natural selection but also partly by mutation, migration, and random genetic drift. In a simplified sense, evolution amounts to gradual changes in allele frequencies in different lines of descent.

3. Mutations provide the raw material for evolution. Dominant mutations are revealed by finding the abnormal trait in the offspring of normal parents. Recessive mutations may not be phenotypically expressed for many generations.

4. Measuring the rates of mutations in humans is imprecise because of the rarity of mutational events for a given gene. A typical human gene mutates at the rate of about one mutation per million gametes per generation.

5. Selection—natural or artificial—occurs when various genotypes in a population differ in fitness, which is defined as the ability to leave offspring. Changes in environmental conditions can lead to changes in the way selection acts.

6. Selection against a rare recessive phenotype decreases the frequency of the recessive allele only very slowly, because most rare recessive alleles are hidden in heterozygotes. Similarly, the medical treatment of recessive diseases will increase their frequency only very slowly.

7. Selection in favor of the heterozygote stabilizes the allele frequencies at intermediate levels. The sickle-cell allele in malarial environments is the only well-understood example in humans.

8. Migration may lead to gene flow, the introduction of new alleles into a population. Estimates of the amount of mixture can sometimes be obtained by allele frequency analysis.

9. Drift refers to haphazard changes in allele frequencies that become more likely the smaller the population. The relative importance of drift in evolution is disputed.

10. Human races, more or less distinctive groups of humankind, have arisen by evolutionary processes. Allele frequencies in one race may differ, to varying degrees, from the frequencies in others.

11. The out-of-Africa theory and the multiregional theory compete as explanations for the origin of human races. The analysis of fast-evolving mitochondrial DNA has sharpened the debate.

12. The genetic differences within human races are large—larger in general than the average differences between races. Each individual has a unique genotype.

KEY TERMS

artificial selection	genetic drift	multiregional model	race
Darwinian fitness	genetic equilibrium	mutation rate	recessive lethal gene
evolution	migration	natural selection	selection
fitness	mitochondrial DNA	*On the Origin of Species*	species
gene flow	(mtDNA)	out-of-Africa theory	superior heterozygote

QUESTIONS

1. If most mutations occur during DNA replication, then sperm would be expected to carry more new mutations than eggs. Why?

2. In families with both normal-height and achondroplastic children, the fathers tend to be younger when they beget the normal-height babies than when they beget the dwarfs. Give a possible explanation.

3. Assume that a rare dominant allele has incomplete penetrance. Thus, some affected children might have normal-appearing parents even when a parent possesses the dominant allele. Would the calculated mutation rate by the direct method then be too high or too low?

4. Assume that all human genes stopped mutating completely. How would you know that this unusual phenomenon occurred?

5. What is meant by the relaxation of natural selection? Is it happening? With what consequences?

6. Explain why selection against a recessive lethal becomes less and less effective as the allele becomes rarer and rarer.

7. The effectiveness of selection against a recessive lethal could be speeded up if we could identify heterozygotes and could counsel them about the personal and social risks of procreation. Why would this procedure be potentially much more effective in reducing the recessive trait than just non-reproduction by affected homozygotes?

8. In some areas of malarial Africa (where heterozygotes for sickle-cell anemia are favored), the equilibrium frequency of the sickle-cell allele is about 15%. Why is this figure much less than the frequency of the normal allele (85%)?

9. Give two reasons why the frequency of the sickle-cell allele among American blacks is less than that among their African ancestors.

10. For the Fy^a allele of the Duffy blood group system, Reed (1969) reports:

$P = 0.0941$ = frequency among Oakland, California, blacks

$P_1 = 0$ = frequency among West Africans

$P_2 = 0.4286$ = frequency among Oakland, California, whites

Calculate M_1 and M_2. Express this result in words.

11. Recall from Chapter 13 that insulin, with amino acid chains labeled A and B, is formed from a longer polypeptide called proinsulin when a segment (called C) is removed and discarded. Kimura (1979) compared the amino acid differences between species for insulin and for the C segment. He found far fewer differences for the A and B chains than for the C segment. Does this make evolutionary sense? Why?

12. (a) Give a genetic definition of race. (b) What are the advantages of using blood groups or other molecular phenotypes to delineate races?

13. Why is mitochondrial DNA useful in studying evolution?

FURTHER READING

The autobiographies by Darwin (1969 reprint) and by Wallace (1908) provide a unique introduction to evolutionary thought. Crow (1987) gives brief comments on the development of population concepts. General accounts of evolution include the informative and masterful books by Dawkins (1986) and Futuyma (1983) and the more traditional textbook by Strickberger (1990). Gould (1991) provides entertaining, stimulating, and authoritative essays on evolutionary topics. The nonscientific viewpoint of creationists is explored succinctly by the Committee on Science and Creationism (1984) and in more detail by Berra (1990). Shreeve (1990) contrasts for lay persons the traditional-fossil and molecular-genetic views on the origin of modern human races. Jameson (1977) has collected important original papers in evolutionary theory. Of many *Scientific American* articles on evolutionary topics, those by Kimura (1979), Stebbins and Ayala (1985), Stringer (1990), Ross (1992), Thorne and Wolpoff (1992), and Wilson and Cann (1992) are of current interest. Cherfas (1991) gives a brief report on analyzing DNA from long-dead organisms.

19 Detecting Genetic Disease

Carol Terry's troubles seemed to begin suddenly at age 25 when she cut her wrists, although she said she didn't especially want to die. She had just wanted to take some action, but she hadn't known what action to take or why any action had been needed. Then, over the course of two years, she saw a series of psychiatrists who moved her in and out of mental hospitals, put her into group therapy sessions, and gave her antidepressants, tranquillizers, and electroshock treatments. To no avail. Her hands became progressively more shaky and clumsy, her mouth became twisted, and she drooled. All this, the psychiatrists said, was the result of hysterical neurosis—her way of seeking attention and showing resentment toward her husband. After the divorce, her symptoms worsened: She had trouble swallowing and lost 25 pounds; also, her right leg went limp, her balance was unsteady, and her speech became badly slurred and squeaky. A series of physical tests revealed an enlarged and damaged liver, the result—it was thought—of all the drugs.

When her insurance ran out, Mrs. Terry entered a community mental health center as a welfare patient. In that unpromising situation, her frightening experience came to an end. David Reiser, a staff physician with a good memory, thought that her appearance and behavior was characteristic of a rare autosomal recessive disorder called *Wilson disease*, whose frequency is about 1 in 50,000 people. He had seen one case in medical school. In patients with Wilson disease, copper accumulates in the body in large amounts, for reasons that are not well understood. An essential trace element, copper is needed for the proper functioning of several enzymes. But excess copper gets deposited in the liver and brain, where it causes serious damage. It also forms golden-brown circles, called Kayser-Fleischer rings, that hide the outer portion of the iris of the eyes (Figure 19.1). The rings are definite marks of Wilson disease, and those in Mrs. Terry's eyes were easily visible.

Relatively few genetic diseases can be effectively treated, but Wilson disease is one of them, if caught early enough. The drug penicillamine binds to copper and causes it to be excreted in the urine. With this treatment, Mrs. Terry's psychiatric symptoms quickly disappeared, her liver slowly returned to near normal, and the Kayser-Fleischer rings faded. She completed a university degree, went to work as an accountant, and remarried. The physical reminders of her brush with death were a slight limp, an occasional hesitation in speech, and a somewhat crooked smile (Roueché 1979).

Wilson disease was first described in 1912. At about the same time, Reginald Punnett (1911), the inventor of the checkerboard for solving genetics problems, wrote: "Increased knowledge of heredity means increased power of control over the living thing, and as we come to understand more and more the architecture of the plant or animal we realize what can and what cannot be done towards modification or improvement." Punnett was thinking of the application of Mendel's laws to farm production, but knowledge is power in human genetics as well.

Recent advances in molecular genetic technologies have greatly expanded our understanding of the human condition, especially in regard to health and disease (Chapter 13). Because it touches the very basis of human life, biotechnology is viewed in different ways by different people. Pessimists predict that the new knowledge will generate a host of problems: damage to the environment or to human health, unwarranted intrusions into private affairs, unethical or irreligious experimentation on animals or human embryos, perversions of conventional methods of human reproduction, or corruption of human individuality. Optimists, meanwhile, expect that genetic engineering will contribute greatly toward alleviating hunger and disease throughout the world. Especially in the last decade, researchers have developed powerful molecular methods for probing human disease—methods that may ease the burden of hereditary disorders.

In this chapter and the next, we examine the implications of advances in human genetics for individuals, families, and society. We begin by surveying the growing field of genetic counseling, including prenatal diagnosis and genetic screening, areas of both promise and controversy. Although few genetic diseases can now be treated, there is nothing inherent in genetic disorders that make them untreatable. Indeed, as many infectious diseases are being brought under control, disorders that are wholly or partly genetic have been receiving increasing attention. We note the limited use of *gene therapy*—the ultimate application of molecular genetic technology. Finally, we ask if anything can (or should) be done to reduce the frequency of genetic conditions in future generations.

The problems inherent in many of the genetic technologies are out of the ordinary, and publicity on the technical, ethical, legal, and social issues has been widespread. For each technique, you might want to consider the following questions:

1. *Is it feasible?* Is a newly announced achievement now applicable only to microorganisms? To laboratory animals? How will it be tested for use in humans? Is the technique so complex, the cost so high, or the morality so debatable that the advancement would only be utilized by a handful of persons? Is *my* family likely to benefit? Sensational pronouncements to the contrary, no human babies are likely to be decanted from flasks as in Aldous Huxley's *Brave New World* (1932).

2. *Who decides?* This question arises in several contexts. Who decides what type of research shall be supported by public funds, and who selects the investigators that receive support? Some decisions on the treatment of malformed newborns or on the choices for recipients of transplantable organs are painfully difficult. Who shall be the primary decision makers? Doctors? Parents? Clergy? Judges? Administrators? Legislators? "Experts" of one sort or another?

3. *Who pays?* In a world of limited resources, we must decide not only who receives an expensive treatment but also who pays the bill. Should scarce medical processes or facilities be available on a random basis, or on the basis of one's ability to pay, or on the basis of one's ability to mount a television appeal for funds? What procedures should or should not be covered by health insurance?

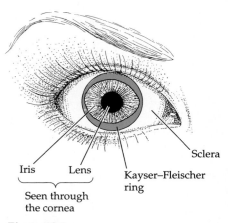

Iris   Lens   Kayser–Fleischer ring   Sclera

Seen through the cornea

Figure 19.1
A Kayser-Fleischer ring that is characteristic of Wilson disease. The golden-brown circle of copper is deposited in the cornea. The cornea is the transparent portion of the outer layer of the eyeball, covering just the iris and the lens behind the iris. Kayser-Fleischer rings hide the periphery of the iris and are wider on the top and bottom than on the sides. (The white of the eye is called the sclera, which is continuous with the cornea.)

4. *Is it ethical*? Is a particular technology good or bad? Does it increase or diminish human freedom? Does it enhance our sense of human dignity or does it dehumanize us? Paul Ramsey (1970), who was a theologian at Princeton University, discussed two types of ethical considerations: the morality of the perfected procedure and the morality of the experimentation in humans that would certainly be needed to perfect the procedure. A fly or a frog that develops badly can be discarded, but what will be done with a defective baby arising from experimentation?

GENETIC COUNSELING

Every year, thousands of families are affected by the birth of an abnormal child. About 0.5% of newborns have a chromosomal abnormality with moderate to severe phenotypic effects. Another 0.5–1% suffer the consequences of single-gene defects (dominant, recessive, or X-linked), and about 2% have a malformation that may be due in part to heritable factors. Altogether, about 4% of newborns have a serious defect that is recognized at birth or within their first year. It has been estimated that about one-third of all children in pediatric hospitals are being treated for conditions that have a genetic component. Thus, there can be no doubt about the need for increased genetic information and counseling services for the many families that require it.

The aim of **genetic counseling** is to convey medical and genetic facts to an affected or potentially affected family in a way that can be understood. Counselors must try to see that the families are provided with a full range of medical and social services while they adjust to their genetic situation emotionally and intellectually. Counseling requires professionals who are thoroughly grounded in both genetics and medicine. To this basic education must be added patience, sensitivity, respect, and the ability to talk easily with people who are likely to be deeply troubled. Illness and disability inevitably bring tremendous stresses to family life, and many broken marriages are found among the parents of children with genetic defects. Counselors must be prepared to deal with people who are experiencing denial, shock, anger, despair, and guilt and who may turn against the partner supposedly "at fault."

All this may require the joint efforts of a team of counselors: a clinician with genetic training, a geneticist with a mathematical background, laboratory personnel, plus a public health nurse, social worker, or genetic associate* skilled in working with families in their home environments. The first heredity clinic employing a team approach to genetic counseling was established at the University of Michigan in 1940. Today, several hundred centers worldwide—usually associated with medical schools—provide advice and guidance on genetic matters (Lynch, Hoden, and Paul 1991).

The Procedure

The need for counseling often arises when a child is born with a possibly hereditary disorder and the parents or other family members are concerned with the well-being of future children. In addition, couples may be troubled by repeated miscarriages, or they may have heard about the greater risk of certain birth defects with advanced maternal age. People without any history of genetic disorder may also wonder whether they

---

* Several dozen training programs in the United States offer a master's degree in genetic counseling. The first one was established at Sarah Lawrence College in 1969 (Scott et al. 1988).

harbor a potentially harmful gene. This situation might occur if a husband and wife are related to each other or if they belong to an ethnic group in which certain genetic diseases are more frequent than usual. In any event, the counseling process will often include the following steps.

Medical Diagnosis. Genetic counseling requires precise diagnosis, because a number of conditions have multiple causes (Chapter 7). Sometimes these variants may be phenotypically indistinguishable from one another, but in other cases an experienced clinician may detect slight differences (such as hemophilia A versus hemophilia B). In addition to examining the propositus, the counselors may want to see other members of the family or possibly even family photographs. The clerical work involved in corresponding with relatives and searching for pertinent medical records and autopsy reports can be very time-consuming. Laboratory work often includes karyotyping; biochemical analyses of blood, urine, or cultured cells; or molecular analysis with restriction enzymes and DNA probes. But a correct diagnosis is the cornerstone of useful genetic counseling, for a wrong diagnosis may have devastating consequences.

Pedigree Analysis. Not only must a complete three- to four-generation family pedigree be obtained, but the reliability of the collected information must also be assessed. As pointed out in Chapter 5, some important clues—miscarriages, stillbirths, or mildly affected relatives—may be entirely missing. But even if the data are complete, more often than not there will be complicating factors of the types discussed in Chapter 7. Thus, deciding among various possible modes of inheritance may not be a simple matter.

Estimating Recurrence Risks. How likely is it that a given condition will recur in a subsequent birth? The answer is straightforward in the case of simple Mendelian traits: 25%, for example, if both parents are known to be heterozygous for a completely penetrant, recessive allele.

With the advent of molecular techniques, more and more disease-causing Mendelian genes can be analyzed prenatally using fetal cells obtained through amniocentesis or chorionic villus sampling (see later). Predictions of risk may then depend on the closeness of genetic linkage to a marker RFLP. Direct detection of a disease-causing allele can also be done in some cases by using DNA probes complementary to one or the other allele of the gene in question. In these cases the risk to a tested fetus would be either 0% or 100%. The number of hereditary diseases that can be analyzed in these ways is growing rapidly, and any list that is given will soon be out of date (Table 19.1).

If the trait in question is multifactorial, counselors cannot establish a risk figure based on rules of transmission. In these cases they may make use of empirical risk figures that rely on the statistics of prior experience with the particular phenotype in question. For many congenital malformations, such as spina bifida, cleft lip, or clubfoot, the risk of recurrence of the particular malformation after the birth of one affected child to normal parents is about 2–5%. For certain traits, the geographical or racial background of a couple may be an important element of risk estimation. For trisomic conditions such as Down syndrome, it is important to distinguish between the presence of a translocation in a parent (relatively high risk of recurrence) and a nondisjunctional cause (relatively low risk).

Whether clear-cut or not, the risk figures must be conveyed in such a way that they can be understood. In particular, it should be emphasized

Table 19.1

Some diseases whose causative genes can be directly detected with DNA probes or mapped to nearby RFLPs or other DNA markers.

Disease	Rough Chromosomal Location	Direct Detection or Approximate Map Units from DNA Marker
Adrenal hyperplasia	6 short	Direct
Agammaglobulinemia	X long	Marker, very close
Alzheimer disease, familial	21 long	Marker, very close
	14 long	Marker, 2–5 map units
Chronic granulomatous disease	X short	Marker, very close
Color blindness (protan and deutan)	X long	Direct
Cystic fibrosis	7 long	Direct
Fragile X mental retardation	X long	Direct
Hemochromatosis	6 short	Direct
Hemophilia A	X long	Direct
Hemophilia B	X long	Direct
Huntington disease	4 short	Direct
Hypercholesterolemia, familial	19 short	Direct
Lesch-Nyhan syndrome	X long	Direct
Muscular dystrophy, Duchenne/Becker	X short	Direct
Myotonic dystrophy	19 long	Direct
Neurofibromatosis	17 long	Direct
Phenylketonuria	12 long	Direct
Retinitis pigmentosa	X short	Marker, 8 map units
Retinoblastoma	13 long	Direct
Sickle-cell anemia	11 short	Direct
Thalassemia, α	16 short	Direct
Thalassemia, β	11 short	Direct
Wilms tumor	11 short	Direct
Wilson disease	13 long	Marker, 3–7 map units

Source: Data updated from Cooper and Schmidtke (1989).

that chance has no memory. The risk of recurrence for a simple Mendelian trait is the same for each birth, regardless of prior outcomes. For example, if normal parents produce one child with a recessive disorder, this does *not* mean that the next three children will be normal. Rather, the risk of an affected child on each successive birth continues to be 1/4.

Options. When the genetic prognosis is unfavorable, a couple may refrain from childbearing or risk a defective birth. If abortion is an acceptable alternative, however, prenatal diagnosis followed by possible termination of pregnancy is now available in a growing number of cases. Alternatively, a couple may choose adoption or an appropriate reproductive technology using either a donor egg or donor sperm (Chapter 6).

One's perception of risk and willingness to accept it are highly subjective matters, depending on one's personality, experiences, moral convictions, and especially on the burden of care imposed by the condition in question. Some newborns with serious birth defects (whether genetic or not) may die within a few months. Others may require constant attention from their family throughout childhood and beyond or may require expensive medical treatment. We do not mean to minimize the grief of parents upon the loss of an infant, but we want to emphasize here the difference between short-term and long-term care. A couple may decide to take a chance on a relatively mild disorder, or on a more serious disorder of short duration, or on one that can be ameliorated by treatment. But they may not be willing to accept the same percentage of risk on a disorder that imposes greater physical, financial, and emotional burdens on themselves and their family.

The humanistic aspects of counseling are extremely important (Applebaum and Firestein 1983). Because a couple may need time to talk it out between themselves and with their friends and relatives, the process should not be hurried. For example, comprehension and decision making by the couple may be especially difficult if counseling follows too closely the birth of a child with a severe defect. The counselor must be prepared to understand and help a mother who blames herself for any one of many different actions taken during pregnancy (e.g., smoking or drinking) or a father who may at first refuse to believe that the child is his biological offspring. The setting of counseling sessions must also be considered; a hospital environment may be intimidating, even if the counselor is skillful in personal interactions.

Follow-Up and Supportive Services. Because a counselor's spoken words may be misinterpreted or forgotten, the relevant medical and genetic information is often put in writing. Such a letter usually includes the names and addresses of specialists and appropriate social services, as well as written answers to the major questions raised by the clients. Families should be informed of useful new research results and be allowed to decide on the value of participating in experimental trials. Services can also be extended to relatives who may unknowingly be at risk, but giving unsolicited information raises a new set of ethical questions. Overall, the counselor must provide expert and caring assistance to persons whose self-esteem has been hurt and whose lives may have suddenly been complicated by the birth of a child with a genetic disorder.

Ethical Concerns

The goal of counseling is to enable couples under stress to plan their families according to their own values and with the maximum amount of information and support, but it is not always clear how this may best be done. Opinion varies among counselors as to the degree of advice that should be given. At one extreme are those who specifically suggest a course of action based on the counselor's own values or experiences (as is common in a doctor-patient relationship). Others may provide advice, if asked, but with the proviso that it is impossible for the counselor to fully appreciate all the variables that enter into an intensely personal situation. Most counselors, however, are nondirective—providing accurate information, much empathy, and complete support, but no specific advice about a course of action (Harris 1988).

Another difference in outlook involves a couple's responsibilities to the human species as a whole. Medical personnel are trained to act primarily for the well-being of their patients, but individual reproductive decisions affect the average welfare of future generations. Whatever our collective obligations to descendants are, however, the views of counseled couples, not those of the counselors, ought to prevail.

The question of withholding certain information also arises (Minogue et al. 1988). For example, in one case told to the authors by a genetic counselor, amniocentesis revealed a twin pregnancy with 46,XY and 47,XYY fetuses. The prognosis for double-Y males is uncertain, although most develop within the range of normal variation of XY males (Chapter

10). Although the existence of the two karyotypes was revealed to the prospective parents, they preferred not to be told which twin had which karyotype. Even if a counselor believes that withholding knowledge is in the best interests of the person being counseled, telling a lie might erode the trust necessary to the counseling relationship. (Also, counselors who withhold information may be subjected to lawsuits if the clients later learn of the secret information.) Generally, counselors tell only the truth—all of it, or virtually all of it.

The conflict between telling or withholding information is especially acute with regard to relatives who have not specifically sought counseling. Although many persons want to know such genetic information, some might not, especially if the "bad news" involves a serious late-onset condition such as Huntington disease. When no treatment is currently available, is a humanitarian purpose served by informing a person that a progressively debilitating neurological disease might strike with such and such a probability? If this person is still in childbearing years, does the counselor bear any obligation to help prevent the conception of additional affected persons? There are no pat answers to such questions.

PRENATAL DIAGNOSIS

The ability to detect prenatally virtually all major chromosomal aberrations and many gene-controlled biochemical defects has been a tremendous aid to genetic counseling. (See partial list in Table 19.1.) If a disorder is diagnosed early in pregnancy, the couple has the option of aborting the fetus and beginning again. Fortunately, in about 97% of prenatal analyses, the fetus is found *not* to have the disorder in question. This finding relieves anxiety for the parents, who can then anticipate a newborn no more at risk to abnormality than a random birth. Many couples at substantial risk of having a child with a serious defect would refrain from further pregnancies unless assured that the risk could be reduced. Thus, *because* of the availability of prenatal diagnosis, life is given to children who would otherwise never be conceived.

Procedures for Obtaining Fetal Cells

There are two major techniques for obtaining a sample of fetal cells for prenatal diagnosis; both are done on an outpatient basis with minimal discomfort for the woman. In **amniocentesis**, which has been performed since the 1960s, a physician removes a small sample of the amniotic fluid that surrounds the fetus (Figure 19.2). A long, thin needle is inserted through the abdominal wall, uterus, and fetal membranes. The position of the needle with respect to the fetus can be accurately seen by ultrasonic scanning, which makes major surface features of the fetus visible through reflected sound waves. The needle must not, of course, penetrate the fetus, the umbilical cord, or the placenta. Amniocentesis is often done at about the sixteenth week of pregnancy, when the 5-inch fetus floats in about a half pint of amniotic fluid (Fuchs 1980).

Suspended in the amniotic fluid that is withdrawn (about 2 tablespoons) are living cells that have sloughed off from the fetal skin, the lining of the respiratory or urinary system, or the amnion. A disadvantage of amniocentesis is that the fetal cells obtained are not dividing. Thus, they must be put in culture medium and allowed to multiply for two to four weeks before cytogenetic analyses can be done. There is a small risk

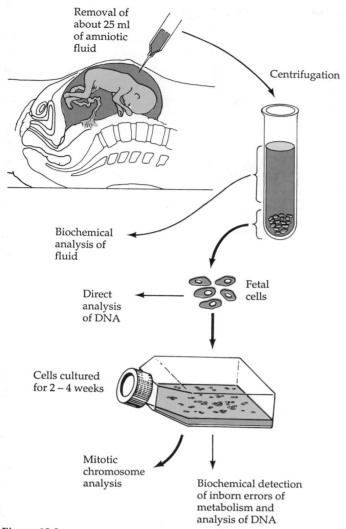

Removal of about 25 ml of amniotic fluid

Centrifugation

Biochemical analysis of fluid

Direct analysis of DNA

Fetal cells

Cells cultured for 2 – 4 weeks

Mitotic chromosome analysis

Biochemical detection of inborn errors of metabolism and analysis of DNA

Figure 19.2
Steps in the prenatal diagnosis of genetic disease by amniocentesis, which is done at about the sixteenth week of pregnancy. The bold arrows show the most common pathway of analysis—karyotyping to detect Down syndrome or some other chromosomal abnormality.

of needle injury to the mother or fetus (bleeding, infection) and a small risk (above the naturally occurring rate) of abortion of the fetus. Amniocentesis is about 99% safe when done by an experienced obstetrician.

Another method of obtaining cells for analysis is **chorionic villus sampling (CVS)**, a procedure that has been performed in the United States only since about 1983. The major advantage of CVS over amniocentesis is that it is done much earlier in pregnancy—from the ninth to twelfth week of gestation. Thus, weeks of anxiety are avoided, the pregnancy can remain a private matter, and abortion (if chosen) is performed during the first trimester, when it is medically safer for the mother.* In one variation of this process, the physician inserts a catheter guided by ultrasound imaging through the cervical canal (Figure 19.3). A few bits of tissue, consisting of branched projections from the chorion (the chorionic villi), are sucked off from the developing placenta. After removal, the fetal villi

* Each trimester is approximately 13 weeks in duration. The fetus becomes viable (able to survive outside the womb) near the end of the second trimester, but it needs technological assistance to survive if born at that time.

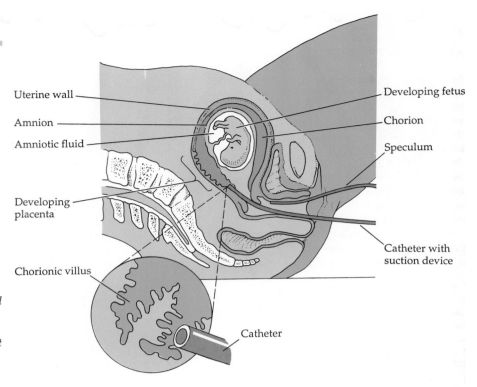

Uterine wall

Amnion

Amniotic fluid

Developing
placenta

Chorionic villus

Developing fetus

Chorion

Speculum

Catheter with
suction device

Catheter

Figure 19.3
The transcervical method for obtaining fetal cells by chorionic villus sampling (CVS), which is done at the ninth to the twelfth week of pregnancy. The projecting villi that anchor the fetal sac to the uterine wall have been likened to a shaggy carpet.

are separated from any attached maternal tissue with the aid of a low-power microscope.

The safety of chorionic villus sampling continues to be evaluated, but the procedure seems to be a few percent more risky and less accurate than amniocentesis according to some (but not all) studies (Anonymous 1991; Jackson et al. 1992). In the *natural* course of events, about 2% of pregnancies that are viable at ten weeks subsequently miscarry, so it has been difficult to evaluate the cause of some abortions following CVS. Another problem of CVS is the risk of Rh immunization of the mother because of possible disruptions in the developing placenta, where maternal and fetal circulations are separated by only a few cell layers (Chapter 16).

Indications and Results

Prenatal diagnosis using fetal cells or amniotic fluid is usually offered in three types of situations.

Increased Risk of a Chromosomal Abnormality. The major use of prenatal diagnosis is for the detection of Down syndrome or other chromosomal aberration (see bold arrows in Figure 19.2). A large percentage of all prenatal diagnoses—as much as 85% at some centers—is done in this situation. The risks of chromosomal abnormalities in fetuses increase gradually with maternal age, but more steeply when a pregnant woman is 35 years or older. At age 35, the risks associated with amniocentesis are more or less the same as the risks of a major chromosomal abnormality in the fetus. Thus, routinely offering amniocentesis at 35 has become the standard of practice, but some younger women also have the test (Box A). Chromosomal analysis of fetal cells is also indicated when a couple (of any age) already have one child with a chromosomal abnormality or when one of the parents is a carrier of a translocation.

BOX A

AMNIOCENTESIS FOR YOUNGER WOMEN?

Although pregnant women under the age of 35 are not routinely offered amniocentesis to detect Down syndrome, an analysis of chemical markers in their blood can aid in deciding whether or not they should consider the test. Between the fifteenth and twenty-first week of gestation, investigators determine the concentrations *in maternal serum* of three chemicals: α-fetoprotein, estriol, and chorionic gonadotropin.* Compared with normal pregnancies, the average values for women who are carrying a fetus with Down syndrome are reduced about 25% for the first two chemicals and increased about 250% for the last one. It is not known why a Down syndrome pregnancy changes the chemical concentrations in these ways. A recent large-scale study (Haddow et al. 1992) discovered 760 pregnant women with significantly altered values; 20 of the pregnancies (about 3%) were later found by amniocentesis to involve fetuses with Down syndrome. In addition, seven other chromosomal disorders were found, including trisomy 13, Klinefelter and Turner syndromes, and the triple X karyotype. The investigators estimated that measuring these maternal serum markers in a random sample of women younger than 35 (followed by amniocentesis when judged to be appropriate) detects about 60% of fetuses with Down syndrome.

* The chemical α-fetoprotein is made in the fetal liver but its function is unknown; it is also used as a marker for neural tube defects. Estriol is a breakdown product of the active form of estrogen. Chorionic gonadotropin, a major pregnancy hormone, is made in the placenta and acts to maintain pregnancy; it is also the basis of home pregnancy tests using urine samples.

There is still no satisfactory explanation why advancing maternal age increases the likelihood of some aneuploid conditions in fetuses and subsequent newborns. We have previously noted that the frequency of Down syndrome increases gradually to about 1 in 100 births for 40-year-old women and becomes more frequent for yet older women (Chapter 10). Other chromosomal abnormalities that increase with maternal age include Klinefelter syndrome, trisomy 13, trisomy 18, and triple X, so that the total risk of chromosomal abnormality in the live-born offspring of a 40-year-old woman is about 1 in 65.

A new technique allows researchers to detect many chromosomal aberrations just one or two days (rather than two to four weeks) after cells are obtained by amniocentesis. Because it relies on *interphase* cells, no cultures are required to obtain the metaphase spreads needed for standard karyotyping. The process uses **fluorescence in situ hybridization**, or **FISH** (noted also in Chapter 13). Very simply, fluorescent probes that are specific for a region of a given chromosome are used to bathe a sample of amniotic cells. Although it is not possible to see interphase chromosomes under a standard light microscope, the probes hybridize and light up regions of DNA on specific chromosomes (Figure 19.4).

Increased Risk for a Single-Gene Disease. Although most inherited diseases are individually rare, medical geneticists are concerned with thousands of them. For about 350 Mendelian disorders, researchers have characterized the molecular defect either as an alteration in the encoded enzyme (or other protein) or as a specific mutational change in DNA (McKusick 1990). A sampling of disorders that have been analyzed at the DNA level was presented in Table 19.1. A sampling of disorders that can

(A)   (B)

Figure 19.4
Quick prenatal diagnosis of the triple X karyotype by fluorescence in situ hybrid-ization (FISH). Researchers used a fluorescent probe that was specific for DNA sequences in the centromere region of the X chromosome, indicated by the bar in (A). They bathed nondividing, interphase amniotic cells with the probe. Although chromosomes are not visible in interphase cells, the probe finds and binds to the DNA of X chromosomes. The three bright spots seen in each of the amniotic cells in (B) indicate three X chromosomes. (From Klinger et al. 1992.)

be detected in a fetus on the basis of the encoded protein is presented in Table 19.2.

Prenatal diagnosis at the protein level can be made only if the en-coding gene is expressed in the particular cells obtained by amniocentesis or CVS. (Many genes are only expressed in specialized tissues that are not obtained by CVS or amniocentesis. An example is the PKU gene, whose product, phenylalanine hydroxylase, is expressed only in liver cells.) Genes that are transcribed and translated in prenatal samples in-clude those responsible for the basic housekeeping functions that occur in virtually all cell types. Tay-Sachs disease is an example of an inborn error of metabolism that can be detected prenatally by an enzyme assay (Chapter 14). Prenatal diagnosis is offered to couples when a prior child is affected or when both parents are otherwise determined to be carriers.

Table 19.2
Some inherited diseases that can be detected prenatally by analyzing for levels of the encoded protein.[a]

Disease	Abnormal Enzyme or Other Protein	Type of Metabolism Affected
Ehlers-Danlos syndrome	Collagen	Connective tissue
Fabry disease	Galactosidase A	Fat
Galactosemia	Galactose-1-phosphate uridyl transferase	Sugar
Hunter syndrome	Iduronate sulfatase	Complex carbohydrate
Hurler syndrome	Iduronidase	Complex carbohydrate
Ichthyosis	Steroid sulfatase	Hormone
Maple syrup urine disease	Branched chain keto acid decarboxylase	Amino acid
Marfan syndrome	Fibrillin	Connective tissue
Porphyria (variegate)	Protoporphyrinogen oxidase	Heme
Tay-Sachs disease	Hexosaminidase A	Fat
Wolman disease	Lipase	Fat

[a]Some of these diseases may also be diagnosed by DNA analysis.

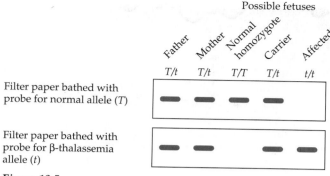

Figure 19.5

The use of radioactive oligonucleotide probes for prenatal diagnosis of β-thalassemia (geno-type t/t) in Sardinia. One probe is specific for the normal allele; the other probe is specific for the thalassemia allele, which differs by a single base. Fetal DNA is cut with a restriction enzyme, then electrophoresed, Southern blotted onto filters, and bathed with one or the other probe. Note that heterozygotes yield DNA fragments that bind both probes, while the two homozygous types bind either the one probe or the other, but not both.

There are many experimental protocols for prenatal diagnosis based on fetal DNA. The diseases that can be determined by a DNA test allow prenatal diagnosis whether or not the gene is expressed and whether or not the protein product of the gene is even known. Some of these tests depend on linkage to RFLPs and other molecular markers. Because there are so many RFLPs scattered throughout the entire genome, soon virtually no gene that is investigated will be without an RFLP to which it can be linked more or less closely. Even if the linkage is not very close, finding RFLP sites on both sides of a disease-causing gene (called *flanking markers*) allows for almost error-free diagnosis if the necessary heterozygosity exists.

Some methods of molecular prenatal diagnosis—generally applicable for diseases caused by known base substitutions—depend on **oligonucleotide probes**, synthetic DNA segments of about 20 nucleotides. These short sequences can be made complementary to the portion of a gene that includes a single base mutation; near the middle is the one base that corresponds to either the normal allele or the mutant allele. Thus, the *normal* probe exactly matches the *normal* allele but has a one-base mismatch with the mutant allele; the *mutant* probe exactly matches the *mutant* allele but has a one-base mismatch with the normal allele. Under carefully controlled conditions of hybridization, each probe will bind only to its corresponding allele. For example, oligonucleotide probes have been used in the prenatal diagnosis of β-thalassemia in Sardinia. Here, the disease-causing change is a G-to-A mutation that makes a stop triplet near the beginning of the gene. Bathed with the radioactive probes, the fetal DNA is hybridized by either the one probe, the other, or both probes in the case of heterozygotes (Figure 19.5).

Oligonucleotide probes can also be used in the prenatal diagnosis of cystic fibrosis. Recall from Chapter 7 that the most common mutation (accounting for about 70% of carriers in the United States) is a three-base deletion known as ΔF508. One oligonucleotide probe is made that binds only to the normal allele, and a second probe binds only to the ΔF508 allele. These and a third probe, common to both alleles and which binds nearby, are used as primers for the polymerase chain reaction. The outcome is that DNA from a normal fetus generates an amplified DNA fragment exactly 63 bases long, and DNA from a CF fetus generates a DNA fragment 60 bases long. Heterozygotes yield fragments of both lengths. The fragments are then identified by gel electrophoresis, which separates them according to size.

Increased Risk for a Neural Tube Defect. In addition to possible chromosomal abnormalities or Mendelian disorders, **neural tube defects (NTDs)** constitute another indication for prenatal diagnosis. By the first month of embryonic development, the furrow of neural tissue that forms along the back closes upon itself to make a tubelike structure, the forerunner of the brain and spinal cord (see Figure 6.11). When the sides of the neural fold fail to close near the head of the embryo, the brain tissue remains exposed and becomes disorganized, resulting in *anencephaly* (literally, "no brain"). Such fetuses are stillborn or survive only a few days. When the failure to close occurs lower down, the result is one or another form of *spina bifida*, often characterized by paralysis or weakness below the level of the spinal opening, incontinence of bowel and bladder, and hydrocephalus (abnormal accumulation of fluid in the brain, producing head enlargement). Some patients with less severe types of spina bifida may live many years with milder disabilities. In the United States, NTDs occur at a frequency of 1 or 2 per 1,000 births, but they are more frequent (up to 7 cases per 1,000) in some other countries, including Ireland, Scotland, and Wales.

Because the recurrence rate for couples with one child with an NTD is about 2% in the United States and about 5% in England, and because the prospect is usually so bleak for affected children and their families, prenatal diagnosis is very useful. The determination of an NTD depends on the concentration of **α-fetoprotein**, a substance of unknown function that is present in fetal blood plasma. Small amounts are also present in the amniotic fluid of normal fetuses, but concentrations in the amniotic fluid of fetuses with NTDs are increased manyfold (especially in the more serious cases). It is thought that plasma with α-fetoprotein leaks into the amniotic cavity from the exposed neural tissue. Ultrasound scanning, done at the time of amniocentesis, can also directly detect many of the deformities associated with neural tube defects.

It is disheartening, of course, that it usually takes the birth of one child with an NTD to initiate the prenatal diagnosis that could prevent a second one. Screening of *all* fetuses might be desirable if amniocentesis carried no risk. It has been found that most (but not all) women carrying a fetus with an NTD have elevated levels of α-fetoprotein in their *own* circulation. Thus, amniocentesis and ultrasonography might be done on the group of pregnant women with elevated levels of serum α-fetoprotein. But this type of screening has presented some problems that are discussed later.

Sex Prediction

Specific prenatal diagnosis is not yet possible for many X-linked recessive disorders that lead to serious disabilities in hemizygous males. Although individually rare, they collectively constitute a significant problem. An example is the *Aldrich syndrome*, in which abnormal blood platelets cause internal bleeding, and immune system defects lead to recurrent and usually fatal infections in childhood. Another example is *anhidrotic ectodermal dysplasia*, which is severe in males, harming especially the skin and mouth, and less severe in heterozygous females, who exhibit only patches of affected skin because of X chromosome inactivation (Figure 7.13).

When a woman is heterozygous for a disease-causing X-linked gene (and her mate is hemizygous normal), one way to prevent the birth of affected boys is to abort *all* male fetuses, even though half of them are expected to be phenotypically normal. The dilemma posed by the abortion of some normal fetuses to avert the birth of possibly abnormal children is not easy to resolve. However, several accurate methods are available for the requisite sex determination of the fetus. These include traditional karyotyping of fetal cells obtained by CVS or amniocentesis, which reliably reveals the Y chromosome. A quicker method is to analyze fetal DNA using the polymerase chain reaction. The primers that are used specifically hybridize to repetitive sequences found only on the Y chromosome.

Currently in an experimental stage is a way to determine the sex of very early embryos prior to implantation. This process could be useful to a couple who wish to avoid a male pregnancy and who are already making use of *in vitro fertilization*. Using a microscope and fine micropipets, a skilled researcher teases out one or two cells from an eight-celled embryo. Then the polymerase chain reaction is run on the cell contents (Chapter 13). The DNA primers are specific for Y chromosome sequences, so that the DNA is amplified only if the extracted cells are from a male embryo. Meanwhile, the rest of the embryo, now with six or seven cells, continues what appears to be regular development (at least in some studies) in its nurturing culture medium. (Recall that identical twins normally originate by the splitting of a few-celled embryo.) Only if the embryo is female would it be transferred to the woman's uterus. Several normal baby girls have been born after such preimplantation diagnosis.

Nor may this technically difficult and costly procedure be practical on a large scale, since in vitro fertilization currently has a low success ("take-home-baby") rate under the best circumstances. Problems are also presented by the male embryos, which may be frozen to avoid the ethical questions posed by a more permanent disposal. Keep in mind, too, that half of the male embryos would not even have inherited the abnormal X-linked gene from a heterozygous mother. Clearly, it would be much better to test for the presence or absence of a *specific* abnormal allele, and researchers are pursuing this goal for several X-linked and also autosomal genes (Simpson and Carson 1992). Indeed, one normal baby has already been born after preimplantation embryos were tested and found not to be homozygous for the $\Delta F508$ cystic fibrosis mutation.

GENETIC SCREENING

Genetic screening is the widespread search of populations for persons having a particular genotype or karyotype. The term refers specifically to genetic testing of those people *without* a family history of the disease in question. Screening serves two general purposes. The first is to detect a serious genetic condition before the onset of debilitating symptoms. This form of screening, typically on a population of newborns, is consistent with an increasing emphasis on preventive medicine. It is similar in intent to *non*genetic screening—for example, periodic mammography or cholesterol determinations. The second general purpose of genetic screening is to identify unaffected carriers in order to counsel them about the risk of producing affected children, thereby helping them to make more informed reproductive decisions.

Although few would disagree with these goals, screening programs initiated with the best of intentions have sometimes caused ill will among the people who were supposed to be helped. Planning and execution of screening require that careful attention be paid to the following concerns (President's Commission 1983).

1. *Testing*. To be useful for screening large populations, a diagnostic test should be convenient, inexpensive, safe, and reliable. Identifications can often be made by biochemical tests involving the protein product of a gene or by molecular methods based directly on DNA. **False negatives** (persons who have the genotype but are not detected) and **false positives** (persons who are said to have the genotype but do not) must be minimized. State legislatures that provide for screening programs are also concerned with cost-effectiveness: The money spent in testing and subsequent services should be less than the money that state agencies would spend on dealing with affected persons. This is likely to be true: The cost of long-term care for institutionalized patients with burdensome genetic conditions is often very great in measurable resources and beyond calculation in terms of human suffering.

2. *Treatment*. Unless the disease in question can be effectively treated, there may be little point in identifying it in a potentially affected person before symptoms begin. Yet in the case of severe diseases with late onset, decisions about marriage and parenthood might be altered by early knowledge of the disease.

3. *Counseling*. Expert counseling services must be made available to guide affected families through difficult times. Considerable psychological harm—low self-esteem and feelings of worthlessness—can result when a person is told that he or she is a carrier of a recessive genetic disease.

4. *Safeguards*. Sometimes it is hard to decide whether screening should be mandatory or voluntary. Although the genetic health of its citizens can be considered a proper governmental concern, genetic diseases do not pose a danger to others in the same way that contagious diseases such as tuberculosis and AIDS do. Although most newborn phenylketonuria screening laws require participation, the constitutionality of this has never been tested in the courts. In any event, most laws provide for a right of refusal on religious grounds.

Strict confidentiality of medical records is another area of concern, especially with the rapid expansion and centralization of computerized records (Holtzman and Rothstein 1992). Although many people fear the loss of health insurance due to genetic or other factors, no insurance companies currently require any kind of genetic testing. They may, however, have access to a doctor's medical data when applicants routinely give permission for such scrutiny. And as genetic tests become more available, more predictive, and less expensive, health insurance companies may change course. Discrimination in employment is also worrisome; but perhaps as a result of the Americans with Disabilities Act of 1990 and efforts of the federal Equal Employment Opportunity Commission, there have been few publicized examples of genetic discrimination in the workplace.

Screening for Early Detection of Genetic Disease

The diagnosis of phenylketonuria (PKU) in newborns was the first large-scale screening program undertaken by government agencies and is still the most widely used. By the late 1950s, it was clear that the most severe abnormalities of PKU could be controlled by diet, but only if treatment was begun in the first few weeks of life. With the development of the so-called **Guthrie test**, the routine screening of newborns prior to their leaving

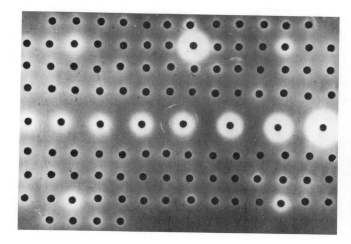

Figure 19.6
A Guthrie test plate for phenylketonuria screening. Filter paper disks of dried blood from 100 babies have been placed on top of bacteria that only grow if phenylalanine is present. The disks in the middle row are controls that contain increasing concentrations of phenylalanine; they support increasingly larger halos of bacterial growth. The test disk near the top center with a wide halo is from an infant with phenylketonuria. (From Levy 1973.)

the hospital became practical. In this procedure, a technician pricks a newborn's heel and prepares a small filter paper disk of dried blood. It is placed on top of an agar surface inoculated with a bacterial strain that does not reproduce *unless* phenylalanine is added to the nutrient medium. After incubation (along with control disks having known amounts of phenylalanine), a halo of bacterial growth around a test disk means that excess phenylalanine is present, and this is usually diagnostic of PKU (Figure 19.6). Confirmatory tests using more refined techniques are performed to rule out false positives. For example, oligonucleotide probes for the several base substitution mutations within the PKU gene can detect heterozygotes as well as homozygotes, but these methods are currently too expensive for a general newborn screening program.

In 1962 Massachusetts became the first state to require that newborn children be screened for PKU; today, all states and several dozen foreign countries have PKU legislation. Tens of millions of newborns have been screened and thousands of affected infants have been detected. Unfortunately, few states provide for financial aid for counseling services or for treatment of children found to have PKU. The PKU screening program has proved to be cost-effective: Tens of thousands of dollars are spent to detect one case of PKU, but considerably more would be spent on institutionalization of affected individuals in the absence of testing.

Some Other Screening Tests. Some states make multiple use of the blood sample obtained from newborns to test for PKU. For example, the autosomal recessive disorder, *galactosemia*, with an incidence of about 1 in 75,000 births, can be diagnosed from dried newborn blood by a direct assay for the missing enzyme (galactose-1-phosphate uridyl transferase). Without treatment, these babies suffer from severe vomiting, diarrhea, and malnutrition—symptoms that may be evident even without screening. The progress of the disease can usually be halted if the infant is *quickly* identified and placed on a diet that includes specially formulated milk substitutes and eliminates all other sources of the sugar galactose, which the babies cannot metabolize.

Screening can also be done during fetal life. As noted in Box A, chemical markers in maternal serum can help detect Down syndrome. In a similar way, general prenatal screening for *neural tube defects* (NTDs) can be done by testing the blood of pregnant women for α-fetoprotein. Recall that *higher* than normal levels of α-fetoprotein are present in the amniotic fluid of fetuses with spina bifida and anencephaly. The α-fetoprotein in the amniotic fluid may diffuse into the maternal plasma, so about 80% of fetuses with open NTDs cause elevated levels in their mother's blood. Testing programs can screen *all* pregnant women in a defined population rather than just those mothers who have had a prior child with an NTD. The justification for general screening is that most newborns with NTDs occur in families with no prior history. This amounts to about 4,000 NTDs per year in the United States.

These screening programs must be carefully managed to minimize uncertainties for the pregnant women. Their blood is tested for α-fetoprotein at the sixteenth to eighteenth week of pregnancy, but high levels do not necessarily—or even usually—mean a defective fetus. There are many benign reasons for an elevated level of maternal α-fetoprotein. A retest for α-fetoprotein plus ultrasonography and amniocentesis are needed to yield a more definitive diagnosis, but these additional procedures take two to four weeks. Since about 90% of women with elevated serum α-fetoprotein deliver *perfectly normal* babies, expert counseling and emotional support must be provided to offset the initially upsetting diagnosis.

Screening for Heterozygotes

It is important to detect the carriers of recessive diseases so that counseling information can be provided to couples when both are heterozygous. For many autosomally inherited inborn errors of metabolism, heterozygotes show only half the level of enzyme activity exhibited by normal homozygotes. Two heterozygote screening programs were initiated in the 1970s, one marked by confusion (at least in the beginning) and one fairly successful from the start. We look at these projects briefly and then at a third very recent screening proposal based on DNA analysis.

Sickle-Cell Anemia. Recall that homozygotes for the sickle-cell allele (about 1 in 500 black Americans) suffer from severe anemia and painful crises. The symptoms are variable, however, and although many affected persons die young, other homozygotes live long, active lives. Heterozygotes (about 1 in 10 black Americans) have normal phenotypes. A simple slide or solubility test can clearly separate persons with at least one sickle-cell allele from homozygous normals (see Figure 12.22), and electrophoretic analysis can easily and reliably distinguish all three genotypes.

The history of screening for sickle-cell phenotypes is intertwined with politics. The civil rights movement of the 1960s, combined with the appreciable frequency of the sickle-cell allele among blacks and the availability of good screening tests, led to a wave of state legislation to try to stem the tide of a "neglected disease." These laws were conceived in good faith and were often passed without debate, having been sponsored by black politicians and supported enthusiastically by community leaders. Unfortunately, later experience showed that some of the laws were too hastily drawn and poorly planned. For example, rather than being *voluntary*, some of the statutes *required* the testing of either prospective marriage partners (who could use the information) or schoolchildren (who could not).

In the first place, the benefits of sickle-cell testing were never very clear. Most homozygotes (i.e., those with sickle-cell *anemia*) already know the state of their ill health from early age; for some milder cases without prior symptoms, the new-found knowledge of homozygosity is not usually medically useful, because effective therapy is not generally possible. Heterozygotes (those with sickle-cell *trait*) might use this information for reproductive decisions, but prenatal detection of sickle-cell anemia through amniocentesis or chorionic villus sampling was not then possible. (It is now feasible by DNA analysis.) Thus, the options available to a couple who were both heterozygotes were either a 25% risk of affected children or no children at all. Neither choice was easily accepted. "As the screening programs were being promoted at a time of rising racial tension in the United States, the implication that certain Blacks should not have children was seen by some members of the community as having hidden racist motivations" (President's Commission 1983). Making matters worse, the early legislation seldom mandated either medical confidentiality or counseling services. Furthermore, community-based educational programs were inadequate, causing needless fear because of confusion of the burdensome *anemia* with the harmless *trait*. As a consequence, some employers and life insurance companies discriminated unfairly against heterozygotes.

The National Sickle Cell Anemia Control Act of 1972 addressed the inequities of the then-existing state laws by providing federal funds to states and private groups that planned voluntary programs incorporating much more community representation, effective education, counseling services, and strict safeguards. Thus, the initial screening difficulties have gradually been replaced with well-planned, comprehensive testing programs that are available to those who request it.

The value of screening *newborn* populations for sickle-cell genotypes has received attention recently, with the recognition that early identification of sickle-cell anemia can reduce mortality by about 15%. Infants with sickle-cell anemia may die quickly from overwhelming *Pneumococcus* infections, which often progress from the onset of fever to death in less than 12 hours. If diagnosed early, however, the frequency and severity of infections can be much reduced by daily administration of oral penicillin.

Tay-Sachs Disease. The opportunity for screening young adults for the recessive allele causing Tay-Sachs disease (TSD) was unusually favorable. The disease is extremely rare in almost all populations, but it has a moderate frequency (about 1 in 4,000 births) in Ashkenazi Jews of eastern and central European descent. Furthermore, the initial screening test for heterozygosity (an assay for hexosaminidase A in serum) is convenient, reliable, and discriminating. When both partners are confirmed as carriers, the option of amniocentesis for all pregnancies and selective abortion of affected fetuses can be offered. Thus, couples who are tested can assure themselves of a family without fear of the emotional trauma imposed by the slow, inevitable death of a TSD child.

After a year of careful planning and organization, a pilot program was undertaken in Jewish communities in Baltimore and Washington, D.C. Information about genetics in general and Tay-Sachs in particular was widely spread through newspapers, radio, television, and special pamphlets. About 7,000 persons volunteered for testing in the initial 1971–1972 period. Although some couples at first expressed shock, anger, or anxiety on learning that one or the other of them was a carrier, these reactions usually abated with further counseling. Nearly everyone was glad to have been tested, the implications were freely discussed with friends and relatives, and in the end, the carrier state was not generally regarded as a stigma (Kaback 1977).

Nevertheless, the feelings of those couples identified as heterozygotes, and the feelings of their (perhaps untested) relatives, cannot be fully known. The problem is particularly acute in some Orthodox Jewish communities, where abortion is prohibited and where the revelation of a TSD allele in one person may adversely affect the marriageability of a

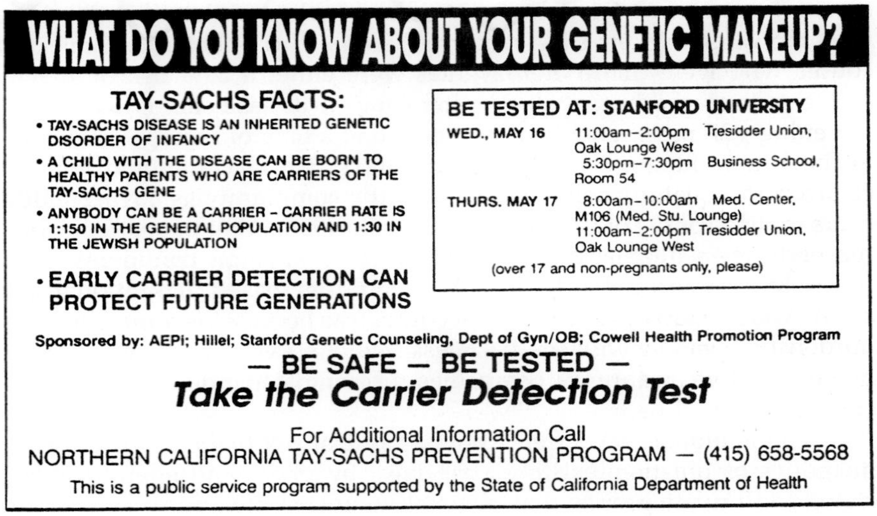

Figure 19.7
Advertising a Tay-Sachs prevention program in California. As part of a public education campaign, this poster promoted the screening program for carriers of Tay-Sachs disease. (Courtesy Alpha Epsilon Pi.)

whole kindred. Some ultra-Orthodox New York Jews participate in an interesting program in which confidentiality is assured by the traditional matchmaker. A person does not learn of his or her carrier status unless matched with another carrier; in this case the families can say that the pair broke up for other reasons (Merz 1987).

Jewish communities throughout the United States and elsewhere have started TSD screening programs (Figure 19.7). Hundreds of thousands of young Jewish adults have been voluntarily tested, with a carrier detection rate of about 1 in 25. Hundreds of couples without a prior history of affected children have been identified as *both* being carriers, and their pregnancies have been monitored. These efforts have contributed to a decline of over 70% in the incidence of TSD among Jewish populations.

Cystic Fibrosis. Testing for cystic fibrosis carriers among people *with a family history* of the disease makes good sense. But controversy surrounds recent screening proposals to detect cystic fibrosis heterozygotes in the *general* U. S. white population. Carriers are present in a substantial frequency, about 1 in 25 people.

A large part of the problem stems from the multitude of different mutations that lead to the disease (Chapter 5). We have noted that the most common mutation (ΔF508) is present in about 70% of carriers. But among the remaining 30%, there are over 200 mutations that can also produce cystic fibrosis when homozygous or when "heterozygous" with a different cystic fibrosis mutation (Table 19.3). Screening tests may use oligonucleotide probes and the polymerase chain reaction on DNA from cells from the inside of the cheek. It is important to note, however, that *one test* only detects *one mutation*. Thus, it may be practical to screen for ΔF508 and a few other specific alleles (at a cost of about $200 per person), but mass screening for *all* the mutations that may be present in a particular population is usually impractical.

There are bound to be many *false negatives*—people who are told they are not carriers, but who nevertheless possess one of the untested cystic fibrosis mutations. In a mating between two false negatives, one-quarter of the children will unexpectedly be affected with cystic fibrosis. What level of carrier detection justifies a screening program for the general population? The question has economic, social, and personal ramifications and is not easily answered. The current level (80–90% of carriers detected) seems to some investigators to be on the borderline. Note that the detection of 85% of carriers means that only 72%, that is $(0.85)^2$, of carrier × carrier marriages will be discovered.

Another problem involves the perceived need for cystic fibrosis screening. CF patients have variable symptoms, some being more mildly affected than others. In addition, modern medical treatments continue to improve the quality and length of life. The stated average life span of 26 years is for a cohort of people born decades ago, not for people born more recently, when treatment is more effective. Some argue that screening for heterozygotes and prenatal diagnosis for fetuses at risk is not needed at all. Although many patients suffer for extended periods of time, advocates for disabled persons (among others) note that some cystic fibrosis patients live productive and independent lives, even into middle age. Thus, they conclude that all people with cystic fibrosis deserve the chance for life.

Cystic fibrosis screening has been done on a limited basis, and undoubtedly population testing will increase gradually as the percentage of detectable mutations rises and the cost falls (Anonymous 1992; Wilfond and Fost 1990). To offer screening to all Americans, however, would be a massive job requiring an unprecedented educational effort. Experienced genetic counselors would be especially in demand to deal with people in uncertain situations—people who must confront the possible stigma of carrier status and who must deal with slippery probabilities involving the concept of false negatives. It is unclear where the expertise and resources would come from, and the American Society of Human Genetics does not believe that mass population screening for cystic fibrosis carriers is justified at this time (ASHG Ad Hoc Committee 1992).

Table 19.3

Nineteen different cystic fibrosis mutations detected in a French Celtic population in western Brittany.

Code Name of Mutation	Frequency (%)	Amino Acid Change	Nucleotide Change
ΔF508	81	One amino acid deletion	Three-base deletion
1078 delT	5	Many changes (frameshift)	One-base deletion
G551D	4	One amino acid substitution	Base substitution
1717−1 G to A	1	Many changes (splicing error)	Base substitution at end of intron
W846X	1	Shortened polypeptide	Base substitution to stop triplet
G91R	1	One amino acid change	Base substitution
1221 delCT	1	Many changes (frameshift)	Two-base deletion
G542X	1	Shortened polypeptide	Base substitution to stop triplet
4005+1 G to A	1	Many changes (splicing error)	Base substitution at end of intron
Ten other mutations	Each <1		

Source: Férec et al. (1992).

473

SUMMARY

1. Inherent in the various procedures for genetic testing and counseling are some troublesome technical, ethical, legal, and social problems.

2. Genetic counselors determine relevant medical and genetic facts for a family confronting genetic disease. They must convey this information, as well as available options, in a sympathetic and understandable way. While sensitive to the emotional needs of affected families, most counselors are nondirective.

3. Prenatal diagnosis usually relies on analyses of fetal cells obtained by chorionic villus sampling or amniocentesis. The former is done at about the tenth week of pregnancy, and the latter (perhaps a bit more safely) at about the sixteenth week. If the fetus is found to be abnormal, abortion is an option.

4. Prenatal diagnosis is usually offered if there is an increased risk of a chromosomal abnormality. In some cases, the time needed for analysis of fetal cells can be reduced by DNA-based methods: fluorescence in situ hybridization, RFLP linkage, or oligonucleotide probes. When there is an increased risk of a metabolic disorder, the concentration of the relevant enzyme is measured in fetal cells.

5. Diagnosis of neural tube defects can be aided by determining α-fetoprotein concentrations in amniotic fluid. Fetal sex can be determined by karyotyping, by Y chromosome probes, or (experimentally) by using one cell excised from an eight-celled in vitro fertilization embryo.

6. Genetic screening is the systematic search of populations to detect a genetic abnormality prior to overt symptoms or to determine heterozygosity for recessive disorders. Programs should be carefully and sympathetically planned to ensure acceptability in the community.

7. The most widely used screening program tests newborns for phenylketonuria. Carrier screening is done among Jewish populations for the Tay-Sachs allele and among black populations for the sickle-cell allele. Prenatal screening for neural tube defects has been undertaken in some populations by detecting high levels of α-fetoprotein in maternal blood.

8. Screening for cystic fibrosis carriers has many problems, including the occurrence of false negatives.

KEY TERMS

α-fetoprotein
amniocentesis
chorionic villus sampling
   (CVS)

false negative
false positive
fluorescence in situ
   hybridization (FISH)

genetic counseling
genetic screening
Guthrie test

neural tube defect
oligonucleotide probe
recurrence risk

QUESTIONS

1. In parts of Africa, Asia, and Latin America, diarrheal diseases, respiratory infections, tuberculosis, malaria, hepatitis, AIDS, and a half dozen other diseases kill tens of millions of people annually. Are the concerns raised in this chapter equally applicable to these populations? What funding should be given to research on tropical diseases versus genetic diseases?

2. In a so-called wrongful life case, an infant with Tay-Sachs disease sought damages for negligence when a medical laboratory failed to find that both her parents were carriers. Had the laboratory made a correct diagnosis, there would have been no child to endure the suffering, because it would have been aborted or not conceived. Would you award damages to the Tay-Sachs family in this case? (Courts in some states have recognized wrongful life actions, while others have rejected them.)

3. Can there be any purpose to amniocentesis or chorionic villus sampling if the parents do not agree beforehand to abort a seriously abnormal fetus?

4. How do you feel about using prenatal diagnosis for the sole purpose of determining sex in advance of birth? Assuming that you were using in vitro fertilization to achieve pregnancy, how do you feel about using preimplantation diagnosis for sex determination?

5. One reason that chorionic villus sampling is a bit less accurate than amniocentesis is that the CVS cell sample comes from fetal tissue that will form the placenta rather than from the embryo itself. Why should this matter, since all the tissue comes from one zygote?

6. Why is chorionic villus sampling not useful for the prenatal diagnosis of neural tube defects?

7. One aspect of genetic screening that has received attention is whether it should be mandatory or voluntary. Do you think that the widespread compulsory screening for phenylketonuria (in which you were probably included) represents an unwarranted government intrusion into personal lives, or is it a legitimate concern of a state's public health service? What about other screening programs?

For the following questions, assume a population in which the frequency of cystic fibrosis carriers is $1/20 = 5\%$. Assume also that CF screening detects 70% of carriers.

8. Show that slightly more than half of carrier $\times$ carrier marriages would be *missed*.

9. Assume that one parent is shown to be a carrier, and the other is not tested. What is the probability that this marriage has a carrier $\times$ carrier status?

10. Assume that one parent is shown to be a carrier, and the other parent is tested and said to be free of disease-causing CF alleles. What is the probability that this marriage has a carrier $\times$ carrier status?

FURTHER READING

A good source of information on screening that includes considerable basic genetics is Office of Technology Assessment (1992). Several chapters on applied genetics in *Principles and Practice of Medical Genetics* (Emery and Rimoin 1990) could be helpful to laypersons. These include articles by Skinner (1990) on genetic counseling, Charrow, Nadler, and Evans (1990) on prenatal diagnosis, Kaback (1990) on heterozygote screening, and Shaw (1990) on legal considerations in applying genetic knowledge to individuals and families. Fraser (1988) gives a general account of genetic counseling, Lewis (1991) treats topics in prenatal diagnosis, and Bonnicksen (1992) discusses the social implications of testing preimplantation embryos. The choice of CVS versus amniocentesis is currently hard to evaluate; balanced views of advantages and disadvantages are provided in Rosenthal (1991) and Anonymous (1991). Finally, the article by Roueché (1979) on Carol Terry's experience with Wilson disease is gripping.

20 Altering Genetic Traits

"I am the mother of five children, three of whom have sickle cell anemia," writes Ola Mae Huntley (1984). Noting that genetic counseling was not offered in the 1960s when her children were born, she continues:

> After our second child was diagnosed as sickle-cell-anemic, a well-meaning but uninformed doctor assured me that, since we already had one normal child and one with sickle cell anemia, our future children would be normal, purely on the basis of numerical chance. . . .
>
> Our sickle-cell-anemic children are now young adults. Only rarely can we all be home together as a family. Usually one or two, sometimes all three, are in hospitals being treated for acute disease crises or for the debilitating effects of the disease. This is now a way of life, or I should say, a way of existence. . . .
>
> In my twenty-five years as a mother very little has changed. The current state of medical art is to treat the symptoms, but little has been done about treating the cause.

She writes of the mental anguish brought about by poor self-esteem and little hope for a better life and likens the relief of physical pain to putting a small bandage on a large wound. Her wish list includes new genetic engineering research that would lead to a more permanent cure or a way to prevent the disease in the first place. Although her specific interest is sickle-cell anemia, Huntley notes that her concerns are relevant to other severe genetic diseases as well.

The treatment possibilities for heritable disorders—various therapies that give a range of benefits—are discussed in this chapter. We include recent attempts at *gene therapy*, that is, adding a new gene to the somatic cells of the body to correct the consequences of a defective one. In theory, gene therapy could be extended to altering or replacing a gene in eggs or sperm, thus affecting future generations. Such considerations call forth the controversy over *eugenics*—that is, attempts to improve the genetic endowment that we leave to our descendants.

TREATING GENETIC DISEASE

In principle, genetically influenced conditions *can* be treated. Effective therapy usually requires an understanding of the cascade of causes and effects. In only a very small percentage of the thousands of genetic diseases, however, can anything like a normal phenotype be restored so that treated individuals can pursue life-styles unhampered by their handicaps. The limitations arise, in part, because the products of mutant genes often interfere with fundamental processes within cells; therefore, it may be difficult to intervene in a substantive way without significant side effects. In addition, most genetic diseases are individually uncommon, and money for research may be limited. Yet even some of the better-known genetic or chromosomal disorders with a long history of well-funded research (e.g., muscular dystrophy, albinism, Huntington disease, sickle-cell anemia, and Down syndrome) remain largely untreatable. Although the results have been generally disappointing, we can hope that recent developments in molecular genetics will open up new avenues for intervention.

Actually, the most effective treatment is prevention. For genetic diseases, this means first of all eliminating some of the mutations to abnormal alleles by reducing exposure to unnecessary ionizing radiation and mutagenic chemicals (Chapter 12). Once new mutations have been transmitted to later generations, prevention becomes a matter of genetic screening and counseling. These may include prenatal diagnosis, selective abortion, and alternative reproductive technologies.

Some Current Practices

To improve everyday functions, many of us alter our private environment by using corrective devices that act at a level far removed from the primary effect of any mutant gene. The list would include eyeglasses, hearing aids, artificial limbs, wheelchairs, and similar devices. In extreme cases, a few children with immune deficiencies of genetic origin have been confined to environments that seal out infectious agents (Chapter 16). Such external management of the physical environment may involve ingenious biomedical engineering, but it can be cumbersome and of restricted value.

Treatments that alter the body rather than the environment are extremely varied (Desnick 1991; Friedmann 1991). Diet modifications may restrict the intake of a toxic substance or supplement a deficient metabolite. Therapy may involve drugs, transplantation, other surgery, or, much closer to the gene, replacement of a polypeptide that the mutant allele is unable to supply. Research at the level of the gene itself, which theoretically could bring about a complete cure, is in experimental and trial stages. Some tactics that can help persons with genetically influenced diseases are listed in Table 20.1 and discussed here.

Surgical Repair and Removal. Alteration of the phenotype by surgery can improve appearance or basic health. The disfigurement, speech difficulties, and swallowing problems of *cleft lip* and *cleft palate* (occurring by themselves or accompanying other genetic and chromosomal syndromes) can be restored to near normal by skillful surgery. The extra digits of *polydactyly* can also be easily removed. In the autosomal dominant disorder *spherocytosis*, the red blood cells assume rounded shapes because of increased permeability of the cell membrane. Although the rounded cells carry oxygen reasonably well, their fragility leads to hemolysis and chronic anemia. The fragile cells are entrapped in the spleen, whose phagocytic cells normally remove aging red blood cells from the circulation. Removal of the spleen in patients with spherocytosis apparently allows the rounded red blood cells to survive longer and relieves most of the disease symptoms.

In the next three categories of treatment, researchers often make use of knowledge of the relevant gene-controlled metabolism, discussed in Chapter 14 and summarized briefly as:

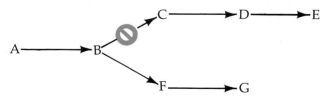

Table 20.1
Some genetic diseases that are treatable.

Method of Treatment	Disease
SURGICAL REPAIR AND REMOVAL	
Surgical repair	Cleft lip and cleft palate
Removal of spleen	Hereditary spherocytosis
Removal of colon	Familial polyposis of colon
DIETARY RESTRICTION OF A PRECURSOR	
Phenylalanine	Phenylketonuria
Galactose	Galactosemia
Leucine, isoleucine, valine	Maple syrup urine disease
Lactose	Lactase deficiency
Fava beans	Favism (G6PD deficiency)
ADDITION OF A DEFICIENT END PRODUCT	
Uridine	Orotic aciduria
Cortisol	Congenital adrenal hyperplasia
DEPLETION OF EXCESSIVE SUBSTANCE	
Copper (by penicillamine)	Wilson disease
Cholesterol (by bile binders)	Familial hypercholesterolemia
Iron (by bloodletting)	Hemochromatosis
Uric acid (by several drugs)	Gout
REPLACEMENT OF A MISSING GENE PRODUCT	
Insulin	Juvenile-onset diabetes
Growth hormone	Pituitary dwarfism
Clotting factor VIII	Hemophilia A
Adenosine deaminase	Adenosine deaminase deficiency
Various enzymes	Lysosomal storage diseases
ORGAN AND TISSUE TRANSPLANTATION	
Bone marrow	Severe combined immunodeficiency
Bone marrow	Thalassemia
Bone marrow	Lysosomal storage diseases
Liver	α_1-antitrypsin deficiency
GENE THERAPY	
ADA gene (into white blood cells)	Adenosine deaminase deficiency
Several other genes	(Experimental stages)

Recall that a metabolic block—here in the conversion of B to C and represented by the "prohibited" symbol—is due to a defective enzyme. The blocked chemical reaction can result in the accumulation of the precursor (beginning) substance, B, to toxic levels. Damage to the body might also result from the buildup of the substances F and G in the alternative pathway. Lack of the vital end product, E, is yet another way for damage to arise as the result of the metabolic block.

Dietary Restriction of a Precursor. The diet therapy for *phenylketonuria* detailed in Chapters 1 and 14 remains the prototype for this method of treatment. The recommended diet drastically reduces the intake of phenylalanine, which is a normal component of digested proteins. Because it can no longer be converted to tyrosine, phenylalanine would otherwise build up in the body to toxic levels. Several dozen genetic diseases are treated with various degrees of success by diet restrictions or diet exclusions. In these cases, the end products that are cut off because of the blockage may be available from food or from alternative metabolic pathways.

Addition of a Deficient End Product. Sometimes the body can tolerate an abnormally high concentration of a precursor but cannot do without a missing end product. One example is the very rare disorder *orotic aciduria*, which is caused by an autosomal recessive allele and characterized by severe anemia plus mental and physical retardation. In affected persons, the biochemical steps from orotic acid to uridine (uracil joined to ribose) are blocked. The accumulation of orotic acid causes no problem, but the deficiency of uridine is harmful because it is utilized in the synthesis of RNA, DNA, and some coenzymes.* In the dozen or so cases of orotic aciduria that have been studied, a daily dose of uridine alleviates the severe symptoms, especially if the diet therapy is begun early in life.

Depletion of Excessive Substance. The rare recessive *Wilson disease* (Chapter 19) results from an error in copper metabolism. In affected persons, the abnormal deposition of copper causes liver and brain damage; if untreated, it leads to death after years of suffering. In some persons, the symptoms are primarily psychological: grossly inappropriate social behavior and bizarre personality changes, which can be mistaken for schizophrenia or manic depression. The age of onset varies from 6 to 50. Although the chain of causation of the symptoms is obscure, drugs that bind copper (so-called chelating agents) and lead to its excretion in the urine can provide dramatic improvement. One such drug is penicillamine (which is derived from penicillin but has no antibiotic activity). Early diagnosis and treatment can completely prevent the serious consequences of Wilson disease.

Recall from Chapter 5 that heterozygotes for *familial hypercholesterolemia* (having a frequency of about 1 in 500 persons) develop atherosclerosis and may have heart attacks beginning in their 30s. The symptoms are due to excessive LDL (low-density lipoprotein), which is the major cholesterol transport system in plasma. Elevated cholesterol levels result from a deficiency of cell receptors (especially in the liver) that take up LDL from the blood. Reductions in serum cholesterol levels are possible by an unusual route involving the bile acids. These substances, rich in cholesterol products, are made in the liver and transported to the intestines where they help digest fats. Normally, the bile acids are reabsorbed in the intestines and used again, but drugs are available that cause them to be excreted in the stool instead. This action forces the liver cells to take up more cholesterol from the serum to manufacture the missing bile acids. Blood cholesterol levels are thereby reduced by about 15–30%.

Replacement of a Missing Gene Product. *Diabetes, pituitary dwarfism,* and *hemophilia* can be fairly well treated with insulin, human growth hormone, and clotting factor VIII, respectively. Recombinant DNA techniques have been used to make the therapeutic proteins. As described in Chapter 13, this manufacture involves inserting the cloned human gene into the DNA of cells growing in culture. Researchers then isolate the gene product, which must undergo years of testing. Factor VIII, of recombinant DNA origin, has only recently received approval for general medical use. The protein has been obtained from pooled human blood (at some risk of acquiring blood-borne viral diseases like hepatitis or AIDS).

---

* A coenzyme is a molecule that combines with a number of different enzymes to activate them. Some coenzymes are synthesized from the vitamins we eat.

Because proteins are broken down by digestive enzymes, oral administration is not usually possible. Thus, intravenous injections are required, often over a lifetime. Although annoying and painful, such treatments allow many hemophiliac and diabetic patients to lead productive lives.

Attempts to replace defective or missing enzymes (rather than nonenzymatic proteins) have generally not been applied at the clinical level. Such treatment would be very helpful to people with albinism, Tay-Sachs disease, phenylketonuria, or hundreds of other enzyme defects. Two major difficulties stand in the way: the short time that the injected enzyme persists in the body, and the inability to deliver the enzyme to the site where it is needed while protecting it from degradation.

A novel method of delivery has been used in the treatment of the rare autosomal recessive disease *adenosine deaminase (ADA) deficiency*. The normal ADA enzyme affects the metabolism of purines, and its deficiency affects white blood cells in particular. The functions of T cells are completely impaired, and antibody production by B cells is also reduced. As a result, ADA-deficient children succumb to overwhelming viral, bacterial, and fungal infections of the skin, airways, and digestive tract. In the new therapy, investigators inject children with the ADA enzyme isolated from cows. But first they coat the enzyme with a chemical (polyethylene glycol, or PEG). Apparently, the life of the coated enzyme is significantly prolonged, since the dozen or so patients treated with PEG-ADA have usually shown some improvement in their immunological functions. ADA-deficient patients have also been treated by bone marrow transplants and by gene therapy (discussed shortly).

Organ and Tissue Transplantation. An interesting procedure for providing the correct genetic information to patients with a hereditary disease is to transplant an organ or tissue from a normal individual. Such a graft may provide a missing enzyme within the patient's body *on a continuing basis*. Within the graft, the enzyme might operate on circulating substrates. Or the enzyme might be secreted by the cells of the transplant to work elsewhere in the body. Most research in this area has been with the transplantation of bone marrow.

We noted in Chapter 16 that David, the "bubble boy" with *severe combined immunodeficiency disease (SCID)*, received a bone marrow transplant from his sister. Although he did not long survive the transplant, over 100 other patients with SCID (including the ADA deficiency type) have improved after receiving new bone marrow. Matching for all HLA antigens is not absolutely required in transplants to infants with SCID, since they lack the functioning B cells and T cells that mount immunological attacks on the transplanted tissue. However, the reverse reaction, graft-versus-host disease, remains a problem in mismatched transplants. Recent techniques to rid the donor marrow of mature T cells have sometimes prevented this problem.

The use of bone marrow transplants to treat *thalassemia* and other hemoglobin diseases is more difficult because immunosuppressive agents must be used to prevent the rejection of the transplanted bone marrow cells. Nevertheless, in one study, about three-fourths of patients survived free of the disease for at least one to three years. About one-fourth died from graft-versus-host disease or other complications of marrow transplantation. Alternative management of thalassemia is also available based on frequent blood transfusions. Because these more conservative techniques can provide almost all affected children with 15 to 20 years of life, it is not clear what therapy is to be preferred.

The impressive variety of therapeutic approaches outlined here should remove any pessimism over the possibility of treatment for genetic diseases. Still, many of the maneuvers require sophisticated medical or technical expertise and the accompanying expense. Furthermore, some are halfway measures that cause significant side effects or discomfort and must be continued for a lifetime. It is clear that knowing the cause of a disease in profound detail does not guarantee effective therapy.

Gene Therapies

Investigators and patients alike look forward to the day when it will be possible to correct the effects of abnormal alleles by supplying the normal alleles. In such **gene therapy,** the hope is that the transferred genes will cure the basic defect rather than merely alleviating its damage. We use the word *cure* to mean a limited course of treatment that restores a person's well-being, just as penicillin cures pneumonia or an appendectomy cures appendicitis. In such procedures, treatment is stopped once health is regained.

We consider here what is called **somatic gene therapy**, in which the body cells, but not the germ cells, of the patient receive the corrected gene. Of course, if the germ cells were also changed from mutant to normal, the cure could extend through future generations. However, objections have been raised to this **germinal gene therapy** because of unsavory overtones—the possibility of unwarranted or unwise tampering with another generation's genes. Therefore, investigations of human gene therapy have been limited to the correction of disease within individuals. This is, of course, *the traditional aim of medicine,* and there is general consensus that somatic gene therapy is an ethical option as long as proper safeguards are in place. Proposals for human gene therapy must pass many levels of review by government oversight committees. In addition to *ethical* issues, these committees examine the *safety* of the procedure for the patient and the general public, as well as the possible *benefit* to the patient in relation to the possible risk (Anderson 1990, 1992; Miller 1992; Verma 1990).

ADA Deficiency. The first federally approved trial of gene therapy to treat a genetic disease was begun in September 1990 at the National Institutes of Health (NIH) and was led by three researchers: W. French Anderson, R. Michael Blaese, and Steven Rosenberg. It involved a four-year-old girl (and later another young girl) with the fatal immunological disease ADA deficiency. Recall that ADA deficiency is treatable to a limited degree by weekly injections of the bovine ADA enzyme coated with polyethylene glycol. In the gene therapy trial, the girls (still on a PEG-ADA regimen, but not doing very well) were injected with their own T cells that had first been removed and manipulated so that they contained the normal ADA gene. Preparation for these injections involved years of research and laboratory trials, not to mention years of scrutiny of the proposed treatment protocols by seven different government agencies.

Initially, it was thought that β-thalassemia, sickle-cell anemia, or some other hemoglobin disease would be the best candidate for initial gene therapy trials. It is easy to get bone marrow (including the cells that differentiate into red blood cells) from a patient and return them to the patient after adding, say, a normal β-globin gene. The major problem here is one of gene regulation. Hemoglobin is composed of equal amounts of α and β polypeptide chains (see Figure 12.19), and the two proteins are normally produced in *exactly* equal amounts in red blood cell precursors (by genes on different chromosomes). An excess of either chain leads to

481

cell damage. Since the addition of a new gene to a human chromosome is generally a hit-or-miss process, the proper genetic controls that regulate gene expression may not be in place. It would be better, it was thought, to choose an enzyme disease where even a small amount of new, normal enzyme would likely be advantageous and an extra large amount would not be harmful.

ADA deficiency seemed to be a good choice, since the PEG-ADA treatment had been shown to increase the level of ADA enzyme a small amount, and patients showed some improvement in immunological functions. Furthermore, it was found that cultured white blood cells deficient in ADA grew and divided slowly, whereas the ADA-corrected cells multiplied faster. Thus, when injected into patients, the latter would likely have a competitive advantage as long as they persisted in the bloodstream.

Getting sufficient amounts of the normal gene to use in the treatment was not a stumbling block. Hundreds of human genes have been cloned and characterized by the recombinant DNA techniques described in Chapter 13. Getting the exact nucleotide sequence of the gene itself is not enough, however. It must be combined with whatever promoters or enhancers are required to achieve transcription and translation in the proper cells.

The major problem is putting the normal DNA into patients' cells so that it is stably integrated and expressed. A variety of techniques have been investigated. The most direct is to simply inject the cloned DNA into recipient cells, where it may integrate into chromosomal DNA. But microinjection is not feasible for somatic human gene therapy; not nearly enough cells can be injected to be effective. Rather, researchers package the correct genetic material into a **retrovirus**. The major advantage is that retroviruses infect target cells with very high efficiency, approaching 100%. Then, in the normal course of their life cycle, they cause a DNA copy of their genetic material (i.e., the provirus) to be inserted into the chromosomal DNA of the host cell (as shown in Figure 15.2). One retrovirus available for human gene therapy is the Moloney murine (mouse) leukemia virus. Before using it, researchers remove most of the genetic material of the virus to render it incapable of further reproduction. The empty space in the viral RNA is then replaced with suitable human RNA (Figure 20.1). This substitution does not affect the ability of the virus to (1) infect human cells and (2) insert a DNA copy of its genetic material into the cells' DNA.

Gene therapy of the ADA patients involves taking a blood sample, treating the isolated T cells with the derived retrovirus, and returning the treated T cells (Figure 20.2). The girls have now received injections of their (presumably) gene-corrected T cells at one- or two-month intervals for over a year. The injections must be repeated periodically because the T cells have a limited life span in the body. Both girls have shown great improvement in their immunological functions. In fact, "both attend regular public school and now have no more than the average number of infections. There have been no significant side effects from the cell infusions" (Anderson 1992).

Additional Comments about the Technology. One worrisome aspect of the gene-corrected T cell therapy for ADA deficiency hinges on a basic concept of the immune system: A single T cell or B cell is programmed to respond to just one antigen (Chapter 16). The range of T cells that is gene-corrected may well have a limited repertoire; that is, the lymphocytes that are treated, injected, and then repopulate the girls' immune systems are not necessarily a complete set. Some competent immune cells that are needed to mount an attack against some of the antigens on disease-causing

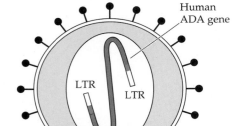

Normal retrovirus

Derived retrovirus

Figure 20.1
Schematic diagram of a normal retrovirus compared with a derived retrovirus used for gene therapy of ADA deficiency. In the derived retrovirus the human ADA gene (actually, its complementary RNA) and some control sequences replace viral genes. LTR stands for long terminal repeat, *part of the normal viral genome required for insertion of the provirus into the host chromosome. (See also Figure 15.2.) The structures other than the genetic material are various types of proteins.*

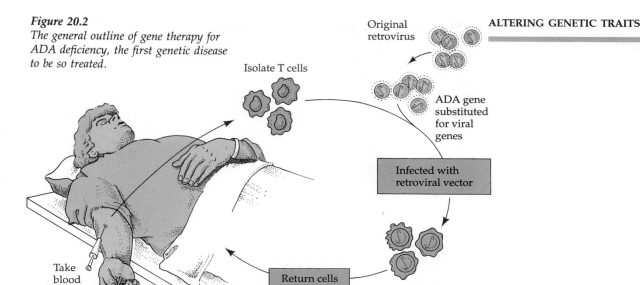

Figure 20.2
The general outline of gene therapy for ADA deficiency, the first genetic disease to be so treated.

Isolate T cells

Original retrovirus

ADA gene substituted for viral genes

Infected with retroviral vector

Gene–corrected T cells

Return cells to patient

Take blood sample

microbes may be missing, because they were not among those that were gene-corrected.

To address this problem, scientists hope to correct and administer, not the mature lymphocytes that are already committed to a particular antigen, but the **stem cells** from which these lymphocytes are derived. The primordial stem cells exist in the bone marrow and serve as a source of both red and white blood cells throughout a person's life (Figure 20.3). When a primordial stem cell divides, it produces one stem cell like itself and one cell that differentiates in succeeding divisions into mature blood cells (Golde 1991). Using marrow stem cells in ADA gene therapy has a further advantage: The injection may have to be given just once, thus achieving a real cure. Since the population of stem cells continually renews itself, their cell divisions produce both B and T cells that then become programmed to respond to particular antigens. The problem is that out of a million bone marrow cells, only a dozen are actually stem cells, and they are hard to identify and collect. Whether they can be effectively gene-corrected and whether they will work in vivo as hoped are still open questions.

Another worry is that insertion of the retrovirus-borne gene into the T cell appears to be at random places in the genome rather than at the locus of the abnormal gene. There is the risk of interruption of some normal gene activities or death of the cell. Researchers also worry that a cellular oncogene may be activated or that infectious viruses may be generated (by recombinational processes) and spread to other cells. Addressing these possible dangers is a major effort of current investigations.

Several protocols for gene therapy involving a total of about 30 patients are now in progress, and more than a dozen other protocols have been tested in animals and are awaiting approval. Several of these gene therapy procedures are listed in Table 20.2. The methods for getting the normal gene inside human cells include the use of retroviruses on extracted cells (as with ADA deficiency), introducing the gene-carrying virus directly to the body (as with cystic fibrosis), or simply injecting the DNA itself (as with Duchenne muscular dystrophy). Other targets of gene therapy include hemophilia, phenylketonuria, β-thalassemia, and sickle-cell

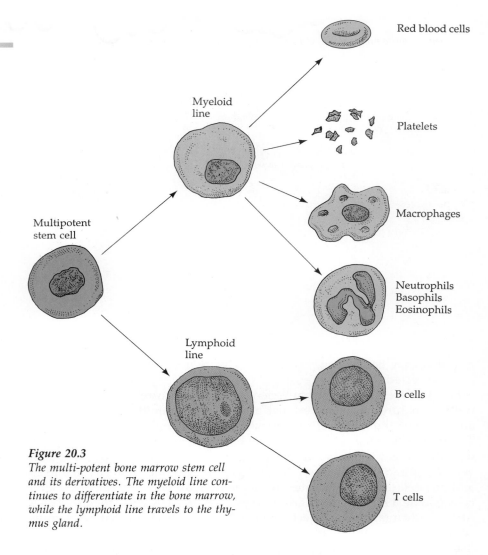

Figure 20.3
The multi-potent bone marrow stem cell and its derivatives. The myeloid line continues to differentiate in the bone marrow, while the lymphoid line travels to the thymus gland.

Red blood cells

Platelets

Macrophages

Neutrophils
Basophils
Eosinophils

B cells

T cells

Myeloid line

Multipotent stem cell

Lymphoid line

anemia. Note that the last two treatments in Table 20.2 are against cancer and heart disease—for the most part acquired rather than inherited disorders. AIDS, too, may eventually be subjected to gene therapies that protect cells from infection by HIV or limit reproduction of HIV in invaded cells. Investigators are optimistic about the widespread use of gene therapy against human disease in the future.

Additional Comments about Ethics. Up to now, gene therapy proposals have been aimed at correcting specific diseases in specific individuals. Newspapers, magazines, and television have reported the progress, and the public seems to accept these "high-tech" applications as a reasonable part of medical practice. Two potential uses of gene therapy, however, involve areas of great ethical concern. One deals with *germinal gene therapy*—modifying the genes of future generations. A second deals with *enhancement*—improving on characteristics that are generally viewed as being within a normal range.

With regard to the genes of future generations, one worry is that manipulation of the germline could harm rather than help. Our knowledge of the inner workings of cells is still fragmentary, and many biological examples could be given of well-intentioned interventions producing unexpected side effects. In addition, by what right do we "play God" with the inherited characteristics of our children? We cannot get their consent

Table 20.2
Some gene therapy proposals in early clinical trials or near approval stages.

Disease	Description of Therapy	Research Facility
ADA deficiency	T cells extracted from patient, gene-corrected (by retrovirus vector), and returned	NIH
ADA deficiency	Same as above but using bone marrow stem cells	NIH
Familial hypercholes-terolemia	Liver cells from patient corrected for LDL receptor gene (by retroviral vector) and reintroduced	Univ. of Michigan
Cystic fibrosis	Adenovirus (cold virus) carrying normal *CFTR* gene introduced directly into airways of patient by an aerosol spray	Three sites: National Heart, Lung & Blood Inst.; Univ. of Wisconsin; Univ. of Iowa
Duchenne muscular dystrophy	Direct injection of dystrophin gene into muscle cells	Univ. of Wisconsin
Cardiovascular disease	Cells lining blood vessels altered to stick to artery walls and produce clot-dissolving protein	National Heart, Lung & Blood Inst.
Skin cancer	Melanomas injected with a gene (in a plasmid) that produces a protein to stimulate the immune system to attack the cancer (no cells removed from the body)	Univ. of Michigan

prior to making deliberate changes. While we all wish for physically and mentally healthy offspring, it is unclear to what extent we should go to ensure that end. Opinion varies. What steps should be taken to foster fetuses that are normal? Is it right to abort fetuses that are not normal? What is normal?

Using gene therapy to try to alter a characteristic such as size would probably not be controversial for, say, a 3-foot pituitary dwarf. Many people would argue, however, that gene therapy merely for enhancement would be wrong, promoting discriminatory attitudes and leading to greater inequalities. But at the borderline between abnormal and normal, the question is muddy. For the undersized youngster whose life is made miserable by continual taunts and bullying, would a new growth-promoting gene be enhancement or therapy? The quandary applies to intelligence as well. A number of specific genes, when mutated, can lead to forms of mental retardation, and these might reasonably be targets for gene correction. But what about a person whose measured IQ is 80, or 70, or 60, or lower? Of course, at present, no specific genes are known that affect intelligence within the normal range, so this form of gene therapy remains completely speculative.

What will be the impact of gene therapy in the future? Just a few years ago, the whole idea seemed bizarre. Now, it is much less so. But will gene therapy ever affect you, your family, or your descendants? If the procedures remain highly specialized, cumbersome, expensive, and applicable only to rare inherited diseases, the impact of gene therapy will remain limited. But millions of people could be affected if researchers develop vectors that deliver genes in a simple way to correct common disorders such as cancer or heart disease.

CHOOSING A CHILD'S SEX

Although not related to genetic disease (except for X-linked inheritance), fetal sex has always been a matter of scientific interest, as well as casual discussion and speculation. It has now become a matter of possible manipulation as well. Suppose, for example, that samples of pure X-bearing sperm or pure Y-bearing sperm were available for artificial insemination. Or suppose that other more natural (as well as safer, more effective, and less expensive) techniques could predetermine a child's sex. Reliable schemes of these types are not now at hand, although they may become available someday. Reports of partial success by several investigators need to be confirmed by other groups.

Do you think that using such technology is ethical? What proportion of people might avail themselves of the opportunity? Would the ability to choose a son or a daughter at each birth alter the composition of families, or the population sex ratio, or overall fertility? What societal changes might ensue? In one study of the consequences of sex preselection, over 7,000 married American women were asked their preferences for the sex of their next child. Although many more women wanted the firstborn to be a boy rather than a girl, the desire for a balanced family was very strong (Figure 20.4). If the firstborn was male, over 80% of women who expressed a preference wanted the second to be female—that is, $72.3/(72.3 + 15.0)$ from Figure 20.4. For a third child, boys and girls were about equally desired. Thus, the major consequences of sex selection would derive from more firstborn children being male. Holmes (1985) points out the many studies that show the greater achievements of the eldest child compared with those of later siblings and notes the benefits that would increasingly accrue to the male sex.

Depending on assumptions about family size and the proportion of women using the techniques, we would expect the overall sex ratio in a population having the opinions expressed in Figure 20.4 to be 52–56%

Figure 20.4
The preference of American married women for the sex of their next child. The figures within the bars express the percentages of women preferring a boy (solid bar), expressing no preference (open bar), or preferring a girl (hatched). (Data from Pebley and Westoff 1982.)

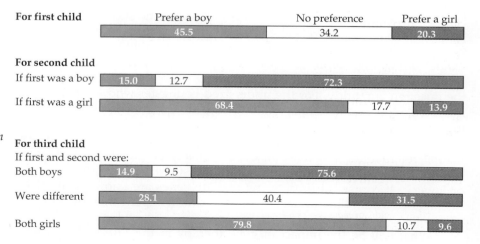

For first child	Prefer a boy	No preference	Prefer a girl
	45.5	34.2	20.3

For second child

If first was a boy	15.0	12.7	72.3
If first was a girl	68.4	17.7	13.9

For third child
If first and second were:

Both boys	14.9	9.5	75.6
Were different	28.1	40.4	31.5
Both girls	79.8	10.7	9.6

male.* This range of values is not too much higher than the actual value of about 51–52% male births. But the role of fathers in implementing their sex preferences, which were not sampled in this study, would probably favor yet more males. A still larger male excess might be expected in countries that have social or religious traditions of male inheritance. In some cultures, females are perceived as family burdens because of poor employment opportunities or the need for large marriage dowries. To counter the widespread practice of abortion of female fetuses, the parliament of India in 1991 outlawed (under most circumstances) prenatal diagnoses that are primarily meant to reveal fetal sex.

Available Techniques (Postconception)

It is now technically possible for a couple to choose the sex of their children with virtually 100% success. But the required methodology carries some small risk and is morally offensive to many persons because it involves terminating unwanted pregnancies solely on the basis of sex. Fetal cells obtained by amniocentesis or chorionic villus sampling can be reliably karyotyped for sex chromosomes or sexed with DNA probes specific for the Y chromosome. It may be that the need for an invasive technique to obtain fetal cells can be avoided: Some research centers claim that the sex of a fetus can be determined by examining the DNA from the few fetal cells that circulate in the blood of the pregnant woman. Successful amplification of Y-specific sequences by polymerase chain reaction signifies a male. Even abortion of the "wrong" sex can be avoided by sexing eight-celled embryos and transferring to the woman only embryos of the desired sex (Chapter 19). Currently, however, this option is limited by the need for special technical expertise.

Although these types of procedures can be useful in the prenatal diagnosis of genetic diseases, only a fraction of American physicians would agree to perform prenatal diagnosis for the *sole* purpose of choosing the sex of a child. Nevertheless, they might agree to abort all male fetuses that are carried by women heterozygous for hemophilia or other serious, recessive, X-linked disorders, even though half the aborted fetuses are expected to be normal. In the United States, abortion before about 24 weeks is currently *legal* for any reason. Parents need to ask whether it is *ethical* to do so to fulfill their desires for a boy or a girl (Box A) (Wertz and Fletcher 1989).

Contemplated Techniques (Preconception)

In an area of abundant scientific and popular publications, one fact stands out about planning a child's sex *prior* to conception: Even a useless technique succeeds about half the time! The ancient who advised tying off the left testicle to produce a son could easily become a seer on the basis of small samples. Over the centuries, hundreds of theories of sex determination were constructed, all of which worked in about 50% of the trials. More modern proposals would alter the behavior of the two sperm types in vivo or separate X-bearing from Y-bearing sperm in vitro for artificial insemination (Levin 1987). We give a few examples.

In Vivo Techniques. Many investigators have looked at the timing of intercourse relative to ovulation. One theory from the 1960s holds that

---

* Following a widespread practice, we express the sex ratio as percent males. This figure, however, is not really a ratio.

BOX A

SHOULD PARENTS BE ABLE TO CHOOSE THE SEX OF A CHILD? YES OR NO?

Yes

Being a desired sex furthers the happiness and well-being of both parents and child. What about a couple with, say, four sons who desperately want a daughter to balance their family? Being wanted is important.

Medical resources are often used to satisfy patients, even if not related to their physical health. (Cosmetic surgery, for example, is commonplace, and can improve a patient's mental outlook.) It is not a doctor's job to make moral judgments or to judge the motives of his or her patients.

No

Sex is not a genetic disorder and should not merit the use of scarce medical resources.

Successful parenting requires the acceptance of a child's individuality, including strengths and weaknesses. Choosing a child's sex (to the point of abortion) is at odds with these values.

It is sexism, pure and simple. The preference for boys, especially for the firstborn, will benefit further the male sex and skew the sex ratio.

Y-bearing sperm swim faster but are shorter-lived than X-bearing sperm (Shettles and Rorvik 1984). If this conjecture were true, then intercourse at the time of ovulation would favor boys because the faster Y-bearing sperm would find a waiting egg. Intercourse that stopped two to three days before ovulation, on the other hand, would favor the birth of girls because the X-bearing sperm, according to the theory, have greater staying power.

Other ideas about sex selection predict just the opposite result. One theory hypothesizes that high levels of maternal gonadotropins at conception are associated with female offspring (James 1990). Since these hormones peak as the egg matures, intercourse at the time of ovulation would favor girls, while intercourse either earlier or later in the cycle would favor boys.

The practice of artificial insemination ought to provide relevant information, because the sperm is transferred as close as possible to the time of ovulation. A summary of reports on artificial insemination published between 1973 and 1984 and involving about 30,000 births found about 57% male births using fresh sperm and 50% using frozen sperm (Zarutskie et al. 1989). Compared to a general sex ratio of 51–52% male, these figures suggest little change due to insemination at the time of ovulation. Whether *artificial* insemination (with its attendant levels of stress) also applies to *natural* insemination is unclear. Even if it does, the amount of the shift is of very limited practical value to a couple desiring a son or a daughter using "old-fashioned" baby-making methods.

In Vitro Techniques. Over the years, many researchers have tried to enrich sperm samples (from humans or farm animals) for X-bearing or Y-bearing sperm. Basic work was slow until recent times, because the only sure way of knowing whether the separation had been achieved was to use the putatively X- or Y-enriched samples in artificial insemination and then wait nine months (for both cows and people). Several methods are now available, however, that are said to identify quickly whether sperm cells carry an X or a Y chromosome. One of these involves the use of quinacrine, which causes the Y chromosome to fluoresce brightly, thus producing a spot within sperm heads. This method is unreliable, however. A better method, at least on a population sample, involves DNA probes that bind to differently sized RFLP fragments derived from X or Y chromosomes.

Once these and other methods were available, many investigators tried to separate X-bearing from Y-bearing sperm using centrifugation, swimming tubes, electrophoresis, cell-sorting cytometry, or immunological techniques. The most remarkable claims of success are those of R. J. Ericsson and colleagues, who place sperm at the top of a column containing progressively more viscous solutions of albumin. It is said that sperm cells that swim downward and accumulate in the most viscous layer at the bottom are predominantly Y-bearing. The Y-enriched samples are then used for artificial insemination. Does the method work? The latest journal article we have found that reports on this method notes the birth of 271 males and 87 females—76% male (Ericsson and Beernink 1987). Considering the number of years that the method has been in use, however, these numbers are relatively skimpy, and we are unsure what to make of them. Other methods are used to produce an X-enriched fraction, also with a claimed 70–80% success rate. If these methods of producing the desired sex work some of the time, it is unclear why they do. At the moment, the matter is unresolved.

CHANGING THE FREQUENCY OF GENETIC TRAITS OVER GENERATIONS

In 1883 Francis Galton (Box A in Chapter 8), coined the word **eugenics** (literally, "well born") to mean the genetic improvement of the human species over time. Eugenic goals included the elimination of genetic disease and the enhancement of desirable traits. Galton wrote (1905):

> What Nature does blindly, slowly, and ruthlessly, man may do providently, quickly, and kindly. As it lies within his power, so it becomes his duty to work in that direction; just as it is his duty to succour neighbors who suffer misfortune. The improvement of our stock seems to me one of the highest objects that we can reasonably attempt.

In 1908 Galton organized the Eugenics Society in London to investigate human heredity and carry out social action programs. Similar societies were established at about the same time in the United States and Germany. Although some people who joined these groups were motivated by altruism, others were concerned with race, class, and privilege. Based, as it was, on classifications of superiority and inferiority, eugenics appealed to both idealists and cranks, and to moderates in between. One American genetic researcher of the early 1900s was Charles Davenport, director of the Eugenics Record Office at Cold Spring Harbor, Long Island, in New York. Davenport, along with others of his time, tried overly hard to fit virtually every human trait into a Mendelian frame, including poverty, moral degeneracy, insanity, and feeblemindedness—as well as a number of bona fide genetic diseases. A small part of their work, including Davenport's analysis of skin color, described in Chapter 8, contributed to the early objective study of human genetics. But much of eugenics "research" was scientifically shoddy, influenced by racism and class consciousness. In fact, the systematic development of human genetics was set back because some serious researchers were reluctant to enter the field of racial and ethnic prejudice (Allen 1983; Kevles 1985).

Although he advocated only voluntary actions, Galton's eugenic aspirations and plans were themselves heavily tinged with nationalism. He noted that a "high human breed" was especially important for the English because they colonized the world and planted the seeds of future millions of the human race. In the United States, eugenicists influenced many aspects of social policy, including enactment of discriminatory immigration laws in the 1920s after their lengthy "scientific" testimony before Congress. Adolf Hitler carried eugenic notions to cruel and bizarre extremes with sterilization programs that championed Aryan elitism. The horrors of the Holocaust soon followed.

Besides overestimating the role of hereditary factors in human behavioral traits, Galton naively believed that eugenic proposals would be easily and voluntarily accepted. Few persons, however, give more than passing thoughts to the overall quality of the genes that are committed to future generations. Indeed, some are even offended by the very idea that methods used daily to improve agricultural plants and animals might be relevant to humans. Yet certain eugenic ideals are involved in traditional medical practices and new reproductive technologies that help individuals and families improve their prospects for health and happiness. In this section we look at social action schemes that seek to prevent the spread of "disadvantageous" alleles (negative eugenics) and encourage the transmission of "advantageous" ones (positive eugenics).

Negative Eugenics

Lessening the incidence of hereditary disorders through the prevention of childbearing, i.e., **negative eugenics**, has a particularly unsavory history related to the enforced sterilization of persons considered "unfit" or likely to have "socially inadequate" offspring. Most infamous was Hitler's policy of "racial hygiene." But even in the United States, tens of thousands of involuntary operations were performed, mostly between 1930 and 1960. Although 19 states still retain statutes allowing for the eugenic sterilization of institutionalized retarded persons after due process of law, its implementation is minimal in this country today (Reilly 1985).

Involuntary Sterilization. The first compulsory sterilization law in the United States was enacted by Indiana in 1907. Although this and other early statutes were declared unconstitutional, a more carefully drawn law in Virginia was upheld by the U.S. Supreme Court in the 1927 decision *Buck* v. *Bell*. The case involved an institutionalized patient, Carrie Buck, who was declared feebleminded, as were her mother and Carrie's seven-month-old daughter. The trio was found to be genetically defective by a eugenics "expert" who never examined them and who misrepresented the daughter. (She died at age eight of an infection, but her schoolteachers considered her very bright.) Furthermore, Buck's ineffective defense lawyer was in collusion with judicial and legislative proponents of the new Virginia sterilization law, so the case was a sham (Lombardo 1985). Yet the patriotic rhetoric of Justice Oliver Wendell Holmes in *Buck v. Bell* (1927) rings out:

> We have seen more than once that the public welfare may call upon the best citizens for their lives. It would be strange if it could not call upon those who already sap the strength of the State for these lesser sacrifices, often not felt to be such by those concerned, in order to prevent our being swamped with incompetence. It is better for all the world, if instead of waiting to execute degenerate offspring for crime, or to let them starve for their imbecility, society can prevent those who are manifestly unfit from continuing their kind. The principle that sustains compulsory vaccination is broad enough to cover cutting the Fallopian tubes. Three generations of imbeciles are enough.

Not only was the case factually incorrect, but the judgment has been questioned on legal grounds: Are vaccinations and sterilizations equivalent? In addition, the genetic suppositions of the decision are groundless, for mental retardation is not a single entity. Some cases are due to the presence of a major gene with differing modes of inheritance, some to the chance combinations arising from polygenic inheritance, and some to various chromosomal aberrations. Other types of mental deficiency are entirely due to environmental damage occurring prenatally, to birth trauma, or to infectious diseases. If the various types of people with mental problems did not reproduce, the subsequent decrease in frequency would vary and would usually be slight. Thus, it is silly to suggest that Carrie Buck and her "kind," whatever that is, would populate the world with criminals and imbeciles. *Buck* v. *Bell* has been soundly censured by judges and others. Such a Supreme Court decision today is unthinkable, but Justice Holmes undoubtedly reflected the public opinion of the time (Box B) (Rafter 1988).

However misguided past practices have been, involuntary sterilization can be of value in specific families to reduce or prevent unhappiness and suffering. Persons with mental retardation are entitled to the full array of constitutional rights that others enjoy. Exercising those rights, however,

THE KALLIKAKS

The Kallikak Family (Goddard 1912) was a much publicized book that purported to be a scientific study of heritable mental defects among the descendants of one New Jersey man. We are told that Martin Kallikak, a young Minuteman, strayed from the path of virtue with a "nameless feeble-minded girl." Martin neglected mother and child, but through this line of descent he became the progenitor of 480 persons, among whom were 143 feebleminded souls, 36 illegitimate children, 33 sexually immoral persons, 24 alcoholics, 8 madams, 3 epileptics, and 3 criminals.

Martin later straightened up and married a respectable girl of good family. On this "control" side, he became the proud ancestor of 496 persons, all with good credentials: "doctors, lawyers, judges, educators, traders, and landholders." From this dual family history, it was concluded that good alleles were perpetuated on one side and bad alleles on the other. In fact, the invented name Kallikak comes from the Greek words meaning "beauty" (*kallos*) and "bad" (*kakos*).

If it is to believed at all, the book is a moralistic tract that almost completely ignores the role of environmental factors and even the role of Martin himself. Far from proving anything about heredity, the Kallikak yarn could constitute a plea for upgrading the social conditions that perpetuate poverty. Gould (1981), for example, made the interesting discovery that the facial features in the book's photographs of persons on the bad side of the family were crudely altered to make them look sinister or stupid. Curiously, the Kallikak author criticizes an earlier analysis of crime, pauperism, and disease in the *Jukes* family as being inconclusive because of unknowable interactions of heredity and environment. In fact, as Rafter (1988) points out, many investigators of "bad-gened" degenerate clans simply found that which they assumed they would find.

raises complex legal issues. After proper procedural safeguards, courts have occasionally decided that sterilization may be in their best interests, allowing them, for example, to live outside an institution without fear of becoming a parent.

Voluntary Sterilization. Many normal adults have chosen voluntary sterilization as the surest contraceptive. Although primarily employed to prevent the birth of a child of *any* phenotype, rather than as eugenic measures, sterilization would be appropriate for persons carrying alleles for genetic diseases. For example, Fraser (1973) tells of a Montreal woman who gave birth to two hemophiliac sons despite the use of contraception (rhythm method). Although she then requested sterilization, she became pregnant once more before her application was approved by a hospital committee:

> By that time . . . the baby was moving, and she just couldn't go through with [an abortion], quite understandably. The result was that she now has three hemophilic sons. She loves the little boy, and she is a good mother, but she is quite sure that she would have preferred that he had not been born. (Fraser 1973)

A common method of female sterilization is *tubal ligation*, in which the oviducts are cut and tied so that eggs cannot pass from the ovaries to the uterus. Although the operation is relatively routine, it requires hospitalization, and the effects of surgery are felt for several weeks. The procedure can now be done using a *laparoscope* (literally, "looking into the abdomen"), a pencil-like instrument with a glass fiber optical system. This tool is inserted through a small incision near the navel, along with an instrument for cutting and sealing the oviducts. In males, sterilization is more easily achieved by *vasectomy*: cutting and tying off both sperm ducts leading from the testes. The operation does not require hospitalization. In neither sex do these sterilizations involve the gonads themselves, and except for blocking the passage of gametes, sexual functions are unimpaired. Millions of voluntary sterilizations have been performed for the purpose of individual birth control, about three-fourths of them on men.

Voluntary sterilizations, then, provide a means for reducing the im-

491

pact of genetic diseases on future generations. But as calculated in Chapter 18, the change in frequency of recessive disease genes is very slight over the course of a moderate number of generations. This is true even when all affected persons do not reproduce. More effective eugenic programs for recessive diseases depend on the detection and counseling of *carriers* prior to the birth of any affected children. The screening program for Tay-Sachs disease has, in fact, dramatically lowered the occurrence of the disease among Jewish groups (Chapter 19).

Positive Eugenics

Rather than advocating the sterilization of the unfit (however defined), Galton stressed the need for increased propagation of men of talent and genius, those with superior health, moral strength, and high "civic worth." **Positive eugenics** could be accomplished, he felt, by educational programs that would influence popular opinion and by wealthy persons interesting themselves in and befriending poor but promising lads. He specifically recommended that exceptionally worthy young couples be provided convenient housing at low rentals.

Although most American eugenicists were obsessed with negative eugenics, the intellectual heir to the positive idealism of Galton was Hermann Muller (a 1946 Nobel laureate for mutation studies). Through his 1935 book, *Out of the Night: A Biologist's View of the Future*, Muller hoped to stimulate interest in positive eugenics and other aspects of social reform. He was an early advocate of sexual freedom, including birth control, legalized abortion, and public child care, especially for women who wanted to work outside the home or further their education. Muller proselytized for positive eugenics, primarily through "germinal choice," that is, artificial insemination of women with semen freely selected from sperm banks to which eminent men of known identity made contributions. Muller (1961) recommended that the sperm be frozen for decades before use in order to better view the individual worth of donors and reduce the danger of hasty choices based on fads and fashions of the day. It was hoped that these methods would enhance basic values distinctive to humans, among which Muller included intelligence, curiosity, creativity, "genuineness and warmth of fellow feeling," and "joy in life and in achievement."

Although there has been no stampede to follow Muller's proposals, more than 70 sperm banks have been established in the United States (many of which are listed in telephone Yellow Pages). Most of these facilities, however, are not intended to fulfill the eugenic purposes envisioned by Muller. For example, some men deposit sperm prior to vasectomy so they can still father a child. One sperm bank with eugenic goals is the Repository for Germinal Choice, which was established in 1981 in southern California by the wealthy inventor Robert Graham. The bank accepts sperm from only a few men, initially just Nobel prize winners. When only three prize recipients donated sperm samples, and when no women chose them, the criteria were loosened to include other eminent scientists and exceptional athletes, including an Olympic gold medalist. Many people (including Muller's heirs) have ridiculed or castigated these blatantly elitist and narrow criteria. Nevertheless, according to Gorman (1992), this bank has "fathered" 156 children.

The net eugenic effect of these births on the human species is miniscule—and would no doubt remain so even if the number of such births was much, much larger. Because of meiotic segregation, polygenic inheritance, and especially environmental effects, the specific characteristics of a parent are rarely passed on intact to a child. For example, the selection of Babe Ruth as a sperm donor would only slightly increase the probability of producing a slugger. (Indeed, none of Ruth's 15 descendants has shown any particular talent for baseball.) We wish to emphasize that phenotypic variability is influenced by differences in both genes and environment. Rather than trying to control the genes with which a person is born, we might be better served by providing the enriching environments that foster optimal expressions of the genotypes that currently exist. The potential benefits of a multitude of social programs for betterment have never been fully realized.

Still, there is some eugenic potential in currently feasible technologies like artificial insemination, proved safe and perhaps to some small extent effective (depending in part on the heritabilities of the traits in question). That these programs are voluntarily undertaken with only minimal government regulation would undoubtedly have pleased Galton and Muller.

CONCLUSION

How can we eliminate the pseudoscience that supported—and in some places still supports—the bigotry of eugenics programs? Perhaps this type of aberration is less likely to occur if society and its leaders know something about the methods of science, the nature of scientific evidence, and the motivations of scientists. Public debate is important to counter prejudice and injustice in any form. Yet in many countries around the world—including democratic ones—racism and other forms of discrimination are almost daily affronts.

Over the years, the positive impact of genetics on the fields of medicine, public health, and agriculture has been impressive. Increased knowledge of genes and increased skill in applying genetic technologies promise to contribute further to humankind. Scientists should, of course, make every effort to consider the ramifications of their own work and convey to the public at large what it is they hope to accomplish. But a precise accounting of future benefits and risks from genetic or any other scientific research is not possible. Scientists can predict neither the outcome of specific experiments nor the many ways their findings might be used. For example, the biochemists who discovered restriction enzymes could hardly have imagined all the applications to gene mapping, prenatal diagnosis, genetic screening, gene therapy—and who knows what to come!

As we learn more about our genetic heritage, it is also to be hoped that people will come to view themselves and *all* the world's fellow creatures in a clearer and more sympathetic light. Human genetics, a science of differences, emphasizes the uniqueness of each individual.

> It is astonishing that man, a species which displays such a range of variety and which lives in a world populated by tens of thousands of other species, should be so conformist and so intolerant of diversity. If knowledge can help to counter this limitation, the study of the origins of human individual differences is very important. Some understanding of the genetic determinants of behavior—their biological qualities, their extent, and their distribution—might . . . give people an enhanced sense of their uniqueness, as well as acceptance, perhaps even tolerance, of their kinship with others. It might lead, above all, to a more charitable view of those foibles, frailties, peculiarities, and eccentricities which cause many people, their virtues notwithstanding, to be set apart. The headlines of the newspapers on any day proclaim the urgency of the need for this understanding. (Barton Childs et al. 1976)

SUMMARY

1. Genetic diseases are treatable in principle, although only a relatively few are really effectively managed. Physicians must generally understand the chain of events extending from mutant gene to altered phenotype.

2. Some treatments rely on diet modifications to restrict a toxic substance or to supply a missing metabolite. In other cases, a metabolite in excess can be depleted by drugs. Injecting a missing gene-encoded protein or transplanting an organ or bone marrow may also be effective.

3. Gene therapy cures a genetic disease by supplying a normal allele to correct the consequences of an abnormal allele. Altered retroviruses can be used to carry the normal gene into the appropriate somatic cells.

4. The first approved gene therapy trial was for a fatal immune disorder, adenosine deaminase deficiency. After more than a year's experience with two patients, the therapy seems to be successful. Many other gene therapy protocols have now begun or are contemplated.

5. Choosing the sex of a child is controversial. Americans who might use such a technique often want a balanced family overall but with a male as firstborn. It is possible to use abortion to achieve the desired sex, but many persons would not do this.

6. Many investigators have sought a method for choosing the sex of the fetus before conception. Methods based on the timing of intercourse do not seem to be effective. Methods for selecting X-bearing or Y-bearing sperm for artificial insemination have been promoted commercially, although it is unclear to what degree these methods work.

7. Eugenics means improving the human species over generations through human genetics research and social action. Despite altruistic aims, the research has often been poor and the actions bigoted.

8. Negative eugenics has sometimes meant enforced sterilization to curtail the transmission of presumed undesirable traits. Voluntary sterilization, however, can be used for similar purposes or simply for birth control. Some proposals for positive eugenics are based on the voluntary use of artificial insemination with banked sperm of selected men.

9. The study of human genetics should lead to a greater appreciation of the uniqueness of each individual.

KEY TERMS

gene therapy	negative eugenics	retrovirus	stem cell
germinal gene therapy	positive eugenics	somatic gene therapy	

QUESTIONS

These questions are meant to stimulate discussion, and we do not provide answers.

1. Ethical considerations aside, gene therapy is not likely to raise the intelligence of persons already within a normal range, but it might someday be used to overcome some specific causes of severe mental retardation. Why is this so?

2. Do you approve or disapprove of the following processes for your own family? For other families? What conditions would you impose on their use?

Sterilization of sexually mature Down syndrome patients

Banking of sperm from distinguished and identified men

Abortion to avoid the birth of a boy or a girl

Artificial insemination with samples of X- or Y-enriched sperm

Timing of intercourse to conceive a child of a given sex

3. Is there a law in your state that allows sterilization of some people without their consent? Is the law utilized?

4. Should people with certain genetic diseases be prevented from marrying? If so, which diseases?

5. If enriching environments and opportunities were available to everybody on an equal basis, would the apparent differences between people disappear?

6. Are Muller's criteria for selecting sperm for a bank of frozen samples the same as yours? Comment on the heritability of these characteristics.

7. If a democratic government decided that some line of research should be stopped, what could the government do to implement such a ban?

8. With regard to the label "playing God" for genetic engineering research, the President's Commission (1982) wrote:

> At its heart, the term represents a reaction to the realization that human beings are on the threshold of understanding how the fundamental machinery of life works. . . . In this view, playing God is not actually an objection to the research but an expression of a sense of awe—and concern.

Is this your interpretation of "playing God"?

FURTHER READING

Although not meant for lay people, Desnick (1991) and Friedmann (1991) are complete accounts of treatments for genetic diseases. The articles on gene therapy by Anderson (1992) and Miller (1992) are worth reading for their non-technical content. Less difficult articles on gene therapy are found in Verma (1990) and Anderson (1990). Zarutskie et al. (1989) have summarized the multitude of confusing claims about choosing the sex of a child. Allen (1983) and Kevles (1985) are authoritative accounts of the eugenics movement, and Bajema (1976) reprints many original papers relating to the history of eugenics. Rafter's (1988) large collection of turn-of-the-century studies of purportedly degenerate clans makes fascinating reading. The President's Commission text on human genetic engineering (1982) is a worthwhile summary of basic information and views on ethical issues.

Appendix 1

Some Common Units of Measurement

Quantity	Unit	Abbreviation	Equivalents
Length	kilometer	km	10^3 m, 0.621 mile
	meter	m	39.4 in, 3.28 ft, 1.09 yd
	centimeter	cm	10^{-2} m (10^2 cm = 1 m), 0.394 in
	millimeter	mm	10^{-3} m (10^3 mm = 1 m)
	micrometer	μm	10^{-6} m (10^6 μm = 1 m)
	nanometer	nm	10^{-9} m (10^9 nm = 1 m)
	Angstrom unit	Å	10^{-10} m (10^{10} Å = 1 m)
	mile	mi	1.61 km
	yard	yd	0.914 m
	foot	ft	0.305 m, 30.5 m
	inch	in	2.54 cm
Mass	kilogram	kg	10^3 g, 2.20 lb
	gram	g	0.0353 oz
	centigram	cg	10^{-2} g (10^2 cg = 1 g)
	milligram	mg	10^{-3} g (10^3 mg = 1 g)
	microgram	μg	10^{-6} g (10^6 μg = 1 g)
	nanogram	ng	10^{-9} g (10^9 ng = 1 g)
	picogram	pg	10^{-12} g (10^{12} pg = 1 g)
	ton	ton	2,000 lb, 907 kg
	pound	lb	454 g, 0.454 kg
	ounce	oz	28.3 g
Volume (liquid)	liter	l	1.06 qt
	deciliter	dl	10^{-1} liter (10 dl = 1 liter)
	milliliter	ml	10^{-3} liter (10^3 = 1 liter)
	microliter	μl	10^{-6} liter (10^6 μl = 1 liter)
	nanoliter	nl	10^{-9} liter (10^9 nl = 1 liter)
	picoliter	pl	10^{-12} liter (10^{12} pl = 1 liter)
	quart	qt	0.946 liter
	pint	pt	0.473 liter
	tablespoon	tbsp	15 ml
	teaspoon	tsp	5 ml
Temperature	Fahrenheit	°F	0°C = 32°F
	centigrade	°C	20°C = 68°F
	(Celsius)		40°C = 104°F
			60°C = 140°F
			80°C = 176°F
			100°C = 212°F

Some Basic Chemistry   Appendix 2

Here we present a few chemical concepts that underlie our discussions of molecules and chemical reactions throughout the book.

ATOMS AND MOLECULES

The materials around us are all composed of tiny particles called **atoms**. One simple model likens the atom to an ultraminiature solar system, with the positively charged atomic nucleus corresponding to the sun and the negatively charged **electrons** corresponding to the orbiting planets. Just as our solar system is mostly space, so are atoms mostly space.

The smallest atom is that of the element *hydrogen*, symbolized by H. It has just one electron spinning about an atomic nucleus containing one **proton**. These two particles carry exactly the same electric charge, but the electron is negative and the proton is positive. The *mass* of a proton, however, is almost 9,000 times greater than that of an electron. The number of protons (or electrons) is called the *atomic number*, which is the same for all atoms of a particular element. The number and arrangement of the electrons about the atomic nucleus largely account for the chemical reactions of an element.

The four most common elements in living matter are (starting with the most abundant) hydrogen (H), oxygen (O), carbon (C), and nitrogen (N). These and other elements join together to form **molecules**. The simplest molecules consist of just a few atoms of the same kind (e.g., molecular oxygen, O_2) or of different kinds (e.g., water, H_2O; carbon dioxide, CO_2; or ethyl alcohol, C_2H_6O). The subscripts give the number of atoms of each element that are present in the molecule.

The atoms in any molecule are held together by **chemical bonds**. Because energy input is often required for the formation of chemical bonds, a molecule, such as the sugar glucose, contains more energy than the sum of its separated atoms. For this reason, the bonds within a molecule represent *stored chemical energy.* This energy gets into our food because green plants convert radiant energy from the sun into the chemical energy of sugars and other molecules. Animals, in turn, extract energy from plant molecules for use in their own bodies. This is why animals need plants, and plants need the sun, in order to survive.

The breakdown of bonds in energy-rich compounds within our body releases some of the sun's energy that was put there when the bonds were originally formed. Complex networks of reactions occurring in all cells release this bond energy in *small steps*, often storing it temporarily in intermediate compounds for future use in other reactions. What remain afterward are simpler, energy-poor molecules.

The most common type of bond in living organisms is the **covalent bond**. This strong and stable bond occurs when two atoms *share* electrons between them. In a molecule of water, for example, one *shared pair* of electrons connects oxygen to each hydrogen. Each connection, a covalent bond, is represented in a structural formula by a short line, as in H—O—H. Hydrogen is always connected by one covalent bond to other atoms; oxygen is connected by two covalent bonds. Nitrogen atoms always form three covalent bonds, and carbon atoms four. Two covalent bonds may be directed toward one atom, as in carbon dioxide (Figure A.1). This arrangement is called a *double bond*.

Figure A.1
Covalent bonding of some simple substances.

Molecule	Structural formula	Shorthand formula
Water	H—O—H	H_2O
Carbon dioxide	O=C=O	CO_2
Ethyl alcohol	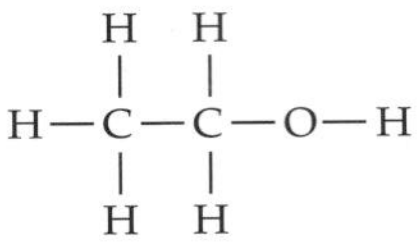 H—C—C—O—H (with H atoms above and below the two carbons)	C_2H_5OH

Another type of connection, much weaker than a covalent bond, is also exemplified by water molecules. The oxygen atom in H_2O attracts the shared electrons more strongly than the hydrogen atoms do, so this part of the molecule carries a partial negative charge. Conversely, because the shared electrons are rather far from the hydrogen nuclei, the hydrogen atoms carry a partial positive charge. Although the molecule as a whole is neutral, with no net charge, there is a separation of charge within the molecule. Such molecules are called *polar molecules*. With both positive and negative parts, they tend to attract other polar molecules. In liquid water, for example, the hydrogens tend to orient toward the oxygens of *other* water molecules. The weak attractions *between* molecules of water are called **hydrogen bonds**. Very little energy is needed to make or break such bonds.

Although hydrogen bonds have only about 1/20 the strength of covalent bonds, they are important in forming the structure of many biological molecules. For example, the helical formation of DNA is stabilized by hydrogen bonding between different parts of the molecule, and proteins owe their three-dimensional conformation (a critically important property) in part to hydrogen bonding. In these cases, some of the hydrogen bonds link a partial positive charge on hydrogen with a partial negative charge on a nitrogen atom.

A third type of bond occurs when an acceptor atom attracts an electron so strongly that the electron's connection with a donor atom is severed completely. In this way, the donor atom loses an electron, becoming a *positively charged ion*; the acceptor atom becomes a *negatively charged ion*. The oppositely charged ions then attract each other to form an **ionic bond**. In table salt, for example, the chlorine atom has pulled an electron away from the sodium atom. Ionic bonds are important in the chemistry of many salts, acids, and bases.

ORGANIC MOLECULES

Molecules that contain carbon are called **organic molecules**. Here we discuss some carbon compounds that are important in the structure and function of living things.

Hydrocarbons contain only hydrogen and carbon. These compounds might have been the first organic molecules on earth, predating the ap-

pearance of anything that could be called living. The simplest hydrocarbons, constituents of natural gas, are methane, ethane, and propane:

Methane   Ethane   Propane

$$H-\overset{\displaystyle H}{\underset{\displaystyle H}{C}}-H \qquad H-\overset{\displaystyle H}{\underset{\displaystyle H}{C}}-\overset{\displaystyle H}{\underset{\displaystyle H}{C}}-H \qquad H-\overset{\displaystyle H}{\underset{\displaystyle H}{C}}-\overset{\displaystyle H}{\underset{\displaystyle H}{C}}-\overset{\displaystyle H}{\underset{\displaystyle H}{C}}-H$$

Note that each carbon atom is involved in four covalent bonds, and each hydrogen is involved in one. Other hydrocarbons have longer chains of carbon atoms, sometimes branched, sometimes in circles of six carbon atoms. Compounds with double bonds between adjacent carbon atoms are *unsaturated*, containing fewer hydrogen atoms than they would if these bonds were single (and the freed electrons were shared with hydrogen instead). Hydrocarbons play a key role in our lives. Examples include synthetic and natural rubber, petroleum, gasoline, fuel oil, asphalt, plastics, and many other petrochemicals.

Many classes of organic compounds have a hydrocarbon skeleton to which is attached a small group of atoms that enter readily into reactions with other compounds. Such attachments, replacing one or more hydrogens, are called *functional groups*. A set of compounds with the same functional group engages in similar reactions. About a dozen functional groups are important in living systems. Of these, we mention three:

1. *Alcohols* have a *hydroxyl* (—O—H) group that replaces one of the hydrogens in a hydrocarbon.

2. *Organic acids* have a *carboxyl group*: $-C\!\!\overset{\displaystyle O}{\underset{\displaystyle O-H}{}}$

3. *Amines* have an *amino group*: $-N\!\!\overset{\displaystyle H}{\underset{\displaystyle H}{}}$

Many important organic compounds possess more than one functional group. **Amino acids**, for example, have both a carboxyl group and an amino group. Many organic molecules are very large, being composed of long chains of similar subunits. Four types of large molecules are extremely common in living organisms: proteins, nucleic acids, lipids, and carbohydrates.

Proteins are chains of dozens to thousands of *amino acids*. The carboxyl group of one amino acid reacts with the amino group of the next one to make the links. Protein structure is coded by genes. The average mammalian cells produces over 10,000 different proteins; most of these are *enzymes*, which catalyze all the chemical reactions within an organism. Other proteins include *hemoglobin*, the oxygen-transporting protein of red blood cells; *antibodies*, the immunity-providing protein of blood plasma and other body fluids; *myosin*, the contractile protein of muscles; *collagens*, the fibrous protein of bones and tendons; and *keratin*, the structural protein of hair and skin.

Nucleic acids consist of long chains of repetitive subunits called *nucleotides*, whose structure and function are discussed in Chapter 12. *Deoxyribonucleic acid (DNA)*, the genetic material of all organisms except certain viruses, contains four different nucleotides. A slightly different set of four nucleotides is found in the *ribonucleic acid (RNA)* molecules that aid in the gene-decoding process. DNA and RNA are *informational* molecules:

The information contained in a sequence of DNA nucleotides is translated by RNA intermediates into a sequence of amino acids — that is, into proteins.

Lipids are fatty substances that do not dissolve in water. One type consists of three long-chain organic acids attached, one on one, to the three hydroxyl groups of a compound called glycerol (Figure A.1). These *triglycerides*, or *neutral fats*, pack a caloric wallop. In *phospholipids* and *glycolipids*, the third hydroxyl group of glycerol is replaced, respectively, by a phosphate group and by a short chain of sugars. Both of these lipid types are major components of membranes. Another kind of lipid, structurally quite different, is the group of *steroids* containing several six-member rings of carbon atoms. The steroids include the *sex hormones* (*estrogen* and *testosterone*) as well as *cholesterol, cortisone,* and *vitamin D.*

Carbohydrates include simple sugars and long chains of these sugars. The breakdown of the simple six-carbon sugar *glucose* is the primary pathway yielding the chemical energy that drives other cellular reactions. In animal cells, excess glucose may be temporarily stored as *glycogen,* which links together hundreds of glucose subunits. In plant cells, the corresponding storage carbohydrate is *starch.* The major component of plant cell walls, *cellulose,* is also composed of long chains of glucose subunits, but these are attached to each other differently than in the storage compounds. About half of all the carbon atoms in living things are in cellulose. We cannot utilize cellulose as food because we lack the enzymes necessary to break the bonds between its glucose subunits. Cows manage the trick by harboring in their largest stomach compartment bacteria that do the job for them.

BIOLOGICAL REACTIONS

Many chemical reactions can be generalized this way:

$$\text{reactant(s)} \longrightarrow \text{product(s)} + \text{energy}$$

In any chemical reaction, energy is conserved; that is, when the chemical bonds joining the atoms of the reactants contain more energy than in the rearranged bonds of the products, this excess energy appears on the product side of the reaction. Part of this extra energy may be picked up as chemical bond energy in other compounds. For example, consider the breakdown of glucose, which occurs in all cells. Although it involves several dozen separate small steps, the overall reaction is:

$$\text{glucose} + \text{oxygen} \longrightarrow \text{carbon dioxide} + \text{water} + \text{energy}$$

The chemical energy within the glucose molecule and oxygen is very much greater than the energy of the corresponding amounts of carbon dioxide and water. Most of this excess energy appears as heat or is simply unavailable for any kind of useful work in the cell. About one-third of the energy, however, is transferred to the chemical bond energy of a molecule called **adenosine triphosphate (ATP)**. The formation of ATP is *coupled* with each of several of the small steps in the breakdown of glucose. That is, generation of ATP occurs simultaneously with the glucose reactions. This is diagrammed as follows:

As indicated, the ATP is made from *ADP,* or *adenosine diphosphate.*

These two molecules differ by just a single phosphate group (which is present in solution in the cell). ATP is a *high-energy compound* that plays a crucial role in the networks of cellular reactions. As an *energy transfer molecule,* it picks up energy from an energy-yielding reaction and delivers it to a different, energy-requiring reaction. Without ATP and other molecules with similar functions, *metabolism* (the totality of chemical reactions in cells) would be extremely slow and cumbersome.

Most chemical reactions in the body are *energy-requiring* and may be generalized as follows:

reactant(s) + energy ⟶ product(s)

All cells must synthesize from simpler substances literally thousands of biochemicals, including all DNA and proteins and many of the necessary lipids and carbohydrates. Almost all of these reactions need energy, which is usually supplied by ATP. When ATP loses a phosphate group, becoming ADP, some of the energy that was put into the ATP molecule in its formation becomes available to drive other reactions. Thus, the energy transfer circle is completed:

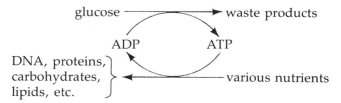

The reactions are far from 100% efficient, however. Thus, the energy content of the final products — DNA, proteins, and so on — is only a small fraction of the energy available in glucose.

ENZYMES

Most chemical reactions, regardless of whether they produce energy or need it, will not proceed unless given an initial nudge. For example, paper burns fiercely, yielding large amounts of energy, but only if it is first brought to its kindling temperature. In living things (with fairly constant, low temperatures), the activation of reactions is accomplished by **enzymes.** These are large protein molecules that *catalyze* (speed up) a reaction by bringing the reactants together in one spot and also by straining the molecular bonds that get broken and remade. Without enzymes, random molecular movements might bring the reactants together, but so infrequently as to be negligible.

Each enzyme is highly specific, catalyzing only one reaction (or sometimes a few reactions of similar type). For example, the enzyme *hexokinase* catalyzes the first of many small steps in the breakdown of glucose. This is a coupled reaction involving ATP. Hexokinase brings together glucose and ATP and causes a phosphate group from ATP to be transferred to the number 6 carbon atom of glucose:

glucose ——hexokinase——▶ glucose-6-phosphate
          ATP   ADP

Glucose-6-phosphate has more energy than does glucose, the difference (as well as the phosphate group itself) being supplied by the high-energy molecule ATP. Although the total breakdown of glucose to carbon dioxide and water ultimately yields considerable energy, this particular starting step requires it.

Enzymes are able to catalyze a reaction partly because of their three-dimensional shape. The folding of the long chain of amino acids leaves an inpocketing of the surface, into which the reactants fit. Hydrogen bonding also helps to anchor the reactants to the enzyme and to orient them to each other. After the reaction, the products are released, but the enzyme is unchanged in the process. Because they are recycled, enzymes are required in only small amounts. They do eventually wear out, however, so new enzymes must be synthesized as needed.

Figure A.2
An outline of the synthesis of the hormone adrenaline from the amino acid phenylalanine, which is present in food. (See also Figure 14.6.)

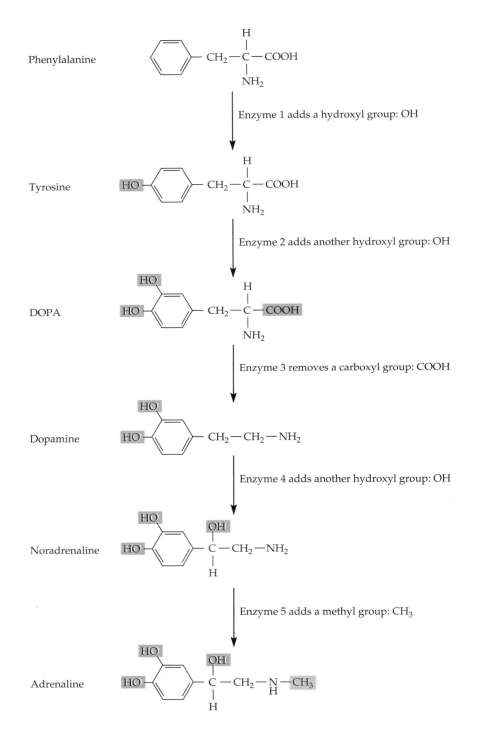

A sequence of several dozen steps is required for the complete breakdown of glucose. Some steps involve just the splitting or the making of one bond or the transfer of a functional group from one molecule to another. So, too, are other molecules constructed or degraded, each small reaction step balanced for energy and catalyzed by a specific enzyme. Figure A.2 shows how the amino acid *phenylalanine* is converted, in five steps, into the hormone *adrenaline*. Note that each step involves the addition or removal of a single functional group.

Within just one cell, thousands of enzymes, involved in many dozens of metabolic pathways, are at work simultaneously. These reactions are dependent on each other, so that an intermediate compound in one reaction may be the starting compound of another. For example, the intermediate compound DOPA in Figure A.2 is the precursor of the melanin pigments found in skin, eyes, and hair. The next two compounds, dopamine and noradrenaline, are themselves neurotransmitters that chemically transfer an electrical impulse from one nerve cell to the next. The biochemical pathways are thus linked, forming networks of diverging and converging patterns (Figure A.3).

In normal cells, each substance appears in the proper concentrations for the correct functioning of the body and its parts. And because the structure of each enzyme depends on one gene (or a few), DNA provides the blueprint for all cellular activity. These networks of chemical reactions help determine the final appearance of an organism.

Figure A.3
Some of the chemical reactions that take place in a cell. About 500 common conversions of small molecules are diagrammed here, with each type of molecule represented by a dot. The breakdown of glucose appears in boldface. (From Alberts et al. 1989.)

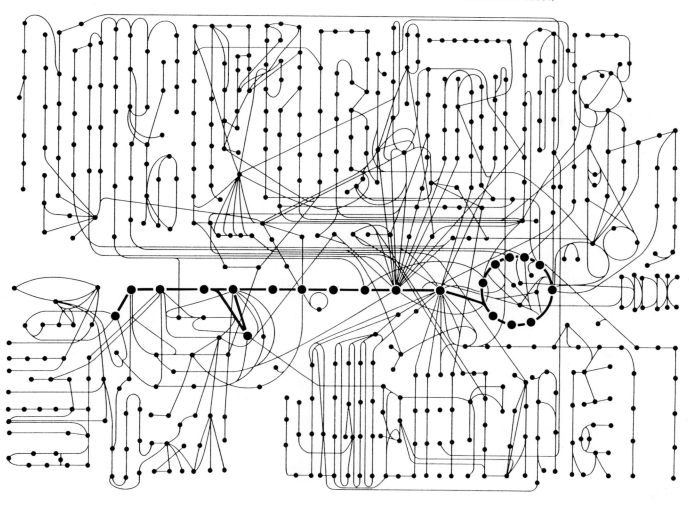

Answers to Questions

CHAPTER 2

2. -G-C-C-T-A-C-A-T-T-G-G-G-A-

3. (a) 47,XXY   (b) 45,X

4. (a) 47,XX,+21   (b) 47,XY,+13   (c) 47,XX,+18

5.

New order	Detectable in unbanded spreads?	Detectable in banded spreads?
(a) *a b c ◆ d*	no	yes
(b) *a b d ◆ c*	no	no
(c) *e f ◆ h g*	yes	yes
(d) *e f g h ◆*	no	no
(e) *j ◆ l k m*	no	yes
(f) *j l k ◆ m*	yes	yes

CHAPTER 3

1. *Embryo/fetus until birth*: Oogonia become primary oocytes and enter a very long meiosis I (leptotene, zygotene, pachytene). This period takes roughly 3 to 7 months. *Birth until beginning of oocyte maturation*: Arrest in meiosis I (diplotene). This period takes roughly 12 to 50 years. *Oocyte maturation until ovulation*: End of meiosis I (diakinesis), metaphase I, anaphase I, telophase I, cytokinesis (to produce secondary oocyte and first polar body), prophase II, metaphase II. This period takes about 1 month. *Fertilization until first cleavage division*: Anaphase II, telophase II, cytokinesis (to produce egg and second polar body). This period takes 24 to 36 hours.

2. At metaphase the chromosome arms extend out to the sides of each centromere, but at anaphase they are pulled behind the centromere.

3. (a) Mitosis   (b) First meiotic division
   (c) Either

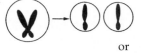

or

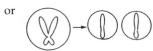

4. One convenient way to systematize such combinations of sets of things is by setting up a branching diagram like the one that follows. This method works not only for homologous chromosomes but also for gene pairs or any other combinatorial data.

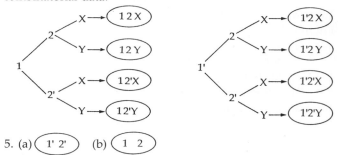

5. (a) ⟨1' 2'⟩   (b) ⟨1 2⟩

6. 46 chromosomes, 92 chromatids

7.

	Chromosomes	Chromatids
(a)	46	—
(b)	46	92
(c)	23	—
(d)	23	—
(e)	23	—

8. $3(23) = 69$; $4(23) = 92$

9. $(64/2) + (62/2) = 32 + 31 = 63$

CHAPTER 4

1. $C/c ♀ \times c/c ♂$ produces 1/2 normal (C/c) and 1/2 albino (c/c).

2. Presence of the albino child shows that she is heterozygous (C/c). No other genotype is possible, barring mutation of C to c in a C/C woman (or adoption of the child).

3. Most likely she is homozygous (C/C), in which case only normally pigmented children are expected. She could, however, be heterozygous (C/c), in which case half of her children are expected to be albino. The probability is low (but definitely not zero) that a heterozygote produces no albinos among several children. For 10 children, this probability is $(3/4)^{10} = 0.056$. [*Note*: This situation exactly parallels Mendel's method for progeny testing his tall F_2 plants (see page 56), running about a 5% chance of mislabeling a heterozygote as a homozygote.]

4. (a) $(3/4)^3(1/4)^2 = 27/1024$ (b) $(3/4)^5$ (c) $1 - (3/4)^5$
   (d) 1/4. Each birth is considered separately and is not influenced by the outcomes of previous births.

5. $B/b \times B/B$ ⎫
   $B/b \times B/b$ ⎬ 50% of progeny expected to be B/b
   $B/b \times b/b$ ⎭

   $B/B \times b/b$ 100% of progeny expected to be B/b

6. (a) $B/B \times B/B$ (b) $B/b \times B/B$ (c) $B/b \times B/b$
      $B/B \times b/b$ $B/b \times b/b$
      $b/b \times b/b$

7. 2/3 Q/q, 1/3 q/q

8. 1/4 AB, 1/4 ab, 1/4 Ab, 1/4 aB.

9. 1/8 each of the gametes carrying the following sets of genes (with only superscripts used for the Hb gene):

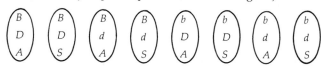

Can you see how the sets of three follow from a tree branching diagram?

10. Sixteen; the 8 kinds above, with Xg^a or Xg added to each.

11. 9/16 normal, type B; 3/16 affected, type B; 3/16 normal, type O; 1/16 affected, type O. This is the Mendelian 9:3:3:1 dihybrid ratio for two independent genes with dominance.

12. (a) $P(B/b\ P/p\ S/s) = (1/2)^3 = 1/8$
    (b) $P(\text{all three dominant traits}) = (3/4)^3 = 27/64$
    (c) $P(b/b\ p/p\ s/s) = (1/4)^3 = 1/64$

13. Four homozygous (A/A, S/S, E/E, C/C,) and six heterozygous (A/S, A/E, A/C, S/E, S/C, and E/C).

14. (a) 1/2 A/A, 1/2 A/S
    (b) 1/4 A/A, 1/4 A/S, 1/4 A/C, and 1/4 C/S
    (c) 1/4 A/C, 1/4 A/S, 1/4 E/C, and 1/4 E/S

15. (a) 1/2 $g^1/(Y)$ = green-shifted male; 1/2 G/g^1 = normal (carrier) female.
    (b) 1/4 $G/(Y)$ = normal male; 1/4 $g^2/(Y)$ = missing-green-pigment male; 1/4 G/G = normal female; 1/4 G/g^2 = normal (carrier) female.
    (c) 1/4 $G/(Y)$ = normal male; 1/4 $g^1/(Y)$ = green-shifted male; 1/4 G/g^2 = normal (carrier) female; 1/4 g^1/g^2 = green-shifted female.
    (d) Parents: g^1/g^2 female $\times$ $G/(Y)$; Offspring are as follows: 1/4 $g^1/(Y)$ = green-shifted male; 1/4 $g^2/(Y)$ = missing-green-pigment male; 1/4 G/g^1 + 1/4 G/g^2 = 1/2 normal (carrier) female.

CHAPTER 5

1. The terms *homozygous* and *heterozygous* imply the existence of two alleles (same or different) in an individual. X-linked genes in males are present only once.

2. (b) It appears that everybody produces and excretes a rather musky-smelling intermediate product (methanethiol) of asparagus metabolism, but not everybody can smell it (Lison et al. 1980). The ability to smell this substance is probably an autosomal dominant trait.

3. Autosomal recessive. Note that the parents of the affected individual are first cousins.

4. (a) Parents: Gl/gl. Child: gl/gl.
   (b) 1/4. No, the same probabilities exist independently for each child, regardless of the outcome of previous pregnancies.
   (c) 1/300; that is, the chance of marrying a heterozygote (1/150) times the chance of then producing an affected child (1/2).

5. (a) Queen Victoria; her children, Alice of Hesse and Beatrice; her grandchildren, Irene, Alix, Alice of Athlone, and Victoria of Spain.
   (b) 10 affected to 10 normal. This appears to be exactly the mathematical expectation, but it is not, for reasons given in Chapter 7.
   (c) No, insofar as descent from Victoria is concerned. The father of Elizabeth II, being unaffected with hemophilia, must have inherited (and passed on to her) an X chromosome lacking Victoria's hemophilia allele.

6. Autosomal dominant. If this were an X-linked dominant: in family 1, male 3 should not be affected; in family 2, males 4 and 7 would be severely affected (as is male 13); and in family 4, female 8 would be affected moderately (like 1 and 4) rather than severely. The condition represented here is familial hypercholesterolemia.

7. (a)

	Among daughters	Among sons
$G/G\ ♀ \times G/(Y)\ ♂ \longrightarrow$	all G/G	all $G/(Y)$
$G/G\ ♀ \times g/(Y)\ ♂ \longrightarrow$	all G/g	all $G/(Y)$
$G/g\ ♀ \times G/(Y)\ ♂ \longrightarrow$	1/2 G/G, 1/2 G/g	1/2 $G/(Y)$, 1/2 $g/(Y)$
$G/g\ ♀ \times g/(Y)\ ♂ \longrightarrow$	1/2 G/g, 1/2 g/g	1/2 $G/(Y)$, 1/2 $g/(Y)$
$g/g\ ♀ \times G/(Y)\ ♂ \longrightarrow$	all G/g	all $g/(Y)$
$g/g\ ♀ \times g/(Y)\ ♂ \longrightarrow$	all g/g	all $g/(Y)$

   (b) All but the first and last matings listed in part (a). All the daughters will be carriers when the mother is homozygous and the father is hemizygous for the other allele (second and fifth matings).
   (c) Those in which the mother is heterozygous. For sons, the genotype of the father is irrelevant with regard to X-linked genes.
   (d) $g/g\ ♀ \times G/(Y)\ ♂ \longrightarrow$ all daughters normal, all sons green-shifted.
   (e) $G/g\ ♀ \times g/(Y)\ ♂$

8. X-linked recessive.

9. Daughters. He transmits a gene-rich X chromosome to daughters but a gene-poor Y chromosome to sons; as a result, daughters carry about 5% more of his genes than do sons.

10. A female receives no sex chromosome from her father's father, whereas a male receives no sex chromosome from his father's mother.

11. It does not skip generations. It is expressed only in females and transmitted only through females. Affected females produce only half as many sons as daughters, but all sons are normal. All but the first feature distinguish this

pattern from an autosomal dominant. This mode of inheritance is rare, but one example is oral-facial-digital (OFD) syndrome type I. Affected females have cleft jaw and tongue plus other malformations of the skull, face, and hands, plus mental retardation.

12. (a) Autosomal dominant or X-linked dominant. (b) In addition to (a), could also be autosomal recessive or X-linked recessive if the mother is heterozygous. These ambiguities point up the importance of having extensive and reliable family data, and also knowing something about the frequency of the trait.

13. (a) The rare allele would have had to be present in spouses I-2 and III-2. Because these persons appear to be genetically unrelated to I-1 and III-1, it is unlikely that a rare allele would occur this way.
(b) The rare allele would have had to occur in I-2.
(c) Individual III-1 would have had to exhibit the trait.
(d) Individuals I-2, II-1, III-1, III-3, and III-4 would have had to exhibit the trait.
(e) III-2 and III-6 should exhibit the trait. There are two simple possibilities. It could be a common X-linked recessive trait (affected = a/a or a/Y), with I-2 as a carrier; III-1, III-3, and III-4 would also be carriers who inherited the mutant allele from their fathers. It could also be a common autosomal recessive trait (affected = a/a), with I-2, III-1, and III-2 (as well as III-3, III-4, and III-6) being carriers.

CHAPTER 6

1. What criteria should be used in defining sex? Although nearly all males are indeed XY and females XX, exceptions do exist, and it cannot be assumed that chromosomal sex always leads to a given pattern of adult sexual development and behavior. Testicular feminization accounts for most cases of XY females who are barred from competition, but confers no athletic advantage over XX women. And some male pseudohermaphrodites brought up as females (with appropriate surgery and hormone treatment from birth) would also be unfairly excluded. For discussions of sex testing for Olympic athletes, see Ljungqvist and Simpson (1992), Ferguson-Smith et al. (1992), de la Chapelle (1986), and Turnbull (1988).

2. Absence or inactivity of the Müllerian-inhibiting substance.

3. All human embryos tend to develop as females unless deflected from this path by the action of testosterone at several critical steps. "If any one of your sexual systems fail, you can coast down the female road, but not down the male" (Money and Ehrhardt 1972).

4. Sex linkage; mitochondrial inheritance; chromosome imprinting; ordinary gene linkage (i.e., the presence in the same chromosome region of two or more genes of interest).

5. The simplest possibility is X-linked recessive inheritance, but it could also be an autosomal dominant gene expressed only in males.

6. It could be homozygosity for some rare recessive "small stature" gene on chromosome 7. But it may be that the small size in these two patients resulted from genetic imprinting. The latter is known to occur in mice that are homozygous for a mouse chromosome that is partly homologous to human chromosome 7.

CHAPTER 7

1. It is easier first to calculate the probability of a normal child and then get the probability of a deaf child by subtraction:

$$P(\text{normal child}) = P(D/- \ E/-) = (3/4)(3/4) = 9/16$$

$$P(\text{deaf child}) = 1 - P(\text{normal child}) = 1 - 9/16 = 7/16$$

2. All sons become bald (B_1/B_2) and all daughters are nonbald (B_1/B_2).

3. Yes, except for female II-5, who should have received the B allele from her father and been nonbald. As noted in the text, however, environmental factors may have contributed to her baldness.

4. Use a checkerboard or the gene × gene method to get:

9/16	$B/-$ $C/-$	= black
3/16	b/b $C/-$	= brown
3/16	$B/-$ c/c	
1/16	b/b c/c	= 4/16 albino

5. Use a gene × gene method to get:

1/2	$-/-$ c/c $-/-$	= albino
1/8	B/b C/c D/d	= black
1/8	B/b C/c d/d	= gray
1/8	b/b C/c D/d	= brown
1/8	b/b C/c d/d	= light chocolate

6. $C/c \times c^b/c$.

7. The parents were homozygous for different autosomal recessives leading to albinism (or deafness). For example, assume that c/c is albino at one locus and that w/w is albino at another locus. (The c is allelic to its normal dominant allele C; likewise, w is allelic to W.) One parent could be c/c W/W, and the other parent C/C w/w. All their children would then be double heterozygous, C/c W/w, and phenotypically normal.

8. (a) 1; (b) 0; (c) 0; (d) 0; (e) 0; (f) 1; (g) 0; (h) 1.

9. As illustrated in Figure 7.13, genotypically identical twin females may differ in the phenotypic expression of heterozygous X-linked genes. Because of the random inactivation of X chromosomes, it is unlikely that mosaic patches in twin females would be identical in size and distribution. Identical twin 46,XY males, however, must express all the same genes on their single X chromosome.

10. Each tumor must have arisen from a single cell (either A or B) rather than from a group of cells. The cell that gave rise to a tumor had *either* an active A allele *or* an active B allele of the *G6PD* gene. This active allele remained the active allele in all descendants of that original tumor cell.

11. All phenotypically normal carrier daughters, and all phenotypically normal sons.

12. (a) No children are expected to be affected, although half are expected to receive allele D.
(b) Half of the children are expected to receive and express allele D.

CHAPTER 8

1. The events in their lives that were different—perhaps very different—are not specified. Did either one like something other than math—or dislike something other than spelling—during all their school years? Did they both drive only blue Chevrolets during their lives? Did either one vacation elsewhere than on the same stretch of Florida beach? We are not told.

2. Perhaps the specific language one first learns to speak. Of course, being able to speak at all is not gene-independent, nor, we guess, is facility with foreign languages.

3. Probably the genetic variability in maze-running ability was exhausted. After seven generations of selection, virtually all the alleles that tended toward maze-brightness or maze-dullness were already concentrated in the one group or in the other.

4. (a) Higher heritability when genetically heterozygous.
   (b) Higher heritability when environmentally uniform.

5. The full explanation is complicated and depends on the effects of dominance, epistasis, and environmental factors. With regard to the environment, however, we can note that heavy persons are often that way because of a "weighty" life-style, in addition to "weighty" alleles. Whatever genes they transmit, however, the environmental contribution to weight gain among their children is not likely to be repeated to the same extreme degree. (A similar argument applies to lightweights.)

6. $G/G\,H/H\,i/i$ $\quad$ $G/G\,h/h\,I/I$ $\quad$ $g/g\,H/H\,I/I$

   $G/G\,H/h\,I/i$ $\quad$ $G/g\,H/H\,I/i$ $\quad$ $G/g\,H/h\,I/I$

7. (a) Hazel × blue-green: $G/g\,H/h\,I/i \times G/g\,H/h\,i/i$, for example. (b) Light brown.

8. (a) True. In addition, no progeny can be darker than medium. (b) False. The child can be medium when the other parent is light, and the child can be dark when the other parent is medium.

9. The paternal contribution to the unusual pair is similar to that for DZ twins, but the maternal contribution is similar to that for MZ twins. Note that in the absence of crossing over, the second polar body is genetically identical to the egg.

10. The range of differences—up to 24 points—shows that environmental factors are important in IQ variation. Other studies also show that being a twin reduces IQ an average of about 5 points below that for single births.

11. The sibs of the severely retarded would generally have IQ scores in the normal range: they have only a 25% chance of being homozygous for the postulated abnormal recessive, and a much smaller chance of abnormality through repetition of a possible developmental error. On the other hand, the sibs of the moderately retarded would tend to have relatively low IQ scores because of polygenic inheritance of the same additive alleles, or perhaps because of similar environmental factors.

12. Within both countries, heritability of IQ is 100%. There would almost certainly be an average difference in IQ scores between persons from A and B, caused in part by the obvious environmental factors. However, there is nothing in this situation as stated that rules out genetic factors contributing to the difference.

CHAPTER 9

1, 2.

	Percent of gametes				Percent recombination
	AB	ab	Ab	aB	
1(a)	25	25	25	25	50
1(b)	50	50	0	0	0
2(a)	49	49	1	1	2
2(b)	46	46	4	4	8
2(c)	39	39	11	11	22

3. To distinguish a recombinant from a parental assortment. If either gene is homozygous, crossing over produces no new assortment.

4. (a) $\quad$ 9% $\quad$ Eh/eh $\quad$ dominant for E, recessive for H

   $\quad\quad$ 9% $\quad$ eH/eh $\quad$ recessive for E, dominant for H

   $\quad\quad$ 41% $\quad$ EH/eh $\quad$ having both dominant traits

   $\quad\quad$ 41% $\quad$ eh/eh $\quad$ having both recessive traits

   (b) $\quad$ 9% $\quad$ EH/eh $\quad$ having both dominant traits

   $\quad\quad$ 9% $\quad$ eh/eh $\quad$ having both recessive traits

   $\quad\quad$ 41% $\quad$ Eh/eh $\quad$ dominant for E, recessive for H

   $\quad\quad$ 41% $\quad$ eH/eh $\quad$ recessive for E, dominant for H

Note that the chromosome eh is received from the second parent in every case in both (a) and (b).

5. Max is ab/ab. His daughters are AB/ab. Among the daughters' sons, 8 out of 40 are recombinants = 20% recombination = 20 map units.

6. 46.5%. The two recombinant types add to 7%, and the two parental types add to the rest, 93%. One of the two parental types has both the G gene and the R gene. Thus, the answer is half of 93%.

7. $(.465)^2 = 22\%$. The probability that the daughters are gr/GR is 0.465. From these daughters, the probability that a gamete carries GR is another 0.465.

8. $(.035)(.465) = 1.6\%$. The probability that the daughters are gr/GR is 0.035. From these daughters, the probability that a gamete carries GR is 0.465.

9. Use a 4×4 checkerboard in which both rows and columns are labeled as follows:

0.4 DR $\quad$ 0.1 Dr $\quad$ 0.1 dR $\quad$ 0.4 dr

Then fill in each of the 16 checkerboard squares with the genotype (or phenotype) determined by the headings of its row and column. The fraction for each of the squares is the product of the row and column fractions. Summing up squares that represent the same phenotype gives:

$\quad$ 0.66 dominant for both traits (sum of 9 squares)
$\quad$ 0.09 dominant for D, recessive for R (sum of 3 squares)
$\quad$ 0.09 recessive for D, dominant for R (sum of 3 squares)
$\quad$ 0.16 recessive for both traits (1 square—lower right)

10. (a) $A\,np/B\,Np$. (b, c) The four types of gametes from the father are indicated below. If a gamete contains A, it will contain the normal (np) allele 90% of the time (45/50).

If a gamete contains *B*, it will contain the normal allele 10% of the time (5/50).

Gamete	Percent
Anp	45
ANp	5
Bnp	5
BNp	45

11. (a) I-1 = *GH*/(Y)   III-1 = *gH*/(Y)

II-1 = *GH*/*gh*   III-2 = *gh*/(Y)

II-2 = *gH*/(Y)   III-3 = *gH*/*gH* or *gH*/*gh*

(b) III-1, getting *gH* from his mother and a Y chromosome from his father.

(c) P(H/H) = 0.03; P(H/h) = 0.97; P(h/h) = 0. Individual III-3 got *gH* from her father. Since III-3 is green-shifted, we know that she got a chromosome with *g* from her mother. Given *g*, was this a parental chromosome with *h* (high probability) or a recombinant with *H* (low probability)?

12. Double heterozygotes who are *male* would make it easier to establish linkage. The less that genes recombine (as in males), the clearer it becomes that they are linked. (Zero percent recombination is the most obvious.) The more that genes recombine (as in females), the harder it becomes to distinguish linkage from independent assortment.

CHAPTER 10

(A) Chromosome loss at meiosis I

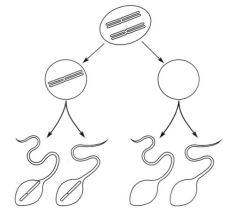

2 normal:2 nullisomic sperm

(B) Chromosome loss at meiosis II

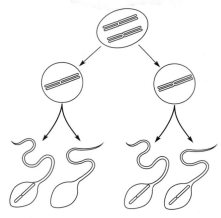

3 normal:1 nullisomic sperm

(Some cytologists see chromosome loss as a form of nondisjunction; others distinguish between the two.)

2. The zygote will be euploid, although both chromosomes 21 came from the mother.

3. Of the two dozen or so recorded cases in which females with Down syndrome have given birth, about half of their children were normal and half had Down syndrome.

4. If the split occurred at the two-cell stage and then mitotic nondisjunction occurred in one of these cells, the result would be one normal twin and one with Down syndrome.

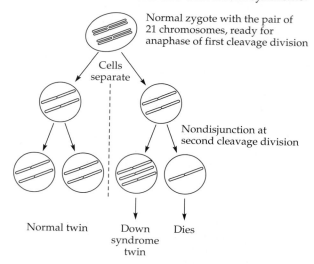

Normal zygote with the pair of 21 chromosomes, ready for anaphase of first cleavage division

Cells separate

Nondisjunction at second cleavage division

Normal twin   Down syndrome twin   Dies

5. Females with Turner syndrome will express color blindness whenever the gene occurs on their single X chromosome, as is true for 46,XY males. Males with Klinefelter syndrome, however, need two doses of the mutant gene to be color-blind, as is the situation with 46,XX females. Thus, the frequencies of X-linked genes in females with Turner syndrome and males with Klinefelter syndrome should parallel the frequencies usually found, respectively, among males and females in the general population.

6. The allele for color blindness must have been present in the heterozygous mother, *G*/*g*, whereas the father was hemizygous normal, *G*/(Y). The females with Turner syndrome must have resulted from fertilization of a *g*-bearing egg by a sperm lacking a sex chromosome; nondisjunction in the father could have occurred at meiosis I or II. The males with

Klinefelter syndrome must have resulted from fertilization of a *g/g* egg by a Y-bearing sperm, with nondisjunction in the mother occurring at meiosis II.

7. The twins were derived from a single 46,XY zygote. If the presumptive embryo split in half at the two-cell stage and one of the two cells lost a Y chromosome, then one twin would be a normal male and the other a female with Turner syndrome.

8. Nondisjunction in females in division(s):

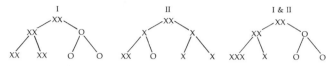

Nondisjunction in males in division(s):

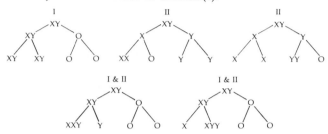

9. Of the eight possible ways, three require only one non-disjunctional event:

		Nondisjunctional events		
Egg	Sperm	In mother	In father	Total
XX	Y	Meiosis I	—	1
XX	Y	Meiosis II	—	1
XX	Y	Meiosis I	Meiosis I & II	3
XX	Y	Meiosis II	Meiosis I & II	3
X	XY	—	Meiosis I	1
X	XY	Meiosis I & II	Meiosis I	3
O	XXY	Meiosis I	Meiosis I & II	3
O	XXY	Meiosis II	Meiosis I & II	3

10. Six types of sperm: XY_1; Y_2; XY_2; Y_1; Y_1Y_2; X

11. Mitotic loss of one X chromosome from a cell that was originally XXY, or nondisjunction of an X chromosome during mitotic division of an XY cell (in any division after the first cleavage) to give an XXY cell plus a Y cell that dies. These two possibilities are diagrammed as follows:

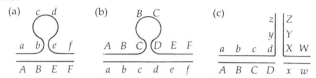

12. Endocrinology clinics would attract those males with Klinefelter syndrome who seek treatment for failure to develop secondary sexual characteristics (i.e., undergo normal puberty); these cases would be more seriously affected. Fertility clinics would attract married males with Klinefelter syndrome—that is, those who are masculine enough to find mates but who later discover they are infertile.

13. *Advantages:* Researchers can learn more about physical

and behavioral phenotypes—including the lack of abnormalities—as well as their frequencies in XYY individuals as they develop. If physical or behavioral problems do turn up, early identification and counseling might help the subjects and their parents. Some genetic conditions are treatable, but it is impossible to develop treatments without first studying the phenotypes.

Disadvantages: Publicity about the 47,XYY karyotype makes it difficult to do prospective studies; subjects might be stigmatized or treated differently by their family and develop problems as a self-fulfilling prophecy. There is no point in studying XYY individuals if no treatment is available for possibly abnormal traits.

CHAPTER 11

1. Both (a) and (b) will show a looping out of one chromosome:

(a)
a b (e f) —
A B E F

(b)
A B C (D E F) —
a b c d e f

(c)
a b c d | X W
A B C D | x w

2. Neither (a) nor (b) require any unusual pairing configurations, since the homologues match:

(a)
A B C (E D) F G
a b c (e d) f g

(b)
A B C X Y Z — U V W D E F
a b c x y z — u v w d e f

3. (a) Half the gametes would contain the (21q21q) translocation chromosome (giving rise to progeny with Down syndrome), and half would contain no chromosome 21 at all (causing lethal monosomy 21). (b) No.

4. Because both chromosomes 15 were stuck together, all her eggs contained either *two* attached chromosomes 15 or *no* chromosome 15. Fertilized by normal sperm, all the resultant zygotes were therefore either trisomic or monosomic for a medium-sized chromosome and thus inviable. One abortus was in fact recovered and found to be trisomic for chromosome 15.

5. By an inversion that includes the centromere:

A B C ◆ D E F { G   to   A B C F E D ◆ G

The reverse of this phenomenon is probably what accounts for the few metacentric Y chromosomes that have been reported in human males.

6. A triploid could arise (1) if the egg and the polar body nuclei fused to form a diploid egg nucleus that was fertilized by a haploid sperm; (2) if two sperm fertilized a haploid egg; or (3) if nondisjunction in the first or second meiotic division of spermatogenesis gave rise to a diploid sperm, which fertilized a haploid egg.

7. Only if the two aneuploidies were complementary, so that the duplicated chromosomes or chromosome segments present in one were balanced by deletions of the same genetic material in the other, and vice versa (i.e., fertilization

involving the pair of gametes in Figure 11.8C or D). This event is highly unlikely, even if the two parents were related and heterozygous for the same translocation.

8. One possibility starts with a normal XX zygote. Before the first cleavage division, both X chromosomes replicate, and the two chromatids of one X will segregate normally. But the centromere of the other replicated X misdivides (as shown in Figure 11.7B) to form an isochromosome of the long arm, with loss of the replicated short arm. Thus the dividing zygote will contain one isochromosome plus two copies of the normal X. Following mitosis, each daughter cell contains one normal X chromosome, but the isochromosome will be present in only one of the two cells. A mosaic results in which one cell line has 46 chromosomes (including the isochromosome of Xq) and the other cell line is 45,X.

9. The son must have inherited both his father's X and Y chromosomes following nondisjunction in meiosis I of spermatogenesis in the father. The XY sperm must have fertilized a no-X egg that resulted from maternal nondisjunction. Replication of the affected father's X chromosome in the zygote resulted in uniparental disomy for that chromosome.

10. If crossing over occurred before nondisjunction, part(s) of the recombinant disomic chromosome pair inherited from one parent would be identical (i.e., derived from just one member of that parental pair), and the other part(s) of the disomic chromosome pair would be different (i.e., derived from both members of that parental pair).

11. Especially for individuals exhibiting rare recessive disorders whose chromosomal locations are totally unknown, the occurrence of uniparental disomy can—when combined with linkage tests for heterozygous molecular markers present in the parents—help to locate the gene on a particular chromosome.

12. The severely affected child was homozygous for the mutant chromosome, having inherited a *CMT1* duplication from each parent.

CHAPTER 12

1. P is present in DNA but not in protein; S is present in virtually all proteins but not in DNA. These differences make possible some elegant experiments in which *either* DNA *or* protein is radioactively labeled.

2. Chargaff's rules only apply to double-stranded nucleic acids. The DNA molecules in some viruses and most RNA's are single-stranded.

3. 1/4 intermediate, 3/4 light; that is, of the eight double helices, two would be intermediate and six would be light.

4.

	(a)	(b)	(c)
A	33%	21%	27%
T	21%	33%	27%
G	18%	28%	23%
C	28%	18%	23%

5. mRNA: GGGUUU
   DNA: CCCAAA

6. Three pieces having amino acids 1–22, 23–29, and 30 only.

7. The β polypeptide has 146 amino acids and so is coded for by (146)(3) = 438 bases. 438/1,606 = 27.3% (see Figure 12.17).

8. The number of nucleotides required to code for the five polypeptides is (5)(438) = 2,190 bases. 2,190/50,000 = 4.4%.

9. Only mutation combinations (b) and (e) produce frameshifts.

10. Ala-Gly-Cys-Lys-Asn-Phe-Phe-Trp-Lys-Thr-Phe-Thr-Ser-Cys

11. (a)   ...Lys-Lys-Lys-Lys...

    (b)   ...Ser-Ser-Ser-Ser...;
          ...His-His-His-His...;
          and ...Ile-Ile-Ile-Ile...

    (c)   ...Phe-Leu-Pro-Ser... (this foursome repeated in tandem)

12.   ACC: Trp   CAC: Val   CCA: Gly (still)

      GCC: Arg   CGC: Ala   CCG: Gly (still)

      TCC: Arg   CTC: Glu   CCT: Gly (still)

13. Enzymes from mammalian liver are sometimes added to the growth medium for whatever effect they might have in metabolizing the tested chemicals.

14. Although radiation can cause mutations, this particular man has no case. He transmitted a *Y chromosome* to his son; the mutant hemophilia gene in question was inherited on an *X chromosome* from the mother.

15. It is estimated that it would take 100–400 rem of chronic radiation to yield the observed rate of spontaneous mutations in humans. The average exposure of a human is much less than this amount.

16. The α chains are a major component of hemoglobin both pre- and postnatally, but β chains are synthesized in quantity only after birth.

CHAPTER 13

1. 20 cycles. That is, 2^{20} (1,048,576) just exceeds 1 million. At 5 minutes per cycle, this will take (in theory) 1 hour and 40 minutes. In practice, it will take longer, since later PCR cycles require more time.

2. Reading the base sequence of one stand to the right is the same as reading the other strand to the left. Almost all restriction enzymes recognize palindromes. Other examples are *Sau*3A (GATC), *Bam*HI (GGATCC), *Hind*III (AAGCTT), and *Not*I (GCGGCCGC).

3. Treat the recombinant plasmid with *Eco*RI. This treatment will yield DNA segments of two sizes: 1 kb for the human gene and something else (one hopes) for the rest of the plasmid. The segments can be physically separated by electrophoresis.

4. $10,000 = (3 \times 10^9 \text{ total bases})/(3 \times 10^5 \text{ bases per YAC})$

5. *Eco*RI lane: 3, 7, and 12 kb

   *Hin*dIII lane: 9 and 13 kb

   "Both" lane: 3, 6, and 7 kb

6. Let **E** = *Eco*RI site, **B** = *Bam*HI site:

```
      5 kb      2   1            14
   ————————————E———B-E————————————————
```

Reversing left for right above is the same map. All other maps that you might draw can be ruled out by trial and error.

7. AGCTGTTTCC

8. 99%. $(30 \times 10^6$ bases sequenced)/(3×10^9 bases total) = 0.01 = 1% will have been sequenced.

9. The reading frame that starts with the triplet AAG cannot represent a gene since triplet 12 is a stop (ATT). The reading frame that starts with AGG also cannot represent a gene since it has stops (ATC) at triplets 9 and 10. The reading frame that starts GGG might represent a gene since it is an ORF (open reading frame), having no stops.

10.

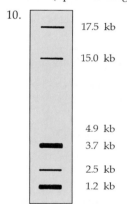

17.5 kb
15.0 kb
4.9 kb
3.7 kb
2.5 kb
1.2 kb

11. In DNA fingerprinting, the fragment-length variation depends on different numbers of repeated segments between two unmutated restriction recognition sites. In the RFLP mapping of the HD gene, the restriction recognition sites themselves have mutations (so that different alleles may be cut or not cut).

12. Chromosome 11. This chromosome is the only one present in all three "plus" lines and absent in all three "minus" lines.

13. No, but mouse genes could be assigned to mouse chromosomes.

14. No; genes on the same chromosome that are more than about 50 map units apart (without intervening marker genes) will appear to be unlinked.

CHAPTER 14

1. Essential amino acids are those that our bodies cannot make "from scratch" (i.e., from simple sugars), because we lack the appropriate enzymes; thus, they must be obtained directly from the diet. Nonessential amino acids are those that our bodies can produce from other substances.

2. The intermediate may be metabolized via a different pathway.

3. This allele is more common among the Irish and their descendants, many of whom live in Massachusetts. In contrast to this, so few PKU cases were found in Washington, D.C. (most of whose citizens are black) that the screening program there was discontinued.

4. The error might occur during the process of polymerization of dopaquinone to form black and brown pigments (left fork at bottom of Figure 14.6).

5. Random inactivation of one X chromosome in each cell during early development would lead to a variety of phenotypes—some with a preponderance of cells in which the mutant X-linked gene is expressed and others with a majority of cells expressing the normal allele.

6. It makes little sense. Milk can induce serious illness among the children and great discomfort in the adults; such experiences may lead recipients of this aid to reject other, more useful, food products.

7. It makes the task of restricting offending foods much more difficult, considering that even products such as processed meats, soups, fruits, salad dressings, and instant coffee are likely to contain lactose.

8. The "cross-correction" studies show that the two mutants are nonallelic, because each excreted a normal substance that was deficient in the other (in this case, two different acid hydrolase enzymes). If they were allelic, neither could provide the normal substance missing in the other. Complementation also occurs in cell hybrids formed from two phenotypically abnormal parental strains. If the hybrid cells show a normal phenotype, the two mutants must involve different genes.

9. Nobody is sure of the answer to this question, but there are two main hypotheses: (1) random inactivation occurs in the precursors of the blood cells, but those expressing the mutant allele are at a disadvantage and become overgrown by their normal counterparts; (2) X chromosome inactivation is nonrandom (favoring the chromosome with the normal allele) in the blood cell precursors but random in other tissues.

10.

```
Metabolites
in minimal   —8→  Ant  —2→  IGP  —3→  Indole  —6→  Trp
medium
```

CHAPTER 15

1. About 30. Each generation doubles the number of cells, so we need to know the power of 2 that gives 10^9. Use the to-the-power key on a hand-held calculator, or the approximation, $2^{10} \approx 10^3$.

2. That all 100 billion members of the tumor have the same active X chromosome suggests, first of all, that the tumor started from a single cell. The X chromosome that was the active one in this original cell remained the active one throughout all cell divisions, as postulated in the Lyon hypothesis.

3. The cultured mouse cells may have already undergone most of the changes on the pathway to malignancy, so only the one final step remained. In fact, the mouse cell line commonly used for such experiments is known to be unusual in several respects.

4. Repeat the experiments with DNA extracted from *normal*, noncancerous cells. Transformation then does not occur.

5. The oncogenic form of the receptor might behave as if it were *always* joined to the growth factor. The cell is thus fooled by its malfunctioning receptor, being stimulated to divide continuously whether growth factor is present or not.

6. There are not many more oncogenes to be discovered.

7. Transcription factors are proteins that interact with DNA and thereby regulate the rate at which genes are transcribed. Figure 15.8 shows a transcription factor directly.

8. The cells of solid tumors are more difficult to culture and are slow to divide. Furthermore, the accumulation of secondary chromosomal defects (seen at the time solid tumors are analyzed) can obscure a specific aberration common to all cells of a given cancer type.

9. Pedigrees of the disease show the standard characteristics of dominant inheritance; that is, about 50% of children of an affected parent get the disease. But in persons carrying the dominant gene, cells become cancerous only if they become homozygous for the genetic defect.

10. When inherited in the germline, *p53* mutations produce a variety of cancers in many members of the same family, as shown in the pedigrees of the Li-Fraumeni syndrome in Figure 15.9. In other people whose body cells become homozygous for *p53* mutations, these same cancers of the breast, brain, bone, etc. can occur. But these cases occur sporadically, that is, in unrelated individuals.

11. Usually dying before age 20, patients with ataxia-telangiectasia (the homozygotes) do not usually survive to ages characteristic of the onset of breast cancer.

CHAPTER 16

1. During proliferation of a specific, stimulated B cell, some cells of the clone do not fully mature to produce antibody, but remain available as memory cells for further quick proliferation.

2. The large amount of a single antibody species in each patient provided a nearly pure substance for investigation. Antibodies from different patients led to the analysis of different variable domains.

3. Papain splits the antibody molecule somewhere along the heavy chains between the bonds holding the heavy chains together and the bonds connecting the heavy and light chains.

4. The columnist had a point. Note that blood grouping can only exclude paternity; any man of type B or AB could have been the father.

5. The babies probably had the right labels but were given to the wrong mothers. The one noteworthy fact is that Mr. and Mrs. W could not have had the baby with the Y label. (You may want to list systematically the possible offspring versus the impossible offspring from each of the ten matings based on ABO phenotypes.)

6. Anti-A and anti-B may be immune antibodies like any others, produced in response to foreign antigens. For a person of type A, substance B, if taken up somehow by the circulatory system, would stimulate the production of anti-B.

7. A, Rh pos: 3/8; A, Rh neg: 1/8; B, Rh pos: 3/16; B, Rh neg: 1/16; AB, Rh pos: 3/16; and AB, Rh neg: 1/16

8. Maternal-fetal incompatibility for an antigen other than Rh. This phenomenon is known to occur sometimes with ABO.

9. I-1: *O/O R/r*   II-1: *O/O r/r*   III-1: *A/O R/r*
   I-2: *B/O R/r*   II-2: *A/− R/r*   III-2: *A/O r/r*

10. $150 \times 12 \times 4 = 7{,}200$.

11.

Subject	Haplotypes
Father	5 2/40 9
Mother	5 3/8 11
First child	5 2/5 3
Second child	5 2/8 11
Third child	40 9/8 11

12. The first child would be better, possessing only one antigen (A3) that is foreign to the recipient.

13. 35 4 23/44 2 30

14. 7 4 3/44 2 3   35 4 23/44 2 3
    7 4 3/58 4 30   35 4 23/58 4 30

15. The HLA system is more polymorphic, so it is more likely to reveal differences between unrelated persons. Of course, DNA fingerprinting (Chapter 13) is more useful yet.

16. T cells, the agents of immunological rejection, do not reach the site of the graft.

17. He had no effective immune system to attack any transplant.

18. Rejection of foreign tissue is a by-product of the general protective business of the immune system.

19. The sooner a vital organ can be removed from a dead person, the more likely its success as a transplant. Therefore, it is important to define the moment of death.

CHAPTER 17

1. freq(A) = $0.1 + (1/2)(0.4) = 0.3$
   freq(a) = $0.5 + (1/2)(0.4) = 0.7$

2. freq(A/A) = p^2 = $(0.3)^2$ = 0.09
   freq(A/a) = $2pq$ = $2(0.3)(0.7)$ = 0.42
   freq(a/a) = q^2 = $(0.7)^2$ = 0.49

3.

Population	Phenotype		
	M	MN	N
Pima	0.49	0.42	0.09
Bushmen	0.36	0.48	0.16
Amish	0.25	0.50	0.25
Polynesians	0.16	0.48	0.36

4. (a) 0.81, 0.18, 0.01. (b) $(0.99)^{40} = 0.67$. (Use the to-the-power key of a hand-held calculator.)

5. The genotype frequencies are the same as those given in question 4. (a) $2(0.99)(0.01) = 0.0198 = 1.98\% \approx 2\%$. (b) $(0.99)(0.01) = 0.0099 = 0.99\% \approx 1\%$.

6. (a) freq(M) $= p = (82 + 38)/200 = 0.6$;
    freq(N) $= q = (42 + 38)/200 = 0.4$;
    Hardy-Weinberg law not assumed.
   (b) freq(type M) $= p^2 = 0.36$;
    freq(type MN) $= 2pq = 0.48$;
    freq(type N) $= q^2 = 0.16$.

7. $p = $ freq(A) $= (1{,}210 + 390)/2{,}000 = 0.80$;
$q = 1 - p = 0.20$.

Genotype	Expected frequency	Expected number
A/A	$p^2 = 0.64$	640
A/S	$2pq = 0.32$	320
S/S	$q^2 = 0.04$	40
Total	1	1,000

8. The only way to make allele calculations is to assume the Hardy-Weinberg law. Assuming the Hardy-Weinberg law, $q = $ freq(c) $= \sqrt{1/10{,}000} = 1/100$; $p = $ freq(C) $= 1 - q = 99/100$. Fraction of population that is heterozygous $= 2pq = 198/10{,}000$.

9. $(1/10)^2 = 1/100 = 1\%$

10. $p = $ freq(A) $= $ frequency of men with trait $= 0.6$. Among women, freq($A/-$) $= p^2 + 2pq = 1 - q^2 = 84\%$.

11. freq(type A) $= $ freq($A/A + A/O$) $= (0.3)^2 + 2(0.3)(0.6)$
$= 0.45$

freq(type B) $= $ freq($B/B + B/O$) $= (0.1)^2 + 2(0.1)(0.6)$
$= 0.13$

freq(type AB) $= $ freq(B/B) $= 2(0.3)(0.1) = 0.06$

freq(type O) $= $ freq(O/O) $= (0.6)^2 = 0.36$
Sum $= 1.00$

12.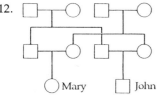

13. (a) Half first cousins, because the fathers of the spouses were half brothers. (b) Double uncle-niece, because C-III was the niece of P-IX through P-VII and also through C-II.

14.

Relationship between parents	Coefficient of inbreeding of their children $= f$
Sibs	1/4
Half sibs	1/8
Parent-child	1/4
Grandparent-grandchild	1/8
Uncle-niece or aunt-nephew	1/8

15. (a) $q^2 = 0.0049$. (b) freq(e/e) $= qf + q^2(1 - f) = (0.07)(1/16) + (0.0049)(15/16) = 0.00897$. (c) $897/490 = 1.8$.

16. (a) 10^{-6}. (b) 63.4. When the allele is rare, the disease is relatively more frequent among the children of first cousins.

CHAPTER 18

1. The cells giving rise to the multitude of sperm continue to divide through a male's middle age, whereas the cells giving rise to the relatively few eggs stop dividing prior to a female's birth. Thus, mature sperm result from many more divisions than mature eggs.

2. Sperm cells from older fathers are more likely to carry a mutated gene because the precursor cells from which they are derived have continued to divide. (See also question 1.)

3. The calculated rate would be too high because some affected children would be tallied as new mutations when, really, an "old" mutant gene, unexpressed in a parent, was transmitted to them.

4. The first consequence might be a sharp reduction in the appearance of isolated (so-called sporadic) cases of dominant diseases. All such patients would have an affected parent (except for instances of incomplete penetrance).

5. Relaxation refers to the increase in reproduction by previously limited genotypes as a result of modern, largely medical, technology. It is happening for a small number of genetic diseases. As a consequence, the frequency of these diseases, which are then treatable, increase very slowly.

6. As the allele becomes rarer, the proportion of alleles present in heterozygotes, where they are masked, becomes greater.

7. Because there are so many more heterozygotes than homozygous recessives.

8. Each generation, a much greater percentage of homozygotes for Hb^S succumb than do homozygotes for Hb^A. Therefore, the frequency of Hb^S would be expected to be less than the frequency of Hb^A.

9. (1) Selection acts against the detrimental sickle-cell homozygote (without favoring the heterozygote as happens in malarial regions of Africa). (2) Some admixture with white populations has occurred.

10. $M_1 = \dfrac{P - P_2}{P_1 - P_2} = \dfrac{0.0941 - 0.4286}{0 - 0.4286} = 0.78$
$M_2 = 1 - M_1 = 0.22$

The ancestry of Oakland blacks is about 78% West African and about 22% white.

11. When a molecule of insulin is formed, the C segment is cut out; thus, its amino acids might be expected to be less constrained—more free to change without affecting critical functions. Mutational changes would therefore be more likely to be found here than in the A or B sections.

12. (a) A subdivision of humankind that has formed by evolutionary processes and that differs from others in the aggregate of its allele frequencies. (b) The correlation between phenotype and genotype is clear; the allele frequency differences are sometimes large, thus allowing for a cleaner analysis of data; the molecular subtypes are sociologically neutral, signifying neither better nor worse.

13. Mitochondrial DNA accumulates mutations at a faster rate than nuclear DNA. It is therefore a more sensitive indicator of evolutionary changes.

CHAPTER 19

Note that questions 1, 2, 4, and 7 have no single correct answer.

3. Prenatal diagnosis may be undertaken to prepare parents for a possibly adverse outcome, without any consideration given to abortion.

5. Mosaicism is possible; that is, mutations or chromosomal aberrations could occur in the embryo itself that do not occur in cell lines giving rise to extraembryonic membranes, or vice versa.

6. The detection of neural tube defects is based on the amniotic *fluid*, which is not obtained by CVS.

8. Carrier $\times$ carrier marriages that would be detected = $(.70)(.70) = 0.49 = 49\%$. Missed marriages = $100 - 49 = 51\%$.

9. $5\% = 0.05$, the probability that the untested parent is a carrier.

10. The probability is the product of 0.05 (the frequency of carriers in the population) and 0.3 (the frequency of nondetection): $(0.05)(0.3) = 0.015 = 1.5\%$.

Glossary

We briefly define here many of the terms, concepts, and processes associated with the study of genetics. For more detail, or for information about specific disorders, genes, or gene products, please consult the index and text of this book or a dictionary of genetics (King and Stansfield 1990; Rieger et al. 1991).

acentric A chromosome or chromosome fragment that lacks a centromere.

acrocentric A chromosome with a centromere near one end and thus having one very short arm and one long arm.

agglutination The clumping together of cells (or viruses) in the presence of a specific immune serum.

allele An alternative form of a gene at a particular locus.

allele frequency The proportion of a particular allele in a particular population.

alleles identical by descent Two alleles that have originated from one particular allele in a common ancestor.

***Alu* sequence** A short (about 300-base-pair), repeated DNA sequence that is dispersed among structural genes and makes up about 5% of human DNA.

Ames test A rapid, simple test (developed by B. Ames) that uses bacteria to screen chemical compounds for mutagenicity.

amino acid A small organic molecule with a carboxyl group and an amino group, which links to other amino acids to form polypeptide chains.

amniocentesis A method of obtaining fetal cells for prenatal diagnosis (to detect chromosomal abnormalities and some biochemical disorders), usually done about the sixteenth week of pregnancy and involving the removal of amniotic fluid from the sac surrounding the fetus.

anaphase A stage of mitosis or meiosis, during which chromatids of chromosomes move to opposite poles of the spindle.

aneuploid Having too many or too few chromosomes or chromosome segments, compared to the normal genotype.

antibody A protein (immunoglobulin) produced by plasma cells in response to a particular foreign substance (antigen) and capable of binding specifically to that antigen.

anticipation The decreased age of onset and/or increased severity of an inherited disorder as it passes from one generation to the next.

anticodon In a transfer RNA molecule, a nucleotide triplet whose base sequence is complementary to that of a particular messenger RNA codon, thereby allowing recognition and binding to the appropriate codon during the process of translation (protein synthesis).

antigen A foreign substance that stimulates the production of a particular antibody.

artificial selection The development of special breeds of organisms through the selective mating of individuals of particular phenotypes.

ascertainment bias A deviation in the observed proportion (compared to an expected proportion) of affected offspring, occurring when sampled families are not randomly chosen — for example, when two heterozygous parents with *no* affected children are excluded from a study.

autoimmune disorder A disease that results when affected individuals make antibodies against their own cells or tissues.

autoradiograph An image produced on a photographic film that was placed in close contact with an electrophoresis gel or tissue section, and which shows the positions of radioactive molecules in the gel or tissue.

autosome Any nuclear chromosome that is not a sex chromosome.

Barr body The sex chromatin; a dark-staining body (discovered by M. Barr) representing a condensed, inactivated X chromosome found in the nuclei of somatic cells in female mammals.

base pair A pair of hydrogen-bonded DNA bases (one purine and one pyrimidine) located between the two backbones of a DNA double helix.

base substitution A type of mutation in which a single DNA base is replaced by a different base.

B cell A lymphocyte (white blood cell) that matures in the bone marrow and responds strongly to bacterial antigens by differentiating into antibody-producing plasma cells.

bivalent A pair of homologous chromosomes synapsed during prophase of meiosis I.

blood group A blood type defined by genetically determined antigens present on the surface of red blood cells; alleles of a gene coding for such a set of antigens.

bottleneck effect A change in gene frequency that results when a population is greatly reduced in size but then expands again with an altered gene pool.

carcinogen A substance that causes cancer.

carrier   An individual who is heterozygous for a recessive mutant allele.

cell cycle   The sequence of events from one eukaryotic cell division to the next, including all stages of interphase (G_1, S, G_2) and mitosis (prophase, prometaphase, metaphase, anaphase, telophase).

centriole   A tiny, self-reproducing structure located just outside of the nuclear membrane in animal cells; it replicates during interphase, and daughter centrioles migrate to opposite spindle poles during nuclear division.

centromere   The indented region of a chromosome that divides it into two arms; also, the region of spindle fiber attachment during mitosis or meiosis.

chain termination (stop) codon   A codon (UAA, UAG, or UGA) that does not specify any amino acid; instead, it stops the translation process and allows the translated polypeptide to be released from the ribosome.

chiasma (plural, chiasmata)   The physical manifestation of crossing over; a cross-shaped connection between synapsed nonsister chromatids that is seen during prophase I of meiosis.

chorionic villus sampling   A method of obtaining fetal cells for prenatal diagnosis (to detect chromosomal abnormalities and some biochemical disorders), usually done about the tenth week of pregnancy, in which bits of chorionic tissue are removed from the developing placenta.

chromatid   One of the two daughter replicas (strands) of a duplicated chromosome, which separate during anaphase of mitosis or meiosis II.

chromatin   The complexed substances (DNA, histones, and nonhistone proteins) that make up eukaryotic chromosomes; the structural unit of chromatin is the nucleosome.

chromosome   A self-reproducing structure in eukaryotic nuclei that consists of DNA and proteins and contains a set of linked genes. Its appearance varies during different stages of the cell cycle.

chromosome arms   The two segments of a chromosome, called p (short) and q (long), that are defined by the position of the centromere.

chromosome banding   Staining techniques (such as G-banding, Q-banding, R-banding, or C-banding) that give rise to a unique pattern of lateral bands along the length of each chromosome and thus provide a means of identification and analysis.

chromosome loss   The disappearance of a chromosome during mitosis or meiosis, often because of lagging on the spindle during anaphase.

clonal selection theory   A model for explaining how the immune system can respond so specifically to vast numbers of antigens. It proposes that a lymphocyte is preprogrammed to recognize and respond to just one antigen, whether or not that antigen ever appears.

clone   A group of genetically identical cells descended by repeated mitoses from one ancestral cell. In molecular biology a clone means multiple replicas of a particular piece of DNA produced by recombinant DNA methods.

codominant alleles   Alleles (e.g., of blood group genes) that are both expressed in a heterozygote.

codon   A group of three nucleotides in messenger RNA that specifies one particular amino acid.

coefficient of inbreeding   The probability that two alleles in one person are identical by descent from precisely the same allele in an ancestor of both parents.

combining site   The pocket, at the upper tip of a Y-shaped antibody molecule, whose shape is complementary to the shape of an antigen.

complementary bases   Nucleotide bases that can pair, purine to pyrimidine (i.e., adenine with thymine or uracil, and guanine with cytosine), by hydrogen bonding.

complementary DNA (cDNA) library   A collection of DNA sequences copied from messenger RNA (using reverse transcriptase) and thus containing all exon coding sequences but no introns or other DNA between the genes.

compound heterozygote   A heterozygote having two different mutant alleles at the same locus, and whose phenotype mimics a homozygote.

concordant twins   A pair of twins who both exhibit a particular trait.

consanguineous mating   Mating between genetically related individuals.

coupling phase   The situation in a double heterozygote when two linked but nonallelic mutants (_a b_) are present on one chromosome and their two wildtype alleles (_+ +_) are present on the homologous chromosome.

crossing over   The exchange of chromosome parts between the chromatids of synapsed homologues during prophase I of meiosis.

cytokinesis   Division of the cytoplasm (as opposed to nuclear division), which occurs during telophase of mitosis or meiosis.

cytoplasm   The protoplasm that surrounds the nucleus of a eukaryotic cell.

Darwinian fitness   A measure of the average number of offspring left by an individual or by a class of individuals.

degeneracy of code   The occurrence of several codons that specify the same amino acid.

deletion   Loss of a chromosome (or DNA) segment of any length.

denaturation   Separation of the two strands of DNA in a double helix by breaking of the hydrogen bonds between base pairs.

deoxyribose   The sugar found in the sugar-phosphate backbones of DNA.

dicentric   A chromosome or chromatid having two centromeres rather than one.

diploid   Having two representatives of every chromosome, as occurs in most somatic cells of higher organisms. The diploid (2n) number in humans is 46.

discordant twins   Twin pairs in which only one member exhibits a given trait.

disomic   Having two representatives of a particular chromosome.

dizygotic (DZ) twins   Nonidentical (fraternal) twins, which arise from two different zygotes (i.e., two eggs fertilized by two sperm).

DNA Deoxyribonucleic acid, the genetic material; a double helix consisting of two deoxyribose-phosphate backbones joined together by hydrogen-bonded purine-pyrimidine base pairs.

DNA fingerprinting A method for identifying individuals by using the unique electrophoretic banding patterns generated by DNA probes for highly polymorphic repeated sequences.

DNA ligase An enzyme that seals together fragments of newly replicated DNA or repairs nicks in one strand of a DNA double helix.

DNA marker A variable site on a DNA molecule, which can be used as a "signpost" for mapping or detecting closely linked genes. DNA markers include RFLPs (restriction fragment length polymorphisms) and VNTRs (variable number of tandem repeats).

DNA polymerase An enzyme that forms new DNA by linking together a string of deoxyribonucleotides, using single-stranded DNA as a template.

dominant The allele or phenotype that is expressed in a heterozygote.

dosage compensation The equalization of expression of X-linked genes in the two sexes, despite a 2:1 difference in their normal dosage.

double-blind study A study in which neither the researchers nor the subjects know whether a given individual belongs to the control group or to the experimental group until after all the data are recorded.

double heterozygote An organism or individual that is heterozygous at two loci.

doubling dose The amount of ionizing radiation needed to double the rate at which mutations occur spontaneously in a given species.

duplication An extra segment of chromosome or DNA, resulting in excess dosage of the genes present on that segment.

ecogenetics The study of genetic predisposition to the toxic effects of chemicals.

egg The female gamete, a haploid product of meiotic cell division.

electrophoresis A method for separating a mixture of molecules (in filter paper or in a gel placed in an electric field) according to their electric charge, size, and shape.

enzyme A protein molecule that, in small amounts, speeds up the rate of a specific biochemical reaction without itself being used up in the reaction.

epistasis The interaction of nonallelic loci, whereby one gene alters or masks the expression of a different gene.

equatorial plate The midregion of the spindle, where chromosomes congregate during metaphase of mitosis or meiosis.

euchromatin The chromatin that is relatively uncondensed during interphase, and which stains differently from the more tightly condensed heterochromatin; it contains nearly all of the known genes.

eugenics Attempted improvement of the genetic endowment of human populations by encouraging the matings of people with supposedly beneficial genes (positive eugenics) and discouraging the matings of people with supposedly harmful genes (negative eugenics).

eukaryote A cell (or organism made of such cells) that contains a membrane-bounded nucleus with chromosomes, and which undergoes mitosis and meiosis.

euploid Having a chromosome number that is an exact multiple of the haploid number characteristic of that species.

evolution Changes in gene frequency in a population over long periods of time that may produce better-adapted organisms and new species.

excision repair A cut-and-patch process for repairing damaged DNA molecules; enzymatic removal of an abnormal piece of DNA, and resynthesis of that piece by using its intact complement as a template.

exon An expressed coding sequence; the part of a split gene whose complementary mRNA sequences remain after the primary transcript is processed to remove the introns.

expanding gene A gene whose DNA content increases (usually because of overreplication of certain triplets), and whose expression may become more severe and show earlier onset from one generation to the next.

F_1 generation The first generation of offspring from a mating between two parental individuals or inbred lines.

false negative An error of testing in which an affected individual is said to be unaffected.

false positive An error of testing in which an unaffected individual is said to be affected.

fluorescence in situ hybridization (FISH) A cytological technique for locating specific gene loci in a metaphase chromosome spread by using fluorescent DNA probes that hybridize to and "light up" the appropriate chromosomal region.

founder effect A change in allele frequencies that occurs when the founders of a new population are not genetically representative of the original population from which they came.

fragile site A heritable constriction or gap that occurs at a specific chromosomal site, and which may cause chromosome breakage.

frameshift mutation A reading frame error caused by the deletion or addition of one or more nucleotide bases (but not multiples of three).

gene A hereditary unit or segment of DNA that occupies a specific site on a chromosome and contains genetic information that is replicated, transcribed into messenger, transfer, or ribosomal RNA, and (if a structural gene) translated to form a polypeptide.

gene flow The introduction of new alleles into a population owing to the migration of individuals.

gene pool All the genes present in a population of organisms.

gene therapy Treatment of a genetic disorder by inserting normal alleles into abnormal cells.

genetic code The 64 triplet nucleotide bases in DNA and mRNA that specify the 20 amino acids found in polypeptides, as well as the chain termination signals.

genetic counseling A process of clinical diagnosis, risk assessment, explanation of options, and provision of social services for families who are affected or potentially affected with genetic disorders.

genetic drift Random and haphazard changes in allele frequencies, occurring especially in small populations.

genetic equilibrium   The maintenance of constant allele frequencies in a population over many generations.

genetic heterogeneity   Multiple genetic causes of the same, or nearly the same, phenotype.

genetic isolate   A community whose members intermarry for generations, thus separating themselves genetically from the general population.

genetic screening   The systematic search of the general population for persons having a particular genotype (or karyotype) that might cause later trouble for them or for their children.

genome   The sum total of all the genes (or DNA sequences) within a single gamete or within a cell's mitochondria.

genotype   The precise allelic makeup of an organism or cell.

germ cell   A gamete (egg or sperm).

germinal mutation   A mutation in a cell that produces gametes (as opposed to a somatic cell mutation).

germ line   Specialized cells (as opposed to somatic cells) that give rise to gametes.

glycoprotein   A protein combined with a carbohydrate.

Guthrie test   A test (devised by R. Guthrie) that screens newborn babies for phenylketonuria, using bacteria that grow only when phenylalanine is present in a dried sample of a baby's blood.

haploid   The chromosome number of a normal eukaryotic gamete, including one representative of each chromosome type.

haplotype   A unique array of closely linked alleles, such as the HLA complex, that is usually inherited as a unit.

Hardy-Weinberg law   A formula that predicts constant gene and genotype frequencies within a large, randomly mating population where there is no selection, mutation, or migration.

hemizygous   Present in a single dose—either in a haploid organism or gamete or in a male, who has just one X chromosome.

heritability, broad-sense   The proportion of phenotypic variation that is due to the genotypic differences among members of a population.

heritability, narrow-sense   The proportion of variation in a population that is due to just the additive portion of the genetic differences among the members.

hermaphrodite   An intersex; an individual who has both male and female sex organs.

heterochromatin   Highly condensed, darkly staining, late replicating regions of chromatin that contain repetitive DNA but few known genes.

heterozygous   Having two different alleles at the same gene locus (or at more than one locus) on homologous chromosomes.

histocompatibility antigen   A genetically encoded tissue or cell surface antigen, which can influence whether a transplanted organ will be accepted or rejected.

histones   Small DNA-binding proteins found in nucleosomes of chromatin, where they regulate chromosome compaction and perhaps gene activity.

HLA system   The major complex of cell surface histocompatibility molecules, the *human leukocyte antigens*, and the cluster of highly polymorphic genes that encode them.

homeobox   A sequence of about 180 nucleotide base pairs found within many developmental genes; it encodes a DNA-binding protein that acts as a gene regulator.

homeotic genes   Developmental genes that regulate pattern formation within body segments; mutants may have extra, misplaced body parts.

homozygous   Having the same alleles at a particular gene locus.

Human Genome Project   An international organization of research teams set up to map and sequence all the DNA in the human genome.

hybridoma   A fused cell (or its clone), derived from an antibody-secreting B cell and a cancer cell, that produces just one kind of antibody and can be indefinitely maintained in tissue culture.

hydrogen bond   A chemical attraction, weaker than a covalent bond, between polar organic molecules that share a hydrogen atom; found in the bases linking two strands of a DNA double helix.

hypervariable region   Within the variable domains of an antibody molecule, amino acid sequences that form the lining of the antigen-binding cavity.

imprinting   The differential expression of genes, depending on whether they are inherited from the male or female parent.

inactive X hypothesis   The Lyon hypothesis of dosage compensation. In female mammalian cells, one or the other X chromosome is randomly inactivated during early development, so that females heterozygous for X-linked genes show mosaic expression for these loci, and gene expression is equalized between the sexes.

inborn error of metabolism   A metabolic disorder caused by the mutation of an enzyme-producing gene and the consequent block of a specific step in a biochemical pathway.

inbreeding   Mating between closely related individuals.

incomplete dominance   The situation in which a heterozygote has a phenotype intermediate between those of the two homozygotes.

incomplete penetrance   Nonexpression of a gene that usually gives rise to a characteristic phenotype.

independent assortment   The random and independent distribution of each pair of chromosomes (or each pair of unlinked alleles) during gamete formation, as stated by Mendel's second law.

interphase   A period of growth and metabolism, including the G_1, S, and G_2 phases of the cell cycle, that occurs between cell divisions.

intron   A noncoding intervening sequence within a gene; a nucleotide sequence that does not code for amino acids, and whose complementary mRNA sequence is excised during mRNA processing.

inversion   A chromosome abnormality formed when two breaks occur in one chromosome, and the intervening segment is inverted before the broken ends rejoin.

isochromosome   A chromosome abnormality that occurs when a centromere splits crosswise rather than lengthwise during mitosis or meiosis; the result is a metacentric chromosome with two identical arms.

karyotype   A photomicrograph of stained metaphase chromosomes, which may be arranged by size and chromosome number, representing an individual or species.

kinetochore A proteinaceous structure, formed during late prophase at the centromere region of a mammalian chromatid, to which spindle fibers attach.

linkage group A group of genes found on the same chromosome.

linked genes Two or more genes located on the same chromosome and showing substantially less than 50% recombination.

locus (plural, loci) The place on a chromosome where a particular gene is found.

lymphocyte A white blood cell, either a B cell or a T cell, that responds to foreign antigens in one of several ways.

lymphokine A glycoprotein, secreted by T cells in response to foreign antigens, that affects various lymphocytes in the immune reaction.

lyonization X chromosome inactivation, as described by M. Lyon's inactive X hypothesis.

lysosome A tiny, membrane-bounded cytoplasmic structure containing enzymes that digest large molecules no longer needed in the cell.

major histocompatibility complex The HLA complex, a closely linked cluster of genes on human chromosome 6 that code for histocompatibility antigens and control activities of immune cells.

map unit A unit of genetic distance on a chromosome, equivalent to 1% recombination between two loci.

maternal effect genes Maternal genes that affect early embryonic development in offspring by producing gradients of gene products in the egg cytoplasm.

meiosis A type of cell (nuclear) division that occurs in sexually reproducing organisms, whereby diploid (2n) cells give rise to haploid (n) gametes; one chromosome replication is followed by two nuclear divisions, called meiosis I and meiosis II.

messenger RNA (mRNA) Ribonucleic acid containing genetic information that is transcribed from DNA and (after processing to remove introns) translated to produce polypeptides.

metabolic block A block in one step of a biochemical pathway, which is caused by an absent or defective enzyme and which may result in an abnormal phenotype.

metabolism The sum of all chemical and physical changes that take place in a cell or organism, allowing it to function and grow.

metacentric A chromosome with a centromere at or near its midpoint and two arms of about equal length.

metaphase The stage of cell (nuclear) division during which the chromosomes are most tightly condensed and lined up on the equatorial plate of the spindle.

migration The movement of individuals from one place to another, possibly resulting in altered allele frequencies between populations.

mitochondrion A self-reproducing, maternally inherited cytoplasmic structure in which the process of cell respiration yields energy in the form of ATP molecules. The DNA found in mitochondria evolved from DNA of aerobic bacteria and contains a slightly different genetic code.

mitosis Cell (nuclear) division that gives rise to two daughter cells with identical chromosomes and genotypes.

modifying gene A gene that alters the degree of expression of another gene at a different locus.

monoclonal antibody A pure and uniform antibody type produced by a clone of hybridoma cells.

monosomic Having just one representative of a particular chromosome, as occurs normally in a gamete or abnormally in an otherwise diploid cell or individual.

monozygotic (MZ) twins Two genetically identical sibs who are derived from one zygote that splits into two during very early development.

mosaic An individual who has two or more genetically different cell lines.

mRNA processing The changes occurring between transcription and translation, that is, conversion of an original transcript to mature mRNA through the removal of introns and other modifications.

multiple alleles The existence of more than two alternative forms of a gene in a population.

mutagen A substance that produces mutations above the rate at which they occur spontaneously.

nested gene A gene that contains another gene within its coding sequence.

nondisjunction The failure of two homologous chromosomes or sister chromatids to separate (disjoin) during anaphase of meiosis or mitosis, giving rise to daughter cells with either a missing chromosome or an extra chromosome.

nonhistone proteins A heterogeneous mix of chromosomal proteins that are not histones.

nucleolus organizer region The region of a chromosomes containing genes that produce ribosomal RNA and thus give rise to a nucleolus.

nucleolus A dark-staining nuclear structure, formed by a nucleolus organizer region of a chromosome and full of ribosomal RNA.

nucleosome A beadlike complex, consisting of a histone core wrapped by DNA and capped by H1 histone, that forms a repeating structural unit of chromatin.

nucleotide A deoxyribose or ribose sugar attached to a phosphate and to a purine or pyrimidine base, and which forms the basic building block of nucleic acids.

nucleus The membrane-bounded structure in eukaryotic cells that contains chromosomes and is surrounded by cytoplasm.

nullisomic A cell or organism that has no representative of a particular chromosome.

oligonucleotide primer A short sequence of bases, complementary to a particular DNA sequence, that is used as a primer in the polymerase chain reaction.

oncogene A gene that can convert normal cells into cancer cells by causing uncontrolled cell division.

oogonium A diploid ovarian cell that divides to form more oogonia or enlarges to form a primary oocyte.

open reading frame (ORF) A coding sequence of DNA nucleotide triplets that does not contain any stop codons.

pattern formation genes Homeotic genes that act during early development by specifying the structures unique to each embryonic region.

pedigree A diagram of two or more generations of a family's lineage, often showing the expression of genetic traits among its members.

pharmacogenetics The study of genetic variation leading to differences in the ways that individuals metabolize drugs, foods, and other substances.

phenocopy An environmentally induced mimic of a genetic condition in an individual who lacks the usual causative gene.

phenotype The observed attribute(s) of a cell or individual, brought about by the interaction of genotype and environment.

Philadelphia chromosome A shortened chromosome 22 resulting from reciprocal translocation between human chromosomes 9 and 22. It is found in bone marrow cells of most people with chronic myelogenous leukemia.

plasma The clear, straw-colored fluid in which blood cells are suspended; it contains specialized blood proteins, salts, nutrients, hormones, intermediate metabolites, and many other substances.

plasma cell An antibody-secreting cell derived from a B cell in response to an antigenic stimulation.

plasmid A tiny, circular molecule of nonchromosomal DNA found in bacteria and used as a vector for transferring genes from one cell to another.

pleiotropy The situation in which a single gene has multiple and seemingly unrelated phenotypic effects, owing to a cascade of reactions stemming from the original gene product.

polar body A tiny, cytoplasm-poor cell that is formed by the meiotic division of a primary or secondary oocyte and which usually degenerates.

polygenic inheritance Quantitative inheritance; the determination of a particular trait by several genes, each with a small effect.

polymerase chain reaction (PCR) A technique for quickly amplifying a particular DNA sequence, starting with just a tiny amount of double-stranded DNA.

polymorphism The presence in a population of two or more relatively common alleles of a particular gene (or forms of a chromosome)—the more alleles, the greater the polymorphism.

polypeptide An unbranched string of amino acids linked together by peptide bonds that are formed during the process of gene translation.

polyploid Having three or more complete sets of chromosomes.

primary oocyte A diploid ovarian cell, which is formed from the enlargement of an oogonium and which undergoes meiosis I.

primary spermatocyte A diploid testicular cell, which is formed from the enlargement of a spermatogonium and which undergoes meiosis I.

proband The affected individual through whom a pedigree is discovered; also called the **propositus**.

probe A substance (such as a radioactively labeled single-stranded DNA or RNA nucleotide sequence or a monoclonal antibody) that is used to identify or isolate a gene or a gene product.

programmed cell death The nonrandom elimination of certain cells at certain stages of development, as controlled by specific cell death genes.

prokaryote A cell or organism (such as bacteria or blue-green algae) that lacks a membrane-bounded nucleus and complex chromosomes and does not undergo mitosis or meiosis.

prometaphase The stage of nuclear division, occurring just before metaphase, when the nuclear envelope breaks down, kinetochores form at the centromere regions, and chromosomes become attached to spindle fibers.

promoter A specific sequence of DNA nucleotide bases to which RNA polymerase binds to initiate gene transcription.

prophase The first stage of mitosis or meiosis, during which chromosomes condense, the nucleoli disappear, and the spindle forms.

protein A gene product; a molecule consisting of one or more polypeptide chains, each one an unbranched string of amino acids.

proto-oncogene A cellular gene that normally regulates cell division, but which can mutate to a dominant oncogene that causes cancer.

pseudoautosomal region A tiny region of homology at the tip of the X and Y chromosome short arms, which normally pairs and crosses over during meiosis in males; genes within this region, being present in duplicate in males, are inherited as though they were autosomal rather than sex-linked.

pseudogene A length of DNA that is similar in nucleotide sequence to a normal gene (at a different locus) but is not expressed, having one or more mutations that render it nonfunctional.

purine (A, G) The larger of two types of nucleotide bases found in DNA and RNA; includes adenine (A) and guanine (G).

pyrimidine (C, T, U) The smaller of two types of nucleotide bases found in DNA and RNA; includes cytosine (C), thymine (T); and (in RNA only) uracil (U).

quantitative variation The situation in which a trait (such as height, weight, or behavioral activity) varies continuously along a range of values, rather than being observable as discrete, nonoverlapping phenotypic groups; involves the interaction of the environment with several or many genes.

race A genetically or geographically distinct subgroup of a species.

random mating Mating, within a population, that occurs without regard to an individual's phenotype or genotype.

reading frame Successive groups of three in a sequence of nucleotide bases. Depending on the starting position, there are three possible reading frames for a particular sequence.

recessive allele An allele that is expressed when homozygous but not expressed when heterozygous.

reciprocal matings Matings in which the sexes representing the two differing parental genotypes or phenotypes are reversed.

recombinant DNA A hybrid DNA molecule formed by the fusion of pieces of DNA derived from different species.

recombination The reassortment of parental genes to form new combinations of alleles in the offspring of a multiple heterozygote, a result of independent assortment or crossing over.

repetitive DNA   Highly repeated or moderately repeated multiples of DNA nucleotide sequences, making up a major fraction of eukaryotic DNA and found in heterochromatin, ribosomal RNA genes, and transfer RNA genes.

restriction enzyme   A bacterial enzyme that can cut across a DNA molecule. Each enzyme recognizes a unique target sequence of bases, often producing staggered cut ends that can join to complementary sequences of any foreign DNA — thus being an important tool of molecular biology.

restriction fragment length polymorphism (RFLP)   Variation in the lengths of DNA fragments cut by a specific restriction enzyme, due to mutations in the target sequences.

retrovirus   An RNA-containing virus that can enter a cell, use reverse transcriptase to reproduce a DNA copy of itself, and integrate into the host's genome. If the retrovirus carries an oncogene, the host cell can be changed into a cancer cell.

reverse transcriptase   A retrovirus polymerase enzyme for "backward" transcription, that is, making DNA from an RNA template.

ribosomal RNA (rRNA)   A type of RNA that becomes part of the ribosomes in the cytoplasm.

ribosome   A cytoplasmic structure composed of proteins and ribosomal RNA, on which protein synthesis (messenger RNA translation) occurs.

RNA   Ribonucleic acid, a single-stranded molecule transcribed from DNA, containing the sugar ribose and the bases adenine, uracil, guanine and cytosine. The three types of RNA, all involved in protein synthesis, are messenger RNA, ribosomal RNA, and transfer RNA.

RNA polymerase   An enzyme that forms RNA by linking together a string of ribonucleotides in a sequence complementary to that of a single-stranded DNA template.

RNA splicing   The processing of messenger RNA by the removal of all intron sequences from the primary transcript and the joining together of all the exons.

secondary oocyte   A diploid female cell that undergoes meiosis II to form an egg and a second polar body.

secondary spermatocyte   A diploid male cell that undergoes meiosis II to form two spermatids.

segmentation genes   Genes that act in very early development to divide an embryo into a series of primitive segments, which are later acted on by the homeotic genes.

segregation   The separation of the two alleles at any given locus into different daughter cells during meiosis.

selection   The differential reproduction of the various genotypes in a population over a period of time.

sense strand   The strand of a DNA double helix that acts as the template for transcription of a particular gene into RNA.

sex chromatin   A Barr body; a dark-staining body that represents a condensed, inactivated X chromosome in somatic cell nuclei of female mammals.

sex chromosomes   The pair of dissimilar chromosomes (X and Y) found in organisms with separate sexes; chromosomes that are not autosomes.

sex-influenced inheritance   The differential expression of autosomally inherited traits (such as baldness) in the two sexes.

sex-limited trait   A trait (such as milk production) affecting structures or processes that exist in only one sex.

sex ratio   The relative proportion of the two sexes (of a given age range) in a population, often expressed as the percentage of males.

somatic cell   Any cell in a eukaryote that is not in the germ cell lineage.

somatic cell fusion   The fusion of cells from two different species or tissues to form a hybrid cell, used for gene mapping and other genetic studies.

somatic mutation   A mutation occurring in a somatic cell and thus not transmitted to offspring.

Southern blotting   A technique of molecular biology, invented by E. M. Southern, in which pieces of single-stranded, electrophoretically separated DNA are transferred from a gel to a nitrocellulose filter and then treated with a radioactive probe to identify the segment of interest.

species   A group of organisms whose members are capable of interbreeding with one another but incapable of producing viable and fertile offspring with members of other species.

sperm   A haploid male gamete.

spermatid   One of four haploid products of male meiosis, which undergo cytoplasmic changes to become sperm.

spermatogonium   A diploid cell in testes that divides mitotically to form more spermatogonia or that enlarges to form a primary spermatocyte.

spindle   A structure made up of spindle fibers, formed during prophase of mitosis or meiosis, on which chromosomes congregate during metaphase and move to opposite poles during anaphase.

structural gene   A unique sequence gene, which is transcribed and translated to form a polypeptide product.

submetacentric   A chromosome whose centromere is located off-center, creating chromosome arms of somewhat unequal length.

synapsis   The pairing of homologous chromosomes during prophase I of meiosis.

synaptonemal complex   A proteinaceous structure that forms during prophase I of meiosis and that binds homologous chromosomes together during synapsis.

T cell   A white blood cell that is produced in the bone marrow, matures in the thymus gland, and is involved in the immune response.

telomere   The tip of a chromosome, containing repeated DNA sequences.

telophase   A stage of mitosis or meiosis during which the spindle disappears, daughter chromosomes decondense, the nuclear envelope re-forms, and nucleoli appear; the cytoplasm splits too.

template   A mold or pattern; a sequence of nucleotides from which a complementary DNA or RNA strand is made.

termination signal   A DNA sequence that causes transcription to stop.

tetrad The four-chromatid structure that makes up a pair of synapsed homologous chromosomes during prophase I and metaphase I of meiosis.

tetraploid A cell or organism with four haploid sets of chromosomes.

threshold trait A polygenic trait (such as cleft lip) that is either present or absent, rather than varying continuously.

thymine dimer A type of ultraviolet-induced mutation in which two adjacent thymine nucleotides on one DNA strand join to form a double base that distorts the DNA molecule and interrupts both DNA replication and transcription.

transcription The formation of a strand of RNA from a DNA template by complementary base pairing, catalyzed by RNA polymerase.

transfer RNA (tRNA) A type of RNA that acts as an adapter during protein synthesis by binding to specific amino acids and fitting them to appropriate codons of messenger RNA.

transformation The conversion by an oncogene of a normal eukaryotic cell in culture to a cancer cell.

transgenic organism An organism that has a foreign gene transferred and stably incorporated into its germ line.

translation The process of protein formation, occurring in a ribosome, whereby the information contained in a sequence of nucleotide bases in messenger RNA directs the synthesis of a sequence of amino acids in a polypeptide.

translocation A chromosome aberration formed by the exchange of parts between nonhomologous chromosomes.

transposon A transposable element or jumping gene; a piece of DNA that can move from one place to another within the genome.

triplet repeat The repetition of three bases in tandem, involved in several genetic diseases (e.g., CAG in Huntington disease).

triploid A cell or organism with three haploid sets of chromosomes.

trisomic A diploid cell or organism having one too many chromosomes, that is, three (rather than two) representatives of one chromosome.

tumor suppressor gene A gene whose normal, dominant allele checks cell division, but whose mutant allele or deletion releases the cell from growth controls.

unequal crossing over A crossover that occurs between misaligned copies of duplicated loci or repetitive DNA sequences, resulting in recombinant chromatids of unequal size.

uniparental disomy Homozygosity for a particular chromosome or chromosome segment, due to the inheritance of two chromosomes (rather than one) from one parent.

variable expressivity Differences in the observed effects of a given allele or genotype in different individuals.

variable number of tandem repeats (VNTR) Short DNA sequences that are tandemly repeated variable numbers of times at many defined regions throughout the genome. This variation forms the basis of DNA fingerprinting procedures.

vector A plasmid, retrovirus, or yeast artificial chromosome that is used to transfer a DNA segment from one cell or species to another.

wobble The looseness of complementarity in the third position of codon-anticodon pairing during translation.

X-linked gene A gene located on the X chromosome.

yeast artificial chromosome (YAC) A synthetic chromosome that can accept a large piece of foreign DNA, thus acting as a vector for making gene libraries and for transferring genes from one species to another.

zygote A fertilized egg; a diploid cell formed by the union of haploid male and female gametes.

Bibliography

To save space, we have (1) cited papers with six or more authors as the first author et al., and (2) used the following journal abbreviations:

AJHG for *The American Journal of Human Genetics*
CSHS for *Cold Spring Harbor Symposia on Quantitative Biology*
NEJM for *New England Journal of Medicine*
PNAS for *Proceedings of the National Academy of Sciences*
SA for *Scientific American*
TIG for *Trends in Genetics*

ABELSON, P. H. 1990. Uncertainties about health effects of radon. *Science* 250: 353.

ADA, G. L. and G. NOSSAL. 1987. The clonal-selection theory. *SA* 257(August): 62–69.

ALBERTS, B., D. BRAY, J. LEWIS, M. RAFF, K. ROBERTS and J. D. WATSON. 1989. *Molecular Biology of the Cell*, 2nd ed. Garland, New York.

ALLEN, G. E. 1983. The misuse of biological hierarchies: The American eugenics movement, 1900–1940. *History and Philosophy of the Life Sciences* 5: 105–128.

AMES, B. N. 1983. Dietary carcinogens and anticarcinogens. *Science* 221: 1256–1264. (See also the editorial by P. H. Abelson.)

AMES, B. N. 1984. Cancer and diet. *Science* 224: 668ff. A reply to five responses to a prior Ames article.

ANASTASI, A. 1988. *Psychological Testing*, 6th ed. Macmillan, New York.

ANDERSON, W. F. 1990. Genetics and human malleability. *Hastings Center Report* 20(Jan./Feb.): 21–24.

ANDERSON, W. F. 1992. Human gene therapy. *Science* 256: 808–813.

ANDERSON, W. F. and E. G. DIACUMA-KOS. 1981. Genetic engineering in mammalian cells. *SA* 245(July): 106–121.

ANDREWS, L. B. 1991. Chapter 23. Legal Issues in Fertility Management. In *Infertility in the Male*, 2nd ed., edited by L. I. Lipshultz and S. S. Howards, pp. 448–456. Mosby, St. Louis.

ANGIER, N. 1987. Light cast on a darkling gene. *Discover* 8(March): 85–96.

ANGIER, N. 1990. Nature may fashion all cells' proteins from a few primordial parts. *New York Times*, Dec. 11, 1990, p. C1.

ANGIER, N. 1991. New debate over humankind's ancestress. *New York Times*, Oct. 1, 1991, p. C1.

ANNAS, G. J. 1990. DNA fingerprinting in the twilight zone. *Hastings Center Report* 20(March–April): 35–37.

ANONYMOUS. 1833. Westminster Medical Society, 6, 13, and 20 April 1833. *The Lancet* II: 146–147.

ANONYMOUS. 1991. Chorion villus sampling: Valuable addition or dangerous alternative? *The Lancet* 337: 1513–1515.

ANONYMOUS. 1992. Screening for cystic fibrosis. *The Lancet* 340: 209–210.

APPLEBAUM, E. G. and S. K. FIRESTEIN. 1983. *A Genetic Counseling Casebook*. The Free Press, New York.

ARNHEIM, N., T. WHITE and W. E. RAINEY. 1990. Application of PCR: Organismal and population biology. *BioScience* 40: 174–182.

ASHG AD HOC COMMITTEE. 1992. Statement of the American Society of Human Genetics on cystic fibrosis carrier screening. *AJHG* 51: 1443–1444.

ATKINSON, M. A. and N. K. MACLAREN. 1990. What causes diabetes? *SA* 263(July): 62–71.

AUSTIN, C. R. 1989. *Human Embryos: The Debate on Assisted Reproduction*. Oxford University Press, Oxford.

AUSTIN, C. R. and R. G. EDWARDS, Editors. 1981. *Mechanisms of Sex Differentiation in Animals and Man*. Academic Press, New York.

AVERS, C. J. 1989. *Process and Pattern in Evolution*. Oxford University Press, New York.

AYALA, F. J. 1982. *Population and Evolutionary Genetics: A Primer*. Benjamin/Cummings, Menlo Park, CA.

BACA, M. and L. ZAMBONI. 1967. The fine structure of human follicular oocytes. *Journal of Ultrastructural Research* 19: 354–381.

BACETTI, B. 1984. The human spermatozoon. In *Ultrastructure of Reproduction*, edited by J. Van Blerkom and P. M. Motta, pp. 110–126. Martinus Nijhoff, Boston.

BAIRD, P. A. and B. McGILLIVRAY. 1982. Children of incest. *The Journal of Pediatrics* 101: 854–857.

BAJEMA, C. J., Editor. 1976. *Eugenics, Then and Now. Benchmark Papers in Genetics*, vol. 5. Dowden, Hutchinson & Ross, Stroudsburg, PA.

BASERGA, R. 1981. The cell cycle. *NEJM* 304: 453–459.

BASKIN, Y. 1990. DNA unlimited. *Discover* 11(July): 77–79.

BAUER, H. H. 1992. *Scientific Literacy and the Myth of the Scientific Method*. University of Illinois Press, Urbana.

BEADLE, G. W. 1980. The ancestry of corn. *SA* 242(Jan.): 112–119.

BEADLE, G. W. and E. L. TATUM. 1941. Genetic control of biochemical reactions in Neurospora. *PNAS* 27: 499–506.

BEAUDET, A. L., et al. 1989. Genetics and biochemistry of variant human phenotypes. In *The Metabolic Basis of Inherited Disease*, 6th ed., edited by C. R. Scriver et al., pp. 3–53. McGraw-Hill, New York.

BEIR V (Committee on the Biological Effects of Ionizing Radiation). 1990. *Health Effects of Exposure to Low Levels of Ionizing Radiation*. National Academy Press, Washington, DC.

BELL, J. and J. B. S. HALDANE. 1937. The linkage between the genes for colour-blindness and haemophilia in man. *Royal Society* of London, Proceedings, Series B—Biological Sciences 123: 119–150.

BERRA, T. M. 1990. Evolution and the Myth of Creationism: A Basic Guide to the Facts in the Evolution Debate. Stanford University Press, Stanford, CA.

BERRIDGE, M. J. 1985. The molecular basis of communication within a cell. SA 253(Oct.): 142–152.

BIALY, H. 1991. Transgenic pharming comes of age. Bio/Technology 9: 786–787.

BICKEL, H. 1987. Early diagnosis and treatment of inborn errors of metabolism. Enzyme 38: 14–26.

BISHOP, J. A. and L. M. COOK. 1975. Moths, melanism and clean air. SA 232(Jan): 90–99.

BISHOP, J. E. and M. WALDHOLZ. 1990. Genome. Simon & Schuster, New York.

BODMER, W. F. 1986. Human genetics: The molecular challenge. CSHS 51: 1–13.

BODMER, W. F. and L. L. CAVALLI-SFORZA. 1970. Intelligence and race. SA 223(Oct.): 19–29.

BODMER, W. F. and L. L. CAVALLI-SFORZA. 1976. Genetics, Evolution, and Man. Freeman, San Francisco.

BOHMAN, M. 1978. Some genetic aspects of alcoholism and criminality: A population of adoptees. Archives of General Psychiatry 35: 269–276.

BOIS, E., et al. 1978. Cluster of cystic fibrosis cases in a limited area of Brittany (France). Clinical Genetics 14: 73–76.

BOND, D. J. and A. C. CHANDLEY. 1983. Aneuploidy. Oxford Monographs on Medical Genetics No. 11. Oxford University Press, Oxford.

BONGIOVANNI, A.M., W. R. EBERLEIN, A. S. GOLDMAN and M. NEW. 1967. Disorders of adrenal steroid biogenesis. Recent Progress in Hormone Research 23: 375–449.

BONNICKSEN, A. 1992. Genetic diagnosis of human embryos. Hastings Center Report 22(July–August): S5–S11. (See also other articles in this Special Supplement called "Genetic Grammar.")

BOUCHARD, T. J., JR., D. T. LYKKEN, M. McGUE, N. L. SEGAL and A. TELLEGEN. 1990. Sources of human psychological differences: The Minnesota study of twins reared apart. Science 250: 223–228.

BOUCHARD, T. J., JR. and M. McGUE. 1981. Familial studies of intelligence: A review. Science 212: 1055–1059.

BOUÉ, A., J. BOUÉ and A. GROPP. 1985. Cytogenetics of pregnancy wastage. Advances in Human Genetics 14: 1–57.

BOWEN, P., et al. 1965. Hereditary male pseudohermaphroditism with hypogonadism, hypospadias, and gynecomastia (Reifenstein's syndrome). Annals of Internal Medicine 62: 252–270.

BOWER, B. 1991. Same family, different lives. Science News 140: 376–378.

BOYER, S. H., IV, Editor. 1963. Papers on Human Genetics. Prentice-Hall, Englewood Cliffs, NJ.

BROOK, J. D., et al. 1992. Molecular basis of myotonic dystrophy: Expansion of a trinucleotide (CTG) repeat at the 3′ end of a transcript encoding a protein kinase family member. Cell 68: 799–808.

BROWN, M. S. and J. L. GOLDSTEIN. 1984. How LDL receptors influence cholesterol and atherosclerosis. SA 251(Nov.): 58–66.

BROWN, M. S. and J. L. GOLDSTEIN. 1986. A receptor-mediated pathway for cholesterol homeostasis: Nobel lecture, 9 Dec. 1985. Les Prix Nobel 1985. Almqvist & Wiksell International, Stockholm. Reprinted in Science 232: 34–47.

BUCHANAN, A. and D. H. BROCK. 1990. Deciding for Others: The Ethics of Surrogate Decisionmaking. Cambridge University Press, Cambridge.

BUCK v. BELL. 1927. United States Reports: Cases Adjudged in the Supreme Court 274: 200–208.

BUISSERET, P. D. 1982. Allergy. SA 247(August): 86–95.

BULMER, M. G. 1970. The Biology of Twinning in Man. Oxford University Press, London.

BUNN, H. F. and B. G FORGET. 1986. Hemoglobin: Molecular, Genetic and Clinical Aspects. Saunders, Philadelphia.

BURNET, F. M., Editor. 1976. Immunology: Readings from Scientific American. Freeman, San Francisco.

CAIRNS, J. 1985. The treatment of diseases and the war against cancer. SA 253(Nov.): 51–59.

CALDWELL, M. 1991. The immune challenge. Discover 12(Dec.): 56–61.

CALDWELL, M. 1992a. Vigil for a doomed virus. Discover 13(March): 50–56.

CALDWELL, M. 1992b. The transplanted self. Discover 13(April): 62–68.

CAROL ANN with K. DEMARET 1984. David's story. People Weekly 122(Oct. 29): 120–141 and 122(Nov. 5): 107–127.

CAROTHERS, E. E. 1913. The Mendelian ratio in relation to certain Orthopteran chromosomes. Journal of Morphology 24: 487–511.

CARSWELL, H. 1982. "Elephant Man" had more than neurofibromatosis. JAMA 248: 1032–1033.

CASKEY, C. T. 1986. Summary: A milestone in human genetics. CSHS 51: 1115–1119.

CASKEY, C. T., R. G. WORTON and J. D. WATSON. 1991. ASHG Human Genome Committee Report. The Human Genome Project: Implications for human genetics (an exchange of letters). AJHG 49: 687–691.

CAVALLI-SFORZA, L. L. and W. F. BODMER. 1971. The Genetics of Human Populations. Freeman, San Francisco.

CHAKRABORTY, R., M. I. KAMBOH, M. NWANKWO and R. E. FERRELL. 1992. Caucasian genes in American Blacks: New data. AJHG 50: 145–155.

CHAMBON, P. 1981. Split genes. SA 244(May): 60–71.

CHANDLEY, A. C. 1988. Meiosis in man. TIG 4: 79–84.

CHANDLEY, A. C. 1989. Awaiting a fairy godmother. TIG 5: 128.

CHANDLEY, A. C. 1990. Infertility and recurrent abortion. In Principles and Practice of Medical Genetics, 2nd ed., edited by A. E. H. Emery and D. L. Rimoin, pp. 313–319. Churchill Livingstone, Edinburgh.

CHANDRASEKHAR, S. 1990. Science and scientific attitudes. Nature 344: 285–286.

CHARROW, J., H. L. NADLER and M. I. EVANS. 1990. Prenatal diagnosis and therapy. In Principles and Practice of Medical Genetics, 2nd ed., edited by A. E. H. Emery and D. L. Rimoin, pp. 1959–1993. Churchill Livingstone, Edinburgh.

CHEHAB, F. F. and Y. W. KAN. 1990. Detection of sickle cell anaemia mutation by colour DNA amplification. The Lancet 335: 15–17.

CHEN, E. 1979. Twins reared apart: A living lab. New York Times Magazine, Dec. 9, 1979, p. 112ff.

CHERFAS, J. 1991. Ancient DNA: Still busy after death. Science 253: 1354–1356.

CHILDS, B., J. M. FINUCCI, M. S. PRESTON and A. E. PULVER. 1976. Human behavior genetics. Advances in Human Genetics 7: 57–97.

CLARKE, C. A. 1968. The prevention of "Rhesus" babies. SA 219(Nov.): 46–52. (Reprinted in Burnet 1976.)

CLONINGER, C. R. 1991. D₂ dopamine receptor gene is associated but not linked with alcoholism. JAMA 266: 1833–1834.

CO, M. S. and C. QUEEN. 1991. Humanized antibodies for therapy. Nature 351: 501–502.

COHEN, L. A. 1987. Diet and cancer. SA 257(Nov.): 42–48.

COLD SPRING HARBOR SYMPOSIA. 1986. Molecular Biology of Homo sapiens. Vol. 51 of CSHS. Cold Spring Harbor Laboratory, New York.

COLLIER, R. J. and D. A. KAPLAN. 1984. Immunotoxins. SA 251(July): 56–64.

COMMITTEE ON SCIENCE AND CREATIONISM. 1984. Science and Creationism: A View from the National Academy of Science. National Academy Press, Washington, DC.

CONLEY, M. E., et al. 1986. Expression of the gene defect in X-linked agammaglobulinemia. *NEJM* 315: 564–567.

CONNEALLY, P. M. and M. L. RIVAS. 1980. Linkage analysis in man. *Advances in Human Genetics* 10: 209–266.

CONNOR, J. M. and M. A. FERGUSON-SMITH. 1987. *Essential Medical Genetics*, 2nd ed. Blackwell Scientific Publications, Oxford.

COOPER, D. N. and J. SCHMIDTKE. 1989. Analysis and diagnosis of human inherited disease by recombinant DNA methods. In *The Metabolic Basis of Inherited Disease*, edited by C. R. Scriver et al., pp. 55–72. McGraw-Hill, New York.

CORWIN, H. O. and J. B. JENKINS, Editors. 1976. *Conceptual Foundations of Genetics: Selected Readings*. Houghton Mifflin, Boston.

CRICK, F. 1966. The genetic code: III. *SA* 215(Oct.): 55–62. Reprinted in Srb, Owen and Edgar (1970).

CRICK, F. 1988. *What Mad Pursuit*. Basic Books, New York.

CROCE, C. M. and G. KLEIN. 1985. Chromosome translocations and human cancer. *SA* 252(March): 54–60.

CROW, J. F. 1982. Sewall Wright, the scientist and the man. *Perspectives in Biology and Medicine* 25: 279–294.

CROW, J. F. 1983. *Genetics Notes, An Introduction to Genetics*, 8th ed. Burgess, Minneapolis.

CROW, J. F. 1987. Population genetics history: A personal view. *Annual Review of Genetics* 21: 1–22.

CROW, J. F. 1988. A diamond anniversary: The first chromosome map. *Genetics* 118: 103.

CROW, J. F. 1988. Eighty years ago: The beginnings of population genetics. *Genetics* 119: 473–476.

CROW, J. F. and A. P. MANGE. 1965. Measurement of inbreeding from the frequency of marriages between persons of the same surname. *Eugenics Quarterly* (now *Social Biology*) 12: 199–203. (Reprinted 1982, 29: 101–105.)

CUNNINGHAM, C. 1988. *Down's Syndrome: An Introduction for Parents*, rev. ed. Brookline Books, Cambridge, MA.

DARNELL, J. E., JR. 1983. The processing of RNA. *SA* 249(Oct.): 90–100.

DARNELL, J. E., JR. 1985. RNA. *SA* 253(Oct.): 68–78.

DARNELL, J., H. LODISH and D. BALTIMORE. 1990. *Molecular Cell Biology*, 2nd ed. Scientific American Books, New York.

DARWIN, C. 1839. *Journal of Researches into the Geology and Natural History of the Various Countries Visited by H. M. S. Beagle*. Henry Colburn, London. (Often reprinted: e.g., in vol. 2 of *The Works of Charles Darwin*, edited by P. H. Barrett and R. B. Freeman, New York University Press, New York, 1987.)

DARWIN, C. 1859. *On the Origin of Species by Means of Natural Selection*. John Murray, London. A facsimile of the first edition has been reprinted by Harvard University Press (1964). The sixth edition (1872) is also often cited.

DARWIN, C. 1875. *The Variation of Plants and Animals Under Domestication*, 2nd ed., p. 319. John Murray, London.

DARWIN, C. 1969. *The Autobiography of Charles Darwin, 1809–1882*. (With original omissions restored; edited with appendix and notes by his granddaughter, N. Barlow.) Norton, New York.

DARWIN, C. and A. RUSSEL WALLACE. 1858. On the tendency of species to form varieties; and on the perpetuation of varieties and species by natural means of selection. *Journal of the Proceedings of the Linnean Society (Zoology)* 3: 45–62. (Reprinted in Darwin and Wallace 1958.)

DARWIN, C. and A. RUSSEL WALLACE. 1958. *Evolution by Natural Selection: A Centenary Commemorative Volume* (with a foreword by Sir Gavin de Beer). Cambridge University Press, London.

DARWIN, F., Editor. 1887. *The Life and Letters of Charles Darwin*. John Murray, London.

DAVENPORT, C. B. 1913. *Heredity of Skin Color in Negro–White Crosses*. Carnegie Institute of Washington, Washington, DC.

DAVIES, K. E., Editor. 1989. *The Fragile X Syndrome*. Oxford University Press, Oxford.

DAVIS, G. T. and A. J. BRUWER. 1991. Two book reviews: (1) *Health Effects of Exposure to Low Levels of Ionizing Radiation: BEIR V* by the Committee on the Biological Effects of Ionizing Radiation, National Academy Press, Washington, DC, 1990; (2) *Radiation-Induced Cancer from Low-Dose Exposure: An Independent Analysis* by John W. Gofman, Committee for Nuclear Responsibility Book Division, San Francisco, 1990. *NEJM* 324: 497–499.

DAWKINS, R. 1986. *The Blind Watchmaker*. Norton, New York.

DE BOULLE, K., et al. 1993. A point mutation in the *FMR-1* gene associated with fragile X mental retardation. *Nature Genetics* 3: 31–35.

DE DUVE, C. 1984. *A Guided Tour of the Living Cell*, vol. 2. Scientific American Books, New York.

DE GROUCHY, J. and C. TURLEAU. 1984. *Clinical Atlas of Human Chromosomes*, 2nd ed. Wiley, New York.

DE GROUCHY, J. and C. TURLEAU. 1990. Autosomal disorders. In *Principles and Practice of Medical Genetics*, 2nd ed., edited by A. E. H. Emery and D. L. Rimoin, pp. 247–271. Churchill Livingstone, Edinburgh.

DE LA CHAPELLE, A. 1986. The use and misuse of sex chromatin screening for gender identification of female athletes. *JAMA* 256: 1920–1923.

DE LA CHAPELLE, A. 1990. Sex chromosome abnormalities. In *Principles and Practice of Medical Genetics*, 2nd ed., edited by A. E. H. Emery and D. L. Rimoin, pp. 273–299. Churchill Livingstone, Edinburgh.

DEAN, G. 1957. Pursuit of a disease. *SA* 196(3): 133–142.

DEFORD, F. 1986. *Alex: The Life of a Child*. Signet, New American Library, New York.

DELLARCO, V. L., P. E. VOYTEK and A. HOLLAENDER, Editors. 1985. *Aneuploidy: Etiology and Mechanisms*. Plenum, New York.

DENIS, J. 1667. An extract of a letter . . . touching a late cure of an inveterate phrensy by the transfusion of blood. *The Royal Society of London, Philosophical Transactions* 2: 617–623.

DESMOND, E. W. 1987. Out in the open: Changing attitudes and new research give fresh hope to alcoholics. *Time* 130(Nov. 30): 80–90.

DESNICK, R. J., Editor. 1991. *Treatment of Genetic Diseases*. Churchill Livingstone, New York.

DEWS, P. B., Editor. 1984. *Caffeine: Perspectives from Recent Research*. Springer-Verlag, Berlin.

DIAMOND, J. 1984. Making a chimp out of man. *Discover* 5(Dec.): 55–60.

DICKERSON, R. E. 1983. The DNA helix and how it is read. *SA* 249(Dec.): 94–111.

DICKMANN, Z., T. H. CHEWE, W. A. BONNEY, JR. and R. W. NOYES. 1965. The human egg in the pronuclear stage. *The Anatomical Record* 152: 293–302.

DIXON, B. 1984. Of different bloods: 20 discoveries that shaped our lives. *Science 84* 5(Nov.): 65–67.

DOBZHANSKY, T. 1937. *Genetics and the Origin of Species*. Columbia University Press, New York.

DOLL, R. and R. PETO. 1987. Epidemiology of cancer. In *Oxford Textbook of Medicine*, 2nd ed., vol. 1, edited by D. J. Weatherall, J. G. G. Ledingham and D. A. Warrell, pp. 4.95–4.123. Oxford Medical Publishers, Oxford.

DONAHUE, R. P., W. B. BIAS, J. H. RENWICK and V. A. McKUSICK. 1968. Probable assignment of the Duffy blood group locus to chromosome 1 in man. *PNAS* 61: 949–955.

DOOLITTLE, R. F. 1985. Proteins. *SA* 253(Oct.): 88–99.

DORAN, G. H., D. N. DICKEL, W. E. BALLINGER, JR., O. F. AGEE, P. J. LAIPIS and W. W. HAUSWIRTH. 1986. Anatomical, cellular and molecular analysis of 8000-

year-old human brain tissue from the Windover archaeological site. *Nature* 323: 803–806.

DOWN, J. L. H. 1866. Observations on an ethnic classification of idiots. *London Hospital Clinical Records and Reports* no. 3: 259–262.

DRLICA, K. 1992. *Understanding DNA and Gene Cloning: A Guide for the Curious*, 2nd ed. Wiley, New York.

DUCHENNE, G. B. A. 1868. Recherches sur la paralysie musculaire pseudo-hypertrophique ou paralysie myosclerosique. *Archives Generales de Medicine* 11: 5–25.

DULBECCO, R. 1976. Francis Peyton Rous. *National Academy of Sciences, Biographical Memoirs* 48: 274–306.

DUNN, F. G., et al. 1976. Pheochromocytoma crisis induced by saralasin: Relation of angiotensin analogue to catecholamine release. *NEJM* 295: 605–607.

DUNN, L. C. 1962. Cross currents in the history of human genetics. *AJHG* 14:1–13.

DUPRAW, E. J. 1970. *DNA and Chromosomes*. Holt, Rinehart & Winston, New York.

DVORAK, M, J. TESARIK, L. PILKA and P. TRAVNIK. 1982. Fine structure of human two-cell ova fertilized and cleaved in vitro. *Fertility and Sterility* 37: 661–667.

EDWARDS v. AGUILLARD. 1987. *United States Reports: Cases Adjudged in the Supreme Court* 482: 578–640.

EDWARDS, R. G. 1989. *Life Before Birth: Reflections on the Embryo Debate*. Basic Books, New York.

EISENSTEIN, B. I. 1990. The polymerase chain reaction: A new method of using molecular genetics for medical diagnosis. *NEJM* 322: 178–183.

EL-HEFNAWI, H., S. M. SMITH and L. S. PENROSE. 1965. Xeroderma pigmentosum—Its inheritance and relationships to the ABO blood-group system. *Annals of Human Genetics* 28: 273–290.

EMERY, A. E. H. and D. L. RIMOIN, Editors. 1990. *Principles and Practice of Medical Genetics*, 2nd ed. Churchill Livingstone, Edinburgh.

ENGEL, E. 1980. A new genetic concept: Uniparental disomy and its potential effect, isodisomy. *American Journal of Medical Genetics* 6: 137–143.

ENGELHARDT, H. T., JR. 1984. Shattuck Lecture: Allocating scarce medical resources and the availability of organ transplantation. *NEJM* 311: 66–71.

EPEL, D. 1977. The program of fertilization. *SA* 237(Nov.): 128–138.

EPEL, D. 1980. Fertilization. *Endeavor* (New Series) 4: 26–31.

EPHRUSSI, B. and M. C. WEISS. 1969. Hybrid somatic cells. *SA* 220(April): 26–34.

EPSTEIN, C. J. 1986. *The Consequences of Chromosome Imbalance: Principles, Mechanisms and Models*. Cambridge University Press, Cambridge.

EPSTEIN, C. J. 1988. Mechanisms of the effects of aneuploidy in mammals. *Annual Review of Genetics* 22: 51–75.

EPSTEIN, C. J. 1989. Down syndrome. In *The Metabolic Basis of Inherited Disease*, 6th ed., edited by C. R. Scriver et al., pp. 291–326. McGraw-Hill, New York.

EPSTEIN, C. J. 1991. Editorial: The forensic applications of molecular genetics—the *Journal's* responsibilities. *AJHG* 49: 697–698.

ERICSSON, R. J. and F. BEERNINK. 1987. Sex chromosome ratios in human sperm. *Fertility and Sterility* 47: 531–532.

EVANS, J., et al. 1991. *Children and Young Adults with Sex Chromosome Aneuploidy: Follow-up, Clinical & Molecular Studies*. Birth Defects: Original Article Series, Wiley, New York.

EZZELL, C. 1992. Alzheimer's alchemy. *Science News* 141: 152–153.

EZZELL, C. 1992. Does gene hike radiation's cancer risk? *Science News* 141: 4.

FELSENFELD, G. 1985. DNA. *SA* 253(Oct.): 58–67.

FEREC, C., et al. 1992. Detection of over 98% cystic fibrosis mutations in a Celtic population. *Nature Genetics* 1: 188–191.

FERGUSON-SMITH, M. A., et al. 1992. Olympic row over sex testing. *Nature* 355: 10.

FERNANDES, J., J.-M. SAUDUBRAY and K. TADA, Editors. 1990. *Inborn Metabolic Diseases: Diagnosis and Treatment*. Springer-Verlag, Berlin.

FERRIS, T. 1990. The space telescope: A sign of intelligent life. *New York Times*, April 29, 1990, Section 4: 1ff.

FIDDES, J. C. 1977. The nucleotide sequence of a viral DNA. *SA* 237(Dec.): 54–67.

FLEMMING, W. 1879. Contributions to the knowledge of the cell and its life phenomena. *Archiv für Mikroskopische Anatomie* 16: 302–406. (Abridged translation reprinted in Gabriel and Fogel 1955.)

FLETCHER, R. 1991. *Science, Ideology, and the Media: The Cyril Burt Scandal*. Transaction Publishers, New Brunswick, NJ.

FRANCIS, D. P. and J. CHIN. 1987. The prevention of acquired immunodeficiency syndrome in the United States. *JAMA* 257: 1357–1366.

FRANCKE, U. 1983. Gene mapping. In *Principles and Practice of Medical Genetics*, vol. 1, edited by A. E. H. Emery and D. L. Rimoin, pp. 91–110. Churchill Livingstone, Edinburgh.

FRANCKE, U. 1992. Chromosome banding:

Methods, myths, and misconceptions. *Cell* 68: 1005–1006.

FRANCKE, U., et al. 1985. Minor Xp21 chromosome deletion in a male associated with expression of Duchenne muscular dystrophy, chronic granulomatous disease, retinitis pigmentosa, and McLeod syndrome. *AJHG* 37: 250–267.

FRANKLIN, D. 1989. What a child is given. *New York Times Magazine*, Sept. 3, 1989, p. 36ff.

FRANKS, L. M. and N. M. TEICH, Editors. 1986. *Introduction to the Cellular and Molecular Biology of Cancer*. Oxford University Press, Oxford.

FRASER, F. C. 1973. Survey of counseling practices. In *Ethical Issues in Human Genetics*, edited by B. Hilton et al., pp. 7–13. Plenum, New York.

FRASER, F. C. 1988. Genetic counseling: Using the information wisely. *Hospital Practice* 23(6): 245–266.

FREIFELDER, D. 1978. *The DNA Molecule: Structure and Properties. Original Papers, Analyses, and Problems*. Freeman, San Francisco.

FRIEDBERG, E. C., Editor. 1986. *Cancer Biology: Readings from Scientific American*. Freeman, New York.

FRIEDMANN, T., Editor. 1991. *Therapy for Genetic Disease*. Oxford University Press, Oxford.

FUCHS, F. 1980. Genetic amniocentesis. *SA* 242(June): 47–53.

FUJITA, T. 1975. Abnormal spermatozoa and infertility (man). In *Scanning Electron Microscopical Atlas of Mammalian Reproduction*, edited by E. S. E. Hafez, pp. 82–87. Springer-Verlag, New York.

FUTUYMA, D. J. 1983. *Science on Trial: The Case for Evolution*. Pantheon Books, New York.

FUTUYMA, D. J. 1986. *Evolutionary Biology*, 2nd ed. Sinauer, Sunderland, MA.

GABRIEL, M. L. and S. FOGEL, Editors. 1955. *Great Experiments in Biology*. Prentice-Hall, Englewood Cliffs, NJ.

GAJDUSEK, D. C. 1977. Unconventional viruses and the origin and disappearance of kuru. *Science* 197: 943–960.

GALINAT, W. C. 1977. The origin of corn. In *Corn and Corn Improvement*, edited by G. F. Sprague, pp. 1–47. American Society of Agronomy, Madison, WI.

GALLO, R. C. and L. MONTAGNIER. 1988. AIDS in 1988. *SA* 259(Oct.): 40–48. The entire issue is devoted to AIDS.

GALTON, F. 1905. Eugenics: Its definition, scope and aims. In *Sociological Papers*, pp. 45–50. Macmillan, London. (Reprinted in Bajema 1976.)

GALTON, F. 1909. *Memories of My Life*, 3rd ed. Methuen & Co., London.

GARDNER, R. J. M. and G. R. SUTHER-LAND. 1989. *Chromosome Abnormalities and Genetic Counseling*. Oxford University Press, New York.

GARROD, A. 1909. *Inborn Errors of Metabolism*. Oxford University Press, Oxford. Reprinted (1963) with a supplement by H. Harris. Oxford University Press, London.

GEDDA, L. 1961. *Twins in History and Science*. Charles C. Thomas, Springfield, IL.

GEHRING, W. J. 1985. The molecular basis of development. *SA* 253(Oct.): 153–162.

GELINAS, D. J. 1983. The persisting negative effects of incest. *Psychiatry* 46: 312–332.

GILBERT, S. F. 1991. *Developmental Biology*, 3rd ed. Sinauer, Sunderland, MA.

GILBERT, S. G. 1989. *Pictorial Human Embryology*. University of Washington Press, Seattle.

GODDARD, H. H. 1912. *The Kallikak Family: A Study in the Heredity of Feeble-Mindedness*. Macmillan, New York.

GOLDE, D. W. 1991. The stem cell. *SA* 265(Dec.): 86–93.

GOLDSTEIN, J. L. and M. S. BROWN. 1989. Familial hypercholesterolemia. In *The Metabolic Basis of Inherited Disease*, edited by C. R. Scriver et al., pp. 1215–1250. Mc-Graw-Hill, New York.

GOLEMAN, D. 1988. An emerging theory on blacks' I.Q. scores. *New York Times, Education Life* (Sec. 12), April 10, pp. 22–24.

GOLENBERG, E. M., et al. 1990. Chloroplast DNA sequence from a Miocene *Magnolia* species. *Nature* 344: 656–658.

GOODWIN, D. W. 1991. The genetics of alcoholism. In *Genes, Brain, and Behavior*, edited by P. R. McHugh and V. A. McKusick, pp. 219–226. Raven Press, New York.

GORDIS, E., B. TABAKOFF, D. GOLDMAN and K. BERG. 1990. Finding the gene(s) for alcoholism. *JAMA* 263: 2094–2095.

GORDON, J. W. and F. H. RUDDLE. 1983. Gene transfer into mouse embryos: Production of transgenic mice by pronuclear injection. *Methods in Enzymology* 101: 411–433.

GORMAN, T. 1992. Spawning a new era? *Los Angeles Times*, April 12, p. A3ff.

GOULD, S. J. 1981. *The Mismeasure of Man*. Norton, New York.

GOULD, S. J. 1986. Reflections from an interior world. *Nature* 320: 647–648.

GOULD, S. J. 1991. *Bully for Brontosaurus*. Norton, New York.

GOWERS, W. R. 1879. *Pseudo-hypertrophic Muscular Paralysis—A Clinical Lecture*. J and A Churchill, London.

GRANDJEAN, P., Editor. 1991. *Ecogenetics: Genetic Predisposition to the Toxic Effects of Chemicals*. Routledge, Chapman, & Hall/World Health Organization, New York.

GRAVES, J. A. M. 1987. The evolution of mammalian sex chromosomes and dosage compensation: Clues from marsupials and monotremes. *TIG* 3: 252–256.

GREEN, E. D. and R. H. WATERSTON. 1991. The Human Genome Project: Prospects and implications for clinical medicine. *JAMA* 266: 1966–1975.

GREENBERG, D. S. 1986. What ever happened to the war on cancer? *Discover* 7(March): 47–64.

GREENBERG, D. S. 1991. A sober anniversary of the "War on Cancer." *The Lancet* 338: 1582–1583.

GRIFFIN, J. E. and J. D. WILSON. 1989. The androgen resistance syndromes: 5α-reductase deficiency, testicular feminization, and related disorders. In *The Metabolic Basis of Inherited Disease*, 6th ed., edited by C. R. Scriver et al., pp. 1919–1944. McGraw-Hill, New York.

GROBSTEIN, C. 1979. External human fertilization. *SA* 240(June): 57–67.

GUSELLA, J. F., et al. 1983. A polymorphic DNA marker genetically linked to Huntington's disease. *Nature* 306: 234–238.

HADDOW, J. E., et al. 1992. Prenatal screening for Down's syndrome with use of maternal serum markers. *NEJM* 327: 588–593.

HAFEZ, E. S. E. and P. KENEMANS, Editors. 1982. *Atlas of Human Reproduction by Scanning Electron Microscopy*. MTP Press, Boston.

HAGERMAN, R. J. and A. C. SILVERMAN, Editors. 1991. *Fragile X Syndrome*. Johns Hopkins University Press, Baltimore.

HALDANE, J. B. S. 1932. *The Causes of Evolution*. Harper & Brothers, London. Reprinted (1966) by Cornell University Press, Ithaca, NY.

HALL, J. G. 1990. Genomic imprinting: Review and relevance to human diseases. *AJHG* 46: 857–873.

HALL, J. G. 1992. Genomic imprinting and its clinical implications. *NEJM* 326: 827–829.

HAMERTON, J. L. 1971. *Human Cytogenetics*, vol. 1. Academic Press, New York.

HARDY, G. H. 1908. Mendelian proportions in a mixed population. *Science* 28: 49–50. (Reprinted in Peters 1959, Morris 1971, and Jameson 1977.)

HARDY, J., M.-C. CHARTIER-HARLIN and M. MULLAN. 1992. Alzheimer disease: The new agenda. *AJHG* 50: 648–651.

HARNDEN, D. G. 1974. Ataxia telangiectasia syndrome: Cytogenetic and cancer aspects. In *Chromosomes and Cancer*, edited by J. German, pp. 619–636. Wiley, New York.

HARPER, P. S. 1990. Myotonic dystrophy and related disorders. In *Principles and Practice of Medical Genetics*, 2nd ed., edited by A. E. H. Emery and D. L. Rimoin, pp. 579–597. Churchill Livingstone, Edinburgh.

HARPER, P. S., H. G. HARLEY, W. REARDON and D. J. SHAW. 1992. Anticipation in myotonic dystrophy: New light on an old problem. *AJHG* 51: 10–16.

HARRIS, H. 1980. *The Principles of Human Biochemical Genetics*, 3rd rev. ed. Elsevier/North-Holland Biomedical Press, Amsterdam.

HARRIS, R. 1988. Genetic counselling and the new genetics. *TIG* 4: 52–56.

HARTL, D. L. 1985. *Our Uncertain Heritage: Genetics & Human Diversity*, 2nd ed. Harper & Row, New York.

HARTL, D. L. 1987. *A Primer of Population Genetics*, 2nd ed. Sinauer, Sunderland, MA.

HASSOLD, T. J. 1986. Chromosome abnormalities in human reproductive wastage. *TIG* 2: 105–110.

HEARNSHAW, L. S. 1979. *Cyril Burt, Psychologist*. Cornell University Press, Ithaca, NY.

HENDERSON, B. E., RONALD K. ROSS and MALCOLM C. PIKE. 1991. Toward the primary prevention of cancer. *Science* 254: 1131–1138.

HEUSSNER, R. C., JR. and M. E. SALMON. 1988. *Warning: The Media May Be Harmful to Your Health! A Consumer's Guide to Medical News and Advertising*. Andrews and McMeel, Kansas City.

HILTS, P. J. 1991. Gene-altered pigs produce key part of human blood. *New York Times*, June 26, 1991, p. 1.

HOFFMAN, M. 1991. How parents make their mark on genes. *Science* 252: 1250–1251.

HOLDEN, C. 1987. The genetics of personality. *Science* 237: 598–601.

HOLMBERG, S. D., M. T. OSTERHOLM, K. A. SENGER and M. L. COHEN. 1984. Drug-resistant *Salmonella* from animals fed antimicrobials. *NEJM* 311: 617–622.

HOLMES, H. B. 1985. Sex preselection: Eugenics for everyone? In *Biomedical Ethics Reviews: 1985*, edited by J. M. Humber and R. F. Almeder. Humana Press, Clifton, NJ.

HOLTZMAN, N. A. and M. A. ROTHSTEIN. 1992. Invited editorial: Eugenics and genetic discrimination. *AJHG* 50: 457–459.

HOOK, E. B. 1985. The impact of aneuploidy upon public health: Mortality and morbidity associated with human chromosome abnormalities. In *Aneuploidy: Etiology and Mechanisms*, edited by V. L. Dellarco, P. E. Voytek and A. Hollaender, pp. 7–33. Plenum, New York.

HOOK, E. B. and G. M. CHAMBERS. 1977. Estimated rates of Down syndrome in live births by one year maternal age intervals for mothers aged 20–49 in a New York state study—implications of the risk figures for genetic counseling and cost–benefit analysis of prenatal diagnosis programs. *Birth Defects: Original Article Series*, vol. 13, no. 3A, pp. 123–141.

HOPPE, K. 1992. Brushing the dust off ancient DNA. *Science News* 142: 280–281.

HORGAN, J. 1992. D_2 or not D_2: A barroom brawl over an "alcoholism gene." *SA* 266: 29ff.

HORGAN, J. 1992. Early Arrivals. *SA* 266(Feb.): 17ff.

HORGAN, J. 1992. Profile: Francis H. C. Crick. *SA* 266(Feb.): 32–33.

HOSTETLER, J. A. 1980. *Amish Society*, 3rd ed. Johns Hopkins University Press, Baltimore.

HOWARD-FLANDERS, P. 1981. Inducible repair of DNA. *SA* 245(Nov.): 72–80.

HOY, M. A. 1991. Three book reviews: (1) *Introduction of Genetically Modified Organisms into the Environment*, edited by H. A. Mooney and G. Bernardi, Wiley, New York, 1990; (2) *Risk Assessment in Genetic Engineering*, edited by M. A. Levin and H. S. Strauss, McGraw-Hill, New York, 1991; (3) *Assessing Ecological Risks of Biotechnology*, edited by L. R. Ginzburg, Butterworth-Heinemann, Boston, 1991. *Science* 253: 89–90.

HRUBEC, Z. and C. D. ROBINETTE. 1984. The study of human twins in medical research. *NEJM* 310: 435–441.

HSU, T.C. 1979. *Human and Mammalian Cytogenetics: An Historical Perspective*. Springer-Verlag, New York.

HUNTER, T. 1984. The proteins of oncogenes. *SA* 251(August): 70–79.

HUNTINGTON, G. 1872. On chorea. *The Medical and Surgical Reporter* 26: 320–321. Reprinted in 1973 in *Advances in Neurology* 1: 33–35.

HUNTINGTON, G. 1909. Recollections of Huntington's chorea as I saw it at East Hampton, Long Island, during my boyhood. From a report given by George Huntington, M.D., at the New York Neurological Society (Dec. 7, 1909). Reprinted in 1973 in *Advances in Neurology* 1: 37–39.

HUNTLEY, O. M. 1984. A mother's perspective. *Hastings Center Report* 14: 14–15.

HUSON, S. M., P. S. HARPER, D. A. S. COMPSTON and P. CLARK. 1988. Von Recklinghausen neurofibromatosis: A clinical and population study in south-east Wales [UK]. *Brain* 111: 1355–1381.

HUXLEY, A. 1932. *Brave New World*. Harper & Row, New York.

ILTIS, H. 1924. *Gregor Johann Mendel: Leben, Werk und Wirkung*. Julius Springer, Berlin. [English translation by E. Paul and C. Paul (1932). *Life of Mendel*. Allen & Unwin, London.]

IMAIZUMI, Y. 1986. A recent survey of consanguineous marriages in Japan. *Clinical Genetics* 30: 230–233.

ISCN. 1985. *An International System for Human Cytogenetic Nomenclature* (1985). S. Karger, Basel. [Also in *Birth Defects: Original Article Series*, vol. 21, no. 1, March of Dimes Birth Defects Foundation, New York.]

JACKSON, L. G., et al. 1992. A randomized comparison of transcervical and transabdominal chorionic-villus sampling. *NEJM* 327: 594–598.

JACOBS, P. A. 1982. The William Allen Memorial Award Address: Human population cytogenetics: The first twenty-five years. *AJHG* 34: 689–698.

JACOBS, P. A. and J. A. STRONG. 1959. A case of human intersexuality having a possible XXY sex-determining mechanism. *Nature* 183: 302–303.

JACOBS, P. A., K. E. BUCKTON, C. CUNNINGHAM and M. NEWTON. 1974. An analysis of the break points of structural rearrangements in man. *Journal of Medical Genetics* 11: 50–64.

JACOBS, P. A., M. BRUNTON and M. M. MELVILLE. 1965. Aggressive behavior, mental sub-normality and the XYY male. *Nature* 208: 1351–1352.

JACQUARD, A. 1974. *The Genetic Structure of Populations*. Translated by B. Charlesworth and D. Charlesworth. Springer-Verlag, New York.

JAMES, W. H. 1990. Variations of the human sex ratio at birth. *Fertility and Sterility* 54: 956–957.

JAMESON, D. L., Editor. 1977. *Evolutionary Genetics. Benchmark Papers in Genetics*, vol. 8. Dowden, Hutchinson & Ross, Stroudsburg, PA.

JENSEN, A. R. 1969. How much can we boost IQ and scholastic achievement? *Harvard Educational Review* 39(1): 1–123. Reprinted in *Environment, Heredity, and Intelligence* (1969). Reprint Series No. 2 compiled from the *Harvard Educational Review*.

JENSEN, A. R. 1980. *Bias in Mental Testing*. The Free Press, New York.

JENSEN, A. R. 1981. *Straight Talk About Mental Tests*. The Free Press, New York.

JERVIS, G. A. 1947. Phenylpyruvic oligophrenia deficiency of phenylalanine-oxidizing system. *Society for Experimental Biology and Medicine, Proceedings* 82: 514–515.

JOHNSON, H. M., J. K. RUSSELL and C. H. PONTZER. 1992. Superantigens in human disease. *SA* 266(April): 92–101.

JOYNSON, R. B. 1989. *The Burt Affair*. Routledge, London.

JUDSON, H. F. 1979. *The Eighth Day of Creation: Makers of the Revolution in Biology*. Simon & Schuster, New York.

KABACK, M. M., Editor. 1977. *Tay-Sachs Disease: Screening and Prevention*. Alan R. Liss, New York.

KABACK, M. 1990. Heterozygote screening. In *Principles and Practice of Medical Genetics*, 2nd ed., edited by A. E. H. Emery and D. L. Rimoin, pp. 1951–1958. Churchill Livingstone, Edinburgh.

KALOW, W., H. W. GOEDDE and D. P. AGARWAL, Editors. 1986. *Ethnic Differences in Reactions to Drugs and Xenobiotics*. Alan R. Liss, New York.

KAMIN, L J. 1974. *The Science and Politics of IQ*. Wiley, New York.

KAZAZIAN, H. H., JR., C. WONG, H. YOUSSOUFIAN, A. F. SCOTT, D. G. PHILLIPS and S. E. ANTONARAKIS. 1988. Haemophilia A resulting from de novo insertion of L1 sequences represents a novel mechanism for mutation in man. *Nature* 332: 164–166.

KEATS, B., J. OTT and M. CONNEALLY. 1989. Report of the committee on linkage and gene order. Human Gene Mapping 10 (Tenth International Workshop on Human Gene Mapping). *Cytogenetics and Cell Genetics* 51: 459–502.

KEATS, B., S. L. SHERMAN and J. OTT. 1990. Report of the committee on linkage and gene order. Human Gene Mapping 10.5 (Update to the Tenth International Workshop on Human Gene Mapping). *Cytogenetics and Cell Genetics* 55: 387–394.

KEVLES, D. J. 1984a. Annals of eugenics, part I. *New Yorker* 60(Oct. 8): 51ff.

KEVLES, D. J. 1984b. Annals of eugenics, part II. *New Yorker* 60(Oct. 15): 52ff.

KEVLES, D. J. 1984c. Annals of eugenics, part III. *New Yorker* 60(Oct. 22): 32ff.

KEVLES, D. J. 1984d. Annals of eugenics, part IV. *New Yorker* 60(Oct. 29): 51ff.

KEVLES, D. J. 1985. *In the Name of Eugenics: Genetics and the Uses of Human Heredity*. Knopf, New York.

KEVLES, D. J. and L. HOOD. 1992. *The Code of Codes*. Harvard University Press, Cambridge, MA.

KIMURA, M. 1979. The neutral theory of molecular evolution. *SA* 241(Nov.): 98–126.

KING, M.-C. and A. C. WILSON. 1975. Evolution at two levels in humans and chimpanzees. *Science* 188: 107–116. (Reprinted in Terzaghi, Wilkins, and Penny 1984.)

KING, R. A., J. I. ROTTER and A. G. MOTULSKY. 1992. *The Genetic Basis of Common Diseases*. Oxford University Press, New York.

KING, R. C. and W. D. STANSFIELD. 1990. *A Dictionary of Genetics*, 4th ed. Oxford University Press, Oxford.

KLEIN, J. 1980. *Woody Guthrie: A Life*. Knopf, New York.

KLINE, A. H., J. B. SIDBURY, JR. and C. P. RICHTER. 1959. The occurrence of ectodermal dysplasia and corneal dysplasia in one family. *Journal of Pediatrics* 55: 355–366.

KLINGER, K., et al. 1992. Rapid detection of chromosomal aneuploidies in uncultured amniocytes by using fluorescence in situ hybridization (FISH). *AJHG* 51: 55–65.

KONG, A., G. O. BARNETT, F. MOSTELLER and C. YOUTZ. 1986. How medical professionals evaluate expressions of probability. *NEJM* 315: 740–744.

KONOTEY-AHULU, F. I. D. 1991. *The Sickle Cell Disease Patient*. Macmillan, Houndsmills, Hants.

KORENBERG, J. R. 1991. Down syndrome phenotypic mapping. In *The Morphogenesis of Down Syndrome*, edited by C. J. Epstein, pp. 43–52. Wiley-Liss, New York.

KORNBERG, A. 1968. The synthesis of DNA. *SA* 219(Oct.): 64–78.

KOSHLAND, D. E., JR. 1987. Nature, nurture, and behavior. *Science* 235: 1445.

KOSHLAND, D. E., JR. 1989a. The cell cycle. *Science* 246: 545.

KOSHLAND, D. E., JR. 1989b. The molecule of the year. *Science* 246: 1541.

KOSTYU, D. D. and D. B. AMOS. 1989. The HLA Complex: Genetic polymorphism and disease susceptibility. In *The Metabolic Basis of Inherited Disease*, 6th ed., edited by C. R. Scriver et al., pp. 225–249. McGraw-Hill, New York.

KRUSH, A. J. and K. A. EVANS. 1984. *Family Studies in Genetic Disorders*. Charles C. Thomas, Springfield, IL.

KUNKEL, L. M., et al. 1986. Analysis of deletions in DNA from patients with Becker and Duchenne muscular dystrophy. *Nature* 322: 73–77.

KUNZE, J. and I. NIPPERT. 1986. *Genetics and Malformations in Art*. Grosse Verlag, Berlin.

LABIE, D., J. PAGNIER, H. WAJCMAN, M. E. FABRY and R. L. NAGEL. 1986. The genetic origin of the variability of the phenotypic expression of the *Hb^S* gene. In *Genetic Variation and Its Maintenance*, edited by D. F. Roberts and G. F. De Stefano, pp. 149–155. Cambridge University Press, Cambridge.

LAFOLLETTE, M. C. 1990. *Making Science Our Own*. University of Chicago Press, Chicago.

LAFOND, R. E., Editor. 1988. *Cancer, The Outlaw Cell*, 2nd ed. American Chemical Society, Washington, DC.

LAKE, J. A. 1981. The ribosome. *SA* 245(August): 84–97.

LANDER, E. S. 1991. Invited editorial: Research on DNA typing catching up with courtroom application. *AJHG* 48: 819–823.

LANDSTEINER, K. 1901. Uber Agglutinationserscheinungen normalen menschlichen Blutes. *Wiener klinische Wochenschrift* 14: 1132–1134. (An English translation is in Boyer 1963.)

LANE, C. 1976. Rabbit hemoglobin from frog eggs. *SA* 235(August): 60–71.

LASKER, G. W. 1985. *Surnames and Genetic Structure*. Cambridge University Press, Cambridge.

LASKER, J. N. and S. BORG. 1987. *In Search of Parenthood: Coping with Infertility and High–Tech Conception*. Beacon Press, Boston.

LAURENCE, J. 1985. The immune system in AIDS. *SA* 253(Dec.): 84–93.

LAWN, R. M. and G. A. VEHAR. 1986. The molecular genetics of hemophilia. *SA* 254(March): 48–54.

LAX, E. 1984. *Life and Death on 10 West*. Times Books, New York.

LECOS, C. W. 1988. Caffeine jitters: Some safety questions remain. *FDA Consumer* 21(Dec. 1987–Jan. 1988): 22–27.

LEDBETTER, D. H. and W. K. CAVENEE. 1989. Molecular cytogenetics. In *The Metabolic Basis of Inherited Disease*, 6th ed., edited by C. R. Scriver et al., pp. 232–371. McGraw-Hill, New York.

LEDER, P. 1982. The genetics of antibody diversity. *SA* 246(May): 102–115.

LEDERBERG, J. 1986. Forty years of genetic recombination in bacteria: A fortieth anniversary reminiscence. *Nature* 324: 627–628.

LEVIN, R. J. 1987. Human sex pre-selection. In *Oxford Reviews of Reproductive Biology*, vol. 9, edited by J. R. Clarke, pp. 161–191. Clarendon Press, Oxford.

LEVINE, L., Editor. 1971. *Papers on Genetics: A Book of Readings*. Mosby, St. Louis.

LEVITAN, M. 1988. *Textbook of Human Genetics*, 3rd ed. Oxford University Press, New York.

LEVY, H. L. 1973. Genetic screening. *Advances in Human Genetics* 4: 1–104.

LEWIN, B. 1990. *Genes IV*. Oxford University Press, Oxford.

LEWIN, R. 1984. DNA reveals surprises in human family tree. *Science* 226: 1179–1182.

LEWIS, R. 1989. Genetics meets forensics. *BioScience* 39: 6–9.

LEWIS, R. 1990. Antibody applications get a boost. *BioScience* 40: 490–492.

LEWIS, R. 1991. Genetic imprecision. *BioScience* 41: 288–293.

LEWONTIN, R. C., S. ROSE and L. J. KAMIN. 1984. *Not in Our Genes*. Pantheon Books, New York.

LINDER, D. and S. M. GARTLER. 1965. Glucose-6-phosphate dehydrogenase mosaicism: Utilization as a cell marker in the study of leiomyomas. *Science* 150: 67–69.

LINDER, M. E. and A. G. GILMAN. 1992. G proteins. *SA* 267 (July): 56–65.

LIOTTA, L. A. 1992. Cancer cell invasion and metastasis. *SA* 266(Feb.): 54–63.

LIPSHULTZ, L. I. and S. S. HOWARDS. 1991. *Infertility in the Male*, 2nd ed. Mosby, St. Louis.

LISON, M., S. H. BLONDHEIM and R. N. MELMED. 1980. A polymorphism of the ability to smell urinary metabolites of asparagus. *British Medical Journal* 281: 1676–1678.

LJUNGQVIST, A. and J. L. SIMPSON, for the International Amateur Athletic Federation Work Group on Gender Verification. 1992. Medical examination for health of all athletes replacing the need for gender verification in international sports: The International Amateur Athletic Federation plan. *JAMA* 267: 850–852.

LOMBARDO, P. A. 1985. Three generations, no imbeciles: New light on *Buck v. Bell*. *New York University Law Review* 60: 30–62.

LONGO, F. J. 1987. *Fertilization*. Chapman & Hall, London.

LORD, T. R. 1989. Exploring the twinning phenomenon in *Homo sapiens*. *Biology Digest* 15(Jan): 11–22.

LUND, J. C. 1860. Chorea St. Vitus Dance in Saetersdalen. *Report of Health and Medical Conditions in Norway in 1860*. (Norway's Official Statistics, 1862, C. No. 4), p. 137. Partial translation in Orbeck 1959.

LUPSKI, J. R., et al. 1991. DNA duplication associated with Charcot-Marie-Tooth disease type 1A. *Cell* 66: 219–232.

LUZIO, J. P. 1992. Golden oldies. *TIG* 8: 117.

LYNCH, H. T., R. H. HODEN and N. W. PAUL, Editors. 1990. *International Directory of Genetic Services*, 9th ed. March of Dimes Birth Defects Foundation, White Plains, New York.

LYON, M. F. 1961. Gene action in the X-chromosome of the mouse (*Mus musculus L.*). *Nature* 190: 372–373.

LYON, M. F. 1988. The William Allan Memorial Award address: X-chromosome inactivation and the location and expression of X-linked genes. *AJHG* 42: 8–16.

LYON, M. F. 1992. Some milestones in the history of X-chromosome inactivation. *Annual Review of Genetics* 26: 17–28.

MABUCHI, H., et al. 1978. Homozygous familial hypercholesterolemia in Japan. *American Journal of Medicine* 65: 290–297.

MACALPINE, I. and R. HUNTER. 1969. Porphyria and King George III. *SA* 221(July): 38–46.

MACGREGOR, H. C. and J. M. VARLEY. 1988. *Working with Animal Chromosomes*, 2nd ed. Wiley, New York.

MADDOX, J. 1991. The case for the human genome. *Nature* 352: 11–14.

MALKIN, D., et al. 1990. Germ line p53 mutations in a familial syndrome of breast cancer, sarcomas, and other neoplasms. *Science* 250: 1233–1238.

MALTHUS, T. 1798. *An Essay on the Principle of Population as It Affects the Future Improvement of Society*. J. Johnson, London. (Reprinted many times: e.g., *An Essay on the Principle of Population: Text, Sources and Background, Criticism*, edited by P. Appleman, Norton, New York, 1976.)

MANGE, A. P. and E. J. MANGE. 1990. *Genetics: Human Aspects*, 2nd ed. Sinauer, Sunderland, MA.

MANUELIDIS, L. 1990. A view of interphase chromosomes. *Science* 250: 1533–1540.

MARAN, S. P. 1991. Science is dandy, but promotion can be lucrative. *Smithsonian* 22(May): 72–80.

MARANTO, G. 1986. Genetic engineering: Hype, hubris, and haste. *Discover* 7(June): 50–64.

MARGULIS, L. 1982. *Early Life*. Science Books International, Boston.

MARRACK, P. and J. KAPPLER. 1986. The T cell and its receptor. *SA* 254(Feb.): 36–45.

MARSHALL, E. 1991. Breast cancer: Stalemate in the war on cancer. *Science* 254: 1719–1720.

MARX, J. 1992a. Boring in on β-amyloid's role in Alzheimer's. *Science* 255: 688–689.

MARX, J. 1992b. Major setback for Alzheimer's models. *Science* 255: 1200–1202.

MASON, V. R. 1922. Sickle cell anemia. *JAMA* 79: 1318–1320.

MASSIE, R. K. 1985. *Nicholas and Alexandra*. Dell, New York.

MAZIA, D. 1974. The cell cycle. *SA* 230 (Jan.): 54–64.

MAZIA, D. 1987. The chromosome cycle and the centrosome cycle in the mitotic cycle. *International Review of Cytology* 100: 49–92.

McHUGH, P. R. and V. A. McKUSICK, Editors. 1991. *Genes, Brain, and Behavior*. Raven Press, New York.

McKEAN, K 1985. Intelligence: New ways to measure the wisdom of man. *Discover* 6(Oct.): 25–41.

McKUSICK, V. A. 1965. The royal hemophilia. *SA* 213(August): 88–95.

McKUSICK, V. A. 1975. The growth and development of human genetics as a clinical discipline. *AJHG* 27: 261–273.

McKUSICK, V. A., Commentator. 1978. *Medical Genetic Studies of the Amish: Selected Papers*. Johns Hopkins University Press, Baltimore.

McKUSICK, V. A. 1990. *Mendelian Inheritance in Man: Catalogs of Autosomal Dominant, Autosomal Recessive, and X-linked Phenotypes*, 9th ed. John Hopkins University Press, Baltimore.

McKUSICK, V. A. 1992. *Mendelian Inheritance in Man: Catalogs of Autosomal Dominant, Autosomal Recessive, and X-linked Phenotypes*, 10th ed. Johns Hopkins University Press, Baltimore. (For people with a computer and modem, this book is available in electronic form at no charge.)

McKUSICK, V. A., J. A. HOSTETLER, J. A. EGELAND and R. ELDRIDGE. 1964. The distribution of certain genes in the Old Order Amish. *Human Genetics: Cold Spring Harbor Symposia on Quantitative Biology* 29: 99–114. (Reprinted in Morris 1971.)

McKUSICK, V. A. and D. L. RIMOIN. 1967. General Tom Thumb and other midgets. *SA* 217(July): 102–110.

McKUSICK, V. A. and F. H. RUDDLE. 1977. The status of the gene map of the human chromosomes. *Science* 196: 390–405.

McLAREN, A. 1988. Sex determination in mammals. *TIG* 4: 153–157.

McLEAN v. ARKANSAS BOARD OF EDUCATION. 1982. Opinion of U.S. District Court Judge William R. Overton. (Reprinted in *Science* 215: 934–943.)

MEAD, M. 1949. *Male and Female: A Study of the Sexes in a Changing World*. Morrow, New York.

MENDEL, G. 1865. Versuche uber pflanzenhybriden. *Verhandlungen des Naturforschenden Vereines in Brünn* 4: 3–47. (An English translation by W. Bateson is reprinted in Peters 1959 and in Jameson 1977. Another English translation, by E. R. Sherwood, is in Stern and Sherwood 1966.)

MERTENS, T. R., Editor. 1975. *Human Genetics: Readings on the Implications of Genetic Engineering*. Wiley, New York.

MERZ, B. 1987. Matchmaking scheme solves Tay-Sachs problem. *JAMA* 258: 2636ff.

MESELSON, M. and F. W. STAHL. 1958. The replication of DNA in *Escherichia coli*. *PNAS* 44: 671–682. Reprinted in Freifelder (1978) and in Corwin and Jenkins (1976).

MICKLOS, D. A. and G. A. FREYER. 1990. *DNA Science: A First Course in Recombinant DNA Technology*. Cold Spring Harbor Laboratory Press, Cold Spring Harbor, New York.

MIESCHER, F. 1871. On the chemical composition of pus cells. *Hoppe-Seyler's medizinische-chemische Untersuchungen* 4: 441–460. (Abridged translation in Gabriel and Fogel 1955.)

MILLER, A. D. 1992. Human gene therapy comes of age. *Nature* 357: 455–460.

MILLER, J. A. 1990. Genes that protect against cancer. *BioScience* 40: 563–566.

MILLER, O. L., JR. 1973. The visualization of genes in action. *SA* 228(March): 34–42.

MILLS, J. and H. MASUR. 1990. AIDS-related infections. *SA* 263(August): 50–57.

MILSTEIN, C. 1980. Monoclonal antibodies. *SA* 243(Oct.): 66–74.

MILUNSKY, A. 1989. *Choices, Not Chances: An Essential Guide to Your Heredity and Health*. Little, Brown, Boston.

MINOGUE, B. P., R. TARASZEWSKI, S. ELIAS and G. J. ANNAS. 1988. The whole truth and nothing but the truth? *Hastings Center Report* 18(Oct.–Nov.): 34–36.

MITTWOCH, U. 1986. Males, females and hermaphrodites. *Annals of Human Genetics* 50: 103–121.

MLOT, C. 1989. On the trail of transfer RNA identity. *BioScience* 39: 756–759.

MONEY, J. and A. A. EHRHARDT. 1972. *Man & Woman, Boy & Girl: The Differentiation and Dimorphism of Gender Identity from Conception to Maturity*. The Johns Hopkins University Press, Baltimore.

MONMANEY, T. 1989. Iron man. *Discover*(July): 62–65.

MOORE, K. L. 1988. *The Developing Human: Clinically Oriented Embryology*, 4th ed. Saunders, Philadelphia.

MORCH, E. T. 1941. *Chondrodystrophic Dwarfs in Denmark*. Munksgaard, Copenhagen.

MOREL, F. M. M., R. F. BAKER and H. WAYLAND. 1971. Quantitation of human red blood cell fixation by glutaraldehyde. *The Journal of Cell Biology* 48: 91–100.

MORGAN, T. H., A. H. STURTEVANT, H. J. MULLER and C. B. BRIDGES. 1915. *The Mechanism of Mendelian Heredity*. Holt, New York.

MORRIS, J. M. 1953. The syndrome of testicular feminization in male pseudohermaphrodites. *American Journal of Obstetrics and Gynecology* 65: 1192–1211.

MORRIS, L. N., Editor. 1971. *Human Populations, Genetic Variation, and Evolution*. Chandler, San Francisco.

MOTULSKY, A. G. 1960. Metabolic polymorphisms and the role of infectious diseases in human evolution. *Human Biology* 32: 28–62. (Reprinted in Morris 1971.)

MOTULSKY, A. G. 1978. Medical and human genetics 1977: Trends and directions. *AJHG* 30: 123–131.

MOURANT, A. E., A. C. KOPEC and K. DOMANIEWSKA-SOBCZAK. 1976. *The Distribution of the Human Blood Groups and Other Polymorphisms*, 2nd ed. Oxford University Press, London.

MOYZIS, R. K. 1991. The human telomere. *SA* 265(August): 48–55.

MULLAN, M., et al. 1992. A locus for familial early-onset Alzheimer's disease on the long arm of chromosome 14, proximal to the α1-antichymotrypsin gene. *Nature Genetics* 2: 340–342.

MULLER, H. J. 1935. *Out of the Night: A Biologist's View of the Future*. Vanguard, New York.

MULLER, H. J. 1961. Human evolution by voluntary choice of germ plasm. *Science* 134: 643–649. (Reprinted in Mertens 1975.)

MULLIS, K. B. 1990. The unusual origin of the polymerase chain reaction. *SA* 262(April): 56–65.

MURRAY, A. W. and M. W. KIRSCHNER. 1991. What controls the cell cycle. *SA* 264(March): 56–63.

NAGEL, R. L. 1984. The origin of the hemoglobin S gene: Clinical, genetic and anthropological consequences. *Einstein Quarterly Journal of Biology and Medicine* 2: 53–62.

NANCE, W. E. and F. E. McCONNELL 1973. Status and prospects of research in hereditary deafness. *Advances in Human Genetics* 4: 173–250.

NATIONAL RESEARCH COUNCIL. 1988. *Mapping and Sequencing the Human Genome*. National Academy Press, Washington, DC.

NATIONAL RESEARCH COUNCIL. 1992. *DNA Technology in Forensic Science*. National Academy Press, Washington, DC.

NEEL, J. V. 1991. Update on the genetic effects of ionizing radiation. *JAMA* 266: 698–701.

NEEL, J. V. and W. J. SCHULL. 1954. *Human Heredity*. University of Chicago Press, Chicago.

NELSON, K. and L. B. HOLMES. 1989. Malformations due to presumed spontaneous mutations in newborn infants. *NEJM* 320: 19–23.

NEUFELD, P. J. and N. COLMAN. 1990. When science takes the witness stand. *SA* 262(May): 46–53.

NEW, M. I. and L. S. LEVINE (1973). Congenital adrenal hyperplasia. *Advances in Human Genetics* 4: 251–326.

NEW, M. I., P. C. WHITE, S. PANG, B. DUPONT and P. W. SPEISER. 1989. The adrenal hyperplasias. In *The Metabolic Basis of Inherited Disease*, 6th ed., edited by C. R. Scriver et al., pp. 1881–1917. McGraw-Hill, New York.

NEW, M. I., P. C. WHITE, P. W. SPEISER, C. CRAWFORD and B. DUPONT. 1990. Congenital adrenal hyperplasia. In *Principles and Practice of Medical Genetics*, 2nd ed., edited by A. E. H. Emery and O. L. Rimoin, pp. 1559–1591. Churchill Livingstone, Edinburgh.

NEWMAN, H. H., F. N. FREEMAN and K. J. HOLZINGER. 1937. *Twins, A Study of Heredity and Environment*. The University of Chicago Press, Chicago.

NILSSON, L., M. FURUHJELM, A. INGELMAN-SUNDBERG and C. WIRSÉN. 1977. *A Child Is Born*. Delacorte Press/Seymour Lawrence, New York.

NILSSON, L. and L. HAMBERGER. 1990. *A Child Is Born*. Delacorte Press/Seymour Lawrence, New York.

NOMURA, M. 1984. The control of ribosome synthesis. *SA* 250(Jan.): 102–112.

NORA, J. J. and F. C. FRASER. 1989. *Medical Genetics: Principles and Practice*, 3rd ed. Lea & Febiger, Philadelphia.

NORMAN, C. 1985. Virus scare halts hormone research. *Science* 228: 1176–1177.

NOSSAL, G. J. V. 1978. *Antibodies and Immunity*, 2nd ed. Basic Books, New York.

O'BRIEN, S. J., D. E. WILDT and M. BUSH. 1986. The cheetah in genetic peril. *SA* 254(May): 84–92.

OBE, G. and A. T. NATARAJAN, Editors. 1990. *Chromosomal Aberrations: Basic and Applied Aspects*. Springer-Verlag, Berlin.

OFFICE OF TECHNOLOGY ASSESSMENT. 1988. *Infertility: Medical and Social Choices*, [OTA-BA-358]. U.S. Government Printing Office, Washington, DC.

OFFICE OF TECHNOLOGY ASSESSMENT. 1988. *Mapping Our Genes. Genome Projects: How Big How Fast?* Johns Hopkins University Press, Baltimore.

OFFICE OF TECHNOLOGY ASSESSMENT. 1990. *Genetic Witness: Forensic Uses of DNA Tests*. [OTA-BA-438] U.S. Government Printing Office, Washington, DC.

OFFICE OF TECHNOLOGY ASSESSMENT. 1992. Cystic fibrosis and DNA tests: Implications of carrier screening. [OTA-BA-532] U.S. Government Printing Office, Washington, DC.

OHNO, S. 1969. Evolution of sex chromosomes in mammals. *Annual Review of Genetics* 3: 495–524.

ORBECK, A. L. 1959. An early description of Huntington's chorea. *Medical History* 3: 165–168.

OREL, V. 1984. *Mendel*. Oxford University Press, Oxford.

OSBORN, D. 1916. Inheritance of baldness. *Journal of Heredity* 7: 347–355.

OVERZIER, C., Editor. 1963. *Intersexuality*. Academic Press, New York.

OWERBACH, D. G. I. BELL, W. J. RUTTER and T. B. SHOWS. 1980. The insulin gene is on chromosome 11 in humans. *Nature* 286: 82–84.

OWOR, R. and C. OLWENY. 1978. Malignant neoplasms. In *Diseases of Children in the Subtropics and Tropics*, 3rd ed., edited by D. B. Jelliffe and J. P. Stanfield, pp. 605–623. Edward Arnold, London.

PÄÄBO, S. 1985. Molecular cloning of ancient Egyptian mummy DNA. *Nature* 314: 644–645.

PATTERSON, D. 1987. The causes of Down syndrome. *SA* 257(Aug): 52–60.

PATTERSON, D. and C. J. EPSTEIN, Editors. 1990. *Molecular Genetics of Chromosome 21 and Down Syndrome. Progress in Clinical and Biological Research*, vol. 360. Wiley-Liss, New York.

PAUL, W. E., Editor. 1991. *Immunology: Recognition and Response. Readings from Scientific American Magazine*. Freeman, New York.

PEBLEY, A. R. and C. F. WESTOFF. 1982. Women's sex preferences in the United States: 1970 to 1975. *Demography* 19: 177–189.

PENROSE, L. S. 1951. Measurement of pleiotropic effects in phenylketonuria. *Annals of Eugenics* 16: 134–141.

PENROSE, L. S. and C. STERN. 1958. Reconsideration of the Lambert pedigree (ichthyosis hystrix gravior). *Annals of Human Genetics* 22: 258–283.

PERUTZ, M. F. 1978. Hemoglobin structure and respiratory transport. *SA* 239(June): 92–125.

PETERS, J. A., Editor. 1959. *Classic Papers in Genetics*. Prentice–Hall, Englewood Cliffs, NJ.

PLOMIN, R. 1990. The role of inheritance in behavior. *Science* 248: 183–188.

POLANI, P. E., et al. 1958. Colour vision studies and the X-chromosome constitution of patients with Klinefelter's syndrome. *Nature* 182: 1092–1093.

PONDER, B. 1990. Neurofibromatosis gene cloned. *Nature* 346: 703–704.

POOL, R. 1990. Electromagnetic fields: The biological evidence. *Science* 249: 1378–1381.

PRAKASH, G. 1988. Helping scientists map genes. *Yale Scientific* 63(fall): 13–17.

PRESCOTT, D. M. 1987. Cell reproduction. *International Review of Cytology* 100: 93–128.

PRESCOTT, D. M. and A. S. FLEXER. 1986. *Cancer: The Misguided Cell*, 2nd ed. Sinauer, Sunderland, MA.

PRESIDENT'S COMMISSION FOR THE STUDY OF ETHICAL PROBLEMS IN MEDICINE AND BIOMEDICAL AND BEHAVIORAL RESEARCH. 1982. *Splicing Life*. U.S. Government Printing Office, Washington, DC.

PRESIDENT'S COMMISSION FOR THE STUDY OF ETHICAL PROBLEMS IN MEDICINE AND BIOMEDICAL AND BEHAVIORAL RESEARCH. 1983. *Screening and Counseling for Genetic Conditions*. U.S. Government Printing Office, Washington, DC.

PROVINE, W. B. 1971. *The Origin of Theoretical Population Genetics*. University of Chicago Press, Chicago.

PUESCHEL, S. M., Editor. 1990. *A Parent's Guide to Down Syndrome: Toward a Brighter Future*. P. H. Brookes, Baltimore.

PUNNETT, R. C. 1911. Mendelism. In *The Encyclopaedia Britannica*, 11th ed. The Encyclopaedia Britannica Company, New York.

RADETSKY, P. 1991. The roots of cancer. *Discover* 12(May): 60–64.

RADMAN, M. and R. WAGNER. 1988. The high fidelity of DNA duplication. *SA* 259(August): 40–46.

RAFTER, N. H., Editor. 1988. *White Trash: The Eugenic Family Studies, 1877–1919.* Northeastern University Press, Boston.

RAMSEY, P. 1970. *Fabricated Man: The Ethics of Genetic Control.* Yale University Press, New Haven, CT.

REDDY, P. G. 1987. Effects of consanguineous marriages on fertility among three endogamous groups of Andhra Pradesh. *Social Biology* 34: 68–77.

REED, T. E. 1969. Caucasian genes in American Negroes. *Science* 165: 762–768. (Reprinted in Morris 1971.)

REIK, W. 1989. Genomic imprinting and genetic disorders in man. *TIG* 5: 331–336.

REILLY, P. R. 1985. Eugenic sterilization in the United States. In *Genetics and the Law III*, edited by A. Milunsky and G. J. Annas, pp. 227–241. Plenum, New York.

RENNIE, D. 1985. "Bubble Boy." *JAMA* 253: 78–80.

RENNIE, J. 1992. Keeping it in the family. *SA* 267(August): 43–44.

RESEARCH AND EDUCATION ASSOCIATION. 1982. *Behavioral Genetics.* Research and Education Association, New York.

RESNICK, M. A. and B. K. VIG, Editors. 1989. *Mechanisms of Chromosome Distribution and Aneuploidy.* (Progress in Clinical and Biological Research Series) Wiley, New York.

RHODES, D. and A. KLUG. 1993. Zinc fingers. *SA* 268(Feb.): 56–65.

RICCARDI, V. M. 1990. The phakomatoses. In *Principles and Practice of Medical Genetics*, 2nd ed., edited by A. E. H. Emery and D. L. Rimoin, pp. 435–445. Churchill Livingstone, Edinburgh.

RICH, A. and S. H. KIM. 1978. The three-dimensional structure of transfer RNA. *SA* 238(Jan.): 52–62.

RICHARDS, R. I. and G. R. SUTHERLAND. 1992. Fragile X syndrome: The molecular picture comes into focus. *TIG* 8: 249–254.

RIEGER, R., A. MICHAELIS and M. M. GREEN. 1991. *Glossary of Genetics: Classical and Molecular.* Springer-Verlag, Berlin.

RIGGS, A. D. and G. P. PFEIFER. 1992. X-chromosome inactivation and cell memory. *TIG* 8: 169–174.

ROBB, S. A., K. R. E. POHL, M. BARAITSER, J. WILSON and E. M. BRETT. 1989. The "happy puppet" syndrome of Angelman: Review of the clinical features. *Archives of Disease in Childhood* 64: 83–86.

ROBERTS, L. 1984. *Cancer Today: Origins, Prevention, and Treatment.* National Academy of Sciences, Washington, DC.

ROBERTS, L. 1989. New game plan for genome mapping. *Science* 245: 1438–1440.

ROBERTS, L. 1990. Huntington's gene: So near, yet so far. *Science* 247: 624–627.

RODGER, J. C. and B. L. DRAKE. 1987. The enigma of the fetal graft. *American Scientist* 75: 51–57.

ROOT, M. 1988. Glow-in-the-dark biotechnology. *BioScience* 38: 745–747.

ROSE, R. J. 1982. Separated twins: Data and their limits. (Book review) *Science* 215: 959–960.

ROSEN, C. M. 1987. The eerie world of reunited twins. *Discover* 8(Sept.): 36–46.

ROSENBERG, L. E. 1972. Diagnosis and management of inherited aminoacidopathies in the newborn and unborn. In *Genetics and the Perinatal Patient.* Mead Johnson Symposium on Perinatal and Developmental Medicine, No. 1, pp. 45–49.

ROSENBERG, S. A. 1990. Adoptive immunotherapy for cancer. *SA* 262(May): 62–69.

ROSENTHAL, E. 1989. The wolf at the door. *Discover* 10(Feb.): 34–37.

ROSENTHAL, E. 1991. Technique for early prenatal test comes under question in studies. *New York Times*, July 10, 1991, p. C11.

ROSS, P. E. 1992. Eloquent remains. *SA* 266: 115–125.

ROUECHE, B. 1979. Annals of medicine: Live and let live. *The New Yorker* 55(July 16): 82–87.

ROWLAND, L. P. 1988. Dystrophin: A triumph of reverse genetics and the end of the beginning. *NEJM* 318: 1392–1394.

RUDDLE, F. H. and K. K. KIDD. 1989. The Human Gene Mapping Workshops in transition. (Tenth International Workshop on Human Gene Mapping, New Haven Conference) *Cytogenetics and Cell Genetics* 51: 1–2.

RUDDLE, F. H. and R. S. KUCHERLAPATI. 1974. Hybrid cells and human genes. *SA* 231(July): 36–44.

SACHS, B. 1887. On arrested cerebral development, with special reference to its cortical pathology. *Journal of Nervous and Mental Disease* 14: 541–553.

SANDBERG, A. A. 1988. Chromosomal lesions and solid tumors. *Hospital Practice* 23(10): 93–106.

SANDHOFF, K., E. CONZELMANN, E. F. NEUFELD, M. M. KABACK and K. SUZUKI. 1989. The G_{M2} gangliosides. In *The Metabolic Basis of Inherited Disease*, 6th ed., edited by C. R. Scriver et al., pp. 1807–1839. McGraw-Hill, New York.

SAPIENZA, C. 1990. Parental imprinting of genes. *SA* 263(Oct.): 52–60.

SATHANANTHAN, A. H., A. O. TROUNSON and C. WOOD. 1986. *Atlas of Fine Structure of Human Sperm Penetration, Eggs and Embryos Cultured in Vitro.* Praeger, New York.

SAYRE, A. 1975. *Rosalind Franklin and DNA.* Norton, New York.

SCARR, S. 1987. Three cheers for behavior genetics: Winning the war and losing our identity. *Behavior Genetics* 17: 219–228.

SCARR, S. and L. CARTER-SALTZMAN. 1983. Genetics and intelligence. In *Behavior Genetics: Principles and Applications*, edited by J. L. Fuller and E. C. Simmel, pp. 217–335. Lawrence Erlbaum Associates, Hillsdale, NJ.

SCHAUB, J., F. VAN HOOF and H. L. VIS, Editors. 1991. *Inborn Errors of Metabolism.* (Nestle Nutrition Workshop Series, vol. 24.) Raven, New York.

SCHROTT, H. G., J. L. GOLDSTEIN, W. R. HAZZARD, M. M. McGOODWIN and A. G. MOTULSKY. 1972. Familial hypercholesterolemia in a large kindred. Evidence for a monogenic mechanism. *Annals of Internal Medicine* 76: 711–720.

SCHULL, W. J. and J. V. NEEL. 1965. *The Effects of Inbreeding on Japanese Children.* Harper & Row, New York.

SCOTT, J. A., A. P. WALKER, D L. EUNPU and L. DJURDJINOVIC. 1988. Genetic counselor training: A review and considerations for the future. *AJHG* 42: 191–199.

SCRIVER, C. R., A. L. BEAUDET, W. S. SLY and D. VALLE, Editors. 1989. *The Metabolic Basis of Inherited Disease*, 6th ed. McGraw-Hill, New York.

SEEMANOVA, E. 1971. A study of children of incestuous matings. *Human Heredity* 21: 108–128.

SEIBEL, M. M. 1988. A new era in reproductive technology: In vitro fertilization, gamete intrafallopian transfer, and donated gametes and embryos. *NEJM* 318: 828–834.

SELIKOWITZ, M. 1990. *Down Syndrome: The Facts.* Oxford University Press, Oxford.

SELKOE, D. J. 1991. Amyloid protein and Alzheimer's disease. *SA* 265: 68–78.

SHAW, M. W. 1990. Legal considerations in the delivery of genetic care. In *Principles and Practice of Medical Genetics*, 2nd ed., edited by A. E. H. Emery and D. L. Rimoin, pp. 2017–2023. Churchill Livingstone, Edinburgh.

SHETTLES, L. B. and D. RORVIK. 1984. *How to Choose the Sex of Your Baby.* Doubleday, Garden City, NY.

SHINE, I. and S. WROBEL. 1976. *Thomas Hunt Morgan, Pioneer of Genetics.* The University of Kentucky Press, Lexington.

SHREEVE, J. 1990. Argument over a woman. *Discover* 11(August): 52–59.

SIBLEY, C. G. and J. E. AHLQUIST. 1984. The phylogeny of the hominoid primates, as indicated by DNA–DNA hybridization. *Journal of Molecular Evolution* 20: 2–15.

SIMMONS, K. 1984. Physicians continue to study cause(s) of 'bubble' boy's death. *JAMA* 251: 1929–1931.

SIMPSON, J. L. 1990. Disorders of gonads and internal reproductive ducts. In *Principles and Practice of Medical Genetics*, 2nd ed., edited by A. E. H. Emery and D. L. Rimoin, pp. 1593–1617. Churchill Livingstone, Edinburgh.

SIMPSON, J. L. and S. A. CARSON. 1992. Preimplantation genetic diagnosis. *NEJM* 327: 951–953.

SINISCALCO, M., G. FILIPPI and B. LATTE. 1964. Recombination between protan and deutan genes: Data on their relative positions in respect of the G6PD locus. *Nature* 204: 1062–1064.

SKINNER, R. 1990. Genetic counselling. In *Principles and Practice of Medical Genetics*, 2nd ed., edited by A. E. H. Emery and D. L. Rimoin, pp. 1923–1933. Churchill Livingstone, Edinburgh.

SKINNER, R. 1990. Unifactorial inheritance. In *Principles and Practice of Medical Genetics*, 2nd ed., edited by A. E. H. Emery and D. L. Rimoin, pp. 95–105. Churchill Livingstone, Edinburgh.

SLOAN, H. R. and D. S. FREDRICKSON. 1972. G$_{M2}$ gangliosidosis: Tay-Sachs disease. In *The Metabolic Basis of Inherited Disease*, 3rd ed., edited by J. B. Stanbury et al., pp. 615–638. McGraw-Hill, New York.

SOLTER, D. 1988. Differential imprinting and expression of maternal and paternal genomes. *Annual Review of Genetics* 22: 127–146.

SPENCE, J. E., et al. 1988. Uniparental disomy as a mechanism for human genetic disease. *AJHG* 42: 217–226.

SRB, A. M., R. D. OWEN and R. S. EDGAR, Editors. 1970. *Facets of Genetics: Readings from Scientific American*. Freeman, San Francisco.

ST GEORGE-HYSLOP, P. H., et al. 1990. Genetic linkage studies suggest that Alzheimer's disease is not a single homogeneous disorder. *Nature* 347: 194–197.

ST GEORGE-HYSLOP, P. H., et al. 1992. Genetic evidence for a novel familial Alzheimer's disease locus on chromosome 14. *Nature Genetics* 2: 330–334.

STEBBINS, G. L. and F. J. AYALA. 1985. The evolution of Darwinism. *SA* 253(July): 72–82.

STEITZ, J. A. 1988. "Snurps." *SA* 258(June): 56–63.

STENT, G. S. 1972. Prematurity and uniqueness in scientific discovery. *SA* 227(Dec.): 84–93.

STEPHENS, F. E. and F. H. TYLER. 1951. Studies in disorders of muscle. V. The inheritance of childhood progressive muscular dystrophy in 33 kindreds. *AJHG* 3: 111–125.

STERN, C. 1949. *Principles of Human Genetics*. Freeman, San Francisco.

STERN, C. 1965. Mendel and human genetics. *Proceedings of the American Philosophical Society* 109: 216–226.

STERN, C. and E. R. SHERWOOD, Editors. 1966. *The Origin of Genetics: A Mendel Source Book*. Freeman, San Francisco.

STEWART, A., Editor. 1990. The functional organization of chromosomes and the nucleus—a special issue. *TIG* 6(12): 377–437.

STEWART, G. D., T. J. HASSOLD and D. M. KURNIT. 1988. Trisomy 21: Molecular and cytogenetic studies of nondisjunction. *Advances in Human Genetics* 17: 99–140.

STRACHAN, T. and R. HARRIS. 1990. The HLA system. In *Principles and Practice of Medical Genetics*, 2nd ed., edited by A. E. H. Emery and D. L. Rimoin, pp. 1453–1460. Churchill Livingstone, Edinburgh.

STRANGE, C. 1992. Cell cycle advances. *BioScience* 42: 252–256.

STRICKBERGER, M. W. 1985. *Genetics*, 3rd ed. Macmillan, New York.

STRICKBERGER, M. W. 1990. *Evolution*. Jones and Bartlett, Boston.

STRINGER, C. B. 1990. The emergence of modern humans. *SA* 263(Dec.): 98–104.

STURTEVANT, A. H. 1959. Thomas Hunt Morgan. *National Academy of Sciences, Biographical Memoirs* 33: 283–299.

STURTEVANT, A. H. 1965. *A History of Genetics*. Harper & Row, New York.

SUMMITT, R. L. 1973. Abnormalities of the autosomes. A *Pediatric Annals* reprint. Insight Publishing, New York.

SUTHERLAND, G. R. 1977. Fragile sites on human chromosomes: Demonstration of their dependence on the type of tissue culture medium. *Science* 197: 265–266.

SUTHERLAND, G. R. and R. I. RICHARDS. 1992. Anticipation legitimized: Unstable DNA to the rescue. *AJHG* 51: 7–9.

SUTTON, W. S. 1903. The chromosomes in heredity. *Biological Bulletin* 4: 231–251. [Reprinted in Peters (1959), in Voeller (1968), and in Gabriel and Fogel (1955).]

SVERDRUP, A. 1922. Postaxial polydactylism in six generations of a Norwegian family. *Journal of Genetics* 12: 217–240.

SYKES, B. 1991. The past comes alive. *Nature* 352: 381–382.

TAY, W. 1881. Symmetrical changes in the region of the yellow spot in each eye of an infant. *Transactions of the Ophthalmological Society, UK* 1: 55–57.

TELLEGEN, A., D. T. LYKKEN, T. J. BOUCHARD, JR., K. J. WILCOX, N. L. SEGAL and S. RICH. 1988. Personality similarity in twins reared apart and together. *Journal of Personality and Social Psychology* 54: 1031–1039.

TEMIN, H. M. 1972. RNA-directed DNA synthesis. *SA* 226(Jan.): 24–33.

TERRY, R. D. and M. WEISS. 1963. Studies in Tay-Sachs disease. II. Ultrastructure of the cerebrum. *Journal of Neuropathology and Experimental Neurology* 22: 18–55.

TERZAGHI, E. A., A. S. WILKINS and D. PENNY, Editors. 1984. *Molecular Evolution: An Annotated Reader*. Jones and Bartlett, Boston.

THAYER, A. M. 1991. Biopharmaceuticals overcoming market hurdles. *Chemical and Engineering News* 69(Feb. 25): 27–48.

THERMAN, E. and M. SUSMAN 1992. *Human Chromosomes: Structure, Behavior, and Effects*, 3rd ed. Springer-Verlag, New York.

THORNE, A. G. and M. H. WOLPOFF. 1992. The multiregional evolution of humans. *SA* 266(April): 76–83.

TIWARI, J. L. and P. I. TERASAKI. 1985. *HLA and Disease Associations*. Springer-Verlag, New York.

TJIO, J.-H. and A. LEVAN. 1956. The chromosome number of man. *Hereditas* 42: 1–6.

TONEGAWA, S. 1985. The molecules of the immune system. *SA* 253(Oct.): 122–131.

TRAINER, M. 1991. *Difference in Common: Straight Talk on Mental Retardation, Down Syndrome and Life*. Woodbine House, Rickville, MD.

TREVES, F. 1885. A case of congenital deformity. *Transactions of the Pathological Society of London* 36: 494–498.

TRYON, R. C. 1940. Genetic differences in maze-learning ability in rats. In *The Thirty-Ninth Yearbook of the National Society for the Study of Education*, edited by G. M. Whipple, pp. 111–119 and 154–155. Public School Publishing Company, Bloomington, IL.

TUDOR, A. 1989. Seeing the worst side of science. *Nature* 340: 589–592.

TURNBULL, A. 1988. Woman enough for the games? *New Scientist* 119(15 Sept.): 61–64.

TURNER, H. H. 1938. A syndrome of infantilism, congenital webbed neck, and cubitus valgus. *Endocrinology* 23: 566–574.

UPTON, A. C. 1982. The biological effects of low-level ionizing radiation. *SA* 246(Feb.): 41–49.

VALENTINE, G. H. 1986. *The Chromosomes and Their Disorders: An Introduction for Clinicians*, 4th ed. W. Heineman Medical Books, London.

VAN BROECKHOVEN, C., et al. 1992. Mapping of a gene predisposing to early-onset Alzheimer's disease to chromosome 14q24.3. *Nature Genetics* 2: 335–339.

VAN OMMEN, G. J. B., et al. 1986. A physical map of 4 million bp around the Duchenne muscular dystrophy gene on the human X chromosome. *Cell* 47: 499–504.

VARMUS, H. and R. A. WEINBERG. 1993. *Genes and the Biology of Cancer*. Scientific American Library, New York.

VELLUTINO, F. R. 1987. Dyslexia. *SA* 256 (March): 34–41.

VERMA, I. M. 1990. Gene therapy. *SA* 263 (Nov.): 68–84.

VIDAUD, D., et al. 1989. Father-to-son transmission of hemophilia A due to uniparental disomy. *AJHG* 45: A226.

VIGILANT, L. M. STONEKING, H. HARPENDING, K. HAWKES and A. C. WILSON. 1991. African populations and the evolution of human mitochondrial DNA. *Science* 253: 1503–1507.

VOELLER, B. R., Editor. 1968. *The Chromosome Theory of Inheritance: Classic Papers in Development and Heredity*. Appleton-Century-Crofts, New York.

VOGEL, F. and A. G. MOTULSKY. 1986. *Human Genetics: Problems and Approaches*, 2nd ed. Springer-Verlag, Berlin.

VOGEL, F. and R. RATHENBERG. 1975. Spontaneous mutation in man. *Advances in Human Genetics* 5: 223–318.

VON BOEHMER, H. and P. KISIELOW. 1991. How the immune system learns about self. *SA* 265(Oct.): 74–81.

VOSS, R., et al. 1989. Isodisomy of chromosome 7 in a patient with cystic fibrosis: Could uniparental disomy be common in humans? *AJHG* 45: 373–380.

WACHTER, R. M. 1992. AIDS, activism, and the politics of health. *NEJM* 326: 128–133.

WAGNER, R., R. L. STALLINGS and M. P. MAGUIRE. 1992. *Chromosomes: A Synthesis*. Wiley-Liss, New York.

WALLACE, A. R. 1855. On the law which has regulated the introduction of new species. *Annals and Magazine of Natural History*, Series 2, 16(Sept): 184–196. (Reprinted in Wallace 1875.)

WALLACE, A. R. 1875. *Contributions to the Theory of Natural Selection—A Series of Essays*. Macmillan, London.

WALLACE, A. R. 1889. *Darwinism: An Exposition of the Theory of Natural Selection with some of Its Applications*. Macmillan, London.

WALLACE, A. R. 1908. *My Life: A Record of Events and Opinions*, new ed. Chapman and Hall, London.

WALLACE, I. and A. WALLACE. 1978. *The Two*. Simon and Schuster, New York.

WALLACE, M. R. and F. S. COLLINS. 1991. Molecular genetics of von Reckling-

hausen neurofibromatosis. *Advances in Human Genetics* 20: 267–307.

WAMBAUGH, J. 1989. *The Blooding*. Morrow, New York.

WARBURTON, D. 1987. Reproductive loss: How much is preventable? *NEJM* 316: 158–160.

WASSARMAN, P. M. 1987. The biology and chemistry of fertilization. *Science* 235: 553–560.

WASSARMAN, P. M. 1988. Fertilization in mammals. *SA* 259(Dec.): 78–84.

WATSON, J. D. 1968. *The Double Helix*. Atheneum, New York.

WATSON, J. D. 1990. The Human Genome Project: Past, present, and future. *Science* 248: 44–9.

WATSON, J. D. and F. H. C. CRICK. 1953. Molecular structure of nucleic acids: A structure for deoxyribose nucleic acid. *Nature* 171: 737–738. Reprinted in Peters (1959) and Freifelder (1978).

WATSON, J. D., N. H. HOPKINS, J. W. ROBERTS, J. A. STEITZ and A. M. WEINER. 1987. *Molecular Biology of the Gene*, 4th ed. Benjamin/Cummings, Menlo Park, CA.

WATSON, J. D. and J. TOOZE. 1981. *The DNA Story: A Documentary History of Gene Cloning*. Freeman, New York.

WEIGALL, A. 1924. *The Life and Times of Cleopatra, Queen of Egypt*. Putnam, New York.

WEINBERG, R. A. 1983. A molecular basis of cancer. *SA* 249(Nov.): 126–1 42.

WEINBERG, R. A. 1988. Finding the anti-oncogene. *SA* 259(Sept.): 44–51.

WEISS, R. 1990. Uneven inheritance. *Science News* 138: 8–11.

WERTH, B. 1991. How short is too short? *New York Times Magazine* June 16, p. 14ff.

WERTZ, D. C. and J. C. FLETCHER. 1989. Fatal knowledge? Prenatal diagnosis and sex selection. *Hastings Center Report* 19(May–June): 21–27.

WETTKE-SCHÄFER, R. and G. KANTNER. 1983. X-linked dominant inherited diseases with lethality in hemizygous males. *Human Genetics* 64: 1–23.

WHITE, R. and J.-M. LALOUEL. 1988. Chromosome mapping with DNA markers. *SA* 258(Feb.): 40–48.

WILFOND, B. S. and N. FOST. 1990. The cystic fibrosis gene: Medical and social im-

plications for heterozygote detection. *JAMA* 263: 2777–2783.

WILFORD, J. N 1992. Critics batter proof of an African Eve. *New York Times*, May 19, 1992, p. C1ff.

WILLARD, H. F. 1990. Centromeres of mammalian chromosomes. *TIG* 6: 410–416.

WILLIAMS, S. A., B. E. SLATKO, L. S. MORAN and S. M. DESIMONE. 1986. Sequencing in the fast lane: A rapid protocol for [α-^{35}S] dATP dideoxy DNA sequencing. *BioTechniques* 4: 138–147.

WILSON, A. C. and R. L. CANN. 1992. The recent African genesis of humans. *SA* 266(April): 68–73.

WILSON, E. B. 1925. *The Cell in Development and Heredity*. Macmillan, New York.

WINN, M. 1990. New views of human intelligence. *New York Times Magazine* (Part 2, Good Health Magazine) April 29, 1990, p. 16ff.

WINTERS, R. W., J. B. GRAHAM, T. F. WILLIAMS, V. C. McFALLS and C. H. BURNETT. 1957. A genetic study of familial hypophosphatemia and vitamin D-resistant rickets. *Transactions of the Association of American Physicians* 70: 234–242.

WRIGHT, K. 1990. Breast cancer: Two steps closer to understanding. *Science* 250: 1659.

WRIGHT, S. 1930. *The Genetical Theory of Natural Selection—a review*. Journal of Heredity 21: 349–356.

YANKNER, B. A. and M.-M. MESULAM. 1991. β-amyloid and the pathogenesis of Alzheimer's disease. *NEJM* 325: 1849–1855.

YUNIS, J. J. 1976. High resolution of human chromosomes. *Science* 191: 1268–1270.

YUNIS, J. J. and O. PRAKASH. 1982. The origin of man: A chromosomal pictorial legacy. *Science* 215: 1525–1530.

ZARUTSKIE, P. W., C. H. MULLER, M. MAGONE and M. R. SOULES. 1989. The clinical relevance of sex selection techniques. *Fertility and Sterility* 52: 891–905.

ZIGAS, V. 1990. *Laughing Death: The Untold Story of Kuru*. Humana Press, Clifton, NJ.

ZOURLAS, P. A. and H. W. JONES. 1965. Clinical, histologic, and cytogenetic findings in male hermaphroditism. *Obstetrics and Gynecology* 25: 768–778.

ZUCKERMAN, H. and J. LEDERBERG. 1986. Postmature scientific discovery? *Nature* 324: 629–631.

Index

Flemming, Walther, 14, 17
Fluorescence in situ hybridization (FISH), 316, 317
  prenatal diagnosis, 463, 464
Fly Room, Thomas Hunt Morgan's research, 197
ΔF508 mutation, cystic fibrosis, 90, 465, 472
Follicle cells, mature egg, 33
Follicle-stimulating hormone (FSH), 110
Fölling, Asbjörn, 1–2
Food, Drug, and Cosmetic Act, carcinogens, 369
Food and Drug Administration (FDA), 294, 295
Food poisoning, immune overreaction, 402
Forks, DNA replication, 255–257
Founder effect, 85, 329, 343, 423
Fragile sites, 26, 99, 155
Fragile X syndrome, 26, 129, 155–157, 236, 458
  behavior, 179
  genetics, 156
  molecular biology, 156–157
Frameshift mutations, 265, 276–277
  cystic fibrosis, 473
  ABO blood groups, 391
Francke, Uta, 231
Franklin, Rosalind, 19, 253
Fraternal twins, *see* Dizygotic (DZ) twins
Fraumeni, Joseph, 361
FRAX, 156
Frequencies, *see also specific disorder*
  altering over generations, 489–493
  population, 407–410
Friedreich ataxia, 93
Frog
  gene splicing with *E. coli*, 291
  homeobox protein, 125–126
Fructose intolerance, hereditary, 326
Fruit flies, *see Drosophila*
FSH (follicle-stimulating hormone), 110
Fusion, somatic cell, 313–315

G, 19, 253, 254
G_1 phase, cell cycle, 36
G_2 phase, cell cycle, 36, 37
Galactosemia, 93, 103, 326, 464
Galton, Sir Francis, 168, 180, 304, 489, 492
Gamete(s), 32–48
  cell cycle, 35–41
  disomic, 210
  fertilization, 32–35
  haploid, 18–19
  law of segregation, 52–53
  meiosis, 41–48, *see also* Meiosis
  monosomic, 210
  nullisomic, 210
  parental, 191, 192
  prenatal tests, 245

proportions, 64
recombinant, 191–192, 193
trisomic, 210, *see also* Down syndrome; Nondisjunction
unbalanced, 234, 238, 241
Gamete by gamete method, *see* Checkerboard method
Gamete fraction, 55
Gamete intrafallopian transfer (GIFT), 119, 120
Gametogenesis, 41–48, *see also* Meiosis
Gamma globulin, 261, 380, *see also* Immunoglobulins (Ig)
Gangliosides
  Sandhoff disease, 330
  Tay-Sachs disease, 329
Garrod, Archibald, 57, 322–323, 335
Gaucher disease, 93
  type I, 327
G-banding, 25, 29
  Down syndrome, 214
Gel electrophoresis, 300–301
  DNA fingerprinting, 306
GenBank, 302
Gender identity, 108–109
  5-α-reductase deficiency, 113
Gene(s), *see also* DNA (deoxyribonucleic acid); *specific genes*
  adhesion molecules, 126–127
  alleles versus, 54, *see also* Allele(s)
  blood group, 387
  candidate, 362
  cell cycle, 41
  cell death, 127
  cholesterol control, 84–85
  components, 19–20
  defined, 20
  dopamine receptor, 187
  expanding, 157–158
  genes within, 160–161
  HLA, 395, *see also* HLA (human leukocyte antigen) system
  homeotic, 124–125
  influencing sexual development, 106, *see also* Sexual development
  jumping, 161, 273
  known versus unknown, 183
  lethal, 72–73
  linked, *see* Gene mapping
  locus, 54
  maternal effect, 124
  mitochondrial, 101
  mobile, 161, 273
  modifiers, 138
  nested, 160–161
  offspring types for two, 64–65
  one with several effects, 142, 143
  organization, 268–273
  pattern formation, 124–126
  segmentation, 124–126

"sex," 110
structural, 272
suicide, 127
tumor suppressor, *see* Tumor suppressor genes
unexpected variations in structure and function, 157–161
X-linked, *see* X-linked genes
Y-linked, 100–101, 247
Genealogy, mtDNA, 449, 450
Gene amplification, cancer, 368
Gene by gene method, 65–67
Gene expression, 257–264, *see also* DNA (deoxyribonucleic acid)
  continuum, 138
  defined, 20
  oncogene products, 355–357
  proteins, 260–262
  RNA, 259–260, *see also* RNA (ribonucleic acid)
  total absence, 141–142
  transcription, 257–259
  translation, 262–264
  variable expressivity, 138–142
  variations, 136–146
Gene families, 270–272
Gene flow, evolution, 441–443
Gene function, pedigree construction, 79–81
"Genelets," immunoglobulins, 383
Gene mapping, 191–206, 316, 318
  chromosomal abnormalities, 247–248
  chromosome 21, 215
  construction and use of maps, 201–203
  crossing over, 192–198, *see also* Crossing over
  deletion, 247
  duplication, 247
  evolution research, 204–205
  Human Genome Project, 205, 302–304
  Huntington disease gene, 311–313
  mapping many genes on X chromosome, 198–201
  recombination concept, 191–192
  RFLPs, 309–313
  road maps of human genome, 316, 318
  significance, 201–205
  translocation, 247–248
Generation(s)
  alteration in frequency of genetic traits, 489–493
  parental, 51
Gene splicing, 291–299
Gene therapy, 455, 481–486
  adenosine deaminase deficiency, 402, 481–483, 485
  cardiovascular disease, 485
  cystic fibrosis, 91, 485
  Duchenne muscular dystrophy, 485
  ethics, 481, 484–486

543